Secret Projects of the Luftwaffe Volume 1

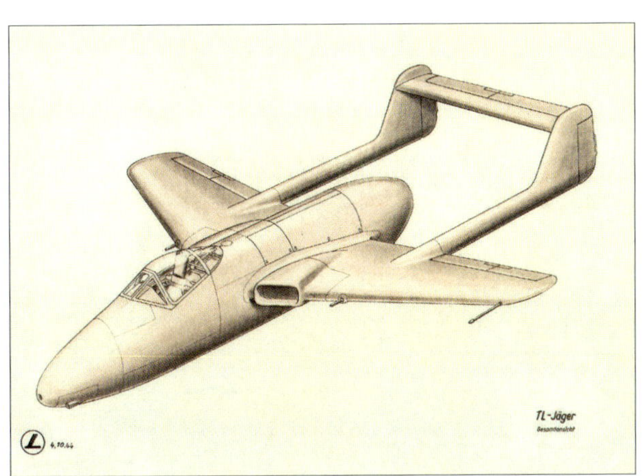

Secret Projects of the Luftwaffe Volume 1

Jet Fighters 1939–1945

Dan Sharp

For Stephen Walton without whom this book could not have been written.

First published in Great Britain in 2020
by Tempest Books
an imprint of Mortons Books Ltd.
Media Centre
Morton Way
Horncastle LN9 6JR
www.mortonsbooks.co.uk

ISBN 978 1 911658 08 5

Typeset by ATG Media

Contents

Preface

German aircraft development during the Second World War has been endlessly studied – but usually only from the perspective of individual designs that were actually built and entered service in one form other another. The deeper story of the military requirements, technical specifications, manufacturer competitions and prototype construction which preceded the delivery of series production model airframes to the Luftwaffe has seldom been explored in any detail.

The result has been much confusion about the sometimes complex processes and discussions underpinning the development of front-line combat machines and the development of the aircraft intended to replace them.

From the beginning of the war to the very end, an ever-changing cast of decision-makers at the Reichsluftfahrtministerium (RLM – the German air ministry) and elsewhere agonised over the likely direction of the air war and attempted to provide the Luftwaffe with materiel which anticipated the enemy's future capabilities – including the introduction of jet aircraft.

This book is an attempt to provide a clear timeline of Germany's attempts to produce high-speed fighter aircraft using innovative new propulsion systems before the Allies could do so. Every effort has been made to base this timeline on surviving contemporary documents and those documents, and other sources, have been cited.

My interest in German Second World War 'secret projects' began in 1991 when film director George Lucas's LucasFilm Games produced a videogame called *Secret Weapons of the Luftwaffe* for the PC. The image on the cover showed a pair of flying-wing aircraft the like of which I had never seen before. They were depicted in Luftwaffe markings attacking a formation of high-flying American bombers. At first I assumed that these aircraft must be fictional but the book accompanying the game included photographs of prototypes indicating that such aircraft were indeed built, albeit in prototype form.

Seven years later, my friend Alexander Power directed me to a website called Luft46.com and I was amazed to discover the vast and bewildering range of 'secret projects' apparently designed by German aircraft companies during the war.

While the RAF's Spitfires and the Luftwaffe's Bf 109s were engaged in a bitter battle to the death above war-torn Europe, Germany's finest engineers had been busily creating outlandish flying wings, rammers, tiny bombers, parasite aircraft, tail-first designs, aircraft with forward-swept wings and aircraft where the wings spun around like the blades of a helicopter – a cornucopia of oddities.

The more I read about the 'projects' the clearer it became that there were discrepancies in the information presented about them. Some were plainly designed at the same time and for the same purpose but they were not linked together. Instead, the projects were grouped by manufacturer. I found this same approach in the few books that existed on the subject.

In the mainstream aviation press of the day, these designs were frequently written off as 'freaks' – products of designers motivated by basic intellectual curiosity, or fear, or boredom, or a toadying desire to please Adolf Hitler. Yet some designs seemed far ahead of their time and some authors of the late 1990s such as Walter Schick and Manfred Griehl were eager to point out similarities they saw between those designs and postwar jets built and flown by the Soviets and the Western Allies.

This idea fascinated me at the time; that the early Cold War jets on both sides had perhaps been based on captured German technology. Certainly, the Allies captured a huge quantity of German aviation industry documents in the immediate aftermath of the war and studied much of it in detail. But the links between specific German wartime projects and postwar aircraft seemed circumstantial in most cases. It was a hypothesis which went nowhere, the once-regular updates to Luft46.com dried up and my interest waned.

More than a decade later I found myself at a loose end and decided to revisit the subject. I discovered that books written about German 'secret projects' remained as confusing and contradictory as ever but this time I resolved to see whether I could discover why those wartime 'oddities' were created and how they might fit together on a straightforward timeline.

The result was *Luftwaffe: Secret Jets of the Third Reich* – published by Mortons in 2015. This was based on a couple of years' research using primary source material and aimed to tell the story of jet aircraft development in wartime Germany with a particular emphasis on fighters. It was a shot in the dark for the company but it paid off and the publication sold out.

I followed up *Secret Jets* with *Secret Bombers* (2016), *Secret Wings* (2017), *Secret Designs* (2018) and *Secret Projects* (2019), all based on my ongoing primary source research. But the first title, *Secret Jets*, was never reprinted – despite repeated calls for Mortons to make it available again in hard copy format.

The opportunity finally arose to reprint Secret Jets as a book in 2019 but by now, after five more years of research, the timeline of jet fighter development in Second World War Germany had become a great deal clearer, as had errors in the original work. I could not allow that publication to be simply reissued. So I promised my publisher that I would completely rewrite Secret Jets – rebuilding it from the ground up, rechecking all the original sources and consulting many new ones, correcting errors and layering on a substantial amount of new detail.

This book is the result. Keen-eyed readers who own or have read the 'Secret' series to date will undoubtedly notice fragments of those works within it – but the overall focus on jet fighters has been tightened and the level of detail now offered is, I believe, unrivalled. I hope too that the inclusion of all sources used in the writing of this volume will allow readers to judge for themselves the accuracy of the account herein presented.

Dan Sharp

ACKNOWLEDGEMENTS AND THANKS

This book would not have been possible without the help of Stephen Walton at the Imperial War Museum and my friend and fellow research enthusiast Steve Coates. I am indebted to them both for their unwavering support and generosity. Elizabeth Borja and David Schwartz at the National Air and Space Museum's archives section and Becky Jordan and Olivia Garrison at Iowa State University Library provided invaluable assistance for which I am most grateful.

Thanks are also due to Jenz Baganz, JC Carbonel, Chris Cocks, Zoltán Csombó, Calum Douglas, Chris Elwell, Paul Fincham, Ian Fisher, Simon Fowler, Chris Gall, Tim Hartley, Pauline Hawkins, Carlos Alberto Henriques, Luca Landino, Scott Lowther, Paul Malmassari, Paul Martell-Mead, Eddie Nielinger, Steve O'Hara, Ronnie Olsthoorn, Martin Pegg, Sean Phillips, Alexander Power, Stephen Ransom, Chris Sandham-Bailey, Dan Savage, Jonathan Schofield, Sven Schultze, J Richard Smith, Kay Stout, Oliver Thiele, Greg Twiner, Daniel Uhr, Gary Webster, Paul Williams and Tony Wilson.

Introduction

German Second World War jet fighter development

Germany gained an early lead in jet fighter development during the Second World War but failed to capitalise on it until it was far too late. The result is a tale of incredible technological advances punctuated by a litany of missed opportunities.

Development of the jet engine in Germany began and progressed in parallel with work on gas turbines taking place in Britain during the mid-1930s. By the end of the decade, however, the Germans had moved more decisively to realise the new powerplant's potential.

The German government encouraged all of the country's major engine manufacturers to begin work on jet engine designs and as a result, despite the technological challenges involved, BMW, Junkers and Heinkel-Hirth all made significant progress during the early part of the Second World War.

Just two aircraft manufacturers were initially entrusted with the task of building a single-seat fighter incorporating this new propulsion technology – Messerschmitt and Heinkel. When the initial specification was issued in January 1939, it called for a single-jet aircraft but both companies ended up building relatively large twin-engine aircraft instead. This was because early turbojets were too weak to power an aircraft loaded with military equipment on their own.

At the time, this detail did not seem to pose any significant problems since turbojet engines were lighter than piston engines and much cheaper and faster to build. A twin-jet fighter was therefore lighter than the equivalent fighter with two piston engines and only slightly more expensive than a fighter with a single piston engine. Messerschmitt designed the Me 262 and Heinkel the He 280. Each had its engines in wing-mounted nacelles allowing easy access for maintenance and negating any loss of thrust that might occur as a result of a long intake duct. In addition, early turbojets were fragile and notoriously unreliable. Having a second engine provided a useful backup in case the other failed.

It was a safe choice of configuration but it had a number of drawbacks. In economic terms, two jet engines meant that building a single Me 262 was still somewhat more

GERMAN JET DEVELOPMENT TIMELINE

• March 13, 1928	Max Valier and Fritz von Opel commission Alexander Lippisch at the Rhön-Rossitten-Gesellschaft (RRG) to build them a rocket-propelled aircraft.
• June 11, 1928	Lippisch's aircraft, powered by solid fuel rockets, is test flown by Fritz Stamer.
• February 1932	Gerhard Fieseler commissions Lippisch to build him a new sporting aircraft – the Delta IV. This aircraft is rebuilt several times before becoming the DFS 39.
• January 30, 1933	Adolf Hitler is named chancellor of Germany and on the same day the RRG is split into the Deutschen Luftsportverband (DLV) or 'German Airsport Association' and the Deutsche Forschungsanstalt für Segelflug (DFS) or German Research Institute for Gliding.
• February 13, 1935	The Heereswaffenamt, RLM and Junkers agree that Junkers should work on rocket propulsion for aircraft.
• September 1935	Heinkel also becomes involved in working on rocket propulsion for aircraft.
• November 9, 1935	Hans von Ohain is granted a patent for Germany's first turbojet design.
• March 1936	Von Ohain is hired by Ernst Heinkel to work on his jet engine.
• April 1936	Dr Herbert Wagner appoints engineer Max Adolf Müller to begin work on turboprop engines at Junkers.
• June 9, 1936	The First World War's second-highest scoring German ace, Ernst Udet, is appointed as head of the RLM's Technisches Amt, in charge of research and development for the Luftwaffe.
• December 1936	Commissioned by the RLM to build an experimental rocket-propelled aircraft, Heinkel begins preliminary work on the P 1033.
• January 16, 1937	A mock-up of the Heinkel P 1033 – an aircraft specifically designed for rocket propulsion – is inspected by the RLM.
• February 16, 1937	A second mock-up inspection of the Heinkel P 1033 takes place – it has now been designated He 176.
• c. February 28, 1937	Von Ohain's first working jet engine, the HeS 1 is completed and installed in a test rig. Heinkel begins working on airframe designs suitable for jet propulsion.
• Early 1938	Junkers begins designing airframes for turbojet-powered aircraft.
• Summer 1938	The Heinkel He 176 V1 experimental rocket aircraft is completed – although it will subsequently undergo a huge number of alterations before its first flight.
• August 1938	Rocket scientist Eugen Sänger reports on the potential of ramjets to the Luftfahrtforschungsanstalt (LFA) or Aviation Research Institute.

ABOVE: Display model of the Lippisch Delta VI photographed and set against a 'sky scene' for promotional purposes. The Delta VI was typical of advanced projects being dreamt up by Germany's aircraft designers as the war neared its conclusion.

expensive than building a single Bf 109 or Fw 190. A larger airframe meant a greater quantity of light metal was required for each Me 262 built.

In development terms, positioning the turbojets under the wings restricted the size of engines that could easily be installed. The He 280 failed partially because its airframe had been designed for the Heinkel HeS 8 and it struggled to accommodate the larger Jumo 004 when required to do so. Furthermore, wings that needed to support engine nacelles had to be strong and therefore thick. It would prove difficult, later, to alter the wing design of the Me 262 without major structural revisions.

Thicker wings had a knock-on effect for performance, as did the use of drag-inducing external nacelles, and most fighter pilots were used to flying single-engine designs, requiring additional training to manage two engines in flight. Despite all this, the twin-jet designs offered the promise of performance beyond anything likely to be possible with piston engines.

The future viability of jet engines for fighter aircraft in general remained in doubt until July 18, 1942, when the Jumo 004-powered third Me 262 prototype was able to fly for 25 minutes without problems. After more than three years of development, it was now clear that turbojets could

• Autumn 1938	Aero engine manufacturers Daimler-Benz, BMW and Brandenburgische Motorenwerke (Bramo) are encouraged to look at turbojet development by the RLM.
• Early 1939	Junkers' Rückstoss-Turbinen-Strahltriebwerk (RTO) or 'Reaction Turbojet Engine' runs up to 6,500rpm on a test stand. It is designed to reach 12,900rpm.
• January 2, 1939	Alexander Lippisch and his team transfer from the DFS to Messerschmitt AG at Augsburg. They form Abteilung L or 'Department L' and develop the DFS 39 as 'Projekt X' which receives the designation Me 163.
• January 4, 1939	Heinkel and Messerschmitt each receive a contract to begin work on the design of single-seat fighter aircraft with single turbojet propulsion – 'Jagdflugzeuge mit Strahltriebwerk'.
• February 1, 1939	Udet is given the new title Generalluftzeugmeister and his area of responsibility is expanded to incorporate all air force equipment including the work of the Luftwaffe's aircraft testing and experimental stations.
• April 1, 1939	Following preliminary studies, Messerschmitt's main project design team set to work in earnest on their first draft of a single-seat twin-jet fighter, project P 65 – later referred to as P 1065.
• April 12-13, 1939	Abteilung L produces the first single turbojet designs in what will become its long-running P 01 jet and rocket fighter series for Messerschmitt.
• c. June 1939	Heinkel completes the He 178 V1 experimental jet aircraft and begins work on two mock-ups of its projected twin-engine turbojet fighter.
• June 7, 1939	The Messerschmitt P 1065 design, intended to be powered by two wing-mounted BMW jet engines, is submitted to the RLM. The RLM asks Messerschmitt to produce a mock-up of the design.
• June 15, 1939	The Heinkel He 176 V1 is flown under rocket propulsion for the first time after completing several dozen short hops earlier in the year.
• July 1939	Bramo is amalgamated into BMW. The jet engine it has been working on is renamed P 3302. BMW's own design, the P 3304, is ultimately abandoned.
• August 27, 1939	An aircraft powered by Von Ohain's latest jet engine design, the HeS 3b, takes flight. The Heinkel He 178 V1 is the first jet-powered aircraft to fly anywhere in the world.
• September 1, 1939	The Second World War begins.
• September 26, 1939	RLM representatives inspect the two completed mock-ups of Heinkel's jet fighter design. After initially being known as the He 180, the aircraft is later redesignated He 280.
• October 16, 1939	Work on fitting the DFS 194 with a Walter rocket motor is completed and ground tests begin.

ABOVE: Much of German aircraft manufacturers' design and development effort during 1944-45 was devoted to finding a way of combatting the increasingly effective USAAF daylight raids over Germany.

be made reliable enough for operational use. But it would be a further 22 months before the Me 262 finally began to enter front-line service with the Luftwaffe.

The rocket-propelled Messerschmitt Me 163 had a similarly tortuous development history – though its shortcomings were even clearer than those of the Me 262. The diminutive aircraft was the culmination of a long line of tailless aircraft experiments carried out by its maverick creator Alexander Lippisch and was intended as a technology demonstrator rather than as a front line fighter.

Its viability was conclusively proven more than nine months earlier than that of the Me 262, when the first

• November 1, 1939	The Heinkel He 178 is demonstrated in front of RLM representatives.
• December 1939	Junkers begins work on the Jumo T1 jet engine design, later known as the Jumo 004.
• December 19, 1939	Messerschmitt's P 1065 mock-up is examined by the RLM.
• Circa spring 1940	Arado is asked to design a bomber/photo-reconnaissance aircraft with either Junkers or BMW jet engines.
• March 1, 1940	It is proposed that 20 prototypes of the P 1065 should be built, to be powered by the BMW P 3302.
• June 3, 1940	The DFS 194 flies under rocket power for the first time.
• September 22, 1940	Glide tests on the first completed airframe of Heinkel's He 280 design begin.
• December 1940	The prototype BMW P 3302 V1 jet engine is run for the first time.
• Winter 1940	Work on the rocket-powered Messerschmitt Me 163 V4 prototype is completed.
• February 1941	Construction of the first prototype Messerschmitt P 1065 begins.
• February 13, 1941	First towed flight of the Me 163 V4.
• March 30, 1941	The Heinkel He 280 V2 flies under its own power for the first time.
• April 8, 1941	Messerschmitt's P 1065 is given the RLM designation Me 262.
• April 18, 1941	The Me 262 V1, PC+UA, makes its first flight, albeit powered by a Jumo 210 G piston engine.
• May 1941	Messerschmitt's projects office comes up with at least 21 designs for a pulsejet-powered aircraft – the P 1079.
• August 13, 1941	First rocket-powered flight of Me 163 V4.
• August 27, 1941	Lippisch proposes the 'Li P 05' interceptor, a scaled-up development of the Me 163.
• October 2, 1941	Test pilot Heini Dittmar becomes the first man to fly faster than 1,000km/h, at the controls of Me 163 V4.
• October 1941	The Horten brothers witness test flights of the Me 163 at Peenemünde.
• October 24, 1941	Arado finally comes up with a concrete proposal for the RLM's bomber/reconnaissance jet requirement – the E.370/IVa. It is examined by the RLM and an initial batch of 50 aircraft is ordered.
• November 1941	Design details of the Me 163 B are finalised. All existing Me 163s become Me 163 As. Two BMW P 3302 prototype engines are delivered to Messerschmitt for installation in the Messerschmitt Me 262 V1 prototype.
• November 17, 1941	Ernst Udet commits suicide. His replacement as Generalluftzeugmeister is Generalfeldmarschall Erhard Milch.
• February 1942	Arado's E 370 design is allocated the RLM designation Ar 234.

prototype was flown at 1,000km/h on October 2, 1941, but rather than following Lippisch's plan to build a much larger interceptor based on the now-established principle of rocket propulsion for a fighter, it was decided that an almost entirely new version of the Me 163 itself should be built – the Me 163 B. This would perform in a similar fashion to the prototype – with all the problems of a dolly-launched, skid-landing small-airframe prototype – but with sufficient room to carry cannon.

Designing, building and testing this new aircraft meant it took even longer to get the Me 163 B into service than it had taken the Me 262, with the Luftwaffe only able to begin limited operations with it in May 1944. And even then it proved to be more of a menace to its own pilots than the enemy – its volatile rocket fuel causing a number of fatal explosions and its unforgiving skid landing gear resulting in several serious back injuries. Operations were soon curtailed when Allied bombing raids knocked out the small number of chemicals factories capable of producing its fuel.

The third jet aircraft to see service with the Luftwaffe was the Arado Ar 234. Designed without advanced features as a simple reconnaissance platform, it perhaps represents the greatest missed opportunity of the three. Arado's brief in 1940 was to create an airborne camera platform able to cover Britain all the way up to the naval base at Scapa Flow. There was no requirement for armament, no requirement for more than one crewman and no particular need for manoeuvrability. The design that emerged was therefore all about achieving a particular range – to the extent that even a wheeled undercarriage was initially sacrificed to keep weight down and provide more room for fuel tanks.

The Ar 234 V1 made its first flight on July 30, 1943, and although Arado had already drawn up various proposals to repurpose the aircraft as a light bomber or even as a fighter, the lack of space within its slender fuselage and its unsophisticated layout made it largely unsuitable for anything other than the task for which it had been created. Its first operational sorties were carried out in August 1944.

ABOVE: With the introduction of the de Havilland Mosquito, the Allies had a means of overflying Germany with near-impunity to take reconnaissance photos. Developing an interceptor that could catch the powerful and high-flying British aircraft was a high priority in 1944.

It is worth noting that the first two developments were begun when Germany was still at peace and neither the resources of the Reichsluftfahrtministerium (RLM – the German air ministry) nor those of the aircraft manufacturers were being taxed by the demands of developing and supplying aircraft for front-line combat. Furthermore, all three developments were begun when the much-maligned

- **Spring 1942** — The Horten brothers begin designing a twin-jet flying wing, the H IX.
- **March 1942** — Drawings and proposals for the Me 163 C and Super 163 Interceptor are drafted.
- **March 25, 1942** — A first flight of the Me 262 V1 is attempted but quickly ends after both of its BMW engines fail.
- **March 31, 1942** — Messerschmitt produces a report emphasising the role of the P 1079, now designated Me 328, pulsejet aircraft as a fighter.
- **April 1942** — The RLM orders six Ar 234 prototypes. Work on the first prototype Me 163 B is completed.
- **May 29, 1942** — The RLM reduces its Me 262 prototype order to just five examples.
- **June 26, 1942** — First towed flight of the Me 163 B V1.
- **July 5, 1942** — Flight testing of the He 280 V3, using two of Von Ohain's HeS 8A jet engines, begins.
- **July 18, 1942** — Me 262 V3 flies for the first time with Jumo 004 engines. It completes 25 minutes of trouble-free flying.
- **September 1942** — Although Me 262 V3 has been wrecked in an accident, its reliable performance convinces the RLM to reinstate the type's formerly cancelled additional prototypes.
- **September 13, 1942** — The first P 11, a two-seater fast bomber powered by two turbojets, is designed by Handrick at Messerschmitt's Abteilung L.
- **November 5, 1942** — Focke-Wulf produces drawings for a jet-engined Fw 190, using a powerplant of its own design.
- **November 6, 1942** — Lippisch presents a talk on his Me 163 A to members of the exclusive Deutsche Akademie für Luftfahrtforschung, including Siegfried Günter of Heinkel and Hans Multhopp of Focke-Wulf.
- **November 18, 1942** — The RLM reduces its He 280 prototype order to just six examples plus one unpowered aircraft for high-speed testing.
- **December 2, 1942** — A second P 11 twin-jet bomber is designed by Handrick. He makes it a single-seater.
- **December 10, 1942** — With the tide of battle in the East turning against Germany, and with no imminent victory in sight, Generalfeldmarschall Erhard Milch orders into effect "an urgent development and production programme under the code word Vulkan". The aircraft given top priority are the Me 163, Me 262, He 280, Me 328 and Ar 234.
- **December 15, 1942** — Messerschmitt produces a new report now emphasising the role of the Me 328 as a bomber, rather than a fighter.
- **December 28, 1942** — The RLM increases its order for Ar 234 prototypes from six to 20.
- **January 4, 1943** — Focke-Wulf designer Julius Rotta sets out how the company will approach future jet design and suggests an aircraft layout similar to what will eventually be produced as the Heinkel He 162.

ABOVE: After D-Day, the Luftwaffe had yet another problem – how to combat powerful and heavily armed Allied fighter-bombers at low level.

ABOVE: The Messerschmitt Me 262 had an incredible turn of speed and awesome firepower but it could not be built fast enough and was unsuitable for combatting low-level enemy fighter-bombers.

Ernst Udet was, as Generalluftzeugmeister, in charge of providing the Luftwaffe's equipment.

When the Second World War began in September 1939, resources that had been available for experimental development were reallocated to more directly supporting the war effort. At Messerschmitt, this meant intensive efforts to upgrade and improve the Bf 109 and Bf 110. Following the suicide of Udet and the appointment of Erhard Milch as Generalluftzeugmeister in November 1941, there was an even tighter focus on existing types to the detriment of all experimental designs.

Milch also proved far less willing than Udet to gamble on unproven technology and his office prioritised the development of new piston-engine aircraft to the extent that between November 1941 and June 1944 – when he left office – there were no new requirements for fighters with turbojet or rocket propulsion. He had, in December 1942, ordered the 'Vulkan' programme into effect which was supposed to give top priority to the development of the Me 163, Me 262, He 280, Me 328 and Ar 234 but this seems to have made little difference to any of them.

The aircraft companies themselves had continued, of their own volition, to half-heartedly work on a handful of new jet and rocket designs, eventually approaching Milch and the RLM in May 1943 with the idea of building lightweight single-jet fighters based largely on existing fighter airframes.

This initiative seems to have been greeted with some enthusiasm at first but that soon tailed off as attention focused once more on the next generation of piston-engine fighters – the Me 209 and Ta 153 as potential replacements for the Bf 109 and Fw 190. With the benefit of hindsight, the

• March 1943	The RLM orders the Horten brothers' Sonderkommando Lln 3 to cease all work.
• March 9, 1943	The RLM decides against series production of the Heinkel He 280 – primarily due to the minimal ground clearance of its low-slung engine pods. This is officially confirmed in a letter to Heinkel dated March 27, 1943.
• March 10, 1943	Arado designs a fighter with a dorsal turbojet arrangement very similar to that drafted by Rotta in January. Focke-Wulf produces two separate reports for a new single-jet fighter – one powered by a BMW P 3302 and another by Jumo's 004 C. Later in the month a third version will be produced, based on the Jumo-engined design, which will become known as the '1. Entwurf'.
• March 18, 1943	Arado designs a pure rocket fighter it calls R-Jäger.
• March 20, 1943	The RLM orders that Lippisch's Department L should be effectively dissolved and absorbed into Messerschmitt. Arado designs a combination turbojet/rocket fighter it calls K-Jäger.
• April 28, 1943	Lippisch officially leaves Messerschmitt. He takes up his new appointment as head of the Luftfahrtforschungsanstalt Wien (LFW) in Vienna, Austria, three days later.
• May 1943	Lippisch presents the RLM with his proposal for an aircraft than can fly 1,000km at 1,000km/h carrying 1,000kg of bombs – the P 11.
• May 22, 1943	General der Jagdflieger Adolf Galland flies the Me 262 V4 at Lechfeld and quickly becomes one of the type's strongest supporters.
• May 28, 1943	A meeting of the RLM's development committee is told that Focke-Wulf and Messerschmitt have suggested creating new lightweight single-jet fighters on the basis of existing designs – the Fw 190 and Me 163. Project sketches for single-jet fighters are submitted by Focke-Wulf, Messerschmitt and Heinkel.
• May 29, 1943	Alexander Lippisch produces his first post-Messerschmitt report on the P 11 '1,000 x 1,000 x 1,000' jet bomber.
• June 1943	Focke-Wulf produces Baubeschreibung Nr. 264 outlining a revised version of its single-jet fighter, also known as '2. Entwurf'.
• June 3, 1943	Arado designs a twin-jet fighter similar to the Ar 234 which it calls simply TL-Jäger.
• June 22, 1943	Messerschmitt's plan for series production of the Me 262 is approved.
• Early summer 1943	The first Ar 234 prototype is completed and taxiing trials begin.
• June 24, 1943	First rocket-powered flight of the Me 163 B V1 prototype.
• July 1943	Lippisch receives a RM 30,000 development contract for his P 11, effectively making it a low-priority project.
• July 30, 1943	The Ar 234 V1 is successfully flown for the first time.

development of a lightweight single-jet single-seat fighter based on the Jumo 004 appears to have been one of the greatest missed opportunities of the war.

Had the Me 262 been cancelled as soon as the Jumo 004 had proven reliable in July 1942, and every effort diverted towards the design and series production of something similar to the He 162, Germany might have had a fleet of cheap high-performance single-jet fighters ready for action by the end of 1943. This might conceivably have altered the course of the war – but as it was the opportunity slipped past unnoticed.

Udet's time in office as Generalluftzeugmeister had been characterised by an explosion in aircraft development – with a huge number of types being commissioned as prototypes either as competitors for a particular requirement or as experiments to study a particular concept. However, where the RLM under Udet had had some success in ruthlessly cancelling projects which failed to meet requirements or which failed to live up to expectations (with a few notable exceptions, such as the Me 210), Milch inherited a large number of aircraft types and projects which, objectively, were surplus to requirements and merely allowed them to continue in the hope that they could somehow be improved.

Even by May 1944, the German aircraft industry was still producing 32 different aircraft types[1] (this having been pared back from some 53 types at one point during 1943),[2] many of them obsolete types that had soldiered on because plans to replace them had come to nothing or which represented the remnants of failed lines of development.

To make matters worse, Germany was facing a critical shortage of all aircraft types at the beginning of the year, with Allied bombing having repeatedly targeted the large factories where they were built. Efforts to resolve this problem resulted in the formation of the Jägerstab or 'Fighter Staff' on March 1, 1944. This new organisation, formed by Albert Speer's Reichsministerium für Rüstung und Kriegsproduktion, had relieved the RLM of its responsibility to oversee aircraft production and set about dispersing it to dozens of small factories and workshops across Germany.

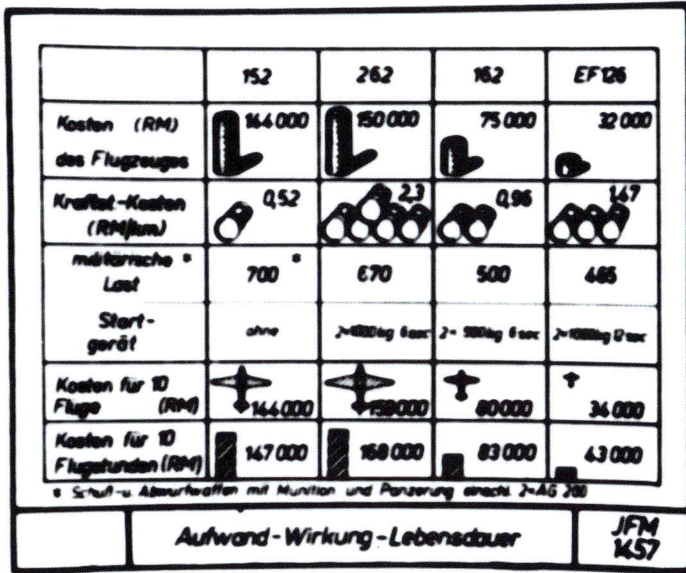

ABOVE: This Junkers-produced slide from December 1944 shows the cost of a twin-jet Me 262 – RM 150,000 – compared to that of a single piston-engine Focke-Wulf Ta 152, RM 144,000 and a single-jet He 162 at just RM 75,000. The economic and production benefits of a single-jet design are clear.

However, it was not until May 25, 1944, that Reichsmarschall Hermann Göring himself outlined plans to reduce the 32 types down to just 16, thereby freeing up vital

• August 1943	The Horten brothers, Walter and Reimar, send their proposal for a 1000 x 1000 x 1000 jet bomber based on their Horten IX design to Reichsmarschall Hermann Göring via Oberstleutnant Ulrich Diesing.
• August 3, 1943	A second round of devastating incendiary attacks on Hamburg prompts Blohm & Voss to consider unorthodox methods of fighting back against Allied bombers.
• August 17, 1943	Me 262 production is delayed when a US bombing raid destroys fuselage construction jigs at Messerschmitt's Regensburg plant, prompting the firm to move key departments to Oberammergau in the Bavarian Alps.
• August 19, 1943	The RLM responds to a proposal by Blohm & Voss to build a glider, the P 186, that can be towed aloft behind a Bf 109 to carry out ramming attacks on Allied bombers.
• September 9, 1943	Dr Paul Karlson produces a paper explaining an idea conceived by his friend Alexander Lippisch to create vertical-launch 'arrows of death' to carry out ramming attacks on Allied bombers.
• September 20, 1943	Jacobs-Schweyer Flugzeugbau (JSF) produces a drawing of the Me 328 redesigned around a single turbojet engine as the Me 328 T.
• September 28, 1943	Göring increases Lippisch's P 11 contract to RM 500,000 and commissions the Horten brothers with the same sum to build the Horten IX – eventually designated Ho 229. Both types will later be considered as heavy fighters.
• October 1943	Oberstleutnant Siegfried Knemeyer, formerly the Luftwaffe's General der Kampfflieger, is appointed head of development (GL/C-E) in the RLM's Technisches Amt. Sänger and his colleague Irene Bredt produce a report on the potential of ramjet engines to power a fighter.
• October 6, 1943	JSF continues to work on a turbojet-powered Me 328 fighter, apparently at Messerschmitt's behest.
• November 1943	Focke-Wulf designs its '3. Entwurf' single-jet fighter.
• November 13, 1943	General der Jagdflieger Adolf Galland expresses an interest in Blohm & Voss's P 186 rammer idea.
• December 1943	Focke-Wulf redesigns its single-jet fighter again, this time to incorporate two rocket motors, resulting in the '4. Entwurf'.
• December 13, 1943	Engineers from Arado and E-Stelle Rechlin discuss the possibility of using the Ar 234 for operations at night.
• December 15, 1943	Blohm & Voss's P 186 rammer, now described as a 'glide fighter', is given the designation BV 40 and 12 prototypes are ordered.
• January 1944	Focke-Wulf engineer Hans Multhopp puts forward a radically different single-jet fighter design which henceforth is known as 'Entwurf Multhopp', although it will later receive many other names including '5. Entwurf', Nr. 279 and Huckebein before eventually receiving the designation Ta 183 in March 1945.

ABOVE: Wind tunnel testing of the Junkers Ju 287 fast bomber. The proliferation of wind tunnels in Germany enabled a huge number of different designs to be tested quickly and rapidly advanced designers' knowledge of aerodynamics. Conversely, it has been argued that the opportunity to test so many designs resulted in too much data - making it difficult and time-consuming to decide which forms and features worked best.

production capacity. Among the aircraft axed immediately or scheduled for expiry over the next 18 months were the Bf 108, Bf 110, Me 210, Me 323, Me 410, Ju 52, Ju 352, Ju 188, Ju 288, Ju 290 and Ju 390. The remaining types included the Ar 234, Do 335, Fw 190/Ta 152, Ta 154, Ju 287, Ju 388, He 277, He 111, Bf 109, Me 163 and Me 262.

Göring explicitly stated that he expected Me 262 production to be substantially increased, that the aircraft should be a fighter (even though, as Milch reminded everyone during the meeting, Adolf Hitler had declared that it should be a bomber) and that units operating the Me 262 should be strategically located to defend important industrial targets.

It had taken years to reach this point and even if the Me 262 had been built as a bomber it would not have been available in time to fly missions over the Normandy beaches when the long-anticipated invasion finally happened on June 6, 1944 – the role Hitler had envisioned for it. Neither was it suitable for combating the low-flying fighter-bombers that would begin ranging across occupied Europe in the wake of D-Day.

Where Germany had been quick to recognise the potential of turbojet propulsion at the end of the 1930s, it had been fatally slow to realise that potential. It was now almost too late to put any completely new jet or rocket-propelled aircraft into production. And it was certainly too late for any such aircraft to have an impact on the remainder of the war. But even in mid-1944, no one involved in aircraft development knew the full extent of Germany's increasingly dire war situation on the ground and no one was planning for defeat.

ADVANCED PROJECTS

With the departure of Milch on June 20, 1944, responsibility for the commissioning and development of new aircraft seems to have briefly entered a period of limbo where no one was directly in charge of development. It was therefore left to Oberstleutnant Siegfried Knemeyer of the RLM's Techniches Amt to tentatively begin working up specifications for a new generation of jet- and rocket-propelled aircraft.

This period of uncertainty came to an end on August 1, 1944, when Speer's ministry assumed full responsibility for Germany's technical air defence and the RLM's

• January 4, 1944	Oberstleutnant Siegfried Knemeyer of the RLM's Techniches Amt tells Willy Messerschmitt to further develop the Me 262 – resulting in the new HG series, P 1099 heavy fighter and P 1100 bomber variants.
• February 1, 1944	Focke-Wulf produces Baubeschreibung Nr. 272 outlining the company's preferred choice of single-jet fighter layout, a twin-boom design with built-in rocket motor that will also be known as 'Flitzer' and '6. Entwurf'.
• February 11, 1944	Focke-Wulf starts working on designs for a twin-ramjet fighter under the designation 'X1'.
• March 1944	Messerschmitt begins work on a new project – P 1101 – comparing two-, three- and four-jet aircraft designs with different armament configurations. The DFS produces a further report on the potential of ramjet engines to propel a fighter. Alexander Lippisch begins working on the ramjet-powered P 12. Focke-Wulf begins work on what will become the Triebflügeljäger triple ramjet vertical take-off fighter.
• March 1, 1944	The Jägerstab (Fighter Staff), a committee of industrialists and RLM officials, including Generalluftzeugmeister Erhard Milch, Reichsminister Albert Speer and Hauptdienstleiter Karl-Otto Saur of the Reichsministerium für Rüstung und Kriegsproduktion is established to rebuild the Luftwaffe's fighter force by whatever means necessary.
• March 5, 1944	The unpowered H IX V1 glider makes its first flight.
• March 16, 1944	Blohm & Voss proposes fitting a rocket motor to its BV 40 glide fighter to turn it into a rocket fighter.
• April 20, 1944	Blohm & Voss is told that General der Jagdflieger Adolf Galland has lost interest in its BV 40 glide fighter.
• April 18, 1944	The DFS produces a drawing showing Lippisch's Delta VI design fitted with a ramjet engine.
• May 1944	Lippisch begins work on the new P 13 ramjet-powered fighter/rammer.
• May 21, 1944	Ten senior engineers from Focke-Wulf meet with representatives of the RLM's C-E 2 department to discuss their single-jet 'Flitzer' design.
• June 1944	The LFW begins building the first prototype of Lippisch's P11, now renamed Delta VI.
• June 6, 1944	Focke-Wulf's X1 twin-ramjet fighter has evolved into the X6.
• June 16, 1944	It is probably on this date that American bombers hit the LFW, killing around 45 of Lippisch's key Delta VI workers.
• June 20, 1944	Reichsmarschall Hermann Göring orders that responsibility for providing the Luftwaffe with new aircraft, including research and development, be transferred from Generalluftzeugmeister Erhard Milch to Reichsminister Albert Speer. Milch leaves the RLM after 11 years to join Speer's ministry as deputy minister.
• Late June 1944	Construction of the Horten IX V2 begins at Göttingen.

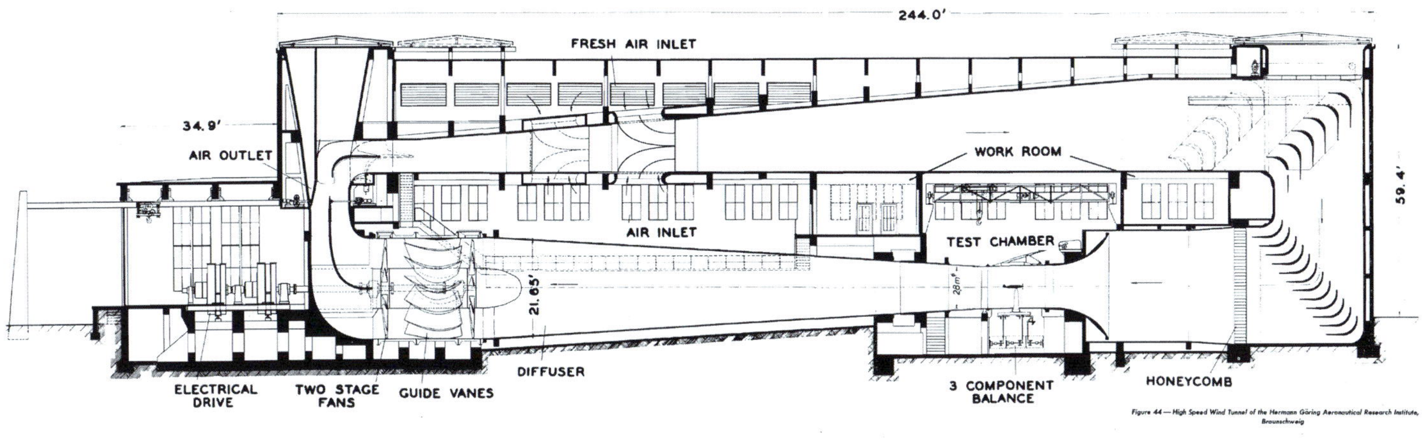

ABOVE: Diagram showing the high speed wind tunnel at the Luftfahrtforschungsanstalt Hermann Göring, Braunschweig-Völkenrode. By the summer of 1943 Germany had access to more modern wind tunnel testing facilities than any other country in the world – having invested huge sums of money in constructing them.

Technisches Amt was replaced by a new organisational structure – the Chef der Technischen Luftrüstung (Chef TLR for short) under Oberst Ulrich Diesing. This effectively gave Knemeyer, on behalf of the Oberkommando der Luftwaffe (OKL – Luftwaffe high command), and with the blessing of Speer's deputy Hauptdienstleiter Karl-Otto Saur, carte blanche to draft ambitious specifications which drew on the full range of technological advances made in Germany since the beginning of the war. Because although Milch had scarcely utilised it, a huge amount of ground-breaking aviation research had been carried out during his time in office.

By the summer of 1943 there were no fewer than 58 German wind tunnel facilities in operation – 10 shared between the Aerodynamische Versuchsanstalt (AVA) and Kaiser-Wilhelm Institut (KWI) at Göttingen, five at the Deutsche Versuchsanstalt für Luftfahrt (DVL) in Berlin, one at the Deutsche Forschungsanstalt für Segelflug (DFS) at Darmstadt, seven at the Luftfahrtforschungsanstalt Hermann Göring (LFA) at Braunschweig, six at the Forschungsanstant Graf Zeppelin (FGZ) in Stuttgart, 19 at various technical schools, one at Blohm & Voss in Hamburg, one at Dornier in Friedrichshafen, one at Focke-Achgelis in Laupheim, one at Focke-Wulf in Kirchorsten, two at Heinkel in Marienehe, three at Junkers in Dessau and one at Messerschmitt in Augsburg. Fifteen of these generated an airflow fast enough for transsonic or supersonic research.[3]

- **July 5, 1944** — Focke-Wulf updates its 'Flitzer' design with a new description – Baubeschreibung Nr. 280. The X6 twin-ramjet fighter is fully described in Baubeschreibung Nr. 283 Strahlrohrjäger.
- **July 6, 1944** — Heinkel produces the first design in a new jet fighter development series – P 1073.
- **July 11, 1944** — A Focke-Wulf memo shows that the company has been given the "urgent" task to prepare for the design, prototyping and series production of a new jet fighter that will be ready to fly on March 1, 1945. It is implied but not explicitly stated that the aircraft should have a single jet engine rather than two.
- **July 16, 1944** — Erich Bachem sketches a tubular vertical-launch rocket-propelled interceptor based on a concept provided by his friend Wernher von Braun.
- **July 17, 1944** — Messerschmitt conducts tests to determine how visible the Me 262's jet engines are at night – a step towards a night fighter version of the aircraft.
- **July 24, 1944** — Messerschmitt's Hans Hornung designs a new single-jet fighter as a follow-on to the P 1101 project.
- **August 1, 1944** — Reichsminister Albert Speer's Reichsministerium für Rüstung und Kriegsproduktion assumes full responsibility for Germany's technical air defence. On the same day, the RLM's Technisches Amt is replaced by a new organisational structure – the Chef der Technischen Luftrüstung (Chef TLR for short) under Oberst Ulrich Diesing, reporting to the Luftwaffe high command (Oberkommando der Luftwaffe or OKL), with Knemeyer in charge of the Fl-E department – aircraft development. Also on August 1, Bachem begins work on what will become the Ba 349 Natter rocket-propelled interceptor.
- **August 4, 1944** — Focke-Wulf records a preliminary technical specification for a new bad-weather day and night fighter.
- **August 9, 1944** — Bachem produces the first Natter project report.
- **August 10, 1944** — Messerschmitt produces the earliest known drawing of the P 1104 towed take-off rocket fighter.
- **August 16, 1944** — Heinkel engineer Wilhelm Benz produces the earliest known drawing of a new vertical launch rocket-powered interceptor known as the 'Julia'.
- **August 18, 1944** — Focke-Wulf produces a description of a turboprop-powered version of the 'Flitzer' in Baubeschreibung Nr. 281. The design's in-house nickname is 'Peterle'.
- **August 24, 1944** — Heinkel executives meet with Knemeyer concerning the P 1073 jet fighter project and come away believing that an order for the design is imminent.
- **August 28, 1944** — The Chef TLR issues a new requirement for a three-seater heavy fighter to combat enemy bombers at night and in bad weather to Blohm & Voss, Dornier, Focke-Wulf, Heinkel and Messerschmitt. Design entries can be powered by any number and arrangement of Jumo 222 E/Fs, AS 413s, DB 603Ls or DB 613s but a piston engine/turbojet combination is also to be tested.

ABOVE: Far from embracing jet types as the future, the RLM under Generalfeldmarschall Erhard Milch tended to prioritise the development of high-performance piston engine types such as the Focke-Wulf Ta 152.

There were even another 13 in occupied countries to which the German firms had access (even though, as the Allies broke through their encirclement on the French coast and drove east, and the Soviets pushed ever westwards, they were about to lose it) – two in Czechoslovakia, four in Poland, two in Holland and five in France – though none of these could be regarded as 'high-speed'. And yet more wind tunnels were built in the Reich itself during the last two years of the war at great expense, such as the 100,000hp Ötztal wind tunnel in Austria which was 70% complete when the war ended and which American investigators later valued at between $60 and $70 million.

Across the industry, the number of personnel involved in development work increased from 7,000 in 1943 to 8,000 by 1945. In addition, the budget for aviation research and development soared from 340 million Reichsmarks in 1943 to RM 500 million in 1945.[4]

In July, Knemeyer had given specifications to Messerschmitt, Heinkel and Focke-Wulf for a new jet fighter to replace the Me 262 and presented Blohm & Voss, Dornier and Focke-Wulf with specifications for 'Hochleistungsjäger' – a high performance piston-engined fighter with a pusher -prop configuration. In August he commissioned former Fieseler technical director Erich Bachem, Messerschmitt and Heinkel to begin work on new rocket-propelled interceptors.

• August 30, 1944	Hornung at Messerschmitt produces a refined version of the P 1101 single-jet fighter design.
• September 1944	Dr Rudolf Göthert at Gotha begins work on a flying wing design with the intention of supplanting the Horten brothers' 8-229, which Gotha has been charged with building.
• September 1, 1944	Messerschmitt is commissioned to develop a two-seater night fighter version of the Me 262 and responsibility for the production and ongoing development of the Me 163 B is transferred to Junkers.
• September 2, 1944	First flight of Messerschmitt's rocket-boosted Me 262 C-1.
• September 8–10, 1944	A project comparison meeting is held at the Messerschmitt facility in Oberammergau for four firms working on single-jet fighter designs. Messerschmitt presents the P 1101, Heinkel the P 1073 and Focke-Wulf the 'Flitzer' but Blohm & Voss offers no project.
• September 10, 1944	Just as the project comparison draws to a close, the Chef TLR invites Arado, Blohm & Voss, Fieseler, Focke-Wulf, Heinkel, Junkers, Messerschmitt and Siebel to tender submissions for a new single-jet fighter requirement soon to be designated 'Volksjäger' – meaning 'National Fighter' or more literally 'People's Fighter'. This calls for a lightweight fighter built from cheap materials and powered by a single BMW 003 jet engine.
• September 12, 1944	The first proposal for an interim night fighter based on the standard Ar 234 B is drafted. Focke-Wulf begins working out how to meet the Chef TLR's August 28 night fighter specification, soon beginning a series of mixed propulsion piston engine/turbojet designs.
• September 14, 1944	A meeting is held at the RLM's offices in Berlin and three of the seven firms invited to tender for the 'Volksjäger' requirement present their designs – Arado, Blohm & Voss and Heinkel.
• September 15, 1944	A new main development committee, the Entwicklungshauptkommission Flugzeuge (EHK), chaired by Luftwaffe chief engineer Roluf Lucht, is established by Germany's minister for war production, Albert Speer, to oversee work on new aircraft types. An initial brief description of the Focke-Wulf Triebflügeljäger is produced.
• September 19, 1944	Another meeting to discuss the 'Volksjäger' finalists is held and the three original entries are reviewed alongside new entries from Fieseler, Focke-Wulf, Junkers and Siebel.
• September 23, 1944	Hitler orders Heinkel's P 1073 design into mass production.
• October 1944	Lippisch works intensively on his ramjet-powered P 13 rammer aircraft design.
• October 3, 1944	Heinkel's P 1073 is given the official RLM designation He 162.
• October 5, 1944	Messerschmitt produces a project description for modifying the two-seater Me 262 B-1 trainer to create an interim night fighter.

The cancellation of more superfluous aircraft types, particularly the He 177, freed up further production capacity and Knemeyer seized upon the opportunity, belatedly, to rush a new single-jet fighter design into production. In July 1945, he told British interrogators[5]: "Finally, in the autumn of 1944, one could afford to use the development and production capacity of the aircraft industry for a short time on a 'shot in the dark' project owing to the fact that production of bomber aircraft had been stopped."

He said that the Me 262 "had the task of combating large bomber formations" but was coming off assembly lines in inadequate numbers, which "led to a demand for a single-jet fighter aircraft especially designed for attacking the enemy low-flying aircraft and this demand went through despite bitter opposition from various leading personalities in the aircraft industry and in particular from Professor Messerschmitt and Director Lusser who did not have the necessary insight of the problem as a whole".

The Volksjäger requirement was issued on September 10, 1944, and resulted in the Heinkel He 162. This single-jet aircraft was built around the BMW 003 turbojet, which had been in development throughout the Second World War. The Jumo 004 B was not specified, despite being more powerful, because every unit was needed for the Me 262. The He 162 was designed, developed and brought to mass production in the space of just seven months (or nine depending on when you start counting). And the end product, though it had its flaws, outperformed every Allied piston-engined fighter by a wide margin.

In addition to these various specifications and requirements, Knemeyer set the ball rolling on several other developments too. Consideration was given to ramjet fighter projects, pulsejet fighters, and a host of other projects – all based on innovative technology and all, as it turned out, far too late to ever see service.

Another key consideration in this switch to aircraft propelled by any power plant other than the piston engine was a severe shortage of 95 octane C3 fuel – required by both the DB 600 series and BMW's 801 – towards the end of 1944. Jumo 213-engined fighters could use the slightly more plentiful 87 octane B4 fuel but even this was running out as the USAAF's bomber fleet systematically attacked and destroyed Germany's fuel production facilities, depots and transport infrastructure.

The Jumo 004 and BMW 003 were run on J2 fuel, essentially light diesel oil, which was cheaper and easier to produce than C3 or B4. Pulsejets were also run on very basic unrefined fuel known as E1 and ramjets could be fuelled with small pieces of coal. If more high-performance fighters could be built with engines which were unaffected by America's relentless fuel industry raids, the Luftwaffe could be kept in business.

On September 15, 1944, responsibility for aircraft development changed hands again, this time being placed with another new organisation known as the Entwicklungshauptkommission Flugzeuge (EHK – the main development commission for aircraft). Its membership included some of the most important men in German aviation, such as Willy Messerschmitt, Kurt Tank of Focke-Wulf, Heinrich Hertel of Junkers, Robert Lusser of Fieseler, Walter Blume of Arado, Richard Vogt of Blohm & Voss and Günther Bock of the DVL but crucially it remained under the auspices of Speer's ministry which restricted its authority somewhat.

December 22, 1944, saw the EHK's members each put in charge of a Sonderkommission or 'special commission' with

- **October 6, 1944** Willy Messerschmitt sends Erich Bachem a project outline of the P 1104, suggesting that he redesign his Natter to make it more like the Messerschmitt design.
- **October 11, 1944** Junkers produces an outline description of the Ju 248 – a new horizontal take-off rocket-powered interceptor designed as an evolution of the Me 163 B.
- **October 15, 1944** Heinkel designs a pulsejet-powered version of its 'Julia' interceptor known as the 'Romeo'. Junkers is working on its own pulsejet design, the EF 126 'Elli' at around the same time.
- **October 21, 1944** Junkers is given a contract to begin development of the Ju 248.
- **October 25, 1944** Work begins on the first He 162 prototypes.
- **October 27, 1944** The earliest known drawing of the Messerschmitt P 1106 single-jet fighter is produced.
- **November 1944** Arado begins work on the E 381 'midget fighter' – designed to be carried aloft by an Ar 234 C. Darmstadt students begin construction the Lippisch P 13 V1 as the 'DM-1'.
- **November 2, 1944** Messerschmitt draws up an equipment plan for the series production version of the P 1101 single-jet fighter.
- **November 8, 1944** Focke-Wulf produces a description of its 'Projekt Multhopp' single-jet fighter design, aka 'Huckebein' – Baubeschrebung Nr. 279.
- **November 9, 1944** The DFS begins work on the towed take-off 'Eber' rocket interceptor.
- **November 10, 1944** Blohm & Voss designs a new pulsejet-powered fighter under the designation P 213.01. Messerschmitt produces definitive drawings for the experimental version of the P 1101.
- **November 16, 1944** Heinkel produces a new report detailing the 'Julia' rocket interceptor.
- **November 21–22, 1944** A meeting, possibly of the EHK, discusses designs for the Objektschutzjäger requirement as well as the Ho 229 and Lippisch P 11. On the 21st, Argus writes to Blohm & Voss explaining in detail how the P 213.01 pulsejet design is a complete failure.
- **November 23, 1944** Focke-Wulf produces descriptions for both a combination piston-engine/turbojet night fighter and a twin-turbojet day fighter.
- **November 25, 1944** Lippisch creates a radically different new version of his P 13 ramjet aircraft – the P 13b. Similar in appearance to the P 11/Delta VI, it is clear that this is a fighter rather than a rammer.
- **November 28, 1944** Blohm & Voss produces its first design outline for the P 212 single-jet fighter.
- **December 2, 1944** Junkers continues to work on getting a pair of Jumo 004s to fit inside the Lippisch Delta VI V2.

ABOVE: The Blohm & Voss BV 155 extreme high-altitude interceptor, originally based on the Messerschmitt Bf 109, was commencing prototype testing as the war came to an end. It remained in development due to uncertainties over jet engine performance above 16,000m.

responsibility for a particular aspect of aircraft development. Messerschmitt was put in charge of day fighters, Tank got night fighters, Hertel was given bombers, Lusser got 'special aircraft' and Bock was put in charge of airframe construction. From this point onwards, the organisation headed by each man would take on the task of assessing new designs and work submitted for his area of responsibility.

Months of indecision followed until the final nail in the coffin of German jet aircraft development was driven home by Hitler himself on March 27, 1945, when he personally stripped Speer's ministry of responsibility for developing jet aircraft, effectively dissolving the EHK in the process, and gave it to SS-Obergruppenführer and General of the Waffen-SS Hans Kammler instead.[6] Four days later, he issued a further edict putting Kammler in charge of managing Diesing "until a suitable replacement can be found".

By now though, several of Germany's major aircraft development centres were about to be overrun by enemy forces – bringing Kammler's brief reign to an end.

- **December 4, 1944** The Chef TLR writes to Blohm & Voss, Focke-Wulf, Heinkel, Junkers and Messerschmitt with guidelines for a revision of their single-jet fighter projects.
- **December 6, 1944** The Heinkel He 162 M1 prototype flies for the first time.
- **December 10, 1944** The He 162 M1 is destroyed during testing, killing its pilot.
- **December 15, 1944** Single-jet fighter designs produced by Blohm & Voss, Focke-Wulf, Heinkel, Junkers and Messerschmitt were discussed and compared. No agreement can be reached on the system of calculation that will be used to compare their projected performances. Junkers produces a new description of the Ju 248, now renamed 8-263.
- **December 16, 1944** Junkers produces an outline description of the new horizontal take-off rocket interceptor EF 127 'Walli'.
- **December 19–20, 1944** At a meeting of the EHK nine special commissions are established with responsibility for different aspects of aircraft development. Willy Messerschmitt is to lead the special commission on day fighters and Focke-Wulf's Kurt Tank is put in charge of night fighters. Robert Lusser of Fieseler is to oversee the development of 'special aircraft' including target-defence types. While this meeting is ongoing, the DVL is carrying out a mathematical assessment of the single-jet fighter designs.
- **December 22, 1944** The EHK issues a statement following the December 19–20 meeting which indicates that the Ju 248/Me 263 is chosen for further development. Development of the rocket-enhanced Me 262 is to continue, work on Junkers' EF 127 'Walli' is to be suspended and both Bachem's Natter and Heinkel's 'Julia' are to be cancelled.
- **December 28, 1944** A full project description for the DFS 'Eber' interceptor is produced.
- **January 5, 1945** Göring cancels the Me 163 B due to its low endurance and a lack of rocket fuel.
- **January 11, 1945** The OKL provides its specifications for both a new single-jet day fighter and twin-jet night fighter. There is no specification for a rocket-propelled interceptor. Blohm & Voss produces an updated description of the P 212 single-jet fighter.
- **January 12–13, 1945** A further comparison of single-jet fighter designs takes place.
- **January 13–16, 1945** Extensive assessment of the single-jet designs is carried out by the DVL.
- **January 17, 1945** Heinkel sends its first description of the single-jet P 1078 to Messerschmitt's special commission for day fighters.
- **January 18, 1945** Messerschmitt draws up a project description for a dedicated Me 262-based night fighter – the Me 262 B2.
- **January 24, 1945** Willy Messerschmitt presents a three-hour briefing on the single-jet designs during a meeting of the EHK in Potsdam. The various night fighter designs submitted to Tank's special commission are discussed and it is decided that a new pure-jet specification be issued to manufacturers.

'FREAKS OF IRRATIONALITY'

It has been stated time and again during the postwar period that the German aircraft designers, particularly after the departure of Milch in June 1944, were deluded – that the war was clearly lost and that working on a great quantity and diversity of 'secret project' designs was a waste of valuable time and resources. There should have been a concerted effort to simply produce the Me 262 in greater numbers, it is argued. Indeed, this particular point was advocated by many ex-Luftwaffe pilots after the war.

Alternatively, some have speculated that the German designers, realising that the war was lost, came up with high-tech designs which they knew could never be realised by their own nation in the hope of buying themselves a degree of favour with the victorious Allies, or in order to ensure that they were not drafted and forced to fight on the front line with the Volkssturm auxiliary units.

Yet even by the beginning of 1945 the Allies had not penetrated the borders of Germany itself – the final collapse, when it did come, came swiftly. There was hope that Germany's remaining armies could rally in defence of the Fatherland and keep both the Soviets and the western Allies at bay. In addition, if propaganda was to be believed, new weapons were being developed which might bring these enemies to a halt and stave off final defeat.[7]

Some Germans also believed that the western Allies would, sooner rather than later, realise that the Soviet Union with its aggressive and expansionist communist ideology was the real threat and join forces with them to oppose the Red Tide sweeping in from the East. If some sort of deal to oppose Stalin could be worked out, the Luftwaffe would need new aircraft to oppose the communists.

Beyond this, it should be remembered that very few of the 'secret projects' worked on actually reached the stage where any significant amount of manpower would be required to make them a fully functioning reality. Surviving documents suggest that a lot more time was spent working through complicated equations relating to aerodynamics and performance, and conducting tests on new construction processes, than was spent actually putting pen to paper and drawing pictures of oddly configured aeroplanes.

The German aircraft companies functioned much as their counterparts in other countries did during wartime: they attempted to meet the requirements set down by the air ministry within an allotted timescale. In the process,

ABOVE: It has been suggested that the German aircraft designers had a fair idea that the war was lost by the beginning of 1945 but few had any real tactical awareness of the unfolding situation on the ground – as the Allies approached – and many hoped that technological advances, such as the V-1 flying bomb, would yet succeed in slowing or stopping the Allied advance.

• **January 27, 1945**	New technical specifications are issued for the 'Schlechtwetter- und Nachtjäger' or 'Bad Weather Day- and Night-fighter' requirement. Submitted designs must now be powered by jet engines only.
• **January 31, 1945**	Sänger produces a report on fitting the Me 262 with a pair of ramjet engines for improved performance.
• **February 2, 1945**	The Horten IX V2 makes its first flight.
• **February 7, 1945**	Focke-Wulf outlines the first in its final series of night fighter designs.
• **February 8, 1945**	The Ju 248 V1 makes its first gliding test flight.
• **February 9, 1945**	The first five production He 162s are completed.
• **February 12, 1945**	Messerschmitt issues a description of an HeS 011-powered Me 262 B2 night fighter.
• **February 16, 1945**	The Horten 8-229 and Gotha P-60 are compared at a DVL conference in Berlin. The EHK demands that work on the Heinkel 'Julia' be stopped.
• **February 18, 1945**	The Horten IX V2 is destroyed in a crash, killing test pilot Erwin Ziller.
• **February 19, 1945**	Final gliding flight of the Ju 248 V1.
• **February 22, 1945**	Hermann Göring rules that no further piston-engined fighters are to be developed.
• **February 26, 1945**	Further discussions take place regarding submissions to the 'Schlechtwetter- und Nachtjäger' requirement. The DVL publishes its second assessment of the single-jet fighter designs.
• **February 27-28, 1945**	The EHK meets to discuss both the single-jet fighter designs – with the expectation that one or more designs will be chosen for further development – and jet night fighters. It is decided that Focke-Wulf should receive a development contract for its Nr. 279 aircraft aka 'Entwurf Multhopp' aka Huckebein as an 'immediate solution' and Messerschmitt should also receive a contract, for its tailless P 1111 design as a longer-term 'optimal solution'. In addition, the technical specifications for the night fighter requirement are altered, making the designs submitted so far inadequate.
• **March 1945**	Blohm & Voss produces a fully illustrated description for its twin-jet P 215.02 night fighter design.
• **March 1, 1945**	Reimar Horten produces a report describing the Ho 229 as a multirole two-seater aircraft for use as a day and bad weather fighter, heavy fighter, light bomber, reconnaissance aircraft and night fighter. In addition, new technical specifications for the 'Schlechtwetter- und Nachtjäger' requirement are sent out based on the demands of the new General der Jagdflieger, Gordon Gollob.
• **March 5–15, 1945**	Between these dates it is decided that Junkers' EF 128 and Blohm & Voss's P 212 should join Focke-Wulf's Nr. 279 as 'immediate solutions' while a design from Henschel, the P 135, should join the P 1111 as an 'optimal solution'.

ABOVE: Huge technological leaps during the early 1940s, such as the invention of effective ground-to-ground long-range missiles, convinced many Germans that the war might still be won, even at the beginning of 1945.

• **March 7, 1945**	Having abandoned attempts to create a new Ar 234-based night fighter, Arado starts work on an entirely new series of twin-jet night fighters.
• **March 11, 1945**	Gothaer Waggonfabrik produces a report outlining its three P-60 twin-jet fighter designs.
• **March 14, 1945**	Focke-Wulf's technical liaison department reports that the OKL has chosen the designation 8-183 A-1 for the company's new jet fighter design. There will be 14 prototypes and two stress-testing airframes.
• **March 15, 1945**	Dornier produces a description for its Do 335-based Do P 256/1 twin-jet design.
• **March 17, 1945**	Messerschmitt publishes its last known Me 262-based project description of the war, for a three-seater night fighter version of the aircraft.
• **March 19, 1945**	Focke-Wulf produces its last report on jet night fighters, describing its Entwurf II-V designs.
• **March 20–24, 1945**	The EHK meets at Focke-Wulf's Bad Eilsen offices to discuss the ongoing aircraft development programmes. It is decided that the night fighter specification of February 27 is too severe and that the requirement should revert to the spec of January 27. Junkers is awarded a development contract for its EF 128 design.
• **March 23**	Skoda produces a description of the P14 ramjet fighter.
• **March 28, 1945**	Heinkel is still attempting to get the EHK's decision on the 'Julia' reversed – without success.
• **March 30, 1945**	Heinkel produces a report on two versions of the He 162 powered by Argus pulsejets.
• **March 31, 1945**	Heinkel's technical design department is evacuated from Vienna.
• **April 8, 1945**	British forces occupy Focke-Wulf's Bad Eilsen offices.
• **April 14, 1945**	American forces capture the incomplete Ho 229 V3 and V4 prototypes at Friedrichroda, Thuringia.
• **April 22, 1945**	Soviet forces occupy Henschel's offices at Schönefeld, Berlin.
• **April 24, 1945**	American forces occupy Junkers' Dessau headquarters.
• **April 29, 1945**	American forces occupy Messerschmitt's Oberammergau design facilities and Dornier's München-Oberpfaffenhofen plant.
• **Early May 1945**	Arado's Brandenburg facility is overrun by Soviet troops.
• **May 3, 1945**	British forces occupy Blohm & Voss's headquarters at Finkenwärder, Hamburg.

RLM JET FIGHTER REQUIREMENTS WITH COMPETING DESIGNS

JAGDFLUGZEUGE MIT STRAHLTRIEBWERK (JET FIGHTER)
As of late 1939
Messerschmitt P 1065 (Me 262)
Heinkel He 280

1-TL-JÄGER
As of September 8-10, 1944
Heinkel P 1073.01-14
Messerschmitt P 1101 (drawing XVIII/113 version)
Focke-Wulf Nr. 280 ('Flitzer')
Blohm & Voss (no proposal)

As of December 15, 1944
Heinkel 'He 162 development'
Messerschmitt P 1106
Focke-Wulf Nr. 279 ('Huckebein')
Blohm & Voss P 209.02
Blohm & Voss P 212.02
Junkers EF 128

As of January 12, 1945
Heinkel P 1078
Messerschmitt I (P 1106)
Messerschmitt II (P 1110)
Focke-Wulf I (Nr. 279)
Focke-Wulf II (Nr. 30)
Blohm & Voss P 212.03
Junkers EF 128

As of February 27, 1945
Heinkel P 1078
Messerschmitt P 1101
Messerschmitt P 1110
Messerschmitt P 1111
Focke-Wulf I (Nr. 279)
Focke-Wulf II (Nr. 30)
Blohm & Voss P 212.03
Junkers EF 128

VOLKSJÄGER (PEOPLE'S FIGHTER)
As of September 14, 1944
Arado E 580
Blohm & Voss P 211
Heinkel P 1073 (simplified version)

OBJEKTSCHÜTZER
As of November 21-22, 1944
Bachem BP 20 Natter
Heinkel P 1077 'Julia'
8-263 (Junkers Ju 248)
Messerschmitt Me 262 with supplementary rocket propulsion

As of December 19-20, 1944
Bachem BP 20 Natter
Heinkel P 1077 'Julia'
Junkers EF 127 'Walli'
8-263 (Junkers Ju 248)

SCHLECHTWETTER UND NACHTJÄGER (LATER 2-TL-JÄGER)
As of January 20, 1945
Arado Ar 234 C-3N
Arado Ar 234 P1
Dornier Do 335 A-6
Dornier Do 335 B-6
Messerschmitt Me 262 B2

As of February 26, 1945
Blohm & Voss P 215.01
Dornier Do P 252/1
Dorner Do P 254/1
Focke-Wulf Entwurf II
Me 262 two- or three-seater with HeS 011

As of March 20, 1945
Arado I
Arado II
Blohm & Voss P.215.02
Dornier Do P 256
Focke-Wulf Entwurf II
Focke-Wulf Entwurf III
Gotha P-60 C

LORIN JÄGER
Heinkel P 1080
Skoda P14

however, they created some fascinating jet fighter designs which still have the power to fascinate aviation enthusiasts today.

In his highly-acclaimed book *The Warplanes of the Third Reich*, published in 1970, William Green wrote: "Other than in cases where they were actually under construction and attained an advanced stage of development (e.g., the Ju 488, the Messerschmitt P 1101), I have ignored projects. Such were legion and undue importance has been attached by some writers to the more exotic of these project studies; freaks of irrationality on the part of their creators which, for the most part, were no more than paper doodles and not to be taken seriously."

While some of the designs produced in Germany during the war years were certainly outlandish, each was intended to provide solutions to very particular problems. What sort of aerodynamic form would work best, in 1944, for a fighter capable of 1,000km/h but still able to handle in a dogfight? How might a fighter be launched when every concrete runway in Germany is covered in bomb craters? What would be the best way to inflict the maximum possible damage on a well-defended formation of enemy bombers? Could a high-performance fighter be constructed without the use of light alloys (because there aren't any) or petrol for its engine (because there isn't any)?

ABOVE: The Aerodynamische Versuchsanstalt (AVA) at Göttingen worked on a wide variety of different layouts for jet aircraft during the war as shown in this drawing produced for the Allies shortly afterwards. The technology was so new and untested it was unclear which configuration would perform best – leading to a profusion of alternatives.

For the most part, German Second World War jet fighter developments were much more than Green's 'paper doodles' – some of the world's finest engineering minds created them with the goal of achieving aerial supremacy for Germany against increasingly overwhelming odds.

Chapter 1

Early developments

Jet and rocket fighters to 1940

When the victorious Allies shattered Germany's aircraft industry after the First World War, they inadvertently encouraged a new wave of innovation in aviation – leading to the design and manufacture of the world's first jet- and rocket-propelled fighters.

ABOVE: A flight of low-wing rocket-propelled fighters drawn by Alexander Lippisch to illustrate a talk he gave to Hauptmann Oskar Dinort at Döberlitz airfield in December 1934 on the military potential of tailless aircraft.
IOWA STATE UNIVERSITY LIBRARY SPECIAL COLLECTIONS AND UNIVERSITY ARCHIVE

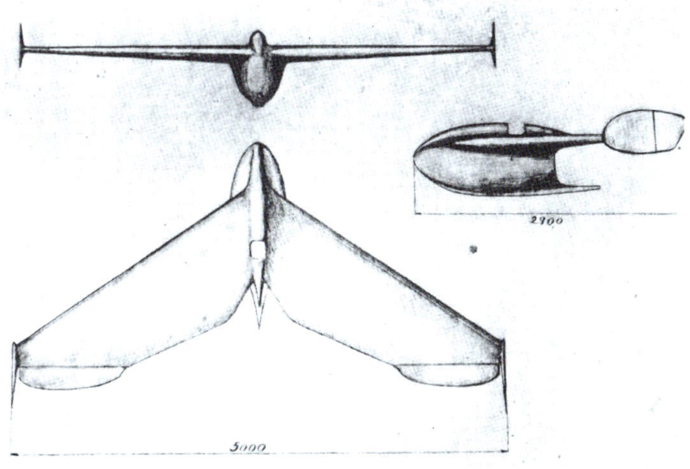

ABOVE: After his tailless Ente aircraft was used as a testbed for Fritz von Opel's solid-fuel rockets in 1928, Lippisch went on to design this rocket-propelled aircraft for him in 1929.

ABOVE: The Delta I, built in 1931, was powered by a 30hp Bristol Cherub engine which came from Theo Croneiss, a director of Bayerische Flugzeugwerke. It was the beginning of an association between Lippisch and the company that would come to fruition eight years later.

RRG AND DFS, 1924–1935

A passion for gliders took hold in Germany soon after the war and can be directly attributed to the imposition of the onerous Treaty of Versailles signed in June 1919. Among its many terms and conditions was a ban on the manufacture and importation of all aircraft and aircraft parts for six months. All German "military and naval aeronautical material" had to be handed over to the Allies – effectively crippling the German aviation industry and leaving thousands of military aviators without aircraft to fly. And the next generation, brought up on tales of German pilots' wartime heroics, had no way of following in their footsteps.

The ban was lifted on building small powered aircraft in 1922 and on larger aircraft in 1926 but Germany's dire economic situation meant large established manufacturers such as Junkers and Albatros struggled to remain afloat. Fokker had left Germany and been re-established in the Netherlands while Gothaer Waggonfabrik and Bayerische Flugzeugwerke both abandoned aviation for several years. A handful of new companies were formed – including Arado in 1921, Heinkel and Rohrbach in 1922 and Focke-Wulf in 1923 – but they remained relatively small.

With powered aircraft in short supply, ex-military aviators and young men eager to fly for the first time turned to gliders instead. Individual enthusiasts and clubs were able to construct their own gliders using widely published plans and a new organisation, the Rhön-Rossitten-Gesellschaft (RRG), was set up in 1924 to regulate and promote the burgeoning glider community. The RRG, led by Professor Walter Georgii, also worked to develop new gliders and to improve existing designs – conducting research on everything from aerodynamics, lightweight materials and control systems to cockpit design, pilot training and the effects of flight on the human body.

At the same time there was growing interest in the possibilities of rocket propulsion. First World War veteran

ABOVE: Lippisch built the twin-engine push-pull Delta IV in 1932 for Gerhard Fieseler, who called it the Fieseler F 3 Wespe. He found it almost uncontrollable in flight, however, and handed it back to the RRG without paying.

and rocket enthusiast Max Valier discussed the idea of using powder rockets to propel an aircraft in correspondence with his friend and fellow enthusiast Hermann Oberth in July 1924. This idea later received widespread publicity in lectures and books, with Junkers becoming interested.

A meeting was held between Valier and Junkers personnel on September 7, 1925, but the company would not consider the viability of rockets for aircraft without a working prototype of the motor Valier had in mind and the idea stalled.

Just over two years later, Valier signed a deal with manufacturer Opel to build a rocket-powered car. Tests of the rocket motor in a standard Opel car commenced in January 1928 and on March 12 it managed to reach 75km/h. The following day, Valier and Fritz von Opel, grandson of the company's founder, visited the RRG and explained to the head of the organisation's technical department, Alexander Lippisch, that they wanted to commission a glider with a canard or 'tail-first' layout. Although the pair were initially rather cagey about exactly why this configuration was necessary, Lippisch soon established that they wanted to fit solid-fuel rockets to the aircraft. The unusual layout would ensure that there would be no tail structure behind the rockets when they were firing.

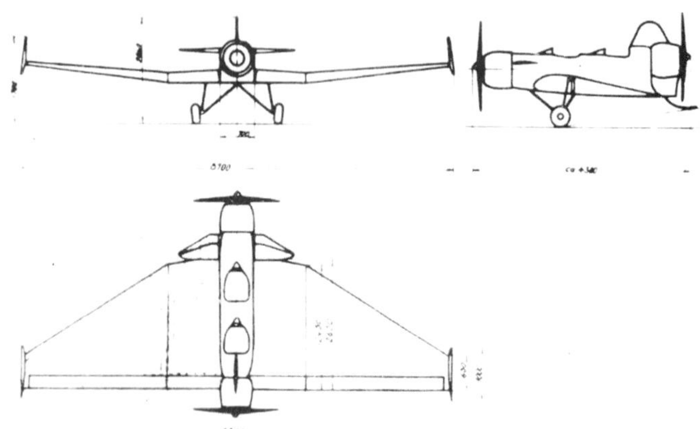

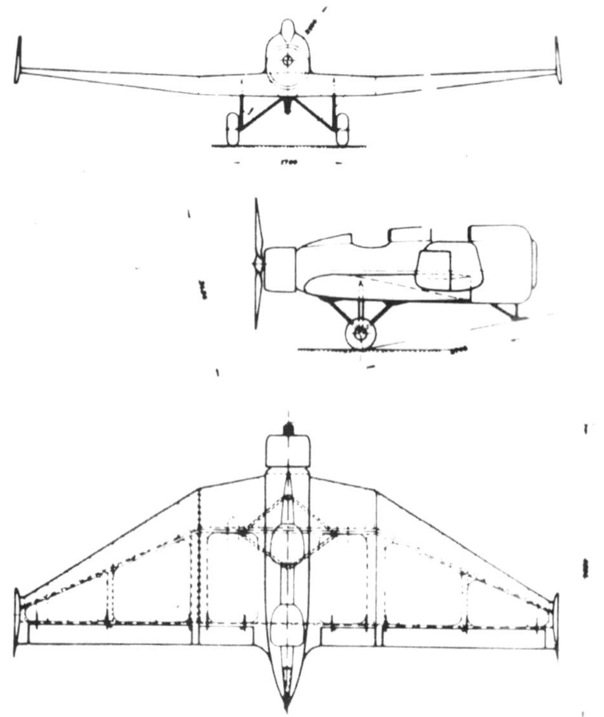

ABOVE: Plan views of the Delta IVa and the aircraft as it was rebuilt after a crash in December 1935 – the Delta IVb.

Lippisch duly designed and had built the straight-winged Ente or 'Duck', which was test flown under rocket power on June 11, 1928,[1] by Fritz Stamer – manager of the RRG's training section and Lippisch's brother-in-law. It was the first rocket-powered manned flight in history but would prove to be the last for some time. Stamer had a lucky escape when one of the rockets blew up and the glider caught fire, von Opel lost interest soon after and Valier was killed in May 1930 when one of his alcohol-fuelled rocket motors exploded.

Lippisch continued to design gliders for the RRG and towards the end of 1929 was commissioned to build a new tailless aircraft for aviation pioneer Hauptmann Hermann Köhl. The result was a high-winged experimental glider with swept leading edges and an entirely straight trailing edge, a wing shape like the triangular Greek letter 'delta'. Unsurprisingly, Lippisch called it the Delta and it was completed in mid-1930.[2]

Testing of the Delta convinced Lippisch that the powered version, which was what Köhl wanted, would perform better with a low wing and work therefore began on another new aircraft, the Delta I, incorporating a Bristol Cherub engine donated by Theodor Croneiss, a director of the Bayerische Flugzeugwerke (BFW).[3] This would be the first in a series of controversial powered Delta designs – controversial because although they offered high manoeuvrability they were reportedly very difficult to control in flight.

In February 1932, First World War fighter ace and stunt pilot Gerhard Fieseler commissioned the RRG to build him three examples of a new tandem-seat delta aircraft, to be powered by a pair of British Pobjoy R engines with one pushing and one pulling, which the RRG knew as the Delta IV but Fieseler himself called the F 3 Wespe. Work on it commenced in April and the first machine was completed in June but Fieseler was deeply disappointed in it – finding it almost uncontrollable.[4]

The Fieseler company itself made numerous alterations to the aircraft, under Lippisch's direction and at its own expense, in an attempt to resolve its problems but eventually Fieseler managed to get out of his contract with the RRG without having to actually buy the aircraft thanks to a technicality. The single completed Delta IV was left with the RRG. Later that year, Ernst Heinkel of the Heinkel company invited Lippisch to a meeting to discuss the potential further development of his delta aircraft but then pulled out of the project.

Following the seismic political shift that occurred when Adolf Hitler became the German chancellor on January 30, 1933, the RRG was reorganised into two separate companies – the Deutschen Luftsportverband (DLV) or 'German Airsport Association' and the Deutsche Forschungsanstalt für Segelflug (DFS) or 'German Research Institute for Gliding'. With fresh funding for aviation research now

ABOVE: The Delta IV stripped to its bare frame after being converted from the IVb into the IVc.

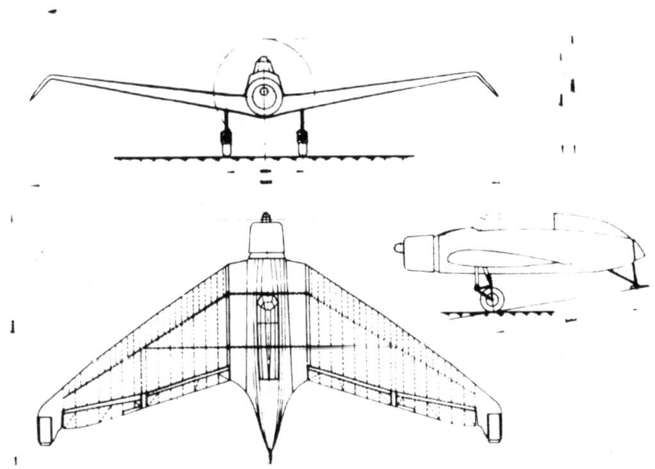

ABOVE: Lippisch planned to rebuild the DFS 39 again with more sharply swept wings as the Delta IVd but this design would remain on the drawing board.

ABOVE: The Delta IV had changed beyond all recognition by the time it became the IVc – subsequently redesignated DFS 39 and seen here in flight.

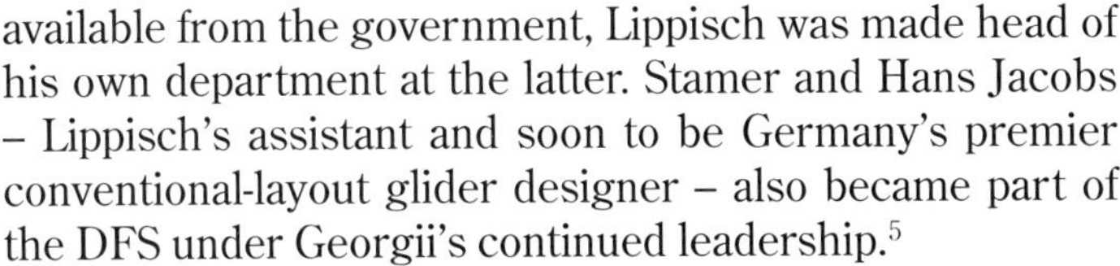

available from the government, Lippisch was made head of his own department at the latter. Stamer and Hans Jacobs – Lippisch's assistant and soon to be Germany's premier conventional-layout glider designer – also became part of the DFS under Georgii's continued leadership.[5]

In December 1934, Lippisch presented a lecture to Hauptmann Oskar Dinort and his men at Döberitz airfield on the military potential of tailless aircraft, illustrating the presentation with a drawing of futuristic-looking rocket-powered fighters.[6] Just over a year later Dinort would become the commanding officer of another flying wing pioneer – Walter Horten. In 1935 the technical department of the DFS was split in two, with Jacobs taking control of everything to do with conventional gliders and Lippisch being given a separate experimental delta and all-wing division.[7]

With the much-altered Delta IV airframe now under his control, and with the DFS's expanded research budget available, Lippisch had it rebuilt without its pusher engine as the Delta IVa. This was found to have significantly improved handling characteristics.

Towards the end of the year the Delta IVa was damaged in a crash and a technical commission was formed from representatives of the Deutsche Versuchsanstalt für Luftfahrt (DVL) or 'German Institute for Aeronautical Research' and the technical office of the RLM to review the performance of Lippisch's delta-wing aircraft. Their report concluded: "The delta type shows no promise and there is no prospect that this type of design will lead to a practical, serviceable aircraft."[8]

However, Georgii came to Lippisch's defence. According to Lippisch himself in his posthumously published 1976 autobiography *Erinnerungen*, or *Memories*: "This situation, which would almost have led to the death of the Delta flight development, was turned around by Professor Georgii who found that he could still make a small contribution to the reconstruction of the Delta IVa."[9]

The Delta IVa was rebuilt as the Delta IVb – although how much of the original remained is questionable since it had entirely new 30° swept-back wings, now with a swept trailing edge too, and a new fuselage made from steel tubing. Trial flights made by test pilot Heini Dittmar resulted in the wings being given a less severe sweep of 23° and the rear fuselage was widened, resulting in the Delta IVc. This was put through official testing as a sports aircraft and approved, receiving the new designation DFS 39.[10]

HEINKEL, 1935–1938

Military research on liquid-fuelled artillery rockets had commenced at the Heereswaffenamt (HWA), the German Army Ordnance Office, in 1931 and less than 12 months later practical experiments were being carried out at the Kummersdorf test range about 30km south of Berlin – the rocketry programme having been joined by 20-year-old mechanical engineering graduate Wernher von Braun.

In 1933, Hans Joachim Pabst von Ohain, aged 22, began theoretical work on gas turbine propulsion, unaware that similar research had been started by Frank Whittle in Britain three years earlier. And in 1934, Dr Herbert Wagner of Junkers undertook some private research into the possibility of a turboprop engine and thereby began work on his own independent turbojet design. Von Braun was by now testing his ground-breaking Aggregat 2 rocket – the development of which would eventually result in the Aggregat 4, otherwise known as the V-2 missile.

ABOVE: Ernst Heinkel Flugzeugwerke founder Ernst Heinkel, right, with his company's chief designer Siegfried Günter. Heinkel was willing to invest heavily in new technology during the late 1930s and regularly attempted to recruit new talent in the field.

ABOVE: Rocket scientist Wernher von Braun joined the Heereswaffenamt, the German Army Ordnance Office, as a civilian employee in 1932. Within three years he would be designing liquid-fuelled rocket motors to power aircraft.

ABOVE: Hans Joachim Pabst von Ohain invented his first turbojet design in 1933 and was recruited by Heinkel in 1936. His engine would power the first jet aircraft to fly – the Heinkel He 178.

In 1935, as the DFS was being reorganised, the chief of development at the RLM's Technical Office Major Wolfram Freiherr von Richthofen became aware of rocket research secretly being conducted at Junkers following an accident at the Dessau factory. He consulted the HWA about this on February 6, 1935, and Junkers was visited by von Braun and RLM representative Hauptmann Zanssen on February 13, 1935.[11]

It was agreed that under strictest secrecy Junkers should work on rocket propulsion for aircraft and missiles for seaplanes. During a meeting on June 27, 1935, von Richthofen echoed Lippisch's lecture of seven months earlier, suggesting that a rocket-propelled fighter could be designed, capable of reaching 12–15km in just one minute to intercept enemy bombers. It was agreed that before such an aircraft could be created, however, an experimental flying testbed was needed.[12]

In September a further agreement was reached between von Braun, the RLM, Junkers and Heinkel that both companies would provide an airframe for testing the HWA's motors. Junkers gave a Junior A 50 ci and Ernst Heinkel himself pledged to provide the wingless fuselage of a He 112 fighter for trial fittings. Heinkel technical director Heinrich Hertel appointed engineer Walter Künzel to lead a small team working on the project at Kummersdorf. The Walter company of Kiel, which was building its own liquid-fuelled rocket motors, was brought on board too.[13]

Meanwhile, von Ohain's turbojet received a German patent on November 9, 1935, and the first model of it was constructed. The inventor met Dr Ernst Heinkel at his home in Warnemünde, a district of Rostock, in March 1936 and the aviation firm boss agreed to both give him a job and pay him royalties on his design.

The rocket team's efforts initially focused on fitting a rocket motor to the A 50 ci but towards the end of 1936 efforts were made to fit one of von Braun's significantly more powerful motors into the He 112 fuselage. It was decided that two complete He 112s – the V3 and V4 – would be fitted with a Walter motor and a von Braun motor respectively. At around the same time, Heinkel was finally commissioned by the RLM to build an airframe specifically designed for rocket propulsion. The company designated this project P 1033 and four aircraft were to be built – two with von Braun motors and two with Walter motors.

The first P 1033 mock-up was inspected on January 16, 1937. It was to be an extremely small aircraft just 6m long and with a wingspan of 5m. The 1.5m-long nose section, containing a cockpit that would be designed around the dimensions of the chosen pilot, was a detachable escape capsule. Work on converting the He 112 V4 to rocket power had also been largely completed by early 1937 and test firings commenced in February. The P 1033 dedicated rocketplane had by now been given the formal RLM number

ABOVE: Having been designed for piston-engine propulsion, the DFS 194's structure was not strong enough to cope with speeds above 550km/h but fitted with a rocket motor it nevertheless contributed valuable test data.

He 176.[14] That same month von Ohain's first demonstration jet engine, the HeS 1, ran successfully for the first time. The realisation that the jet engine was a viable powerplant prompted Heinkel's projects team to commence work on a variety of single- and twin-jet project designs.

The Luftwaffe experimental station at Rechlin chose Erich Warsitz as the rocket He 112 pilot and he made his first flight in the He 112 V4 with rocket thrust on June 3, 1937.[15] By this time, it had been decided that the Heinkel jet aircraft should have only a single engine positioned in the centre of its fuselage, with a nose intake and a single exhaust outlet beneath the fin at the rear of the fuselage. Testing of the He 112 V3 commenced on November 13, 1937.[16] Completing von Braun's HWA engine took longer so efforts were largely concentrated on the He 112 V4 and He 176 with their simpler Walter motors.

By the end of 1937, Heinkel was deeply invested in both jet and rocket technology. The company was in the ascendant – working alongside both the HWA and the RLM on a range of rocketplanes and pressing ahead with its highly innovative turbojet aircraft design. And the RLM seems to have been content, at this stage, in allowing all this pioneering work to remain concentrated at a single aircraft manufacturer.

DFS, 1936–1938

During the winter of 1936–37 the RLM placed a contract with the DFS for an experimental mid-engined pusher-prop tailless design that was expected to form the basis for a fighter – the DFS 194. With work on this project well under way, Alexander Lippisch was approached during the autumn of 1937 by RLM representative Dr Hermann Lorenz, who technically had some oversight of work carried out by the DFS, and Dr Joseph Jennissen, an RLM technical liaison officer. Lippisch would later recall:[17] "Dr Lorenz and I did not get on very well. I was always doing something that was not on the agenda. Actually, it was not my fault: I was used to applying newly acquired knowledge immediately in order to get ahead quickly and, of course, this did not fit into the well-ordered scheme of things in the research department.

"In the autumn of 1937, Dr Lorenz, accompanied by Dr Jennissen, came to Darmstadt to make an inspection. I showed the gentlemen into the workshop where work was being carried out on the new DFS 194. The DFS 39 was also standing there. Dr Lorenz asked me which aircraft had the better flying characteristics. 'That's unquestionably the DFS 39, Herr Dr Lorenz.' 'Tell me, Herr Lippisch, could you build us a second aircraft of this sort with exactly the same wings but with a different fuselage?' 'Certainly we can do that.' 'We want to test a new engine and this must be installed in the rear fuselage with the cockpit forward, not like in the layout of the DFS 39.'

"As I had already heard similar talk from Fritz von Opel back in 1928 the penny dropped immediately and I replied dryly, 'You want to try out a new rocket engine?' Lorenz would have liked to have strangled me for revealing this state secret. 'What on earth made you shout that out here among all these people and how do you know anything about it anyway? This is quite outrageous!' 'I don't know anything about it at all, but we were flying with rocket power ten years ago,' I managed to gasp when I had got my breath back. 'I must ask you not to use the word "rocket" in public again,' Dr Lorenz said very sharply and firmly. I felt very sheepish as we retired to the privacy of my office.

"After all the doors and windows had been barred, I was informed in a whisper that it was planned to award a contract for a research aircraft for testing 'jet' propulsion at high speeds. The RLM development department had ordered such a bird from Heinkel. I was to design an aerodynamically improved DFS 39 for the research department. This design was designated Project X with the security classification 'burn after reading' and was the start of work on the project which was to become the Me 163."

The fuselage was to be constructed by Heinkel, since it already had experience of installing rocket motors in airframes, while the wings would be designed and built by the DFS. Wind tunnel testing and free-flying model tests were carried out by Lippisch, his department's aerodynamicist Josef Hubert and engineer Rudolf Rentel. The design drawings were made by Fritz Krämer. According to Lippisch's own account "ample" funds were made available for this work.

The DFS 194 airframe was completed during the winter of 1937/38 but installing its Argus petrol engine proved to be a "headache" due to its complicated cooling system. Project X was a design in progress.

DFS AND MESSERSCHMITT, 1938–1939

Just a few months into 1938, events were set in motion that would result in Lippisch leaving the DFS and joining Messerschmitt AG at the beginning of 1939. Several different versions of exactly why Lippisch made this surprising move but by far the most detailed and nuanced is offered by *Erinnerungen*. In order to understand what happened, it is necessary to backtrack a little.

By 1936 all was not well between Lippisch, Jacobs and Stamer. Jacobs felt that Lippisch was wasting the DFS's time and money on the dead-end field of tailless aircraft design and Stamer was dissatisfied with the way Lippisch treated his wife. By his own admission, thanks to his workaholic lifestyle, Lippisch seldom had much time for his wife Katharina – Stamer's younger sister. Neither Jacobs nor Stamer, caught up in the wave of nationalism then sweeping Germany, was impressed when Lippisch began to actively pursue a professional relationship with aircraft designers abroad – particularly in England. He even used his influence within the DFS to set up an office where foreign works on aerodynamics could be translated into German.

Matters came to a head in March 1938, just as the installation of the DFS 194's piston engine was being completed, when Katharina died aged 34.[18] Lippisch was 43 and now had his 10-year-old son Hangwind and one-year-old Jürgen-Günther to look after. But rather than spending more time with his family, and mourning his wife, he redoubled his efforts at work and threw himself into the development of new smoke tunnel technology. He also became increasingly friendly with a young female colleague – 23-year-old Gertrude Knoblauch. For Stamer in particular, this appears to have been the last straw. During the summer of 1938, together with Jacobs, he persuaded Professor Georgii that it was time to put Lippisch in his place.

Unaware of what was going on behind his back, Lippisch continued to refine the aerodynamic form of Project X. At the same time, it was decided that the whole DFS organisation should be moved east to Ainring near the Austrian border. With war in the west becoming an increasing possibility, the DFS's base at Griesheim, west of Darmstadt, was strategically vulnerable.

Now Stamer, Jacobs and Georgii saw an opportunity. They drew up plans for a restructuring of the DFS, coinciding with the upheaval of the move, which would see Lippisch's role dramatically decreased. According to *Erinnerungen*: "Thus a shift and a reorganisation of the DFS was worked out, but I was not a little attracted to it because my former friends Stamer and Jacobs now convinced Professor Georgii that my all-wing division would no longer be of any interest at the outbreak of the war.

"They had stripped me out of the organisation and left me only a small model building department. This plan was to be concealed from me, but the secretaries still had so much character that they could not take part in this ignoble contempt for my work. So I learned through back-channels of all these machinations and had nothing more to do than to put myself in my car and drive the 1,000km to Rechlin in one go, to get in touch with my friends from the Darmstadt group and from the gliding flight and I then went to Berlin, where the attitude was 'thank God for your work', something that I thought was quite different from the DFS organisation.

"At first I contacted Heinkel, but soon found out that he was really interested only in the order to test the high-speed aircraft, not in me and my co-workers. So I went to Messerschmitt, whom I knew quite well from the years on the Rhön. There was also director Croneiss, who was very much in our favour and helped us with the Delta l.

"So I agreed with Messerschmitt to move from the DFS to Messerschmitt AG in Augsburg at the beginning of 1939, and to take a number of my best employees into a newly founded department. All this I did without the knowledge of the clique at the DFS, and I then submitted the final plan, sanctioned by Berlin, to them. There were long faces, for such a move had not been expected on my part."[9]

Lippisch had neatly sidestepped an attempt to derail his career with the help of Messerschmitt director Theo Croniess and chairman Willy Messerschmitt himself. By this time Messerschmitt was riding high on the success of the Bf 109, which was becoming the Luftwaffe's standard fighter. The company was in line for some very lucrative

ABOVE: Lippisch, right, believed that his brother-in-law Fritz Stamer, left, was behind efforts to deprive him of his influence at the DFS in the wake of his wife's – Stamer's sister's – death. It was these machinations which evidently prompted his move to Messerschmitt, rather than anything to do with his work.

contracts and could afford to make an investment in some additional research and development muscle.

Georgii allowed Lippisch to take his DFS 39 and DFS 194 away with him. He recalled: "On January 2, 1939, at 8 o'clock in the morning, I arrived at Messerschmitt with my car from Darmstadt, where a group of my co-workers were already waiting for me. So we were first stamped in as Messerschmitt employees with many signatures on forms, all possible examinations, and finally we were correct members of Messerschmitt AG in Augsburg-Haunstetten.

"Now we were a sort of secret department, because we were supposed to build and test the rocket-driven experimental aircraft, whose aerodynamics we had already worked on at DFS. Of course the rocket propulsion was strictly secret, so that the whole department had to be housed in a special part of the building and secured by appropriate positions at the doors. Since no other room was available at the moment, we moved into a large screening room on the top floor of the main building."[0]

The newly formed Abteilung L or 'Department L' was given special treatment at first too: "In the morning a waitress with a large tray would be admitted to our secret room. We would then have the Bavarian-style quite rich breakfast. This went on so well until the commercial director, Mr Kokothaki, accidentally rode in the same elevator as the waitress, with the large tray full of treats for Abteilung L. This resulted in a somewhat excited discussion and my gentlemen then had to eat their breakfast in the canteen."

Lippisch was also pleased that he no longer had to deal with Dr Lorenz at the RLM, now being allocated Hans-Martin Antz, an advisor seconded to Section LC 2/III of the RLM's technical office.

HEINKEL, 1938–1939

Lack of space within the He 176's fuselage caused difficulties for Heinkel throughout the early months of 1938 while a 180mm-long steel model of the aircraft was tested in the DVL's wind tunnel up to Mach 0.9.[21] The V1 was completed during the early summer and sent to the Aerodynamische Versuchsanstalt (AVA) at Göttingen for full-scale wind tunnel testing from July 9 to July 13. Ground trials commenced during the autumn at the Peenemünde test centre.

Design work on the jet-propelled He 178 V1 had been completed by now and a mock-up was then made and inspected on August 29 before construction of the first prototype began.[22]

Antz produced a memo entitled 'Present status and future development work in the area of jet engine rapid flight' on October 14, 1938,[23] which offers an overview of Heinkel's rocket and jet aircraft up to that point.

It states: "According to our current knowledge there is a limit to the performance increases possible for aircraft with internal combustion engines and propellers, partly due to engine size and weight, and partly due to the deterioration of propeller efficiency at high Mach numbers. This limit may be in the vicinity of 800–850km/h.

"In addition, increases in performance up to and above the range of the speed of sound are possible when using special thrusters according to the jet principle. Therefore only aircraft with this special drive are discussed in this report."

Antz then reviews progress on the He 176: "Present status. Ernst Heinkel Flugzeugwerke [EHF] is the first and so far only company to have been entrusted with the development of a fast single-seater which has a special thruster according to the jet principle. Four prototype aircraft of the model He 176 were ordered: He 176 V1 with Walter engine, initially 600kg, later 1,000kg thrust at a maximum velocity of 1,000km/h. Flight duration about two minutes. He 176 V2 with Braun's engine. He 176 V3 and V4 with Walter engine.

"Work on the He 176 V1 has progressed so far that the aircraft will be ready for flight after the arrival of the engine in about one and a half to two months. On the fuselage side, some changes have to be made for stability reasons; for example lowering the tailplane (initially on the V2–V4, later also on V1) and bringing forward the centre of gravity.

"Furthermore, the testing of the droppable cockpit with parachute is still ongoing. It must be emphasised here that these four prototypes can only be experimental aircraft – gathering information on stability and flight characteristics of the aircraft itself and on the engine side testing this type of engine in flight.

"The poor efficiency of the engine with a fuel consumption of 8g per kg of thrust per second shows that this pattern with the current motor cannot be used for a front-line application. Fitment of weapons was therefore ignored. The only use which could be considered would be that of an interceptor – but only because of the large climbing speed (up to 10,000m in two minutes).

"However, the work done so far and wind tunnel investigations have provided a number of insights that can be used in the construction of future special-purpose fast aircraft. Also under development is a tailless project of the DFS (DFS V3), which has not yet gone beyond project-related investigations and some wind tunnel measurements. In the context of this project, the second part of this report is followed by a specific proposal."

Under a heading of 'future development', Antz explains that little is known about flight conditions close to the speed of sound. He says developing an engine suitable for high-speed flight is going to be easier than developing a suitable airframe to house it. He goes on: "For EHF, for example, under internal development (Dr von Ohain) on the test stand is an engine, which at 600kg of thrust has a fuel consumption (gasoline) of 0.5g per kg of thrust per sec at a forward speed of approx. 250m per sec = 900km/h. This engine weighs 300kg and has an outside diameter of 700mm. In order to achieve a speed of 900km/h with a normal combustion engine an output of approx. 3,000hp would be required.

"The jet engine, which is still in the early stages of development, still suffers from poor internal efficiency, i.e. high specific consumption. If, however, the above data is used as the basis of a project for a fast single-seater, the result is a flight duration of one hour."

Antz then lists the "major benefits" of using jet propulsion: "1) Small dimensions of the engine, which is also installed behind the pilot's seat and only requires an air intake from the fuselage tip to the actual engine. 2) The related and for the fighter highly desirable full vision canopy without the disturbance of the engine housing and propeller. 3) Concentration of firearms in the nose is facilitated, which gives a better fire effect. 4) The elimination of the torque of the propeller, which is very considerable in a normal engine of the corresponding power, improves stability and flight characteristics. 5) The high percentage of fuel weight in the total aircraft weight brings a significant reduction in landing weight, i.e. a reduction in landing speed."

As a drawback, Antz gives limited range due to high fuel consumption. But he adds: "The above-mentioned engine is merely an imperfect example of the possibilities that will be offered for high-speed flight in the future. Due to the development starting with LC 8 an increase of the jet velocities in and beyond the sound range is to be expected. Unfortunately, it must be said that the conditions for the correct use of such an engine are not yet given on the side of the airframe."

Ten problems to be overcome in the area of airframe design and testing are then listed, ranging from a lack of high-speed wind tunnel facilities to the question of whether a landing skid or full landing gear would be necessary, given the need to keep wing profiles thin and therefore free of bulky wheels. He says the flying wing form with the smallest possible surface area would be advantageous, that a prone pilot position might be preferable and that a pressure cabin escape capsule would be essential.

As if attempting to convince himself, he says: "It should be emphasised again that these are not utopian ideas or probabilities lying in the future, but based on our current knowledge about already possible developments in this fast-moving field."

ABOVE: One of only two known photographs showing the Heinkel He 176 V1 rocketplane. The aircraft's precarious centre of gravity, diminutive size and narrow track undercarriage made it difficult to land – hence the nosewheel and protective loops at the wingtips.

Antz then outlines a series of 'next steps' – expediting completion of wind tunnels, conducting drop tests of aerodynamic forms, increasing the development priority of high-speed research, construction of test stands for jet engines at the Peenemünde test centre and "extension of the development of high-speed aircraft to another aircraft factory". On this latter point, he says: "So far, only the EHF firm was entrusted, partly for reasons of secrecy. However, it now seems appropriate to entrust another aircraft factory with the development of fast single-seaters. This is due to the preliminary work at Messerschmitt AG.

"In this connection, it is further proposed to remove the Lippisch group from the DFS and to add or integrate it into Messerschmitt. Within DFS, further development of the tailless, fast single-seaters is not possible. Lippisch has already made his decision to go into industry. Negotiations are pending with Prof. Heinkel – on the other hand, Prof. Messerschmitt has repeatedly invited Mr. L. to come to him. Mr. L. asks for a decision by the RLM in this matter. Since it is still completely unclear which type of aircraft in the speed-of-sound and faster-than-sound range is most suitable, and there is much to be said for the tailless low-wing aircraft with low aspect ratio and large wing depth, it seems appropriate that a parallel development with tailless monoplanes (Proj. DFS 39 V3) be held in another factory. An extension of the development to a second aircraft plant seems moreover appropriate in order to avoid a monopoly of EHF in this area."

The He 176 V1 underwent numerous further detail changes during its ground testing and efforts were made to test the aircraft's rudder by towing it behind a supercharged 7.6 litre Mercedes car. Unfortunately, it was only possible to reach 100km/h on Peenemünde's grass airfield. The tow car and aircraft were then taken to the beach at Usedom, where 155km/h was reached before the car became bogged down in wet sand and had to be recovered using a tractor.

Among the modifications made during this time appears to have been the addition of a nosewheel. The aircraft was designed as a tailsitter with a retractable tail skid but it became clear that the airframe's awkward centre of gravity made it prudent to allow for a nose-down landing. According to Warsitz more than 100 short hops and flights were made in the He 176 V1 between the autumn of 1938 and the spring of 1939.[24]

JUNKERS, 1938–1939

Wagner was made head of special developments at Junkers in May 1935 and was now in a position to further his own interest in turboprops. In April 1936, he appointed engineer Max Adolf Müller to lead a small turboprop team at Junkers' Magdeburger Werkzeugmaschinenfabrik subsidiary. They began by investigating heat-resistant materials, combustion chambers and axial flow compressors and by the summer of 1938 were working on an experimental turbojet engine with a 12-stage axial flow compressor and two-stage turbine, the turboprop, a stationary gas turbine and other designs.

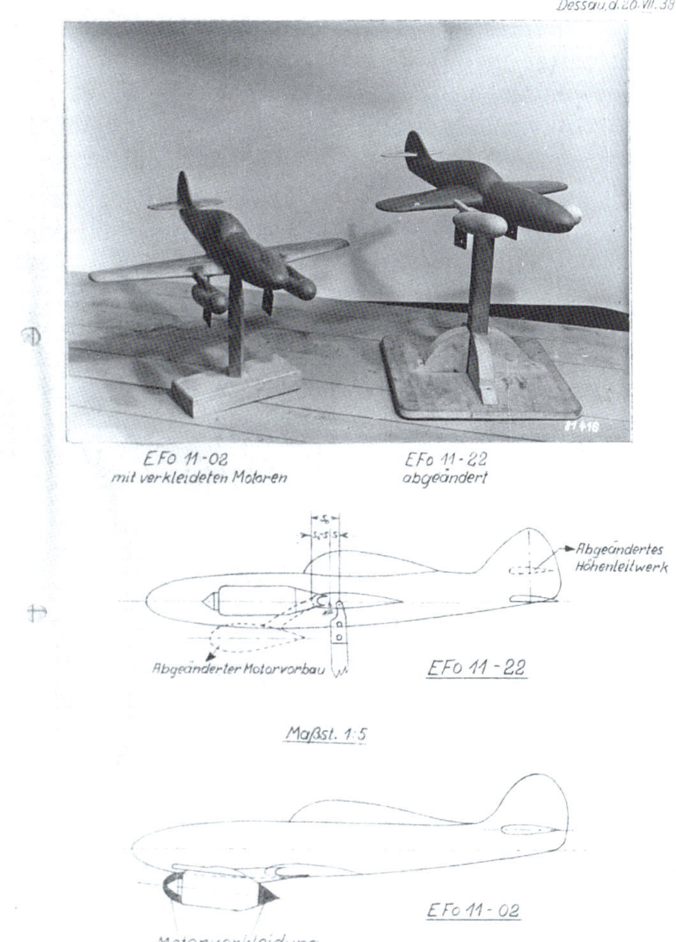

ABOVE: The Junkers EFo 11-02 and EFo-22 from a document dated July 20, 1938. The designs are similar but their wings are differently positioned and the -02's cockpit canopy is more blended into the rear fuselage.

Around the beginning of 1938 – perhaps nine or 10 months after von Ohain's first successful engine test – Junkers began designing aerodynamic forms for aircraft that could be powered by jet engines. A variety of wind tunnel models were made and tested as part of the 'EFo' series. Judging by Hans-Martin Antz's October 1938 report, the RLM was not made aware of this work.

While the full extent of the EFo series is unknown, it appears to have encompassed numerous designs for jets fighters, bombers and racers intended for attempts at breaking the world air speed record. Surviving drawings and associated documents date from between July 20 and December 16, 1938,[25] but the series evidently began earlier in the year and continued on into 1939.

The earliest known drawing from the series is EFo 08 – which would appear to be a large aircraft with two engines under each wing. EFo 12, 13 and 15 have the same basic configuration with variations in how the engines are positioned, wing size and undercarriage arrangement.

EFo 09 is a bizarre design with ten turbojets clustered around its fuselage forward of large deep-chord wings. No details are known for certain about its role or even its scale. Nothing is known of EFo 10 but EFo 11 covers a range of different twin-engine designs. The EFo 11-02 and -22 are similar in having their engines projecting forwards from their straight wings and each has a streamlined but otherwise conventional-looking fuselage. But while the -02 has a low-wing arrangement with its tailplanes in a median position, the -022 has a mid-wing and low tailplane.

The EFo 11-03 has an altogether more radical character with the cockpit apparently integrated into a large fin to the rear, a pointed fuselage and small wings. Unlike preceding models, EFo 11-04 has its engines attached to its fuselage – one on either side of the fuselage forward of the wings. It also has a high wing position. EFo 11-05 of July 29, 1938, has a low wing, mid-tailplane arrangement like the -02 but with a pronounced fuselage 'spine' and a more triangular tailfin.

EFo 18 appears to be another approach to the basic EFo 11 design but with four turbojets positioned beneath its wings rather than two. EFo 17 and 19 are further variations on the twin-engine layout but with their engines integrated into their wings rather than projecting away from them. The EFo 17 has a very narrow fuselage with its engines attached to the underside of its wings while the EFo 19 has a broad flat fuselage with its engines forming a structural part of its wings.

EFo 14/1 is arguably the most radical design in the series. It appears to be a single-jet aircraft built around a large turbojet – certainly larger than the units planned for the other designs. The turbojet's housing makes up most of the fuselage, with the wings attached on either side and the cockpit positioned above the exhaust. The small rear fuselage and tail fin extends backwards from the cockpit.

Assuming this is what it appears to be, it would be perhaps the earliest known German single jet aircraft design after the He 178 itself. Alexander Lippisch's Messerschmitt P 01-116, the next earliest, would not be committed to paper until April 1939.

This early work at Junkers came to an abrupt end just as Lippisch was getting started at Messerschmitt. In April 1939, Wagner was forced to transfer Müller's team to the main Junkers engine works under Professor Otto Mader. Mader was dismayed by the sheer number of different engines Müller was working on and said just one should be chosen for ongoing development. This led Müller and most of his team to resign in May and the following month they joined Heinkel on increased salaries.

Since August 1938, supercharger specialist Anselm Franz had been tasked with reviewing Müller's work and he reported to Mader in July 1939 that none of Müller's designs were worth continuing. Instead, Franz put forward a new design known as the T 1 – which would later become the Jumo 004.

ABOVE: Junkers' work on turbojets commenced in 1936 when a research team was formed by Herbert Wagner, the company's head of special developments.

ABOVE LEFT: The bizarre Junkers EFo 09. Little is known about this design - even its scale. It could be a small aircraft with ten tiny turbojets or it could perhaps be a large aircraft similar in size to the He 111. It is unclear too whether the lines depicted on the nose are an original part of the drawing and if so what they represent.

ABOVE RIGHT: A design thought to belong to the EFo 11 series - though its sub-designation, if it has one, is unknown.

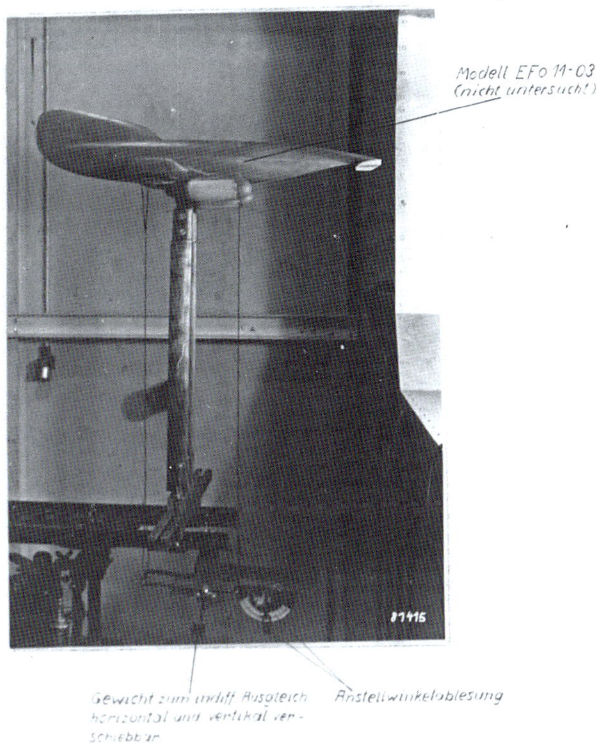

ABOVE: The radical EFo 11-03 of July 20, 1938, was not developed further than this wind tunnel model.

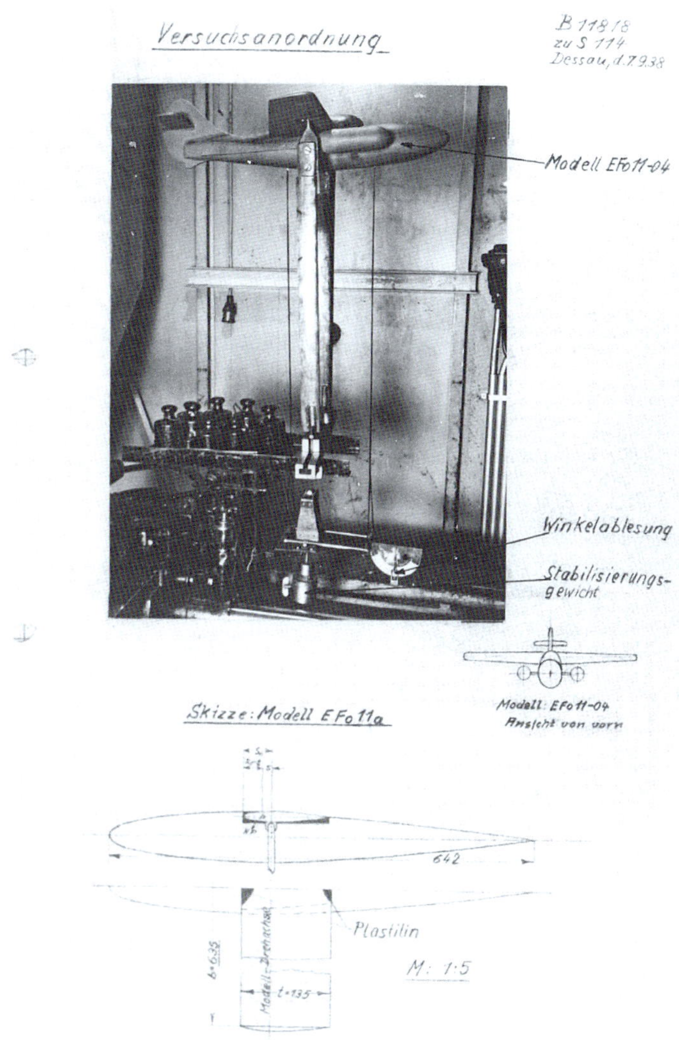

ABOVE: A later development from the EFo 11 series – the 11-04 – had a high wing, a reshaped fuselage and fin, and its engines attached to the fuselage at the front. It is dated September 9, 1938 – later than the EFo 11-05.

HEINKEL, 1939–1940

Efforts to complete the He 176 V2 with its von Braun engine continued and Warsitz made the first full flight of the V1 on June 15, 1939. He demonstrated the aircraft in front of Adolf Hitler on July 3 at Rechlin-Roggentin.[26]

The He 178 V1 was completed in June 1939 but at the same time Heinkel was already working on the design of its jet fighter, dubbed the He 180.[27] This aircraft was to be powered by a pair of under-wing turbojets and would have a tricycle undercarriage. During a company meeting on June 20, 1939, it was decided that mock-ups of two different He 180 configurations would be built. The main difference was apparently the position of the cockpit. In one version it would be in the aircraft's nose while in the other it would be in the centre of the fuselage. Either way, the aircraft would be armed with three or four MG 151s.[28]

Ground trials of the He 178 V1 now took place at the Heinkel factory airfield of Rostock-Marienehe. Design work on the He 180 was completed on August 10 and Warsitz, having switched over from the He 176 V1, made the world's first flight in a jet-propelled aircraft when he took off in the He 178 V1 on the morning of August 27, 1939.

All work on the He 176 project ceased on September 12, 1939, on the orders of Ernst Udet – the Second World War having begun less than two weeks earlier.[29]

A mock-up of the He 180's forward section, including the cockpit, was inspected on September 26, 1939 though by this time the aircraft had been re-designated He 280. Two days later, the RLM decided that 20 examples of the He 280 should be constructed with the first to be delivered on April 20, 1940.[0]

Work on the He 178 continued until early November when the engine was removed and the aircraft was disassembled. A second prototype, the He 178 V2, was never finished and neither airframe survived the war – but testing of von Ohain's engines continued attached to the fuselages of He 111s.[31]

Weapons trials commenced using the He 280 mock-up on April 30, 1940, and the landing gear for the first

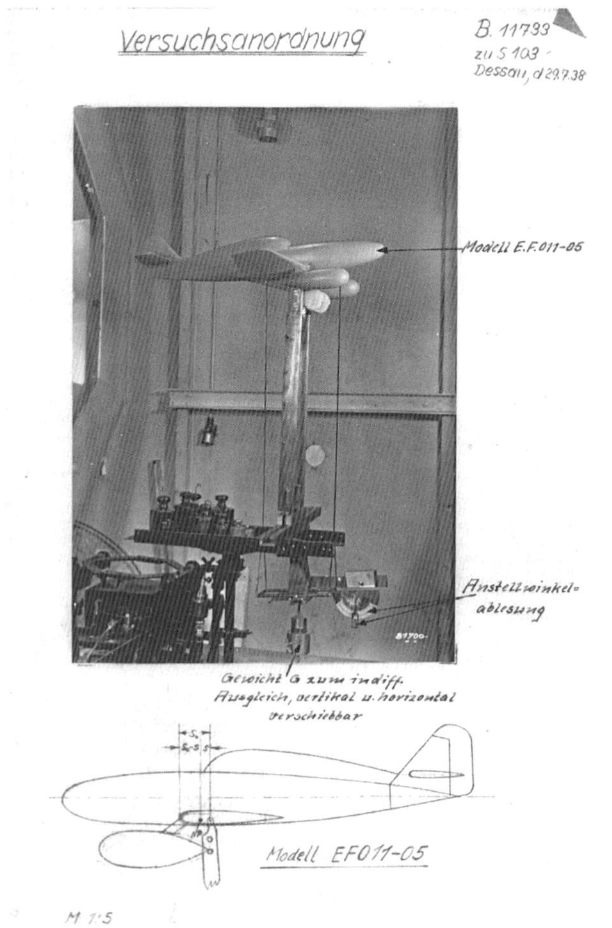

Versuchsanordnung

B. 11733
zu S 103
Dessau, d 29.7.38

Modell E.F. 011-05

Anstellwinkel-ablesung

Gewicht G zum indiff. Ausgleich, vertikal u. horizontal verschiebbar

Modell EF011-05

M 1:5

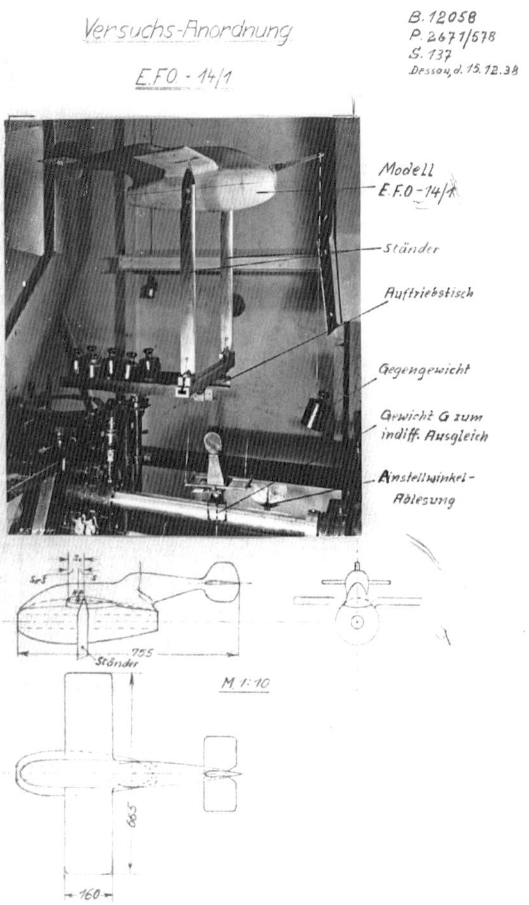

Versuchs-Anordnung

E.F.O. - 14/1

B. 12058
P. 2671/578
S. 137
Dessau, d. 15.12.38

Modell E.F.O-14/1

Ständer

Auftriebstisch

Gegengewicht

Gewicht G zum indiff. Ausgleich

Anstellwinkel-Ablesung

755

Ständer

M 1:10

665

160

ABOVE LEFT: The **EFo 11-05** of July 29, 1938. This design features a taller tailfin than earlier 11-series models and a raised spine extending back from the cockpit. **ABOVE RIGHT:** Dated December 15, 1938, the **EFo 14/1** appears to be a single-jet aircraft. The scale is 1:10 so the fuselage length appears to be 7.55m and the wingspan 6.65m. By way of comparison, the single-jet **He 178** was 7.48m long with a wingspan of 7.2m. Presumably the undercarriage would have folded up into the engine housing. The **EFo 14/1** would appear to have been a contemporary of the **He 178** – making it one of the two earliest-known single-jet aircraft designs in history.

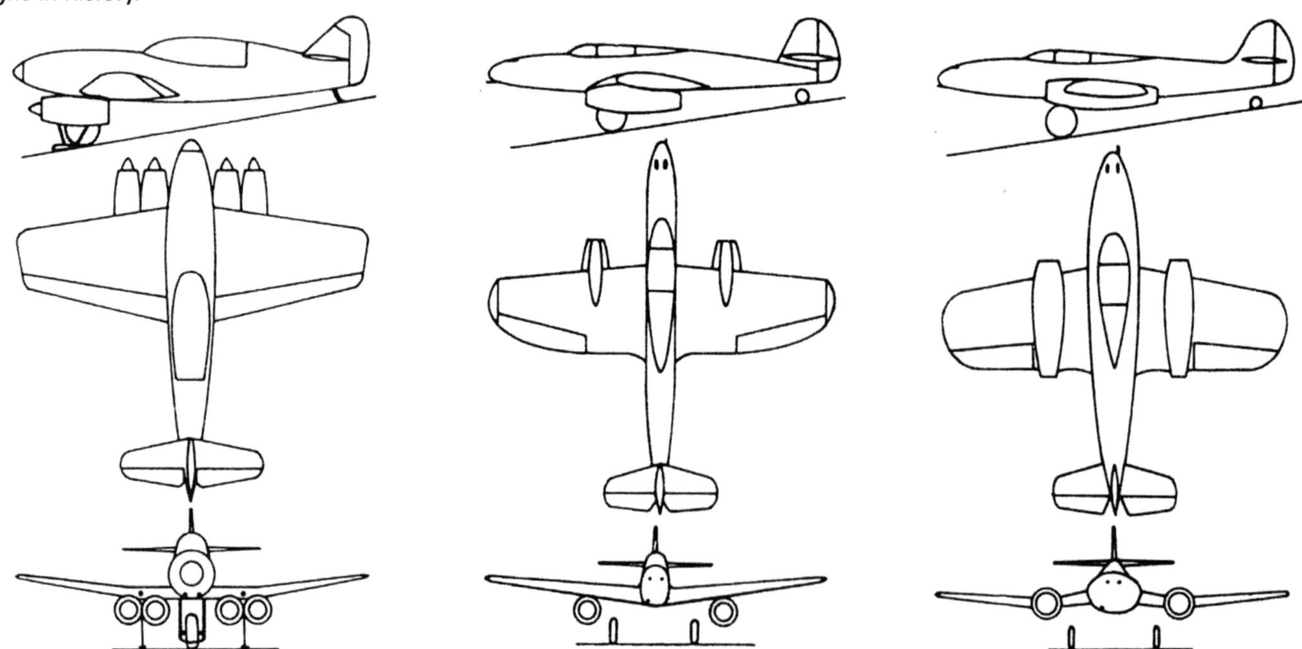

ABOVE LEFT: The **EFo 18** appears to have been yet another variation on the **EFo 11** theme but this time with four turbojets. **ABOVE MIDDLE:** Junkers' **EFo 17** has been regarded as a competitor for the **Me 262** but whether it was or not is unknown. **ABOVE RIGHT:** The **EFo 19** features engines integrated into its wings.

prototype was delivered on June 12 – well behind schedule – and ejection seat testing began on June 15. Fuel tank test installations in the He 280 V1 began on June 21. By August 1, 1940, the RLM had ordered five prototypes and 20 pre-production series examples. The first ground tests on the engineless V1 began on August 24 and its first towed test flight with Flugbaumeister Paul Bader at the controls was on September 22 at Rechlin-Roggentin – with aerodynamic mock-up nacelles fitted to simulate the drag effect of the engines.[32]

On October 18, 1940, it was decided that the number of He 280s on order should be scaled back to just five prototypes until such time as 10-hour endurance testing of its HeS 8 A engines could be satisfactorily completed. During a presentation to the RLM on December 16, 1940, Heinkel's representative stated that the He 280 V1 was ready to fly and that its first powered flight would take place in mid-January. He 280 V2 was undergoing final assembly, V3's wings and fuselage were under construction and the structural frames of the V4 and V5 were being made.[33]

MESSERSCHMITT, 1939–1940

The arrival of a whole new team at Messerschmitt in January 1939, dedicated to working on tailless aircraft designs, meant the company now had two separate project departments – the other being the company's existing projects team headed by Robert Lusser, although he left the following month to rejoin Heinkel and was replaced by Woldemar Voigt.

Lippisch's first order of business in his new role was to scrap all the existing drawings of Project X and start again, since it had been decided that the aircraft's rocket thrust would be increased. Secondly, it was decided that since data on the performance of tailless aircraft fitted with rocket engines was practically non-existent, a test vehicle was needed. The DFS 194 was a suitable candidate for conversion so this work too got under way – though there was no engine yet available to fit. Thirdly, Lippisch began to work on a series of designs for a new fighter.

It seems likely that this third project, designated P 01, was prompted by a study contract Messerschmitt had

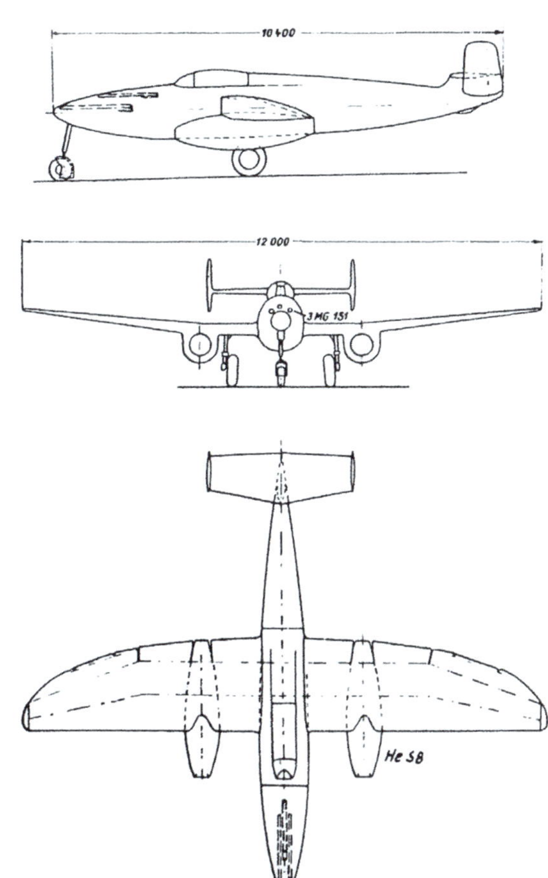

ABOVE: An RLM recognition drawing of the Heinkel He 280. Work on the He 178 was abandoned as attention switched to the twin-turbojet fighter design.

ABOVE: The Heinkel He 178 V2 – which never flew. It was supposed to have been powered by the HeS 6 engine.

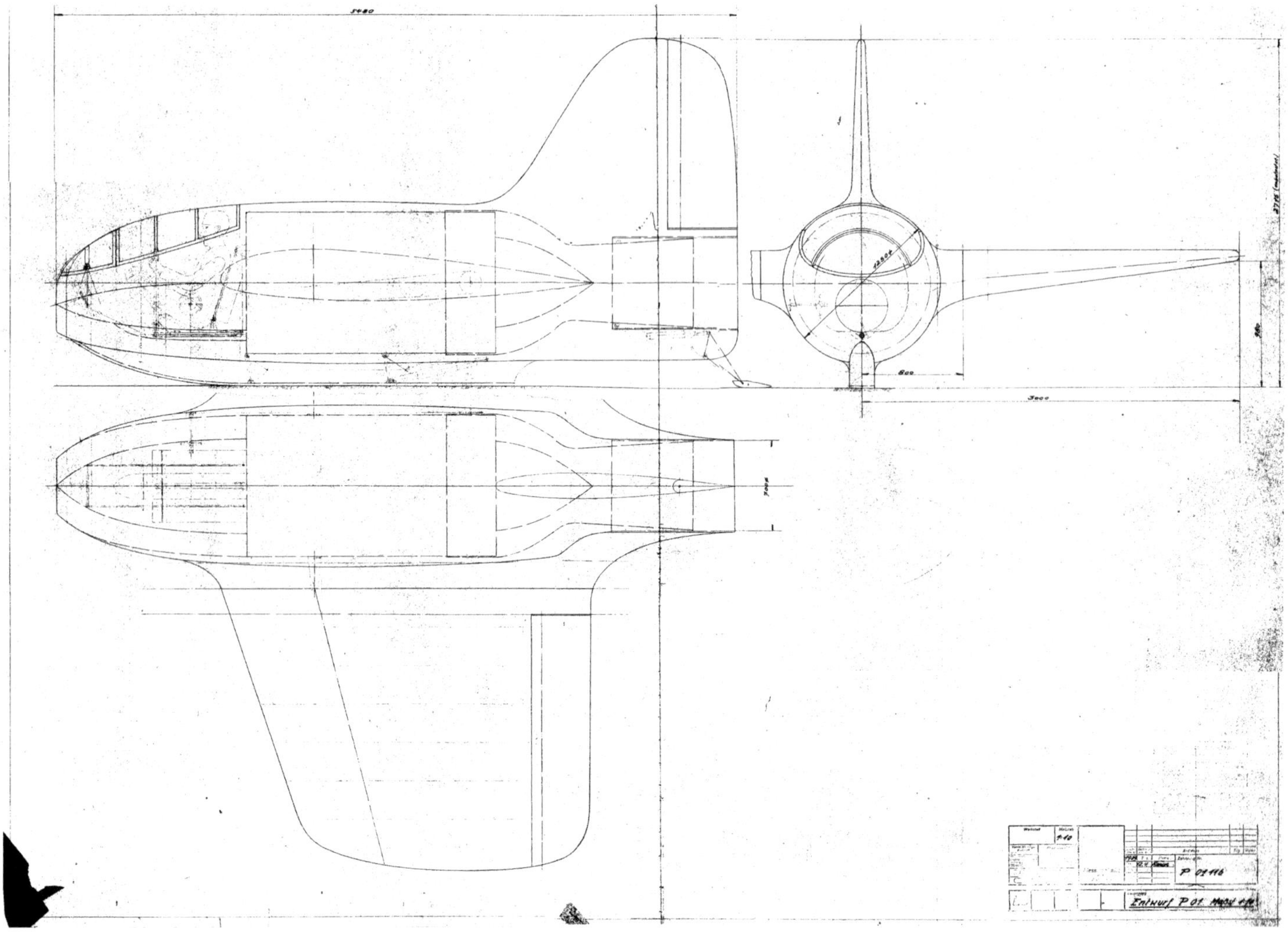

ABOVE: One of the earliest drawings of a jet aircraft ever committed to paper – Körner's P 01-116 of April 12, 1939. *IOWA STATE UNIVERSITY LIBRARY SPECIAL COLLECTIONS AND UNIVERSITY ARCHIVE*

received a month earlier, in December 1938. The actual specification[34] for this was only received on January 4, two days after the arrival of Lippisch and his team. It called for a single-seater powered by a single turbojet engine in two roles – high-speed fighter and interceptor.

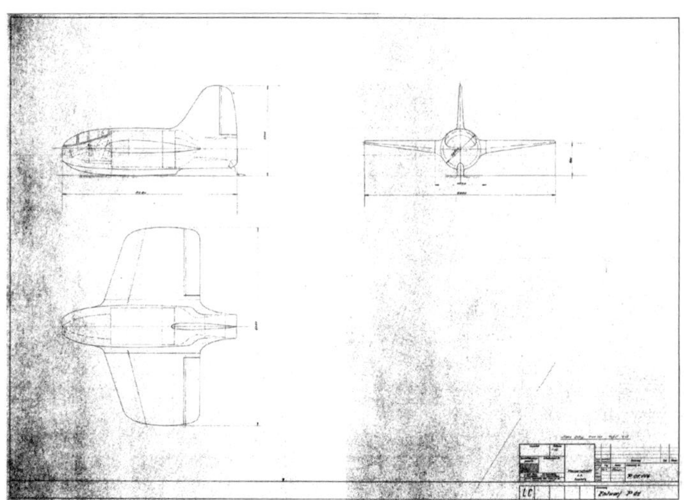

ABOVE: A full three-view drawing of the P 01-116 – the earliest version of the Messerschmitt P 01 despite the out-of-sequence '116' number – dated April 13, 1939. *IOWA STATE UNIVERSITY LIBRARY SPECIAL COLLECTIONS AND UNIVERSITY ARCHIVE*

In either configuration it was to be armed with two MG 17 machine guns and one MG 151/20 cannon. Top speed was to be 900km/h – this figure having previously been cited in Antz's report – and landing speed was to be not more than 120km/h (75mph). Flight duration was one hour as fighter and 30 minutes as interceptor.

It appears that Lippisch began by writing out a list of potential configurations,[35] engine options and other features and allocating each one a different number from 111 to 119. Then, rather than having his department work through the different concepts in numerical order, he started work on the design designated P 01-116. Where many of the other ideas featured rocket engines in one way or another, the initial concept of P 01-116 was that of a single-seat fighter powered by a single turbojet engine. The concept designated 111 was also a single-seat single turbojet fighter but work started with 116 – which might have been seen as the concept most closely matching the January 4 specification.

The first drawings were produced by Körner, one of the less-well-known members of Lippisch's team who had followed him over from the DFS and remained with Abteilung L until it was dissolved. Lippisch's only comment on it was "this design was based on the first available data for a turbojet".

The P 01-116 of April 12 and April 13, 1939, was a remarkably small tailless aircraft[36] – measuring just 5.48m long and with a wingspan of 6m. A large tailfin gave it a height of 2.715m. By way of comparison, the diminutive Me 163 B-1 was 5.98m long and 2.75m high with a 9.33m wingspan. The P 01-116's fuselage was cylindrical and its wings were extremely broad and almost rectangular in shape, with only a slight sweepback to the leading edge. Take-off would have presumably involved a wheeled dolly and landing was to be accomplished on a long skid with a small additional skid at the extreme end of its tail.

It is uncertain precisely which turbojet Lippisch had received data on, although it is unlikely to have been the Heinkel HeS 3 which would power the Heinkel He 178 during its first flight four months later. Unlikely, because Ernst Heinkel was unwilling to offer his jet engine designs for use in other manufacturers' aircraft at this stage.

While Lippisch got started on a single-jet design as P 01, so did Voigt's department under the designation P 65. Their first designs evidently had a similar twin-boom layout to the de Havilland Vampire[37] but by June 1939 it had been abandoned in favour of a twin engine design.

The first towed gliding flights of the DFS 194 took place on July 28, 1939, with further flights taking place in August. Work on the P 01 continued; if there was a P 02 all information about it has been lost and Lippisch himself never seems to have mentioned it. Work was also begun on the P 03 – the intended next-step development of what had now become known as the Me 163.

Projekte 1939

1.)	Entwurf P01 -	113 (2x)		
2.)	" "	114 (2x)		
3.)	" "	115 (2x)		
4.)	" "	116 (2x)	Körner	13.4.1939
5.)	" "	" (2x)	"	12.4.1939
6.)	" "	" (2x)	Rentel	12.6.1941
7.)	" "	" (2x)	"	16.7.1941
8.)	" "	117 (2x)		
9.)	" "	118 (2x)		
10.)	" "	119 (2x)		
11.)	" "	111 (2x)		

ABOVE: A note from Lippisch's personal papers, presumably produced in 1941, appears to show Abteilung L's P 01 schedule. The first design in the series to be worked on was the P 01-116, rather than the P 01-111 as logic might suggest. *IOWA STATE UNIVERSITY LIBRARY SPECIAL COLLECTIONS AND UNIVERSITY ARCHIVE*

In his memoir, Lippisch wrote: "The name '163' was also a camouflage name, because Messerschmitt had previously designed and tested a slow-flying aircraft as competition to the Fieseler 'Storch', which, however, was not good enough and was supposedly to be improved. If then our drawings with the 163 number on them went to the workshops, so did the new parts for the improvement of this slow-speed aircraft. And so the work we did on the 163 and then on the converted airframe of the DFS 194 remained astonishingly secret."[38]

The Bf 163, a two- to three-seater, had been ordered by the RLM on June 13, 1936. The first two prototypes, the V1 and V2, were to be powered by Argus As 10 C III engines while the third, V3, was to be fitted with a Hirth HM 508 E.[39] Therefore, the first prototype of the

ABOVE: An early model of Abteilung L's Projekt X with wingtip control surfaces undergoing wind tunnel testing at Göttingen during the spring of 1939.

ABOVE: A second model of Projekt X was made and tested with a large central fin in place of wingtip surfaces. This configuration was chosen to form the basis of what would become the Me 163.

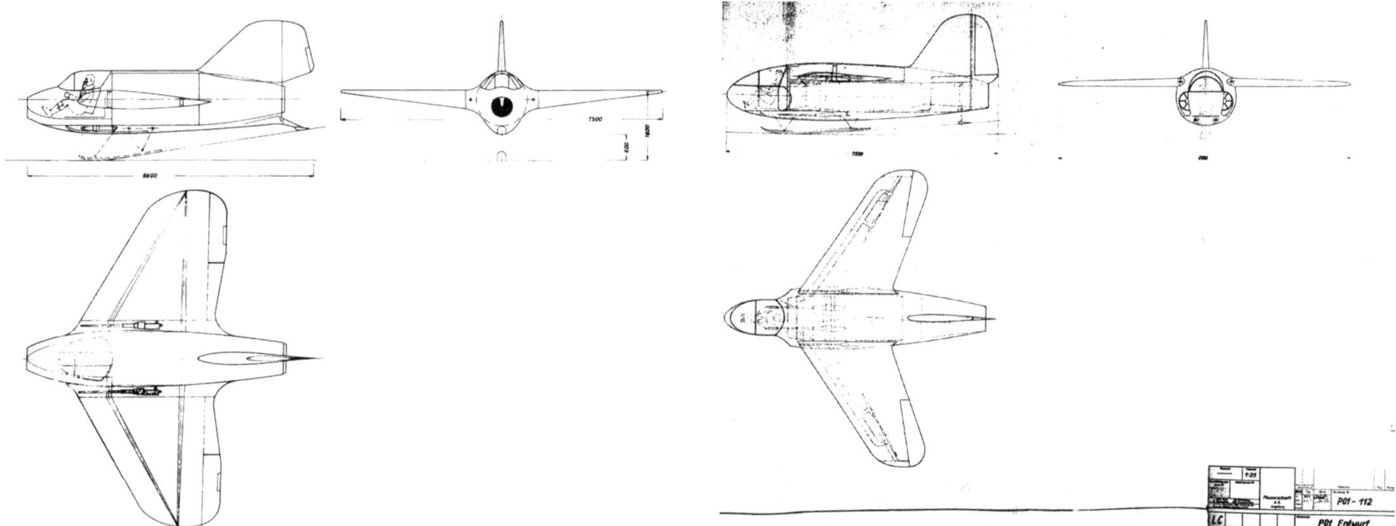

ABOVE LEFT: The single-turbojet P 01-111 was first drafted in November 1939, seven months after the P 01-116. The space left for the engine is large and undefined since the precise dimensions of the turbojet likely to be used were unavailable.
ABOVE RIGHT: The original broad fuselage P 01-112 fighter, drawn up on January 31, 1940.

Me 163 rocket-propelled aircraft was given the designation Me 163 V4.

When the war began in September 1939, the Messerschmitt company came under massive pressure to step up production and development of the Bf 109 and Bf 110. Resources were diverted away from Abteilung L's work on the Me 163 but the engineless DFS 194 was transported to Peenemünde and began to undergo further glide testing on October 16, 1939.

Work also continued in earnest on the P 01. The next design in the sequence, first drafted in November 1939, went back to the numerical beginning – P 01-111.[40]

This was a development of the P 01-116 in every respect, being larger at 6.6m long with a 7.5m wingspan. The pilot sat higher up in a slightly larger cockpit with a nose intake for the single-jet engine positioned in front of his feet. The P 01-111's skid design was improved and a pair of cannon were even shown buried in its wing roots. A sharper sweepback was applied to the wings, including a swept trailing edge, the tailfin was also more swept and the rudder was enlarged.

The 'conventional' Messerschmitt design department had by now been working on exclusively twin-engine layouts for its P 65 jet fighter – which would eventually receive the designation Me 262 – for nearly six months and had reached the mock-up inspection stage by December 1939.

It would seem that further data on the early jet engines was received around this time, since the next Abteilung L design, the P 01-112, featured a pair of turbojets buried within its fuselage. A full brochure on this design was issued in February 1940 with the P 01-112 appearing to 'shadow' the work of the conventional design department.[41] But where the P 65 had thin straight wings, the P 01-112 had a 35° sweepback on its thin wings. And where the P 65's engines were carried in wing-mounted pods, the

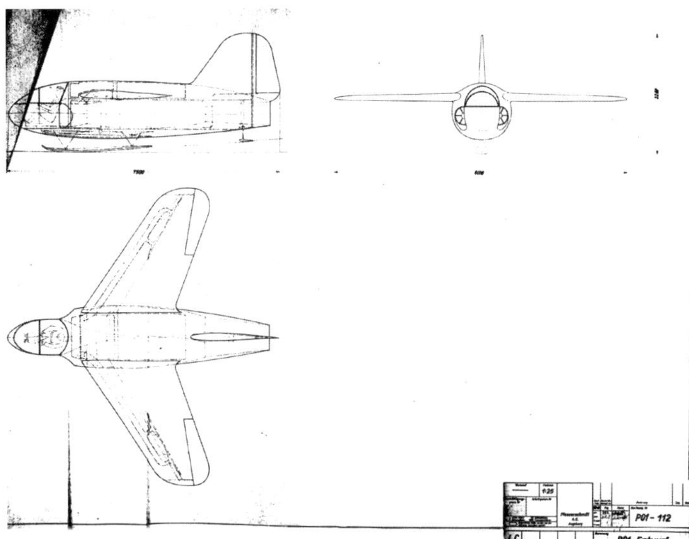

ABOVE: The broad fuselage P 01-112 as an experimental aircraft, dated January 31, 1940.

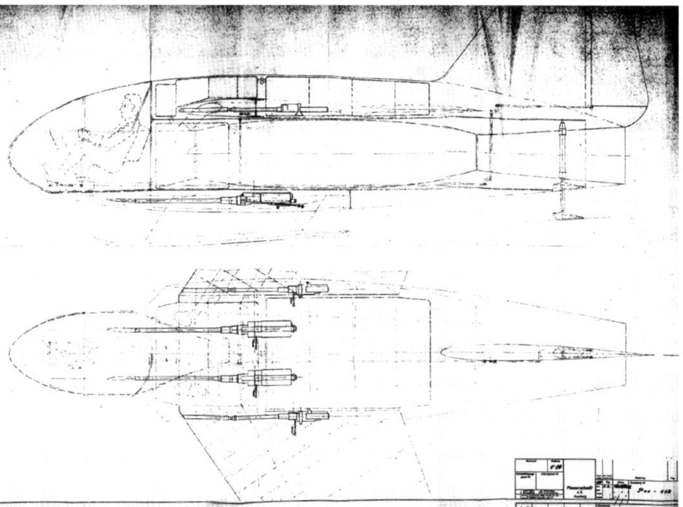
ABOVE: Detailed view showing the P 01-112's armament of two MG 17 machine guns in the upper fuselage and two cannon below the engines. The drawing is dated February 1, 1940.

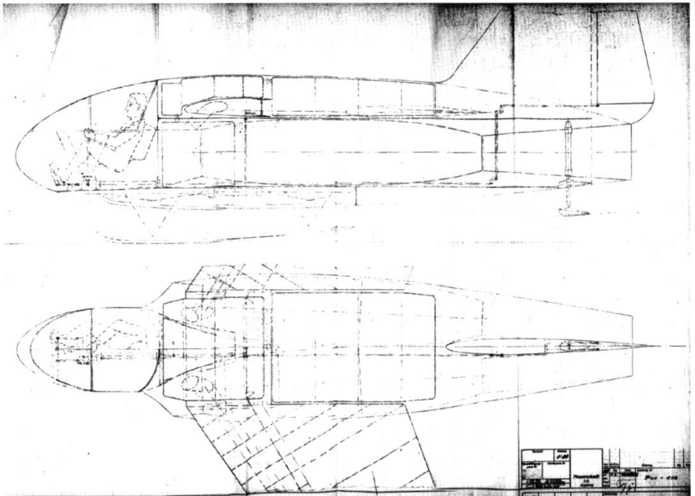
ABOVE: The P 01-112 as an unarmed experimental aircraft, as drafted on February 1, 1940.

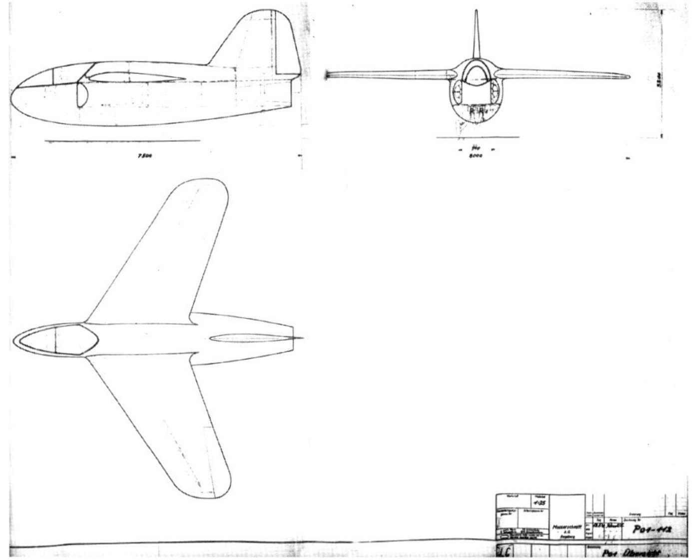
ABOVE: The final-form twin-jet P 01-112 with narrow fuselage, wheeled undercarriage and straight leading-edge wings, dated February 12, 1940.

P 01-112's were fitted within its fuselage – fed by a pair of curved intakes, one either side of the pressurised cockpit.

As the foreword of the P 01-112 brochure notes: "The drive is carried out by two jet engines (Jumo), which are arranged in the fuselage under the wing. This eliminates any negative influence of the engines on the flow on the wing.

"The pilot's seat, arranged in front of the wing and the engines, offers the best visibility and allows the formation of an armoured pressure booth which can be designed independently of the static structure of the fuselage.

"A central runner and spur, both retractable, are provided as the main landing gear. The take-off takes place on a jettisonable trolley. For the initial unarmed version without equipment, fuel is provided for half an hour's flight time. The space for armament and fuel tanks for one hour flight time will be available however."

The fuselage was to be made of a load-bearing light metal shell with the skid in a box under the floor along the centre. The wings, built as a single piece, were to be fitted onto the fuselage over the engines, with the fuel tanks then fitted on top of that. The engines themselves were to be separated by a central partition and could be accessed or removed by taking off the upper portion of the fuselage.

There would be space for weapons both beneath the cockpit and the wing roots next to the fuselage.

Another advanced feature was a system for blowing air over the upper wings to improve lift during landing: "A suction line runs through the wing from the suction chamber in front of the engines, which can be switched on, in particular, during landing and sucks the boundary layer on the outer wing through slots arranged behind the main hoop.

"The installation of a slat is therefore not planned. Also, an extensible slat would interfere with the sensitive upper

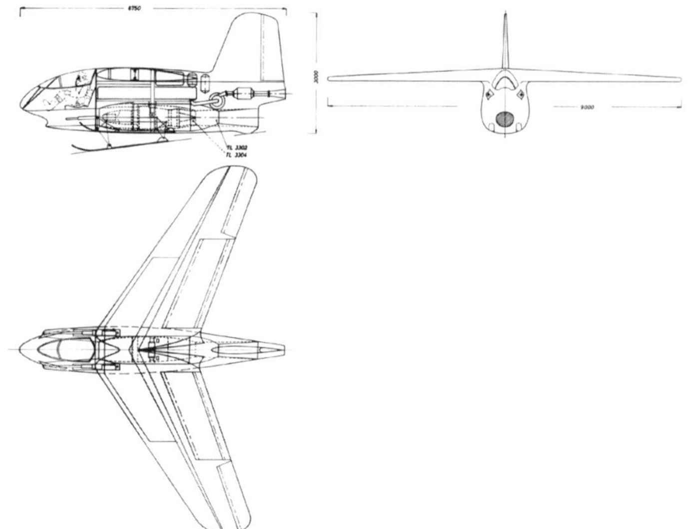

ABOVE: The mixed rocket/turbojet propulsion P 01-113 of July 1940 – five months after the P 01-112. The whole P 01 sequence was spread out over a considerable time period.

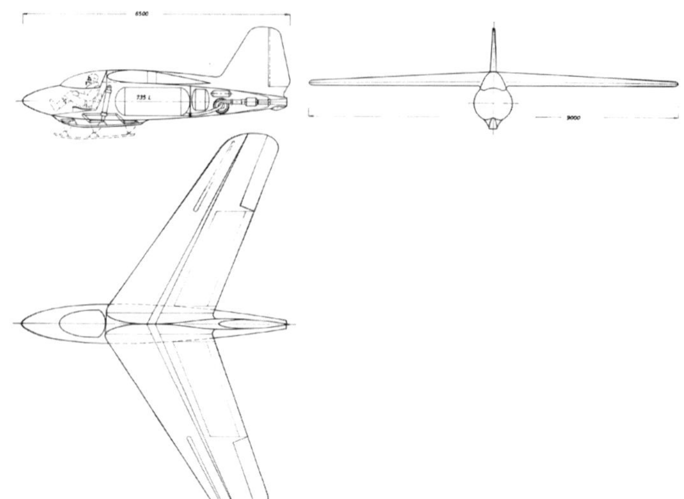

ABOVE: One of Abteilung L's most important designs was the single-rocket interceptor P 01-114. Despite its high-wing layout, it is similar in many respects to the Me 163 B. Drafted in July 1940, it received the designation Me 263 and provided the basis for an extensive wind tunnel test programme.

side of the outer wing at high flight speeds. The thickness ratio of the profiles in the direction of flight is 8%."

Furthermore: "The air sucked in by the engines flows in through large openings in the fuselage. The exhaust line is branched off from the vacuum chamber located in front of the engine to one wing each. The normal air supply to the engine is throttled by flaps, which are attached to the rear edge of the pilot's compartment, so that in particular the engines are used only for suction during landing. The extent to which the exhaust can also be used for take-off and climb requires testing."

This may be the earliest example of a variable geometry jet engine intake in history. Unfortunately, the P 01-112's design was being hampered by the lack of information from Junkers about the engines: "Since details of the engines are not yet available from the manufacturer, they can only be made later. The design provides that the exhaust jet of both engines is discharged through an extension nozzle on the fuselage tail."

Four weapons were to be fitted – two MG 151s under the pressure cabin with 500 rounds each and two MG 17s, one in each wing root, with 1,000 rounds each.

The performance figures provided suggest a clear advantage over those provided in a late September 1939 report on the P 65 – albeit powered by Junkers engines rather than the BMW ones then being considered for that design. Maximum all-up weight of the P 01-112 was to be 3,840kg, compared to 4,325kg for the P 65. Top speed at 4,000m was 1,080km/h compared to 975km/h at 3,000m. Landing speed was 125km/h compared to 150km/h.

By now progress on turning Abteilung L's Me 163 drawings into a working aircraft had slowed dramatically. Lippisch wrote: "Our scientific work on the creation of the high-speed aircraft 'Project X', now called Me 163, was practically stopped by the outbreak of the war and the design office had to provide assistance in all possible places."[42] With practical design and development work on tailless projects all but halted during the early months of 1940, Lippisch busied himself with theoretical work on transonic and supersonic flight, calling on the talents of another recently recruited Messerschmitt employee, mathematician Dr Friedrich Ringleb.

A few months later, "apparently at the Ministry of Aviation was the realization that the planned Blitzkrieg could not be brought to a rapid conclusion and that, as far as possible, one had to resume and carry forward promising developments".[43] Now there was an engine ready for the DFS 194 – a Walter unit capable of producing 400kg of thrust – and it was installed at Peenemünde's own workshops. The DFS 194 made its first rocket-powered flight on June 3, 1940.

According to Lippisch's own account: "In the summer of 1940, a series of experimental flights with a rocket drive began, with Heini Dittmar taking over this task, which was not without danger, and carried out with great success.

"Since the fuselage was comparatively large, we were also able to install an Askania quadruple recorder and the performance measurements that resulted from it gave us very valuable new insights. The original fuselage of the DFS 194 was not designed for high speed, so we had to abort our experiments at 550kph because otherwise the loads exceeded the permissible limit."[44]

With this exciting work now taking place, Abteilung L returned briefly to the P 01 series in July 1940 – designing two more concepts which unsurprisingly each featured rocket propulsion – the P 01-113 with a single turbojet plus a rocket engine, and the P 01-114 which had the same wings as the 113 but a new slender fuselage housing only a rocket engine.

In fact, according to the design brochure,[45] the P 01-114 was intended simply as an experimental vehicle to assess the features of the P 01-112 and P 01-113 designs – rather than being a prototype for a production aircraft in its own right.

It states: "The model P 01-114 is a single-tail tailless aircraft which is used to test the design at very high speeds and high gradients with respect to the patterns P 01-112 and 113. The structure perfectly corresponds to the wing provided for the heavily armed designs, while the fuselage was simplified with regard to simple and fast manufacture. The HWK engine, which is provided for the drive, permits a step-by-step test with different thrusts, so that the properties can be flown out at very high speeds."

The wings were to be adjustable so that different angles of sweepback could be tried and a device was to be fitted which could alter the sensitivity of the elevator control to make it easier to use at very high speeds.

"The fuselage was designed as a rotationally symmetrical body in order to ensure the simplest design of parts, so that only a few devices are required for producing the fuselage shell," it states. "The cockpit is pressure tight. A normal chassis was dispensed with. The take-off takes place on a jettisonable roller train, while the landing takes place on the retractable skid and the spur. The experimental testing of a crawler roller is planned."

On November 26, 1940, Messerschmitt sent a copy of the P 01-114 brochure to Ernst Udet – but with '263' written beneath 'P 01-114' on the cover.[46] The accompanying note to Udet described the design as the 'Me 263' but this seems likely to have been an in-house designation at this stage rather than an official RLM type number.

But now, with work recommencing on the construction of the first two Me 163 prototypes and promising results already having been obtained from flight testing the DFS 194, Abteilung L suspended almost all project design work for nearly a year to carry out what Lippisch later described as "intensive work on the DFS 194 and the Me 163".[47]

Messerschmitt's main project office, meanwhile, continued to work on the P 65.

ABOVE: A receipt for Messerschmitt's short description of the Me 263 from Ernst Udet, dated November 1940.

ABOVE: The DFS 194 made its first rocket-powered flight on June 3, 1940. The Me 163 was designed as an experimental aircraft capable of reaching and exploring speeds that the DFS 194 could not.

Chapter 2

Abteilung L

Messerschmitt Me 163 and projects 1940–1943

While Messerschmitt's projects bureau worked on the Me 262 fighter, Alexander Lippisch's Abteilung L team focused their efforts on designing a rocket-propelled interceptor.

ABOVE: A 1:2.5 scale model of the Me 263 being tested by the AVA during the autumn of 1941.

THE FIRST ME 263

Abteilung L had been part of the Messerschmitt organisation for nearly two whole years by the end of 1940 and the company continued to enjoy large contracts as Germany's main supplier of fighter aircraft – working on substantial upgrades for the Bf 109 while readying the Me 210 to replace the Bf 110.

The final flight tests of the DFS 194 were conducted at Peenemünde on November 30, 1940, and the first towed flight test of the newly completed Me 163 V4 took place at Augsburg on February 13, 1941, with Heini Dittmar at the controls. In the meantime, Abteilung L had been joined by a new recruit – Dr Hermann Wurster had previously been Messerschmitt's chief test pilot, conducting many early tests involving the prototype Bf 109s and the Bf 209 speed record aircraft; now he switched to designing aircraft rather than flying them. His first job was to work on designs for a new tailless piston-engined bomber.

According to Lippisch: "After a lapse of almost a year which was filled with intensive work on the DFS 194 and the Me 163, we renewed our project design work. In the spring of 1941 we conceived a version of the Me 163 with greatly improved aerodynamic properties, the P 03-Me 263.[1] This project was designed as a test aircraft for higher velocities. The angle of sweepback was 32.1° (compared to 23.4° of the Me 163). The wind tunnel model of this design was tested at the AVA in Göttingen and the measurements showed an extremely low friction coefficient. However, work on the project was stopped when the Me 163 passed the 1,000km/h mark."[2]

This was the P 01-114, which had already been given the Messerschmitt in-house designation 'Me 263'[3] by November 1940. By April 1941, the P 01-114 was being tested in the wind tunnels of the AVA at Göttingen, known simply as the 'P 01'. In May, a model was tested with wings set further forward than those of the original design, a lower fin and a small set of tailplanes. The near-cylindrical fuselage was labelled 'Rumpf I'. In June another model was tried with the same high wing layout and tailplanes but with a shorter nose and thinner chord wings. This one had 'Rumpf IIa'.[4]

Having previously been suspended, P 01 now recommenced as a fourth ongoing strand of the department's work alongside the Me 263, testing of the Me 163 and Wurster's bomber designs.

Fittingly, the first design drawing to emerge for the 'new' P 01 series, on June 12, 1941, was given the designation P 01-116 and like its namesake of over two years earlier was a single-jet fighter. This out-of-numerical-sequence reboot of the series left a gap at P 01-115 which was filled on July 2 with the creation of a mixed propulsion single turbojet and rocket-propelled design – the same day that Heini Dittmar made his last unpowered flight in the Me 163 V4 before it

with the creation of a mixed propulsion single turbojet and rocket-propelled design[5] – the same day that Heini Dittmar

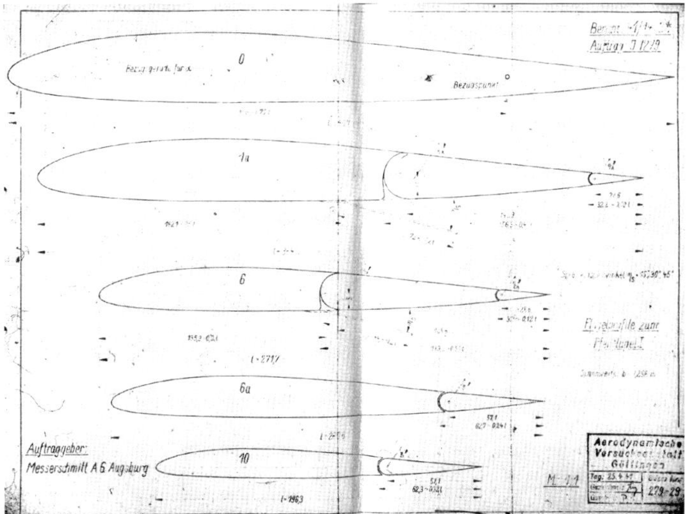

ABOVE: Wing profiles for the P 01 as tested by the AVA on April 25, 1941.

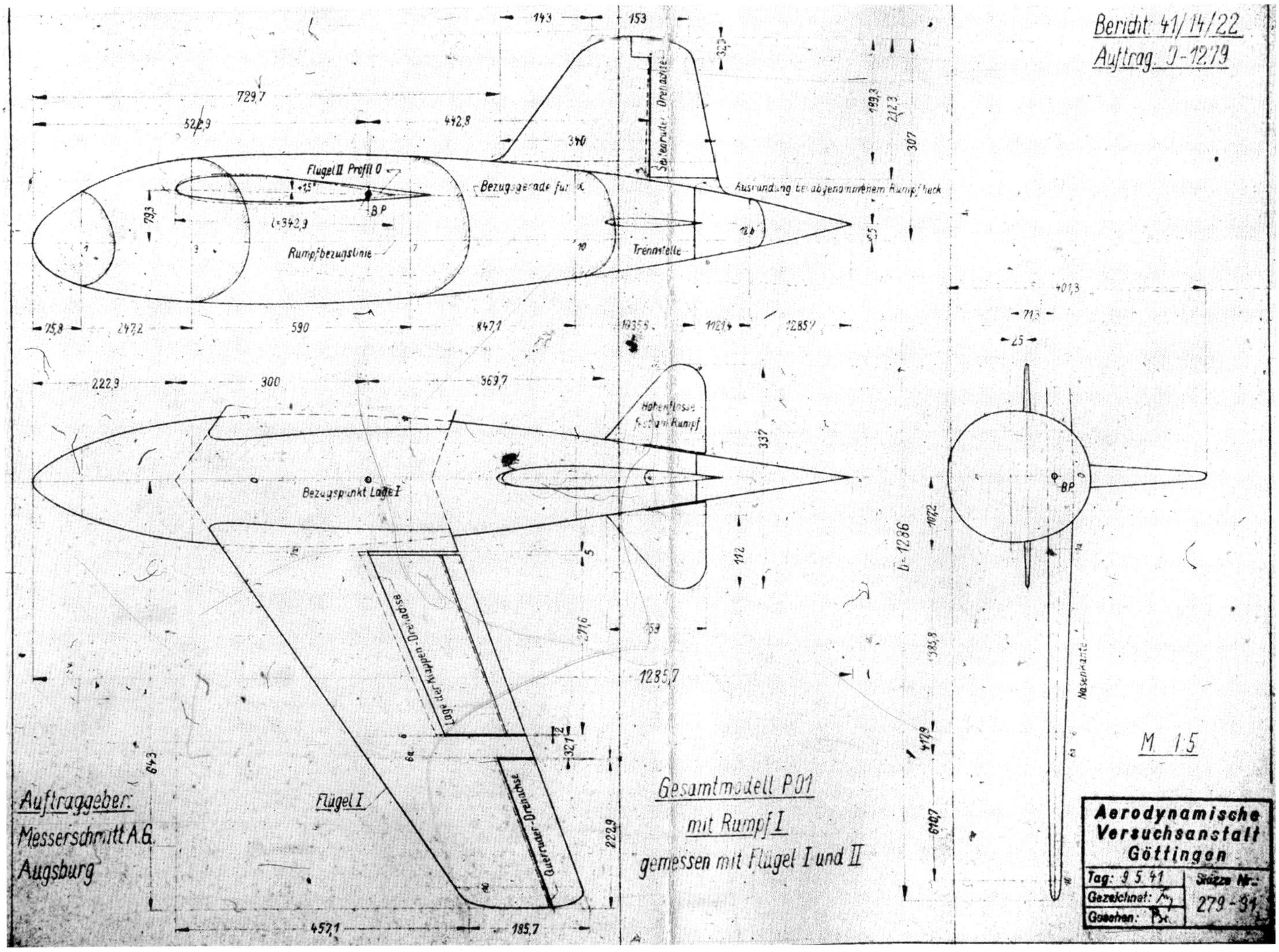

ABOVE: Early model of the Messerschmitt P 01 from May 9, 1941. The fuselage is the basic 'Rumpf I' and the swept wings are 'Flügel I'.

made his last unpowered flight in the Me 163 V4 before it was moved to Peenemünde so that its Walter rocket engine could be fitted.

According to Lippisch: "In the summer of 1941, we moved into our quarters at Peenemünde and the people from Walter installed the engine. Heini was familiar with the aircraft's characteristics because of the gliding tests he had carried out and, therefore, the initial flights got off to a good start. Seeing him fly this bird was a wonderful sight. After take-off, the aircraft stayed close to the ground while gaining speed and then climbed steeply to an altitude of 4,000-6,000m.

"The flying characteristics were excellent and when, more or less as a spectator, I watched the bird disappear into the clouds, trailing smoke, my thoughts went back to the beginning, to our creation's childhood and adolescence, and I was very glad that, in the end, we had managed to overcome the many, almost insurmountable problems."[6]

A couple of weeks after this important phase of the Me 163's development began, on July 16, yet another new P 01-116 was drafted – a pure rocket-propelled design. This meant that the P 01 sequence ran as follows: P 01-116, P 01-111, P 01-112, P 01-113 and P 01-114, then a gap of nine months, followed by a second P 01-116. This 'new' P 01-116 retained the squared-off tailfin of the P 01-114 and the amorphous P 01 wind tunnel models being tested by the AVA, making it a natural next step in the sequence. The P 01-115 which then followed had a new rounded tailfin and this feature was carried over to the third and final P 01-116 – making the latter the only one of the three P 01-116s to fit correctly into the numerical sequence.

Within three weeks, the last trio of P 01 designs were produced – the P 01-117 on July 22, the P 01-118 on August 3 and the P 01-119 on August 4.[7] The first featured a cockpit where the pilot was expected to lie flat on his belly in a prone position, the second boasted a novel tilting seat arrangement to improve the pilot's visibility during a steep near-vertical climb, and the third had a pressure cabin for high-altitude operations. Each of them was to be powered by a single Walter rocket engine for the initial climb to altitude, but the P 01-118 also featured a second rocket engine for cruising to provide greater endurance.

In August 1941, another P 01/Me 263 model was tested with 'Rumpf IIb' and three different wing positions – the usual shoulder wing setting and two lower settings. The little tailplanes remained. At some point copies of the original P 01-114 report dated August 1940 had '263' retrospectively added to them. The original P 01-114 drawing was removed and a drawing of the '8-263' dated August 5, 1941, was added.

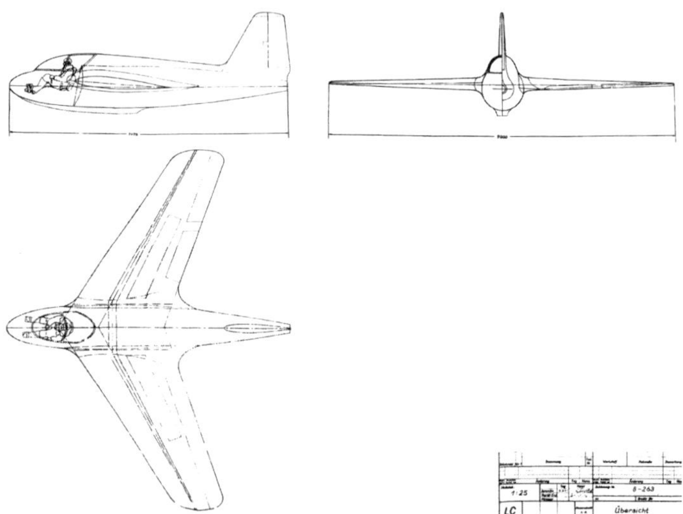

ABOVE: This drawing of the Messerschmitt 8-263, dated August 5, 1941, was inserted into year-old copies of the P 01-114 brochure, reflecting both the change in designation and the change in design.

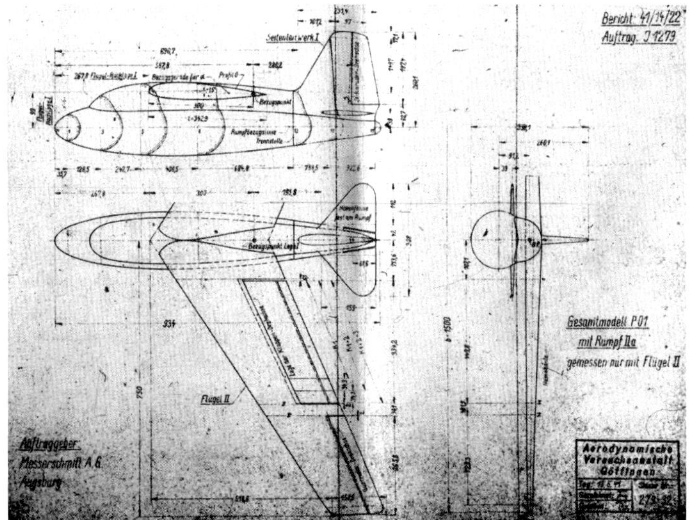

ABOVE: Another drawing of an AVA wind tunnel model, this time composed of 'Rumpf IIa' and 'Flügel II'. It is dated June 13, 1941.

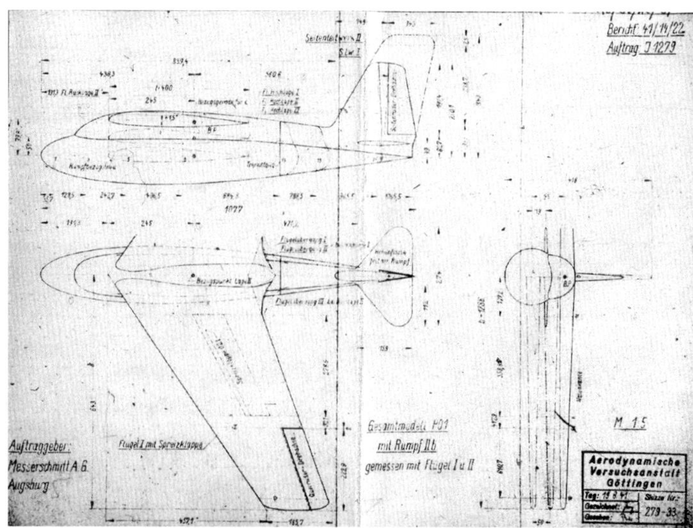

ABOVE: AVA wind tunnel model of the P 01 now being referred to elsewhere as the P 01 (Me 263) or simply the Me 263, from August 19, 1941. This model has the definitive 'Rumpf IIb' fuselage and modified 'Flügel I' wings, though three different options are offered for their positioning. An alternative tailfin form is also ghosted in.

This aircraft was the contemporary 'Rumpf IIb' design with the mid-wing position and at 7.475m it was nearly a metre longer than the original.

On August 13, 1941, eight days after the '8-263' drawing was drafted, Heini Dittmar made the first powered flight in a Messerschmitt Me 163. The outcome of this and further tests seems to have been twofold – firstly it demonstrated to any doubters that the concept of a highly manoeuvrable rocket-powered aircraft was sound and an avenue worth pursuing. Secondly, the start of practical testing freed up Abteilung L's design staff to concentrate on whatever might come after the purely experimental Me 163.[8]

P 05 – THE ENLARGED ME 163

Lippisch then seems to have turned his attention away from the Me 263 and on August 27, 1941, unveiled his first proposal for a military design based on Me 163 flight experience – the armed P 05 Interceptor drafted by Rudolf

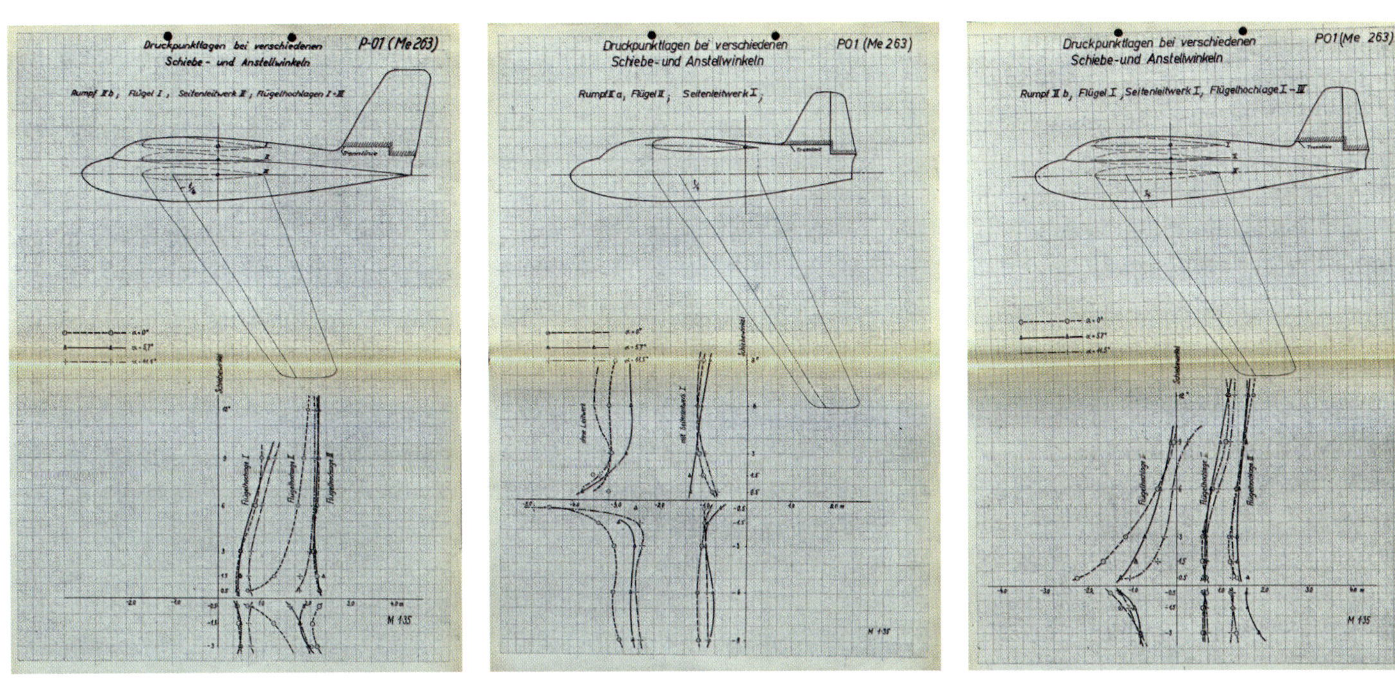

ABOVE LEFT: Undated AVA P 01 (Me 263) drawing showing 'Rumpf IIb', 'Flügel I', 'Seitenleitwerk II' (fin) and 'Flügelhochlagen I+III' (wing positions). ABOVE MIDDLE: The stubbier 'Rumpf IIa' fitted with the lengthy 'Flügel II' wings and tailfin version I. ABOVE RIGHT: The ubiquitous 'Rumpf IIb' with 'Flügel I', 'Seitenleitwerk I' and 'Flügelhochlagen I+III' (wing positions).

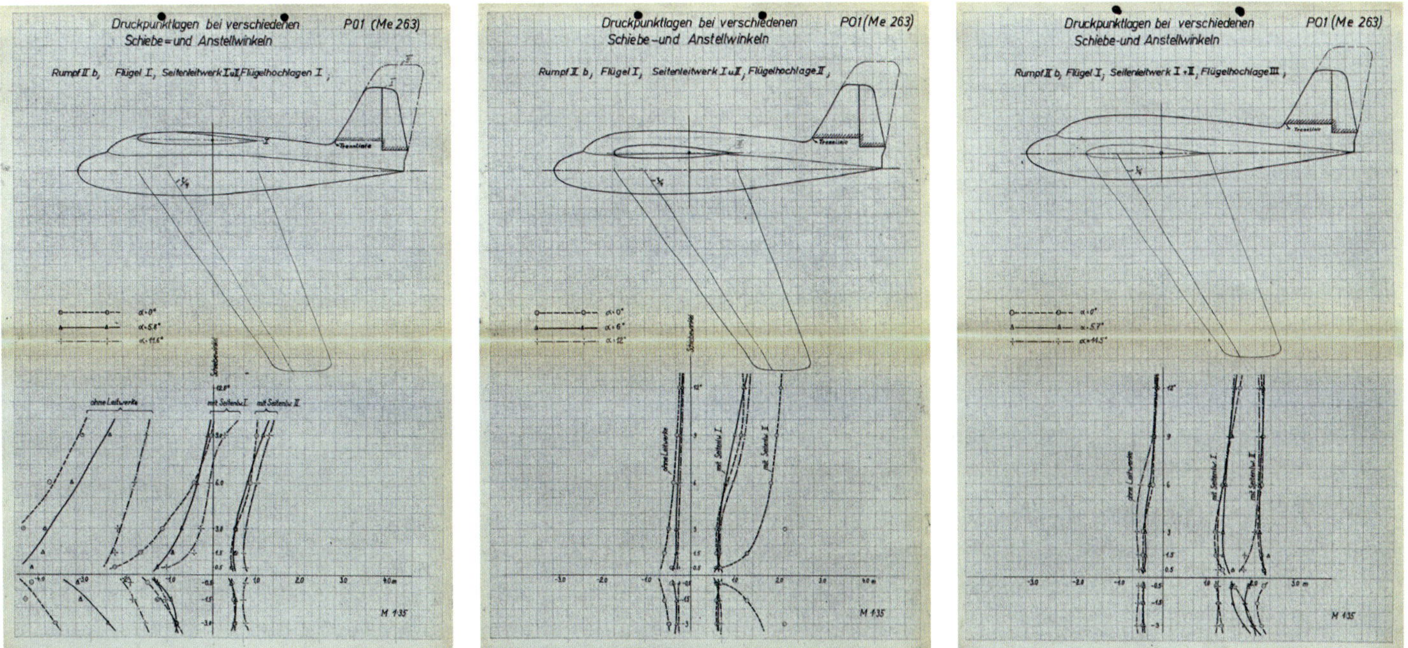

ABOVE: 'Rumpf IIb' again with 'Flügel I' and both fin versions. The wings are set high in position I, in the centre in position II and low in position III.

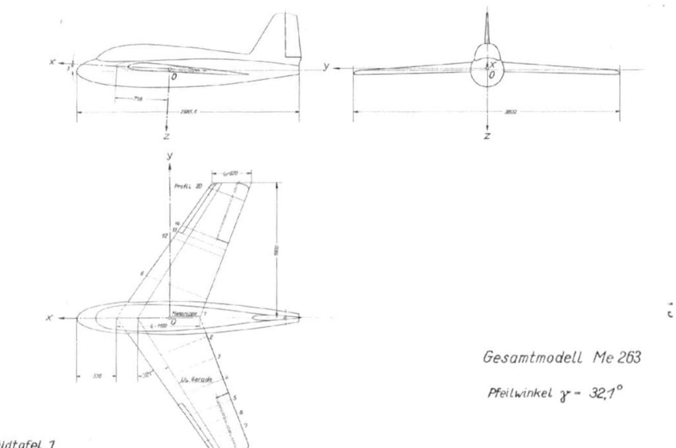

ABOVE: Three-view of the final-form Me 263 from February 1942. The wings now have a slight anhedral.

ABOVE: Side view of the Me 263 fuselage from the February 1942 report.

Rentel. The title page of the P 05 brochure[9], bearing Lippisch's signature, is revealing. It states: "If, during the course of the further flight testing of the Li 163, comparative engine changes are necessary, the manufacturer reserves the right to carry them out. Messerschmitt AG, Augsburg."

Lippisch is now referring to 'his' creation as the Li 163. Indeed, the brochure is headed 'Projektbaubeschreibung Li P 05 Interceptor' even though it is printed on a standard Messerschmitt-headed form.

The foreword states: "On the basis of the flight experiences and characteristics test of the Me 163 V4 the present

ABOVE LEFT AND RIGHT: The large Me 263 model is readied for tests at the AVA.

ABOVE: Views of the Me 263 model in the AVA wind tunnel from the February 1942 report.

project was developed as a massive enlargement of the tried and tested pattern. In accordance with the intended use, the rocket engine propulsion interceptor achieves the shortest climbing time at the application altitude and on the other hand ensures superior speed and climb performance compared to normal combat and fighter aircraft.

"The drive type allows only a short operating time because the fuel consumption of the currently developed engines of this type is many times higher than in Otto engines with a propeller drive. The superior climb and speed performance combined with a strong armament still make the interceptor a weapon effective for air defence."

This last point is reiterated in the technical section of the brochure which describes "usage: daytime use as a local guard with strong armament and high climb and flight speed".

As with most of Abteilung L's other creations, the P 05's wings were to be made entirely from wood except for the rudders, while the fuselage was a steel tube and light alloy construction. The pressure cabin cockpit was to be assembled separately then attached to the fuselage as a single piece.

Unlike the Me 163 with its single 1,500kg thrust HWK rocket engine, the P 05 was to have no fewer than four engines – three of them with 1,500kg thrust each. According to the brochure: "The three climb engines are not controllable, they are started one after the other by means of a common throttle lever, so that a certain thrust graduation of 1,500 to 3,000 to 4,500kg thrust is possible."

The fourth engine was for level flying once operational altitude had been reached and could be throttled back if necessary: "The cruising device is infinitely adjustable within the limits of 200–750kg thrust."

One of the 1,500kg engines was set at the extreme rear of the aircraft and the other two engines were on either side of it but set further back. The 750kg engine was set just above the former.

The P 05's three main fuel tanks were positioned behind the pressure cabin and two of them fed the launch engines. The third tank, armoured and sandwiched between the others, was for the cruise engine. The brochure notes that there is space in the wings available for the installation of further fuel tanks. Landing involved a sprung skid actuated using compressed air and a brake parachute, while armament was four MG 151/20s with 100 rounds each.

The publication of the P 05 brochure on August 27 was followed on August 28 by a pre-arranged meeting between Willy Messerschmitt, Lippisch, Theo Croneiss,

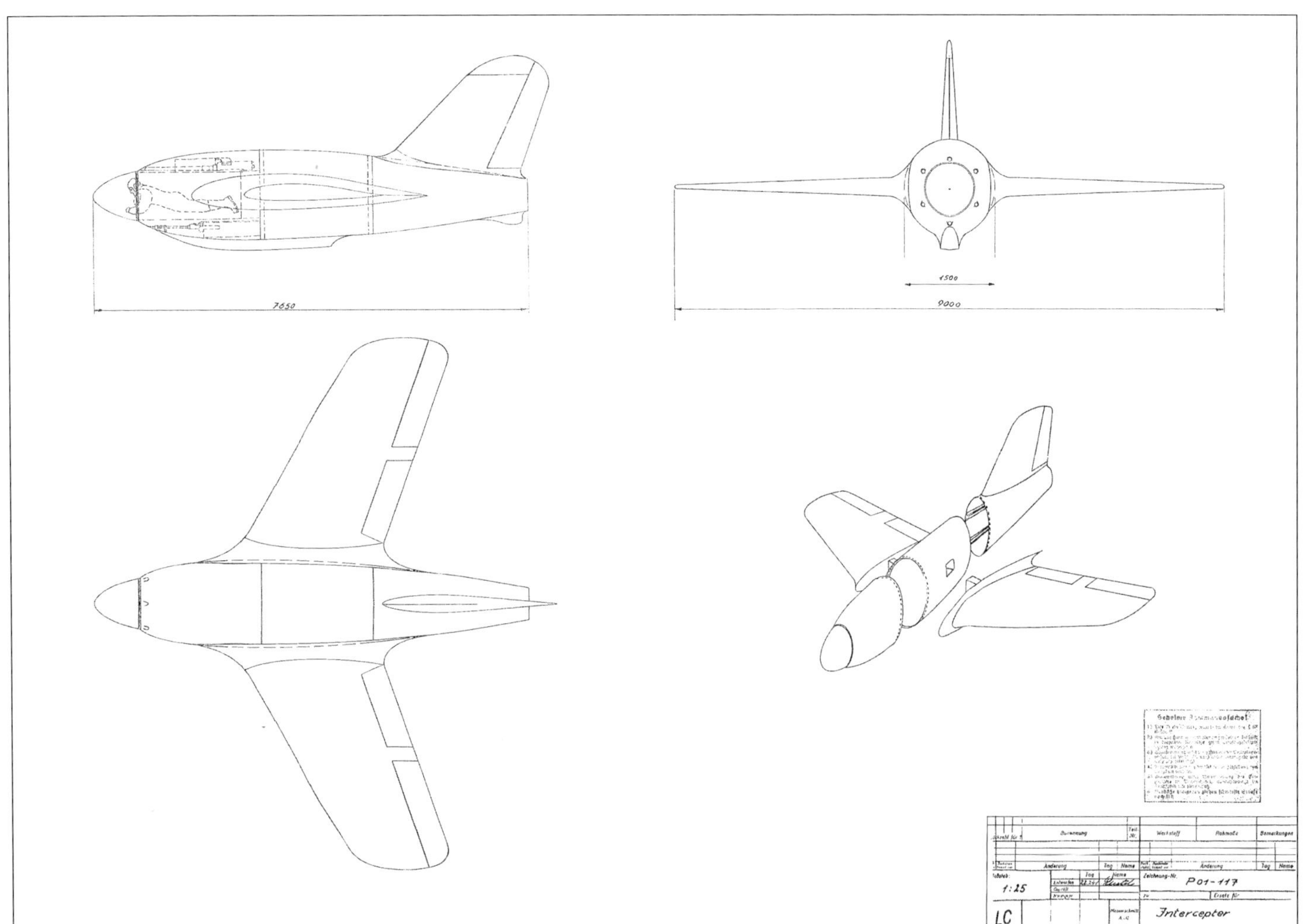

ABOVE: One of the various late-series P 01 designs, the P 01-117, which featured a prone pilot position. The drawing is dated July 22, 1941. *IOWA STATE UNIVERSITY LIBRARY SPECIAL COLLECTIONS AND UNIVERSITY ARCHIVE*

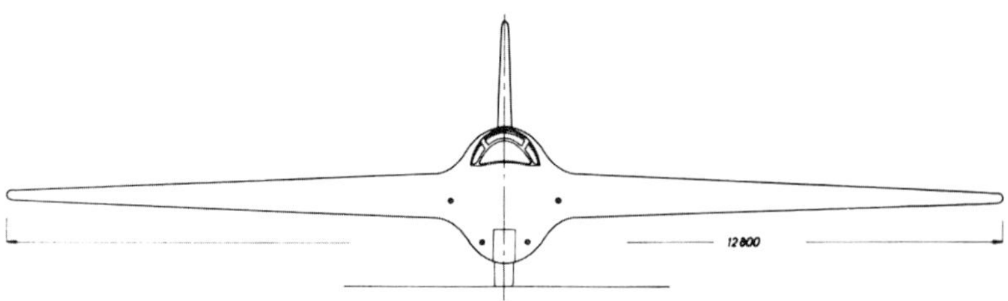

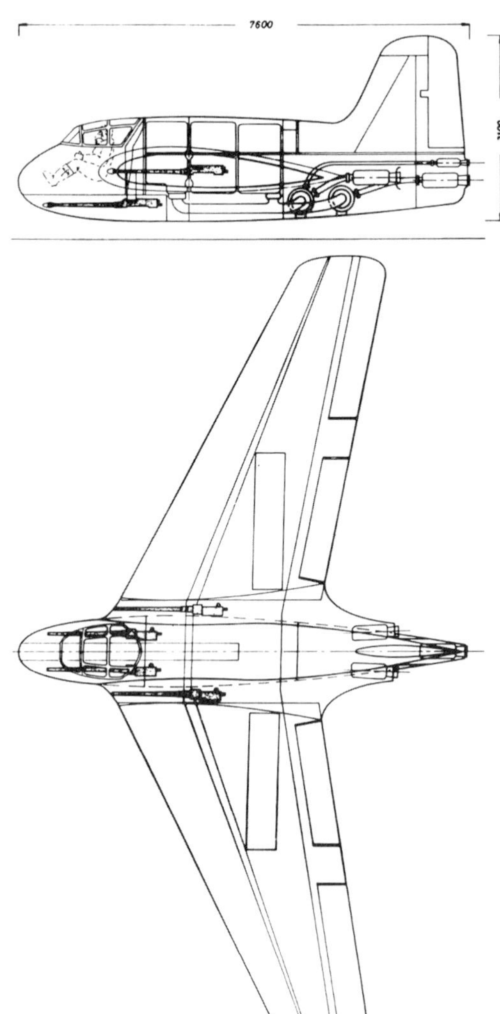

ABOVE AND LEFT: A three-view drawing of the P 05 showing the positioning of its four rocket engines – three for take-off and one for cruising at altitude.

ABOVE: A perspective drawing of the scaled-up Me 163 – the P 05 interceptor – from the project brochure. *IOWA STATE UNIVERSITY LIBRARY SPECIAL COLLECTIONS AND UNIVERSITY ARCHIVE*

Messerschmitt production director Fritz Hentzen and Oberst-Ing Gottfried Reidenbach, head of the RLM's Technical Office.

Up to this point, Abteilung L had spent more than two and a half years as a productive and integrated part of the Messerschmitt organisation but all that was about to change. At the meeting, several points of disagreement concerning the future of the department and its work were discussed.

Evidently Lippisch wanted it clarified that his project office had equal status to that of Woldemar Voigt, while Generalluftzeugmeister Ernst Udet, possibly at Lippisch's instigation, had decided that Abteilung L should be transferred to its own separate facility at Obertraubling and that Lippisch's type designations should bear his 'Li' prefix rather than the Messerschmitt 'Me'.

At another meeting on August 29, again attended by Croneiss, Hentzen and Reidenbach but this time accompanied by Hans-Martin Antz, minutes from the August 28 meeting were read out by Croneiss:[10] "Oberst-Ing Reidenbach accepted Professor Messerschmitt's objections to moving Abteilung L to Obertraubling and agreed that Abteilung L should remain in Augsberg. He asked whether Lippisch and Voigt's project offices would have equal status within the organisation under Professor Messerschmitt and whether the design office and prototype workshop would support both project offices to the same degree.

"Brigadeführer Croneiss and Direktor Hentzen answered both questions in the affirmative. Oberst-Ing Reidenbach said he was satisfied with this solution."

Now it was Reidenbach's turn to read out another set of minutes, from a recent meeting held by Udet and attended by his adjutant Max Pendele, Antz and Heini Dittmar, who had presumably just climbed out of the Me 163 V4's cockpit at the time: "At the meeting, which had taken place after a demonstration of the Me 163 at Peenemünde, General-Oberst Udet had decided that: 1) the Abteilung Lippisch should be transferred to Obertraubling, 2) the aircraft designed by Lippisch should bear his name, 3) the manufacture of the first batch of aircraft should be accelerated so that the aircraft can become operational next year, 4) Lippisch should be provided with the personnel he requires to undertake this task.

"Oberst Ing Reidenbach added the following comments to these minutes: to 1) he would inform the Generaloberst that he should reconsider this decision, since, after re-examining the reason for transferring Abteilung L to Obertraubling, Professor Messerschmitt and Lippisch had reached an agreement on this point at a meeting held on the 28th of this month. To 2) Oberst-Ing Reidenbach did not have the authority to decide this point and recommended that Professor Messerschmitt discuss the matter directly with Generaloberst Udet. Brigadeführer Croneiss and director Hentzen pointed out that an earlier decision made by the RLM Technical Office required aircraft types to be prefixed by the name of the manufacturer. Oberst-Ing Reidenbach did not accept this argument but still thought Professor Messerschmitt should discuss this point personally with Generaloberst Udet.

"To 3) Brigadeführer Croneiss handed over the description of the enlarged version of the Me 163 to Herr Antz for further attention, to 4) this will be the subject of a discussion at Augsburg on September 4."

In short, Lippisch and Abteilung L would be staying put in Augsburg by mutual agreement and Lippisch might get his 'Li' designation unless Messerschmitt could talk Udet out of allowing it.

It has been suggested that the "enlarged version of the Me 163" mentioned here refers to the Me 163 B – but this had not yet been designed, whereas Lippisch had only published his description of the P 05 the day before. It seems likely that this is what was handed to Antz for "further attention". Indeed, graphs exist dated September 10, 1941, which show that a revised version of the P 05 was still being worked on at that time.

Continuing with the August 29 meeting: "Herr Antz informed the meeting that the project office chief Voigt had submitted a project closely resembling that from Abteilung L. He had received a drawing of P 1079/13c,[11] dated August 11, 1941, and an accompany description of an aircraft, which although not tailless, had a sharply swept-back wing and was powered by a jet engine. Herr Antz remarked that when Herr Voigt had handed over this information he had said that if this project were developed it would not be necessary to build the Me 163 and the interceptor.

ABOVE: A view under the starboard wing of Me 163 V5, GG+EA, in high gloss finish. Its leading edge slot, landing and control flaps are visible.

ABOVE: An Me 163 fuselage under construction. Once design work on the experimental Me 163 was completed in early 1941, work on building the first prototypes advanced rapidly.

"Direktor Hentzen remarked that Professor Messerschmitt had considered it necessary to fit high-speed aircraft with sweptback wings after learning of the investigations made by Professor Betz and had, therefore, issued instructions for this project to be so designed. Croneiss added that, at a meeting held on the 28th of this month, Professor Messerschmitt had expressly forbidden Herr Voigt to investigate tailless designs, since these were and would remain the responsibility of Abteilung L."

Voigt's drawing, extracted from the wide-ranging P 1079 series of studies, would end up providing the basis for the later pulsejet-powered Me 328 – albeit in much modified form and without the swept wing.

In the meantime, Lippisch had won the backing of Ernst Udet and clearly had a degree of support from within the RLM, particularly from Antz. His new assertive attitude did nothing to endear him to Willy Messerschmitt, however, and it is likely that Croneiss was acting as an intermediary at this point.

During the summer of 1941, Messerschmitt suffered the first of a series of blows to its reputation as a company – the RLM, while carrying out due diligence checks on the P 1061, now the Me 264 'America aircraft', had discovered that Messerschmitt's numbers did not add up. It was apparent that the aircraft would be unable to fly 15,000km nonstop, let alone the 20,000km Messerschmitt had originally claimed it could manage.

According to a summary written on May 9, 1942, by Generalleutnant Eccard, Freiherr von Gablenz: "Because Messerschmitt lacks experience in building heavy aircraft, and also because, in the opinion of the RLM technical specialist, the Me 264 plan-form was too narrow, the design could not be adopted as the standard long-range bomber for the Luftwaffe and other firms had to be invited to tender."[12]

The order for 30 Me 264s was put on hold and the number of prototypes was reduced to five while a new set of requirements for a re-run of the Fernkampfflugzeug competition were drawn up. Messerschmitt would have to re-tender and now ran the risk of another firm being given the contract. It was a disaster and it resulted in further consternation and conflict within the company.

During September, as planned, discussions took place about the P 05 Interceptor during which it appears that the project was effectively split in two. Rather than having a single large aircraft that could climb extremely fast then cruise along at altitude, engaging in dogfights, two projects emerged which each took up part of the P 05's role. One was a fast-climbing, short endurance rocket fighter and the other a version of the P 05 with a conventional undercarriage and powered by two turbojets.[13]

Lippisch later wrote that the Li P 05 "was a true-to-scale enlargement of the Me 163 V4 to ensure the design of an aircraft of known characteristics, good weaponry (four machine guns), and great range. However, since it was feared that in this aircraft of 12.8m wingspan the control forces would be too high, it was decided to return to the smaller size.

"On September 14, 1941, we produced a smaller design, the Li 163 S (S for series production) with a wingspan of 9.2m that was to be armed with four machine cannon."

The Li 163 S fighter was to be a fairly straightforward development of the experimental Me 163 and was given the designation Li 163 S in the confident expectation that Messerschmitt would be unable to persuade Udet that

ABOVE: Completed Me 163 V4, KE+SW, in July 1941 at Peenemünde – later that year it would be flown to 1,003km/h by Heini Dittmar.

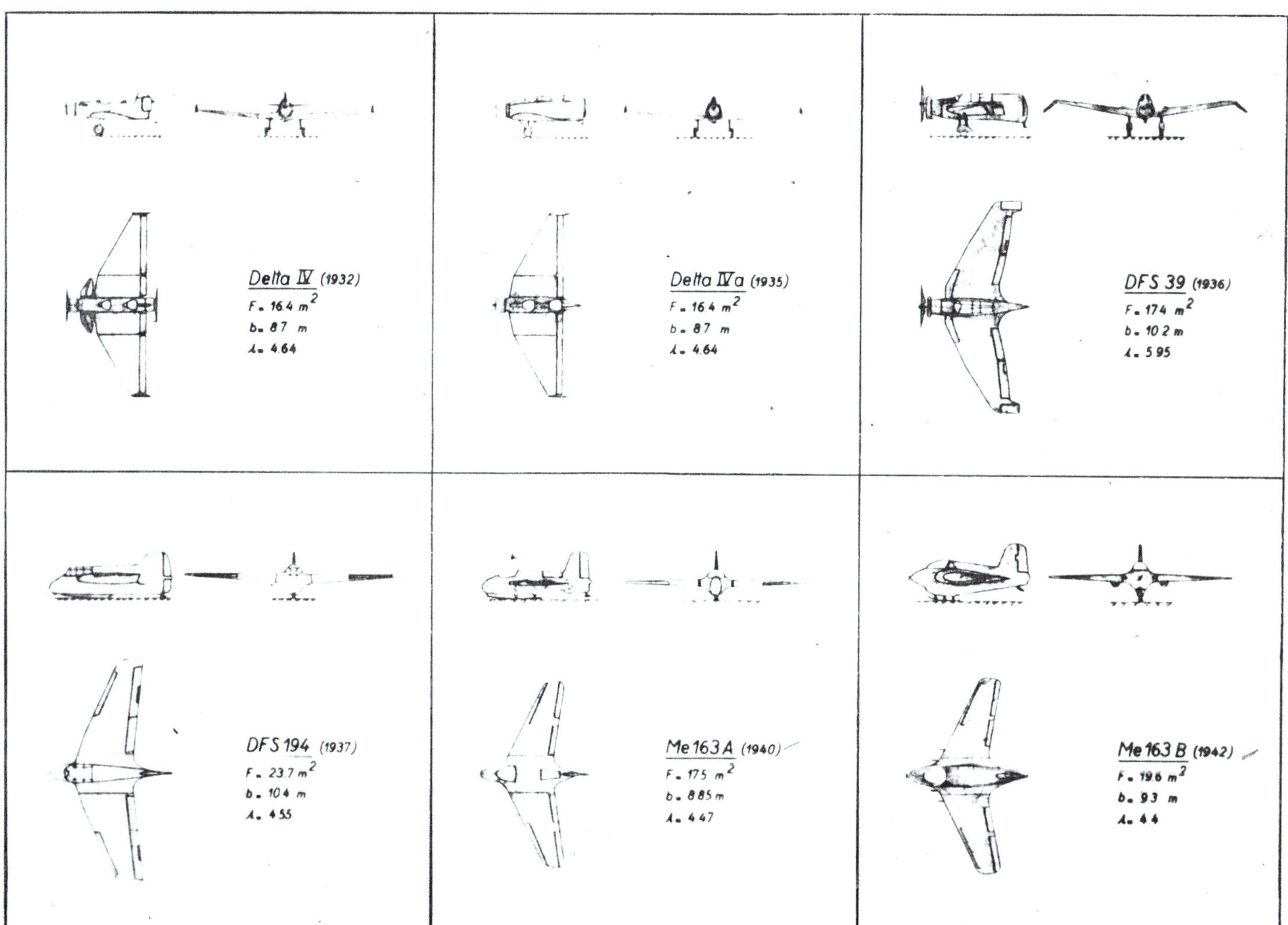

ABOVE: This diagram from a presentation given by Lippisch to the Deutsche Akademie der Luftfahrtforschung in 1943 shows a timeline of designs leading up to the Me 163 B along with dimensions. The design of the Me 163 A was indeed enlarged to provide the basis for the Me 163 B – but not by much.

Lippisch should be denied his 'Li'. Rudolf Rentel took on the task of turning his own rocket-propelled P 05 design into a similarly scaled twin-turbojet fighter – a task which evidently took a little longer than the swift makeover which had resulted in the Li 163 S.

In the meantime, on October 2, 1941, Heini Dittmar became the first man to fly faster than 1,000km/h in the Me 163 V4. Lippisch wrote: "Our success in 1941 was crowned by Heini flying faster than 1,000km/h. At just over 1,000km/h the shock wave caused the outer wing to separate suddenly and the aircraft pitched nose down, the instruments registering a negative acceleration of 11 g. Heini pulled the throttle back quickly, the aircraft slowed down and he was able to take control again. In the evening the Askania theodolite measurements were evaluated: the final result revealed a speed of 1,003km/h!

"After it had been reported to Berlin, the significance of our progress began to be recognised. When I returned to Augsburg, Messerschmitt did not speak to me at first: he first had to digest the fact that my aircraft was faster than his."[14]

TAILLESS PATENTS

Between October 11 and October 13, 1941, he sketched out at least three separate patent applications, each of which seemingly arose from the need to improve the characteristics of tailless aircraft during take-off and landing. The first of these saw him returning to research he had carried out five years earlier while working at the DFS – forward-swept wings.

Lippisch wrote: "We should not omit mentioning a special project of 1936 which was of unusual design. In England Professor Townsend had shown that by sweeping the wing forward and strongly tapering it toward the tip, stalling at the outer wing section could be effectively postponed.

"Even at large angles of attack and low speeds, roll stability could thus be maintained, as opposed to the normal swept-back wing where stalling started at the wing tips. To study this problem we designed an experimental glider Kormoran DFS 42 with forward sweep of the wings, which was built in 1936.

"We observed the flow by means of filaments and clearly observed the Townsend phenomenon. We took a number of photographs of the wing flow on subsequent flights. However, although the flow on the outer wing remained steady, the flow over the midsection of the wing stalled relatively early causing loss of longitudinal stability and lower maximum lift.

"We therefore installed an additional rudder over the midwing section which prevented the incipient loss of lift. On the whole, these experiments were interesting but failed to prove any superiority of the wing with forward sweep, since on normal swept-back wings stalling of the outer wing could be prevented by installing wing slots which produced higher maximum lift. From that time on no further experiments with swept-forward wings were carried out by us."[15]

While there is no evidence to indicate that Lippisch did conduct further practical experiments, he evidently did return to the problem and his employer, Messerschmitt AG, patented what he considered to be a potential solution on his behalf.

Entitled 'Flugzeug mit nach vorne gepfeilten Flügel' or 'Aircraft with forward swept wings', the patent[16] says: "Swept aircraft wings with both positive and negative arrow position (wing ends swept back or swept forward) are made inherently stable by appropriate twisting (setting angle at the leading parts larger).

"The negative arrow position brings greater security against stalling, because in contrast to the positive arrow position no flow can take place from the boundary layer

ABOVE: Drawings from Alexander Lippisch's patent for a W-wing arrangement intended to improve the take-off and landing characteristics of tailless aircraft. Abb. 1 shows a plan view of the unusual wing form, Abb. 2 shows the 'twist' applied to the wing leading edges and Abb. 3 shows a forward view of the design.

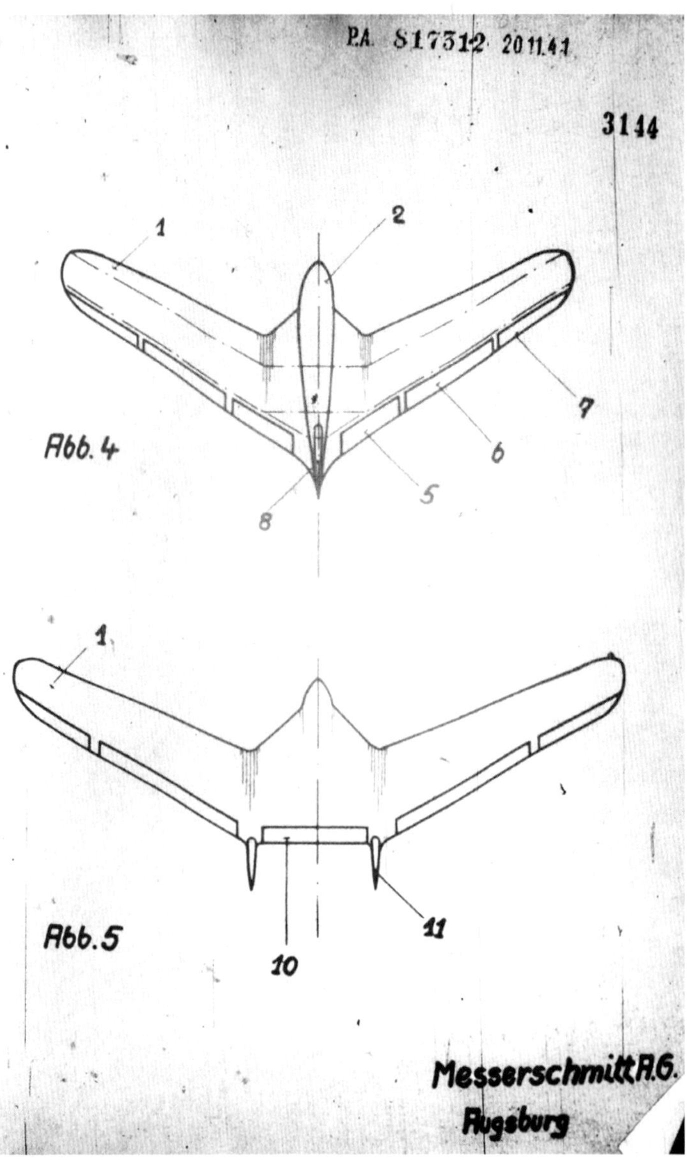

ABOVE: Abb. 4 shows a generic teardrop-shaped fuselage with vertical fin and rudder to the rear with another unusual wing planform. This time only the leading edge has the W-wing shape, while the trailing edge is straight. This results in extremely deep wing roots. Abb. 5 shows a similar arrangement but with twin fins and a straight horizontal stabiliser.

to the outer wings and the boundary layer instead flows inwards towards the middle of the wing. This means, however, that the flow, which has already been disturbed by the influence of the fuselage, is even easier to break off in the middle part of the wing; the result is great additional resistance and loss of lift even at moderately high angles of attack.

"The present invention is based, then, by appropriate shaping of the wing plan to direct the boundary layer where it can do the least damage. According to the invention, in an aircraft with a negative arrow position, with predominantly forward-facing wings and at the front of all wing-ends, which are connected to the hull and, where appropriate, to engine nacelles, with positive sweep. The result is that the destructive boundary layer flow is not only kept away from the tips of the wings, but also from near the fuselage, where there are already favourable flow conditions.

"It is sufficient if the positive sweep to be used according to the invention at certain points extends to the leading edge of the wing; this is enough to divert the boundary layer in the desired sense. The twisting of the wing according to the invention depends on what has been said above: the setting angle becomes larger the farther the relevant wing part is from the fuselage.

"A further improvement of the flight characteristics can be achieved in a further embodiment of the invention in that the wing receives a weak W-shape not only in the top view, but also in the forward view."

The patent is illustrated with five diagrams. The first shows Lippisch's W-shape design from above, the second shows the wing 'twist' he proposes and the third a forward view of the same design, indicating a slight gullwing effect. The fourth drawing shows "another embodiment in plan view; the positive sweep in the area of the fuselage (2) is limited here to the leading edges. This leads to a strong increase in the wing depth against the fuselage and therefore allows a very favourable connection of the wing (1) to the fuselage (2). For the control, three

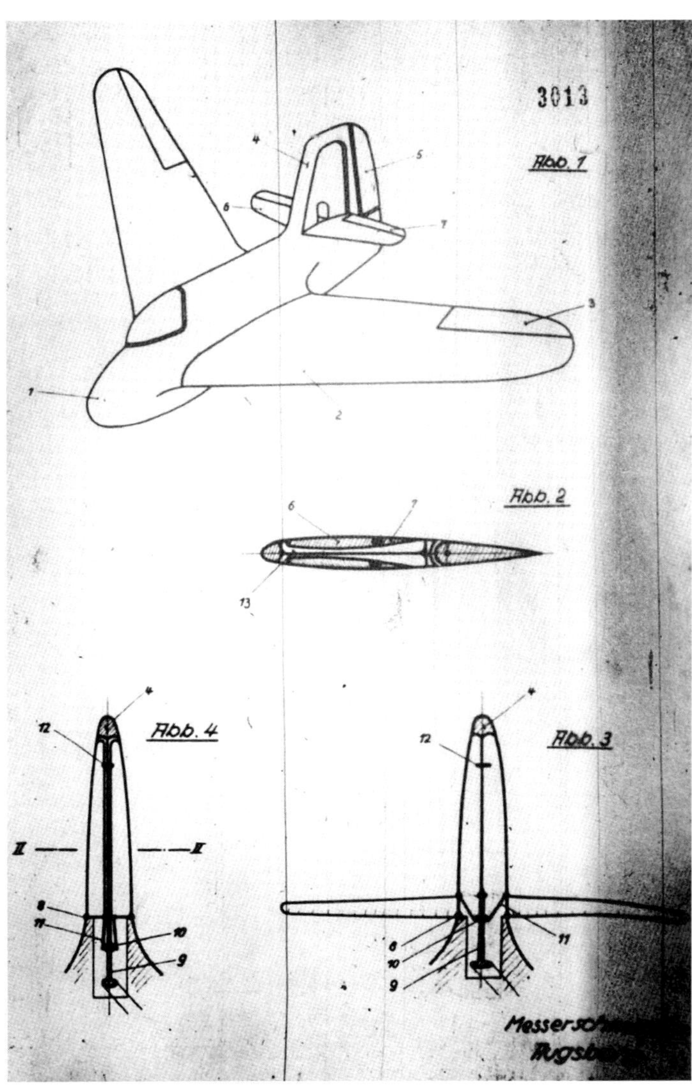

ABOVE: The first page of illustrations from the formal submission of Lippisch's ideas to the Reich patent office by Messerschmitt AG. This shows the flip-down tail planes.

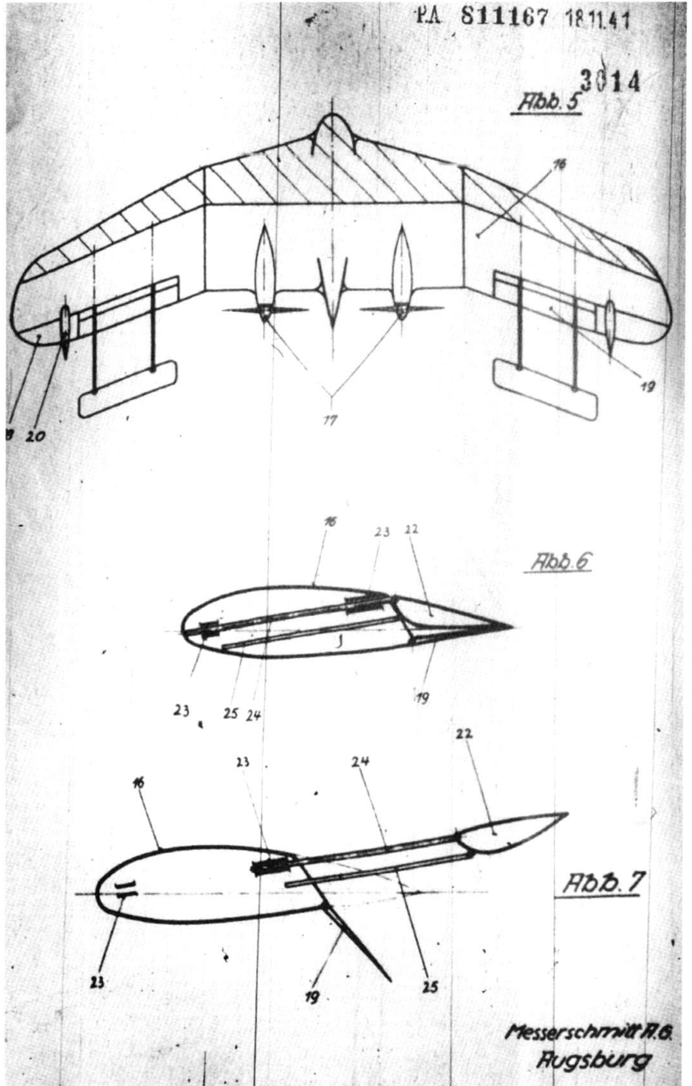

ABOVE: A flying wing with 'tail planes' that can extend from the trailing edge of its wings.

pairs (5), (6) and (7) of wing flaps are attached, as well as a central rudder (8).

"The inner flap pair (5) serves as an aileron and is just as actuated as the elevator of an aircraft with a tail fin. The pair of flaps (6) is used for lift increase during take-off and landing, since the flap neutral axis is almost in the centre of gravity and therefore no tilting moment arises around the transverse axis at flap deflection.

"Adjusted in the opposite direction, the flaps (6) can also be used for trimming around the longitudinal and vertical axes. The outer pair of flaps (7) serve primarily as horizontal stabilisers, also in the same direction adjustment for trimming about the transverse axis. The flap arrangement is thus adapted to the aircraft's layout, which allows the achievement of high lift coefficients at take-off and landing."

The fifth drawing is of a design "basically similar" to the fourth drawing "but here the trailing edge of the wing centre section is placed perpendicular to the direction of flight and equipped with a single continuous horizontal stabiliser (10), while the rudder (11) is divided and relocated to the kinks in the wing".

The formal version of this patent application, filed on November 14, 1941, received PA number 817312 of November 20, 1941. Filed on the same day, evidently having been drafted at the same time, but seemingly dealt with sooner was patent number PA 811167 of November 18, 1941.

This is entitled 'Höhenleitwerk für Flugzeuge' or 'Tailplanes for aircraft'[17] and effectively incorporates four different ideas into a single package. The patent description states: "The invention consists of attaching an additional retractable tailplane unit, which is spatially separated in the operative position from the wing, on an aircraft with horizontal stabilisers which are mounted on the wing itself, and which can be used for altitude control. This should be retracted during the high-speed flight and deployed at take-off and landing, expediently together with the lift flaps or with the landing gear. This tailplane may consist of a fin with an invariable angle of incidence, of a controllable rudder alone or, ultimately, a combination of both.

"One might ask why one does not build this tailplane on the aircraft and use it all the time. Consider not only the additional harmful drag of the separate tailplane but also its unfavourable location for high-speed flight in the area of the fuselage flow, where it is greatly affected by changes in the airflow. These flow disturbances often cause dangerous rudder vibrations, which have often led to severe accidents. In high-speed flight, therefore, the conventional control of tailless aircraft by means of wing flaps is much more advantageous since they lie in undisturbed flow."

The patent states that there are a number of different possibilities for the inclusion of tailplanes – the first illustration shows "an embodiment of the invention in a perspective view, Figure 2 shows a horizontal section through the tail fin, Figures 3 and 4 show the planes in

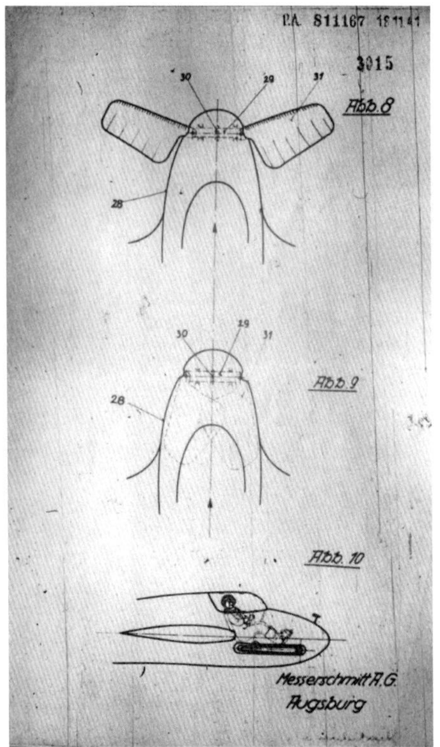

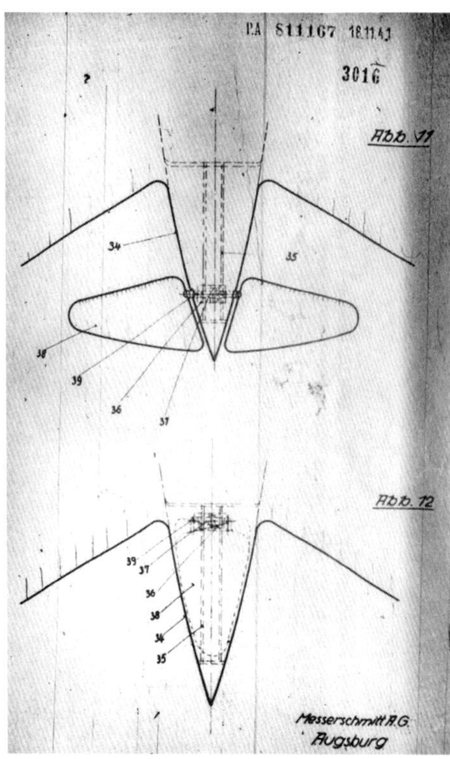

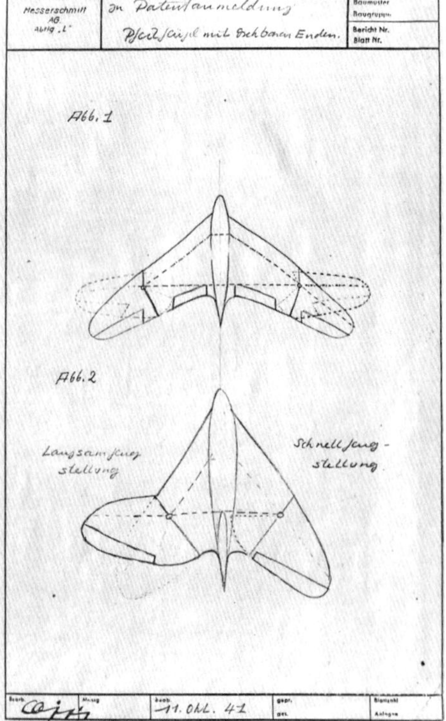

ABOVE LEFT: The formal submission of Lippisch's idea for pop-out canards. **ABOVE MIDDLE:** Patent drawing showing the mechanism employed by Lippisch for his flip-out tail planes. **ABOVE RIGHT:** Lippisch's patent drawing of October 11, 1941, for a tailless swept-wing aircraft with rotatable wingtips. The tips would be swivelled forwards for low-speed flight, then swivelled back for a sharply swept form at high speed.

extended and retracted positions, Figure 5 shows a tailless aircraft with another embodiment of the invention in plan, Figures 6 and 7 show the extendable tailplane section as part of the wing in the extended and retracted position, Figures 8 and 9 show a third embodiment in the extended and retracted position, Figure 10 is a side view, Figures 11 and 12 a fourth embodiment in the extended and retracted position".

While the first means of introducing retractable tail surfaces for landings and take-offs is straightforward enough – tailplanes that pop out of the fin or retract back into it as needed – the second consists of surfaces that extend backwards from the wings of a tailless aircraft, the third is a set of pop-out canards at the front of the aircraft and the fourth is a set of tailplanes that slide horizontally out of the aircraft's fuselage rather than from the fin. All were intended to achieve the same effect.

The third and last-known patent from this fertile interlude in Lippisch's career has the title 'Pfeilflügel mit drehbaren Enden' or 'Swept wings with rotatable ends'.[18] Unlike the other two patents, it is unknown whether this one was actually filed by Messerschmitt. The only known copy comes from Lippisch's own collection of papers and is both handwritten and signed by him.

The description says: "The sensitivity of the wingtips of a swept wing is based on the fact that the slipstream flows outwards and the boundary layer is accumulated on the outer wing. As a result, the swept wing has the property of tilting at maximum lift. The idea of the invention eliminates this property in that e.g. when taking off and landing the wing ends are pivoted forward about a substantially vertical axis, so that these wing sections are approximately perpendicular to the direction of flight.

"The build-up of air coming from the inner wing will then flow backwards at the bend, so that the flow on the outer wing remains healthy. The axis can be placed by slight inclination to the flight level so that rotation simultaneously generates a small pitch change in the sense of a twisting of the outer wing.

"The rotation of the outer wings can also be used during the flight for trimming. The outer wings carry the ailerons, which can serve as rudders at the same time. The control cable goes through the axis of rotation. In a known manner also slats can be attached to the outer wing. In particular, the inventive concept is suitable for high-speed aircraft with a very strong arrow shape.

"With these aircraft, which have a very high-speed range, it is necessary to achieve the high-speed flight by nose-heavy trimming and the slow flight by tail-heavy trimming. The forward pivoting of the outer wing provides besides the tipping safety the desired trim change for slow and fast flight."

In short, the large outer sections of the aircraft's wings could be rotated to provide straight leading edges and therefore additional lift for take-offs and landings and when flying slowly. Perhaps the Messerschmitt company felt that this was the least plausible of Lippisch's three radical proposals for improving the performance of tailless aircraft at low speeds and therefore left it out when putting forward the latest round of patent applications.

ABOVE: Junkers tested a model of the Me 163 B with a new nose and canards later in the war.

The idea deemed most likely to succeed appears to have been the retractable canards, since wind tunnel models with this arrangement were tested by Junkers later in the war.

SUPER 163 AND ME 163 C

Rentel's first completed design for the turbojet P 05, now designated P 09, had been produced on October 28, 1941.[19] The type of engine to be used for this new fighter was unspecified and its wingspan was reduced from the P 05's 12.8m down to 11.6m, while its length was reduced by 50cm from 7.6m to 7.1m. Its undercarriage consisted of two main wheels which retracted into the wings and a tail skid.

Ernst Udet committed suicide on November 17, 1941, and with him died Lippisch's hopes of having his Messerschmitt type designations prefixed with 'Li'. The Li 163 S had, in any case, been superseded by another new design which received the designation Me 163 B. The original Me 163 was now retrospectively renamed the Me 163 A and plans to build the 'B' in series got under way with an order being placed for 70 examples.

The final new Abteilung L design of 1941 was a ground-attack spin-off from Rentel's P 09, given the unusual designation P 010,[20] presumably to indicate that it was more of a P 09.5 rather than the P 10, a designation which would not be applied until May of the following year. The P 010, dated November 26, 1941, was substantially increased in size from the P 09 fighter, having a 13.4m wingspan and being 8.15m long. This resulted in large part from the need to provide space for a bomb bay in the central portion of the fuselage.

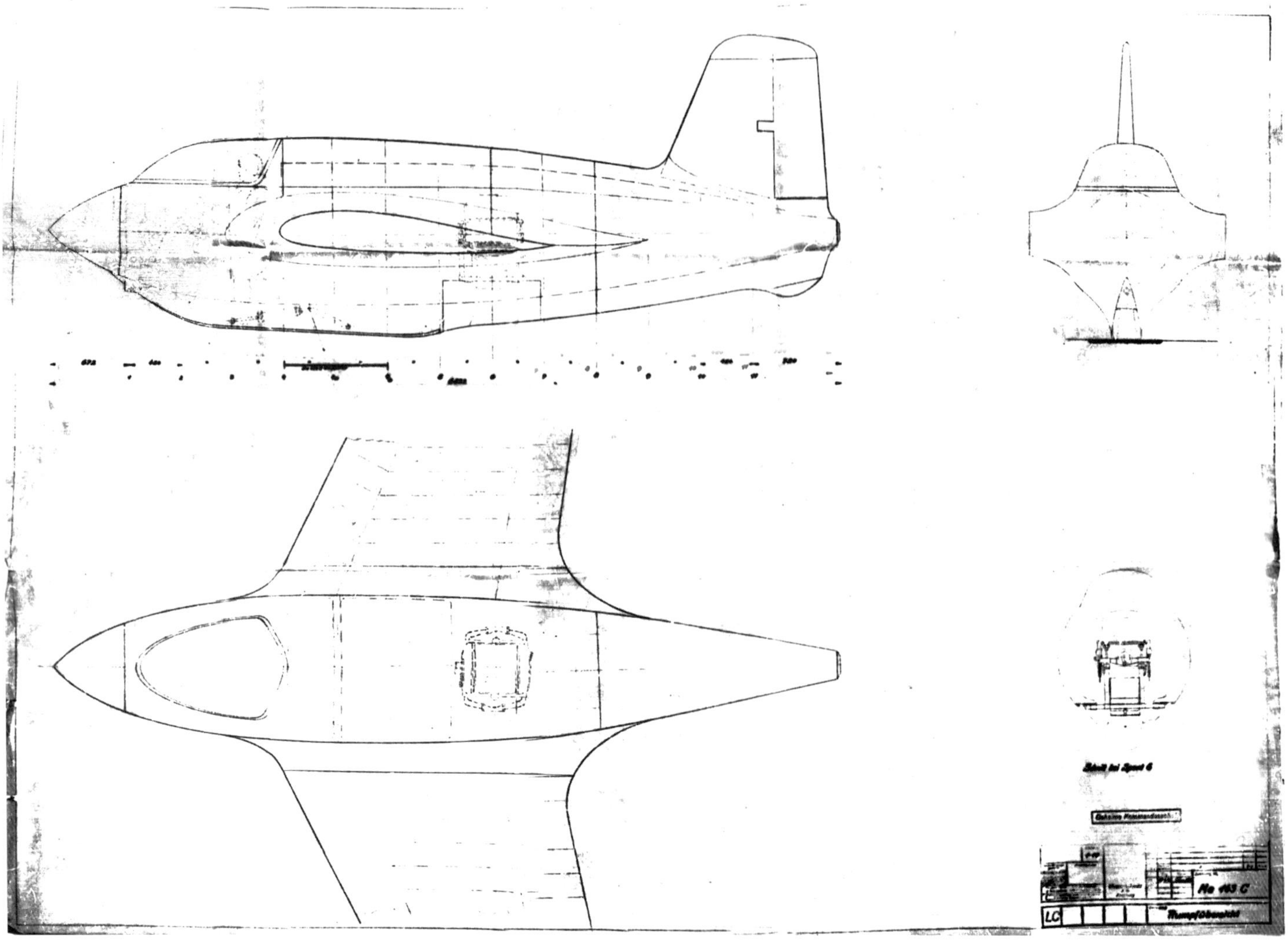

ABOVE: Drawing from the original report on the Me 163 C, dated March 1942. It was to be a reconnaissance version of the Me 163 – with an extended fuselage, camera and small second rocket engine for cruising.

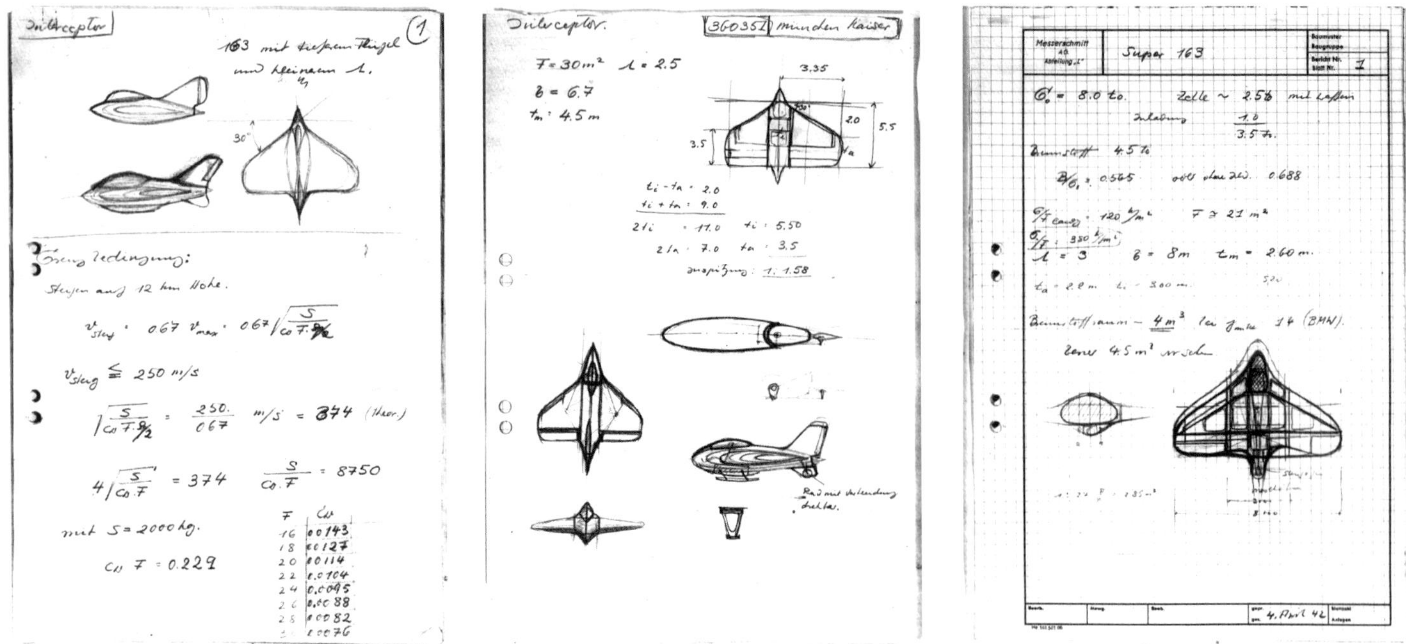

ABOVE: Lippisch's designs for a low aspect ratio interceptor, from a sketchbook dated January/February 1942 but including work up to April. Earlier designs are labelled simply 'Interceptor' but this later becomes 'Super 163'.

On December 1, construction of the first Me 163 B prototype commenced and later in the month the deadline for design submissions to the re-run of the Fernkampfflugzeug competition passed. There were two entries from Messerschmitt – but neither of them was Wurster's flying-wing bomber, the P 08. Instead two different versions of the Me 264 were submitted, one with four engines and one with six.

However, despite all this productivity there were further storm clouds gathering on the horizon for the Messerschmitt company. Work on the long-awaited multi-role Me 210 was not going well and Udet's successor would be less inclined to tolerate Messerschmitt's seemingly endless delays.

Abteilung L spent the first four months of 1942 fully engaged in design work on the Me 163 B. Lippisch himself, however, seems to have had little direct involvement with the detail design of the Me 163 B. One side effect of the Me 163 A's success was an increase in the number of meetings Lippisch needed to attend at the RLM in Berlin. He also became increasingly conscious that despite his tailless aerodynamic ideas having now been 'proven', none of the new aircraft designs being created by Abteilung L were being chosen for development.

He later wrote: "Our advocate, General Udet, had chosen suicide and despair about the leadership of the Luftwaffe and his successor, Milch, had nothing more to do than to reduce our programme. Udet had no longer been able to watch with what true dilettante measures the air war was carried out.

"None of the new machines, the single-seater fighter Me 209, the two-engine destroyer Me 210 of the Messerschmitt AG and the big bomber He 177 of Heinkel, were used," he wrote. "They were complete mis-constructions and better types were not in sight. How should the German Luftwaffe, with its now obsolete types of aircraft, face the newly created types of their opponents?"

He bemoans the delays at rocket engine manufacturer Walter which resulted in the Me 163 failing to reach front-line service until January 1944, then continues: "Leaving the universally enthusiastically acclaimed features of the Me 163, we also made parallel designs for tailless fighters with turbojet and piston engines, so as not to be reliant on the unsafe rocket.

"We also made designs for fast tailless destroyers and bombers with turbojet or piston engines. These, however, often only reached mock-up stage and none of them were built, since comparisons made with other designs were corrupted and tainted for different reasons.

"In addition to these machinations, all sorts of personal differences had taken place which had a profound effect on productive work. The whole mishap of the aircraft and many other things was already apparent in 1941. In 1941 I often had to go to Berlin, where I noticed many things in the RLM, which did not appear in the press under Goebbels, about the fact that we had lost the war and that it was only a question of preventing a total collapse as far as possible."[21]

It seems unlikely that Lippisch was noticing "many things" at the RLM which indicated that the war was lost in 1941, but nevertheless he was spending more and more time in Berlin while his team got on with their work on the Me 163 B at Augsburg.

Lippisch did, however, still find time to muse on what a successor to the Me 163 B might look like. In a notebook bearing his signature and dated January to February 1942,[22] he penned drawings of what he called the 'Super 163', a low aspect ratio interceptor based on the Me 163 with a short and wide delta wing, a longer flatter canopy and a sharply swept tailfin with substantial under-fin. Alongside this, he jotted down calculations on an Me 163 C.

It was the latter that became the subject of a full report on March 23, 1942 – the Super 163 evidently proceeding no further than a handful of calculations and sketches. However, the Me 163 C (Nahaufklärer) or 'Me 163 C (close reconnaissance)' report stood in stark contrast to the Abteilung L reports which had preceded it. Gone was the Abteilung L-headed notepaper, and in its place the standard Messerschmitt AG form complete with title page Messerschmitt logo.

It begins: "Task: To examine the performance of a fuselage of the Me 163 B pattern modified for the purpose of close reconnaissance. In this case, it is assumed that the fuselage is designed without armament and armour and without fuel tank protection. An FK 50/30 camera is to be installed. The aircraft is equipped with a pressurised cabin.

"Constructive measures: The unchanged fuselage of the Me 163 B is unsuitable for the installation of an FK 50/30. By extending the fuselage without changing the wings and tailpiece, i.e. by inserting two sections at the thickest point of the trunk, the space is created."

The aircraft's internal fuel tanks were to be shuffled around to create room for the camera and in addition to the usual 1,500kg thrust rocket engine, in this case to be supplied by "BMW (Zborowski)"; a second much smaller rocket engine of just 100kg thrust was to be provided to extend the aircraft's cruising range.

In operation, the Me 163 C would climb rapidly to 20,000m (65,616ft) before gliding over its target and then back to base. The maximum possible range would be 900km, though launched from a standing start on the ground using only its own internal engine range would be just 500km.

Summing up, Lippisch writes: "The provision of such an aircraft can take place in the autumn of 1942, if the construction work on the Me 163 B, ending in April 1942, is immediately switched over to this pattern." This was the first appearance of the Me 163 C and would be by no means the last but for now the project went no further.

DISSOLUTION

The Me 210 was a huge failure for Messerschmitt The intended successor to the Bf 110 heavy fighter, its development had been subject to a string of delays and the aircraft that had been completed proved to be flawed – fatally so in some cases where their pilots were concerned.

Yet there had been such confidence in the conventional twin-engine design that vast quantities of components had been produced and delivered to Messerschmitt's factories in anticipation of full production being given the green light. The expense involved was immense and the company was required to pay all of its suppliers in full.

Production of the Me 210 was halted on April 14, 1942, and within a month Willy Messerschmitt had been removed as chairman of the board and managing director of his own company. Even his personal aircraft, Bf 108 D-IMTT, was taken off him.

However, he stayed on at the company as chief designer – giving him direct day-to-day responsibility for both Abteilung L and the conventional project office and making him effectively Lippisch's line manager. In the meantime, Abteilung L had been busy working on a tailless piston-engined fast bomber personally commissioned by Milch under the designation P 10. Disagreements over this design and Lippisch's direct 'back channel' to Milch soured his relationship with Messerschmitt still further and another political shift occurred at the company on November 7, 1942, when Theo Croneiss died aged 47. Croniess, though a great friend, colleague and supporter of Messerschmitt himself, had also been a supporter of Lippisch since 1931, as previously related. His loss may have contributed to the deteriorating relationship between the two men.

During early 1943 Abteilung L focused on another design of Wurster's, the piston-engined tailless Me 329, and Lippisch continued to clash with Messerschmitt. After the P 09 and P 010 jet aircraft came another, the P 11, and two further piston-engine bombers, the P 12 and P 13. Although the latter two designs went nowhere, Lippisch seems

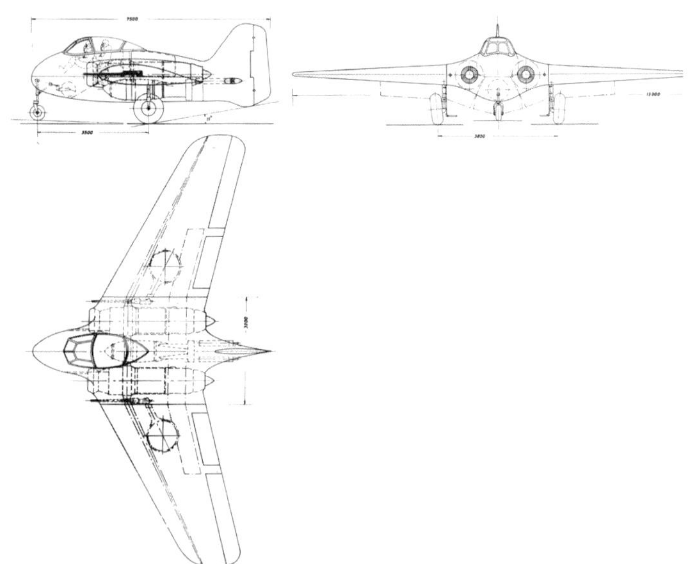

ABOVE: Abteilung L's P 11 two-seater twin-jet fast bomber, dated September 13, 1942.

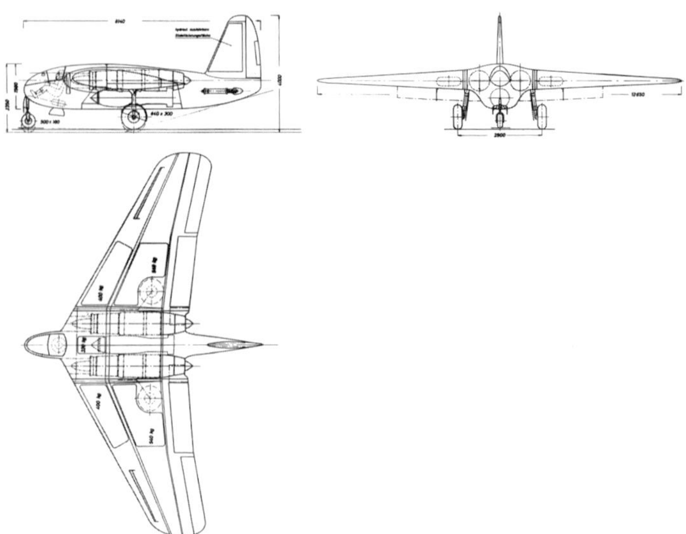

ABOVE: Lippisch evidently realised that the concept of a twin-jet bomber had potential and stuck at P 11, evolving it into this aerodynamically cleaner version, dated December 2, 1942.

ABOVE: Two views of the Messerschmitt Me 163 BV2, VD-EL, photographed at Augsburg in September 1942 before being fitted with four 20mm MG 151s for firing trials in October.

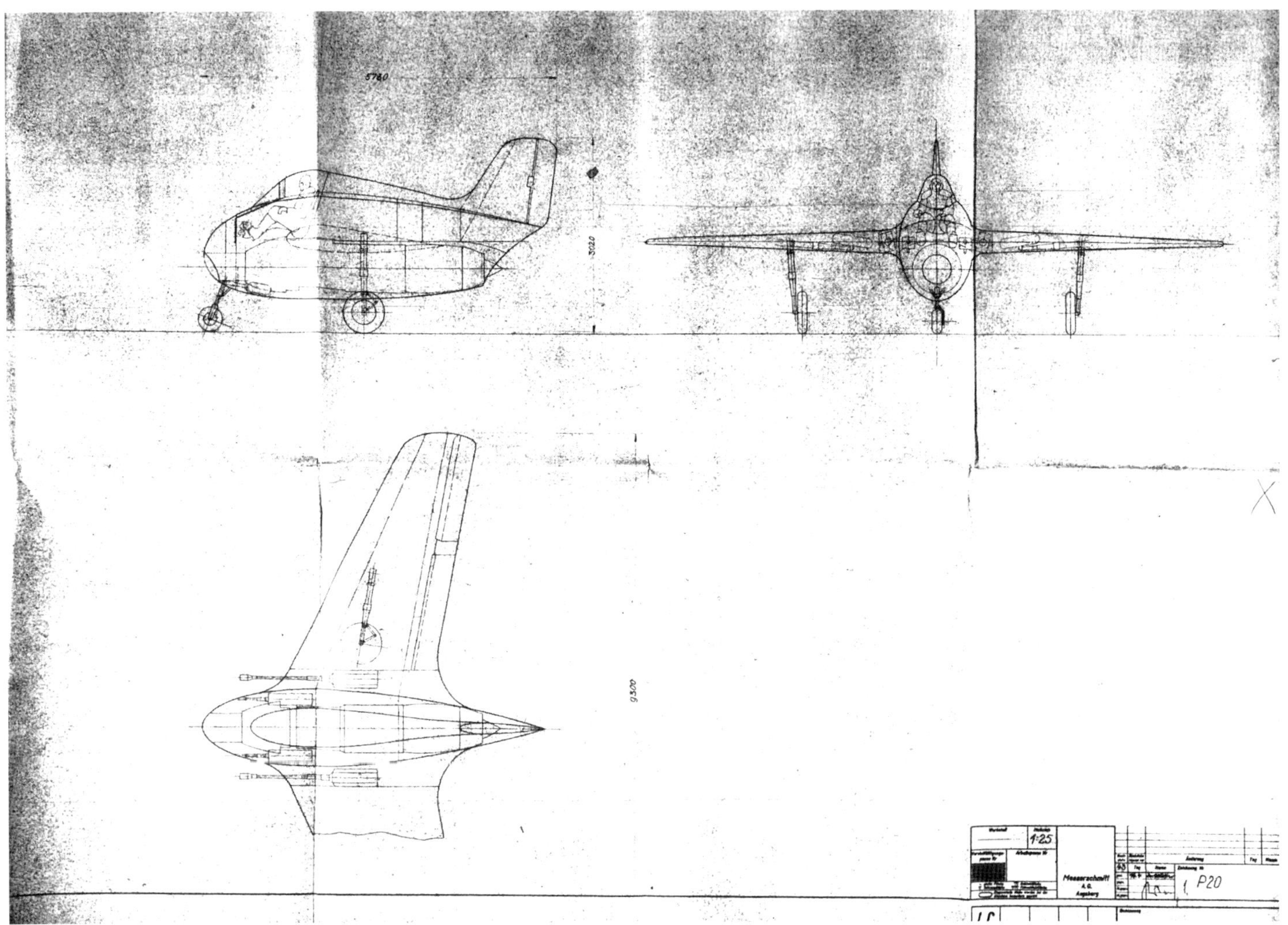

ABOVE: The final Abteilung L design – the Me 163-derived P 20 jet fighter. This drawing is dated April 16, 1943, and shows an early version of the design.

to have become increasingly convinced that the P 11 had potential.

The end of Abteilung L came on March 26, 1943 when it was formally announced that the division would be wound up and its personnel redistributed within the Messerschmitt company. Lippisch himself was to leave Messerschmitt and take up a new position within the Luftfahrtforschungsanstalt Wien (LFW) or 'Aeronautical Research Institute Vienna', where he would have a 100-strong workforce at his disposal.

Abteilung L still had one last card to play, however. On April 19, 1943, a report was produced on the P 20.[23] The design itself was for a stubby single-jet fighter or bomber. The report lacks an introduction – commencing on Blatt 1 or 'Page 1' with a graph showing only the performance of the Jumo 004 C – and is produced on 'Messerschmitt AG Projektbüro' standard forms.

The only text given in a total of 11 pages is almost a series of bullet points: "General comments on the provisional performance specifications. Fuselage: Tailless jet fighter P 20. Originated from Me 163 B by installing a Jumo jet device in the fuselage. Engine: 1 x Jumo special engine 109 004 C.

"Power determination: The indicated flights and weights are to be regarded as provisional and do not constitute guarantee values. The thrusts and consumptions were taken from preliminary Junkers data according to curve sheet 109 004 C.

"The horizontal speed, climbing speed and climbing times were measured with an average speed. Weight = 3,000kg. No external loads were taken into account. All flight performances are without Mach influence!" The drawing attached, dated April 16, is by Wurster.

A memo dated April 27, 1943,[24] shows the 86 remaining members of Abteilung L, including nearly all of the men who had followed Lippisch from the DFS, being divided up between the project office, stress office, design office, flight test department, workshops and administration office of Messerschmitt AG. Lippisch officially left the company the following day, April 28.

Although Abteilung L was no more, several of its projects lived on – including the Me 163 and its derivatives, the P 11 and the P 20.

Chapter 3

The first competition

Messerschmitt Me 262 early versions and Heinkel He 280

The P 65 was Messerschmitt's official competitor for Germany's first jet fighter requirement - regardless of what was happening simultaneously within Abteilung L.

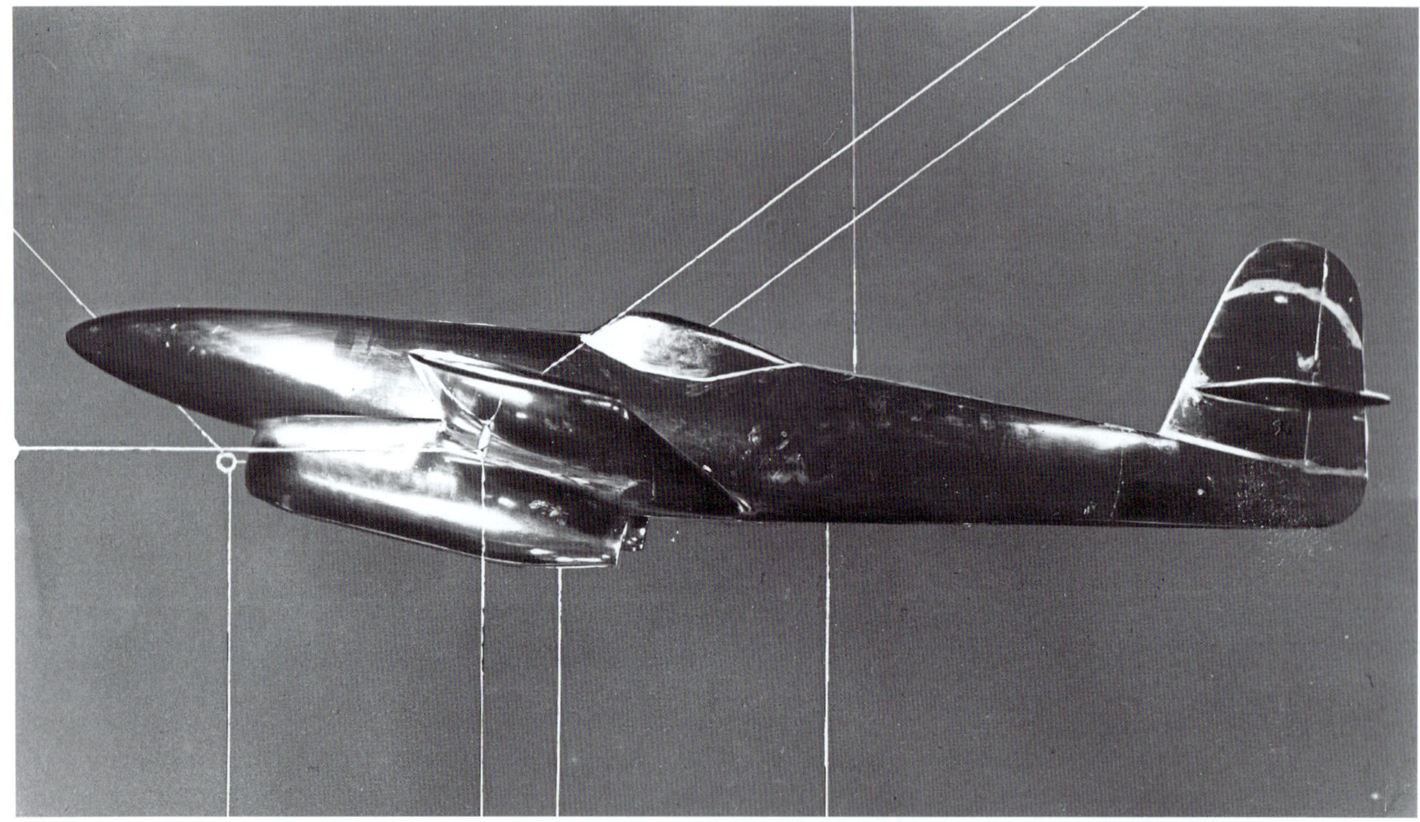

ABOVE: This wind tunnel model of Messerschmitt's P 65 with straight wings similar to those of the Bf 109 E and an unswept tailfin was tested by the AVA during 1940.

ME 262 AND HE 280

Following on from the P 65 mock-up inspection carried out in December 1939, a series of 20 BMW P 3302-powered prototypes was proposed on March 1, 1940,[1] though the production model was intended to get the much smaller and more advanced BMW P 3304. The first full project brochure,[2] issued in March 1940, showed the aircraft as a tail sitter powered by a pair of P 3304s, each mounted mid-wing rather than slung underneath. By this time Messerschmitt was adding '10' to the beginning of its project bureau designs, making the P 65 the P 1065 – and this is what appeared on the front of the brochure. Armament was to consist of a trio of MG 151s in the nose with 250 rounds each, and the whole nose tip could be swung upwards on hinges to allow access for reloading.

A highly ambitious RLM Lieferplan or 'delivery schedule' dated July 1, 1940, showed 20 Messerschmitt P 1065s on order with the first due to be delivered in September 1940, then two more in October, three in November, five in December, five more in January and the remainder in February.[3]

Unfortunately, BMW was struggling to get the P 3304 to a point were it could reliably power an aircraft in flight and by January 1941 the situation was becoming desperate. The rival He 280 had already begun flight testing four months earlier – albeit under tow – whereas the P 1065 did not even exist as a prototype.[4]

The same Lieferplan had showed 20 He 280s on order, with two due in August, three in September and then five per month for the rest of the year.

With the BMW turbojets delayed, Messerschmitt now proposed the use of a Jumo 210 G piston engine, fitted in the P 1065 V1's nose, bolstered by a pair of Walter R II 211 rocket engines. This plan was approved and the P 1065 V1 was built between February and March 1941.[5]

By February 5, 1941, a total of 32 gliding flights had been completed by Heinkel's He 280 V1. It was reported that stress-testing of the He 280 fuselage design had been carried out and tests were ongoing with the wings, tail and landing gear. The V2 was now fitted with its engines and had been ground-tested under power on January 24. The He 280 V3 was also being readied for powered flight and was due for completion on April 1. V4 was due for completion on May 1, 1941; V5 on June 1, 1941. Components for an initial run of five pre-production aircraft were already in the process of being manufactured.[6]

The He 280 V2 made its first powered flight, lasting three minutes, on March 30, 1941. Company test pilot Fritz Schäfer made the flight with the aircraft's engines exposed – because they leaked fuel so badly that housing them in their aerodynamic nacelles would have led to a dangerous build-up of combustible material. Flights were suspended the following month because the HeS 8 A was still too fragile for prolonged testing – primarily because the materials used in its construction simply could not withstand the temperatures involved.[7]

The P 1065 received the official RLM designation Me 262 on April 8, 1941,[8] and 10 days later the newly renamed Me 262 V1 flew for the first time – though only powered by its Jumo piston engine. The He 280 V2 had flown with jet engines nearly three weeks earlier.

The RLM confirmed the Me 262 V1-V5 series and also approved the building of 20 pre-production machines on July 25, 1941.[9] That same month it was proposed that high-speed testing of the He 280 airframe might be continued by fitting Walter rockets into its nacelles but this idea apparently came to nothing.

Two BMW P 3302 prototype engines were delivered to Messerschmitt in September 1941 and it took until December to get them fitted. And by the end of 1941 Heinkel's HeS 8 A remained incapable of sustaining the He 280 in powered flight for any significant length of time. During March 1942, Heinkel approached Argus to discuss the possibility of flying the He 280 using its pulsejet motors. It was suggested that a pair of pulsejet tubes could be fitted beneath each wing, enabling a return to flight testing during May.[10] At the same time, Heinkel also investigated the possibility of obtaining engines from its competitors – Junkers and BMW.[11]

Having taken off 47 times on its piston engine alone, the Me 262 V1 took off again on March 25, 1942, this time with the P 3302 prototypes under its wings. All eyes were on Messerschmitt test pilot Fritz Wendel as he lifted the aircraft off the runway at Augsberg – only for both jet engines to suffer compressor blade failure. Wendel managed to get back on the ground safely with just the Jumo piston engine but the damage to the Me 262's reputation was already done.

The He 280 V1 was retired from towed glider flights on May 22 in readiness for the installation of four Argus engines and the RLM scrapped the order for 20 pre-production Me 262s on May 29, 1942, only keeping the initial order for five prototypes.

Just three days later, a pair of Junkers T1 (Jumo 004) engines were delivered to Messerschmitt and over the next six weeks they were fitted to the Me 262 V3. On July 18, Wendel flew the aircraft for 12 minutes without problems in the morning, then flew it for another 13 minutes at around midday, managing to reach 342mph.[12]

ABOVE: Another wind tunnel model of Messerschmitt's **P 65** – by now known as the **P 1065** – but with the engine nacelles built into the wing, rather than slung beneath it, and a graduated sweepback. The tailfin has also been given its familiar triangular shape.

ABOVE: Drawing from Messerschmitt's **P 1065** project construction description issued in March **1940**. At this point the aircraft was a tail sitter armed with three MG 151s in its nose.

ABOVE AND BELOW: Few photographs of the Messerschmitt Me 262 V1 survive. It was fitted with a Jumo 210 G driving a twin-bladed propeller.

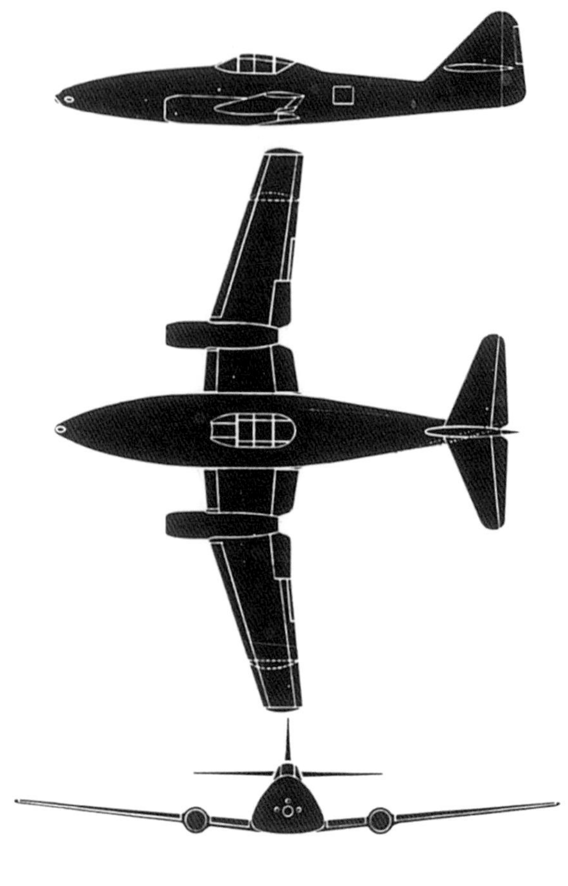

ABOVE: A silhouette drawing of the P 1065 from the March 1940 description.

The Me 262 V3's sustained success with its T1s was the first real evidence that all the time, effort and money invested in jet fighters was going to pay off. The Me 262's series of pre-production aircraft was reinstated and it was decided that the Me 262 V2, which had been built for BMW P 3302 engines, should be converted to T1s by the end of September. The V3 was to be put back together and the V4 and V5, also on T1s, would need to be ready by January and March of 1943.

An He 280 was finally flown with jet engines again on July 5, 1942, but it was too late. The He 280's development priority was downgraded and Heinkel redrew its plans to incorporate the Junkers engine. But the T1 was longer than the HeS 8 and weighed almost twice as much, meaning that

ABOVE: The second He 280 takes off without the cowlings of its engine nacelles. The HeS 8A engine was notorious for minor fuel leaks – which could result in a dangerous build-up of flammable liquid within the cowlings over time.

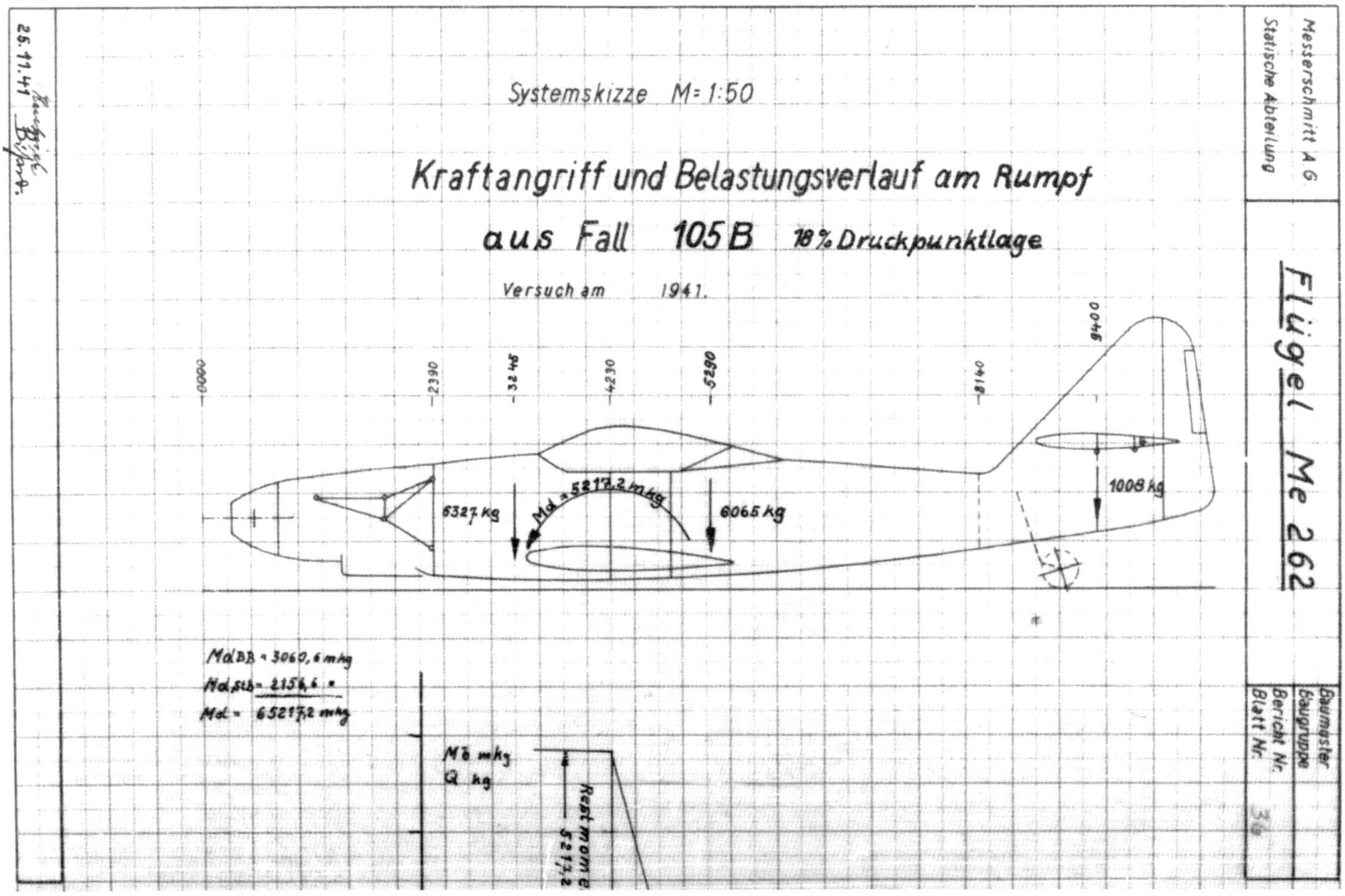

Systemskizze M= 1:50

Kraftangriff und Belastungsverlauf am Rumpf

aus Fall 105 B 18% Druckpunktlage

Versuch am 1941.

Flügel Me 262

Messerschmitt A.G.
Statische Abteilung

25.11.41

Baumeister
Baugruppe
Bericht Nr.
Blatt Nr.

ABOVE: Messerschmitt diagram showing pressure distribution across the Me 262 V1 – its profile distinctive due to the Jumo 210 in its nose.

not only was there a danger of the engines making contact with the ground on take-off, but the aircraft's centre of gravity was also altered.

In September, Milch told Ernst Heinkel that getting the He 177 bomber into front-line service should be his top priority, followed by the He 219 night fighter, developments of the He 111 and then the He 274 high-altitude bomber. The He 280 had the lowest priority[13] and on October 15, 1942, the RLM decided that just 20 examples of the He 280 should be built – the V1 powered by Argus pulsejets, the V2, V4, V6, V8, V9 and V10 powered by Jumo T1s, the V3 and V5 powered by HeS 8s, and the V7 left engineless for flight testing as a pure glider. The V11 to V20 would have either Heinkel or BMW engines.

The V1 was evidently fitted with its pulsejets and towed to the Luftwaffe test centre at Rechlin on October 29, 1942, but a mechanical failure meant that the pilot had to make a belly landing – putting the aircraft out of action while repairs were made.[14]

Plans for the short He 280 production run were changed on November 2, 1942, with the V8 now earmarked for an HeS 8, the V9 for a BMW 003, the V10 a Jumo 004 as before, V11 BMW 003, V12 HeS 8, V13–V18 either BMW 003 or Jumo 004 and V19–V24 BMW 003s.

On November 18 the number of He 280 prototypes was cut back to just nine by the RLM. However, on December 10, 1942, Milch included both the Me 262 and He 280 in his new development and procurement programme codenamed 'Vulkan' (Volcano), alongside the Me 163, Me 328 and Ar 234.[15]

ABOVE: The He 280's troublesome HeS 8A powerplant as viewed from the side.

ABOVE: A rare view of the He 280 V3 in flight, powered by a pair of HeS 8A engines. Flight testing began on July 5, 1942, and continued sporadically for seven months. It was halted on February 8, 1943, when the machine was damaged during a crash landing.

ABOVE: Messerschmitt test pilot Fritz Wendel powers up the Me 262 V3 ahead of his historic flight at Leipheim on July 18, 1942. This image is a still from a cine film.

Test pilot Heinrich Beauvais from Rechlin flew the only airworthy He 280, the V3, on December 11, 1942 and found that while regulating its engines was difficult, it was easier to fly at take-off and landing than the Me 262.

ME 262 ONGOING DEVELOPMENT

At the RLM's development meeting on January 8, 1943,[16] Milch read out a letter from Ernst Heinkel which promised that the He 280, fitted with BMW engines, would have a speed of 800km/h and better armament. Development time would be cut from two years to six months and the aircraft would now have a single central tailfin rather than a twin-tail. Oberst Wolfgang Vorwald, head of the RLM's Technical Office, told the meeting that this latter development would result in a considerable delay and that he had "therefore rejected this already".

ABOVE: It is difficult to underestimate the importance of Messerschmitt Me 262 V3. By the time the aircraft was ready, the German government was losing faith in jet engine development. Heinkel and BMW both seemed unable to deliver anything close to sufficient reliability for a full production model and it was left to Junkers to rescue the Me 262. The V3 flew successfully for 12 minutes on the morning of July 18 using turbojet power. It then flew for another 13 minutes at around midday – reaching 342mph. The Heinkel He 280 was doomed, the Me 262's future was assured and faith was restored in the turbojet programme.

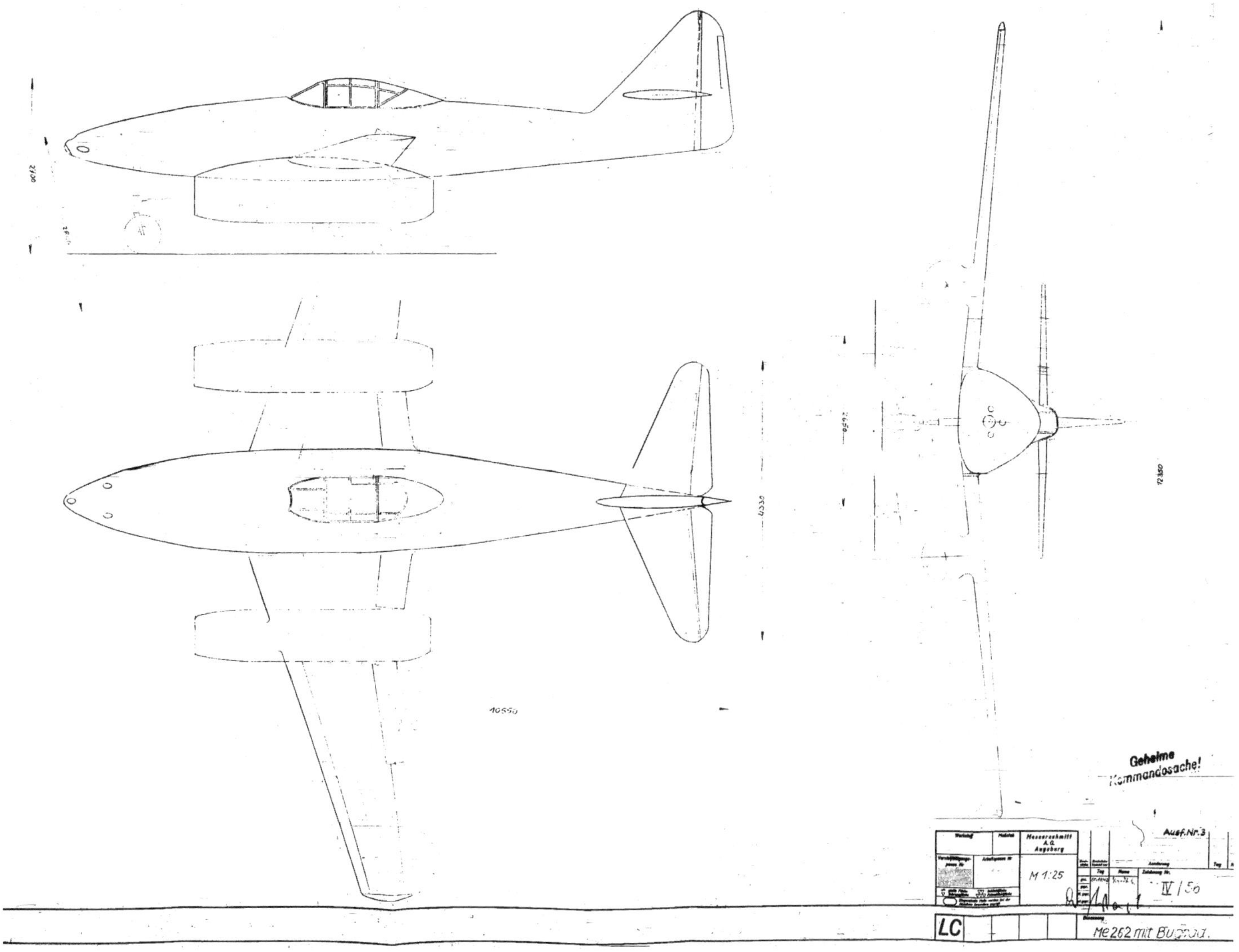

ABOVE: A three-view drawing of the Me 262 from a project description of November 20, 1942. At this point the aircraft's armament was still intended to be a trio of nose-mounted MG 151s.

During a further RLM development meeting on January 21,[17] Vorwald outlined the current state of planning for jet-propelled aircraft: 100 examples of the He 280 0-series (pre-production), Me 163 increasing by 70 prototypes to 190 in total – 50 per month – and the Me 262 increasing from 30 to 50 then 100. Special permission for the latter increase was being obtained from Göring.

The pulsejet-equipped He 280 V1 was wrecked in a crash on January 28, 1943,[18] before it had made a single powered flight and by March 19 the type had been cancelled.[19] An RLM Entwicklungsbesprechung on March 22 heard more detail of its cancellation.

According to the minutes of the meeting,[20] RLM staff engineer Hans-Martin Antz told the assembly officials that "according to Messerschmitt, Me 262 has already flown 850km/h and is equipped with a pressurised cabin for heights over 12km. The armament meets the requirements of the General der Jagdflieger. With some reinforcement of the construction capacity, 140 aircraft

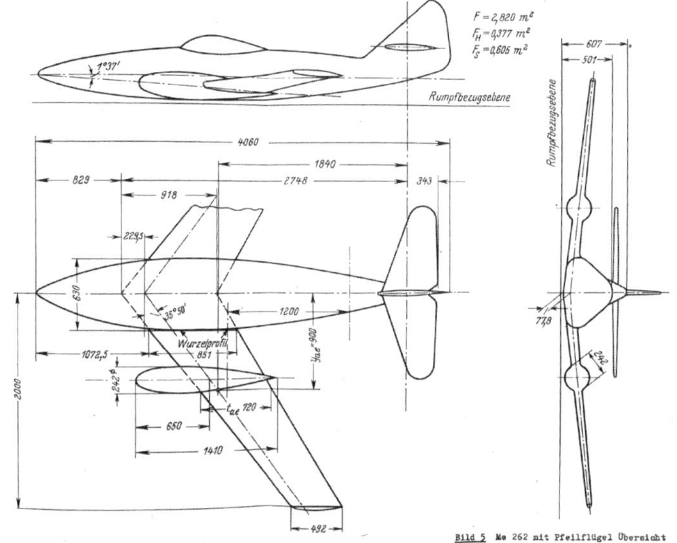

ABOVE: Drawing of an Me 262 wind tunnel model with 35° wing sweepback tested by Messerschmitt. It appears in a report dated July 20, 1942.

ABOVE: General of Fighters Adolf Galland test-flew the Me 262 V4 on May 22, 1943, and famously told Hermann Göring that: "It flies as if there is an angel pushing."

could be built by the end of 1944 according to a 'forceful plan' by Messerschmitt.

"Of the He 280, which has already flown, 20 V-aircraft have been ordered, of which nine are to be delivered by the end of 1943. However, since a continuation of both patterns next to each other is prohibited for reasons of capacity, it is proposed to discontinue He 280 except for the first nine V-patterns and to use the capacity that is released in the processing of the Me 262."

Staff engineer Schelp then "gives a brief overview of the development of the turbojet engines and emphasises the need for extensive flight testing as soon as possible. After both the Jumo and the BMW device have delivered unexpectedly favourable results and are running in smaller pilot tests, the HeS 11 engine will also come into operation on the test stand in the next few weeks.

"In this device, the experience of the past four years is taken into account, so that 1,200 to 1,500kg start thrust can be achieved without significant enlargement and weight increase. By reducing the specific fuel consumption by around 25%, normal gasoline consumption at high altitudes can already be achieved."

The minutes report that "the Generalfeldmarschall [Milch] approved a proposal of C-E Chief [Dr Georg Pasewaldt] for removal of the He 280. The nine V-models under construction are being completed for engine testing. Extensive transfer of the people working at Heinkel to the Me 262 task must be checked.

"The Generalfeldmarschall then discussed the question of placing the entire 'jet-powered aircraft' sector, also with regard to the evaluation of their equipment, under a strong personality familiar with the circumstances. However, the

ABOVE: Me 262 V6 was the first Me 262 prototype to be fitted with a fully retractable tricycle undercarriage.

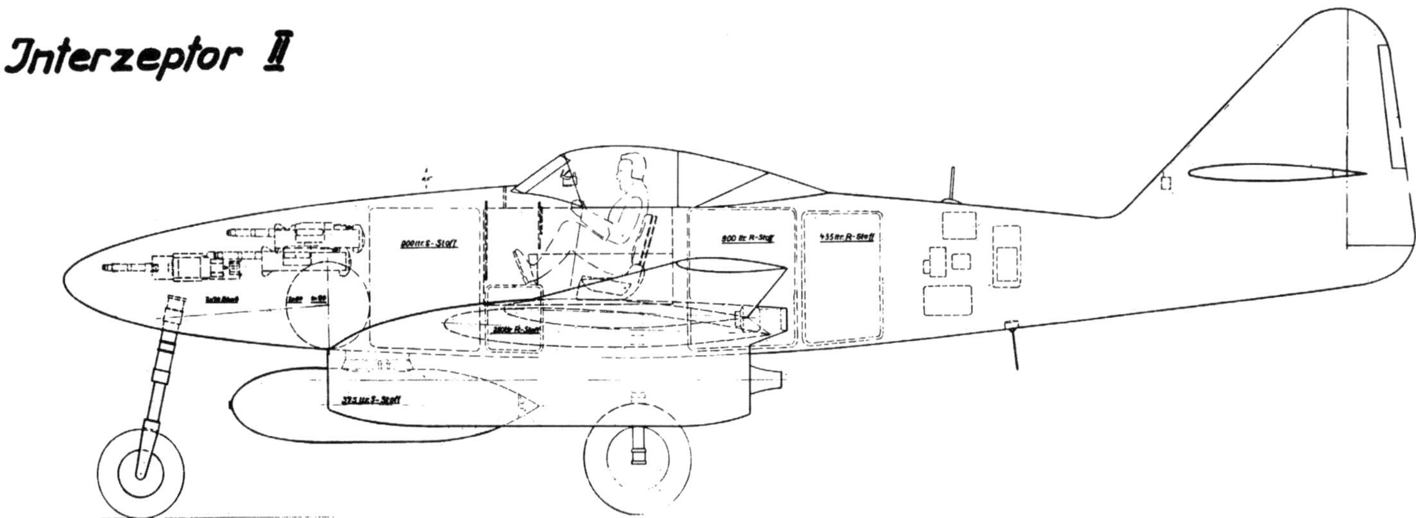

ABOVE: Oddly, given its designation, the Me 262 Interzeptor II design was chronologically the earliest of the drawings featured in a September 11 1943, report on potential developments of the aircraft. This perhaps explains the slightly strange-looking pilot depicted in drawing II/160 of July 6, 1943. The aircraft itself was to be powered by a pair of BMW 003 R composite turbojet/rocket engines. Armament was six MK 108s – the same for all three interceptor designs.

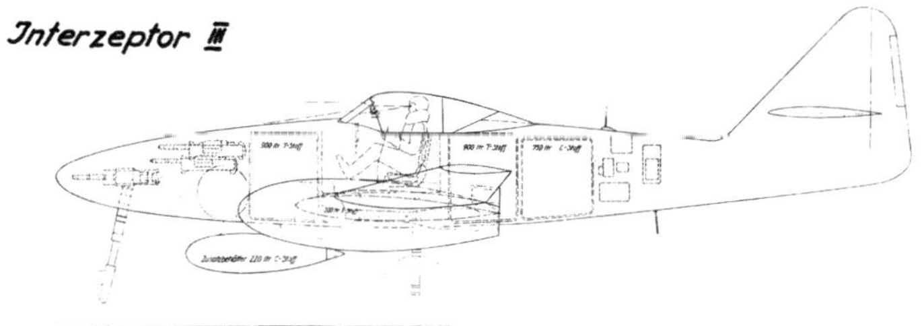

ABOVE: Chronologically the second Me 262 interceptor design, Interzeptor III has no turbojets, just a pair of HWK 509 rocket motors in its wings. It also has an out-of-sequence drawing number – II/181 of July 21, 1943.

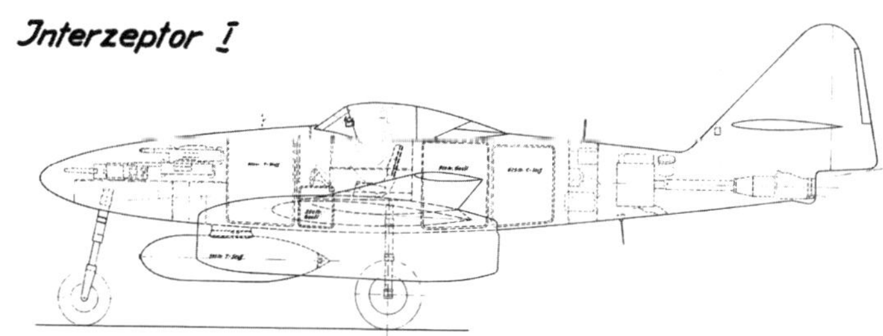

ABOVE: Interzeptor I, drawing number II/169 of July 22, 1943, has a Walter R II/211/3 rocket motor in its rear fuselage to provide additional thrust for take-off. Like Interzeptor II, it carries additional fuel under its forward fuselage.

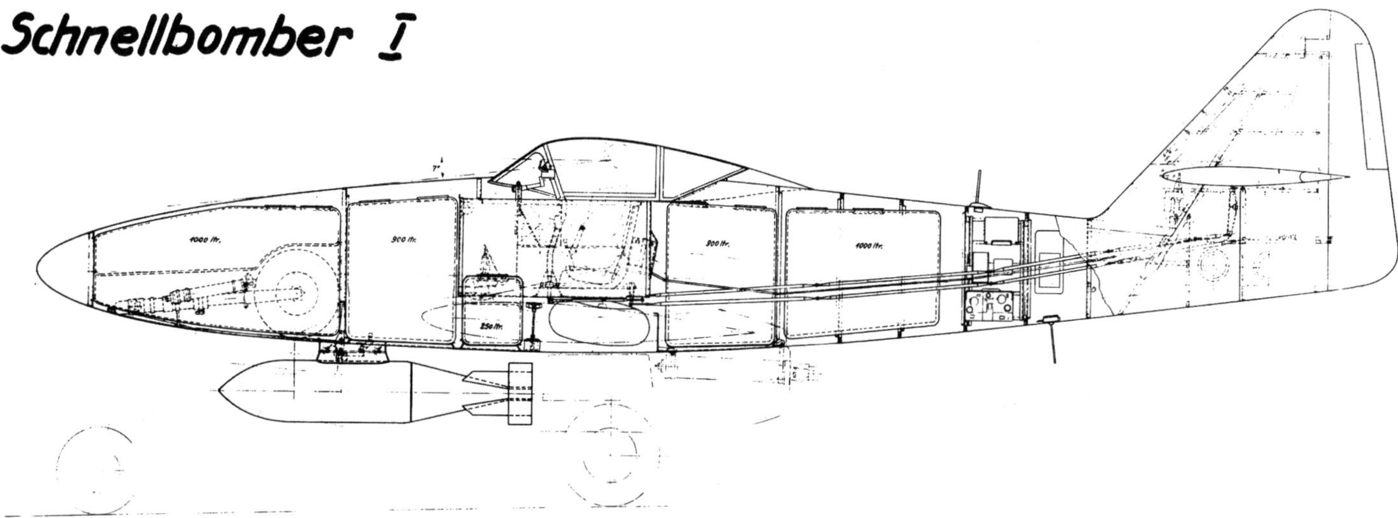

ABOVE: The first and least radical of three designs for a fast bomber version of the Me 262 from the September 1943 report. Every available internal space was filled with additional fuel tanks for a total capacity of 4,050 litres and the payload was slung beneath the forward fuselage. This is drawing number II/167 dated July 22, 1943.

as yet immature stage. The Generalfeldmarschall agrees to this proposal."

On March 25, 1943, a new brochure for the Me 262 was produced[21] – stating in its introduction that it superseded the P 1065 brochure of March 1940. This appears to be the first formal proposal to use the Me 262 as a fighter-bomber, since the brochure presents the Me 262 as both Jäger 'fighter' and Jabo 'fighter-bomber', the latter capable of carrying an SC500 500kg bomb. Armament otherwise remained three MG 151s although options for two, four or even six MK 108s are mentioned. The engines were to be two Jumo 004 Bs and later 004 Cs.

The aircraft is also clearly shown with a nosewheel in the accompanying drawing. Wingspan is 12.56m compared to the production version's 12.6m and length is 10.55m compared to the production version's 10.6m. Wing area is 21.68m² compared to 21.7m² for the finished product.

The engineless He 280 V7, now seconded to the DFS, made its first glider flight on April 19, 1943. This airframe would continue to be used for aerobatic and aerodynamic tests well into 1944.[22] The Jumo 004-powered He 280 V8 first flew on July 19, 1943, and ongoing testing led to the discovery that there were strong vibrations in the tail unit above 550km/h. Information about the V4, V5, V6 and V9 is scant but little further work was done on the He 280 prototypes.

Adolf Galland, General of Fighters, test-flew the Me 262 V4 on May 22, 1943, and quickly became the

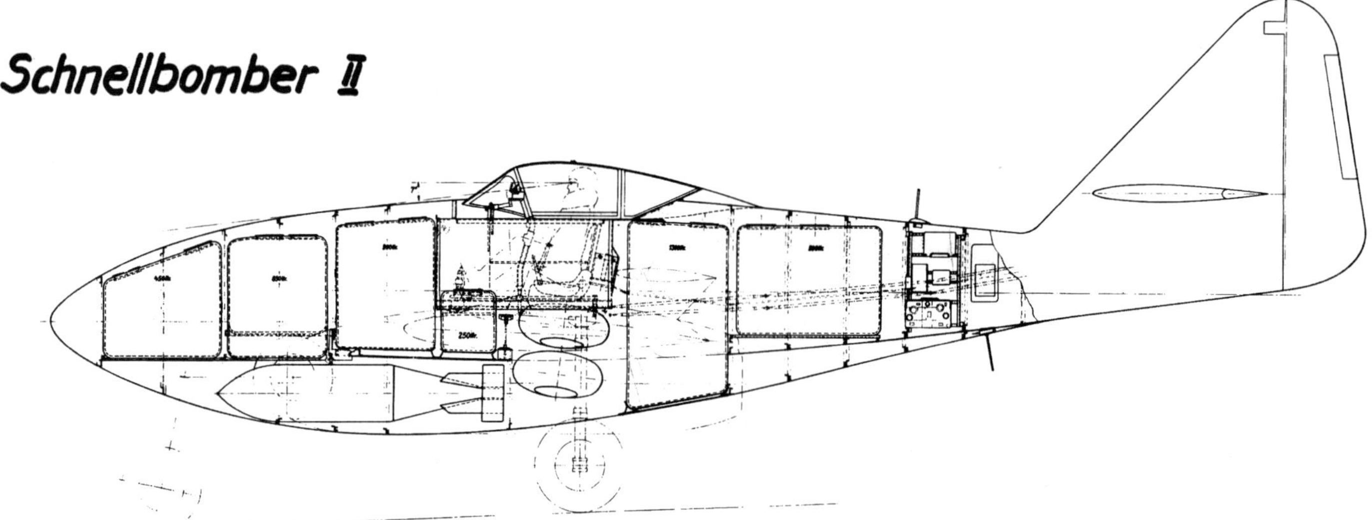

ABOVE: Schnellbomber II, shown in drawing number II/168 of July 22, 1943, involved a new enlarged fuselage for the Me 262 with significantly increased fuel capacity - up to 4,550 litres - and an internal bomb bay. This design, which featured two wheels for each main undercarriage leg and thicker wings - had obvious benefits and Messerschmitt commissioned the AVA to study a wind tunnel model of it, later detailed in AVA report 43/W/37 of July 31, 1943.

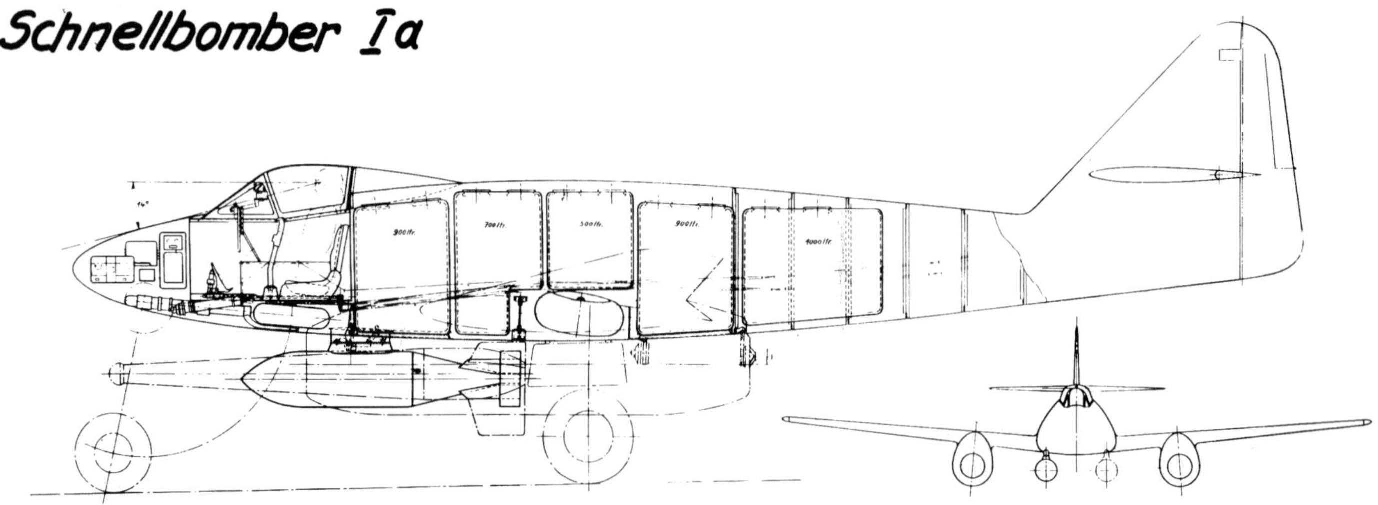

ABOVE: The second fast bomber design from September 1943 saw the cockpit shifted to its nose with the nosewheel now rotating to lie flat within the fuselage. Total fuel capacity was slightly less than that of Schnellbomber I at 4,050 litres. The drawing itself, II/183 of September 10, 1943, was produced nearly three weeks after those of the Schnellbomber I and II.

type's most ardent supporter. He famously reported to Reichsmarschall Hermann Göring that: "It was as though angels were pushing."[23] Me 262 V5, the first prototype to be fitted with a tricycle undercarriage, first flew on June 6, 1943, and demonstrated a marked improvement in take-off performance compared to the earlier 'tail dragger' versions.

Messerschmitt drew up a plan on June 17 showing production of the Me 262 beginning in January 1944 but Milch ordered that the first 100 aircraft should be ready by the end of 1943. Even at this late stage however, during July and August 1943, Messerschmitt's engineers and draftsmen worked on an array of designs showing how the basic Me 262 platform might be significantly altered to improve performance in a variety of different roles. The resulting report[24] published on September 11, 1943, shows the standard Me 262 A-1 fighter compared against a fighter-bomber 'Jabo' version as before, but now with three additional 'Schnellbomber' fast bomber versions, three new 'Aufklärer' reconnaissance versions and three 'Interzeptor' interceptors plus a 'Schulflugzeug' trainer version.

The Schulflugzeug had an extended fuselage enabling the instructor and pupil to sit in tandem with dual controls. A ballast weight of 150kg was installed in the nose in place of armament. The Jäger u. Jabo 'fighter-bomber' version was similar to the normal fighter but with four MK 108s in the nose rather than six, an extra 750-litre fuel tank in the rear fuselage and the option to carry either a single SC500 bomb or two SC250 bombs under its forward fuselage.

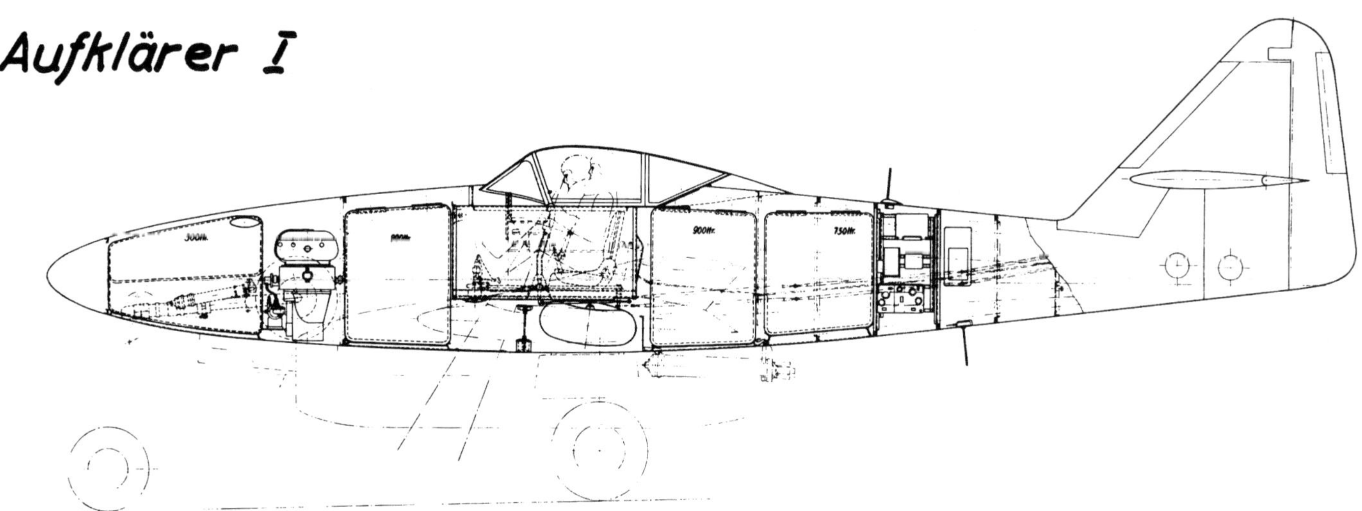

ABOVE: Messerschmitt's simplest reconnaissance version of the Me 262 from September 1943 had two cameras side by side within its nose and a transparent plate in the cockpit floor for the pilot to look through. This is drawing number II/171 of July 22, 1943.

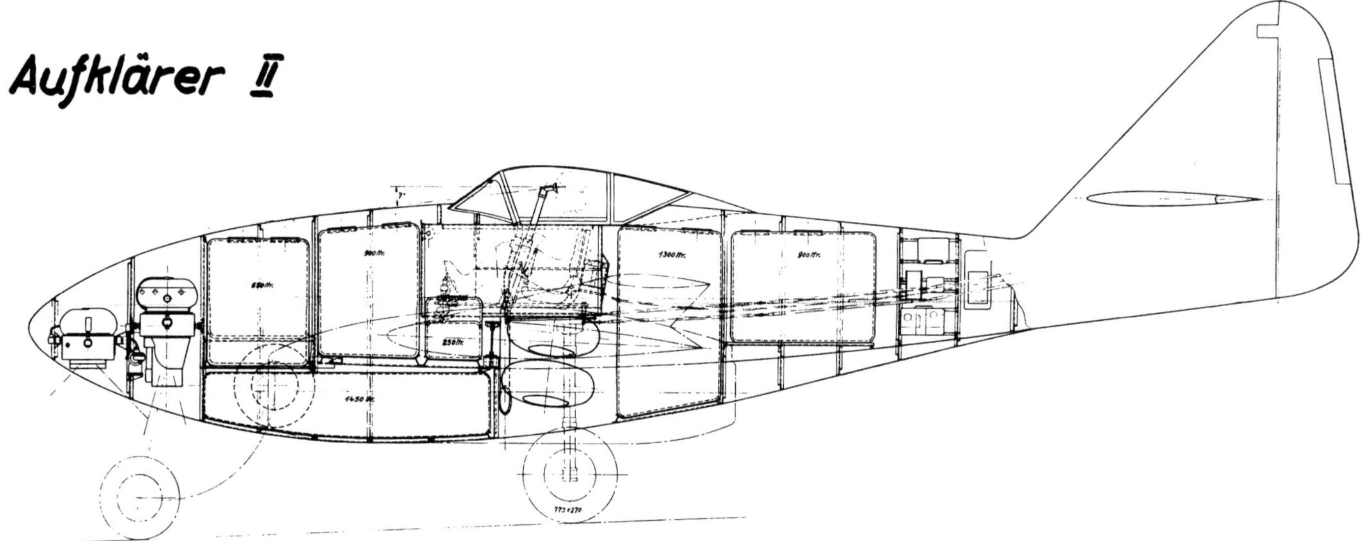

ABOVE: Drawing number II/172 shows the large-body version of the Me 262 fitted with three cameras in its nose - an Rb 20/30 and a pair of Rb 75/30s. The pilot has a periscope for downward visibility.

The Schnellbomber I looked like the A-1 fighter but had no weapons other than its bomb load – one SC1000 bomb, two SC500s, or two SC250s. A fourth option was a single BT 700 torpedo bomb. In place of the fighter version's cannon its nose was fitted with a 1,000-litre fuel tank. A second tank of equal capacity was installed in the rear fuselage. The Schnellbomber Ia had its cockpit in its nose and fuel tanks filling the rest of the fuselage. A pair of MK 108s could be positioned beneath the pilot and bomb load possibilities were the same.

For the Schnellbomber II, the aircraft's forward fuselage was made deeper for creation of an aerodynamic fairing over the bomb load. Again, there was no provision for defensive armament and the same payload options were available. Each undercarriage mainwheel had a second wheel on it, giving the aircraft a total of five wheels on three legs.

The layouts of the three Aufklärer 'reconnaissance' versions followed the same pattern. The Aufklärer I was outwardly identical to an A-1 but had a 500-litre fuel tank inside its nose and a pair of cameras immediately behind that – either an Rb 75/30 and an Rb 20/30 or two Rb 75/30s. The Aufklärer Ia, like its Schnellbomber counterpart, had its pilot sitting in its nose with the option of two MK 108s. The cameras, two Rb 75/30s, would be housed in the rear fuselage.

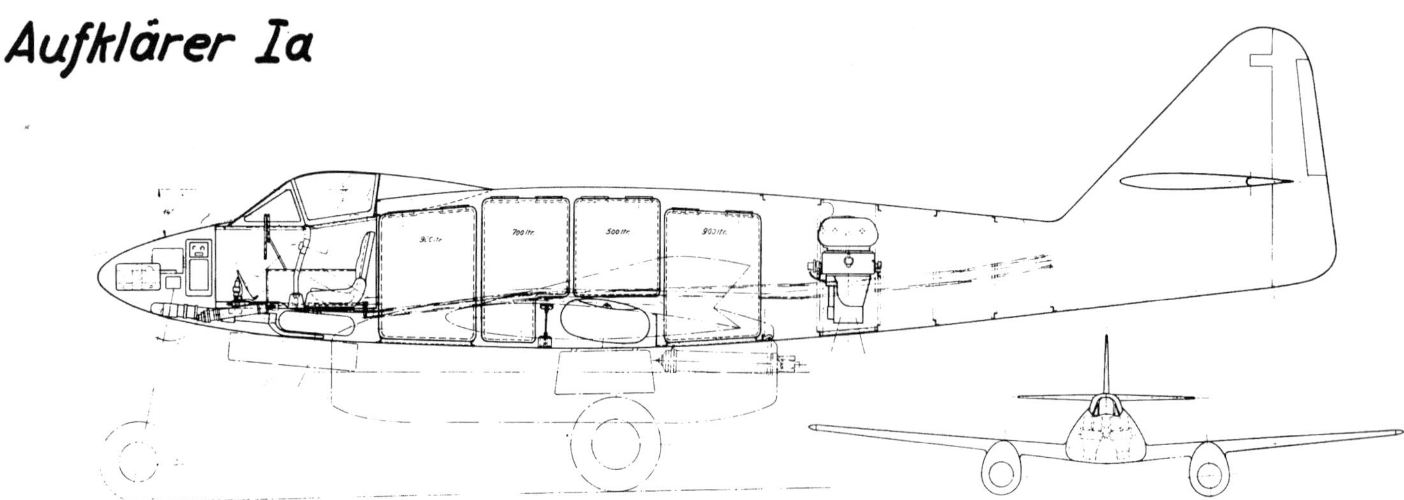

ABOVE: Like the Schnellbomber Ia, the Aufklärer Ia was created slightly later than the designs based on the normal Me 262 and the large-body version. Drawing II/180 shows a design with two Rb 75/30s in its rear fuselage but no real downward visibility for the pilot in its nose.

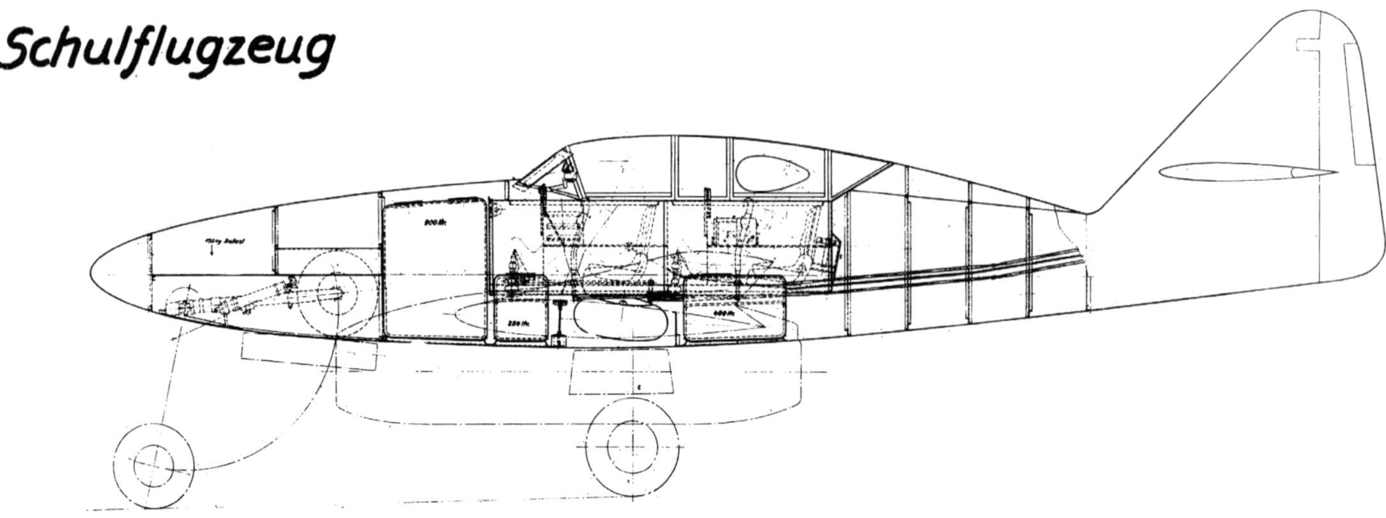

ABOVE: Drawing II/175 of July 22, 1943, shows the proposed two-seat trainer version of the Me 262, labelled Schulflugzeug. Rather than weapons, the nose carried 450kg of ballast. Dual controls were provided and the instructor was to benefit from blisters on his section of the canopy for improved visibility.

The Aufklärer II had the same undercarriage arrangement and expanded fuselage as the Schnellbomber II, allowing it to carry a trio of cameras – an Rb 20/30 and two Rb 75/30s in the end of its nose. The underhanging section of the fuselage would be used to house a 1,450-litre fuel tank.

None of the Interzeptor designs featured the radical layout alterations of the bomber and reconnaissance versions but each was subtly different to the A-1, being configured for maximum speed and rate of climb with the inclusion of rocket engines. Each had four MK 108s in the nose.

Interzeptor I had the Me 262's usual Jumo 004s supplemented by a Walter R II/211/3 rocket motor mounted inside its tail. Interzeptor II had a pair of BMW 003 R engines in place of the Jumo 004s in its wings. The BMW 003 R was essentially a normal BMW 003 turbojet with a BMW P 3395 rocket engine attached to it, the latter running on concentrated nitric acid (SV-Stoff) and monoxylidene oxide/triethylamine (R-Stoff), rather than the methanol-hydrazine-water mixture (C-Stoff) and hydrogen peroxide (T-Stoff) used for Walter-type rockets. Lastly, the Interzeptor III was to be a pure rocketplane – powered by a Walter HWK 509 rocket engine under each wing.

In addition to the normal Me 262 A-1 fighter, the Jäger u. Jabo was built as the Me 262 A-2, the Schulflug-

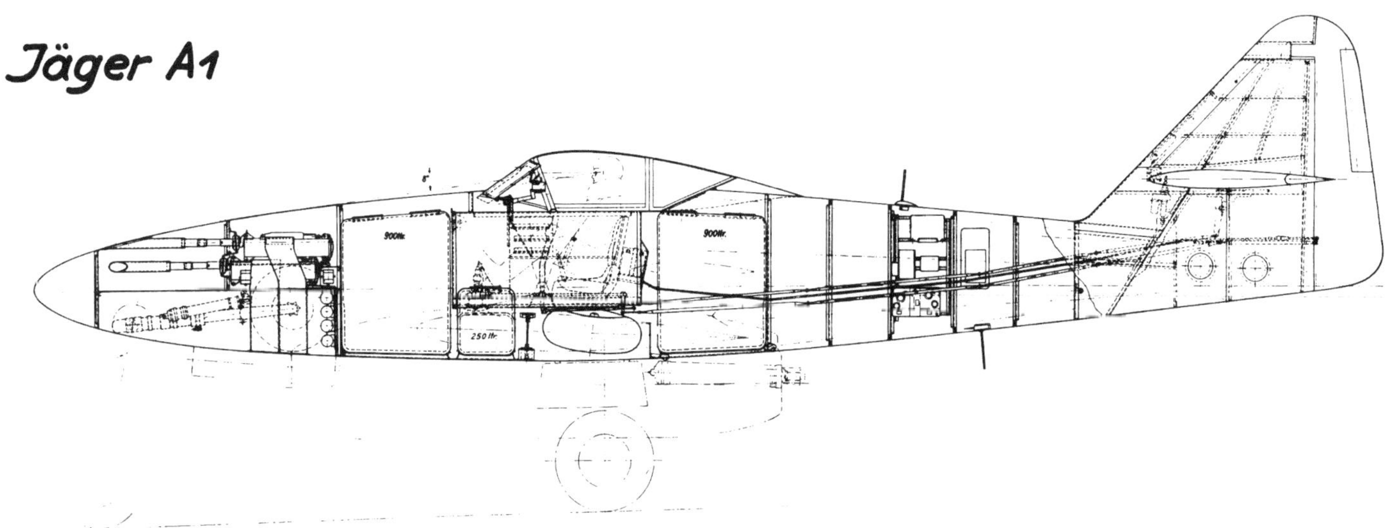

ABOVE: The proposed standard production model Me 262 A1 as shown in drawing II/170 of July 23, 1943, was close to the final production model as built.

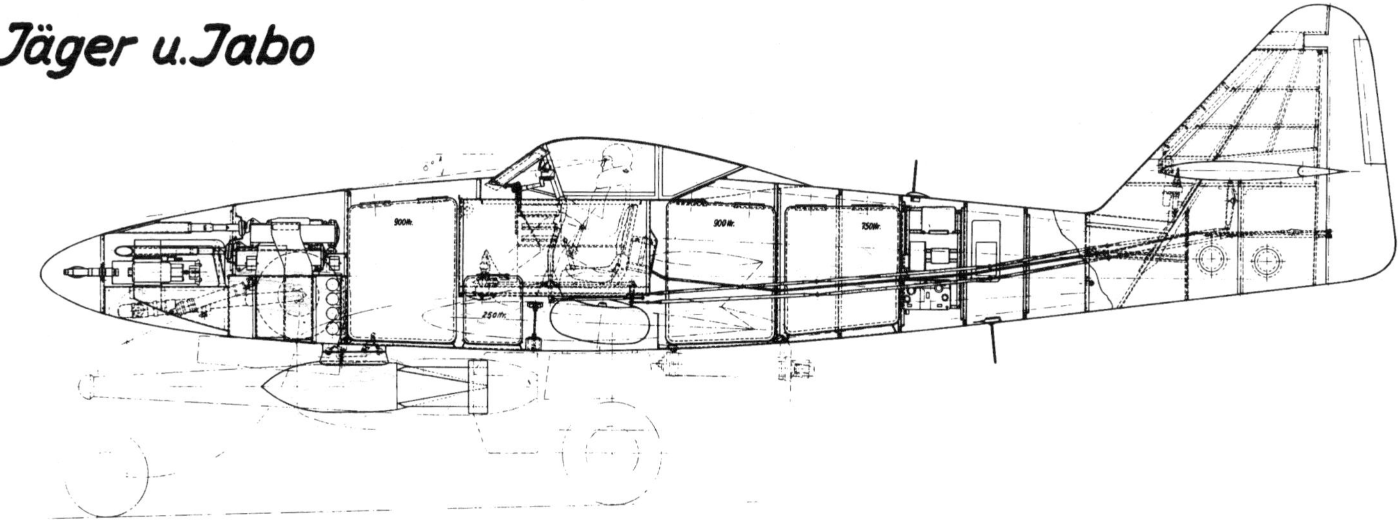

ABOVE: The incredibly heavily laden 'Jäger u. Jabo' or 'fighter and fighter-bomber' version of the Me 262 was to have six MK 108 cannon in its nose and an extra 750-litre fuel tank in its rear fuselage – in addition to whatever payload of bombs it was to carry externally. It is shown in drawing number II/173 of July 22, 1943. The fighter-bomber Me 262 would later be designated A-2.

ABOVE: The engineless Heinkel He 280 V7 as it appeared while performing towed pitch testing with the DFS during late 1943 and into 1944. It made a total of 115 flights and gathered valuable data that helped Heinkel during its work on the He 162.

ABOVE: Close-up of the He 280 V7's tail rigged with camera equipment for pitch testing.

zeug as the Me 262 B-1, the Interzeptor I as the Me 262 C-1 and the Interzeptor II as the Me 262 D-1, later redesignated Me 262 C-2. But this wasn't where development of the Me 262 ended. Far more radical versions were yet to be designed.

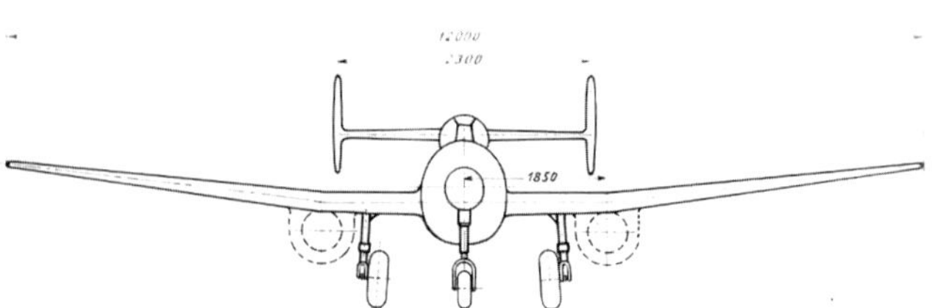

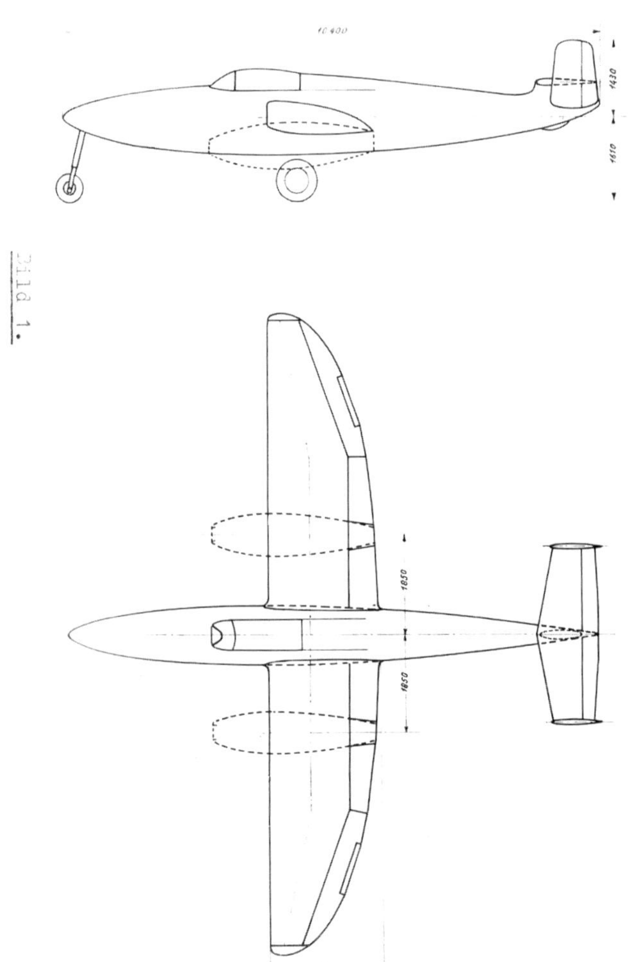

ABOVE: A three-view drawing of the engineless He 280 V7 from a DFS report.

ABOVE: A sequence of photographs showing the He 280 V7 during flight testing.

Chapter 4

Light fighter

Messerschmitt P 1079/Me 328

As work progressed on the Me 262, Messerschmitt's project office was tasked with investigating options for an aircraft propelled by another promising new powerplant – the Argus pulsejet.

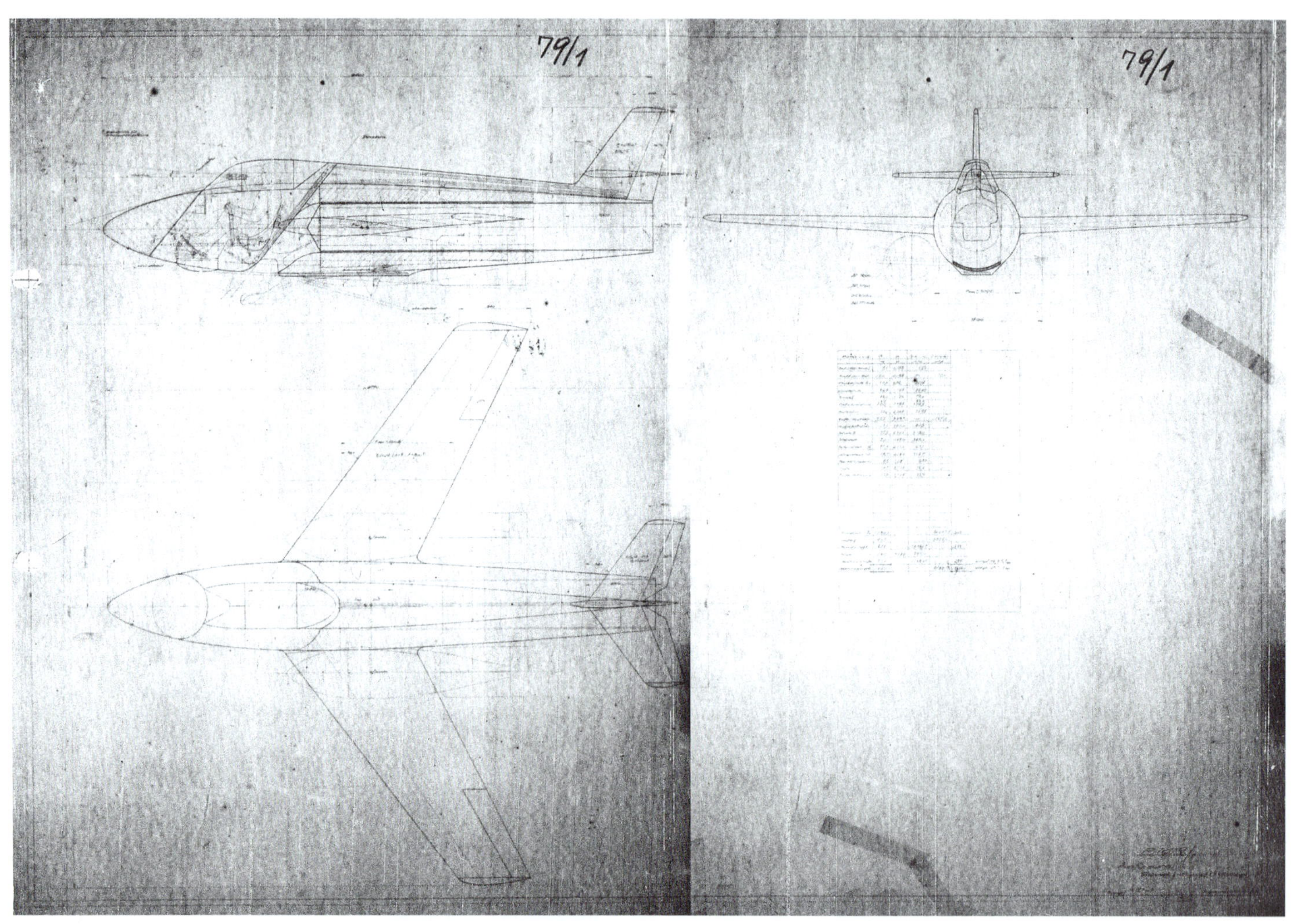

ABOVE: Messerschmitt P 1079/1 – a conventional design but with its main fuel tank wrapped around its hot pulsejet engine.

P 1079

When Messerschmitt project office chief Woldemar Voigt handed the RLM's Hans-Martin Antz a drawing of the pulsejet-powered P 1079/13c in August 1941 and claimed that this project made the Me 163 and P 05 unnecessary, he seems to have genuinely believed it.

Earlier in the year, Messerschmitt had begun talks with engine maker Argus about the pulsejets it was working on with inventor Paul Schmidt. There had been tentative plans to fit the P 65/Me 262 with pulsejets but greater efforts were concentrated on creating a dedicated lightweight pulsejet aircraft similar in size to the Me 163.

During May 1941, at least 21 designs were drafted by the project office under the umbrella designation of P 79, also written as P 1079 owing to the change in Messerschmitt's project numbering system. A report produced at the end of July detailed P 1079/1 to P 1079/17, with /10 and /13 being subdivided into /10a-c and /13a-c.

The report, entitled simply 'Flugzeuge mit Strahlrohren' or 'Aircraft with jet tubes',[1] compared the projected performance of the Argus pulsejet against that of BMW's P 3302 turbojet and concluded that while the former provided only 500kg thrust compared to the latter's 600kg, the pulsejet weighed only 80kg, compared to the turbojet's 600kg. On a very small aircraft, the pulsejet's light weight made a vast difference – although its fuel consumption was almost twice that of the turbojet.

According to the report, under the heading 'design and task': "For the time being only generally required: high maximum speed, good climbing performance with comparatively large loads (bombs up to 2,000kg)." The pulsejet could apparently guarantee a high maximum speed so "consideration of the climbing attributes should be omitted here". The P 1079's swept wings were also designed with high speed in mind.

It was hoped that Argus could develop an 800kg thrust pulsejet but "the largest device currently under development produces 500kg of thrust (it is not yet possible to specify an enlargement of the device)" so the fuselage had been kept as small as possible in the designs examined. It was hoped that even with a pair of 500kg thrust engines, the P 1079 would be able to fulfil five different roles: "1. Fighter – high-speed good climbing ability, range up to 800km. 2. Reconnaissance – good climbing performance and maximum altitude, sufficient ranges to fly over large oceans. 3. Interceptor – high climb rates with low flight times. 4. Bomber – sufficient speed with enclosed bombs to escape today's fighter aircraft. 5. Parasite aircraft (as bomber or fighter) – small dimensions."

The P 1079 document, as it has survived, includes only 11 of the maximum 21 designs. Each is presented with a single drawing and a very brief typed description – often including critical comments.

For P 1079/1, a relatively conventional design but with its fuel tank wrapped right around the pulsejet tube, it says: "Accommodation of the fuel in two tanks; one of them is executed as a ring tank (fire hazard). Fuel quantity 1,200 litres. Unfavourable air intake for the engine. Little opportunity to enclose the bombs. Unfavourable for engine upgrade. The engine is also the tail carrier."

The second design, P 1079/2, was characterised by a very long rear fuselage and a flat horizontal intake on its underside. According to the comments: "Division: Cockpit – fuel tank space – fuselage with engine. Fuel quantity 1,200 litres. Unfavourable air intake."

P 1079/3 looked somewhat similar to P 1079/1, and with good reason: "Further development from 79/1. Slightly improved air intake. Improved visibility. Fuel storage in a tank (ring tank). Fuel 1,200 litres. Unfavourable centre of gravity position."

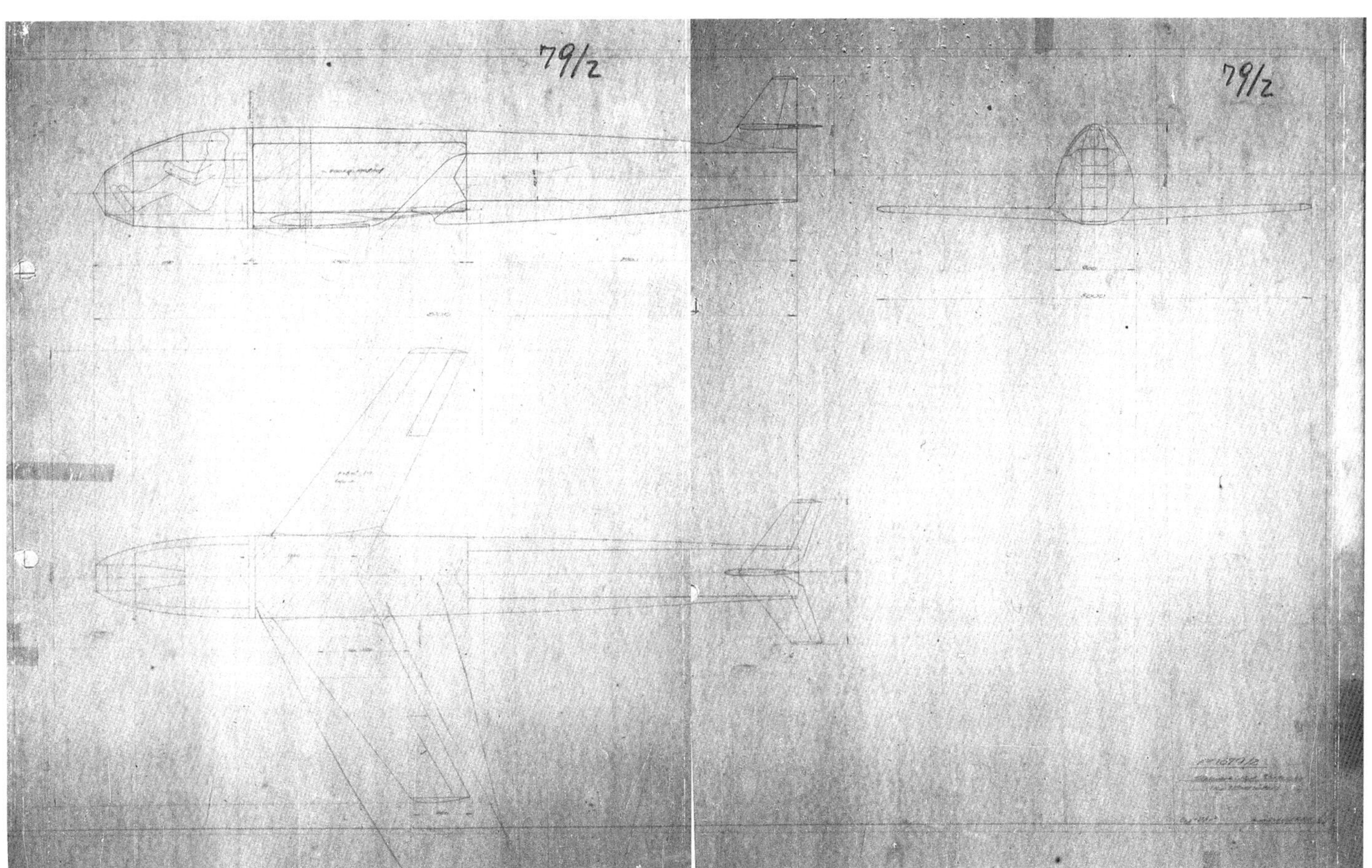

ABOVE: The bizarrely elongated P 1079/2 with its pulsejet at the extreme rear and a flat intake on its underside.

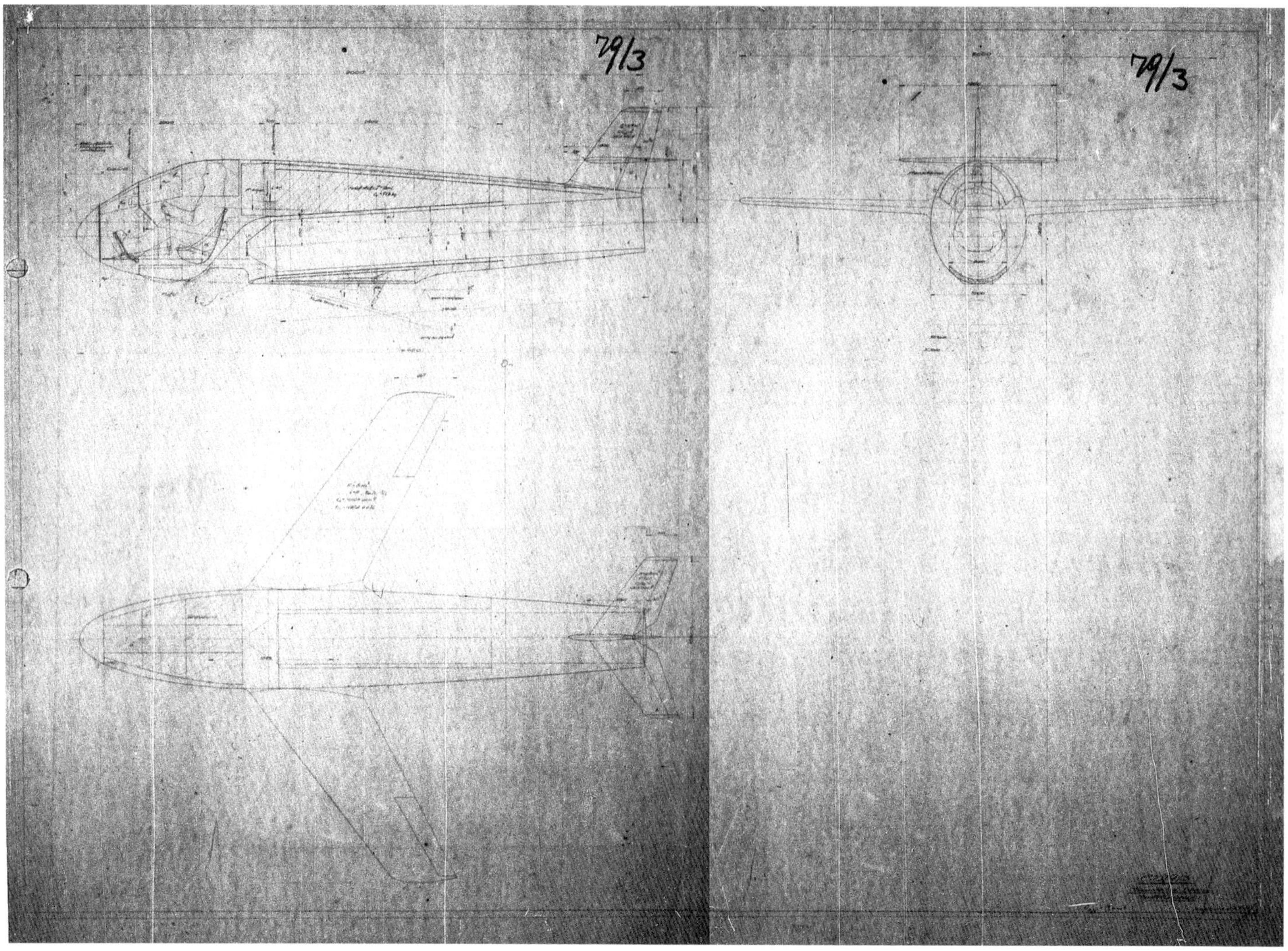

ABOVE: P 1079/3 was a development of P 1079/1 offering better visibility and an improved intake.

ABOVE: Another development of the /1 – the P 1079/4 – this time with side intakes.

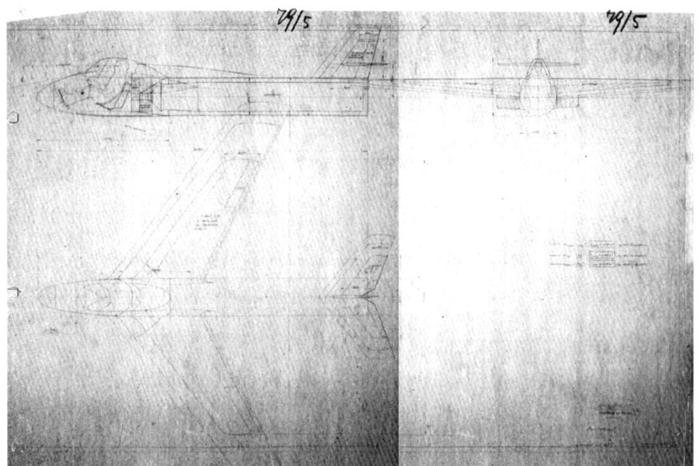

ABOVE: The P 1079/5 was used as a basis for exploring different wing and tailplane lengths.

Although it doesn't say so, P 1079/4 appears to have been another improvement on the same theme: "Fuel storage in a saddle tank. Residual fuel in the tip of the fuselage underside. Engine on the fuselage underside (improved upgradeability). Lateral air intake."

The P 1079/5 again took a similar layout but this time it was an opportunity to try scaling the wings: "Fuel accommodation in the wing. Large wing thickness 10-25% while the following amounts of fuel can be accommodated: wing area 5m^2 approx. 330 litres, wing area 8m^2

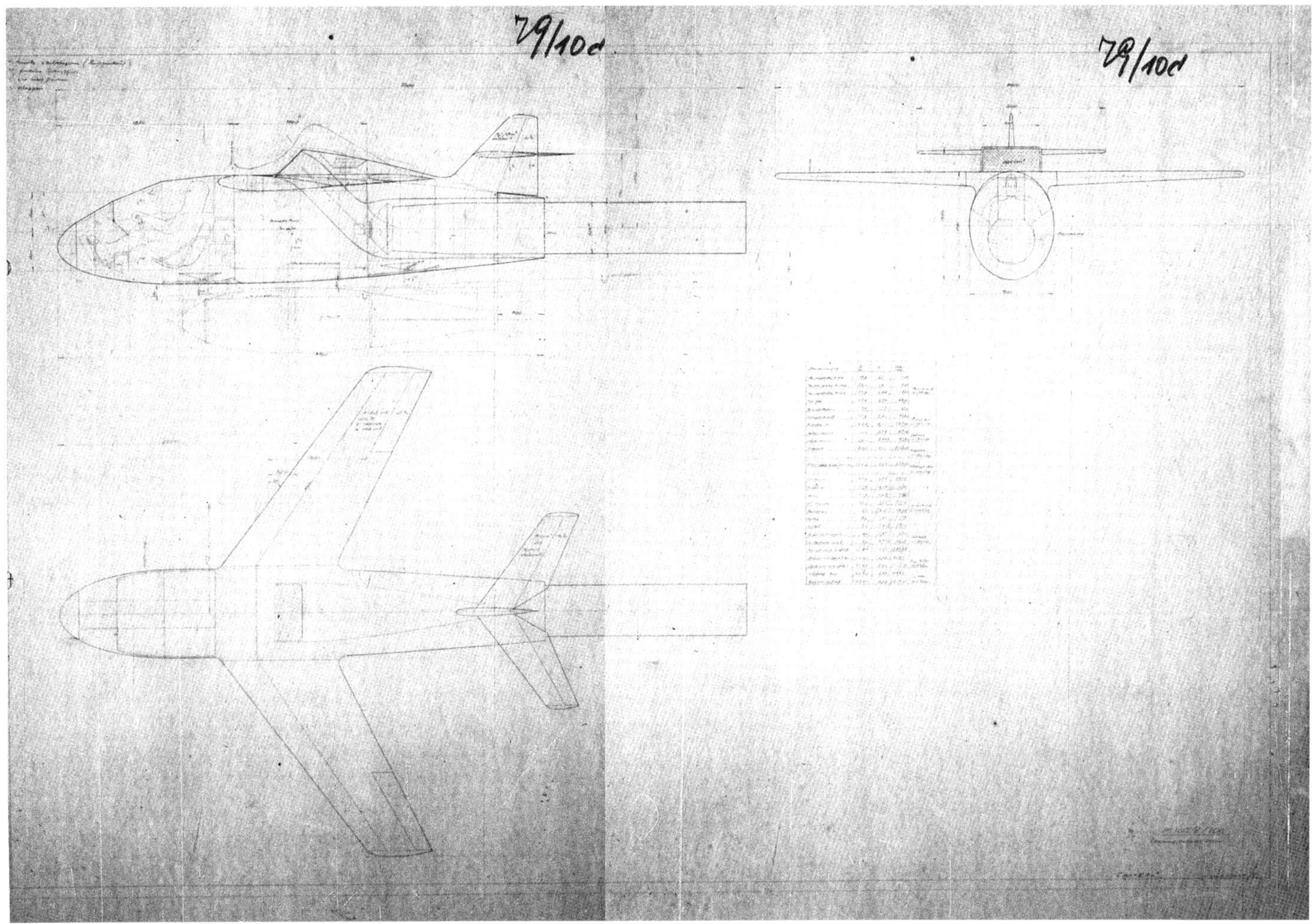

ABOVE: Another /1 development, the P 1079/10c shifted the engine to the rear – allowing it to protrude from the aircraft – and introduced a dorsal intake.

approx. 640 litres, wing area 10m^2 approx. 920 litres. Engine = tail unit."

The sixth design is entirely missing and nothing is known about it. P 1079/7 features a radically repositioned engine – right up on top of the fuselage and to the rear: "The engine is laid free on the fuselage and serves as a tail carrier. Air intake significantly improved. Inadmissible load changes with gas take-off."

Designs eight, nine, 10a and 10b are then missing, with P 1079/10c being next in the series: "Engine installation in the fuselage end, the rear part of the tube protruding freely. Unfavourable air intake. Fuel content 800 litres. Unfavourable bombing, especially with two bombs."

Designs 11, 12 and 13a are missing, then arguably the most important known design is outlined – P 1079/13b. Important because it's the closest the report gets to describing the P 1079/13c, which was the basis for the Me 328. The report provides a slightly longer description of the /13b: "Use of two engines. Air intake significantly improved. Fuel content 900 litres. Unfavourable arrangement of the engines. Clear and cheap construction. Easy interchangeability of the engines. Uncertainty in the stability. Good chance to enclose it even when using two bombs. Relatively simple solution of the dive brake. Better flight performance than single engine design."

P 1079/13c and 14 are missing, then comes the remarkable asymmetrical P 1079/15: "Impeccable air intake for the engine. Large amounts of fuel – 1,540 litres or, when using the entire left fuselage space, 2,375 litres. Possibility of two-man crew. One man crew bad visibility. Possibility to enclose the bombs as in draft 79/13b."

P 1079/16 was similar to 15 but more slender: "Modification of the draft 79/15 (omission of the left fuselage half). Fuel quantity 770 litres. Unbalanced, otherwise like design 79/15." And finally there was the somewhat more conventional P 1079/17: "Modification of the draft 79/16. Improved visibility. Improved stability around the transverse axis. Fuel content 800 litres. Possibility to install fixed weapons in the fuselage tip. Engine side same as draft 79/16. No possibility to reasonably arrange and enclose the bombs."

The P 1079/17 is the only design in the series to have been depicted with the fixed weaponry of a fighter; the others are shown entirely unarmed or carrying bombs.

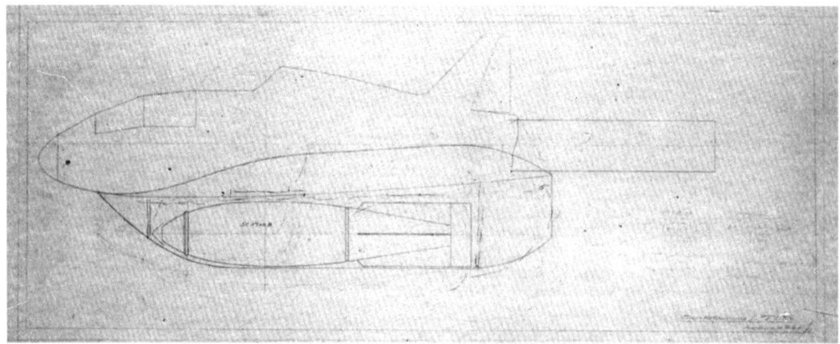

ABOVE: Another view of the P 1079/10c, this time fitted with the bulbous bomb shroud Messerschmitt seems to have intended for the series.

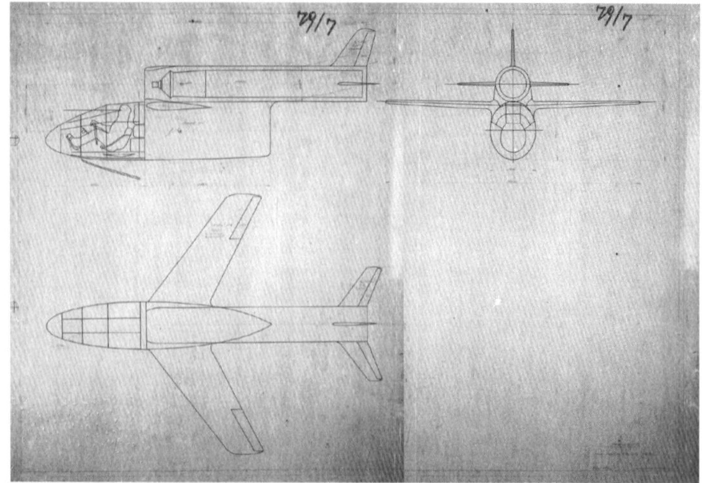

ABOVE: P 1079/7 took the radical step of shifting the engine and tail into a dorsal position, the fuselage tapering off to a point aft of the wings.

Graphs included with the report show various values for P 1079/4 to 10, then 10c-13a, then 14, then 13b and finally 15–17. This was followed by a full construction description of the P 1079 (given as the 'P 79') evidently based on the P 1079/13, dated August 1, 1941.

Under 'usage', this states: "The aircraft is designed as a jet bomber for use on short routes. The deciding factor here is that the aircraft with the highest bomb load still reaches high airspeeds when climbing, which ensure that it escapes the enemy fighter aircraft with certainty. The take-off takes place by being released from a parent aircraft (Me 323, Do 17). By replacing the bombing raid equipment with a weapon, the aircraft can be used as a fighter and interceptor. Because of its small dimensions, it is also possible to accommodate it inside a parent aircraft as an on-board aircraft (fighter or bomber)."

In other words, right from the outset the P 1079 was designed to be launched from another aircraft – although whether it was to be towed aloft, carried under a wing, carried on the back of the 'parent' or even launched from inside it is unclear.

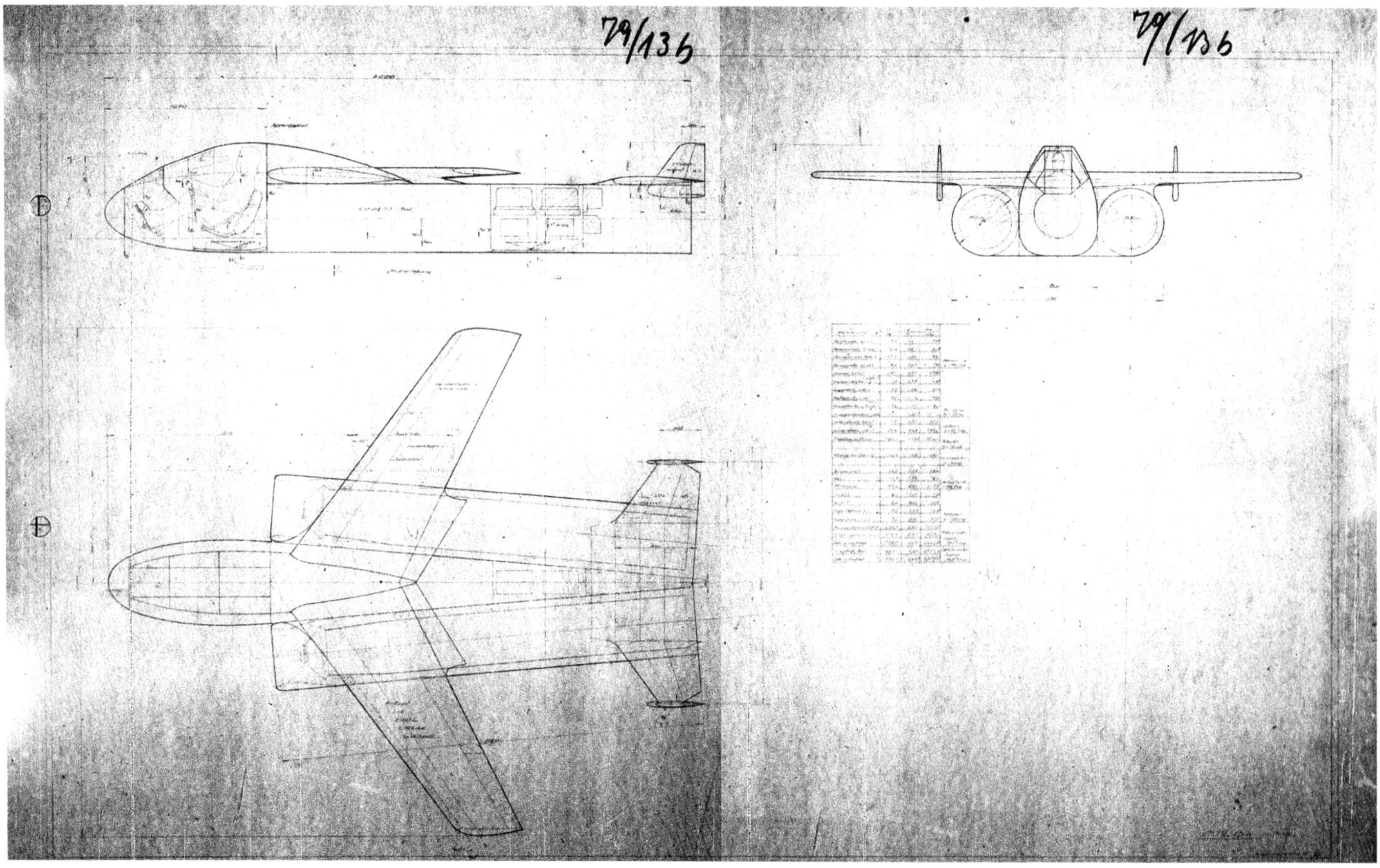

ABOVE: It may not look it, but P 1079/13b was the design chosen for further development and a similar layout, P 1079/10c, would become the very different-looking Me 328.

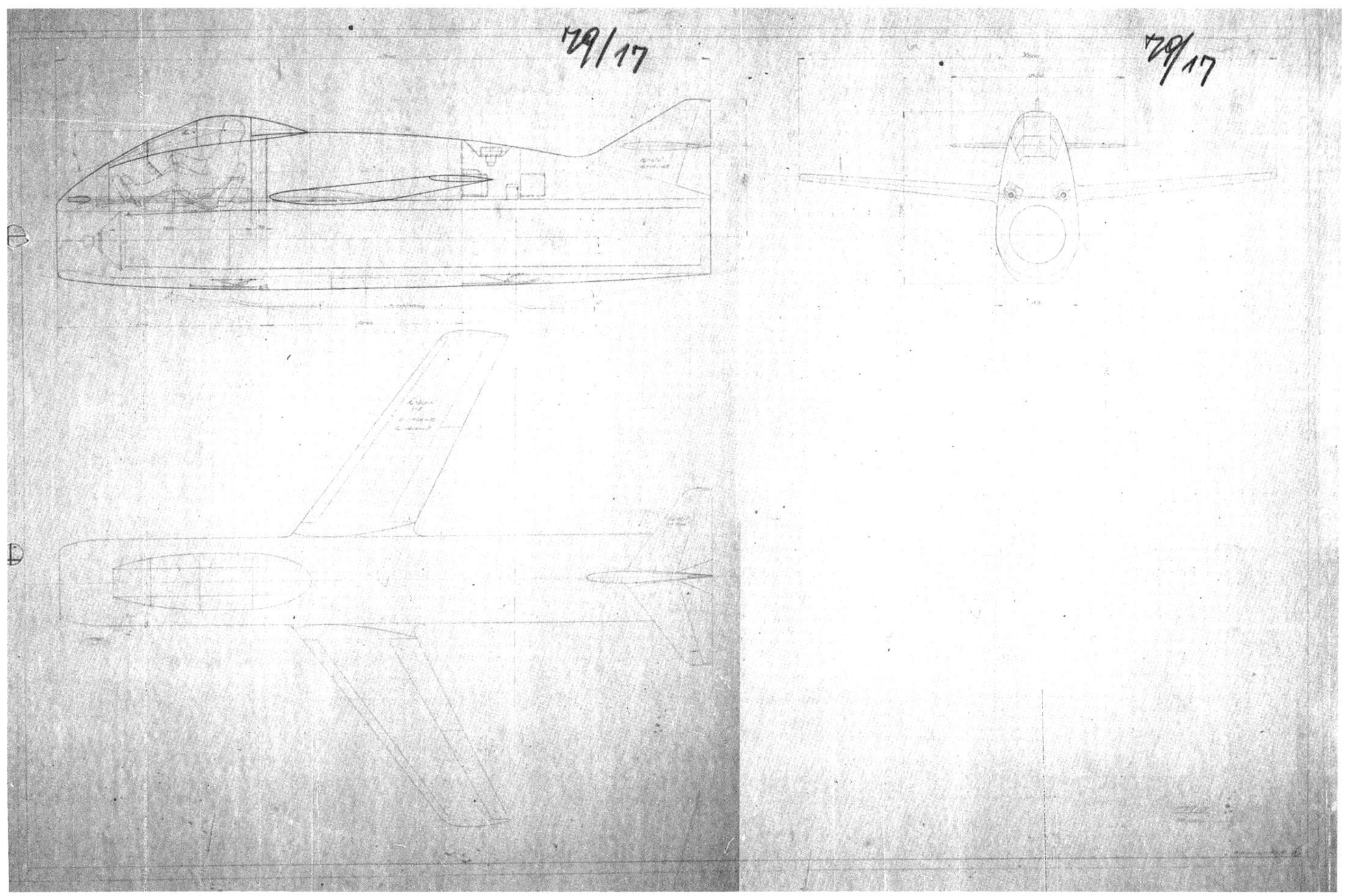

ABOVE: The final design in the sequence, P 1079/17, was created by shifting the cockpit of the /16 up on top of the pulsejet.

In physical design, it was to be a "self-supporting, twin-engine high-wing all-metal construction with retractable landing skids and enclosed pilot's compartment".

Also, under a heading of 'construction': "In the design, the highest priority was given to cheapness and simple construction in view of the early war effort of the aircraft." The airframe consisted of the cockpit, fuel tank mid-section and fuselage tail section. The tail was split with twin end-plate rudders. It seems much of the P 1079's design was intended to help it reach front-line service as quickly as possible: "As immediate solution the standard cabin has primitive equipment which can later be replaced without significant changes to the rest of the aircraft if a fully equipped pressurised cabin is provided. The forwardcanopy is made in bulletproof glass."

The mid-section had a rectangular cross section and was shielded against radiated heat from the 500kg thrust engines on either side. According to the report: "The engines are arranged on both sides of the fuselage converging towards the rear. The installation is carried out by attaching them to the fuselage side walls, thus allowing them to be changed rapidly. The possibility of installing engines of other sizes without any significant change in the airframe is also given."

The fuel tank behind the cockpit could be enlarged by extending the fuselage forward, meaning "it is possible to revert to longer ranges. An additional fuel tank can be housed under the fuselage in the bomb shroud. Four container pumps with a delivery rate of 1,000 litres per hour each are provided. Quick release is considered".

The report gives a considerable level of detail on the P 1079. Each pulsejet was to have its own power lever with a quick-stop function, and other pilot equipment included a pressure gauge, fuel gauge and fuel flow meter on the left-hand side of the cockpit. On the right were controls for power distribution and safety devices as well as a toggle switch for operating the bomb payload. On the dashboard would be an airspeed indicator, variometer, altimeter and turn indicator. And the pilot would also get a flare gun, shoulder and waist belt, parachute, oxygen system, clock, compass or gyro and a map pocket.

There was a wind-powered generator on board, FuG XVI Z and FuG XXV radio equipment with aerial housed in the rear cabin structure, and an intercom system so that the pilot could communicate with the crew of the carrier aircraft.

Evidently "the aircraft can be optionally equipped with one or two 1,000kg to 1,800kg bombs housed under the fuselage in a drop-down fairing". Alternatively the fairing could be

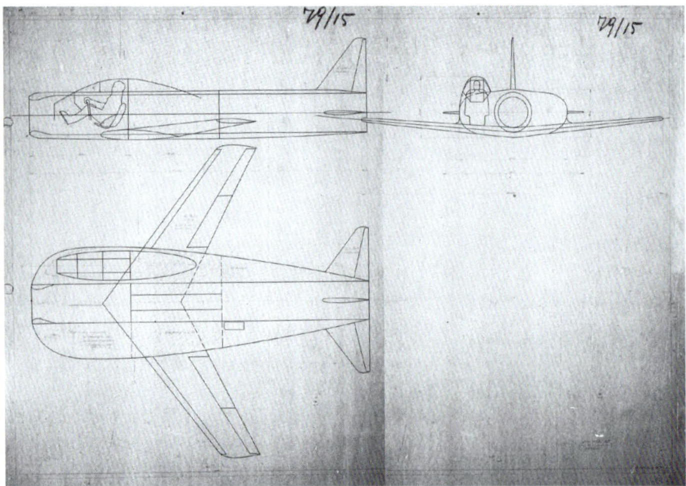

ABOVE: Giving the P 1079/15 a bulbous fuselage meant it had tremendous fuel-carrying capacity but resulted in a very odd-looking aircraft.

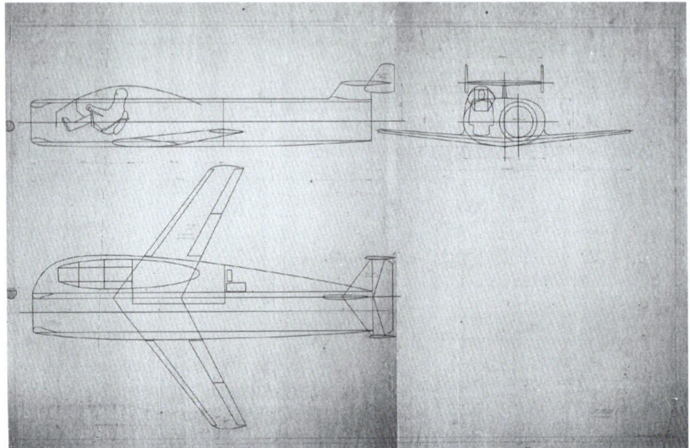

ABOVE: A development of the /15, the P 1079/16 had part of the fuselage removed to create an even more asymmetrical design.

used to house an additional fuel tank. Finally, "to allow for perfect bombing in the dive a dive brake isprovided".

Also appended to the document are a comparison between the P 1079 and the Me 109 D, dated July 31, 1941, and a comparison between the P 1079, the Me 109 F and the Me 262 – this being dated July 28, 1941.

FROM P 1079 TO ME 328

A meeting was held at the RLM on December 17, 1941,[2] to discuss the latest version of the P 1079 and the type's lead designer Rudolf Seitz explained that it was now being considered primarily as a fighter. Stats given were a wingspan of 6.2m, a wing area of 7.5m², equipment weight of 1,250kg and overall weight, including 60kg of ammo, was 2,200kg. It would be fitted with two 150kg thrust pulsejets and later two 250kg pulsejets but "eventually more powerful thrust is planned – 2 x 400kg or 4 x 250kg". With two 250kg pulsejets, horizontal top speed would be around 700km/h or 1,000km/h in a dive. For take-off a four-wheeled trolley would be provided.

A chart drawn up four days later compares the P 79 with a wing area of just 5m² and options of two 150kg engines, two 250kg engines or two 400kg engines with the 'P 79 V1' with a wing area of 7.5m² and either two 150kg engines or two 250kg engines. Also included in the comparison are the 'Me 109 F' (as opposed to Bf 109 F) with a wing area of 16.1m² and the 'Me 109 D' with a wing area of 16.4m².

On February 17, 1942, RLM Technical Office chief General-Ingenieur Gottfried Reidenbach wrote a letter to Willy Messerschmitt headed 'P 1079',[3] saying: "As reported by the DFS, constructional work on the three P 1079 test aircraft will now be carried out by the DFS at Ainring on the basis of a discussion with the Messerschmitt company. I agree with this, but please look after the work in Ainring with the gentlemen of your project office.

"In view of the remaining final uses of the P 1079, I think it is right to make these three experimental aircraft as simple as possible, in tubular steel and wood construction. At the same time, this makes it possible to produce this aircraft with the help of the DFS production and operating resources, and to relieve the experimental design department of your plant. Incidentally, I will try to accelerate the work in Ainring by temporarily presenting some designers who have experience in this type of construction."

Messerschmitt wrote back on February 24, 1942,[4] saying: "We have agreed with the DFS to design and construct as many parts of the aircraft as possible, under the supervision of a master of our project office. With us only the hull and the landing skids are constructed and built. These parts were already too advanced, so switching to DFS Ainring would have led to unnecessary delays."

The following month, the type received the Me 328 designation and work on the three prototypes commenced. A further report was produced on March 31, 1942, entitled 'Flugzeug mit Strahlrohren' or 'Aircraft with jet tubes'[5], which emphasised the suitability of the Me 328 as a fighter.

The foreword stated: "The jet tubes that have been in development for around two years are being used for the first time on the Me 328 aircraft ... The following reasons speak for the construction of aircraft with jet tubes: 1) Considerable increase in flight performance compared to that achievable with Otto [piston] engines. 2) Extraordinary reduction in construction costs (engine and fuselage) and thus savings in work capacity. 3) Due to the low engine weight, small wing area or overall dimensions of the aircraft and thus less material. 4) Possibility of increasing the capacity of the engine production to a high multiple of the current output."

Messerschmitt assumed, perhaps wrongly, that the Allies were already in possession of German pulsejet technology and would be working on their own pulsejet-powered fighter. Getting the Me 328 into production was therefore a matter of some urgency: "5) The structure of the engine is of an astonishing simplicity, which enables it to be built according to the least information; i.e. the use of aircraft with such engines must, if it takes place at all, take place as early as possible and then in bulk in order to deprive the opponent of his experience or knowledge in this area, which he will already have today due to the lack of complete secrecy.

"6) Engine testing has so far proven its usability up to speeds of 550km/h on the ground and up to 5km flight altitude. Higher speeds and altitudes are being flown at the moment. In view of the extraordinarily low cost, the great advantages and also the danger that may arise from the late use of these devices on the German side, the development of a usable aircraft is already being carried out in the current testing stage of the engines.

"These requirements (date and production output) were largely determinative in the design of the Me 328; in order to take the least possible risk with regard to flight characteristics, the shape and arrangement of the wing and tail unit were adopted from known aircraft. In order to avoid cooling problems, the engines were arranged on the wing, the construction of the fuselage was kept as simple as possible, etc."

The Argus pulsejet engines intended for the Me 328 weighed just 47kg but provided a thrust of 150kg – which allowed "the development of relatively small and high-performance, extremely high-quality aircraft. The higher specific fuel consumption of the jet pipes is largely eliminated by the low construction weight of the aircraft."

In addition, each pulsejet took just 100 man hours to make, compared to 8,000 for the Bf 109's DB 601 – not including the time required to make the propeller and its related equipment. And "further advantages of the jet tube are: low sensitivity to shooting, easy maintenance, easy handling".

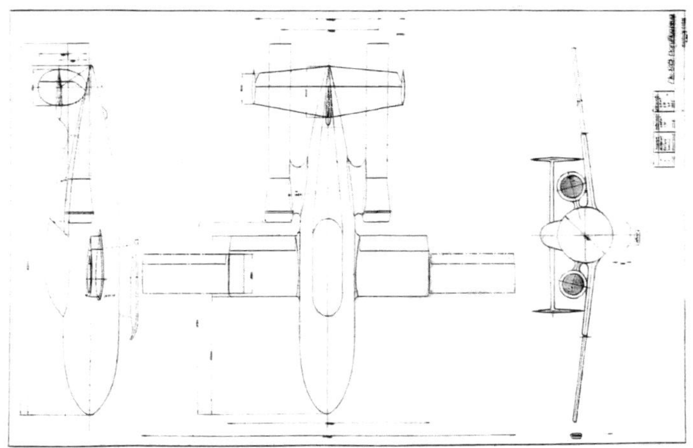

ABOVE: The second of two Messerschmitt Bordflugzeug designs published on July 7, 1942. This slightly less radical design had telescoping wings – providing the aircraft with a small wing area for high-speed flight and a large area for landing.

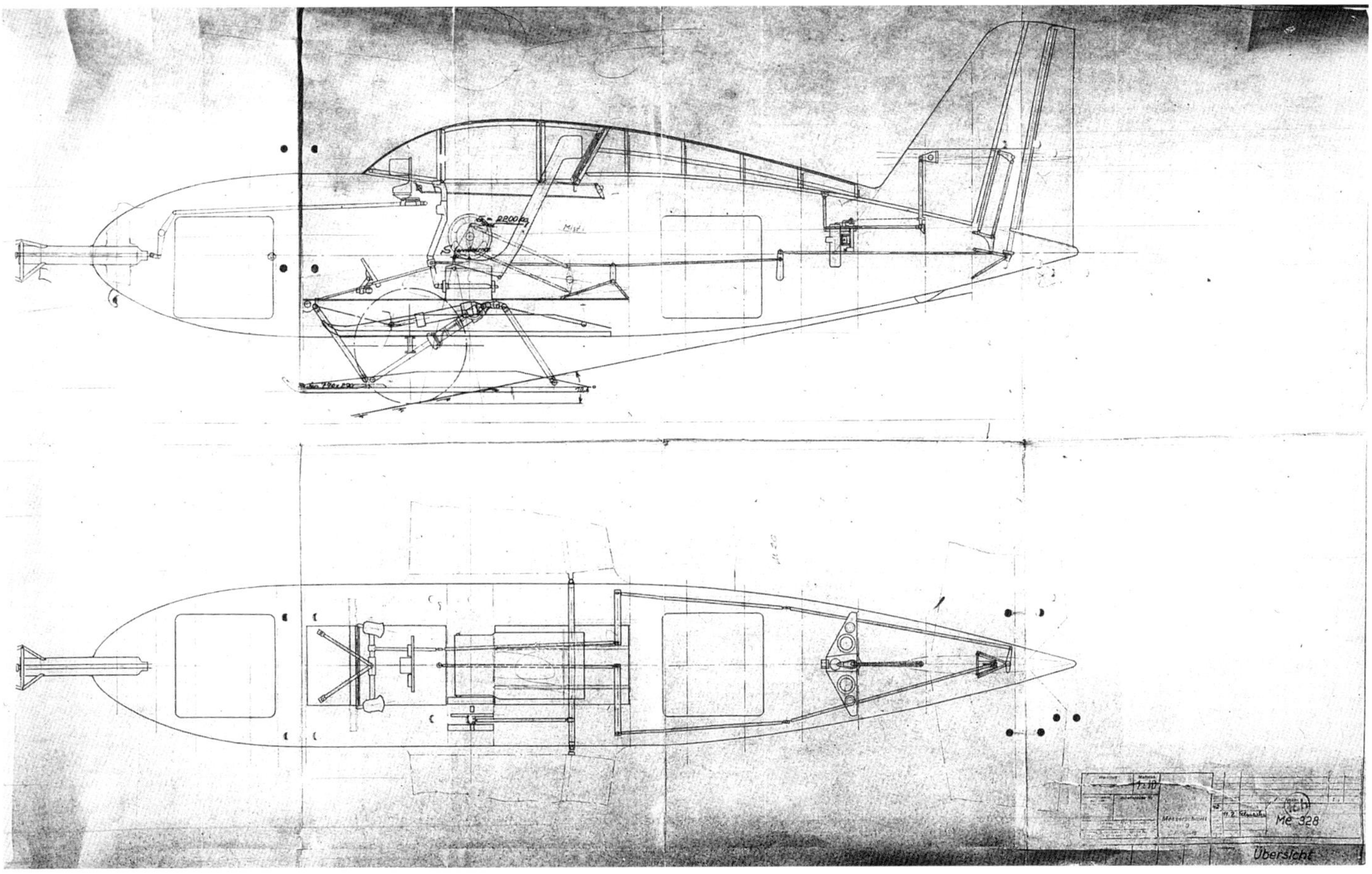

ABOVE: Side view of the Messerschmitt Me 328 dated July 11, 1942.

Four uses for the Me 328 were outlined in the following order: fighter, bomber, parasite aircraft and reconnaissance. As far as the first use went: "Due to its high performance, the aircraft type is particularly suitable as a fighter plane. Version I has a wing area of 7.5m² and a fixed armament of 2 x MG 151; it is designed with 2 jet pipes with a thrust of 300kg each, which will initially be replaced by 4 jet pipes with a thrust of 150kg each on the first prototype machines.

"For the time being, the take-off takes place on a droppable dolly, the landing on a skid. Version II has an interchangeable, enlarged wing area of 12m² with reinforced armament (additionally 2 x MK 105, 1 x MG 151 to the rear), 4 jet pipes each with 300kg thrust are provided. On both versions, the range can be increased significantly by enlarging the tank (lengthening the fuselage)."

Given the Me 328's later switch to being purely a light bomber, surprisingly little detail is given on the bomber version at this point: "The aircraft type can be used in both Versions I and II as a bomber with the heaviest bombs; RS bombs can be used particularly cheaply here. For Version I starting aids are required; for Version II, independent use is possible."

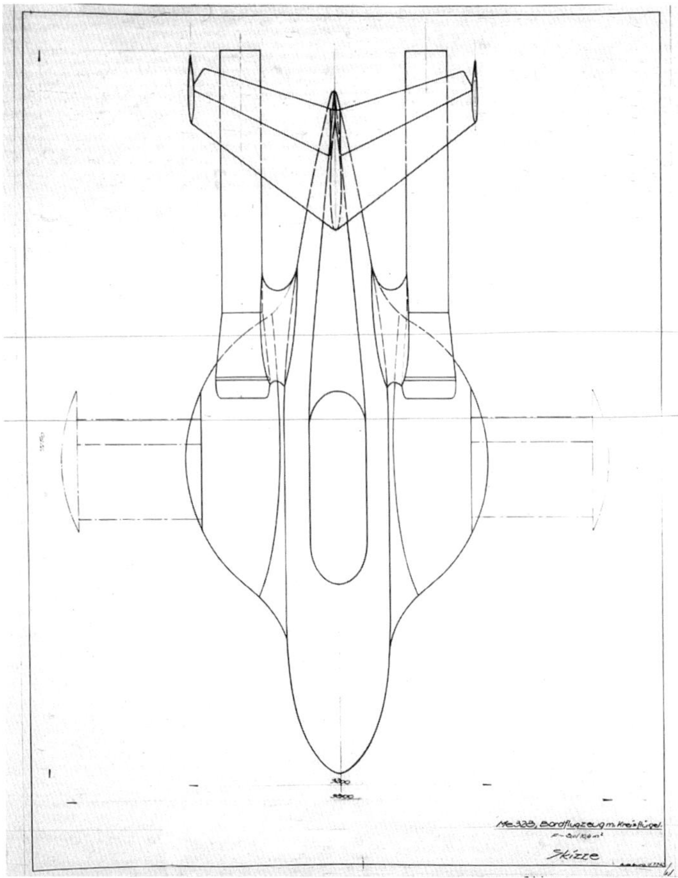

ABOVE: The first of two designs detailed in a Messerschmitt report of July 7, 1942, showing a version of the Me 328 with a 50° sweepback on the leading edge of its wings for high-speed flight but with the ability to extend straight sections to improve lift during low-speed flight and when landing. The small bomber was designed to be carried aboard a large mothership based on the Me 264.

The parasite fighter or 'Bordjäger' version was particularly innovative: "The fighter Version I fulfils the requirements of a Bordjäger due to its small dimensions (about half the size of Me 109) and its low weight. For such an application, the aircraft is provided with a towing device for connection to the mother aircraft (Fw 200, He 177, Me 264). By re-attaching and refuelling in flight (in principle already clarified by tests), multiple use is possible.

"The final state of the fuselage is aimed at an even smaller wing area, decision only after the test results are available with the 7.5m² wing area. The weight of the Bordjäger with fuel for half an hour at full throttle is about 1,650kg. As a bomber, the aircraft can be carried over long distances (150–200km) in tow, which has also been experimentally clarified, can carry out self-contained attacks with the heaviest bombs (e.g. SD 1400) and is then brought home by the mother aircraft.

"In addition, the aircraft can be carried on merchant ships and submarines and, due to its low weight, can be launched from on-board without much effort. The landing takes place on the skid in the water, after which the aircraft is taken on board again."

The only detail of the reconnaissance version given is that it would be based on Version II of the fighter. In terms of construction, the fuselage "is as simple as possible (fuselage with circular cross-section, wing continuous, engine suspended under the wing, etc.). The fuselage is selected so that it is possible at any time to insert a cylindrical fuselage piece without any other changes worth mentioning, thus doubling the fuel tank space. The installation of protected tanks is possible at any time.

"The engines are arranged in such a way that they can be replaced by stronger ones practically without modification. It is also possible to arrange several engines side by side or one above the other. In addition, the aircraft can optionally be equipped with wings of different sizes."

A brief development update was also given: "Originally, the Technical Office commissioned Messerschmitt AG to build 3 test machines in accordance with construction status I of the fighter, whereby construction of the wing and vertical tail were assigned to DFS Ainring. Later, the number of prototype machines was increased to 10, all of which are to be built by DFS Ainring. With these prototype machines, the general characteristics, the towing procedure, re-attaching and refuelling in flight, the influence of extremely high wing loads, swept wings, fixed and launching weapons should be tested."

Messerschmitt was constructing the prototype aircraft fuselages from sheet steel and the DFS was already building the wings, tails and fuselage components. The tailplane was a Bf 109 part and the first prototype was expected to be flight-ready on about May 1, 1942.

Pulsejets with 150kg thrust "can be delivered immediately" but these had very little endurance: "service life (due to valve springs) has been proven to 2 hours. Engines are constantly being tested in flight". Engines with 300kg thrust were "under development, available in a few months".

The testing up to this point had gone smoothly but "of course, it can still be expected that difficulties will arise during testing on a broader basis, which may restrict the use of the engines, postpone them or possibly prohibit them. On the other hand, we believe the mere fact that on this new basis real mass production is possible for the first time for airplanes and engines, it is necessary to advance the development with a view to an early deployment."

According to the appended data sheet, the 'Me 328 Jäger Ausführung I' had a wingspan of 6.4m, wing area of 7.5m^2 and a length of 6.83m. A fuselage extension option increased length to 8.63m. Armament was two MG 151/20s. The 'Me 328 Jäger Ausführung II' had a wingspan of 8.5m, wing area of 12m^2 and was 8.63m long. Armament was three MG 151/20s and two MK 105s.

During the summer of 1942 various designs for the 'Bordflugzeug' version were considered. A report dated July 7, 1942, with the long-winded title 'Me 328 als Bordflugzeug, unter dem Gesichtspunkt der Leistungssteigerung des Flugzeuges, sowie wirtschaftlicher und sparstoffscarender Möglichkeiten für den Tragwerksbau' or 'Me 328 as on-board aircraft, from the point of view of increasing the performance of the aircraft, as well as economic and fuel-saving possibilities for structural engineering'[6] begins by discussing the possibility of an Me 328 with 50° wing sweepback.

This would make landing nearly impossible, so it was proposed that the aircraft could have pop-out straight wings, providing a 0° sweepback when necessary. The alternative was a telescoping straight wing – allowing the aircraft to have short wings for high-speed flight and long wings for landing. It was proposed that the aircraft could be carried within the narrow fuselage of an Me 364 bomber – the earliest known designation for the six-engine version of the Me 264.[7] But by now the Me 328 seems to have been increasingly seen as a bomber itself rather than a fighter. While within the fuselage of its parent aircraft, the Me 328 could be fitted with more bombs and refuelled. Launch involved simply lowering the aircraft from its parent before releasing it. The report gives no details of an Me 328 armed with fixed forward-firing weapons – instead concentrating on its performance while carrying a bomb load.

On December 15, 1942, Messerschmitt produced a high-quality brochure entitled 'Me 328 B Leichtes Schnellkampfflugzeug' – or 'Me 328 B light fast bomber'[8] which described the aircraft as flying coastal artillery for fighting off an invasion. From this point on, very little consideration seems to have been given to the pulsejet Me 328 as a fighter. It mattered little because the aircraft was fundamentally flawed.

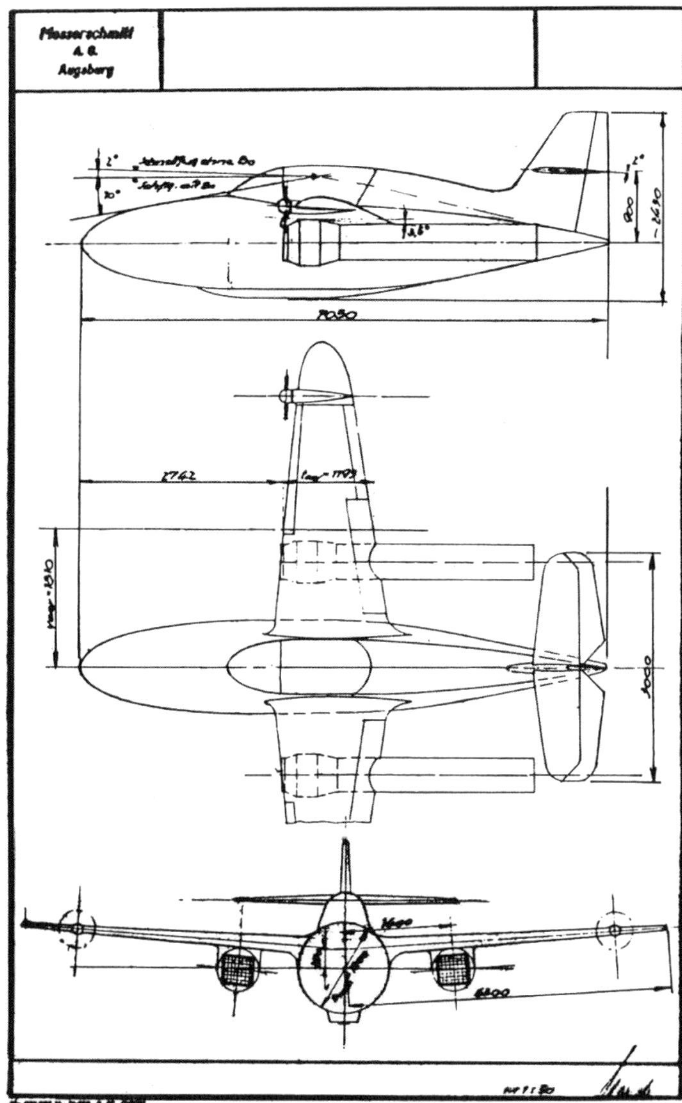

ABOVE: The Me 328 as it appeared circa December 15, 1942. By this time, the pulsejet version of the type was intended exclusively as a bomber.

Voigt explained the type's development to his Allied interrogators on September 7, 1945:[9] "The flight testing of the prototype without engines gave good results. The pilots blamed the low towing speed of the Me 110 and the bad field of view from the 328 at high angles of attack. Special devices were necessary to shorten the landing run on wet lawn.

"With engines no satisfactory results were obtained until the development was stopped; the engine itself did not work satisfactorily and did not give the expected thrust; the noise of the engine was hardly bearable without a sound-proof cabin; the vibrations of the engines were detrimental to the strength of the airframe.

"The Me 328 was not ordered in series; the development was stopped when the turbojet engine proved to be reliable and the Me 262 was given higher priority. The Me 328 was the only model designed and constructed during the war by Messerschmitt which was not ordered in series."

Chapter 5

1-TL-Jäger part I

Single-jet fighters 1943–1944

The first jet to fly had only one engine and as more powerful engines entered development, the thoughts of German aircraft manufacturers turned towards the possibility of a single-jet fighter.

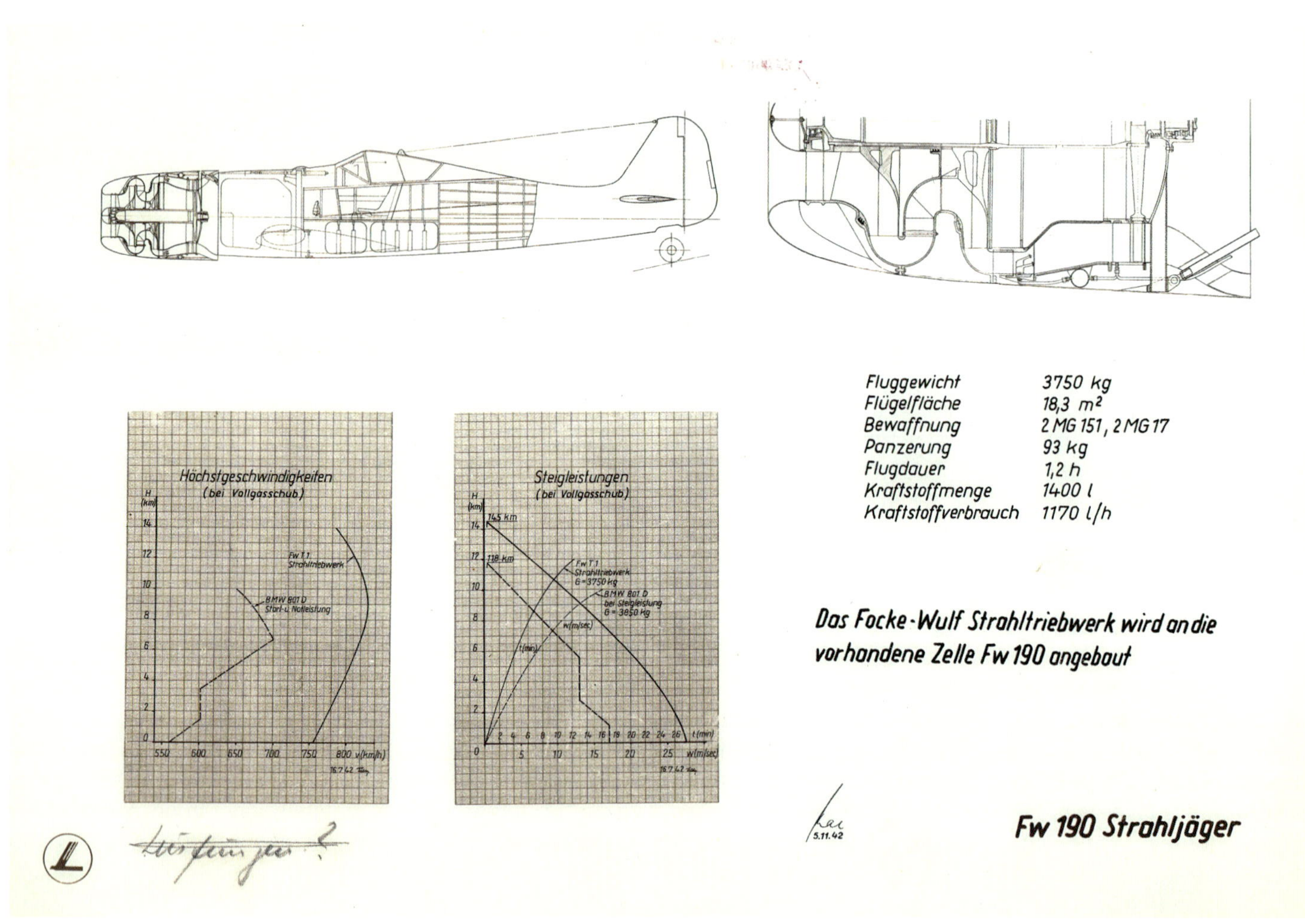

ABOVE: Focke-Wulf's earliest known attempt at designing a 'jet fighter' – a standard Fw 190 airframe with a Focke-Wulf-designed turbojet bolted on in place of the BMW 801.

FOCKE-WULF, NOVEMBER 1942 TO MARCH 1943

With the cancellation of the He 280 and with the pulsejet Me 328 now being regarded as a bomber, the Me 262 with its Jumo 004 B engines was the only jet fighter under development in Germany during the spring of 1943. Its twin-engine configuration had been chosen early on to account for the weakness of contemporary jet engine designs, but two engines did not mean double the power. A larger, weightier, draggier airframe was needed to accommodate both powerplants and double the amount of fuel was required.

It seemed clear at this point, however, that significantly more powerful jet engines would soon be available – making it possible to create a high-performance fighter powered by just one. Such a fighter would have several advantages over the Me 262: it would be cheaper and quicker and less resource-intensive to build, its performance would not be significantly worse and it would be easier to fly since the pilot would have only one engine to manage.

Both Messerschmitt and Heinkel had been investing heavily in twin-jet designs up to this point and although certainly the former and probably the latter had investigated single-jet designs, neither had expended much effort on them. Focke-Wulf, on the other hand, had not been involved in the initial jet fighter competition but had conducted its own turbojet research.

Interrogated after the war, Focke-Wulf engineer Dr Otto Pabst told his British captors that, like Heinkel, Focke-Wulf had attempted to design its own jet engine. The summary report[1] states: "Prior to the war, Dr Pabst had also worked on a gas turbine engine to be constructed by Focke-Wulf, which consisted of a double entry radial compressor and a single stage axial flow turbine with a single annular burner chamber which was expected to produce 600kg thrust at 11km or 2kg thrust at sea level. This power plant was being considered for the Fw 200 (four engine bomber of the B-17 type)."

When efforts were made to either re-engine or entirely replace the Fw 190 with a new design in 1942, among the options considered was a jet-propelled version. This involved simply bolting a turbojet of Focke-Wulf's own design to the standard Fw 190 airframe. The device had an annular thrust nozzle – which meant it exhausted down the sides of the fuselage and underneath it, though not over the pilot's canopy.

This was expected to produce a top speed of 516mph, less than that projected for the Me 262 and He 280. Although it had the advantage of being able to use factory-fresh Fw 190 airframes with very little modification, it appears to have gone no further than a presentation drawing and graphs dated November 5, 1942.[2]

At around the same time, a young self-taught aerodynamicist working in Focke-Wulf's design department, Julius Rotta, was asked to undertake a wide-ranging 'blank sheet' study looking at what sort of airframe might be best suited to housing a single turbojet powerplant and what sort of performance could be expected from it.

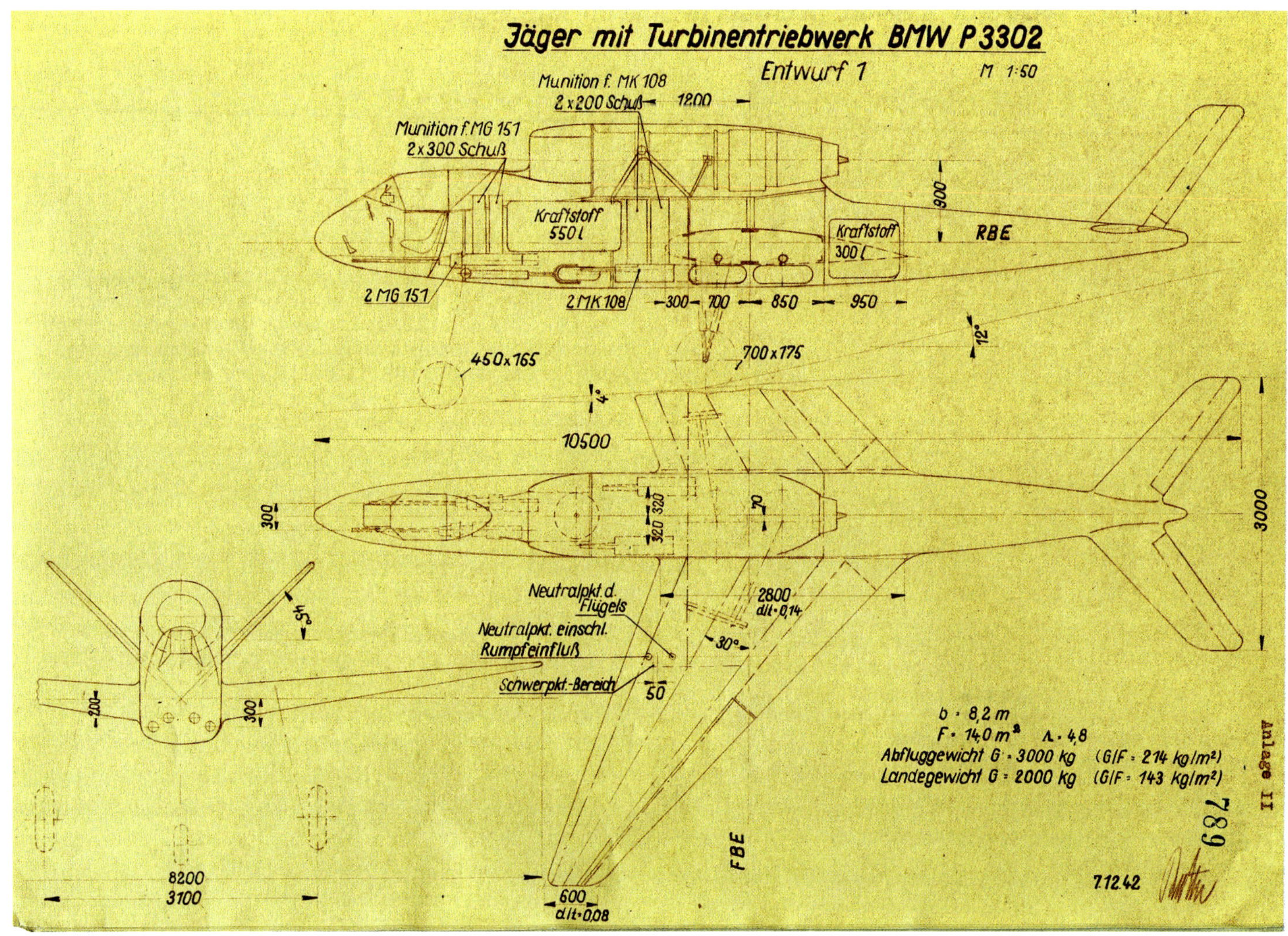

ABOVE: Rotta's single-jet design of December 7, 1942 was remarkably prophetic. A dorsally mounted turbojet, swept V-tail and forward-swept wings would all appear in later designs of other companies but Focke-Wulf itself never really followed up on any of them.

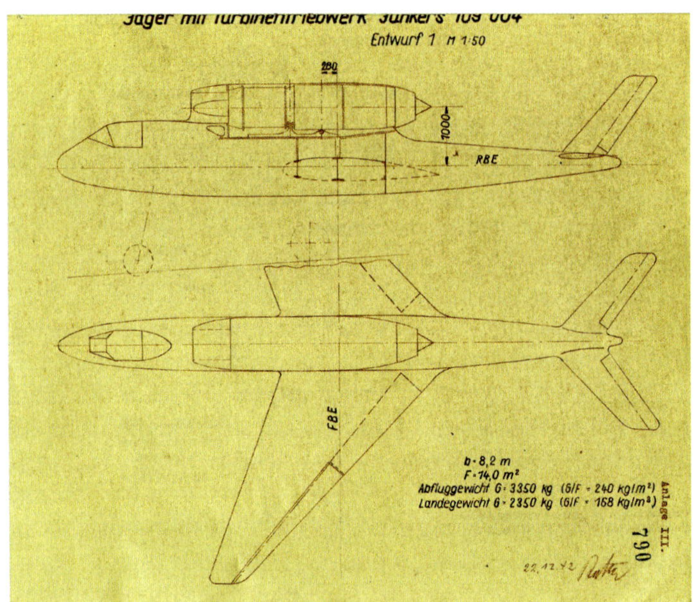

ABOVE: A second version of Rotta's 'Entwurf 1' but now fitted with a Jumo 004 in place of the BMW P 3302. It is dated December 22, 1942. The writing at the top of the page is partially obscured by the way the original document is bound.

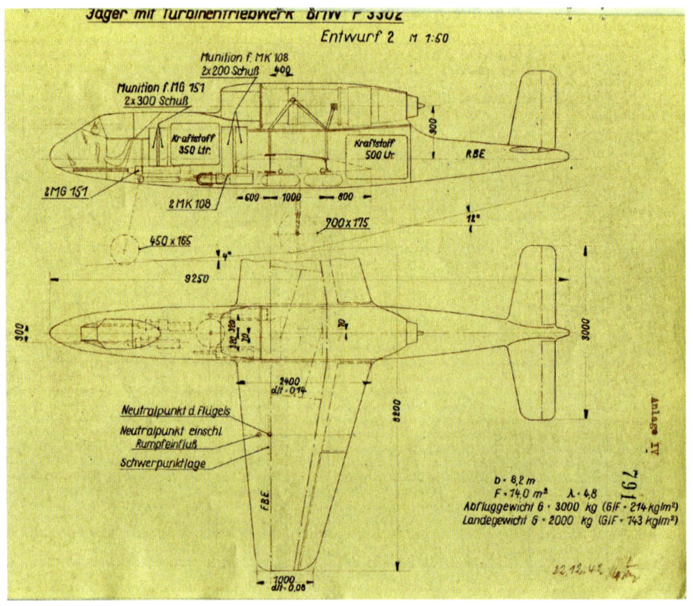

ABOVE: For his 'Entwurf 2', again powered by a single BMW P 3302, Rotta envisioned a less risky design with straight wings.

His report, 'Grundlagen zum Entwurf eines Jägers mit Turbinentriebwerk' or 'Fundamentals For The Design of a Jet Fighter',[3] was published on January 4, 1943, and looked at how large a single-jet fighter ought to be, what sort of shape and layout would be best, what jet engines could be fitted and how, what the advantages and disadvantages of piston engines and jet engines were and aerodynamic issues.

To illustrate his points, Rotta came up with a trio of remarkably foresighted designs: Jäger mit Turbinentriebwerk BMW P 3302 Design 1, Jäger mit Turbinentriebwerk BMW P 3302 Design 2, and Jäger mit Turbinentriebwerk Junkers 109 004.

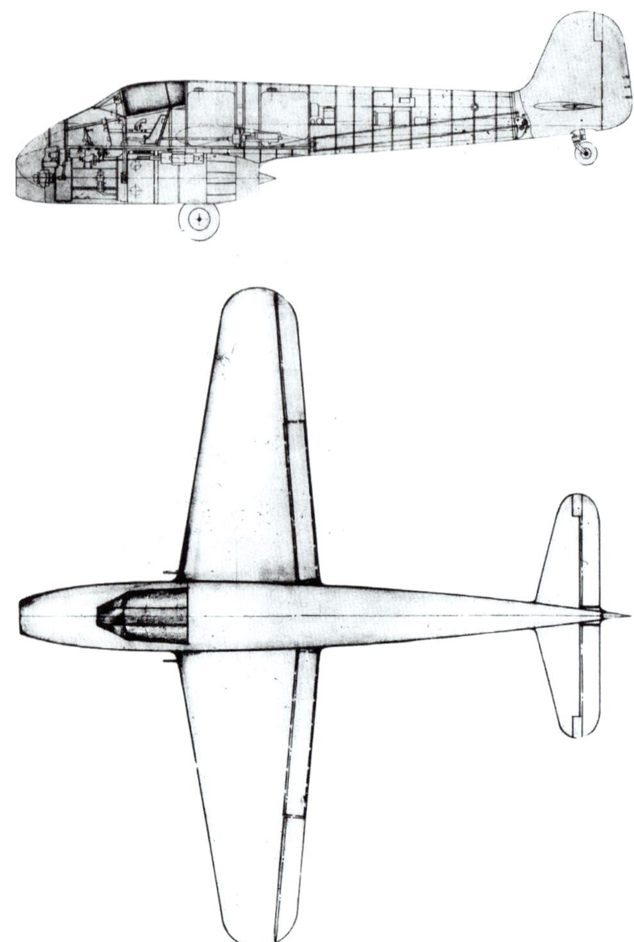

ABOVE: Focke-Wulf's first attempt at a practical single-jet fighter in March 1943. The fuselage and wings owe much to those of the Fw 190 – which seems to have been part of the design brief. Unfortunately, the tail-sitter undercarriage arrangement was less than ideal given the turbojet's position. The design's correct designation is unknown but it was referred to retrospectively in a company report of August 1944 as the '1. Entwurf', the first design.

All three had the jet engine mounted on their backs, pre-dating the next earliest known design with this layout – Arado's E 570/K-Jäger – by two months, the Henschel Hs 132 by perhaps a year and the Heinkel P 1073/He 162 by 18 months. The first and third designs also had forward-swept wings and backward-swept V-tails. The second BMW P 3302 design had unswept wings and an unswept V-tail.

Rotta wrote: "There will now be described the draft of a fighter with the BMW P 3302. The design has the advantage that optionally also the Junkers 109 004 engine can be installed without major changes to the airframe.

"General structure – since the proportion by weight of the engine on the flying weight is quite large, the engine must not be very far away from the centre of the aircraft. This fact and the large size of the engine mean basic aspects of the aircraft's layout come into consideration.

"Engine above or below the hull, engine on the wing next to the fuselage (unbalanced design), engine within the fuselage of the aircraft or an arrangement of two hulls to accommodate pilots and fuel. The latter version is ruled out because with such a small wing caused by three hulls, disturbance of lift distribution creates intolerable conditions. The same also applies for the mentioned asymmetrical design."

He then concludes that the undercarriage must retract into the fuselage, rather than the wings, for reasons of weight distribution – precluding the fitting of a jet engine into or under the fuselage, he says, and also ruling out the asymmetrical design.

"So the only remaining solution is the arrangement of the engine above the fuselage. A suggested armament would be 2 x MK 108 with 200 rounds and 2 x MG 151 with 300 rounds. The fuel quantity is 850 litres. The landing gear is retracted into the hull, so that the two wheels of the main landing gear are located one behind the other. For aerodynamic reasons, a very strong wing sweep was provided to the front. An alternative suggestion of design has also been made without an arrow shape of the wing."

Apparently there was some concern at this time about the hot gases exhausting from jet engines because Rotta writes: "The heating by exhaust plume of the parts behind the engine is safe and is in no case more than 100°C, as can be seen with reference to a chart image supplied by BMW of the temperature distribution in the beam. The installation of the Junkers engine 109 004 in the same airframe is also shown. Since both the electrical connections and operating terminals are provided on the upper side of the engine, in accordance with an arrangement of it being fitted under a wing, a pivoting of the engine by 180° is required for the present case. The integrated oil system would have to be rebuilt. In the airframe, no changes other than minor modifications to the fuselage fairing would be necessary."

Comparing the two engines, Rotta allows for the fact that further development of both designs ahead of full production status is likely to considerably alter their characteristics. Similarly, he states that more research needs to be carried out on the flight performance benefits of forward-swept wings: "How far the theoretically expected improvement occurs practically in the presence of the fuselage cannot be said. There are still a large number of outstanding

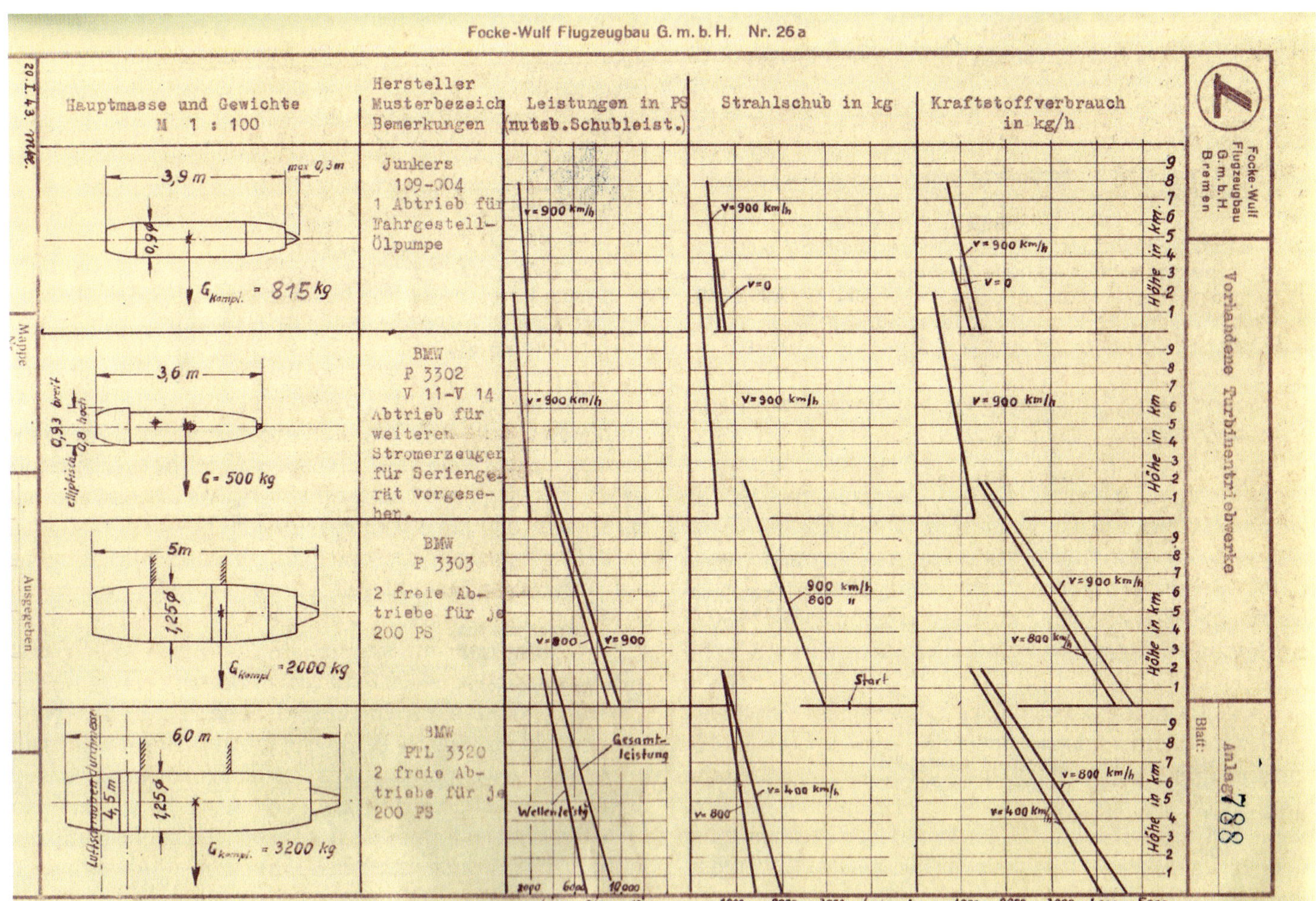

ABOVE: This was the selection of engines facing Rotta when he wrote his January 1943 report on jet fighter configurations for Focke-Wulf – a motley collection of BMW designs and the Jumo 004.

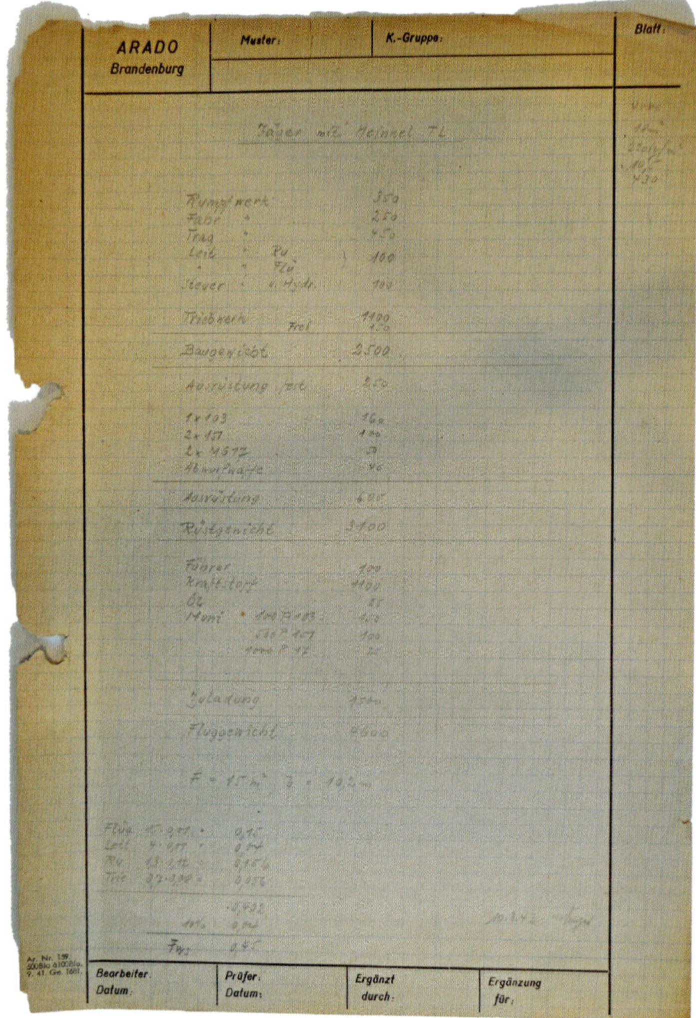

ARADO
Brandenburg

Muster: | K.-Gruppe: | Blatt:

Jäger mit Heinkel TL

Rumpfwerk	350
Fahr "	250
Trag "	450
Leit " Ru / " " Flü	100
Steuer " u. Hydr.	100
Triebwerk / Frel	1100 / 150
Baugewicht	2500
Ausrüstung fest	250
1 x 103	160
2 x 151	100
2 x MG 17	50
Abwurfwaffe	40
Ausrüstung	600
Rüstgewicht	3100
Führer	100
Kraftstoff	1100
Öl	25
Muni	150
	100
	25
Zuladung	1500
Fluggewicht	4600

F = 15 m², b = 10,2 m

Bearbeiter: Datum: | Prüfer: Datum: | Ergänzt durch: | Ergänzung für:

ABOVE: List of weights for a fighter powered by a Heinkel turbojet. The aircraft was to have a wingspan of 10.2m and wing area of 15m². Armament was one MK 103, two MG 151s and two MG 17s. What appears to be the design in question was sketched upside down on the reverse of this sheet.

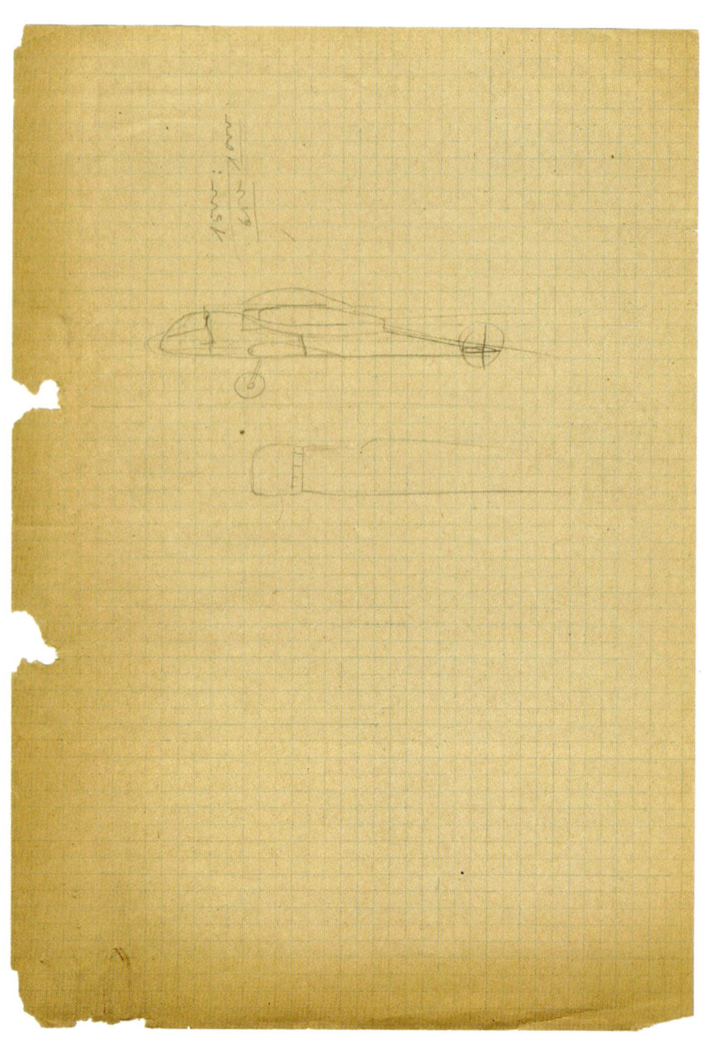

ABOVE: Rough sketch from the reverse of a list of weights for an Arado 'Jäger mit Heinkel TL' dated March 10, 1943. The design features a dorsally mounted turbojet, twin-fin tail and tail-sitter undercarriage.

issues in this area. Therefore, the design was investigated under no sweep."

Rotta's radical designs do not seem to have been very highly regarded within Focke-Wulf however, because the company's first stab at a practical jet fighter during March 1943 rejected nearly all of their features. It was detailed in two separate reports, both dated March 10, 1943: 'Jagdflugzeug mit Turbinentriebwerk mit BMW-P 3302'[4] and the other 'Jagdflugzeug mit Turbinentriebwerk Jumo 109 004 C'.[5] In each case, the aircraft was a tail-sitter like the Fw 190 but with its cockpit moved to the front of the aircraft and its turbojet positioned beneath the cockpit. A simple nose intake fed the turbojet and it exhausted beneath the fuselage. The straight wings appeared similar to those of the 190 in plan view but the undercarriage position must have differed since the mainwheels would have been unable to tuck up into the underside of the fuselage.[6]

The first report, which included two drawings, numbers 0310 226-20 and -21, stated that the design was 8.025m long, had a wingspan of 9.5m and a wing area of 15m². With its BMW P 3302 turbojet, the aircraft stood at 2.3m high. Armament was four MG 151/20s – two in the fuselage and two in the wing roots. Take-off weight was 2,950kg and top speed at an altitude of 8km was 796km/h.

There was just one drawing in the second report, number 0310 226-22, and the aircraft had the same dimensions as before except for height – the larger Jumo 004 C made it 2.6m high. Take-off weight with the bigger engine was 3,220kg but top speed at 8km was 814km/h.

A third version of the design appears to have been created, also in March 1943, which had a 10.2m wingspan but the same wing area. It was this design, evidently based on the Jumo 004 C variant, that would reappear in a timeline of Focke-Wulf jet fighter designs produced by the company in August 1944 under the label '1. Entwurf'.[7] Although very little more of this design appears to have survived, it was an important project for Focke-Wulf and was evidently worked on for several months.

ARADO, FEBRUARY TO AUGUST 1943

While privately owned Messerschmitt and Heinkel had been busy with their twin-jet fighter designs, government-owned Arado had been working on a twin-jet reconnaissance platform with ambitions to become a light bomber – the Ar 234. Before 1943, the company appears to have made no efforts to design a single-jet fighter but this changed on February 18, 1943, when the RLM sent Arado a set of fairly exacting specifications for a single-seat 'Nahkampfflugzeug' or 'close support bomber'.[8]

This would appear to be the specification for a single-seat Junkers Ju 87 replacement with good visibility and an air-cooled 2,000hp engine providing a top speed not below 500km/h with a dive speed, with dive brakes, of 550km/h. Range was 1,000km with an option for drop tanks to achieve 1,400km/h. Several armament options were specified and the aircraft had to be easy to maintain and operate from rough airstrips

After studying the Ju 87, the Hs 123, the Douglas DB 19 (apparently the Douglas SBD Dauntless, based on the stats Arado gives for it) and the Fw 190, Arado began project E 570 for a 'Jabo mit BMW 801 C'. Three different sizes were projected for this aircraft, a wing area of either 35m², 30m² or 22.5m² and each had two MG 151s, one in each inner wing.

But on April 15, 1943, Arado sent the RLM a letter which said: "The work for a single-seat close support aircraft according to the guidelines mentioned in the above letter of February 18 was started in our design office and we searched for favourable versions. However, this had to be cancelled because the capacity in the design office is not sufficient to carry on this work in addition to the project of the multi-role transporter and the Ar 234 B, as well as the design of the Ar 396 and Ar 432. So, to our great regret, we need to refrain from further work, but we do not want to offer nothing. In short, here are the lines of thought that we consider appropriate to solve the problem.

"Since the visual demand in a single-engine normal form with air-cooled radial engine results in very high fuselage superstructures, which contradicts the wishes for great speed, the following routes were investigated: a) Retractable cockpit. As a result, good forward visibility is created in the dive down and at the same time an increased drag, while in the retracted state the fuselage cross-section can be kept small.

"b) Engine behind the pilot's cockpit. As a result, very good visibility was achieved for small fuselage cross-sections. The disadvantage of this arrangement is that probably only one liquid-cooled motor can be installed. c) Jet engine. The engine is a jet-unit, which is mounted in aerodynamically good shape to the fuselage, whereby the pilot's space can be formed as full-vision cockpit. Unfortunately, with these preliminary studies, we had to conclude the work for reasons mentioned above."

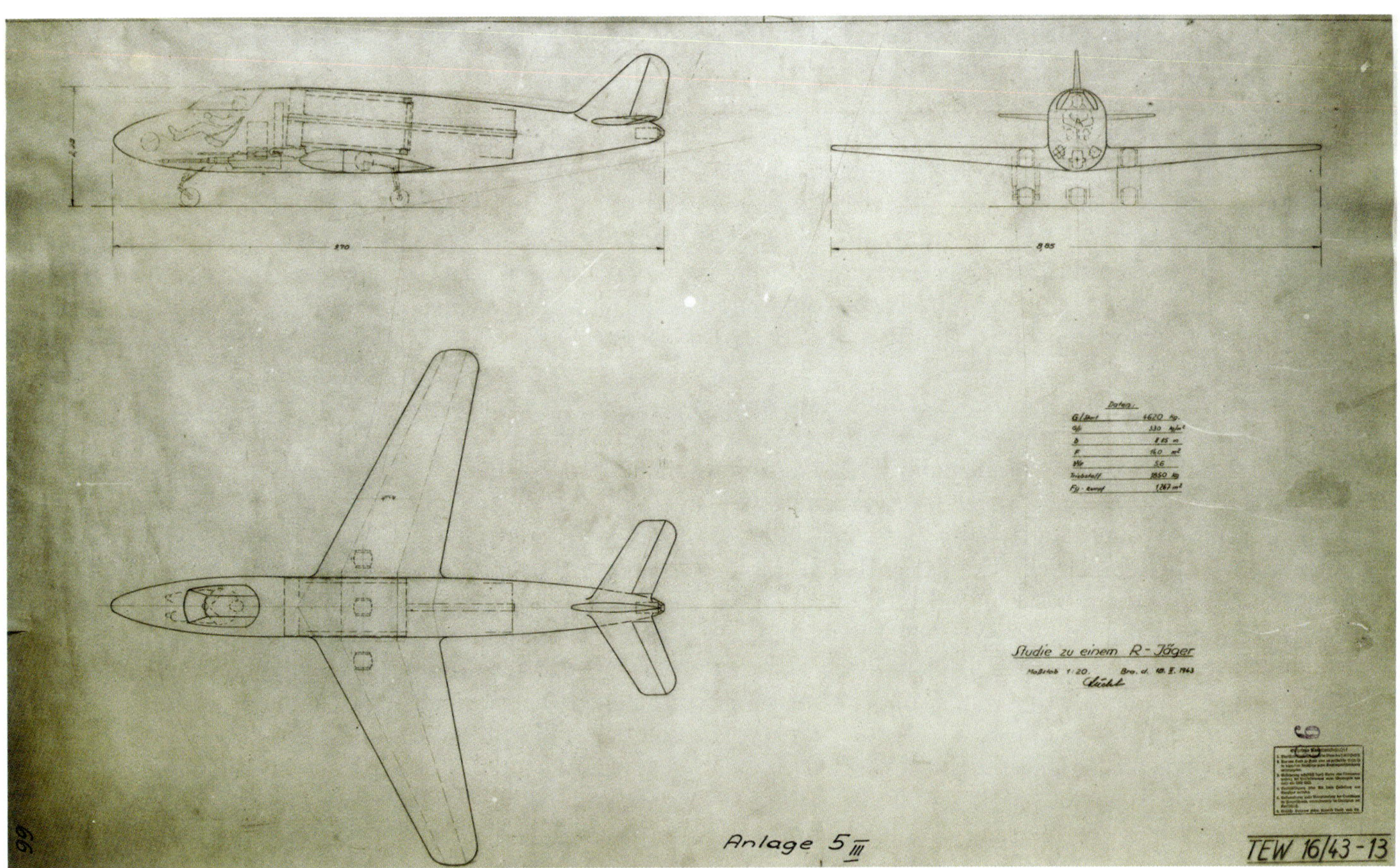

ABOVE: Arado's R-Jäger of drawing TEW 16/43-13, dated March 18, 1943.

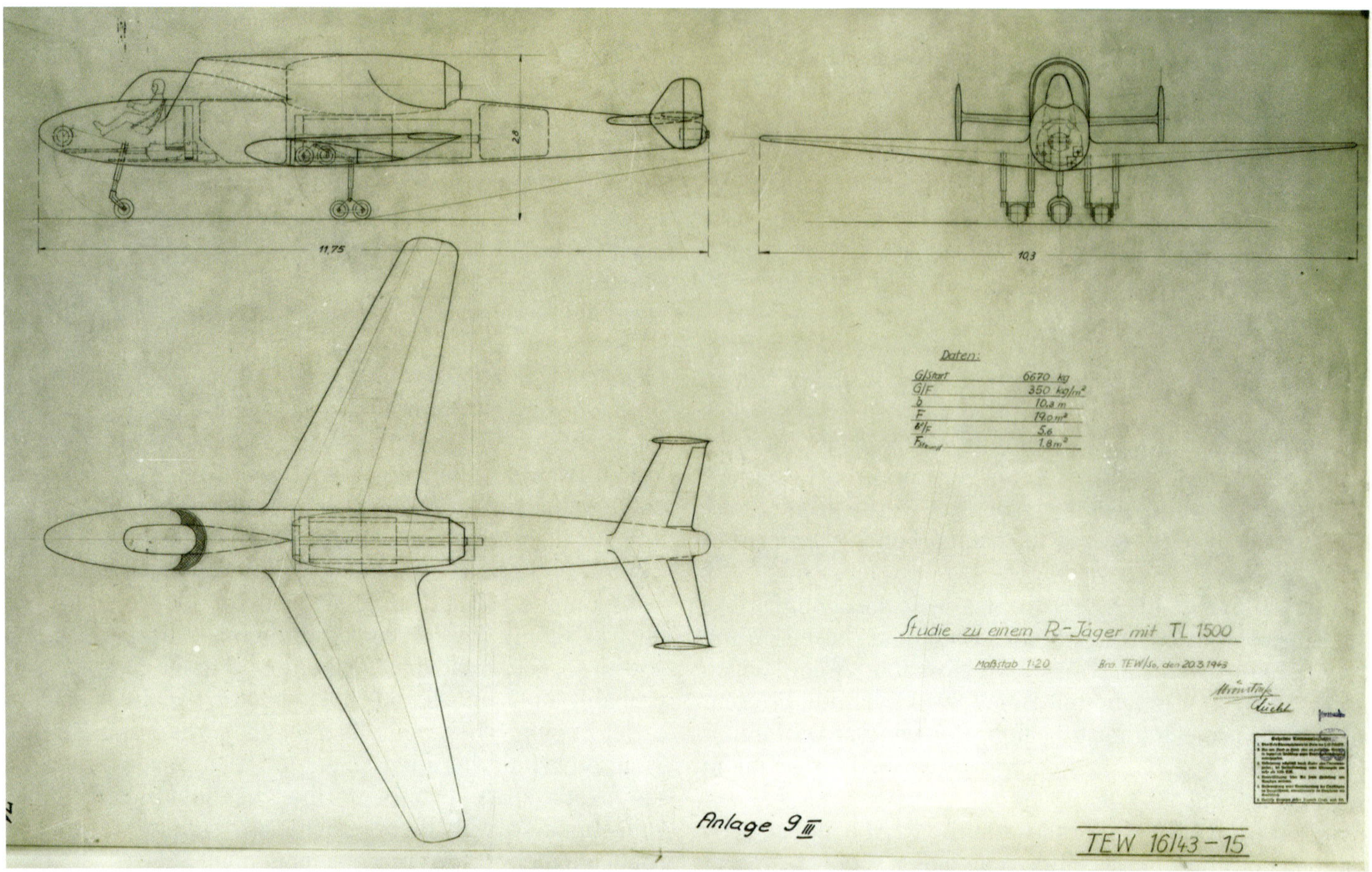

ABOVE: 'Study of a rocket fighter with TL 1500' shown in Arado drawing TEW 16/43-15 of March 20, 1943. The design, which appears to be an evolution of the E 570, combined jet and rocket propulsion and was referred to in a company report of August 1943 as the K-Jäger.

The letter states that three sketches are appended showing the three designs but these are missing. However, all the information known about the E 570 is contained in a single folder marked 'Jabo' and among the various notes is a rough sketch showing what appears to be the 'c) Jet engine' design. It is a low-wing tail-sitter aircraft with twin-fin tail and the turbojet positioned on its back – allowing the inclusion of a 'full-vision cockpit' similar to that of the Ar 240 at the front.

Beneath this drawing is an even sketchier outline showing an aircraft with an air-cooled radial piston engine on its nose and a very low cockpit canopy – presumably the 'a)' design.

On the other side of the same sheet of Arado notepaper, written in pencil, is a list of weights for a 'Jäger mit Heinkel TL' or 'fighter with Heinkel jet'. The list is dated March 10, 1943. Arado's designers next turned their attention to rocket propulsion for drawing TEW 16/43-13 of March 18 with what was dubbed the 'R-Jäger'. This rather basic low-wing fighter was 9.7m long with a wingspan of 8.85m, wing area of 14m² and take-off weight of 4,620kg – 1,850kg of which was rocket fuel. It also featured a novel thick-wheeled undercarriage – with a wheel retracting into each wing, a third into the centre of the fuselage plus the nosewheel, making a total of four wheels. At the extreme rear of the aircraft was a single rocket motor.

Drawing TEW 16/43-15 of March 20 showed the Arado 'K-Jäger'. The 'K' stood for 'Kombinations' since the design combined the basic form of the original E 570 sketch with the R-Jäger's rear fuselage-mounted rocket motor. Fuselage length was 11.75m, wingspan 10.3m and wing area 19m². Take-off weight was 6,670kg. The undercarriage had the same fat wheels as that of the R-Jäger but rather than have a fourth wheel in the centre of the fuselage, each mainwheel leg had two wheels rather than one.

The R-Jäger and K-Jäger were joined by a third design – the unimaginatively named 'TL-Jäger' of drawing TEW 16/43-23 dated June 3, 1943, in a report of August 11, 1943, concerning the development of fast two-seaters.[9]

The section of the report that deals with the three fighters, Appendix 2, begins: "Preview of future technical and tactical possibilities of next-step fighter aircraft. General requirements for a fighter to combat rapid aircraft.

"For rapid aircraft flying at high altitude, despite a well-organized reporting system, defensive fighters have little time to go on the attack. Therefore, a very large rate of climb is absolutely necessary and aircraft with Otto engines must be separated from those with more suitable propulsion. The fighter must be at least equipped with jet engines, but would not be significantly superior to the enemy in this arrangement. Above all, the rate of climb

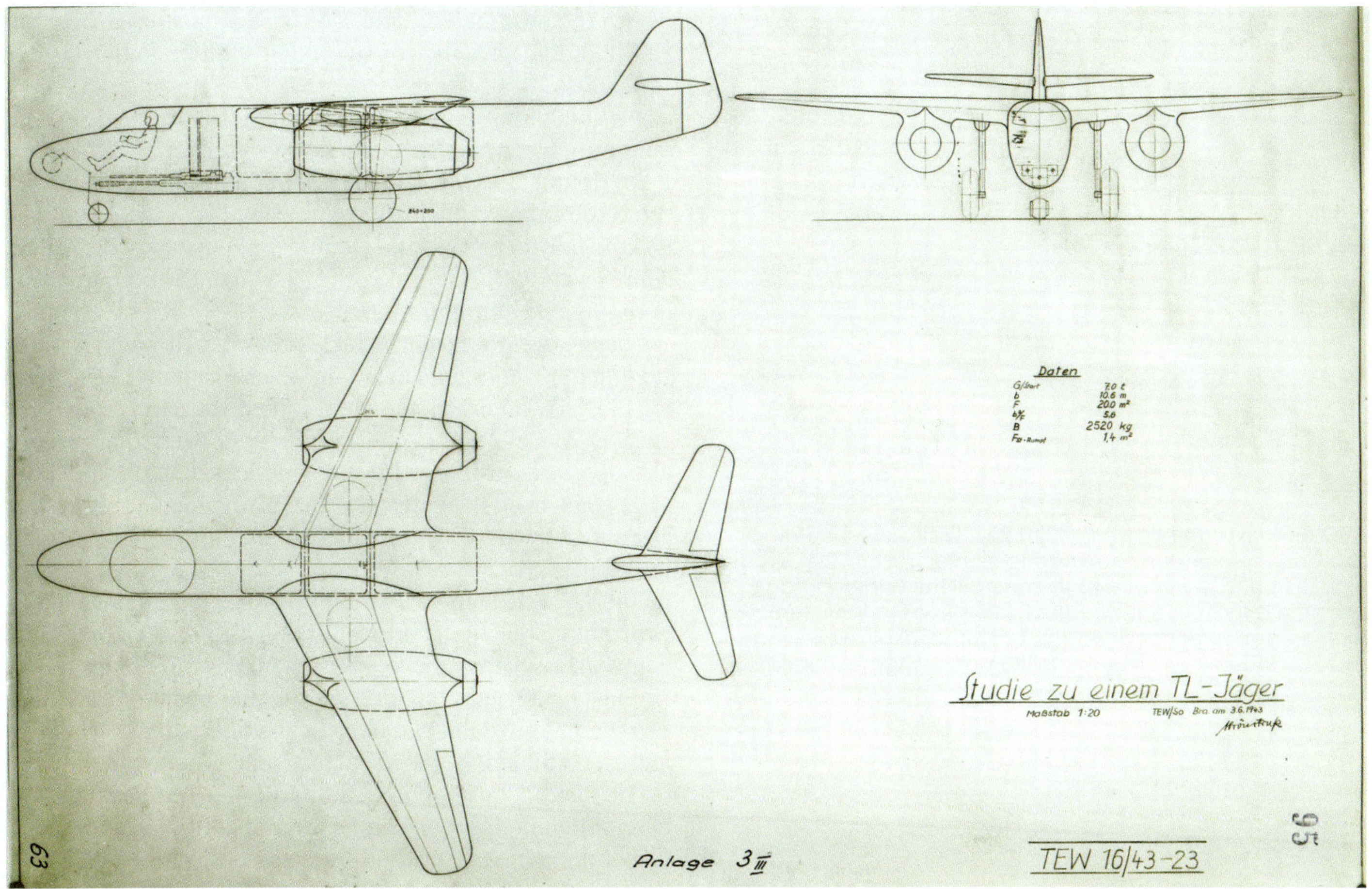

ABOVE: The third of Arado's 1943 jet fighter studies was a straightforward design with two turbojets – the 'TL-Jäger' of drawing TEW 16/43-23, June 3, 1943. The undercarriage retraction appears to be inconsistent between the side and top views.

would still be regarded as insufficient, since the altitude greatly decreases the thrust of jet devices, which means a significant decrease in the rate of climb.

"By contrast, with a rocket-powered fighter (here called R-Jäger) large rates of climb and steep climb angle are possible, but due to the high specific fuel consumption has completely inadequate range, whereby the possibility of a dogfighting battle is called into question. If one wants to succeed with a fighter against a rapid opponent, one must combine high-altitude performance with climbing speed, long range and horizontal velocity. Even under these conditions, a successful interception will be possible only with excellent reporting and guidance to the target.

"These conditions are only applicable when the fighter is equipped both with a turbojet engine and an intermittently connectable rocket engine. This combination fighter (here called K-Jäger) alone has the potential to fight and destroy rapid opponents."

The report then went on to outline the equipment, advantages and disadvantages of the TL-Jäger, R-Jäger and K-Jäger before concluding that the K-Jäger offered the best performance. It is not clear where any of these designs was actually intended as the basis for a real aircraft but it would appear that at least one other company was influenced by Arado's report.

1-TL STRAHLJÄGER, MAY 1943

The third item on the agenda at the RLM's development committee meeting of May 28, 1943,[10] was the '1-TL Strahljäger' or single-jet fighter.

The discussion was opened by RLM staff engineer Walter Friebel and the minutes have him saying that "it was suggested by Focke-Wulf and Messerschmitt to provide an existing fighter (Fw 190 or Me 163) with a jet engine with the least amount of design effort. Working through the design showed that a complete reinterpretation [of the aircraft] is inevitable. The possibilities and prospects of such a 1-TL-Jägers have been investigated".

Friebel reported on the advantages of the single-jet fighter: "Horizontal speeds 150km/h greater than the Otto-Jäger [piston-engined fighter]. In the climb as fast as the Otto-Jäger flying horizontally. A pursuit in the climb would therefore be possible. Central installation of several heavy weapons in the fuselage means strong concentrated firepower as with the heavy 2-TL-Jäger Me 262, which is more of a destroyer, preferably against heavy bombers.

"Special advantages over Me 262: the possibility of fighter against fighter combat (manoeuvrability). Saving of materials by about 20%. Savings in hours spent fabricating the airframe about 25%. Savings on fabricating the engines about 50%. Disadvantages over the Otto-Jäger: fuel

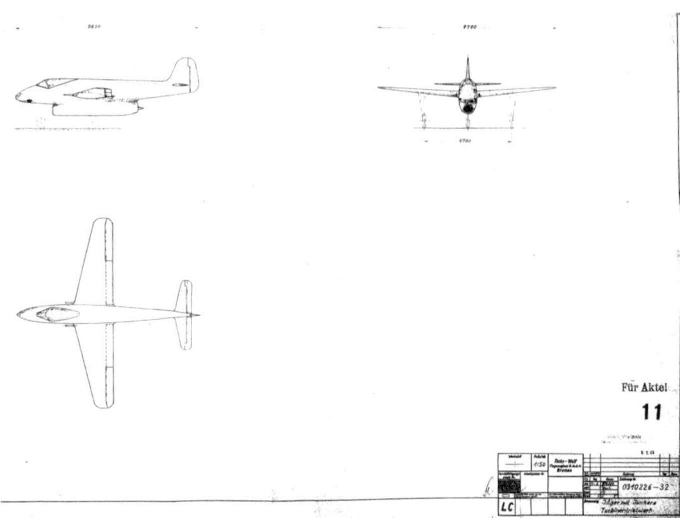

ABOVE: Focke-Wulf drawing 0310 226-32 of June 17, 1943 showing the company's revised jet fighter design. The turbojet was shifted to the centre of the aircraft, the undercarriage became a tricycle arrangement, alterations to the internal fuel tanks allowed for a redesign of the cockpit, new wings were provided and even the tailfin was slightly reshaped.

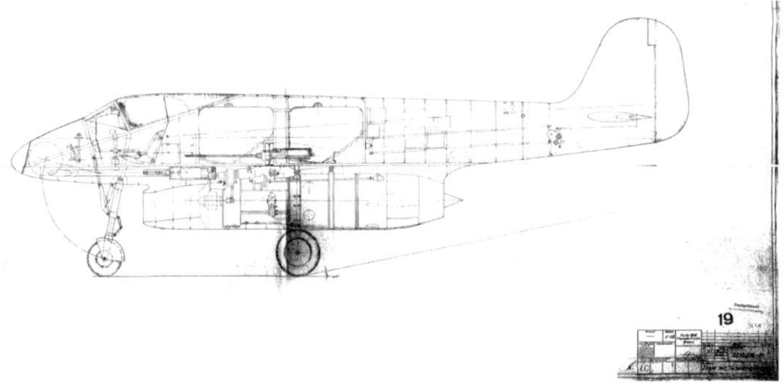

ABOVE: A detailed internal view of Focke-Wulf's jet fighter is shown in drawing 0310 226-31 of June 1943. The aircraft was described in Baubeschreibung Nr. 264.

consumption at least twice as high. For the same route significantly shorter flight duration."

Production capacity for building a single-jet fighter was then discussed. Friebel said: "Although Focke-Wulf has been able to complete the task, the [RLM's Technical] Office believes it is certain that it will have a detrimental effect on Ta 154. This view is confirmed by the Special Representative for the Ta 154. Whether Heinkel from the end of 1943, in addition to taking care of the He 177, He 277 development and He 219-development as well as He 274 can still be charged with developing a jet fighter is being investigated.

"Heinkel is also heavily involved in the jet bomber project. Most likely, Messerschmitt would be suitable for the new task provided that it would be tackled to avoid diversion of effort and only after completion of the Me 262 development (about the end of 1943)."

Project sketches by Messerschmitt, Focke-Wulf and Heinkel were then submitted but "the designs are not even aligned [based on the same mathematical formulas] due to the shortage of time. For performance reasons, the HeS 11 engine is to be considered in addition to the weaker Jumo 004 C".

It is likely that Focke-Wulf's design was the '1. Entwurf' outlined earlier, and Messerschmitt's was the Me 163-based P 20. Heinkel's entry is unknown but it may well have been the P 1069 about which almost nothing is known. Certainly, it is the only design on the Heinkel project list[11] – described simply as "1069 / Jäger / 1 x TL" – which fits both in terms of timing and configuration.

During the discussion which followed, the General der Jagdflieger Generalmajor Adolf Galland said that "The need for the 1-TL Jägers alongside the (expected to be one year earlier) Me 262 is beyond doubt, since with that aircraft air superiority can only be achieved locally and temporally at certain fronts. In the interest of manoeuvrability, this aircraft must be very small and maximum ceiling should be 13km".

Oberst Edgar Petersen, Kommandeur der Erprobungstellen (commander of the Luftwaffe's test centres), said that there was a danger of damage to the single-jet aircraft's tail due to the hot exhaust and that Fowler flaps would be a big advantage when it came to increased lift and reduced surface load for slow flight.

The head of development at the RLM's Technical Office, Oberstingenieur Dr Georg Pasewaldt, who had succeeded Generalingenieur Gottfried Reidenbach in June 1942, said he expected that the danger of the tail being affected by the exhaust was not insurmountable and that tailless construction was not necessarily the answer, since it would introduce new centre of gravity problems.

Friebel said that "the very high strength requirements of the Fowler flap make the steel wing appear desirable. However, housing the Fowlers will be difficult with the thin wing profiles".

Professor Dr Friedrich Seewald of the Luftwaffe's Forschungsführung – its research command – "emphasized the importance of the piston engine for speeds up to 900km/h despite high weights for propellers etc. Since a corresponding engine does not presently exist, the need for the use of TL devices from the Otto [piston engine] point of view is beyond doubt".

At the end of this section of the meeting's minutes, under 'decision', it says: "The Generalfeldmarschal [Milch] orders that the project of the 1-TL Jägers be pursued with the utmost vigour. In the interpretation, the two factors are to be considered a) highest speed, b) a peak service ceiling of at least 13km. Scheduling and performance suggest using the HeS 011 engine. It would be appropriate to entrust a fighter company – like Messerschmitt or maybe Focke-Wulf – with the execution of the test. This will be decided at a later date. Also Heinkel is to be involved in the project work at first."

A date was then set for the next discussion about single-jet fighters – June 25, 1943. While the minutes for this meeting, if it took place, have yet to be discovered, both Focke-Wulf and Messerschmitt set about developing new and improved designs for single-jet fighters. It would appear that Heinkel was too focused on its new P 1068 to pursue the 1-TL-Jäger any further at this stage since the company's next three major projects were the P 1070 flying-wing four-jet bomber, the asymmetrical twin piston-engine P 1071 fighter and the P 1072 piston-engine bomber.

FOCKE-WULF, MAY 1943 TO JUNE 1944

Following the meeting in May, Focke-Wulf concluded that the design later referred to as '1. Entwurf' was inadequate, specifically stating that "with this arrangement, no satisfactory taxiing properties were to be expected and there was also the risk of burning the airfield surface. This design was abandoned".

Work then began on a redesign – outlined in the company's Baubeschreibung or 'construction description' Nr. 264 'Jäger mit Junkers Turbinentriebwerk' dated June 9, 1943.[12] The aircraft was still to be powered by either a Jumo 004 B or C but this was now to be fitted centrally beneath the fuselage. It had a wingspan of 9.7m, wing area of 15m² and length of 9.85m. The undercarriage had become a tricycle arrangement with the nosewheel directly in front of the engine intake. Armament was two MK 108s in the fuselage on either side of the pilot and two MG 151s in the wing roots.

The actual description itself was somewhat threadbare, amounting to a list of the design's features without any further explanation. The aircraft was to be made from dural and steel, weighed 3,350kg, had a Revi 16 c gunsight, was fitted with an FuG 16 and FuG 25 and was not provided with a pressure cabin – although it would be possible to fit one.

It would appear that Focke-Wulf seriously considered building the Nr. 264 design. A company plan labelled 'Strahl-Jäger' and dated July 26, 1943,[13] showed construction commencing in November 1943 and continuing into June 1945. Between the end of February 1944 and January 1945, ten prototypes were to be completed along with a single additional airframe for stress testing.

Eventually though, by Focke-Wulf's own admission, the engine-under-fuselage configuration was "rejected because of the risk of engine damage during belly landings".

Another redesign, today lacking a detailed description or even proper designation, was produced in November 1943. In August 1944 it would be referred to as the '3. Entwurf'.[14] Now the Jumo 004 B (or C) turbojet was positioned directly behind the cockpit with the intakes on either side. It exhausted over the tail, which had a twin-fin layout. Wingspan was 70cm shorter than that of the previous design at 9m but wing area remained 15m². The aircraft was slightly lighter at 3,340kg. Armament appears to have been four MG 151s below the cockpit but further information is unavailable. Repositioning the turbojet mitigated the belly

ABOVE: Focke-Wulf drawing showing the fuel tanks, weaponry and armour protection of the Baubeschreibung Nr. 264 design.

ABOVE AND BELOW: Wind tunnel model of the Nr. 264 design circa June 1943.

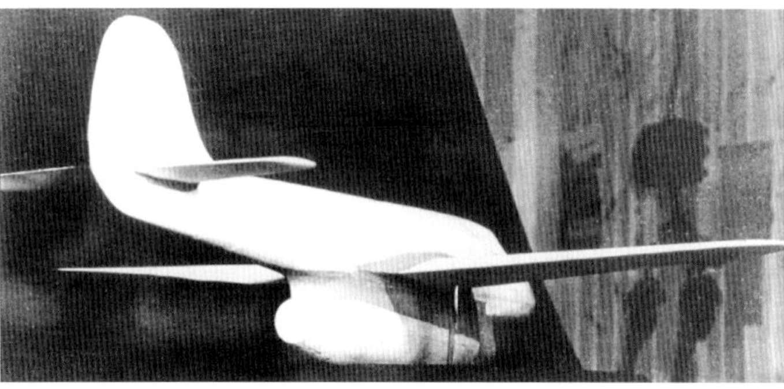

landing problem but "the calculated flight performance was inadequate due to the lateral intakes".

This design was refined the following month, December 1943, with a crucial design change – the addition of two rocket motors, one on either side of the Jumo 004 B or C. It appears that between November and December 1943, Focke-Wulf either became aware of Arado's work on the R-Jäger, K-Jäger and TL-Jäger, published just four months earlier, or simply came to the same conclusion independently: a rocket motor for a short take-off and fast climb combined with a turbojet for cruising at altitude was the best option.

The forward part of the new design's fuselage was similar to that of its predecessor – albeit much broader – but ended abruptly, with the tail being instead composed of twin booms attached to the trailing edge of the aircraft's wings. This '4. Entwurf', its true designation also unknown today, was rejected because "horizontal velocity remained unsatisfactory" – although presumably the twin rocket boosters gave it a more than acceptable climbing speed. Wingspan was the narrowest yet at 8.6m but wing area increased slightly to 15.5m². Another factor restricting the design's horizontal speed may have been its hugely increased weight: 4,900kg.

At this point, January 1944, company aerodynamicist Hans Multhopp suggested a completely different layout to the Focke-Wulf Entwurfsabteilung's (design department's) steadily evolving twin-boom aircraft. He proposed a tall aircraft with the turbojet built into the lower rear part of the fuselage, fed by a single large nose intake. This was the first Focke-Wulf jet design to have swept-back wings and it also featured a long swept T-tail.

Work now took place in parallel on 'Entwurf Multhopp' and the 'Entwurf mit Leitwerkstragern' or 'Design Multhopp' and the 'Design with tail booms'. The latter was considered to be the more promising of the two and the design department created a detailed description of it as Baubeschreibung Nr. 272, dated February 1, 1944.[15] The twin-boom design now featured intakes built into the leading edge of its wings and there were four propulsion options: Jumo 004 B, Jumo 004 C, HeS 011 and HeS 011 with a single Walter 109-509 rocket motor. Length was 9.8m, wingspan was 8m and wing area remained 15.5m². There were three armament options: two MG 151s in the fuselage and two MK 108s in the wings, one MK 103 in the fuselage and two MK 212s in the wings or two MG 213s in the fuselage and two MG 213s in the wings. The gunsight was a Revi 16 c and radio equipment was a FuG 16 and FuG 25. The aircraft had a total of six fuel tanks – a forward fuselage tank of 550 litres and a rear fuselage tank of 380 litres, 240 litres in each outer wing section and 100 litres in each tail boom.

Baubeschreibung Nr. 272 featured a lengthy introduction which strongly echoed Arado's conclusions about a combination of jet and rocket propulsion: "For protection of the homeland, fighters equipped with Otto power plants are currently used. Because of their low and limited climbing speed, a great part of their flying time is used for the approach. The time required for reaching the enemy, the large area to be covered in the warning zone, and the probability that the bombing unit engages in combat before release of the bombs, are factors to be taken into consideration; especially with regard to an attack by the enemy with fast aircraft from a high altitude. Repeat operations of the same fighter units against the same bombing unit are possible on rare occasions only.

"For these reasons it is imperative to develop a fighter with high horizontal speed with the climbing time reduced to a minimum. To meet the demand for high horizontal speeds, the propeller drive must be abandoned because the required power, especially at high altitudes, cannot be produced with sufficient effectiveness when using propellers. Of jet propulsion, only turbine jet propulsion (Junkers, Heinkel, BMW) or rocket propulsion (Walter) enter into consideration.

"Fighters powered with jet propulsion exclusively satisfy as far as horizontal speed, ceiling and also range are concerned, with not too much difference between single and twin-engine types. The disadvantage of these fighters is found in the long take-off run and low climbing speeds, which lie within the same limits as comparable types equipped with Otto power plants; therefore, the prevailing superiority in speed in combat at high altitudes rarely becomes effective.

"Aircraft equipped with rocket propulsion exclusively can easily attain high horizontal speeds. Their climbing speed

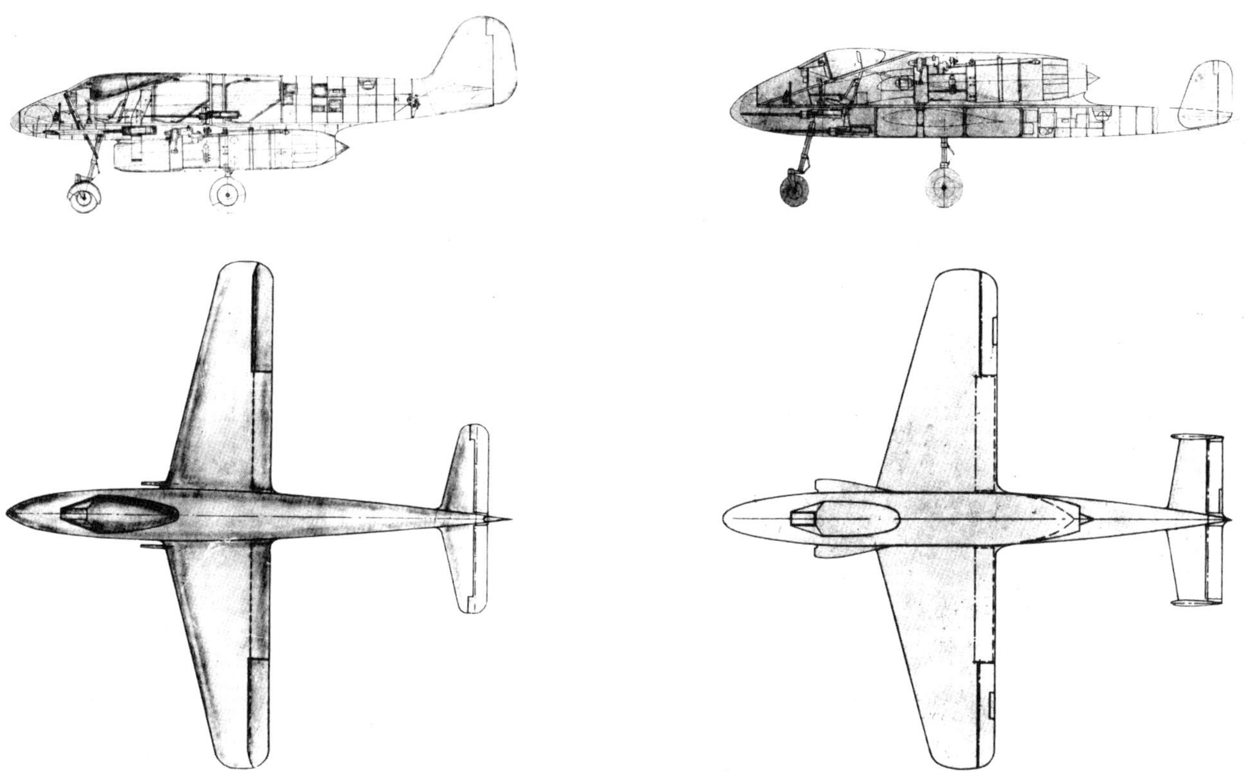

ABOVE LEFT: The drawings included with Baubeschreibung Nr. 264 were not Focke-Wulf's last word on the design and work on the same basic layout appears to have continued. This drawing, used to illustrate a retrospective report in August 1944, shows a slightly different version with a longer undercarriage with larger nosewheel and revised MK 108 positioning.
ABOVE RIGHT: Between June and November 1943, Focke-Wulf revised its jet fighter once again – this time repositioning the turbojet within the rear fuselage behind the cockpit. By now the last vestiges of the Fw 190 airframe had fallen away.

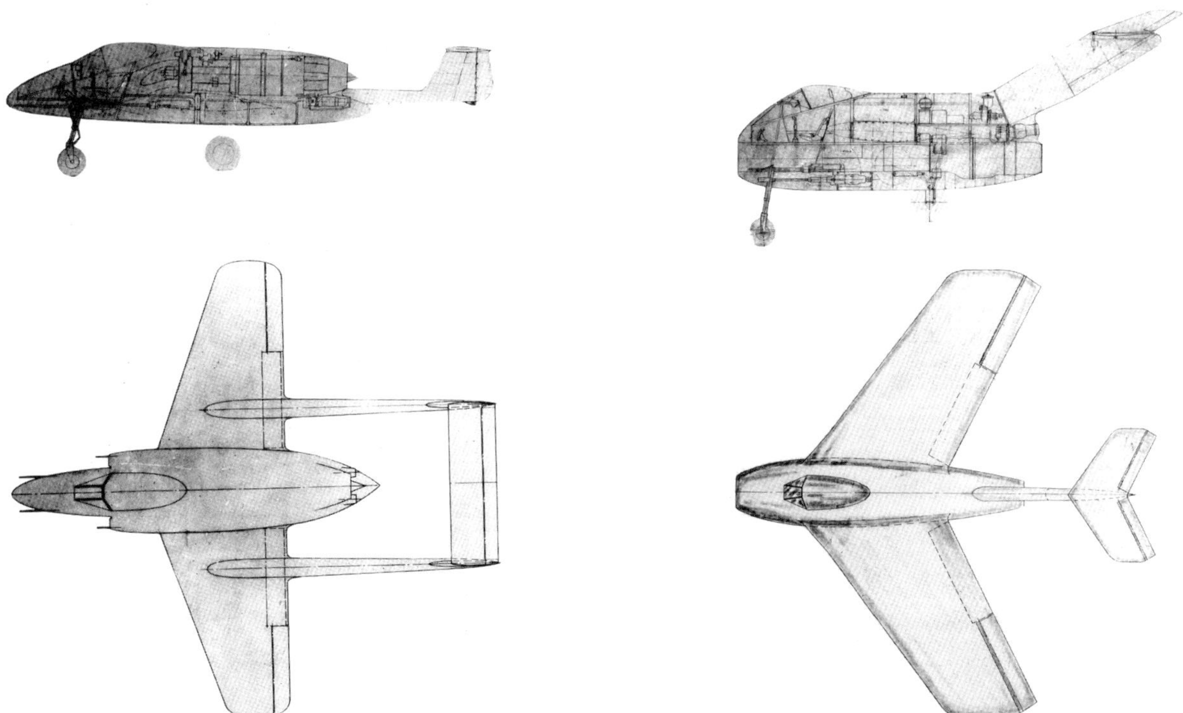

ABOVE LEFT: In December 1943, Focke-Wulf followed Arado's lead in deciding that a built-in rocket motor was essential for a single-jet fighter. The jet fighter design evolved once more to incorporate a pair of rockets alongside its turbojet. This resulted in a bulky fuselage and necessitated the switch to a twin-boom layout – shifting the control surfaces away from the rocket exhaust.
ABOVE RIGHT: Focke-Wulf aerodynamicist Hans Multhopp came up with an alternative to the twin-boom project being worked on by the company's Entwurfsabteilung in January 1944 – a design with deep chord swept wings and a T-tail to keep its control surfaces away from the exhaust of its single-rocket motor. Naturally this design was dubbed 'Entwurf Multhopp'.

takes on proportions which could never be attained with Otto or turbine-jet power plants only. Unfortunately, the consumption of special fuel for rocket devices is so high that an aircraft equipped with rocket propulsion exclusively has a very limited flying time and range.

"For tactical attack purposes an ideal fighter would have to meet the demand, not only for speedy performance and strong armament, but also for shortest time of climb to combat altitude, with a flying time at this altitude of at least half an hour. Therefore the advantages of the jet aircraft (duration of flight, range), and the advantages of the rocket aircraft (time of climb, take-off run), must be combined while retaining the great horizontal speed. This demand can be met only with an aircraft equipped with jet propulsion and supplementary rocket propulsion, in which the rocket propulsion is supplied with fuel for start and climb only."

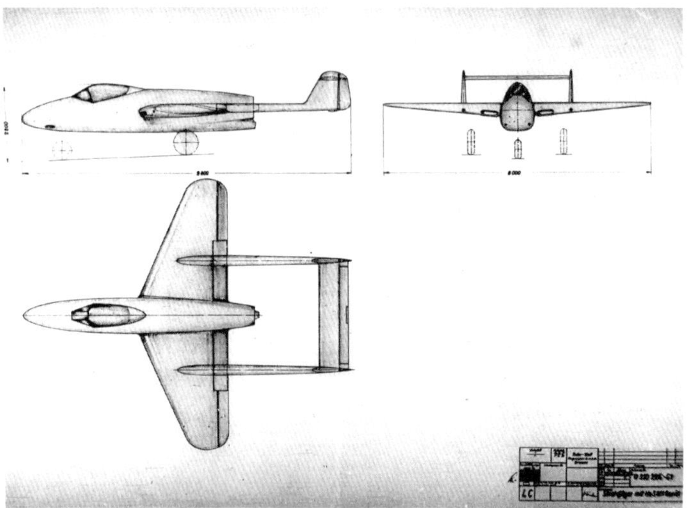

ABOVE: The Focke-Wulf Entwurfsabteilung's refined twin-boom jet fighter was crystalised in Baubeschreibung Nr. 272 of February 1, 1944. Drawing 0310 226-61 shows the aircraft with a slimmed-down fuselage thanks to its single-rocket motor and slot intakes built into its wing roots. A distinctive characteristic of the Nr. 272 aircraft was the dead straight trailing edge of its wings.

In essence, the single-jet fighter's chief drawback – the poor acceleration afforded by early turbojets – could best be overcome by equipping the aircraft with an integral rocket motor. This would provide the fighter with massive thrust for the take-off and climb. Once up to speed it would be able to more than hold its own against enemy piston-engined aircraft. It is worthy of note that Focke-Wulf explicitly states that the aircraft is intended for use against bombers, since this was in February 1944 before American daylight bombing raids had reached their crescendo.

The introduction continued: "During development of a jet-and-rocket fighter, careful consideration should be given to adoption of the single versus the twin-engine type: whether, in contrast to the single engine, the disadvantage of the twin-engine type's larger fuel consumption is offset by its greater flying performances. In this connection it should be remembered that by installing efficiently performing jet power plants, such as the HeS 011, the twin-engine type is subject to a limitation in the critical Mach number similar to that of the single-engine type.

"However, with a sufficiently small shearing load, the only difference in horizontal speeds to be obtained is from the critical Mach number, which is contingent upon the design of the aircraft. Since it does not follow in the principle that a single-engine horizontal type has lower Mach numbers than a twin-engine type, the difference in speeds to be obtained would therefore prove to be of no great consequence. For this reason it is advisable to promote the development of the single engine jet-rocket plane."

In other words, the sound barrier imposed a limit on how fast jet aircraft could fly no matter how powerful or numerous their engines. Therefore, the top speed of a single-jet fighter would be the same as that of a twin-engine design since neither was likely to go supersonic. The introduction continued: "Comparative calculations showed that the proposed design, with an installation of two supplementary rocket units, each of 2,000kg thrust, could make a vertical climb from an altitude of about 4,000m, and thereby bring to realisation the shortest possible time of

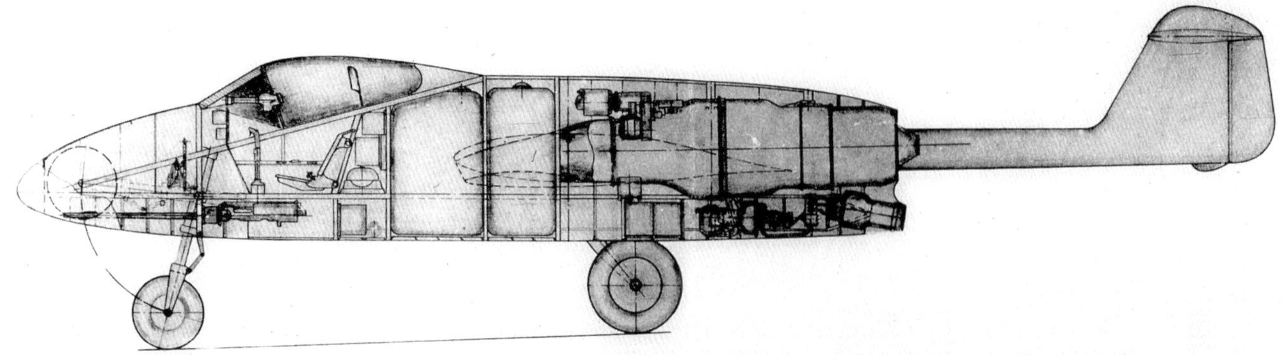

ABOVE: Side view of the Nr. 272 aircraft, drawing 0310 226-62, showing substantial internal alterations from the preceding design. Without the engine's intakes in the way, the fuselage is able to hold far more fuel and the cockpit is less cramped. The tail boom has been raised and the rocket exhaust angled downwards to further separate them.

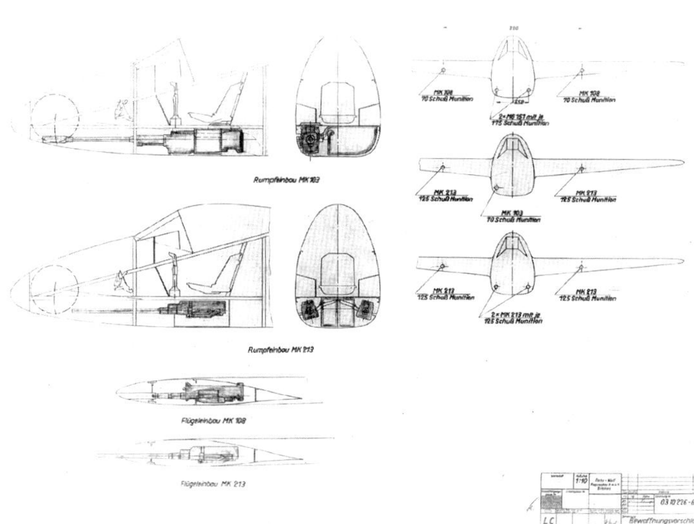

ABOVE: The different weapons options offered by Focke-Wulf for its Nr. 272 fighter are shown in 0310 226-60.

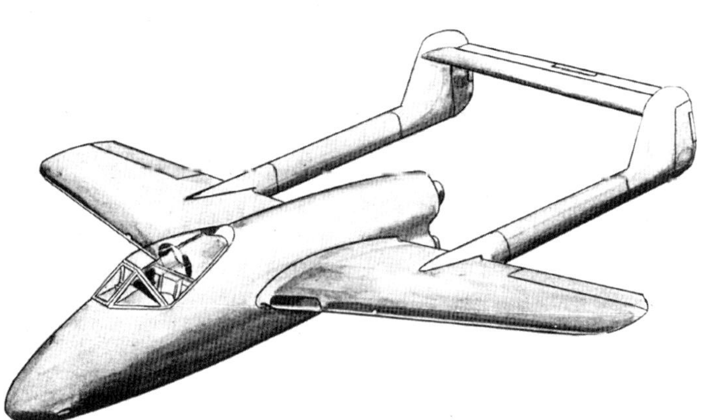

ABOVE: Perspective view of the Nr. 272 aircraft.

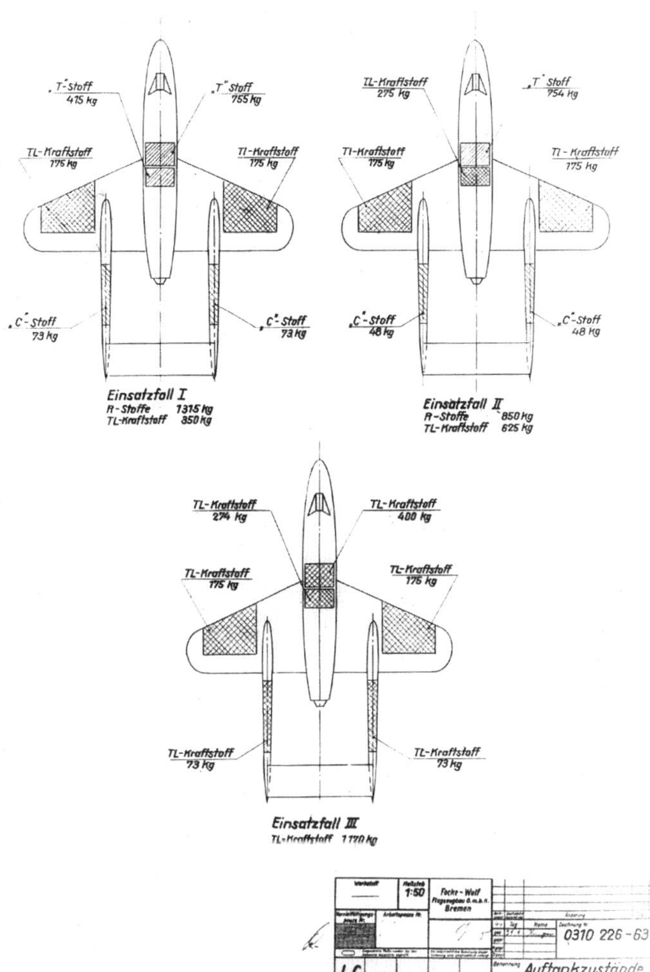

ABOVE: Focke-Wulf drawing 0310 226-63 of January 31, 1944, shows the three different 'cases' of the design. The difference was limited to fuel load - the first had more rocket fuel, the second a balance of jet and rocket fuel and the third carried no rocket fuel.

calculations also showed however, that, with an installation of a single supplementary rocket device of 2,000kg thrust, the vertical climb must be sacrificed, and an increase of from 81 to about 125 sec in the total flying time to 10,000m (including time for take-off and acceleration) taken into the bargain.

"Also, there is no perceptible change in the amount of fuel to be carried for the rocket propulsion. In view of the very short climbing time, an increase of about 44 sec seems to be quite acceptable. An additional advantage is the greater simplification of the system with one rocket device, its easier manipulation, and lessened structural weight.

"As a supplementary thrust device the Walter design has been selected. It is operated with T- and C-Stoff or with fuel derived from sage. A pump unit feeds these fuels to the rocket unit independent of the jet power plant. The installation of a jet-rocket combination is, of course, also possible if the rocket unit with 2,000kg thrust can be placed at the necessary angle below the jet unit. The installation of a jet-rocket unit combination, however, means sacrificing the independence of the rocket unit, and also a larger consumption of rocket fuels because of their joint operation.

"For the proposed design an HeS 011 turbine power plant and a supplementary rocket device of the Walter type 109-509 have been selected for producing thrust. By interchanging the tanks and removing the rocket unit, the aircraft can be re-equipped for three different cases of operation."

The first 'case' involved carrying far more rocket fuel than jet engine fuel – 1315kg compared to 350kg. The take-off run would last 400m and 12 seconds, with time to 12km altitude being 129 seconds at a fairly stunning average climbing speed of 93m/sec. Range at 12km was 470km and flight duration – not counting an unpowered 20-minute glide in to land – was 46 minutes. Take-off weight was 4,750kg.

The second 'case' rebalanced the proportion of rocket fuel (850kg) to jet fuel (625kg). This lighter burden, a take-off weight of 4,560kg, reduced the take-off run to 380m and 11 seconds. Time of climb to 7km was 77 seconds at an average climb speed of 91m/sec. but at this point the rocket fuel would run out – and ascending from 7km to 12km on jet power alone would take 12.8 minutes at an average climbing speed of 15.7m/sec. Range at 12km was now 820km and flight duration was 76 minutes.

The third and final 'case' was entirely without rocket fuel, meaning a take-off weight of just 4,110kg. But the take-off run was now 700m and would take 20 seconds. Time of climb to 12km was 23 minutes at a rate of 8.7m/sec. Range was a healthier 1,520km and duration was 132 minutes.

On February 21, 1944, new pages were drafted for slotting into copies of Baubeschreibung Nr. 272 which saw all mention of the Jumo 004 B or C deleted. Although the Baubeschreibung Nr. 272 aircraft seems to have been considered a safer bet and was much more heavily worked on, 'Entwurf Multhopp' never entirely went away. During the spring of 1944, Focke-Wulf's evolving single-jet development programme was given the nickname 'Flitzer', meaning 'Streaker' or 'Whizzer'. This appears in numerous company memos and notes.

A modified version of the 'Entwurf Multhopp' was drafted on April 7, 1944, in drawing number 0310 226-92.[16] This was 8.85m long and had a wingspan of 9m, with a wing area of 23m². Its take-off weight, including a rocket booster and fuel, was 5,200kg.

But on May 31, 1944, ten of Focke-Wulf's most senior engineers – Cassens, Kosel, Mittelhuber, Mathias, Heinzelmann, Wolff, Plöger, Freyer, Schaaf and Nieten – met two representatives of the RLM's C-E 2 department, Dr Braun and Otto, at their company's Bad Eilsen offices to discuss the benefits of the 'Flitzer'.[17] 'Entwurf Multhopp' was not mentioned in the meeting summary notes published on June 14, 1944.

MESSERSCHMITT, MAY 1943 TO JANUARY 1944

With Heinkel apparently not competing and Focke-Wulf making steady progress, Messerschmitt appears to have been somewhat conflicted about the single-jet fighter since it was potentially a replacement for the Me 262 – upon which huge efforts continued to be focused.

On July 3, 1943, the company produced Technischer Bericht Nr. 90/43,[18] which compared the performance and characteristics of the twin-jet Me 262 to those of the P 20 and a new single-jet design of more conventional

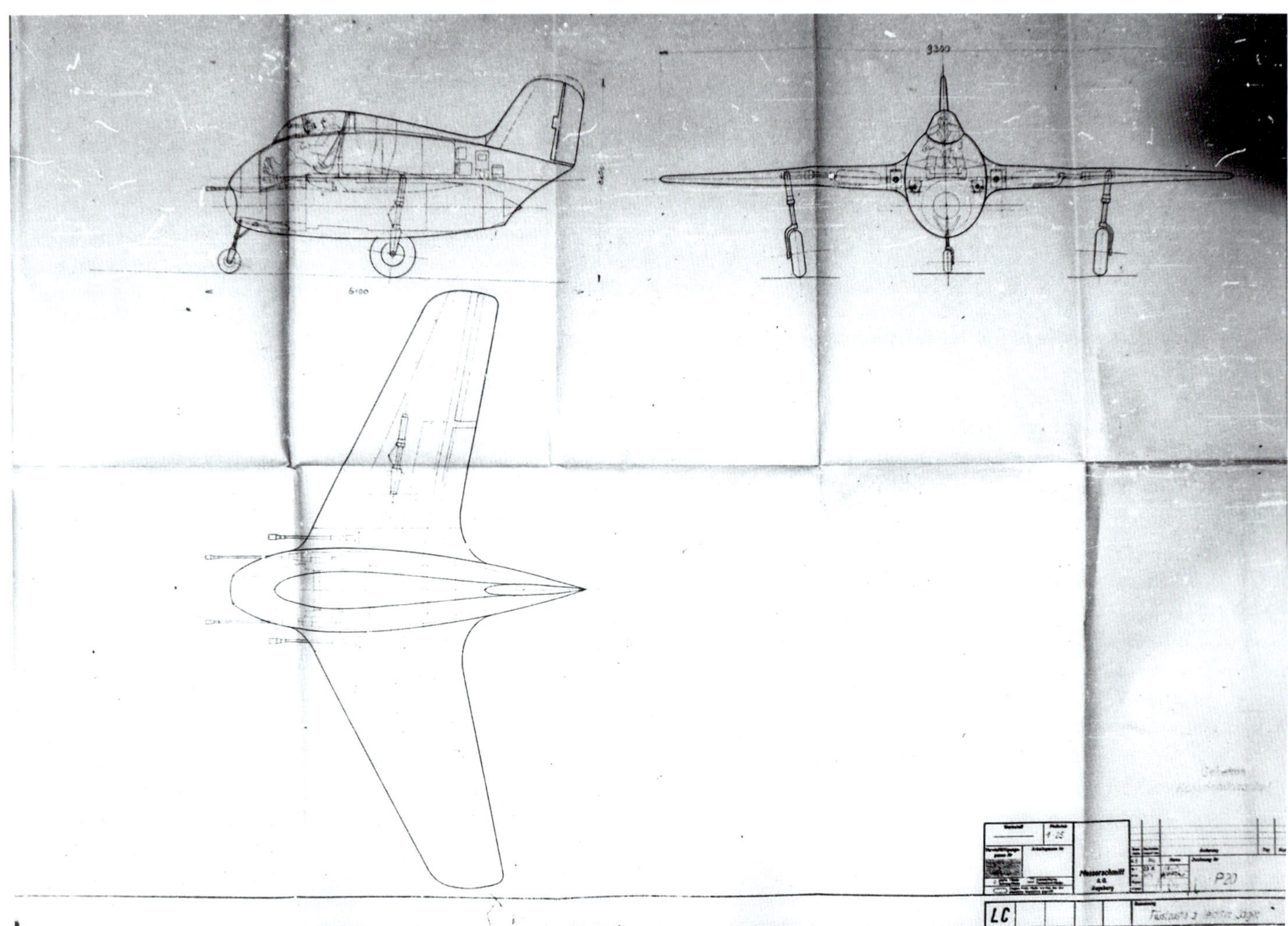

ABOVE: Messerschmitt's P 20 design – a turbojet Me 163 – was slightly revised between April 16 and June 23, 1943. The June version, shown here, was 37cm longer and 18cm taller but its wingspan remained the same.

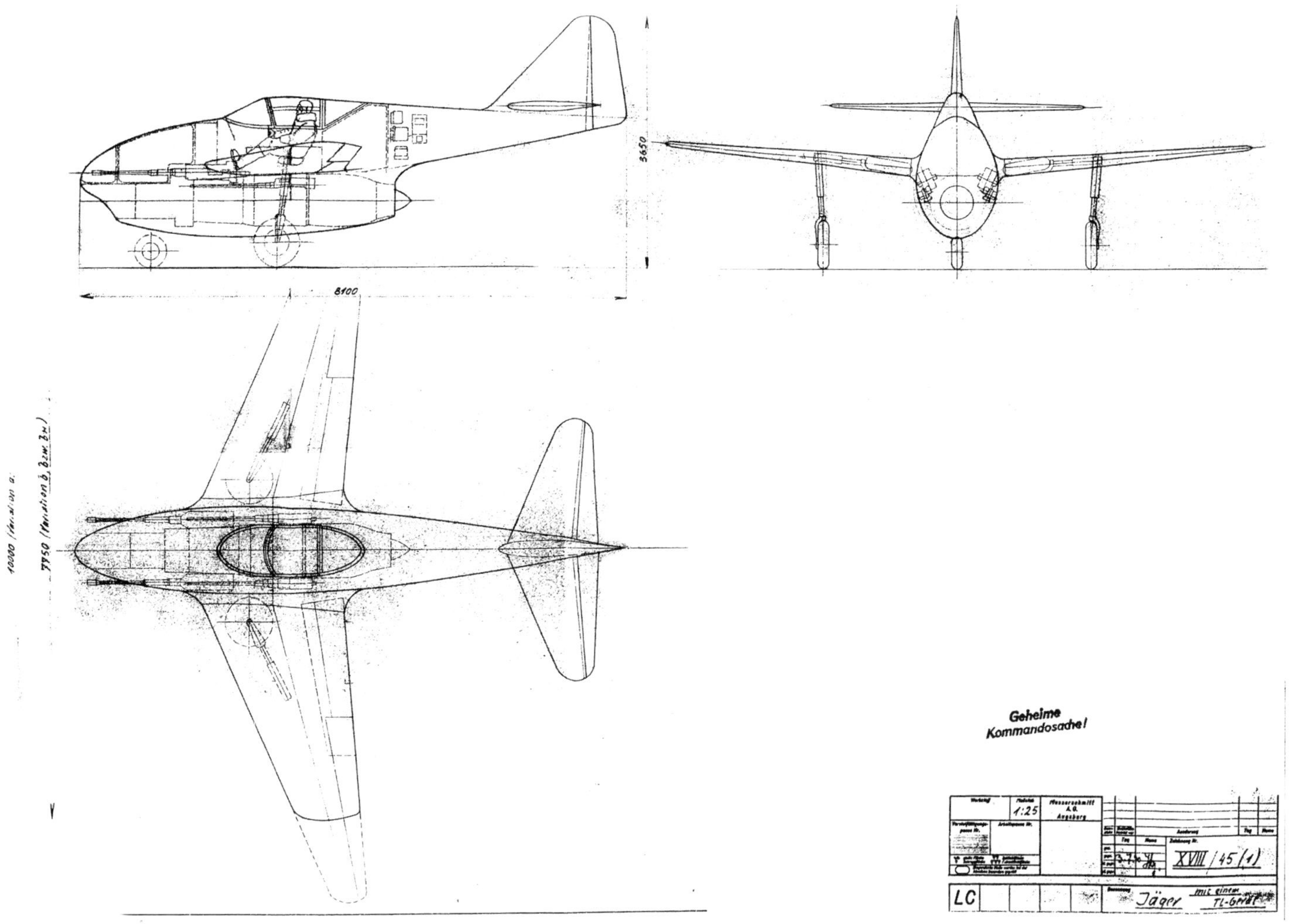

ABOVE: P 1092 of drawing XVIII/45 (1), dated July 3, 1943, was the Messerschmitt Projektbüro's conventional counterpart to the now-defunct Abteilung L's P 20.

layout, the P 1092. During project work on P 1092 a number of different layouts were apparently conceived, including a highly unusual variant with the engine, weapons and undercarriage clustered together at the front of the aircraft and the cockpit suspended above the ground to the rear just ahead of the tailfin. Another version appears to have been designed with twin-pulsejet propulsion in mind.

Three similar P 1092 single-jet fighter designs were presented in Technischer Bericht Nr. 90/43, labelled P 1092/a, P 1092/b and P 1092/bH. The main difference between them was in wing surface area: 12.7m², 14.25m² and 14.45m² respectively, shown in drawing number XVIII/45 (1). Armament for all three designs was described as two MK 103s and two MG 151/15s and all three were powered by Jumo 004 C engines. The Me 262 was assessed with a higher fuel content because "the twin-engine fighter handles a larger weight at take-off".

The landing speed of the Me 262 was higher than that of the other designs by 8km/h but when the twin-jet design was required to fly the same distance as the single-jet designs it would be able to do so carrying 1,000kg of bombs which the single jets would not be able to manage – although it would need a longer take-off run. With the same range and same take-off run it could manage 300kg of bombs. Range and bomb-carrying capacity of the Me 262 would be even further improved if the bombs could be carried internally with "an enlarged fuselage cross-section".

It was calculated that either single-jet fighter would require 45% less material to manufacture and would take 10–15% less time to build but "the single engine fighter is specifically made more expensive by having a less favourable weapon arrangement (divided into two locations)" and a less accessible engine.

In short, the Me 262 could be regarded as the superior design in most respects and neither the P 20 nor the P 1092 went any further. But even so the company revisited the idea of a single jet design later in the year as a way of rescuing the ailing Me 328 programme. Testing of the Me 328 with its twin pulsejets had revealed an unacceptably high level of vibration which threatened both the health of the pilot and the structural integrity of the aircraft.

Precisely when the idea of turning the Me 328 into a single-jet fighter arose is unclear but a drawing of September

20, 1943, shows the 'Me 328 T – V1' – the Me 328 pulsejet fighter reimagined as a single jet aircraft.[19] The drawing is also marked 'Darmstadt', suggesting that it originated with Darmstadt-based Jacobs-Schweyer Flugzeugbau, which had been charged with building the prototypes for the Me 328 B.

The Me 328 T – V1 has an upper profile similar to that of the Me 328 B but slung beneath the fuselage is a single turbojet. The characteristic bullet nose of the Me 328 is replaced with a beaky looking nose housing a nosewheel and the aircraft has a non-retractable mainwheel strut housing – visible in the forward view. There is also a small tail skid. Comically, seated within the cockpit a cartoonish pilot is depicted picking his correspondingly beaky nose.

The date is interesting because Messerschmitt had ordered that all work on the Me 328 pulsejet fighter be stopped on September 3, 1943. However, a letter written by Willy Messerschmitt to Rudolf Seitz on September 29, 1943, suggests that Seitz himself had pushed for a jet-propelled Me 328 to take the place of the pulsejet version, allowing work on the project to continue.[20]

Messerschmitt's letter is headed "Ihre 'Stellungnahme der Entwicklungsleitung Me 328 zum Stoppbefehl vom 3.9.43'" or "Your 'Statement of the development management Me 328 to the stop command of 3.9.43'". It is his response to Seitz's reaction to the 'stop' command.

He writes: "A final decision on whether to accelerate the 1TL [single-jet] aircraft and start it in series depends on the following conditions: 1) The likelihood that the flight characteristics will not cause any difficulty. 2) The time needed for the entire process from the exact planning including series preparation to the time of the production of at least 100 aircraft per month. 3) The performance of the aircraft a) with fixed undercarriage b) with retractable undercarriage. 4) The capacity available for the production of the standard aircraft.

"This is the quickest way to do it, but I would like to point out that it needs to be handled very carefully. If this result shows that in a relatively short time larger numbers of an aircraft are to be created, which will have better flight performance with sufficient defensive armament than the best enemy aircraft with piston engine may have, the flight performance may of course be worse than that of the 262, if the aircraft is cheaper accordingly."

A week later, on October 6, 1943, a number of drawings were produced showing a turbojet-powered Me 328 that was largely the same as the 'Me 328 T' drawing. One of these is marked 'Darmstadt' while another actually appears on official Jacobs-Schweyer Flugzeugbau drawing paper. These are marked 'Me 328 C'. Presumably the designation had been reapplied from the much larger 'Me 328 C' of March 1943.

Although similar to the Me 328 T, the rechristened Me 328 C is armed with a pair of MK 103s which require bulged fairings on either side of the fuselage. A much larger fuel tank now sits above the nosewheel housing, the nosewheel strut itself is now straighter and angled forwards, the fuel tank behind the cockpit is enlarged and the biggest change involves the undercarriage mainwheels, which have been repositioned further away from the fuselage and given full aerodynamic housing, including spats.

It would appear that Seitz was unable to demonstrate the merits of the turbojet Me 328 C and it went no further. It has been suggested that the Me 328 C was derived from the Messerschmitt P 1095 single-jet fighter project or vice-versa. The only known primary source evidence of the P 1095 is an undated memo entitled 'P 1095 Vorbemerkung' or 'P 1095 preliminary note'.[21] This appears to follow the Messerschmitt philosophy by using as many components from existing or apparently soon-to-exist full production aircraft types as possible. It says: "In the following, the proposal is made to develop a 1-TL-Jäger whose components are largely taken from the current or just beginning full production series fighters.

"Investigations conducted so far show that with today's tangible jet engines (up to approx. 1,000kg thrust) superiority of the single TL aircraft over the two TL aircraft (apart from turning and landing behaviour) cannot be fully achieved, but on the other hand only about half the fuel is needed.

"A completely new development for the 1-TL-Jägers with the TL engines of today is hardly worthwhile, but on the other hand, it is urgently necessary in the current air warfare and fuel situation to develop the 1-TL-Jäger, given its great performance superiority over the enemy aircraft with piston engines. A compromise solution suitable for this purpose may be the following proposal: From the series of the Me 262 and Me 209, as many parts as possible are taken over with or without slight changes for the one TL aircraft project P 1095 to be created from them.

"The development and production costs are then reduced to a minimum. In the enclosed short description the parts that are acceptable for the series are listed for the 1 TL-Jäger and their share in the overall production is estimated. It is irrelevant whether the metal version or the later wooden version of the Me 262 is used.

"The following is a summary of the main possible uses: Taking off under its own power (without or with launch rockets) – 1) Light fighter. Ranges and flight duration approaching today's requirements for single-engine propeller-driven fighter. Armament 2 x MK 103 with 60 rounds or 3 x MK 108 with 60 rounds each in the fuselage nose. 2) Fighter with stronger armament. Basic armament the same as 1.) but additionally 2 gondola weapons each with 1 x MK 108 or 2 rocket launchers. 3) Ground-attack aircraft with up to 5 MG 81. 4) Destroyer.

"Taking off under its own power but with the aid of rockets – 5) Fighter-bomber. Basic armament as fighter with 2 x SC 250 bombs or up to 4 rocket launchers. 6) Long-range fighter. Possible by enlarging the fuel system. For catapult, towing or carrying towing start only – fast bomber with up to 2,000kg bomb load."

The P 1095 was to consist of Me 262 wings and forward fuselage, including pressurised cockpit, tail from the Me 209, controls from the Me 262 and/or Me 209, undercarriage from the Me 262 and Me 209, and a single Jumo 004 B-2 turbojet. Wing area was to be 15.3m², wingspan was 9.77m. These figures seem slightly at odds with the rest of the description since the Me 262's wings, with a turbojet each, gave the aircraft a wingspan of 12.6m and a wing area of 21.7m².

Re-drawn pictures of the P 1095 purportedly dated October 19, 1943, do appear to show the aircraft with slightly swept wings similar to those of the Me 262 and with a tail which may be derived from that intended for the Me 209. This would suggest that the P 1095 was separate from the Me 328 C and came about after the Me 328 programme had become the focus of test pilot Hanna Reitsch's attempts to create a suicide weapon.

Going into 1944, Messerschmitt switched its project efforts to further developments of the Me 262. But the company would return to the single-jet layout later in the year.

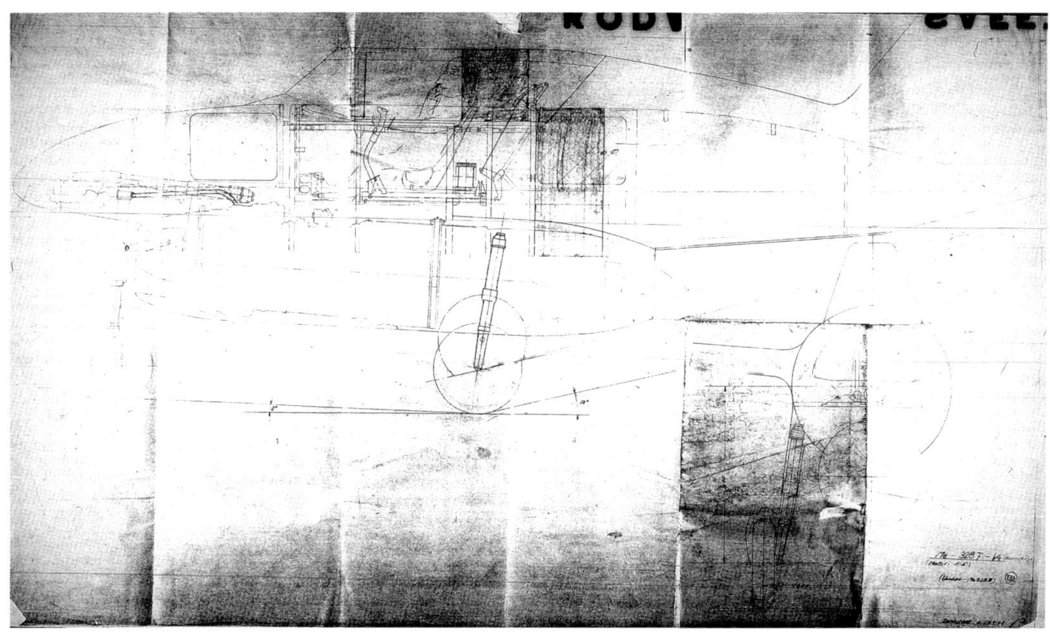

ABOVE: The cartoonish 'pilot' of the Me 328 T appears to pick his long nose in reference to the aircraft's own unique forward profile.

ABOVE: The Me 328 T - V1, dated September 20, 1943, as designed by Jacobs-Schweyer Flugzeugbau. This unarmed turbojet-powered version of the Me 328 has a fixed shock-absorbing undercarriage fitted close to its fuselage, a long nose and small fuel tanks. In brackets after 'Me 328 T - V1' the inscription says 'Umbau Me 328 B' or 'Modification Me 328 B'.

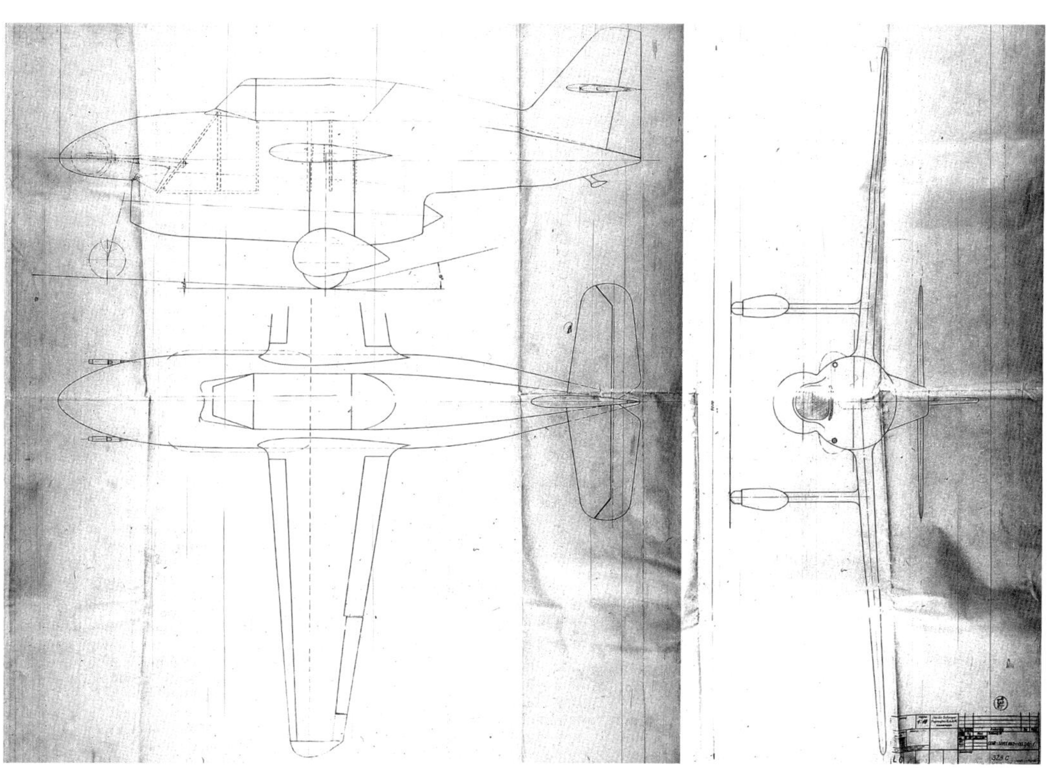

RIGHT: A three-view drawing of the Me 328 C dated October 6, 1943, on Jacobs-Schweyer Flugzeugbau paper. Unlike the Me 328 T, the undercarriage of the 'C' is fixed to the wings, much further away from the fuselage, with large gaiters and spats over the wheels. The top and front views show the bulged fairings necessary to accommodate a pair of MK 103 cannon.

Chapter 6

1,000 x 1,000 x 1,000

Fast bombers turned heavy fighters

Presented with two separate designs for a twin-jet bomber supposedly capable of carrying 1,000kg of bombs for 1,000km at 1,000km/h, Reichsmarschall Hermann Göring gave each designer RM 500,000 and told them to get on with prototype construction...

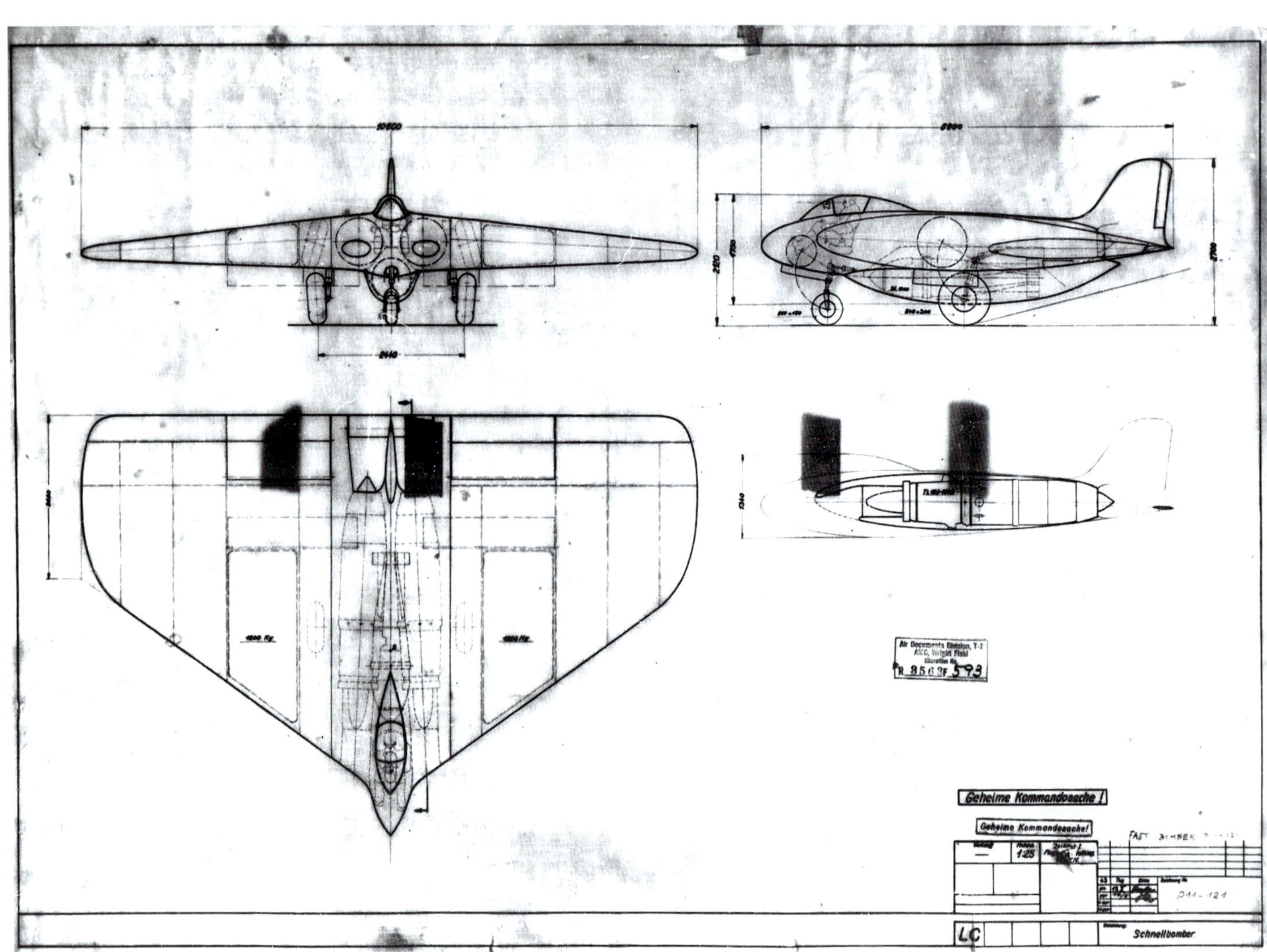

ABOVE: Lippisch took the P 11 project with him when he left Messerschmitt in April 1943 and continued to work on it. This drawing shows the state of the design on May 17, 1943. The aircraft was designed to fly at 1,000km/h for 1,000km carrying a 1,000kg bomb load.

P 11 AFTER MESSERSCHMITT

Messerschmitt designer Alexander Lippisch had left the company at the end of April 1943 and moved to the Luftfahrtforschungsanstalt Wien (LFW) or Aeronautical Research Institute Vienna where he was put in charge of a 100-strong workforce.

Having taken four trusted colleagues with him from Messerschmitt – Dr Friedrich Ringleb, Handrick, Sanders and Dr Volker – he set about continuing his work on the twin-jet fast bomber P 11. Less than a month later, on May 29, 1943, he produced a 13-page report entitled 'Projektbaueschreibung Versuchsflugzeug für

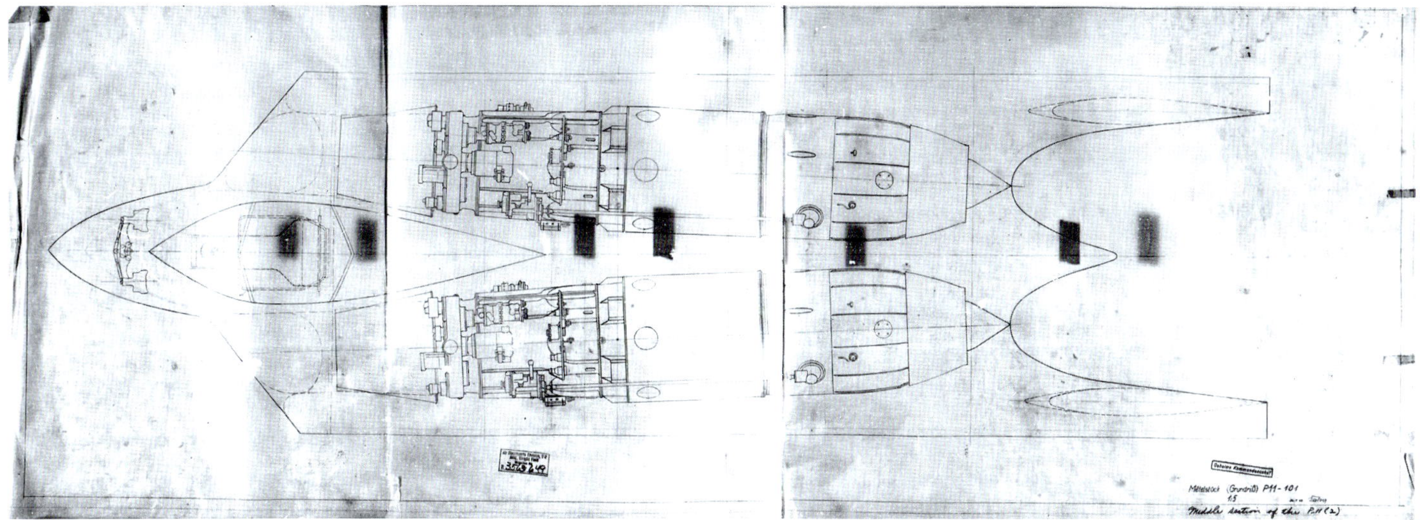

ABOVE: The P 11 soon became a twin-fin design. This drawing from July 26, 1943 shows how the two Jumo 004 engines would sit within the centre section.

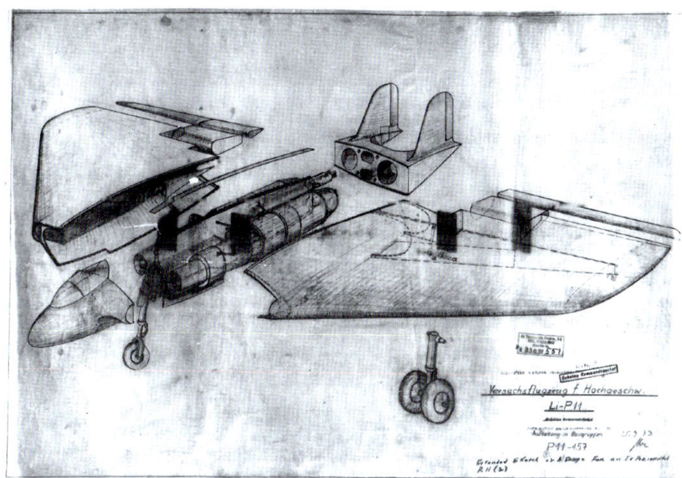

ABOVE: The internal structure of the P 11 glider with empty cylinders where the engines would be. The drawing is dated September 25, 1943 – three days before the design was presented to Göring during a meeting lasting just 15 minutes.

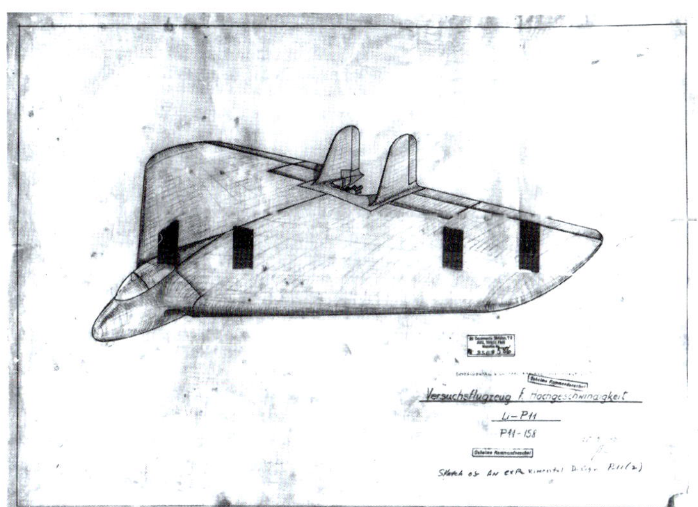

ABOVE: This perspective drawing dated September 25, 1943 also appears to show the P 11 as an experimental engineless glider. It is fitted with a rocket booster to the rear for take-off.

Hochgeschwindigkeit' or 'Project construction description of an experimental aircraft for high-speed'.[1]

The aircraft it outlined, powered by two Jumo 004 B or C jet engines, was intended to fly at 1,000km/h with a flight endurance of two hours – giving it a range of 1,000km. Although described as 'experimental', the design was intended as a stepping stone towards an operational version. Indeed, a drawing of the aircraft dated May 17, 1943, depicted it carrying a single 1,000kg bomb – making it the first German 1,000 x 1,000 x 1,000 design.

While the drawing of May 17 gave the aircraft's wingspan as 10.6m, the May 29 report gave it as 10.8m; the design's development would be characterised by constant detail changes of this nature. It had a single fin at the rear of an enormous 50m² wing with no fuselage to speak of and only the pilot's small cockpit protruding from the leading edge, resulting in an overall length of 6.8m. There was an intake on either side of this, set into the leading edge of the wing which was swept back 40°. The low tricycle undercarriage retracted neatly into the underside.

While the report may have been ready in May, it took Lippisch until September 28, 1943, to get an audience with Reichsmarschall Hermann Göring to present it personally.

THE HORTENS

Meanwhile, tailless aircraft enthusiast siblings Walter and Reimar Horten had been working on their own twin-jet aircraft. Without official sanction, using only their contacts, personal charm and creativity with paperwork, the two Luftwaffe officers had managed to set up their own tiny 'off the books' organisation within the labyrinthine bureaucracy of the Luftwaffe, Sonderkommando LIn 3, for the sole purpose of designing and building their own flying-wing type aircraft.[2]

During the winter of 1941–42, Walter had presented his brother Reimar with rough copies he had made of

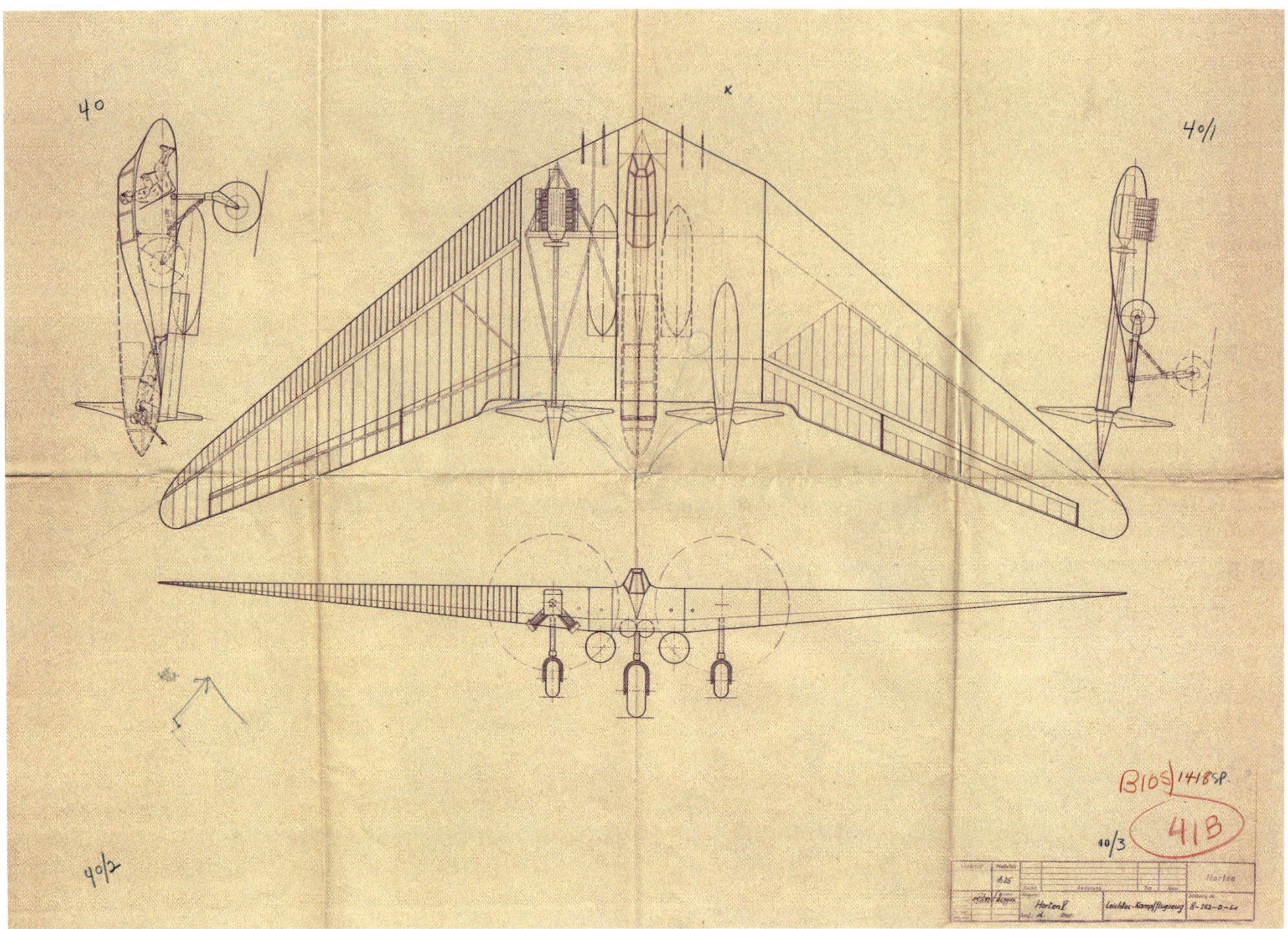

ABOVE: The Horten V d flying-wing design as it stood on March 19, 1942. The Horten brothers had been working on twin-engine tailless designs for years before an opportunity presented itself to use this layout as the basis for a twin-jet fast bomber.

documents showing the projected performance and physical dimensions of the Junkers Jumo 004 turbojet being developed by Dr Anselm Franz at Dessau.[3] And what was more, Walter believed he might be able to lay his hands on some of these engines.

The pair considered what sort of aircraft might be propelled by these engines and initially thought about putting a single turbojet into their piston-engined H VII flying wing design. However, it soon became clear that an entirely new aircraft was needed. The brothers therefore began work on what they designated the H IX. This would be a flying wing, like most of the brothers' other designs, but the question remained – where to put the engines? Eventually it was decided that they should be entirely buried within the wing, with their intakes on the leading edge and exhausting over the trailing edge.

The H IX design evolved during the course of 1942 and into 1943 but disaster struck in spring 1943 when the RLM ordered that Sonderkommando LIn 3 be disbanded. Thinking quickly, the brothers managed to give the impression that the order had been obeyed by 'dispersing' rather than actually disbanding Sonderkommando LIn 3.[4]

In March 1943, Walter attended a speech made to aviation industry leaders by Hermann Göring where he called for an aircraft that could carry 1,000kg of bombs 1,000km with a maximum speed of 1,000km/h.[5] This appeared to be the opportunity that the brothers had been hoping for. If they could massage the H IX design a little to meet this specification, at least in part, they might have an opportunity to secure a development contract.

Unable to meet the 1,000/1,000/1,000 spec, they managed to come up with calculations showing a speed of 950km/h, a range of 700km and a payload of 2,000kg using a pair of BMW 003 turbojets. Walter worked his contacts once again and managed to get a 20-page design proposal to Oberstleutnant Ulrich Diesing. Diesing apparently passed this on to Göring and a meeting was arranged for September 28, 1943.

MEETINGS WITH GÖRING

Göring met Lippisch concerning the P 11 from 3.45pm to 4pm on the 28th, then met the Hortons about their H IX from 4pm to 4.20pm on the same day.[6] The P 11 Lippisch presented had changed somewhat from his May design

– there were now twin fins in place of the single fin and the experimental engineless first prototype was to be launched using a rocket booster installed in the trailing edge of the wing and exhausting between the fins.

According to the minutes, "Prof Lippisch reports that he left the Messerschmitt company and that he is currently working on a new project on behalf of the research leadership of the Luftwaffe in the Vienna Research Institute. Task: Lippisch fast-attack aircraft, shell construction, two Jumo jet engines. Project capabilities: 1,000kg bombs, 1,000km/h speed, 3,000km flight distance [1,000km to the target, 1,000km back, with a 1,000km margin for manoeuvring and safety]. V1 should fly in June 1944. Apart from normal teething troubles Lippisch does not expect major difficulties despite the three fundamental innovations (jet engine, all-wing design, plastic-wood construction). If the level of urgency, which is currently set at RM 30,000, is increased to approximately RM 500,000, Lippisch believes that the test pattern can fly as early as February. Apart from the increased extent of urgency, Prof. Lippisch has no wishes at the moment.

"When questioned by Herr Reichsmarschall why his project did not come to fruition earlier, Lippisch reports that in earlier times there was strong resistance to all-wing projects, especially by General-Ing. Reidenbach and Stabs-Ing. Friebel. Lippisch traces back these resistances ideologically, partly to the old rivalry of the Akaflieg [Akademische Fliegergruppe, groups of aerodynamics students from German technical universities] against the Rhönflieger [non-academic glider builders and designers who flew their creations from the Wasserkuppe in the Rhön mountains]. Herr Reichsmarschall shows an extraordinary interest in the project of Prof. Lippisch and gives his full support to Prof. Lippisch for the construction of his test samples. He wishes that the urgent request of Prof Lippisch be met by the Generalluftzeugmeister [Erhard Milch] after a factual examination as soon as possible."

It is unknown whether Lippisch crossed paths with the Hortens on the way out but the next meeting then started. The minutes state: "Herr Reichsmarschall has Hauptmann [Walter] Horten present his work on the basis of photographic material and asks about the new projects. Hauptmann Horten reports that currently three twin-engine propeller-driven trainer aircraft of all-wing layout are under construction at the Peschke company. On the orders of the general, Awia and the local designer Kaupa have been included [although the meeting transcript says 'Awia' and 'Kaupa', this may well be a reference to Czech company Avia and Austrian engineer Otto Kauba].

"In the opinion of Hauptmann Horten, however, neither this company nor the designer Kaupa is suitable for cooperation, since the construction system Kaupa (crosswise welded tubular spar) espoused by is not suitable for a shell construction of a high-speed aircraft."

Horten told Göring that he would rather have the construction work done at Peschke. Then "at the request of Herr Reichsmarschall, as to whether a fighter or combat aircraft in all-wing construction was being worked on, Hauptmann Horten reports that at present he is running a project for a fighter jet (fighter-bomber), which he is building in a road maintenance centre near Göttingen. It is to have two Jumo jets, 1,000kg bombs, 950km/h speed with a flight distance of 1,500km. The first V-pattern will be finished without engines in February as a towed glider, be equipped with an engine in June and from August 1944 also have full equipment.

"An order from the RLM for this aircraft has not yet been issued. The aircraft is well-suited for a large series in the opinion of Hauptmann Horten because of its simple mixed construction method, which requires only small tooling capacity. At the request of the Reichsmarschall, Hauptmann Horten reports that his work has received little support from the Technical Office [of the RLM], that occasionally their work has been interrupted by order of the Office, and that in the past especially General-Ing Reidenbach and Stabs-Ing. Friebel have pronounced against the all-wing construction.

"Hauptmann Horten, in his opinion, could also considerably speed up his work were he too receive a corresponding urgency to procure the necessary material. In addition, Hauptmann Horten asks for the assignment of his designated soldiers who are currently referred to by the military as indispensable. Hauptmann Horten also asks for his command to be issued with a budget and clear referrals to ensure that soldiers who have made a valuable contribution to his service are not left behind in promotions and awards.

"Herr Reichsmarschall thanks the Horten brothers for their research and development work and assures them of full support for their further work. In conclusion, Herr Reichsmarschall states that, in his view, there has not been sufficient attention in the past in the direction of the development of all-wing aircraft, in particular the swept-wing, and that this area must be more closely respected by the Generalluftzeugmeister".

It is not recorded in the minutes, but speaking years later[7] Reimar said Erhard Milch had given Walter the order for the construction of the H IX, although it had come from Göring: "Milch drew up all the paperwork for the transfer of the RM 500,000 grant from Göring. Milch asked us who should this contract be made out to? Hauptmann Horten or what? Walter paused and told Milch that he was not sure, however he would return in a day or two with the correct information. Before going back to Milch, we incorporated ourselves and called our new company the Horten Company for Aircraft Design or the Horten Flugzeugbau."

H IX/8-229/HO 229

With a million Reichsmarks between them, Lippisch and the Hortens set about building their prototypes. Reimar said that he and his brother did not keep track of how much of the RM 500,000 they had spent, however: "I do not know how much money we had used. Walter spent several days in Dessau to arrange for funds. We never had any problems and the Ministry of Finance said that they would add up all the expenditure after the war... so for now go about and construct the aircraft which the Reich needed."[8]

The unpowered H IX V1 was ready on March 1, 1944 – albeit with a Heinkel He 177 tail wheel assembly as its nosewheel, main landing gear wheels from a Messerschmitt Bf 109 G, components from a captured B-24 Liberator and other assorted bits and pieces from a damaged Me 210. Bad weather delayed its first flight however, so the brothers sent Göring a telegram anyway, telling him that the aircraft had flown, even though it had not.[9]

Eventually, on March 5, 1944, it was towed up to 3,600m by an He 111 before gliding back down to the runway, but on touching down the pilot found he was unable to brake effectively on the icy runway. In order to avoid colliding with a hangar, he deliberately retracted the nosewheel, putting the skidding aircraft's nose onto the ground.

More tests followed on March 23 and April 5. During the latter, the nosewheel failed and repairs led to further delays. There was still no sign of the BMW 003 becoming available for the H IX V2 – the powered version – so the Hortens reverted to their earlier intention of using Jumo 004s and the V2's airframe was designed according to the dimensions Walter had come up with back in the winter of 1941.

These showed that each engine was effectively a tube 60cm in diameter. However, Reimar later discovered that with all the necessary accessories added to their exterior, the engines were actually closer to 80cm. This was a major problem and necessitated a redesign. Reimar said:[10] "I resolved to make the centre section greater by one additional rib on each side – 40cm and 40cm, 80cm in total. And raise the height of the profile about 13% to 15%. With this increase in height, I could fit in the 80cm Jumo 004 turbojet. This change cost us one full month."

The original plan had called for the V2 to be flown for the first time on June 1, 1944, but work was still ongoing as this deadline passed. The RLM placed an order with Horten

ABOVE: The Horten IX V1 flying wing glider under tow after a test flight - its brake parachute is bundled up on the rear section.

ABOVE: The H IX in gliding flight.

ABOVE: The Horten 8-229 V2, with jet engines installed, undergoing ground tests at Oranienburg.

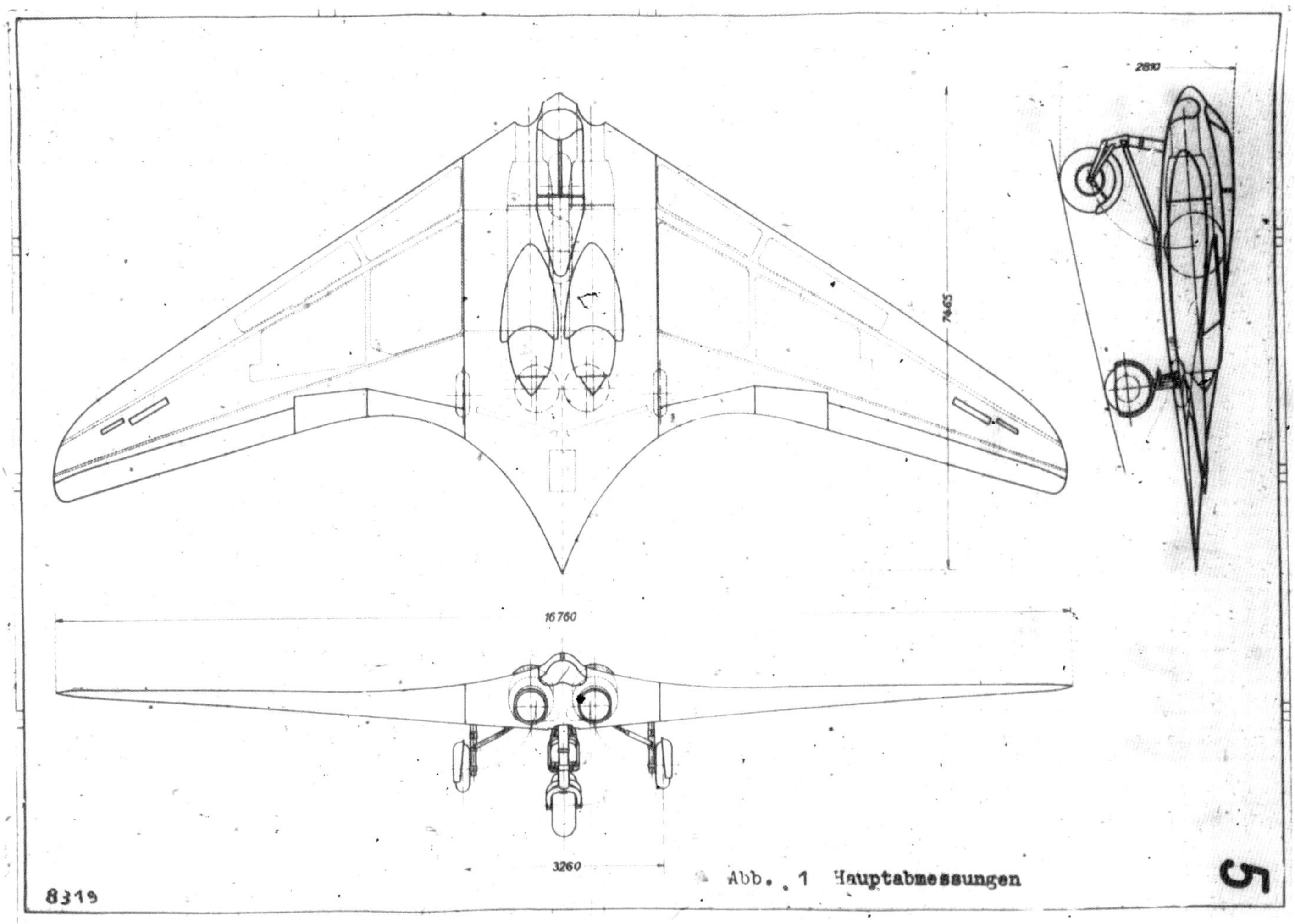

ABOVE: A three-view drawing of the 8-229 from an unusual document published by Gotha on November 22, 1944, entitled 'Vorläufige Baubeschreibung' or 'preliminary building description'. It gives a full description of the 8-229 without naming it once.

Flugzeugbau for a number of further prototype H IXs on June 15 with the new type designation 8-229. The '8-' simply denoting a powered aircraft.

Horten Flugzeugbau, however, did not have the facilities or the manpower to build more than the one airframe it was then working on. A subcontractor was needed: Gothaer Waggonfabrik, commonly known as Gotha, came to mind.

Reimar said: "Gotha had not done any work for us before they received the order to continue production of the H IX. The name of Gotha was mentioned to us about August of 1944. Walter mentioned to me that he was thinking about using Gotha to manufacture the H IX series. I told him it did not matter to me which firm was picked to build it because I was not interested in that part of aircraft production.

"My main interest was in design and development work. Once it got to production I did not want to be involved. I had other things to do. It was now the RLM's responsibility to see our series production should take place and how. Gotha had had some practice in wood construction so they had been selected by Walter and the RLM.

"Klemm had also been selected as a builder of what was now being called the Ho 229. Klemm had had considerable experience in building light planes and Walter and the RLM had selected Klemm at the same time they had picked Gotha. The thing with Gotha is that during the war they had built for DFS their 230 troop glider. They had changed the design, not aerodynamically but in structure to be built in series. With these modifications by Gotha, they had not

ABOVE: The H IX V1 is rolled out of its hangar at Göttingen for another flight.

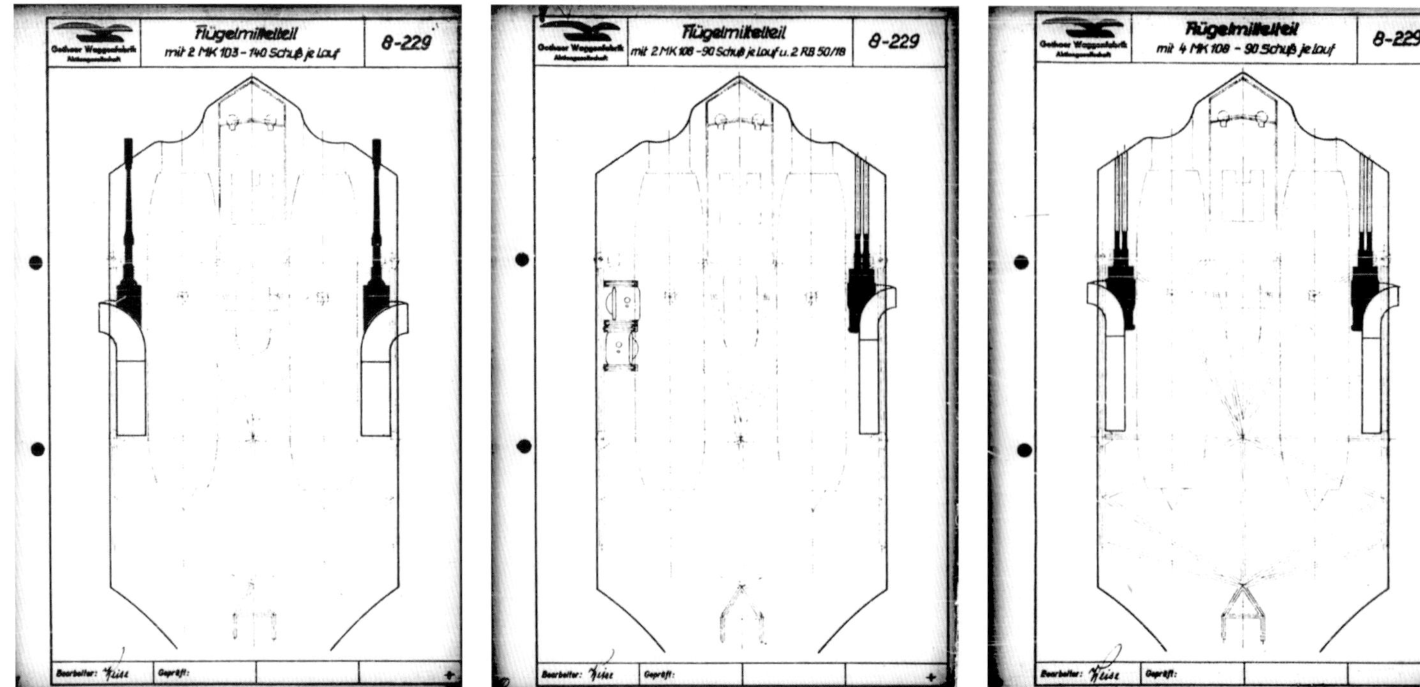

ABOVE: Three different potential armament options for the 8-229 V6 outlined in a Gotha report by Kalb and Weise dated December 15, 1944.

really been wanted by the troops. For the modifications they had made in order to ease construction had been bad and several DFS 230 gliders had broken up in flight.

"So I told Walter, I want you to observe Gotha during the construction of H IX V3 through V6 that they don't make any modifications which might harm the aircraft during flight, just for the sake of quicker construction."[11]

In fact, Gotha aerodynamics specialist Rudolf Göthert, who had worked for the firm since 1942, told his Allied interrogators in 1945 that the company had become involved with the H IX/8-229 project as early as late 1943 or early 1944.[12] The Hortens had little faith in Göthert but nevertheless, Gotha set to work on the four airframes. Horten Flugzeugbau was still struggling to complete the H IX V2 by

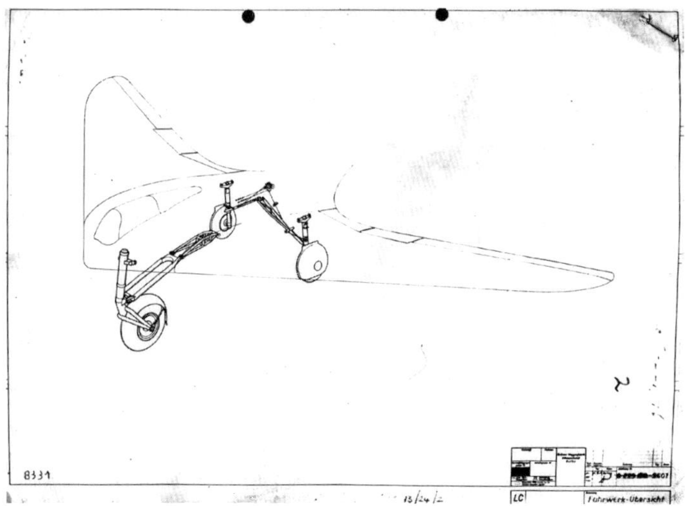

ABOVE: The 8-229's tricycle undercarriage arrangement.

November 1944. The work was briefly discussed during a meeting,[13] probably of the Entwicklungshauptkommission Flugzeuge (EHK), the recently formed chief development commission for aircraft, on November 21–22: "The Ho 229 was to be developed in conjunction with Gotha, and three prototypes of the Horten VII were to be completed."

On November 22, Gotha published a preliminary description[14] of the 8-229 which made it clear that "the aircraft is a single-seat fighter design. It can also be a bomber with a two-ton bomb load or a reconnaissance type". Evidently the type was no longer being considered primarily as a fast bomber.

The V2 was finally completed on December 17, 1944, and transported by rail from Göttingen to Oranienburg, just north of Berlin, for flight testing. The designated pilot, Erwin Ziller, received twin-engine jet flight training in a two-seat Me 262 B-1 on December 29-31 but testing was halted while the brothers travelled back to Bonn to spend Christmas with their parents. During January 1945 Gotha appears to have slowed or even halted its programme of work on the V3, V4, V5 and V6. Göthert had by now decided that the original Horten design was irredeemably flawed and designed his own alternative flying-wing fighter – the Gotha P-60, which had its two crew in prone positions in a glazed cockpit built into the leading edge of the aircraft's wing. The two turbojets were positioned one above and one below the wing/fuselage. The rest of the Gotha company was occupied with gearing up for and then commencing mass production of wings for the Heinkel He 162.

Ziller finally took the 8-229 V2 up for its first powered flight on February 2, 1945. Keeping the undercarriage

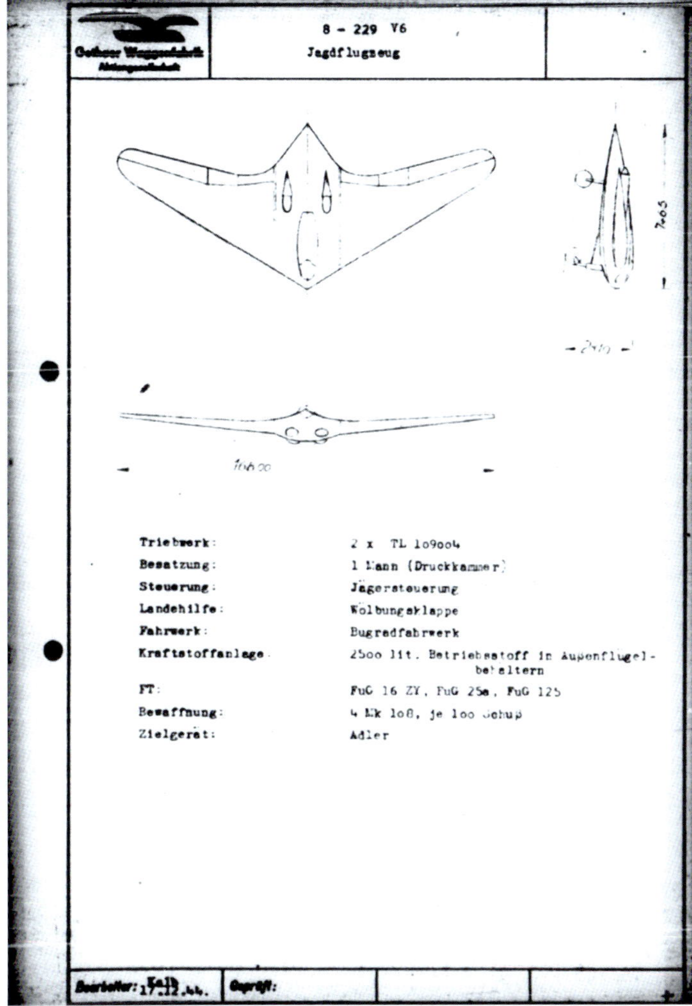

ABOVE: A three-view of the 8-229 V6 from the December 15 report. At this stage it is clear that the aircraft is to be a single seater.

locked in the 'down' position, he reached 300km/h before throttling back and landing again. A second flight is believed to have taken place on February 3 but this time a hard landing due to a too-early deployment of the aircraft's brake parachute resulted in undercarriage damage.

Following repairs, a third flight took place on February 18, 1945. Ziller made three passes over the airfield so that a team from the Rechlin test centre could make speed and altitude measurements – apparently clocking up 795km/h (494mph) below 2,000m. The right engine failed after 45 minutes in the air and Ziller was unable to restart it. He tried to bring the aircraft in to land but the landing gear hydraulic system, powered by the right engine, was inoperative so Ziller had to use the contents of a compressed air bottle, his emergency backup, to lower the gear 400m from the runway. This meant that once it was down, it could not be retracted.

The increased drag slowed the aircraft and Ziller realised he wasn't going to make the strip. He powered up the remaining engine but the drag produced by the gear was too great and its airspeed could not be increased. The aircraft then entered a broad turn to the right which it maintained until it hit the ground. The impact was so great that both engines and Ziller himself were thrown from the aircraft. He hit a tree and was killed instantly.[15]

It has been suggested that Ziller was rendered unconscious by engine fumes entering the cockpit during these final moments, which would explain why he made no further efforts to recover the aircraft after it entered its wide turn. It was found that his harness, though torn open by the force of the impact, had not been unfastened.

HO 229/GO 229/HO 267

Following Ziller's death, the Hortens were distracted by the inclusion of their H XVIII bomber in the Grossbomber/Langstreckenbomber competition – the main conference for which took place on February 20–23.[16] Once this was over, Reimar's thoughts returned to the 8-229 and he prepared a new report[17] evidently intended to help Gotha acquire more resources for the production of the remaining prototypes.

The cover of the report folder was marked 'Horten [underlined] Flugzeugmuster 8-229 Zerstörer' and the first page read: "Preface. The airplane described in the following is a flying wing aircraft, which has neither fuselage nor special control surfaces. It was designed in 1943 on the basis of systematic preliminary work on more than 30 aircraft of the most varied tasks.

"The construction of the aircraft was carried out in two work-pieces by the Luftwaffen-Sonderkommando IX, Göttingen. Both aircraft have expanded our experience with numerous flights with various pilots in such a way that series construction can be realised on this basis. However, a further testing of prototype aircraft is required for the large series."

The page is signed by Reimar Horten, Göttingen, March 1, 1945. The next page is headed 'Description of the flying wing aircraft 8-229 [underlined] according to project "Horten IX"'. The 8-229 is described as a two-engine jet multipurpose two-seater aircraft for use as a) Day fighter and bad weather fighter, b) destroyer, c) light bomber, d) reconnaissance, e) night fighter.

It is clear that, going forward, the 8-229 is to be a two-seater whatever its intended role. The single-seater 8-229 appears, as of March 1, 1945, to have been abandoned.

The report description from that date continues: "The flying wing has 53.6m² wings and a wingspan of 16.8m. It is three-part, the outer wings are made in wood construction, the wing middle part as steel tubular framework. The outer wings are supported by the rudder flaps acting as height and ailerons, as well as the side rudders configured as an air brake at the wing ends.

"On the wing root, drag flaps are sunk as take-off and landing aids. Inside the wings are four fuel tanks for 1,200 litres of fuel. It is intended to glue and conserve the wing later, if necessary, so that approximately 2,000–2,200 litres of fuel tank space is produced per wing.

"The wing is lined with an auxiliary hull and with a 17mm-thick solid wood shell. The dimension of the wing shell is selected to be this large for stiffness reasons, whereby it is at the same time possible to use locking wood of a poor quality without incurring disadvantages. The wing centrepiece with a width of 3.20m carries the landing gear, the two-seated pilot cabin, the engines 109 004 or 109 003 in the interior, as well as the weapons, armouring, bomb hanging devices and navigation equipment.

"The static assembly consists of untreated steel pipes; the claddings of plywood shells. These can be dismantled, so that the accessibility to the engines, weapons, frames and FT installations is ensured at all times.

"The landing gear is designed as a nosewheel gear and gives the aircraft an angle of 9°. By approximately 50% of the total load lying on the nosewheel, it necessarily forced this angle of attack when rolling, which significantly reduces the starting distance, especially in the case of grass and soft ground. All wheels are retractable. The rear undercarriage, consisting of standard bogies from the Me 109 in the case of V-models, can be replaced later by the chassis of the Me 262 without major changes, which makes it possible to increase the weight of the aircraft up to 11.5 tons."

The engines, either Jumo 004s or BMW 003s, were to be installed inside the wing and would be as close as possible to the very centre of the aircraft. They could be installed or removed from the front via their intakes.

At the front of the aircraft "the cabin for the pilot and the observer is provided between the inlets of the engines. This cabin is designed as an armoured pressure cabin. It contains catapult seats for the tandem seated pilot and observer". Armour would be provided to protect the crew and ammunition boxes from bullets up to 12.7mm from the front and back. Weaponry would be four fixed forward-firing MK 108 cannon with 120 rounds each. The gunsight was to be an EZ 42 but also "the possibility to install an additional 24 to 36 R4M rockets is available. Between the engines, a 503 rack is installed for 1,000kg bombs".

No detail is given on exactly how the unguided R4M rockets would be installed but presumably they would be fitted to underwing racks as they were on the Me 262. Equipment in terms of radar units or other electronic devices would be "later determined by OKL", the Luftwaffe high command.

Weighing 7,000kg, the 8-229 would require a take-off run of 520m. Landing speed with the same load was 144km/h but with a reduced fuel load (weight down to 5,450kg) it was just 127km/h. The 'performance' section of the report also mentions a range of 4,040km when fitted out for reconnaissance duties with a pair of 1,000kg drop tanks.

ABOVE: Artwork showing the 8-229 in its final series production form – a two-seater – from Reimar Horten's March 1, 1945, 'Horten Flugzeugmuster 8-229 Zerstörer' report.

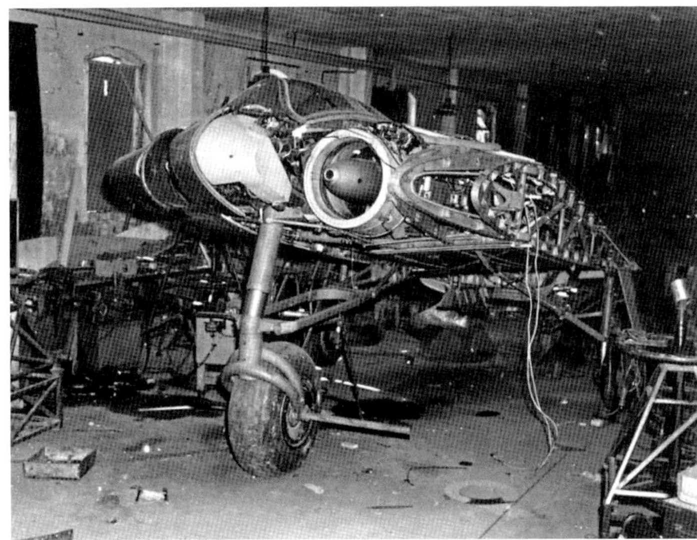

ABOVE: The partially completed 8-229 V3, captured by the Americans at Ortlepp Möbel Fabrik in Friedrichroda.

The drag from these would result in top speed being lowered from around 950km/h to 830km/h. Ceiling was 15–16km although the report notes that the available engine specifications were "not sufficient for more accurate height determination".

Next the report provides a summary of 22 advantages "of the 8-229 in front of today's jet fighters". These are given as "higher speed (950km/h near the ground), higher climbing speed, greater flight time and range, higher payload (2 tons bombs) higher ceiling (15–16km), shorter start run, narrow turn radius (smaller surface load), lower landing speed, jet inlets are higher (lower risk of contamination), airfield suitability, better visibility, better shooting position (side control), better accessibility (without tailplane), fuel is far from the pilot in the outer wing, lighter roll handling (nose wheel and intercepting force), two-seater, construction has less work required (4,000 hours), building material wood and steel (65% wood), better single-engine flight due to closer engines, tactical brake in three stages, all three wheels can be braked, very short run-out due to: small landing speed, screen brake, 3 x wheel brake".

The final written section of Reimar's report is headed 'production' and gives the position as of March 1, 1945: "The company Gothaer Waggonfabrik 20 prototypes have been ordered with DE-urgency starting from November 1944. The output begins with V3. From V3-incl. V5 is the single-seater. As of V6, the aircraft will be built as a two-seater. The standard processing can be carried out by the company GWF.

"In order to achieve the lowest possible risk during a series run, it is necessary that the test aircraft are manufactured as quickly as possible. The current deadline, which provides for the production of one aircraft per month, is sufficient to ensure a satisfactory trial during the course of the year 1945.

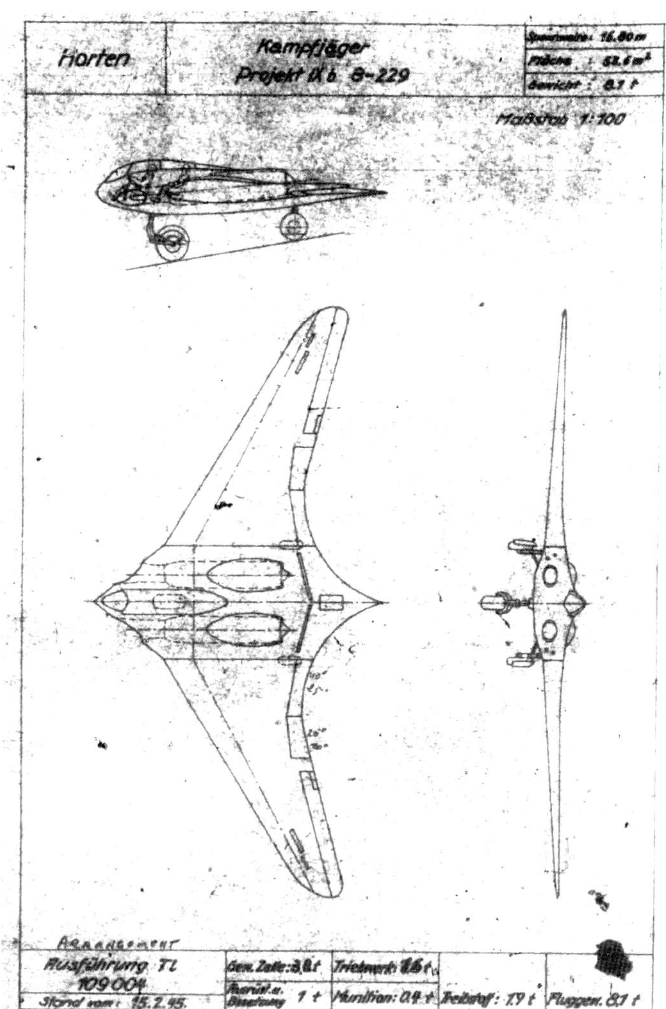

ABOVE: A three-view drawing of the two-seater production model 8-229. Whether it would have been the Ho 229, the Go 229 or even the Ho 267 is debatable.

"The prerequisite for this is that the specified dates are kept. For example, the first aircraft to be deployed by the company GWF has a deadline of three months, a larger support of the company is necessary."

At the end of February, a meeting of the EHK, discussing night and bad weather fighters, had confirmed that work on "another project, fighter variant Horten IX" was still ongoing at Gotha.[18] Junkers produced a brief report six days later, on March 7, 1945, entitled 'Triebwerkeinbau in Go 229 (Horten) (V3 + V5)'[19] which outlined the difficulties that the company was experiencing in fitting the engines into the airframe, primarily due to the lack of available space.

Under a heading of 'Horten IX' the war diary of the Chef TLR for March 16 to April 4 states that "apart from the three V-patterns V3 to V5 in the V2 version, a further 10 aircraft V6 to V15 will be built at GWF".[20]

This suggests that Gotha had been authorised to construct a series of 10 two-seater 8-229 aircraft, presumably some time between March 1 and April 4. There appears to be some evidence that the Ho 229/Go 229 received another the American T-2 Technical Intelligence Glossary of German Aeronautical Codes, Models, Project Numbers, Abbreviations, Etc. Final Edition,[21] published in November 1947, has entries for both the Ho 229 and Go 229. Under 'Ho 229', it states: "Aircraft type number for an experimental flying-wing reconnaissance airplane powered by two Ju 004 turbojet engines redesignated as Ho 267." The entry for 'Ho 267' just below it reads: "Aircraft type number for a Horten flying wing powered by two Ju 004 turbojet engines."

The 'Go 229' entry states: "Aircraft type number for an experimental twin-jet flying-wing reconnaissance airplane developed from the Horten IX of which the Go 229 – V6 modification is known." Under 'Go 267' it simply states: "Aircraft type number for a Gotha airplane."

Evidently work on the V3 to V5 8-229 prototypes had been taking place on Gotha's behalf at the Ortlepp Möbel Fabrik at Friedrichroda – which is where the incomplete airframes were captured by elements of the American Third Army's VII Corps on April 14, 1945. Both the V3 and V4 had their engines fitted, the V3 being near completion. The V5 existed only as a steel frame.

By now both Horten brothers had also been captured and taken into custody as prisoners of war.

GOTHA P-60

Although Göthert apparently started work on his 8-229 alternative, the P-60, during the late autumn or winter of 1944, his earliest known report on the design, entitled 'Go P-60 Hochgeschwindigkeitflugzeug' or 'high-speed aircraft', was not published until March 11, 1945.[22] It gave the P-60's intended roles as heavy fighter, fighter-bomber, reconnaissance aircraft and night fighter. Three different two-seater designs had been drawn up: the P-60 A and P-60 B, in which the crew lay prone in the nose, and the P-60 C, which had conventional upright seating in tandem. The first two had control surfaces built into their wingtips which could be extended or retracted as required, while the 'C' had a pair of fixed fins.

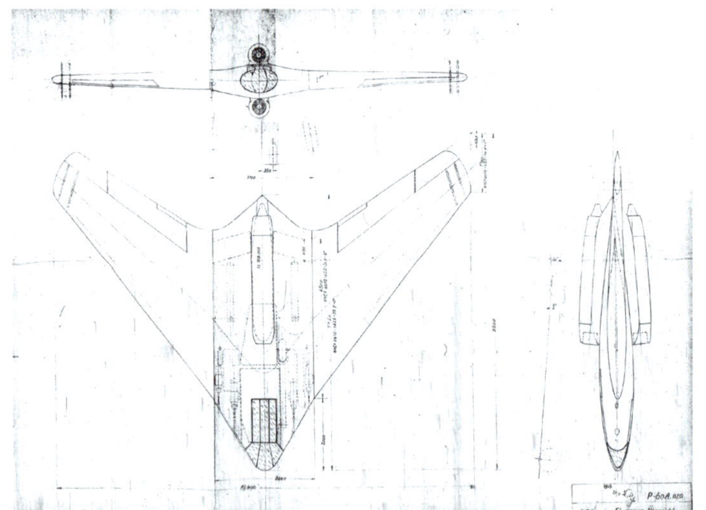

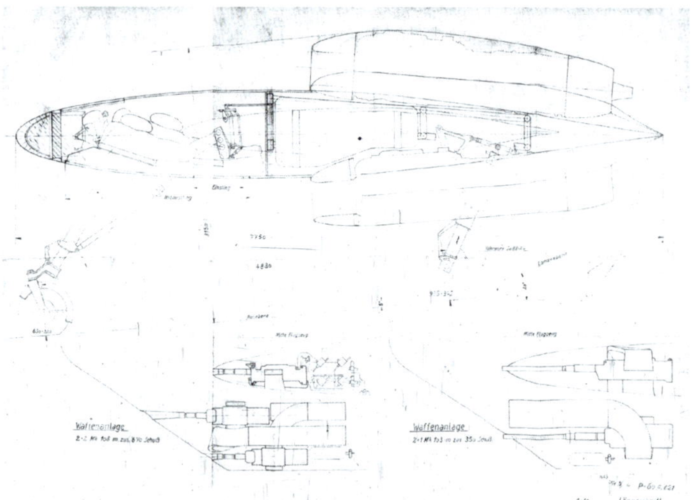

ABOVE LEFT: Gotha P-60 A with BMW 003 engines as shown in drawing P-60 A.020 of January 28, 1945. An alternative arrangement labelled P-60 A.019, drawn up on the same day, shows the aircraft with its turbojets built into the rear section of the wing.
ABOVE RIGHT: Side view of the P-60 A showing the prone crew positions, short undercarriage main legs and weapons loadout.

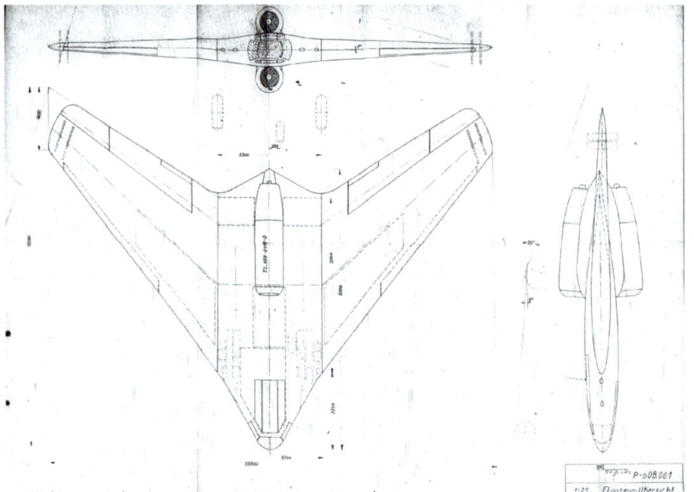

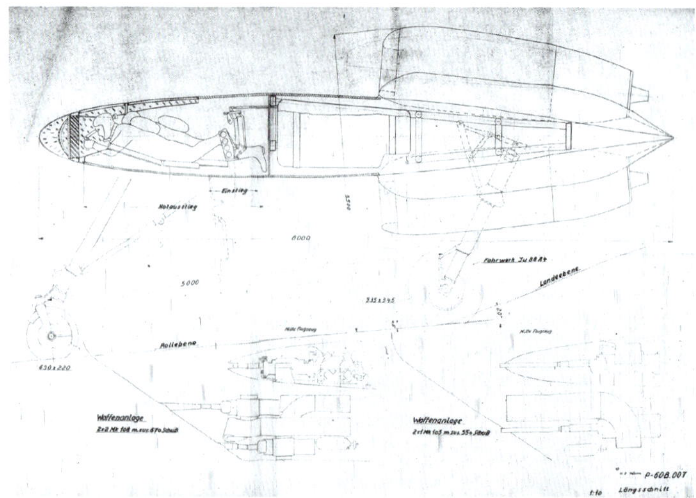

ABOVE LEFT: The HeS 011 was much shorter than the BMW 003 and the difference is evident in this drawing of the Gotha P-60 B. The design is also slightly larger than the P-60 A and features different cockpit glazing. ABOVE RIGHT: A view of the P-60 B from the side.

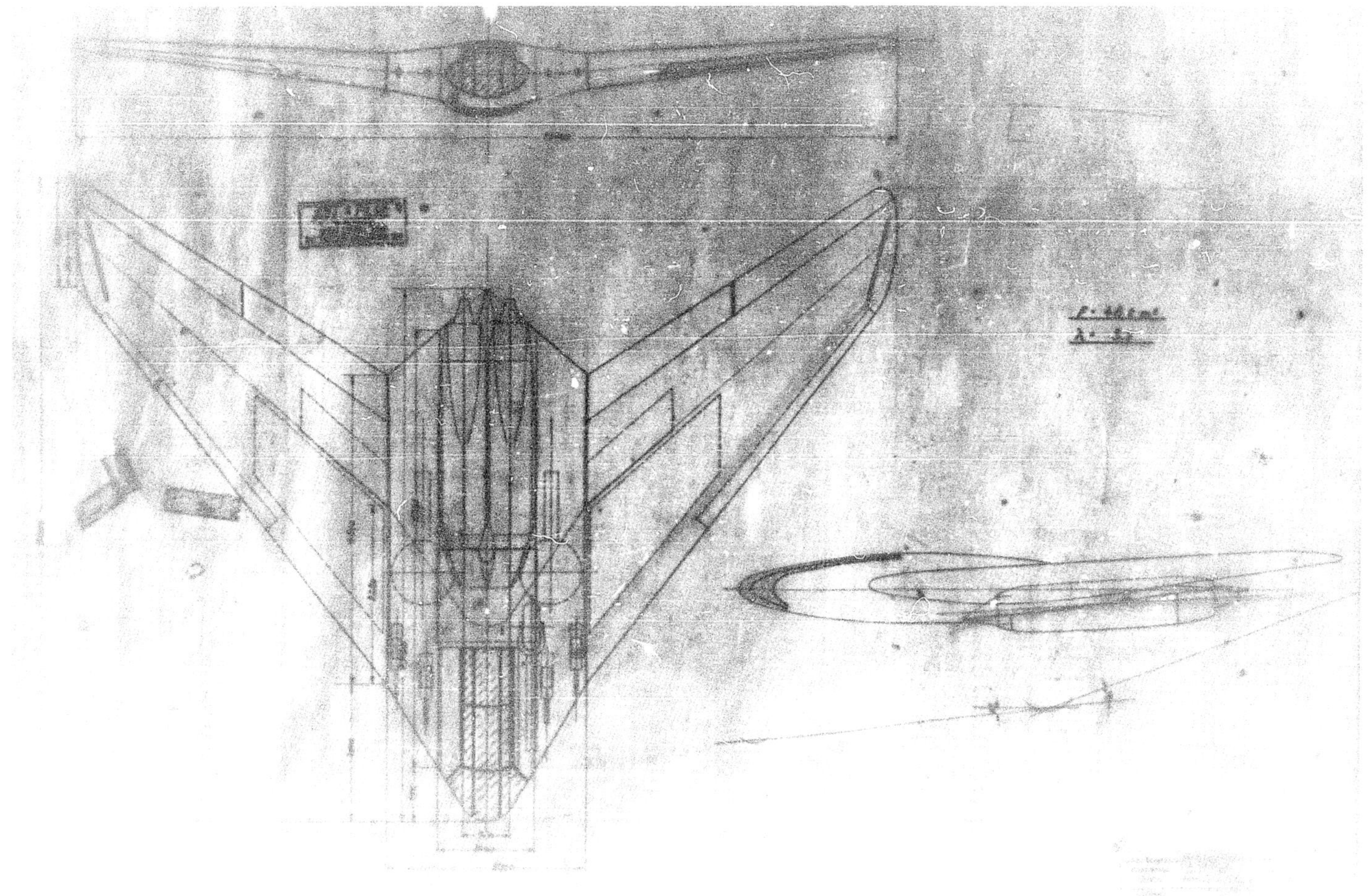

ABOVE: The Gotha P-60.007 was significantly different from the rest of the series in having both of its engines on its underside. It also had a sharper trailing edge profile, a pronounced dihedral and different wingtip control surfaces.

The BMW 003-powered P-60 A, appearing in drawing P-60 A.020 dated January 28, 1945, was based on experience of working on the 8-229 and was a director competitor for it. It had a wingspan of 12.4m – compared to 16.76m for the 8-229 – and was 7.75m long, based on its centreline (compared to 7.465m for the 8-229), or 9.5m from its nose to its wingtips, which unlike those of the 8-229 extended rearwards beyond the centreline. Armament was either four MK 108s or two MK 103s. A distinctive feature was its main undercarriage wheels, due to being sourced from a Ju 88 A-4 or A-7 which tilted inwards.

The P-60 B was an HeS 011-powered replacement for the P-60 A/8-229 and based on a drawing of February 12, 1945, was somewhat larger than the P-60 A, with a wingspan of 13.5m, a centreline length of 8m and a nose to wingtips length of 10.3m. It had a broader three-pane cockpit nose window than the 'A' and its Ju 88 undercarriage mainwheels were vertical. Armament options, however, were the same.

The P-60 C, also HeS 011-powered, was designed to compete against new jet night fighter designs already being worked on by other major aircraft manufacturers as 1944 drew to a close and is outlined in greater detail in Chapter 12.

Göthert's report outlines each version in detail – the P-60 A and P-60 B both had armoured and pressurised crew compartments with air conditioning, adjustable air-cushioned couches for the pilot and radio operator, and 'hanging' foot pedals. He also mentions a fourth P-60 but does not name it beyond calling it a 'variant'. He writes: "In a variant, a twin arrangement, in which the engine units are arranged below the central wing, recessed next to one another, is possible. Current wind tunnel tests will prove whether this type of installation is also suitable."

This suggests that the 'variant' was a work in progress as of March 11, 1945, and a drawing labelled P-60.007 appears to show this design. The most obvious difference from the other P-60s is the two turbojets being recessed deep within the rear of the wing centre section, with only a broad lip intake protruding below. In addition, compared against the P-60 A, the centre section protrudes further rearwards and the point where it meets the wings is sharp and angular rather than the gentle curve shown in the other designs.

The outer sections of the wing have a pronounced dihedral – angling upwards compared to the other versions – and have different control surfaces at their tips.

Another drawing, P-60 A.019 of January 28, 1945, showed an alternative version of the P-60 A design which featured the turbojet arrangement from P-60.007 – with both 003s built into the rear part of the wing and fed by a low-profile underside intake. This allowed for shorter rear undercarriage legs but the design was otherwise largely identical to the P-60 A shown in drawing P-60 A.020, which was included in the final report on the type.

In fact, even apart from the P-60.007 and P-60 A.019 designs, the Gotha report on the P-60 actually lists six potential versions of the design including two with additional rocket motors fitted at the rear of the centre section between the two turbojets – P-60 A mit BMW 003, P-60 A mit HeS 011, P-60 B mit HeS 011, P-60 B mit HeS 011 und R-Gerät, P-60 C mit HeS 011 and P-60 C mit HeS 011 und R-Gerät. There is no mention of a version of the P-60 A with 'R-Gerät' – rocket propulsion – however. This made a total of seven versions of the P-60, but there was still one more to come – a three-seater development of the P-60 C (see Chapter 12). With the end of the war, all work on the P-60 came to a halt.

LIPPISCH P 11/DELTA VI

Not long after his meeting with Göring (perhaps immediately after, since the last known 'P 11' drawing is dated three days before the meeting) Lippisch's tailless P 11 was renamed the Delta VI. His small workforce at the LFW in Vienna diligently laboured over his designs, even as he constantly changed and altered them. Before long, the Delta VI bore little more than a passing resemblance to the P 11 endorsed by the Reichsmarschall.

Work on building the unpowered first prototype had begun by June 1944 when disaster befell the project. According to Technical Intelligence report no. A.424:[23] "The construction of this aeroplane was started in a factory in Vienna which was bombed out in June 1944 by the American Air Force.

"By this event Dr Lippisch lost 43 of his collaborators. The factory was then rebuilt in the Wiener Wald but the first experimental aircraft of this all-wing type, Li P 11, was never finished as the Russians invaded the region of Vienna."

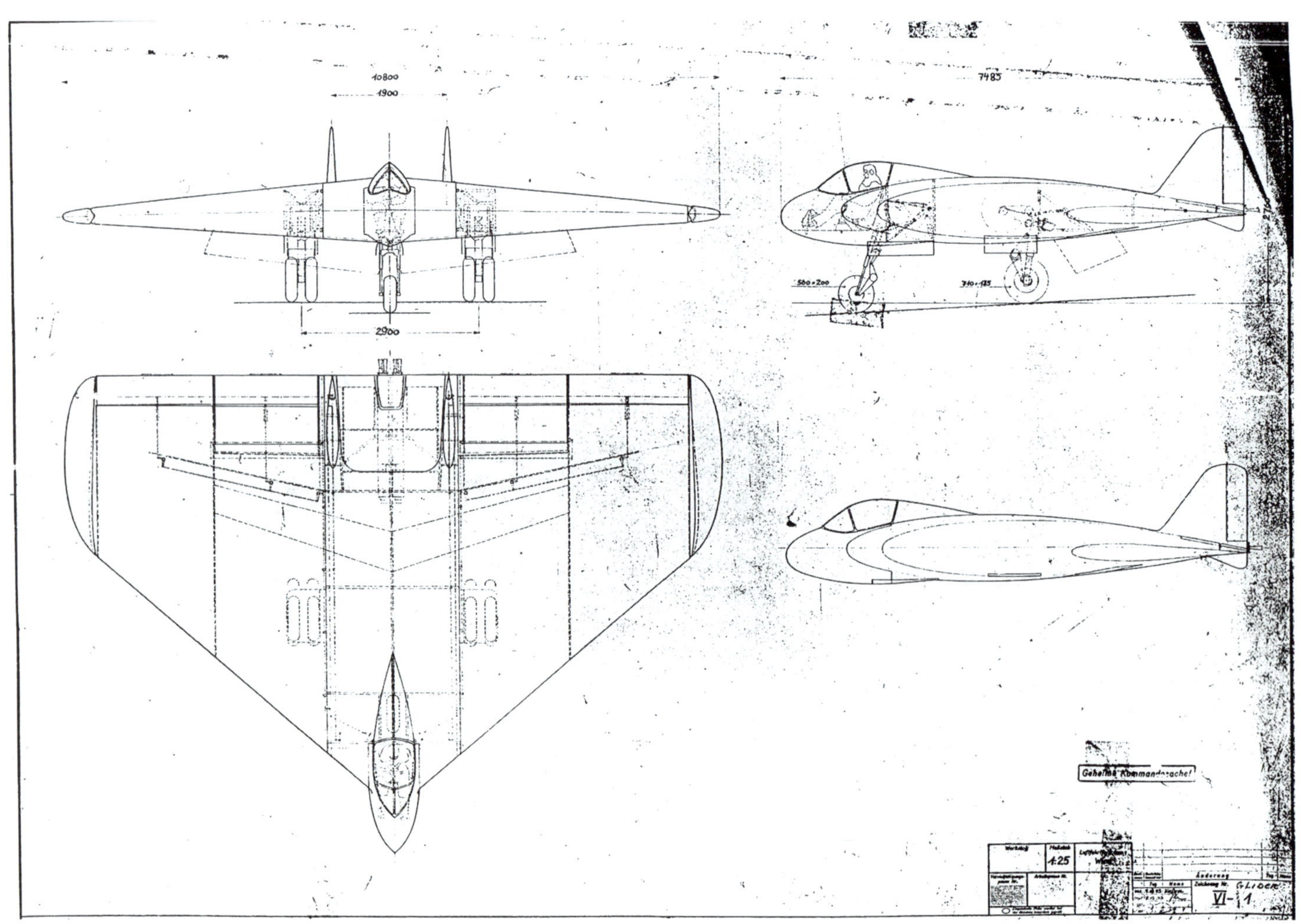

ABOVE: The Lippisch Delta VI V1 rocket-boosted glider as of December 9, 1943. Compared to the version of September 25, the design now lacks mock-up intakes and the cut-out on the trailing edge has been filled in, with the rockets exhausting directly from the trailing edge. Variable wingtips are now also clearly a part of the design. Wingspan is 10.8m and overall length is 7.485m.

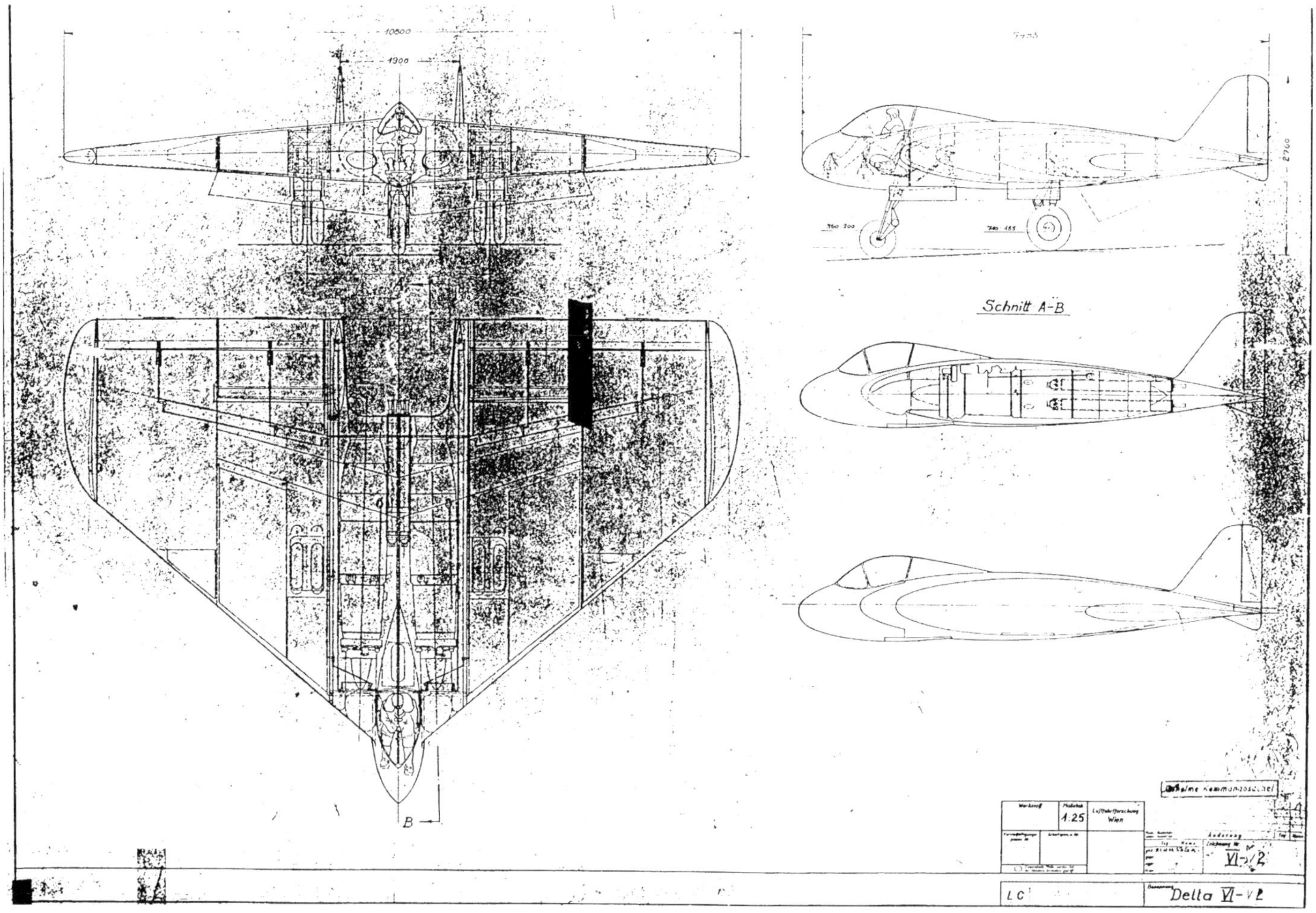

ABOVE: The twin turbojet-powered Lippisch Delta VI V2 as of December 21, 1943. The design has the same overall dimensions as the V1 but the trailing edge cut-out has been retained in this version with the rocket boosters wedged between the turbojets.

Vienna was bombed twice in June 1944 by the USAAF – on the 16th and 24th. The former seems more likely to have hit the LFW.

Writing about the attack in his book *Erinnerungen* some 30 years later, Lippisch put the number of dead slightly higher: "At the time of the highest activity an air attack on aeronautics research in Vienna (LFW) occurred in June 1944, during which severe damage and, above all, 45 deaths were to be lamented; including some of my most valuable employees."[24]

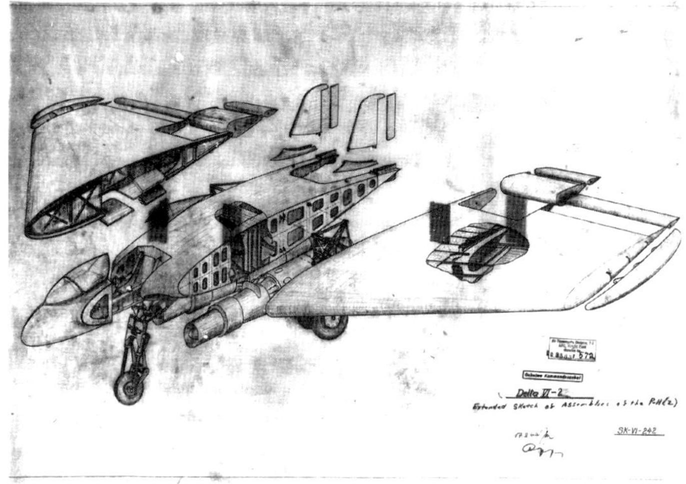

ABOVE: Structural breakdown of the Delta VI V2 dated February 17, 1944.

ABOVE: The Delta VI mock-up under construction at the LFW in Vienna during 1944. *IOWA STATE UNIVERSITY LIBRARY SPECIAL COLLECTIONS AND UNIVERSITY ARCHIVE*

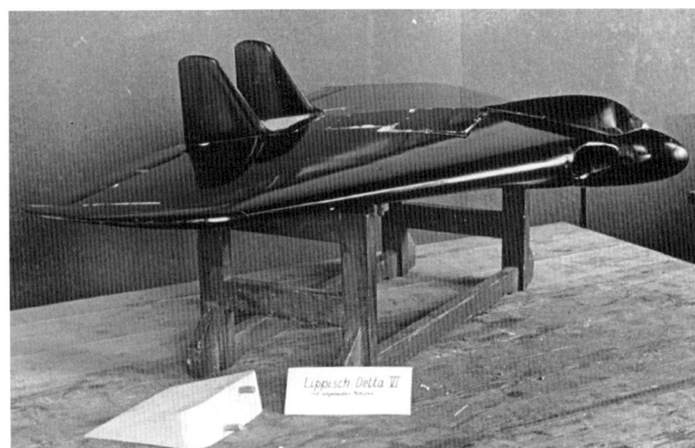

LEFT: Another model of the Delta VI at the AVA. The card below says 'Lippisch Delta VI mit eingebauten Motoren' – the Lippisch Delta VI with built-in engines'. The model itself has a panel on its back allowing access to its inner workings. The cockpit canopy section is also removable.

ABOVE: Wind tunnel model of the Delta VI without upturned wingtips, photographed at the AVA in Göttingen.

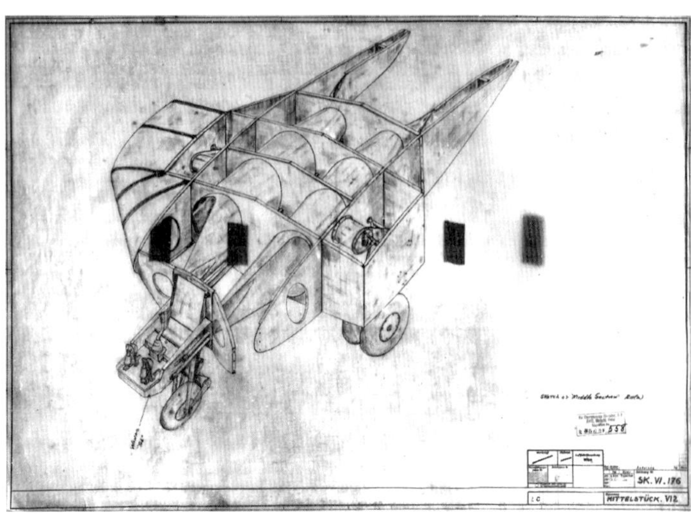

ABOVE: The Delta VI V2 internal structure shown in a drawing dated May 31, 1944. The rocket boosters are apparently absent and the caption near the nose says 'Wanne "Tub"'.
BELOW: Another Delta VI V2 drawing from May 31, 1944. External dimensions remain the same, with the engines being installed at an angle, but although the rocket boosters are now shown it is unclear how they will be installed.

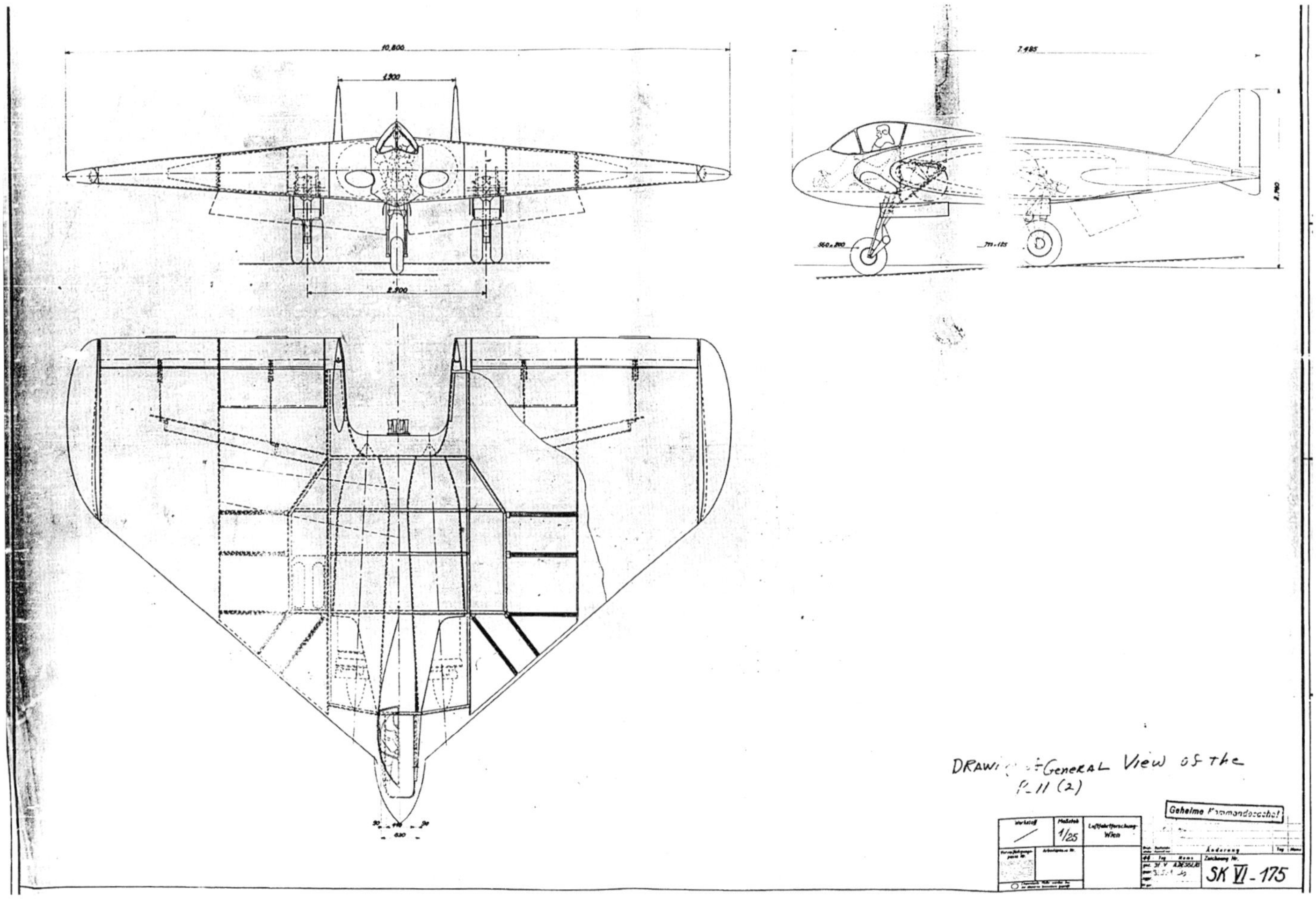

10,800

2,900

2,700

7,485

DRAWING OF GENERAL VIEW OF THE P.11 (2)

Geheime Kommandosache

1/25

SK VI - 175

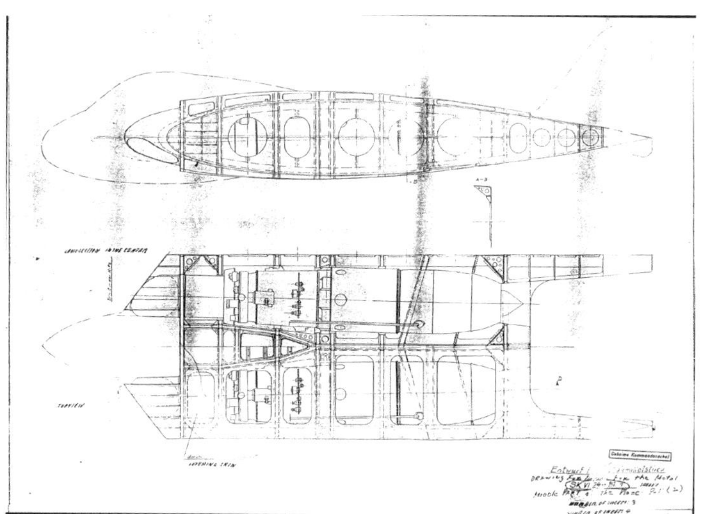

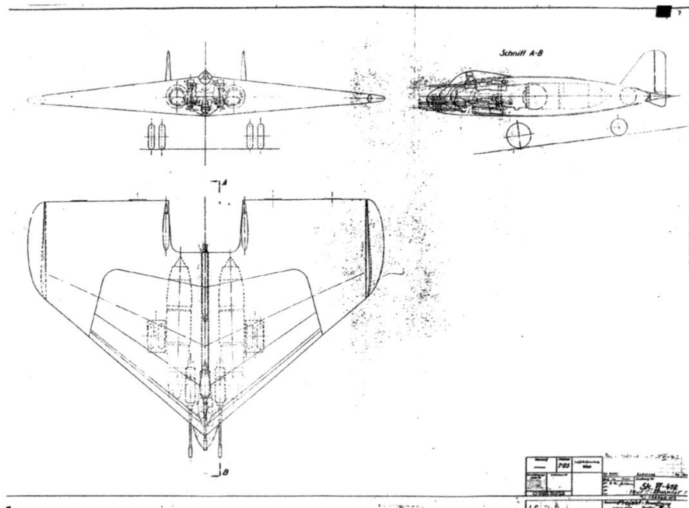

ABOVE LEFT: Undated drawing showing the internal structure of the Delta VI. Based on its lack of rocket boosters, the drawing would appear to date from between May 31 and September 8, 1944. ABOVE RIGHT: This drawing dated September 8, 1944, appears to show a concept for the in-service version of the Delta VI – where every known earlier drawing has shown either the V1 or V2 prototype. The changes are substantial: the engines are still slightly angled, though less so than before, and have been separated to allow room for the rocket boosters. The rear cut-out has consequently been widened, the pilot's seating position has been altered to allow for a lower canopy and the nosewheel has been deleted and replaced with tailwheels. Armament shown is four MK 103s.

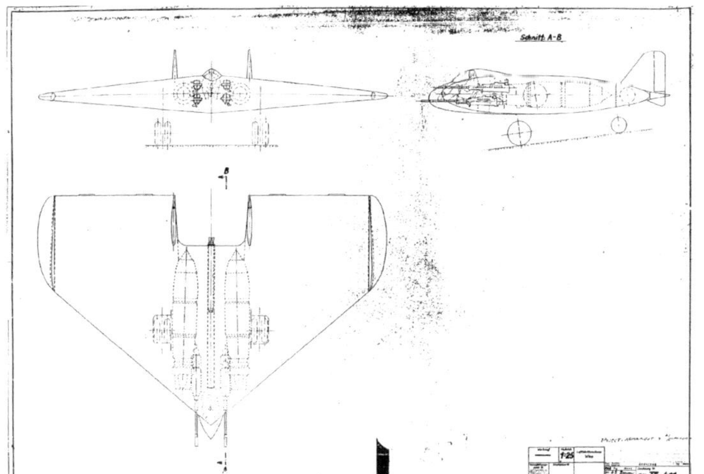

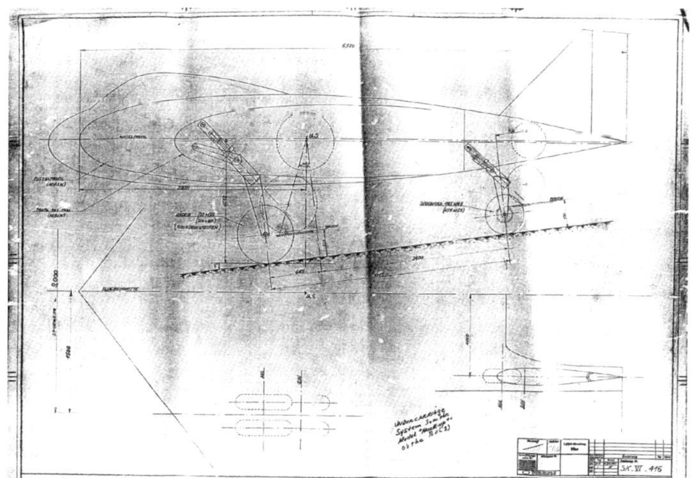

ABOVE LEFT: This undated drawing is similar to the drawing of September 8, 1944, but the cockpit canopy has become even smaller and narrower. The tailfins have become thicker too. Armament is five MK 108s in the nose. ABOVE RIGHT: This drawing shows how the production model Delta VI's tailwheels would retract into a space beneath the leading edge of the tainfins.

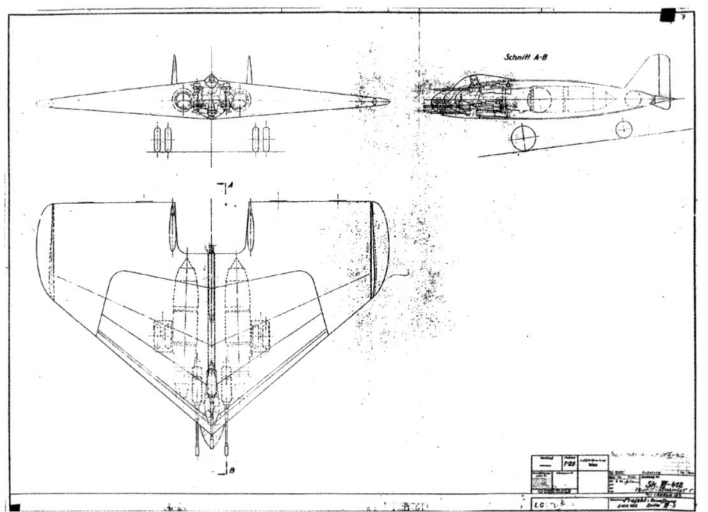

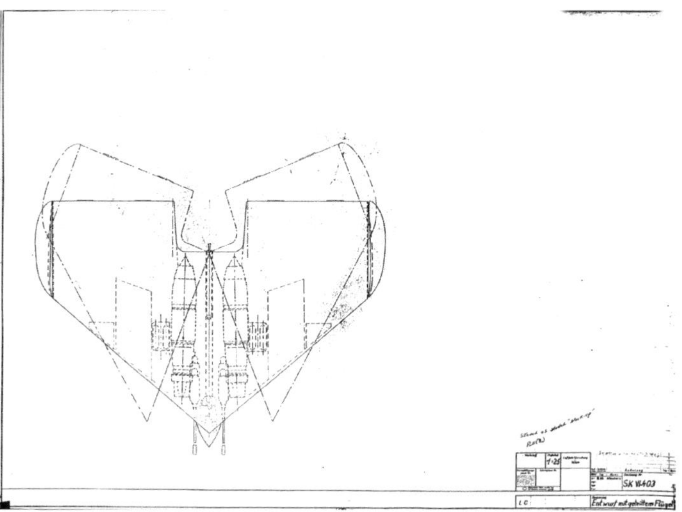

ABOVE LEFT: A third concept drawing for the production model Delta VI, dated October 8, 1944, this time with five MK 103s. ABOVE RIGHT: Split wing concept for the Delta VI – allowing easier access to the aircraft's interior. The drawing is dated October 8, 1944.

Now work on the Delta VI slowed to a crawl as Lippisch became increasingly focused on new projects – the ramjet-powered P 12 and P 13 designs. Yet work did not stop completely. At one point Henschel appears to have become involved, perhaps as a construction partner to help with building the Delta VI. Minutes from a meeting on November 21–22, thought to be a gathering of the EHK, state "the Lippisch P 11, a parallel development with the Ho 229, was to be developed in collaboration with Henschel" and a report from Henschel chief designer Friedrich Nicolaus suggests that this did happen, although to what extent is currently unknown.

Even as late as December 2, 1944, Junkers' special engines division OMW-Kobü Sondertriebwerke was struggling to get a pair of Jumo 004s to fit inside the Delta VI V2 – the powered version that was to follow the V1 glider.

When the Russians reached Vienna at the beginning of April 1945, Lippisch found it necessary to cease work on the Delta VI. The unfinished V1 was abandoned on the edge of a motorway.

ARADO 'DREIECK'

There seem to have been numerous instances of the design team at government-owned Arado being used as a stalking horse for industry. When a requirement was issued to the private companies, such as Heinkel, Messerschmitt and Focke-Wulf, Arado would be quietly asked to see what it could come up with to meet the same requirement.

Sometimes the resulting design made Arado a competitor for those companies, but on other occasions it simply served to represent what might reasonably be achieved. Projects such as the E 470 transport/bomber, E 580 single-jet fighter and E 395 jet bomber seem to have resulted from this process.

In June 1943, Arado project engineer Hans Walland had begun to draw up designs for a twin-jet flying-wing aircraft of a comparable size to Lippisch's P 11. Like the P 11 it had a single central tailfin – but unlike the P 11 its planform was that of an isosceles triangle – a pure delta wing.

It had a wingspan of 20m and was to be powered by a pair of Heinkel HeS 011 turbojets to the rear of the airframe with

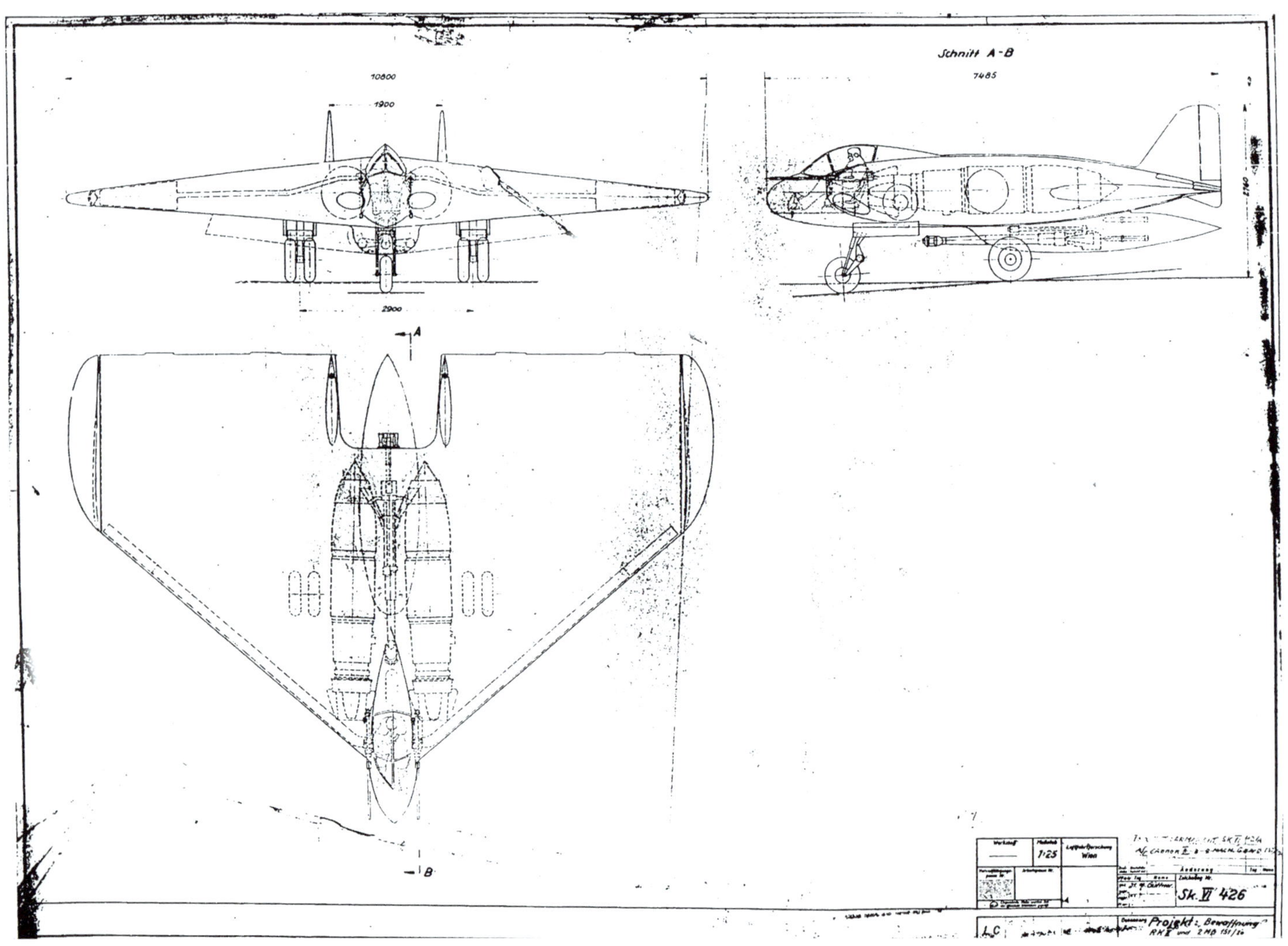

ABOVE: This is the last known drawing of the Delta VI, dated October 31, 1944. It shows the basic airframe of the V2 but fitted with an enormous RK II cannon in an underslung pod and two MG 151/20s in the nose. Presumably by this point Lippisch had abandoned his production model concept with its elaborate tailwheel undercarriage and settled on arming the V2 design as it stood.

its fuel tanks positioned in the middle around a clear space at the centre of gravity. The tanks appear to be positioned directly in front of the turbojets themselves, so whether air from the engine intakes would go above or below them is unclear. Presumably the clear space represents the aircraft's 1,000kg bomb bay.

Walland's triangle sat on a tricycle undercarriage and while the drawing does not show a cockpit, it is reasonable to assume that a cross placed towards the leading edge of the wing in the centre is where it was intended to go.

By September 1943, Arado had already been engaged for several months in designing its series of large flying-wing jet bombers – E 555. The shape of these designs was what the company referred to as 'knickpfeilflügel' or 'creased arrow wing' and was far removed from Walland's regular triangle form. So what could have prompted a design which deliberately stepped away from the curves and refinements of other Arado projects?

During the Second World War, German scientists and engineers tested a wide range of different aerodynamic shapes to determine which might have advantageous characteristics for future aircraft designs.

Some of these shapes must have seemed outlandish even at the time but they were tested anyway to gather baseline data. On September 27, 1943, the DVL published a report entitled 'Prüfbericht über 3- und 6-Komponentenmessungen an der Zuspitzungsreihe von Flügeln kleiner Stockung (Teilbericht: Dreieckflügel)' or 'Test report on 3- and 6-component measurements on a series of tapered wings of small aspect ratio (partial report: triangular wing)' by Lange and Wacke, Untersuchungen und Mitteilungen ('Investigations and Communications' or UM) Nr. 1023/5. This effectively assessed the aerodynamics of pure delta wings in flight and it is not inconceivable that Arado, which received the same aerodynamics reports as every other aircraft manufacturer in Germany via the ZWB (Zentrale für wissenschaftliches Berichtswesen or Central Office for Scientific Reporting), saw and acted upon it.

Evidently Arado designers' notes were filed in date blocks from, say, April 1943 to September 1943, and the only known evidence of Walland's triangle or 'dreieck' project today comes in the form of a few basic sketches and pages that were captured by the Allies in 1945 jumbled up among pages from other reports and projects from that date block.

It seems unlikely that a triangular aircraft could have been made to work given the limited technology available in Second World War Germany, even with a tailfin to add directional stability, but nevertheless it indicates that the '1,000 x 1,000 x 1,000' concept was more significant than has perhaps previously been acknowledged.

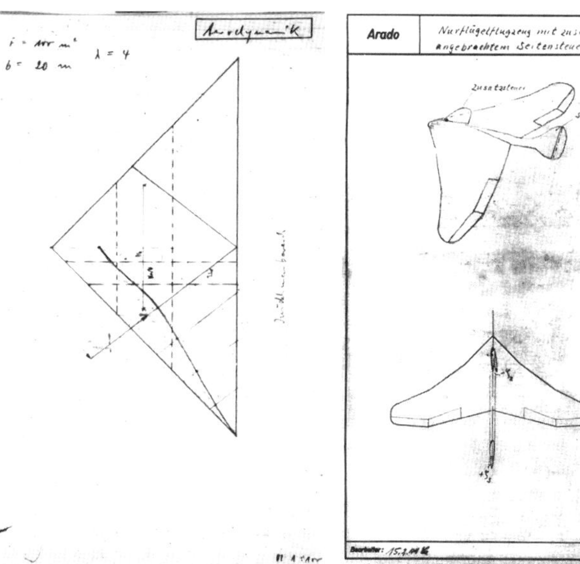

ABOVE LEFT: Sketch by Hans Walland showing the triangular shape of his design. **ABOVE RIGHT:** The state of Arado's flying-wing design by February 1944, from a patent application. The company evolved the wing shape but struggled to find a form of rudder that would provide adequate stability.

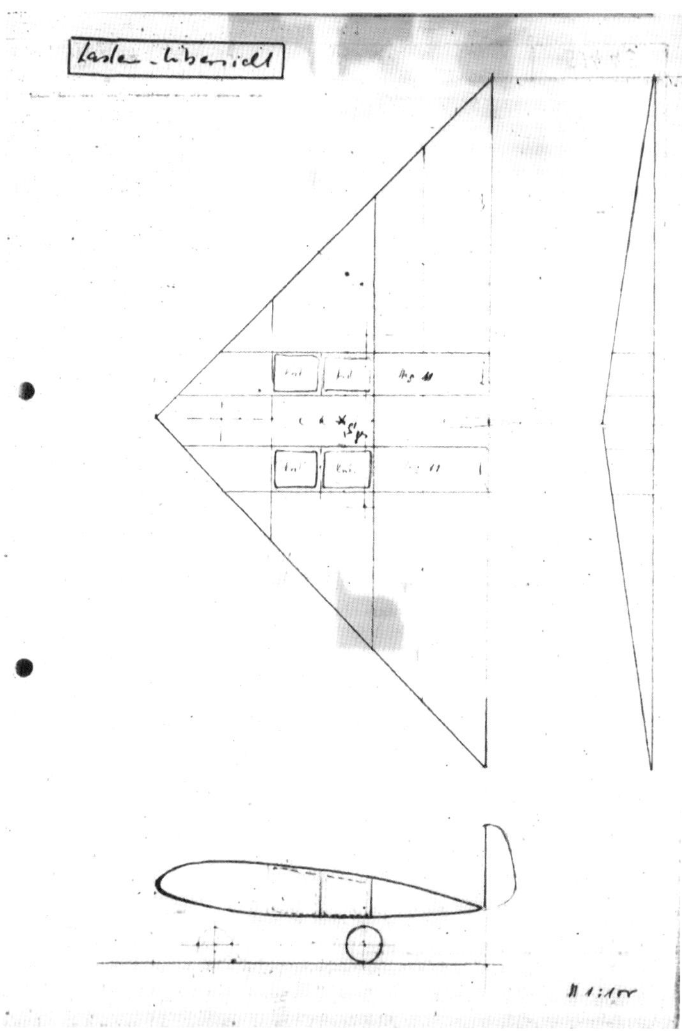

ABOVE: Sketch of Arado's triangular twin-jet aircraft, drafted by Hans Walland in September 1943.

Chapter 7

Messerschmitt Me 262 later developments

P 1099 and the HG series

Further development of the Me 262 beyond its A-1 and A-2 form was instigated at the beginning of 1944 and resulted in new heavy fighter, bomber and experimental high-speed research versions.

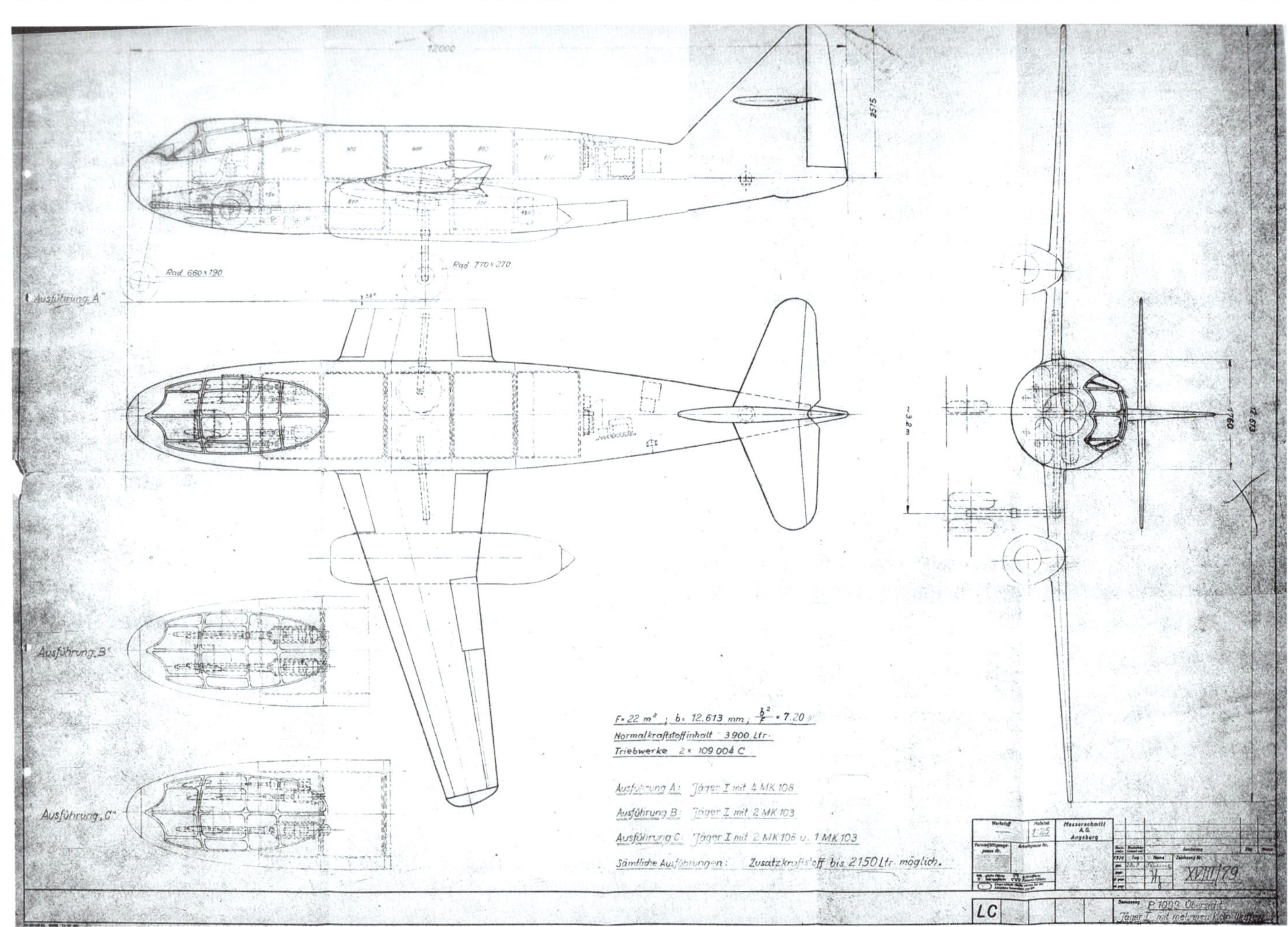

ABOVE: The Messerschmitt P 1099 Jäger I of drawing number XVIII/79, dated March 22, 1944. Three different armament options are illustrated. The drawing, which has the same date as the report it appears in, seems to have been made a month later than other drawings in the XVIII sequence with higher numbers.

P 1099 HEAVY FIGHTER

During a meeting with Oberstleutnant Siegfried Knemeyer, head of development (GL/C-E) in the RLM's Technical Office, on January 4, 1944, Willy Messerschmitt was told that his company's "most important task is to push ahead and develop the jet fighter Me 262 by all means".[1]

Knemeyer went on to say that "an attempt should be made to achieve and maintain a high technical superiority and a time advantage over the opponent under all circumstances with this pattern [Me 262] and its derivatives".

This order of January 4 was evidently exactly what Messerschmitt wanted to hear and the following day the company commenced three new strands of development for the twin-jet fighter.

According to a report of March 22, 1944, entitled 'Baubeschreibungen der Weiterentwicklungen der Me 262 (mit neuem vergrössertem Rumpf) P 1099 (mehrsitzige schwere Jäger) P 1100 (mehrsitzige Bomber)'[2] or 'Construction description of the further developments of the Me 262 (with new enlarged fuselage) P 1099 (multi-seat heavy fighter) P 1100 (multi-seat bomber)', following Knemeyer's order: "On the one hand a great deal of work was done to improve the current basic form of the Me 262, the performance, the flight characteristics, the armament etc., while on the other hand studies were carried out for rapid developments to achieve the hoped for superiority over the opponent by developing new modifications of the Me 262. The extensive thoughts that led to this development are listed in the variants of the two building specifications."

Both the P 1099 heavy fighter and P 1100 bomber were based on the same platform – an Me 262 with a new fuselage. According to the report: "During the investigation the greatest emphasis was placed on confining all modifications to the Me 262 into a single new fuselage assembly, while the other parts of the aircraft are the same as or only slightly different to those used for the series version of the Me 262.

"In this way, starting series production to tight deadlines is made possible even under today's difficult conditions (especially in the equipment sector) by exploiting the series production of many assemblies already under way. On the other hand, care was taken from the outset that this use of Me 262 parts should not lead to a significant reduction in the development potential of the new aircraft types. For this reason, these derived patterns have already been and are being examined for a later development stage with increased weights, enlarged equipment, etc.

"The P 1100 jet bomber is more detailed in the appended building descriptions, as it transpired in the course of investigations that with the unequalled Me 262 wing the creation of a bomber (without and with middle defensive armament) can be guaranteed without further ado. The wing is too small for the heavy fighter P 1099 for extremely strong armament. The employment with normal fighter weaponry is possible with the small wing area.

"The investigations on the wing size for the bomber with strong defensive armament and heavy fighter or heavily armoured destroyer are ongoing. For the fuselage, which is the same for both projects, beyond the present specification there are fairly complete project documents covering all essential points of the design, as well as a first mock-up."

The detailed description[3] of the P 1099 states that "the development of destroyers was prompted by the following ideas: 1) In the conversations with the RLM on the development of the Me 209, the tendency to considerably emphasise the importance of the heavy fighter over the single-seat fighter was quite clear. This direction is based mainly on the fact that the two-seater heavy fighter is much better suited for the bad weather use (or night use) than the single-seater and that on the other hand, the size and the layout of the two-engine nacelle aircraft offers a very great freedom of movement for armament and equipment.

"2) Performance research has shown that the Me 262 is unlikely to achieve the performance superiority in jet fighter against jet fighter and jet bomber combat needed to effectively carry out its attacks ... The reason for this lies in the fact that at the moment the strong drag increase at Mach numbers of 0.7 to 0.8 limits all jet aircraft to the same top speed and may set it. As a technically possible way out of this difficulty, we have the opportunity to use fighters to fight jet bombers. Flying at about the same speed they can sit in front of the bomber group and fight the enemy with a movable weapon that fires backwards."

The report states that this idea of attacking bombers by flying along steadily in front of them was based on the difference in range and projectile travel time between the fighter's weapons firing backwards and the bombers' weapons firing forward against the flow of air. The fighters' projectiles, facing less air resistance, would have greater range and speed than those of the bombers.

"The designs for the fighter Me 262 fuselage were basically carried out so that the same hull with only moderate changes can also be used for a two-seater jet bomber with full-view cockpit. This highlighted the fact that the jet bomber technically represents the more favourable task for the given wing and tail sizes, since the landing speed of the heavy fighter is quite noticeably higher than that of the bomber after it has dropped its load because of the high fixed weapon weights. On the basis of the same landing speeds, the weakly armed or unarmed bomber therefore has a higher maximum speed (disregarding the Mach limit) than the fighter achieves.

"The design work was carried out in two different forms. Both the heavy fighter with a rigid attack armament and the fighter with a movable, rearward attack armament were examined. Both have been designed to have essentially the same fuselage except for weaponry. It is intended to incorporate the HeS 11 engine in a later development stage with significantly strengthened armament and to adapt the wing area for heaviest armament and increased flight weights by inserting an additional section (as with the Me 109 H). The fighter or heavy fighter with defensive armament to the rear is possible for reasons of weight only in the further development stage."

All together seven different designs were presented – three variants of the P 1099 fighter and four of the P 1100 bomber – although at least five P 1099 and seven P 1100 variants had been drawn up. The drawings they appear on are in the XVIII sequence, a continuation from the P 1092 designs of the previous year.

For the P 1099, 'Lösung A' or 'Solution A' Jäger I was a two-seater with either two or four fixed forward-firing 3cm weapons. It was shown in drawing number XVIII/79. Solution A Jäger II of drawing XVIII/87 was also a two-seater but with heavier weapons. P 1099 Solution B of drawing XVIII/85 was a three-seater with a nose turret, dorsal turret and tail barbettes.

The Solution A Jäger I design was 12m long with a wingspan of 12.613m and wing area of 22m². It had standard Me 262 wings and an Me 262 tailfin. It had a tricycle undercarriage with the nosewheel retracting straight back into the underside of the cockpit and the mainwheels retracting into the fuselage mid-section. It was to have two Jumo 004 C engines but "the final solution is the later HeS 11 engine" and its five fuel tanks could hold 3,900 litres. A pressure cabin would be introduced when the HeS 11 engines became available.

The three armament options offered were either four MK 108s, two MK 103s or one MK 103 and two MK 108s. A large empty space in the fuselage below the fuel tanks offered the option of carrying extra fuel or even a bomb load. Take-off weight ranged from 8,762kg to 10,262kg depending on loadout.

Solution A Jäger II had the same dimensions and engines but armament was either one MK 108 and one MK 112 or a single 5.5cm MK 214.

Solution B was a sketchier design without given dimensions. According to the report: "This task is not finished yet. It is only enclosed with a general outline drawing, without any liability, showing how with this project this task could be solved."

The brief text outline of Solution B describes it as a 'Schwerer Jäger mit halbstarrer Angriffsbewaffnung nach hinten' or 'Heavy fighter with semi-rigid weapons to the rear'. It states: "The attack armament consists of 2 x MK 103 in semi-rigid frame with limited movement (5° cone angle). The main firing direction of this aircraft

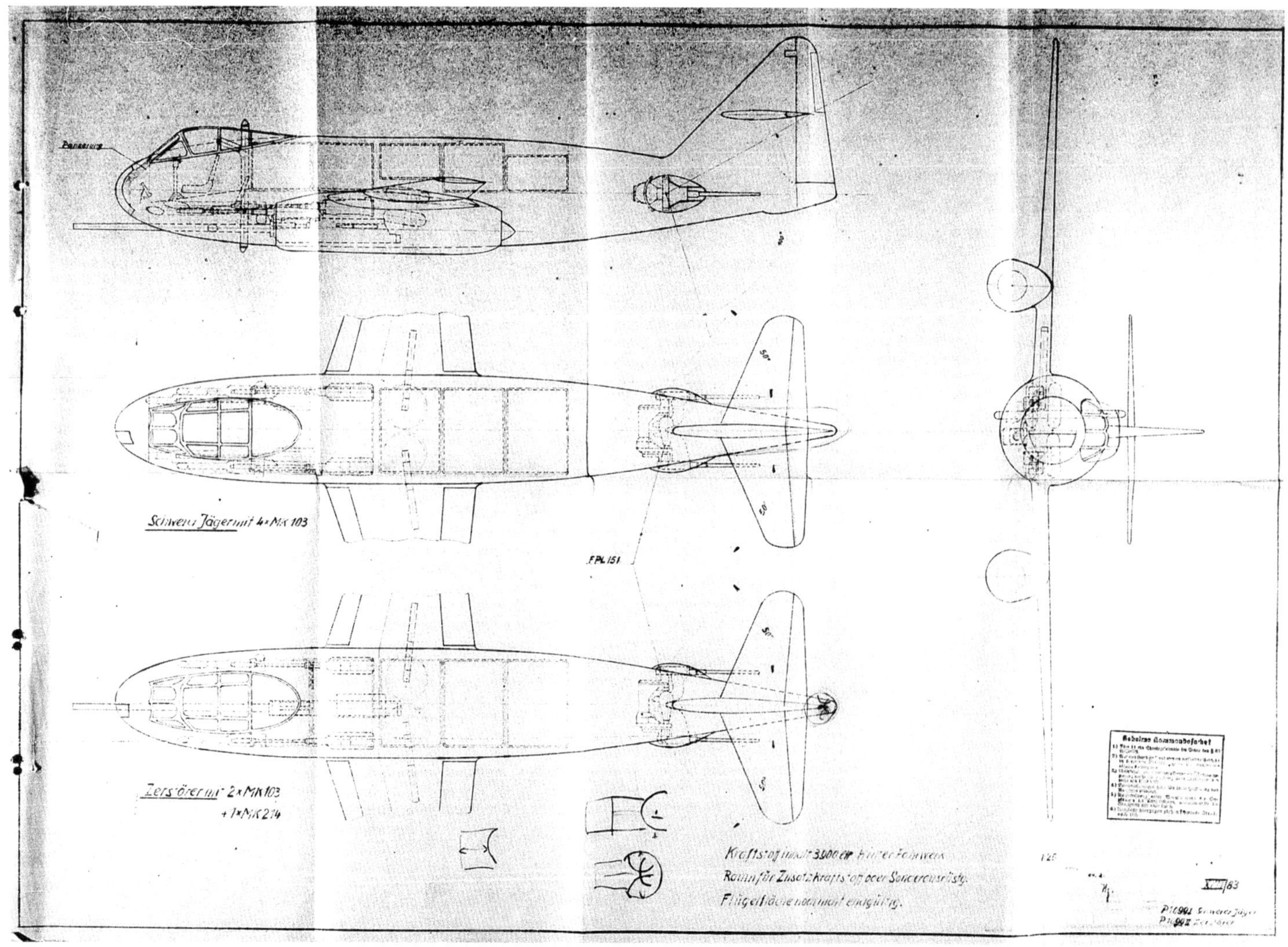

ABOVE: Drawing XVIII/83 of February 22, 1944 shows heavy fighter and destroyer versions of the P 1099. The heavy fighter is armed with four MK 103 cannon while the destroyer has two MK 103s and a single long-barrelled Mauser MK 214 50mm auto-cannon.

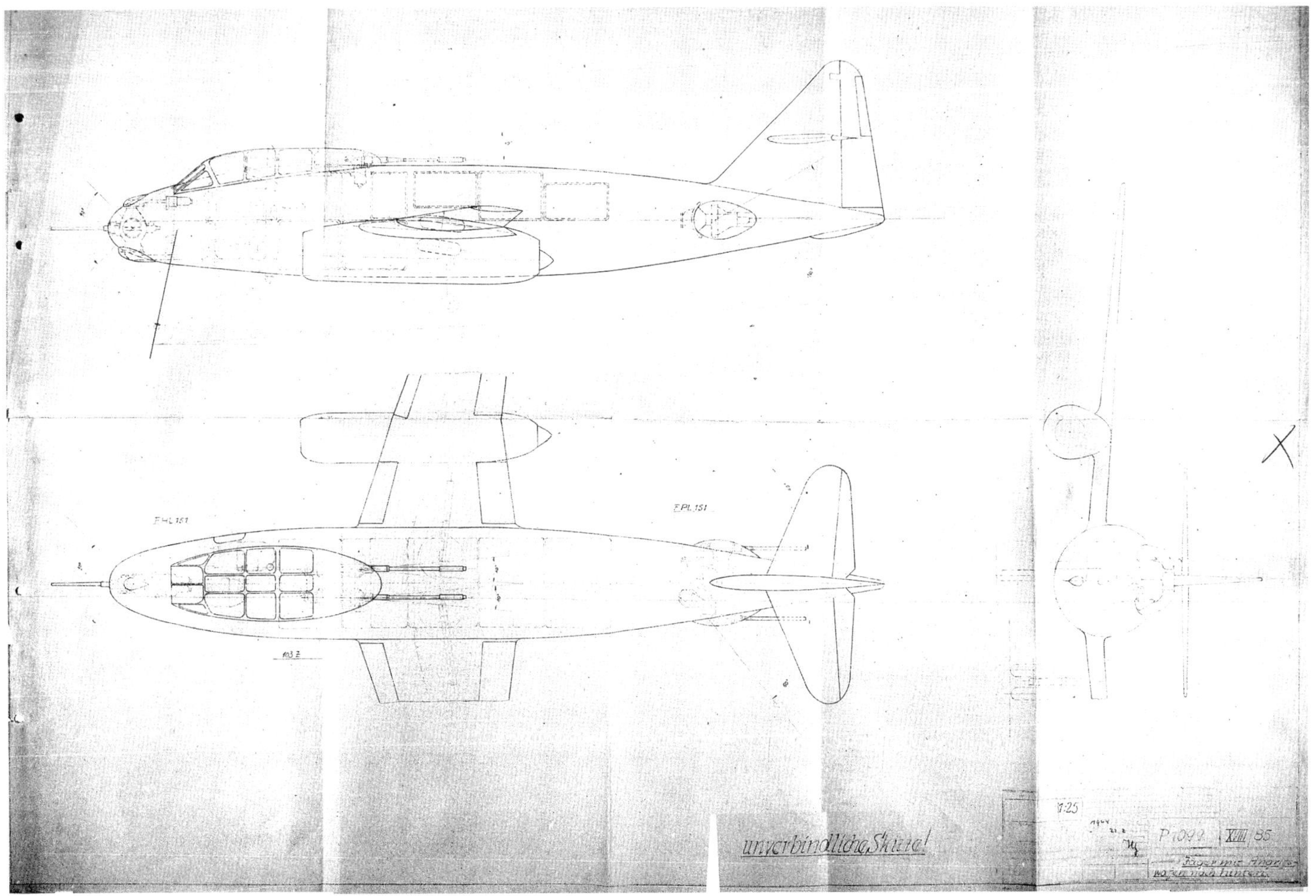

ABOVE: The heavily-armed P 1099 fighter shown in drawing number XVIII/85. It has an FHL 151 turret on its nose, FPL 151 in its tail and twin MK 103s firing over its back. The aircraft was designed to fly ahead of its targets, shooting backwards at them using its MK 103s – the idea being that the enemy's defensive weaponry would lack the range, firing against the flow of air at high speed, to hit the P 1099 ahead.

is to the rear. Attempts would be made to turn the plane around 180° during the fight in order to do justice to the attacking tactics of the rear and the defence.

"In order to keep away attacking fighters and to perform the attack task undisturbed, a movable armament facing forward (FHL 151) and to the rear (FPL 151) is provided. The pilot uses a panoramic telescope to bring the aircraft into shooting range for the gunner (3rd man) on the semi-rigid MK 103 position. The defensive weapons are operated by the 2nd man next to the aircraft pilot. The visor is a periscope.

"The front fuel tanks, replaced by the 3rd man and the semi-rigid weaponry, are repositioned to the lower front fuselage. The possibly remains to use the front space as well as the entire space behind the undercarriage in the lower fuselage for the carrying of bombs."

An early draft of the report shows that Solution A Jäger I was originally to have been a fighter/heavy fighter shown in drawing XVIII/83 while Solution A Jäger II was a night fighter shown in drawing XVIII/84. Neither of these designs made the final cut but Solution B, the XVIII/85 design, remained the same.

A description survives of these designs, with Jäger I (drawing XVIII/83) being a heavy fighter or destroyer for daytime operations and Jäger II (drawing XVIII/84) being a night fighter.

This states that the aircraft were the same as the P 1100 Schnellbomber with only minor changes. As a fighter, the Jäger I would be armed with four MK 103s and as a fighter it would have two MK 103s and an MK 214. "For both cases, the MK 103 installation is the same. The attack weapon installation is located in the lower front fuselage (front bomb bay in the fast bomber).

"Droppable weapons. Between the undercarriage bay and tail in the lower fuselage is a free space (rear bomb bay in the fast bomber) 3m in length, which is to accommodate droppable weapons up to about 1,000kg or can be used for disposable additional fuel tanks or special equipment."

The Solution A night fighter of drawing XVIII/84 "does not need a long-range weapon and the landing weight should be kept as low as possible with regard to night landings. The armament consists of 6 x Mk 108, whereby 2 weapons can possibly be installed upwards for oblique firing'. The only mention of radar was equipment of "additional devices for night fighting" and flight endurance would be increased with the addition of five more fuel tanks in the bomb bay areas. Total fuel capacity would be 5,500 litres.

The P 1099 and P 1100 designs appear to have gone no further after the report of March 22, 1944, but by then Messerschmitt's projects office was already progressing with the second and third strands of development arising from the January 4 meeting.

ME 262 HG SERIES

In keeping with Knemeyer's wishes that Messerschmitt retain technical superiority, at a meeting held on January 5, 1944, Willy Messerschmitt proposed that alongside the P 1099 fighter and P 1100 bomber developments of the Me 262 his company should design and develop a new twin-jet aircraft to explore the potential for flight near the speed of sound. The designer of the Me 328, Rudolf Seitz, was appointed to oversee the project on January 31, 1944.[4]

Theoretical work was carried out during February comparing the potential performance of the 'Höchstgeschwindigkeiten Me 262' when fitted with Jumo 004 C or HeS 011 engines and models were designed for a programme by aerodynamics tests by the DFS. Though based on the Me 262, the test vehicle was to be an almost completely new aircraft. On February 17, 1944, a drawing was produced showing an Me 262 fuselage with completely new short, sharply swept wings of very thick chord – effectively a notched delta wing. The engines remained in their original positions.

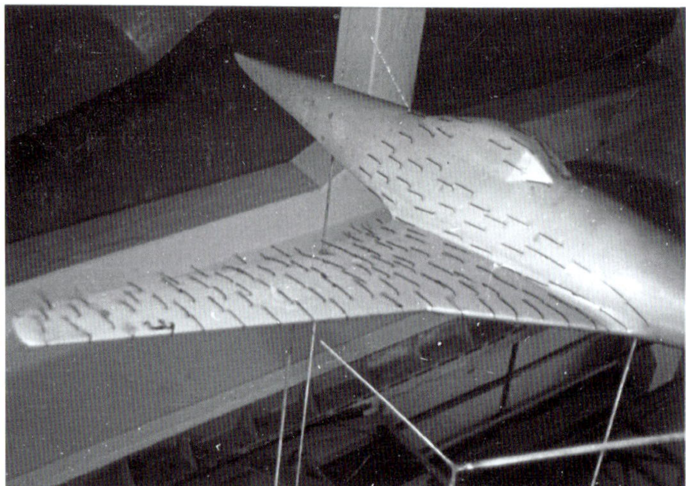

ABOVE: Wind tunnel model of an Me 262 fitted with experimental wing root extensions.

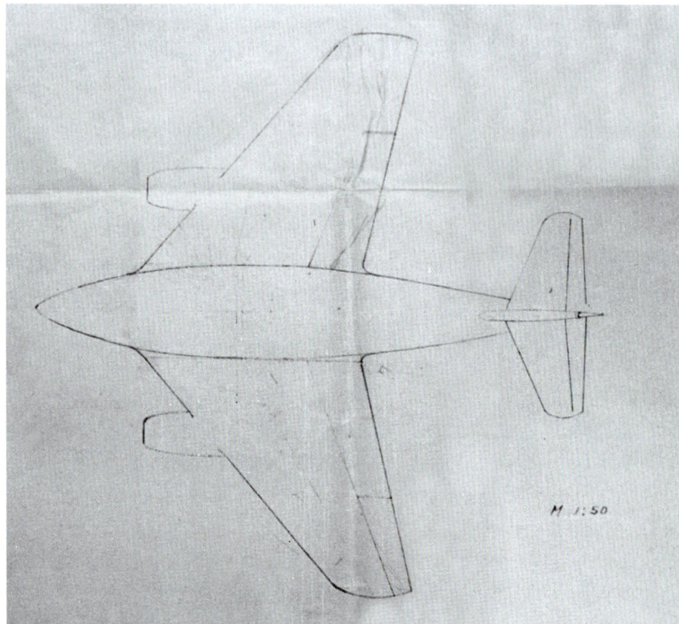

ABOVE: This detail from a drawing dated February 17, 1944, shows an Me 262 fuselage and tail with a notched delta wing. The engines appear to be installed in exactly the same position, relative to the fuselage, as those of the standard Me 262. The overall drawing is much larger but only shows a larger view of the same wing arrangement.

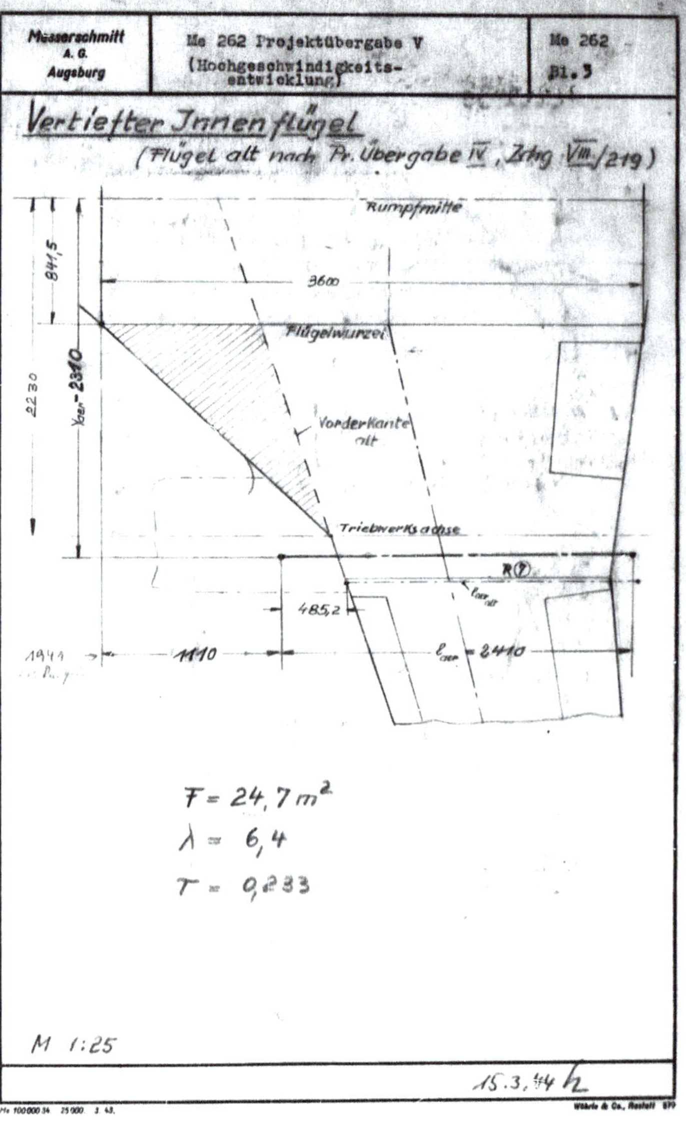

ABOVE: Drawing showing the modified wing roots intended for the Me 262 HG I.

A meeting was held on February 25 and Seitz proposed developing the existing Me 262 rather than beginning with the test vehicle originally proposed. Willy Messerschmitt approved this idea and also approved static tests on an Me 262 airframe. Seitz's project diary then offers a blow-by-blow account of the progress made.

The company's construction office for tails – Kobü-Leitwerk – started preliminary work on a swept tailplane on March 10. Four days later, Messerschmitt was presented with a schedule for progress on the project. Step one would involved wider chord inner wings, swept tail and streamlined 'rennkabine' cockpit canopy. Step two involved 35° swept wings, improved engine nacelles and reprofiled fuselage. Step three was new wings with built-in engines. It was decided later that step three's wings would be swept by 45°.

Only the first of these steps was reasonably well defined by this point – the other two steps existed primarily as concepts. It is unclear whether the three steps became known as the Me 262 HG I, II and III immediately or whether these designations were applied later.

Kobü-Leitwerk and the construction office for wings, Kobü-Tragwerk, received project documents for the swept tail and wider chord wings respectively on March 17. Head office for wings, Stabü-Tragwerk, began working on the calculations for the aircraft's inner wings the following day.

Kobü-Rumpf, the fuselage office, began work on the 'rennkabine' on March 20, based on sketches provided by the design department. The next eight days were spent preparing project documents for the modified fuselage and inner wings.

During this time, a number of sketches were made showing possible layouts for the wings of the second and third stages of development. Two pages of designs from March 25, 1944, show variations on the earlier thick wing but this time also include some designs for wings with built-in engines. Two days later, Willy Messerschmitt himself drew a sketch showing how the Me 262's engines could be repositioned to the trailing edge of the wing.[5]

On March 31, Kobü-Rumpf reported that it would be possible to lower the height of the Me 262's cockpit structure by 150mm, allowing a corresponding lowering of

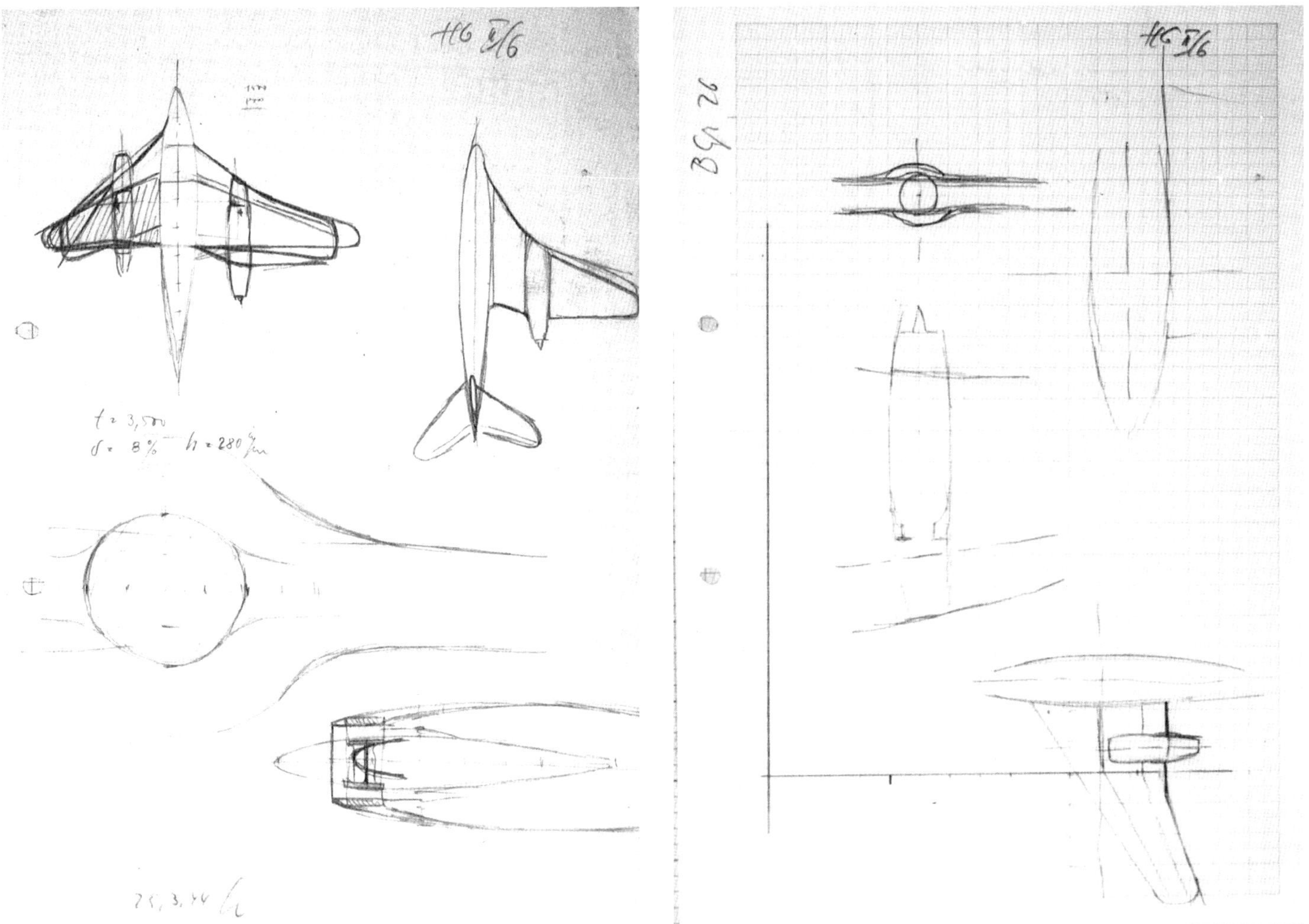

ABOVE: Two pages of sketches from March 25, 1944, showing potential options for delta or other wing forms on future developments of the Me 262 HG series. The note at the top right of each page 'HG II/6' appears to have been added later in a different hand.

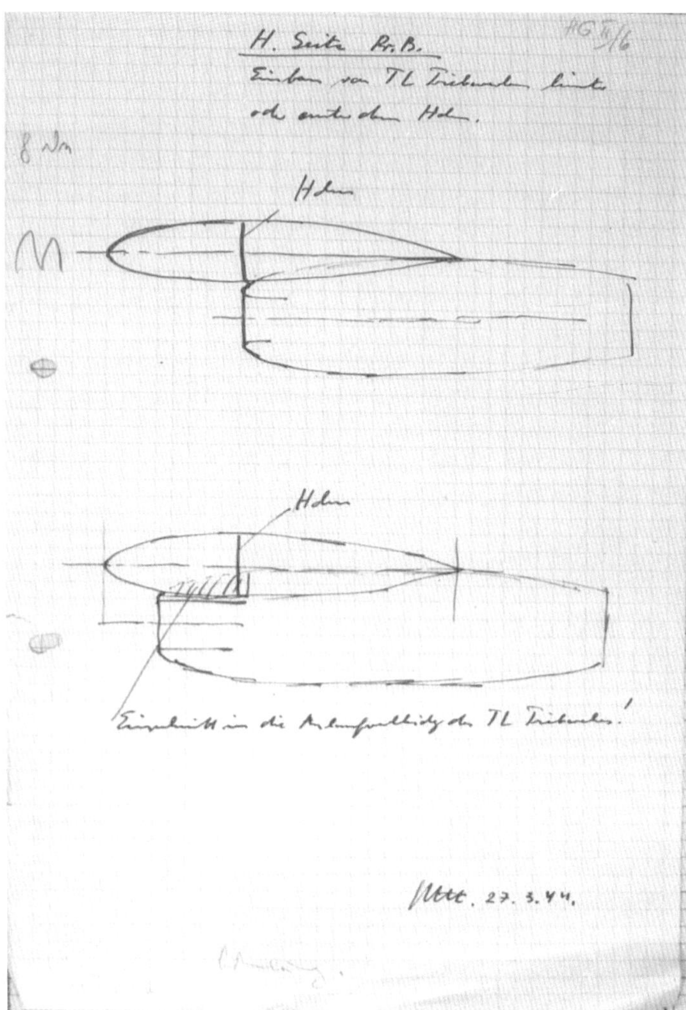

ABOVE: Sketch of March 27, 1944, signed by Willy Messerschmitt and addressed to Herr Seitz, which shows how the HG series could have its engines set back beneath the trailing edge of its wings.

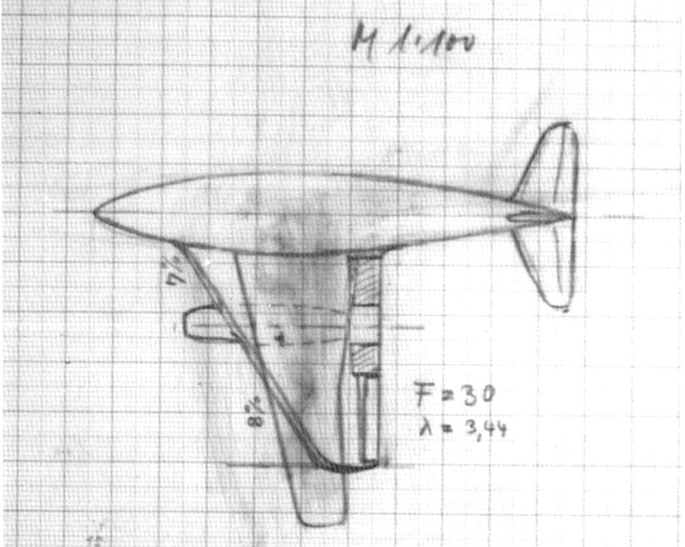

ABOVE: This sketch, apparently produced during the first few days of April 1944, shows both a delta-wing and swept-wing form. Beyond this point the delta would be dropped in favour of the slender swept-wing form.

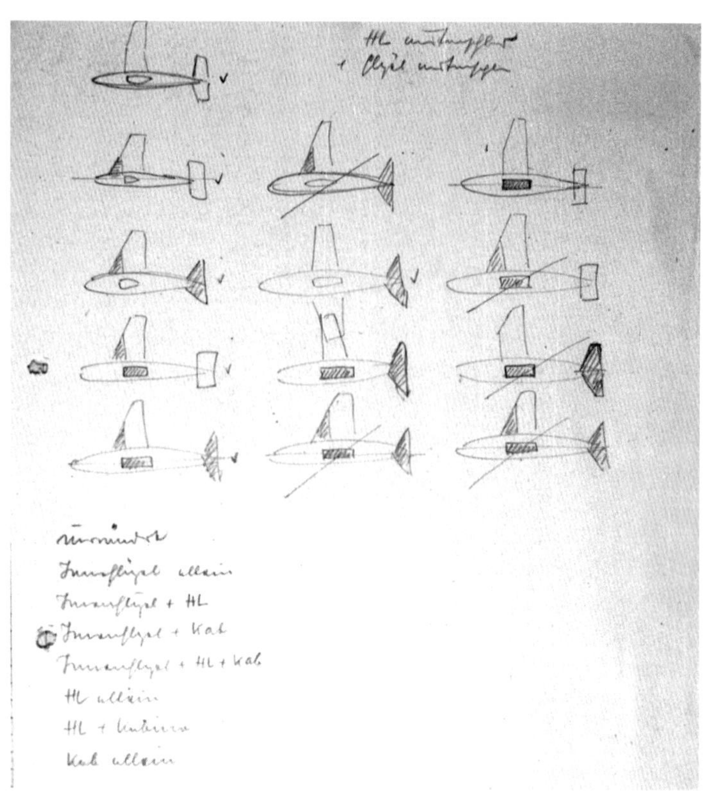

ABOVE AND BELOW: Two pages from the April 3, 1944, draft report showing numerous different potential wing, engine and tail arrangements for future HG series models.

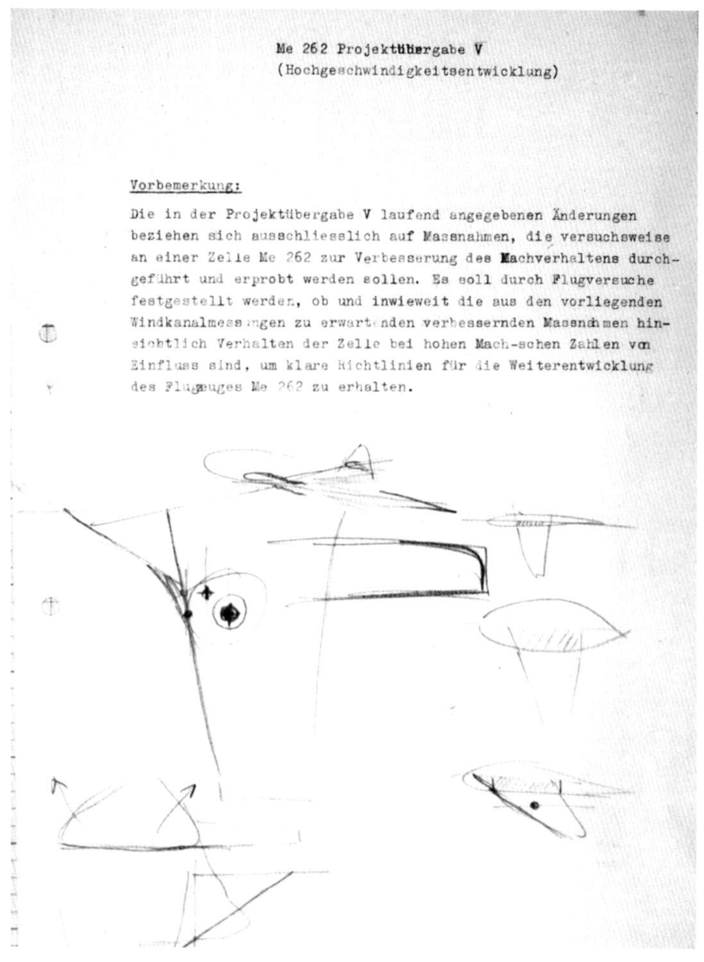

Me 262 Projektübergabe V
(Hochgeschwindigkeitsentwicklung)

Vorbemerkung:

Die in der Projektübergabe V laufend angegebenen Änderungen beziehen sich ausschliesslich auf Massnahmen, die versuchsweise an einer Zelle Me 262 zur Verbesserung des Machverhaltens durchgeführt und erprobt werden sollen. Es soll durch Flugversuche festgestellt werden, ob und inwieweit die aus den vorliegenden Windkanalmess ngen zu erwart nden verbessernden Massnahmen hinsichtlich Verhalten der Zelle bei hohen Mach-schen Zahlen von Einfluss sind, um klare Richtlinien für die Weiterentwicklung des Flugzeuges Me 262 zu erhalten.

the canopy.[6] Work on the inner wings was halted on April 1 and the staff responsible were switched to carrying out urgent work on the Me 109 K instead, as were the staff of Kobü-Leitwerk.

April 3, 1944, saw a report being written on the preparations already under way entitled 'Me 262 Projektübergabe V (Hochgeschwindigkeitsentwicklung)' or 'Me 262 Project Delivery V (High-speed development)'[7] – even though only two men were now working on the '262 HG' including Seitz. This began: "The changes currently mentioned in Project Delivery V relate exclusively to measures that are to be carried out and tested on an Me 262 airframe to improve the performance.

"It is to be determined through flight tests whether and to what extent the improvement measures to be expected from the present wind tunnel measurements are influential with respect to behaviour of the airframe at high Mach numbers in order to obtain clear guidelines for the further development of the aircraft Me 262."

There were three changes to be made: "The inner wing is deepened by attaching a new nosepiece while retaining the wing piece behind the main spar", the tailplane was to be swept back by 40° and the tailfin deepened, and "in order to improve the Mach behaviour of the cabin, its structure is to be reduced to the lowest possible level". There were three stipulations for the latter – the pilot had to be able to open the canopy while in flight, the cockpit seat had to be height adjustable so that the pilot could raise himself up to the position he had in the unmodified Me 262, and a hinged windscreen had to be positioned in front of the pilot.

It was expected that these three changes, which for safety's sake would be made without worrying about adding extra weight, would "shift the critical Mach number of the entire airframe to about 0.83. Furthermore, all three changes would be made at once "and then the improvements are gradually reversed" so it would be possible to work out what worked and what didn't by a process of elimination.

Appended to this report were several pages of sketches showing further alternative layouts for future Me 262 HG wing configurations. One page shows 13 different arrangements of different wings, tails and cockpits – among them a number of them with near-triangular delta wings.

Another undated drawing, apparently produced around this time, shows an Me 262 fuselage and wing arrangement similar to that shown in the February 17 drawing but this time with a sharply swept leading edge and a straight trailing edge – the delta form – but overlaid with a more complex swept-wing form.

Once again, on April 5 the 262 HG group was reduced to just two men – Seitz and his colleague Brutscher.[8] April 6 saw another series of drawings produced showing complex swept-wing designs. The delta wing now disappears from the HG project records and does not seem to have been reconsidered.

Seitz went to Messerschmitt personally on April 7 and told him that the Kobü was dragging its heels. Messerschmitt replied that he wanted more planning work done on the project.[9] Six days later BMW was informed that Messerschmitt was planning to fit jet engines "centrally in the wing behind the spar" of an Me 262 and the company pledged to carry out static testing to see how airflow to the engines would be affected by this arrangement.

For April 15, Seitz noted in his diary: "Leinsinger refuses to schedule the construction work as long as there is no order for it at present (fears influence of the Jägerstabes)".

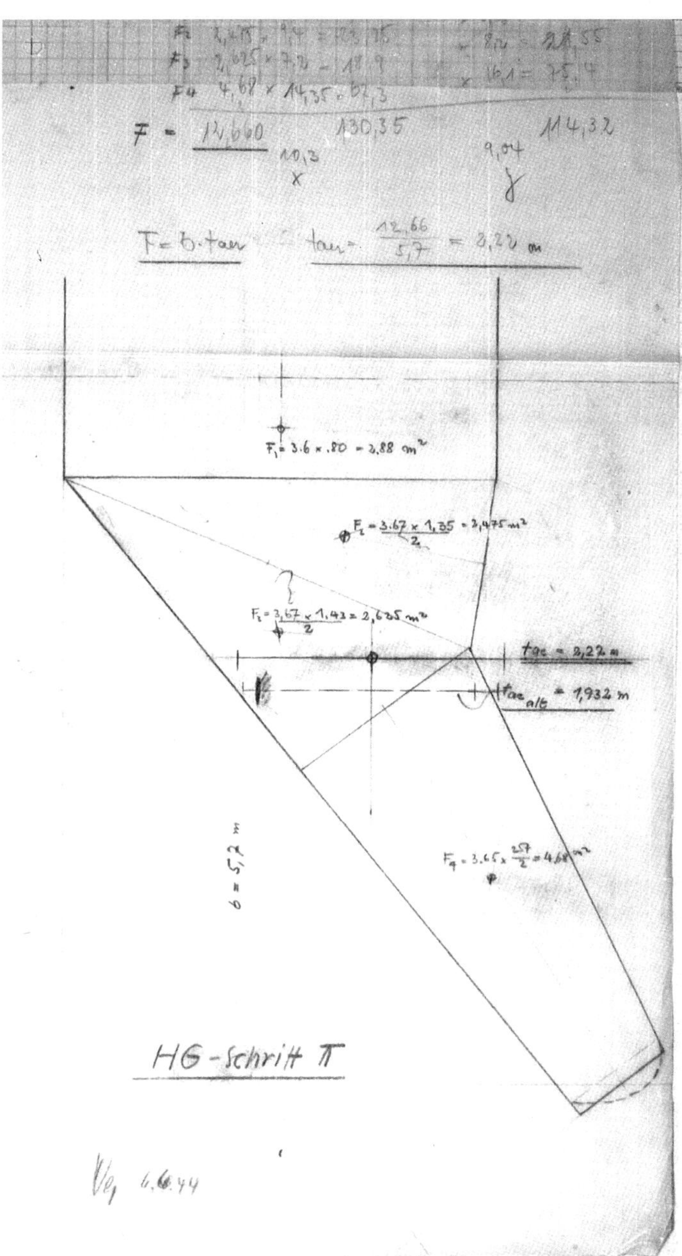

ABOVE: Numerous different swept-wing forms appear in the Me 262 HG files after April 6, 1944, indicating that although the company had not yet firmly decided on the wing shape of the HG II and III, the delta was no longer under consideration.

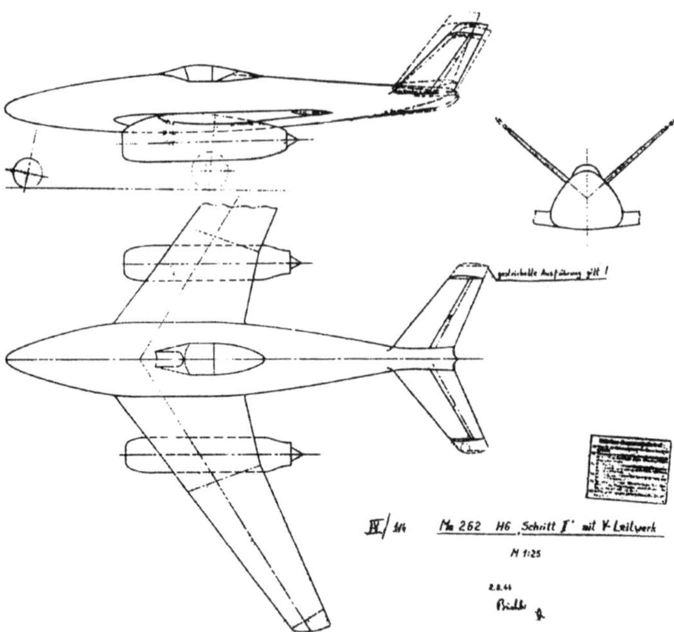

ABOVE: Messerschmitt drawing IV/114 of August 8, 1944 shows the HG II with swept wings, a V-tail and lowered cockpit.

However, the following day, during an official visit to Messerschmitt's Oberammergau facility, "Knemeyer announces high-speed development Me 262. Mtt AG must maintain its lead in the high-speed area".

With Knemeyer's backing Seitz was able to issue construction orders to Messerschmitt's Kobü and Leinsinger agreed on the following dates "a swept tailplane 4.5.44, deepened inner wings 29.4.44, lowered cockpit 30.5.44".

New drawings were made on April 20, showing the Me 262 fitted with HeS 011 engines fitted behind the spar on a modified wing, and the following day projects office chief Woldemar Voigt ordered that installation of BMW engines behind the spar should also be investigated.[10] On April 22 it transpired that the Kobü had mistakenly thought that the HG I was to have accommodation for a second crewmember and with that cleared up, on April 24, work began on a mock-up showing the lowered cockpit canopy.

Further meetings with Willy Messerschmitt were held throughout 1944 as work on the HG designs progressed. In a meeting on July 24, 1944, Messerschmitt bemoaned the poor dissemination of technical reports from the DVL and other research bodies and "occasion was given by a submitted report containing measurements from 1942 that have only now become known". He also asked "in which way the engine nacelles can be optimally accommodated aerodynamically in connection to the swept wings" and said that forward sweep of the outer wing surfaces should be investigated "but probably will not be investigated at first".

He later suggested that a V-tail should be fitted to the HG II in order to overcome unfavourable aerodynamic interference effects between the swept wing and tailplane. This was certainly being worked on by early September 1944.

ABOVE: Wind tunnel model of the HG III circa January 1945.

ABOVE: AVA wind tunnel model of the ultimate Me 262 development – the HG III. After the delta-wing designs were dropped, efforts were concentrated on sharply swept wings and burying the engines in the wing roots.

The HG project continued throughout the remainder of 1944, with the designs for the HG I, II and III becoming more concrete as time went on. The form of all three had been more or less firmly decided by September. In ADI(K) Report No. 1/1946, dated January 7, 1946, an account is given of Woldemar Voigt's interrogation in which he discusses the Me 262 HG series.[11] The Me 262 HG II "was to be a high-speed version of the Me 262. It had the same fuselage as the Me 262, the same wing (or same main parts of the wing), and the same kind of power plant installation. Wing and horizontal tail surface were arranged to have a large sweep back angle. The prototype was completed, but it was destroyed by some other aircraft crashing into it.

"The HG III was similar to the HG II but with a sweep back of 45°. The power plant was installed in the wing roots aft of the main spar, and faired into wing and fuselage. The air intakes were in the wing root leading edge with ducts to the engine. The calculated speed of this machine was 1,000km/h. A wind tunnel model was completed, but it was lost and destroyed during the occupation of Oberammergau."

The third strand of development arising from the Knemeyer's January 4, 1944 meeting with Messerschmitt was the P 1101 – beginning as a development of the P 1099/P 1100 fuselage with new wings, removing all vestiges of the Me 262, then moving on to entirely new designs with neither Me 262 wings nor the P 1099/P 1100 fuselage.

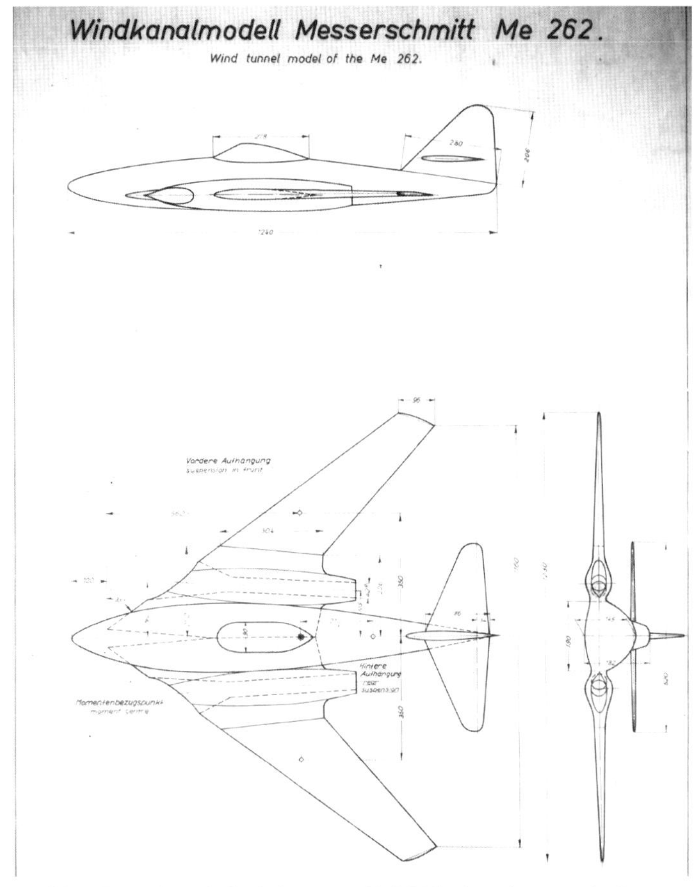

ABOVE: Drawing dated January 5, 1945, showing a wind tunnel model of the HG III.

Chapter 8

1-TL-Jäger part II

Single-jet fighters 1944–1945

A competition to design the Me 262's high-performance single-jet replacement commenced in earnest in July 1944. The contest would continue until the end of the war with no clear winner being decided. It would, however, produce a wealth of astonishing technological innovations.

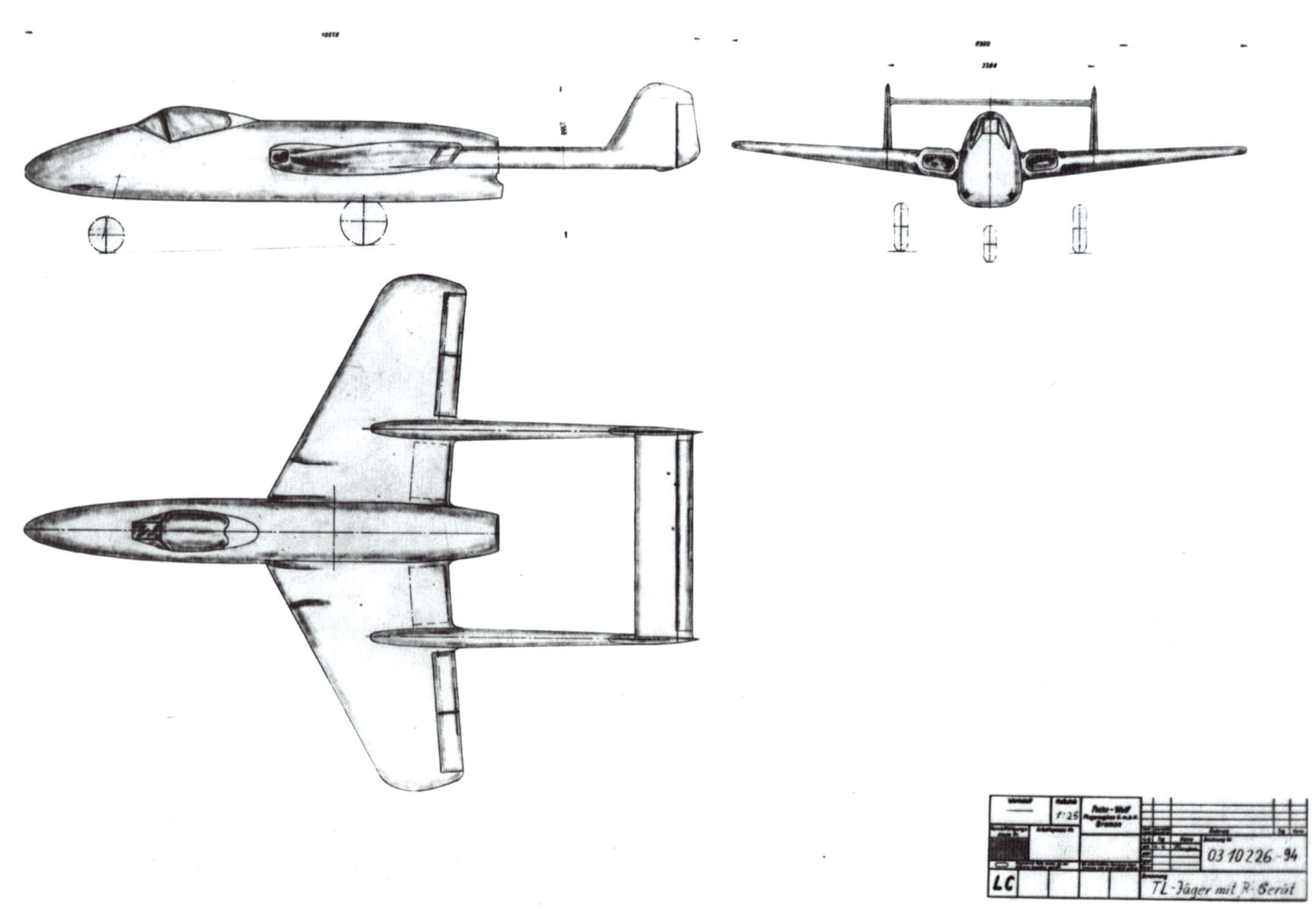

ABOVE: The evolved 'Flitzer' as it appears in drawing number 0310 226-94, included in Focke-Wulf Baubeschreibung Nr. 280 of July 5, 1944.

FOCKE-WULF, JULY TO SEPTEMBER 1944

A new construction description of Focke-Wulf's 'Flitzer' was published on July 5, 1944, and it was clear that the twin-boom 'Entwurf mit Leitwerkstragern' had defeated the swept-wing T-tail 'Entwulf Multhopp'.[1] Baubeschreibung Nr. 280 showed the aircraft with modified wings – featuring swept rather than straight trailing edges – set slightly further back on the central fuselage. The intakes set into the wing roots were substantially enlarged too. Internal alterations had been made to allow for heavier armament in the aircraft's nose, pushing the fuel tanks further towards the rear. The aircraft was now longer overall, from 9.8m to 10.55m, though wingspan remained 8m. Wing area was 17m² compared to the earlier design's 15.5m².

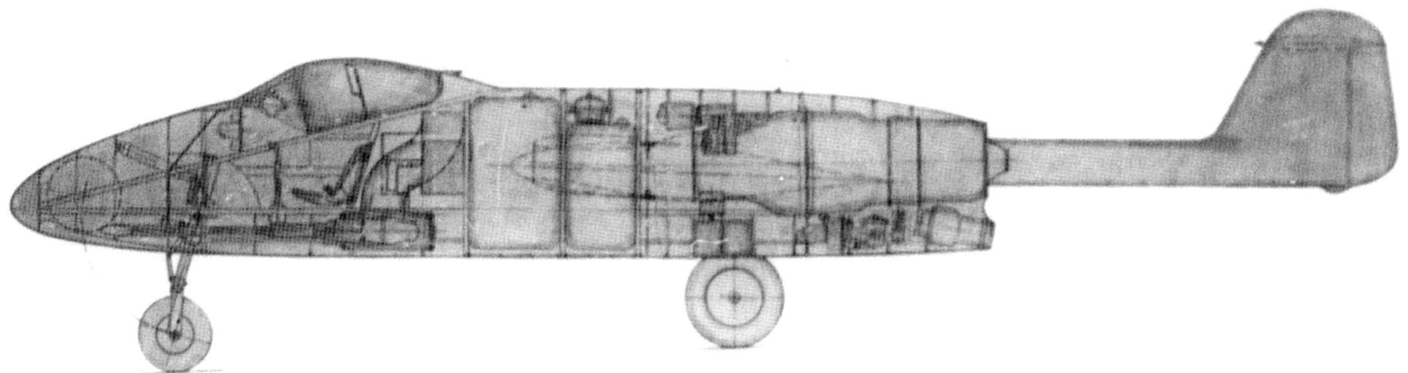

ABOVE: Focke-Wulf drawing 0310 226-87 shows the internal structure of the 'Flitzer'.

The wording of the introduction to Baubeschreibung Nr. 280 was almost exactly the same as that of Baubeschreibung Nr. 272 despite being published five months later. The only difference was in the performance statistics presented for the three different 'cases'. The first 'case', with a greater proportion of rocket fuel than jet fuel, allowed a climb to 11km in 120 seconds – whereas Baubeschreibung Nr. 272 gave 12km in 129 seconds. Take-off weight was 5,105kg compared to 4,750kg before.

The more balanced second 'case' involved a climb to 7.5km taking 90 seconds, compared to the 7km in 77 seconds previously given. Take-off weight was 4,925kg compared to 4,560kg. And the rocket-less third 'case' offered a range of 1,100km with endurance of 100 minutes – compared to the 1,520km and 132 minutes mentioned before. Take-off weight had increased here too, from 4,110kg to 4,455kg.

Two of the three armament options were the same but the third swapped one MK 103 and two MK 212s for one MK 103 and two MG 151/15s.

Overall, Baubeschreibung Nr. 280 ran to 37 pages, compared to just 25 for Baubeschreibung Nr. 272. It was heavily illustrated and showcased detailed design work on the aircraft's wings, undercarriage retraction system, tail booms, engine installation and armament. It was obvious that between February and July 1944 the company had carried out a huge amount of work on the 'Flitzer' – a name generally only used in Focke-Wulf internal company documents.

Six days later, on July 11, 1944, Rodde, the member of Focke-Wulf's Technischer Aussendienst or Technical Field Service responsible for the Ta 154 and Ta 254 under Oberingenieur Ernst Lammel, wrote a memo headed 'Development Communication – TL-Jäger'.[2] This began: "Subject: General definition for the start-up jet fighter. Task: It is extremely urgent to tackle the design of the jet fighter and to carry out both prototyping and serial production in such a way that a first prototype aircraft is ready to fly on 1.3.45."

The note was based on a technical direction meeting with the RLM. It is unclear exactly when the meeting took place, but presumably not long before the note was written. It went on: "Implementation: The design work for the prototype aircraft shall be carried out in such a way that a first aircraft is ready for flight at 1.3.45, taking into account the fact that a series of prototype aircraft will be prepared with a maximum output of 30 aircraft per month. Parallel to this, the completion of the series drawings, where as far as possible steel is to be used."

Rated weight was to be 3,700kg, take-off weight 3,500kg – significantly lighter than Focke-Wulf's 'Flitzer'. Power was to be supplied by an unspecified number of HeS 011 or Jumo 004 C turbojets (although the Jumo 004 C part of Rodde's note is crossed out on the original document).

A wooden dummy HeS 011 was to be "available in a few weeks" and all documents on the Jumo 004 C were being made available through the RLM. Fuel tanks in the fuselage were to be protected against 13mm bullets and "the outer wings are to be tightly riveted from the first machine, also the tail unit is required as container". It was to be clarified which corrosion protection was to be used for the tail containers since these might be required to hold rocket fuel.

Further clarification was needed on the points of heating for the cockpit and weapons. The aircraft was to be made using "mainly normal materials as far as possible" and "high-quality materials must be specially approved". Details on the use and processing of steel for series production of the new fighter were to be provided by the RLM during a follow-up meeting on July 19.

The following month, on August 15, 1944, Focke-Wulf published a new report entitled 'Betrachtungen zum Entwurf eines einmotorigen Jagdflugzeuges mit TL-Triebwerk'[3] or 'Considerations for the design of a single-engine fighter with jet engine'.

It began by stating: "With the existing jet engine HeS 011, it is possible to develop a single-engine jet fighter which has performance that can be achieved with lower material and manufacturing costs and fuel costs than any other fighter aircraft. This aircraft design is described in Focke-Wulf Baubeschreibung Nr. 280. The task of this report is to prove the above surprising statement on the basis of verifiable information."

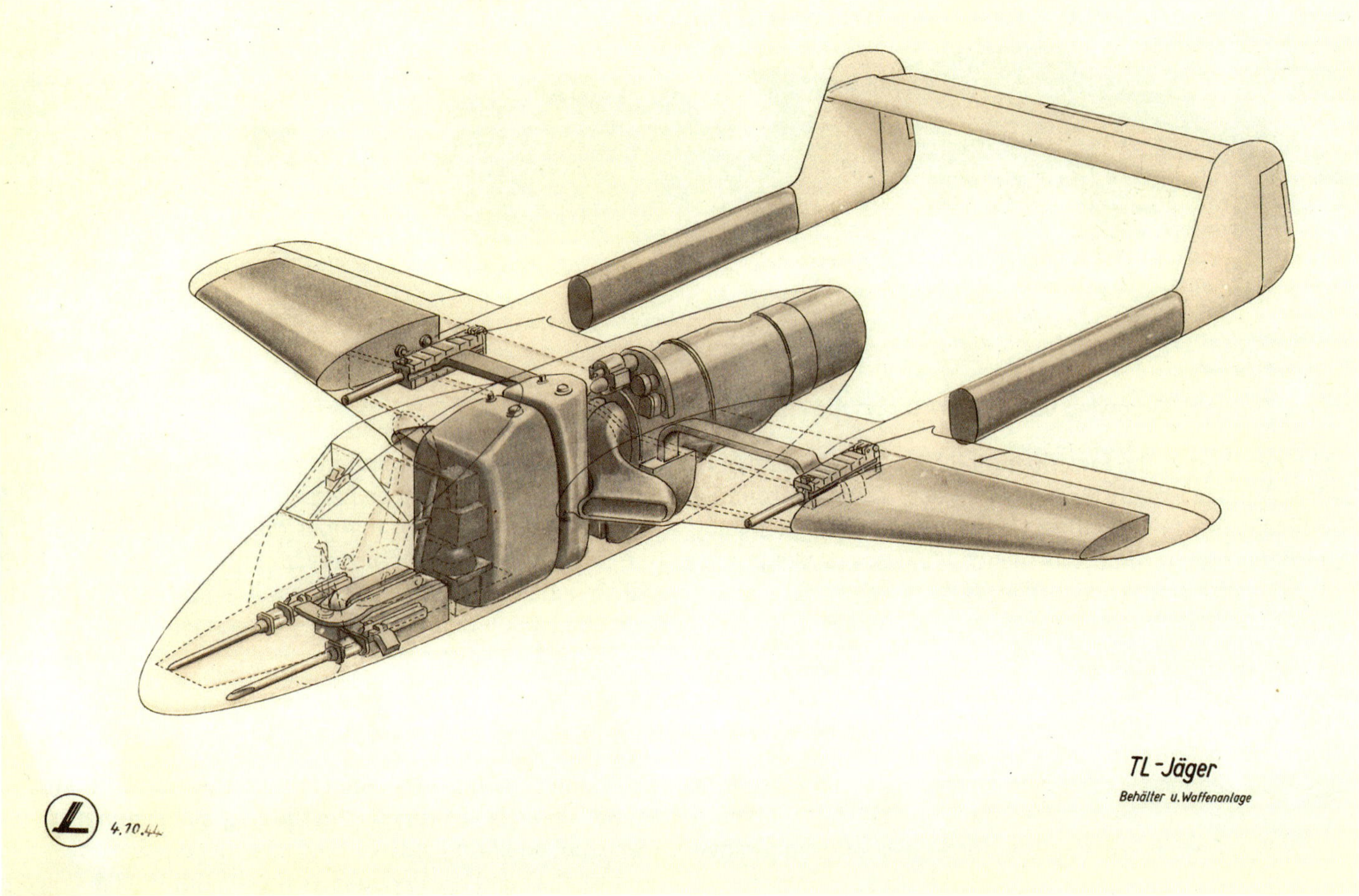

ABOVE: A perspective view showing the weaponry, fuel tanks and engine of the 'Flitzer' design.

A chart was presented comparing the straight line speed of the 'Flitzer' against the de Havilland Mosquito, North American Mustang, Ta 152 H, Ta 152 with Jumo 222 E/F, Dornier Do 335 and Messerschmitt Me 262. According to the report: "It turns out that the speed of the single-engine Fw-TL fighter aircraft is still sufficiently superior to the performance of the existing enemy aircraft with piston engines that are considered to be the fastest possible opponents, even if one considers an increase in this performance.

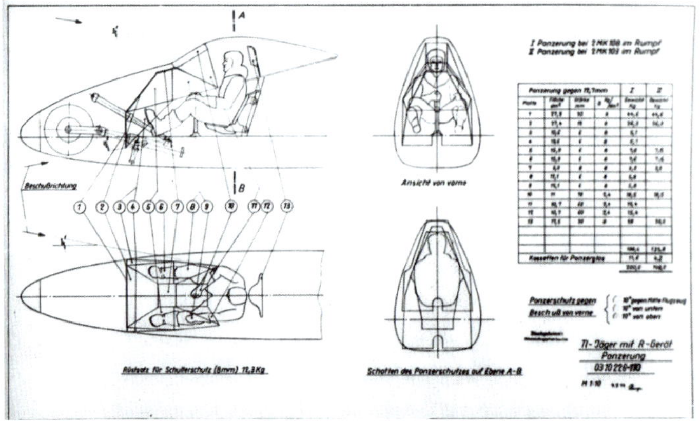

ABOVE: Armour protection for 'Flitzer' is shown in drawing 0310 226-110, dated July 4, 1944.

"In this context, it should be emphasised that the use of this jet fighter aircraft is essentially against piston-engine aircraft, not against jet fighters or jet bombers. At the moment it cannot be determined at all whether it will be possible to successfully combat enemy aircraft with jet engines through jet fighters, as both have the same physical limits (critical Mach number).

"From the graph of the horizontal speeds can be seen that the Me 262, according to information provided by Rechlin, has a lower maximum speed in its present form than the Fw design offers. To the speeds and their prospective further development one can say the following: 1) The Fw-TL-Jaeger can expect significant speed increases from the 1,500kg thrust HeS 011 engine (possibly later 1,700kg). The engine of the Me 262 no longer has great development possibilities. 2) The installation of HeS 011 engines in Me 262 will only be possible or useful after extensive reconstruction, as the current design of the airframe cannot exploit the performance of the HeS 011 engines (critical Mach number, wing form)."

It was noted that without its rocket motor the climbing performance of the 'Fw-TL-Jäger' was comparable to that of the Me 262 – but with its booster the Focke-Wulf design far surpassed it. A comparison then followed between single-engine and twin-engine layouts, comparing the Ta

154 against the Fw 190 and the Me 410 with the Bf 109 – concluding that while the outright performance of a twin-engine design would always be better than that of the single-engine design with the same engine, manoeuvrability suffered and manufacturing costs were significantly greater. The single-engine jet fighter was clearly superior to any piston-engine fighter and was superior to the Me 262 in terms of agility in combat and at a far lower cost in terms of manpower and materials.

The report then stated that "the single-engined jet fighter shown in Baubeschreibung Nr. 280 is the culmination of several years of development work" and proceeded to give a history of Focke-Wulf's single-jet fighter development up to that point. The report concludes: "Finally, the advantages of the single-engine jet fighter according to Baubeschreibung Nr. 280 are summarised again in comparison with the Me 262: 1) The fuel consumption of the Fw-TL Jäger amounts to only about 63% of the fuel required for Me 262. 2) The material cost for the airframe of the Fw-TL Jäger amounts to about 72% of the material cost for an Me 262 airframe. 3) The production cost for airframe and engine is about 69% of the comparable value of an Me 262 in the Fw-TL Jäger. 4) In the case of the Me 262 airframe equipped with HeS 011 engines, a major remodelling of the airframe is required, which would need new construction equipment. 5) The Fw-TL Jäger achieves climbing speeds of more than 100m/sec with the standard rocket motor. 6) The Fw-TL Jäger can be used with max. 1,250kg jet fuel loaded inside the airframe, reaching a full throttle flight time of two hours at a height of 12km, whereby take-off and climb consumption have already been discontinued. 7) When the fuselage is rebuilt, the Fw-TL aircraft can be equipped with the PTL 021. With a take-off weight of 5,000kg climb rates of about 32m/sec and maximum speeds of about 820km/h at ground level."

The latter aircraft – a version of the 'Flitzer' equipped with a Daimler-Benz DB 021 turboprop – was subsequently outlined in Baubeschreibung Nr. 281 of August 18, 1944.[4] It was similar in many respects to the Baubeschreibung Nr. 280 design but measured only 9.9m long and was fitted with a propeller shaft which passed from the turbojet in the rear of the aircraft, beneath the pilot, to a propeller in the nose. Internally, the company sometimes referred to this design as 'Peterle'. However, the pure jet 'Flitzer' of Baubeschreibung Nr. 280 remained Focke-Wulf's main contender going into September 1944.

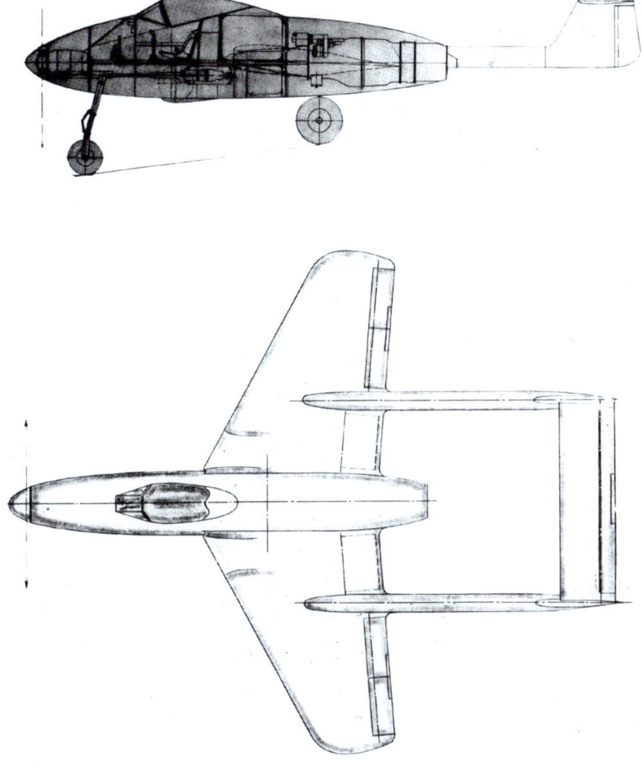

ABOVE: Focke-Wulf's turboprop version of 'Flitzer' was outlined in Baubeschreibung Nr. 281 and nicknamed 'Peterle'. This is the left-hand portion of drawing 0310 226-114.

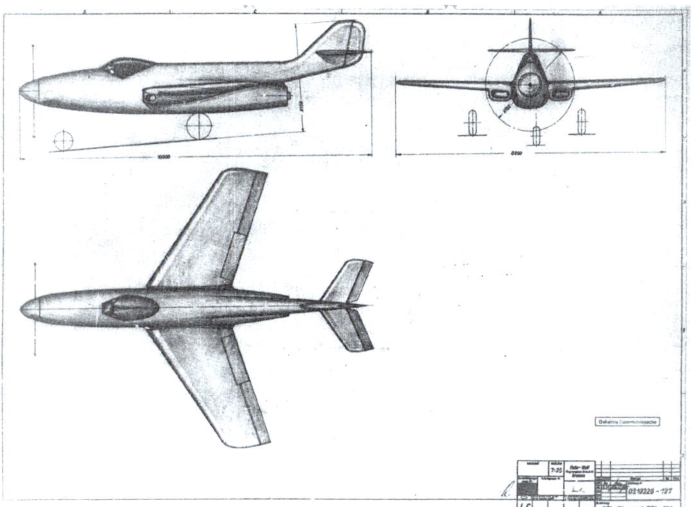

ABOVE: An alternative version of 'Peterle' with a conventional fuselage and tail appears in Focke-Wulf drawing 0310 226-127.

HEINKEL, JULY TO SEPTEMBER 1944

Heinkel's situation looked bleak in early July 1944. Göring had cancelled the He 177 heavy bomber[5] on July 1 and its He 277 derivative, with separate rather than linked engines, was cancelled two days later. This meant the only aircraft bearing the Heinkel name still in production was the He 219 night fighter and that had not been ordered in large numbers.

So when the RLM issued its specification for a new jet fighter, Heinkel grasped the opportunity with both hands. A new project, P 1073, was launched on July 6 – the day after Focke-Wulf's Baubeschreibung Nr. 280 had been published – with three new drawings. Each showed a twin-engine jet fighter with its HeS 011 engines in nacelle attached to the exterior of its fuselage. A similar engine arrangement had been worked on extensively by the company six months earlier during the course of its now largely defunct P 1068 bomber project and here was an opportunity to make practical use of all those calculations and wind tunnel test results.

The designs, numbered P 1073.01-01, -02 and -03, each featured swept wings – forward-swept in the case of P 1073.01-02 – and swept tail surfaces. The -01 had one engine on its back and the other under its nose (and the option to fit Jumo 004s instead of HeS 011s) plus a V-tail while the -02 had one on either side of its forward fuselage and the -03 had one on either side of its back. Four days later, on July 10, 1944, a fourth design was created. P 1073-04 was similar to the first design but with a slightly modified undercarriage arrangement and a short report[6] was drafted to explain its features entitled 'P 1073. Schneller Strahljäger' or 'P 1073. Fast Jet Fighter'.

It stated: "Gaining air superiority depends not only on numerical superiority of single-seaters but also in superior flight performance. If the enemy use single-seat jet fighters, a superiority of the Me 262 is very unlikely due to its conventional design, because its unswept wings and its arrangement of the engine nacelles on the wings give rise to substantial drag resistance. It therefore seems necessary to construct and build a faster aircraft that has a low construction time.

"The construction cost and range are proportional to the military fixed load – the armament, equipment, ammunition and crew. Near the ground, the fuel consumption is very large and the range is small. For this reason, it is necessary to limit the design to a single-seat aircraft with the smallest possible amount of equipment and for fuel to make up a good proportion of the total weight.

"Only a significant improvement in critical speed over the Me 262 justifies a new design. The improvement requires wings as thin and swept as possible as well as a much improved attachment of the engine nacelles.

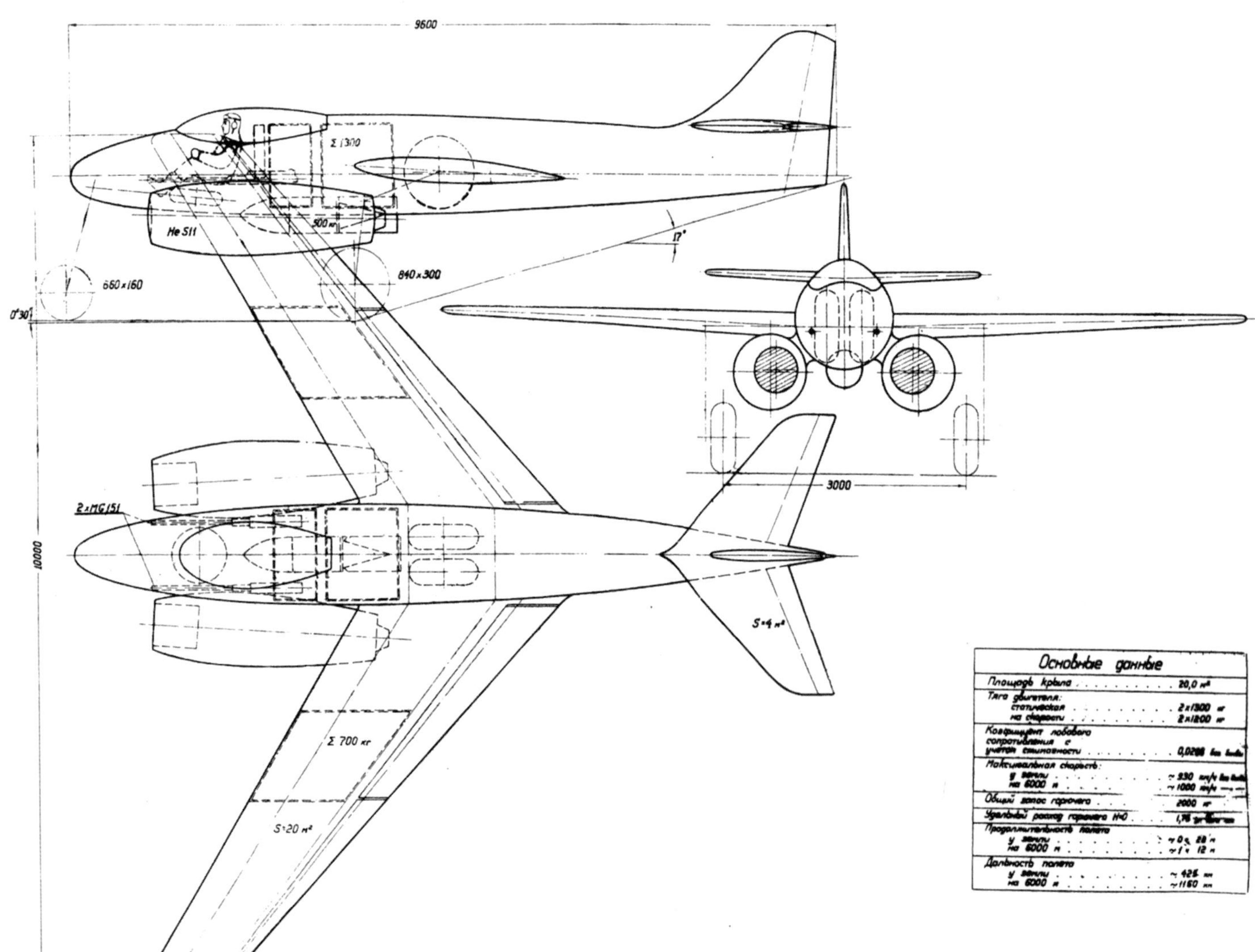

ABOVE: One of the first three P 1073 designs – the P 1073.01-02 of July 6, 1944 – had forward-swept wings and its turbojets were attached to the sides of its cockpit. The drawing has been given a Russian information panel but otherwise retains its original German captions.

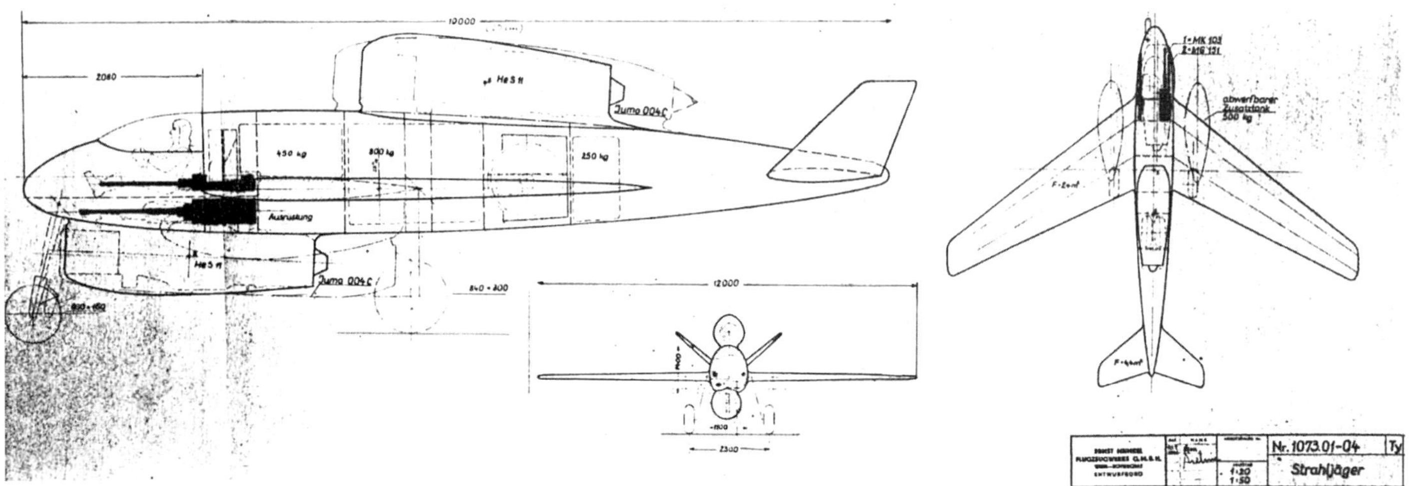

ABOVE: The fourth version of Heinkel's P 1073 – a twin-jet fighter with a layout inspired by the company's work on the P 1068 bomber design. This was the design Heinkel sent to the RLM in July 1944.

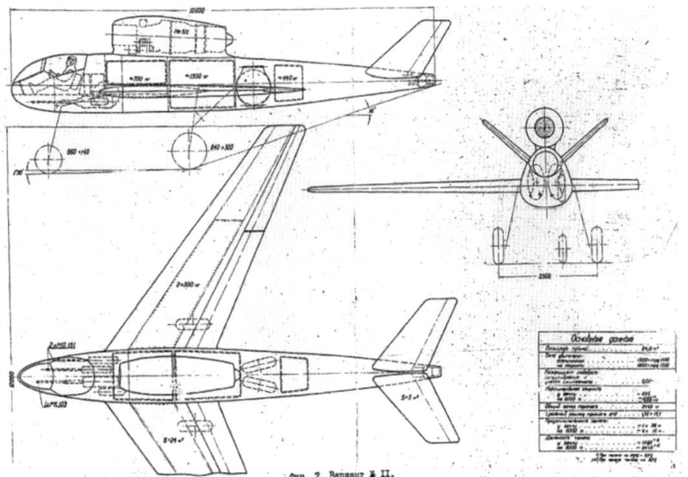

ABOVE: The eleventh P 1073, of August 5, 1944, featured an unusual cockpit canopy. Once again the drawing has been given Russian notes.

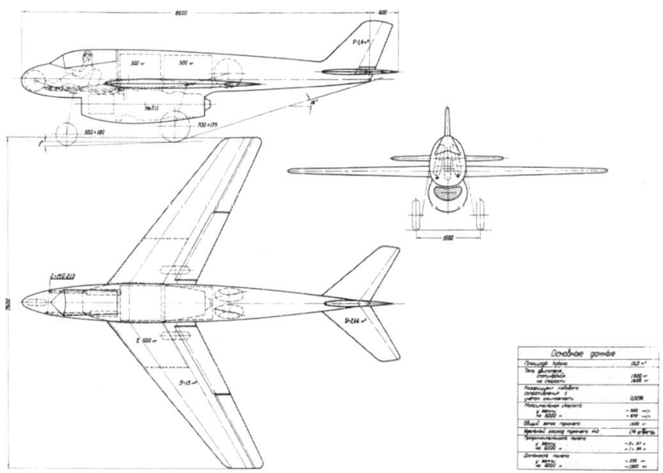

ABOVE: The Heinkel P 1073.01-13 of August 18, 1944 was the last design in the series to feature any configuration other than the dorsal turbojet.

"The numerous drag measurements carried out by Heinkel on jet engine nacelles have shown that in order to achieve high critical velocities, nacelles attached to the wings either protrude very far forward over the wing leading edge or have to be located very far back. However, this arrangement is not suitable for construction. The measurement results, which were obtained with the arrangement of the engine nacelles on the fuselage, resulted in considerable improvements.

"The achievable speed improvement over today's usual construction method (Ar 234 and Me 262) with wing sweep and arrangement of the engines on the fuselage amounts to at least 70km/h. The attachment of the engines above and below the hull is structurally simple. The view to the rear should not be worse even with the engine arranged above the fuselage as, for example, in the Ar 234, since the hidden angle is very small and the aircraft pilot can not look exactly behind anyway.

"The location of the engine under the fuselage is such that no stones can be thrown from the nosewheel into the intake opening. The undercarriage is very simple and requires little space and only small fuselage openings. The jet fighter P 1073 is designed for the HeS 11 engine. Until this engine is operational and in series production, the Jumo 004 is installed. This results in a reduced, but still considerable speed that is above the Me 262's maximum speed (with HeS 11 = 1,010km/h, with Jumo 004 = 940km/h in 6km altitude)."

In sharp contrast to Focke-Wulf's view that a single-jet aircraft with supplementary rocket boost was the best replacement for the Me 262, Heinkel felt that another twin-jet was needed but with advanced aerodynamic features to address those areas where the Me 262's performance was hampered by its late 1930s design. Ernst Heinkel himself sent the report to Göring the following day with a covering letter stating that his firm could have the P 1073 built and ready in a very short time.

No doubt aware of what Focke-Wulf was proposing by now, Heinkel drew up a single-jet version of the P 1073 on July 22, 1944 – P 1073.01-05. It retained the earlier designs' swept wings and V-tail but deleted the dorsal turbojet, making its engine arrangement somewhat similar to that of Focke-Wulf's Baubeschreibung Nr. 264 aircraft from 13 months earlier. On July 23, a twin-jet version of the -05 was drawn up as the P 1073.01-06.

A week and a half later, on August 3, 1944, Heinkel produced three more designs to explore different aspects of the P 1073 layout: P 1073.01-07 had a shorter fuselage than previous iterations and was powered by a single HeS 011 on its back, P 1073.01-08 was identical to -06 except for its long straight wings. P 1073.01-09 was also similar to -06 but offered a refinement of the cockpit and nose section.

Two days later Heinkel appears to have switched from considering a twin-engine direct replacement for the Me 262 to focusing more on a single-jet design. Both P 1073.01-10 and -11 had a single dorsally mounted HeS 011, otherwise differing primarily in armament and cockpit size. Nearly two weeks later, on August 18, 1944, Heinkel produced its final twin-jet design – the P 1073.01-12, which again had one HeS 011 under the cockpit and another on its back. The second design of that day, P 1073.01-13, took another stab at the arrangement Focke-Wulf had dismissed, with a single turbojet under the cockpit but featuring a modified intake form.

On August 19, the P 1073 took on its 'classic' form with a single HeS 011 mounted on its back. P 1073.01-14 retained the swept wings and V-tail of earlier designs but was small and compact with an armament of two MG 151s. Heinkel senior executives met with Knemeyer on August 24 and submitted the P 1073.01-14 to him. They evidently came away from the meeting believing Knemeyer had agreed to give them an order for their '1-TL-Jäger'.[7]

MESSERSCHMITT, MARCH TO SEPTEMBER 1944

Having worked on the P 1099 heavy fighter and P 1100 bomber developments of the Me 262 during the first three months of 1944, Messerschmitt's projects team next took the fuselage designed for those types and looked at what heavier armament and improved equipment might be possible with the addition of entirely new wings. This follow-on project was numbered P 1101.

It evidently soon became apparent that since the P 1099/P 1100 fuselage was mostly new and unbuilt anyway, there was no reason to retain it. The result was a series of what were essentially concept designs based on the use of

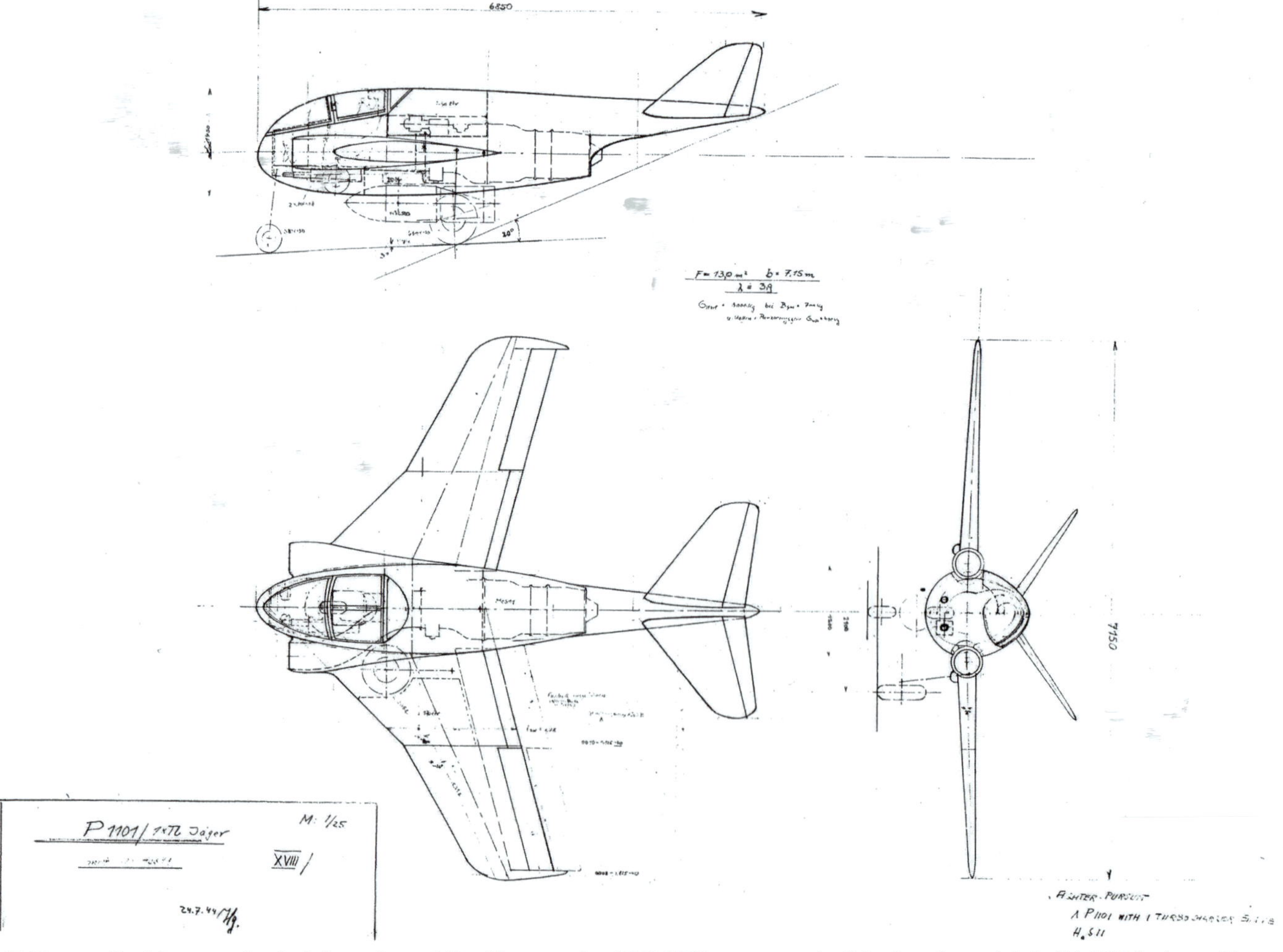

ABOVE: The earliest known single-jet version of the Messerschmitt P 1101 appears in this drawing of July 24, 1944, signed by Hans Hornung.

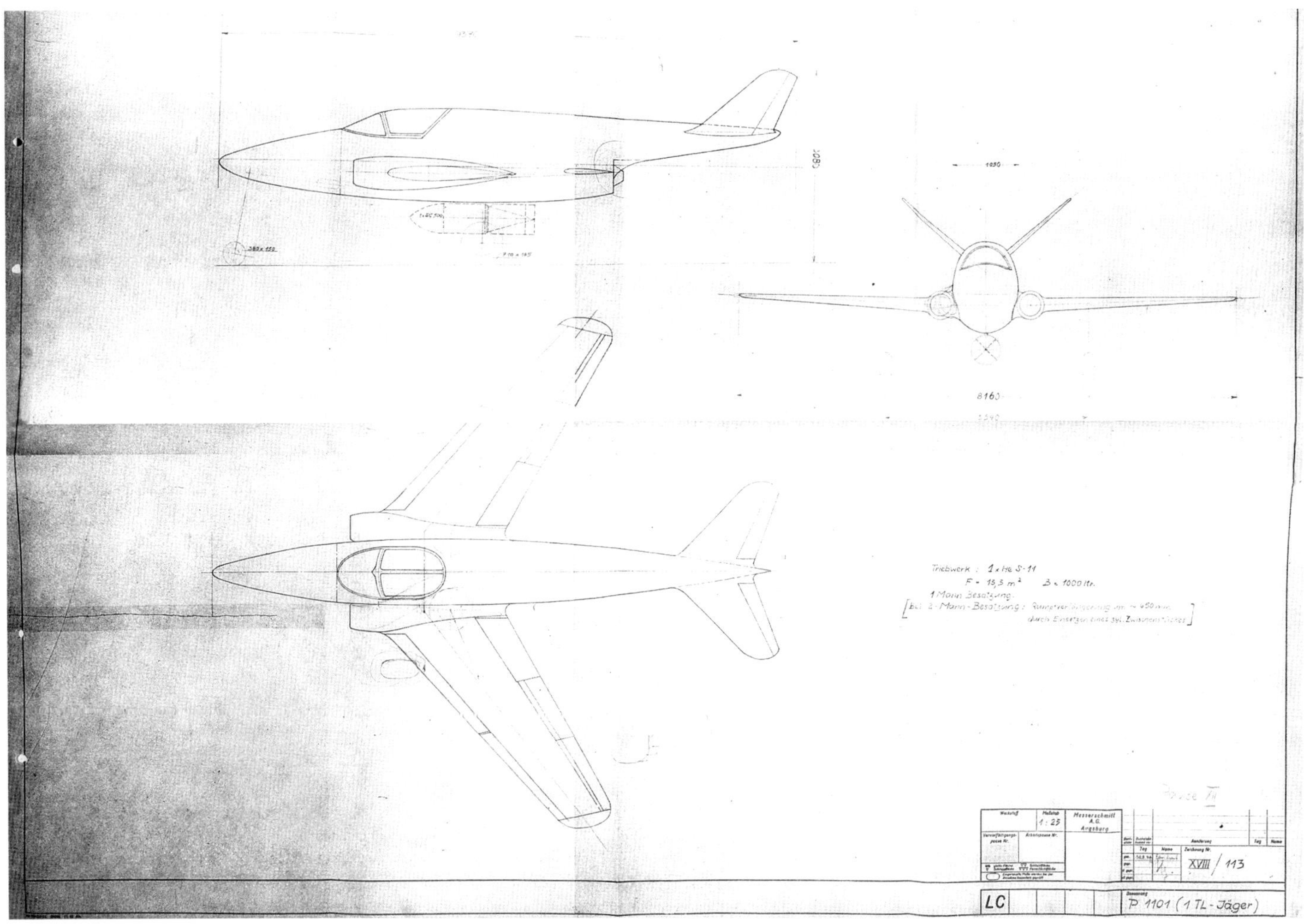

ABOVE: Messerschmitt presented this version of the P 1101, shown in drawing XVIII/113 of August 30, 1944, at the September single-jet fighter comparison meeting. Designer Hans Hornung considered that it could be turned into a two-seater by adding a 65cm-long fuselage section.

two, three or four HeS 011 engines. Each of these three layouts was compared against the Me 262 and although they were substantially heavier, they were also much better armed. Where the Me 262 could carry six MK 108s, the P 1101/2TL could carry one BK 7.5 cannon or three MK 112s, the P 1101/3TL could carry one BK 7.5 or four MK 112s and the P 1101/4TL could manage one MK 7.5 and five MK 112s.[8]

Some of these designs were sketched out with drawing numbers following on directly from the P 1099/P 1100 sequence – XVIII/92 showed the P 1101/2TL design armed with a single BK 7.5 while XVIII/99 showed the P 1101/4TL armed with one MK 7.5 and five MK 112s, one of them firing forwards and the remaining four firing upwards at an oblique angle. This work continued into July 1944 when Messerschmitt evidently received the same new jet fighter specification as Focke-Wulf and Heinkel. The result was a complete about-face for P 1101. Instead of examining larger and larger heavy fighters, the project concentrated instead on the smallest possible advanced single-jet fighter.

The earliest known result was the 'P 1101/1 x TL Jäger mit HeS 011', drawn on July 24, 1944, and signed by Hans Hornung. This fighter measured 6.85m long (just over a metre longer than the Me 163), had swept wings with a span of 7.15m and an area of 13m^2 and a V-tail. It also had a tricycle undercarriage, a pair of MK 108s and was depicted as being able to carry a single SC 500 bomb. The pilot sat right at the front of the aircraft beneath a full-vision canopy and with a large circular engine intake on either side of him. The design was part of the XVIII drawing sequence but its exact number is unclear.

The next known design in the series, however, appears in drawing number XVIII/113. This 'P 1101 (1 TL-Jäger)', also signed by Hornung, was dated August 30, 1944. It was 9.37m long, making it more comparable with the Bf 109's 8.95m length, with an option to extend by a further 650mm to accommodate a second crewman. The swept wings had a span of 8.16m and an area of 13.5m^2 and the V-tail was retained. The pilot's cockpit was shifted to a more central position and the circular intakes remained, though a pencilled-over outline indicates that a more oval form was now being considered. No fixed weaponry appears in the only known drawing of the design but it was nevertheless shown carrying an SC 500 bomb.

1-TL-JÄGER FIRST COMPARISON

A meeting to compare the Focke-Wulf, Heinkel and Messerschmitt designs for an Me 262 replacement was apparently arranged for August 31 but it was decided that Messerschmitt's design was not sufficiently advanced[9] so a three-day assessment meeting was scheduled for September 8 at Messerschmitt's Oberammergau facility.

By this point Blohm & Voss had also been invited to participate but the company was unable to produce a design in time for the meeting. According to the post-meeting report,[10] issued on September 10, in attendance from Blohm & Voss was Hans Amtmann; Focke-Wulf sent Ludwig Mittelhuber, Herbert Wolff, Begandt, Nieten, Voightberger and Tielcke; Messerschmitt's delegates were Woldemar Voigt, Hans Hornung, Wackerle, Horn, Puffert and Narr, and Heinkel sent Walter Hohbach, Schulz and Eberhard.

According to the introduction of the report, "the task was to create comparable fundamentals for the performance calculation and, as far as possible, to carry out a comparison of the present jet aircraft designs".

The Focke-Wulf team submitted the 'Flitzer' of Baubeschreibung Nr. 280, while Messerschmitt offered the P 1101 of drawing XVIII/113 – armament options now being given as two MK 108s, three MK 108s, one MK 103 and a 20mm weapon or a single MK 112 5.5cm cannon – and Heinkel offered a P 1073 with a wing surface area of 14m², a wingspan of 8m and a fuselage length of 9.3m. Wing sweepback was 35° and a V-tail was included. Armament was given as two MG 213s, two MK 108s or one MK 103 plus another 20mm weapon.

The three designs presented such diverse armament options and maximum fuel loads that the RLM assessors decided to calculate weights based on each design being

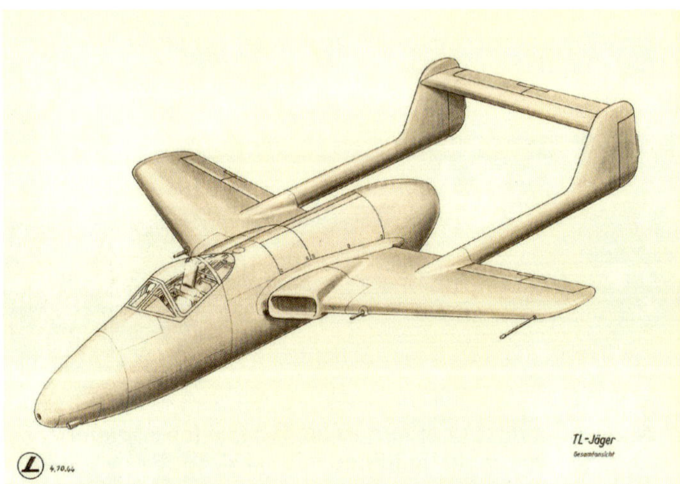

ABOVE: The September single-jet fighter comparison demonstrated to Focke-Wulf that 'Flitzer' was considered inferior to its competitors and during early October 1944 the company reworked the design one last time, without its rocket motor.

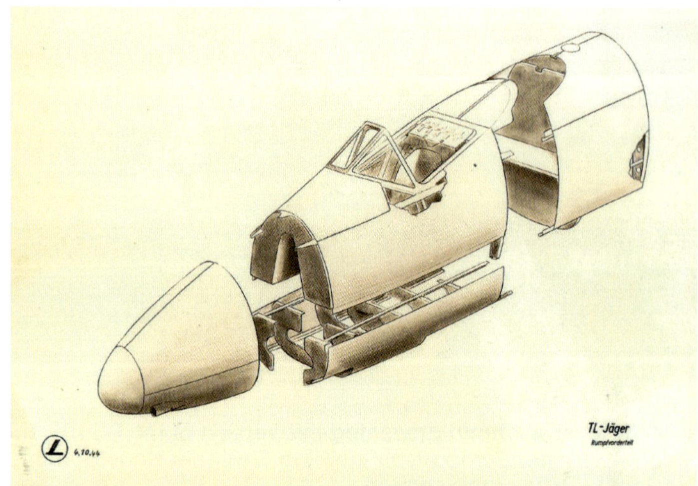

ABOVE: While the 'Flitzer' of Focke-Wulf Baubeschreibung Nr. 280 was very well worked out and practically ready to build, calculations showed that it would be severely lacking in performance compared to the designs tendered by Messerschmitt and Heinkel.

ABOVE: Focke-Wulf was so confident about its 'Flitzer' design it invested in a full scale mock-up.

ABOVE: A rear view of the 'Flitzer' mock-up.

ABOVE: A mock-up of Focke-Wulf's 'Entwurf Mittelhuber' aka 'FW II', left, compared against a mock-up of the design that would eventually become the Ta 183 – the Baubeschreibung Nr. 279 design, aka 'Entwurf Multhopp' aka 'Huckebein'.

ABOVE: Mock-up of the 'Entwurf Mittelhuber' nose section photographed outdoors, presumably during early 1945.

ABOVE: Mock-up of Focke-Wulf's Baubeschreibung Nr. 279 design.

fitted with just two MK 108s with 60 rounds each and a fuel load of 830kg. This made the P 1101 3,554kg, the Flitzer 3,660kg and the P 1073 3,604kg.

The report also notes that while the weights for different components were available in Focke-Wulf's documentation, these had to be estimated for both the Messerschmitt and Heinkel designs since specific values were unavailable. Furthermore, the Focke-Wulf design's given weight had to be reduced by 70kg because the company had specified steel in some components where Messerschmitt and Heinkel had both assumed lightweight alloys. The vagueness of the requirement issued in July had resulted in design details which made accurate mathematical comparison very difficult.

In terms of performance, the report stated that "according to the participating companies, the designs for Heinkel and Messerschmitt have approximately the same top speed. Due to the loss of thrust as a result of the long jet intake, the Messerschmitt design is disadvantaged by about 10km/h compared to the Heinkel design. The Focke-Wulf design should be about 50–100km/h slower compared to the designs of Heinkel and Messerschmitt".

A cautionary note was added with regard to the possible effects of compressibility on each of the designs but "due to a lack of sufficient documentation, it was decided not to use compressibility effects for the fuselage and engine for the purposes of the comparison".

The representatives of Heinkel and Focke-Wulf, it was recorded, thought that the calculated speeds were generally too high, "while Messerschmitt on the basis of the experience with the Me 262 and the Do 335 holds the contrary view and hopes to achieve even higher speeds by further improvements".

ABOVE AND BELOW: Models of both the Nr. 279 and Nr. 280 designs were tested in free flight during the summer of 1944 – with the Nr. 280 'Flitzer' evidently coming out on top.

It was concluded that further general clarification was needed on all points and that a broad exchange of experience was needed. It was also "urgently necessary that all RLM development branches and the Luftwaffe test stations be made available to provide the flight measurement results and related documents necessary for the evaluation".

The report was signed by Amtmann, Wolff, Voigt and Hohbach on behalf of their respective companies. What Amtmann, Wolff and Voigt didn't know, however, was that even as Hohbach was presenting his team's design on the first day of the meeting, September 8, his colleagues back at Heinkel were hard at work on a significantly downgraded and simplified version of the P 1073 with straight wings, a twin rudder and BMW 003 jet engine – the P 1073.01-15. Only on the final day of the meeting, September 10, would the other companies each be sent a telegram with a less demanding requirement for a different 1-TL-Jäger – a cheap fighter, inferior in performance to the Me 262 but quick to build and made from wood and steel, that would become known as Volksjäger (see Chapter 9).

1-TL-JÄGER SECOND COMPARISON

The sudden demand for Volksjäger designs from not just Focke-Wulf, Heinkel, Messerschmitt and Blohm & Voss but also Arado, Fieseler, Junkers and Siebel seems to have temporarily derailed the competition to design a high-performance single-jet Me 262 replacement.

It was put back on track with a letter sent to Focke-Wulf, Blohm & Voss, Messerschmitt, Junkers and Heinkel on December 4 by General-Ingenieur Hermann of Fl. E 2.[11] This was the airframes subsection of what had been the RLM's Technical Office; following a reorganisation on August 1, 1944, a new organisational structure directly responsible to the OKL (Luftwaffe high command) had been formed from the old RLM known as Chef der Technischen Luftrüstung (TLR) led by Oberst Ulrich Diesing.

The letter, of which only a damaged copy is known to survive, said: "Subject: 1 TL-Jäger with HeS 011. After projects were received from a number of different companies ... some guidelines are now being presented as a follow-up [and companies] are asked to revise the projects under these directives. The documents are to be exchanged between the involved companies in good time and to be cleared up with regard to performance ... and submitted to a joint debate in the OKL at TLR / Fl. E 2 room 5474 on 15.12.44".

The revised requirement included 1,200kg fuel capacity – enough for a flight endurance of about two hours at 60% throttle at 9–10km altitude – armour protection against 12.7mm rounds from the front, protection against 20mm rounds from the rear, protected fuel tanks, basic armament of two MK 108s with 100 rounds each and the option to fit an additional one or two MK 108s, or armament variations of two MK 103s or one BK 5.5.

The meeting took place on December 15 as scheduled[12] with Focke-Wulf, Messerschmitt, Junkers and Heinkel each submitting one project while Blohm & Voss submitted two.

FOCKE-WULF NR. 279

Focke-Wulf's well-thought-out Flitzer was gone and in its place was a thoroughly reworked version of 'Entwurf Multhopp' with its nose intake, short but relatively

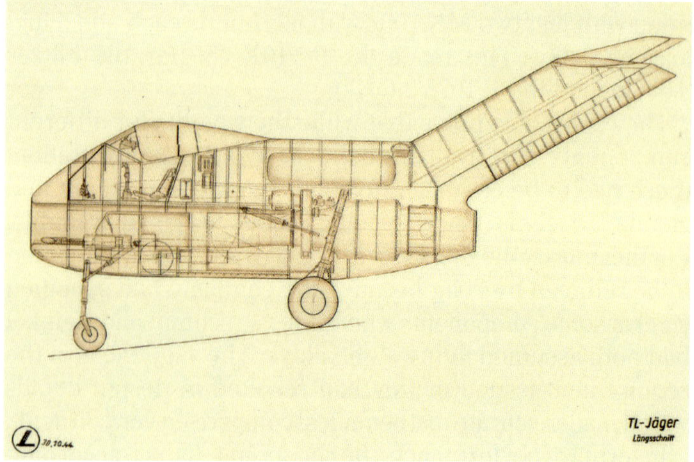

ABOVE: The internal structure of the Baubeschreibung Nr. 279 design on October 30, 1944.

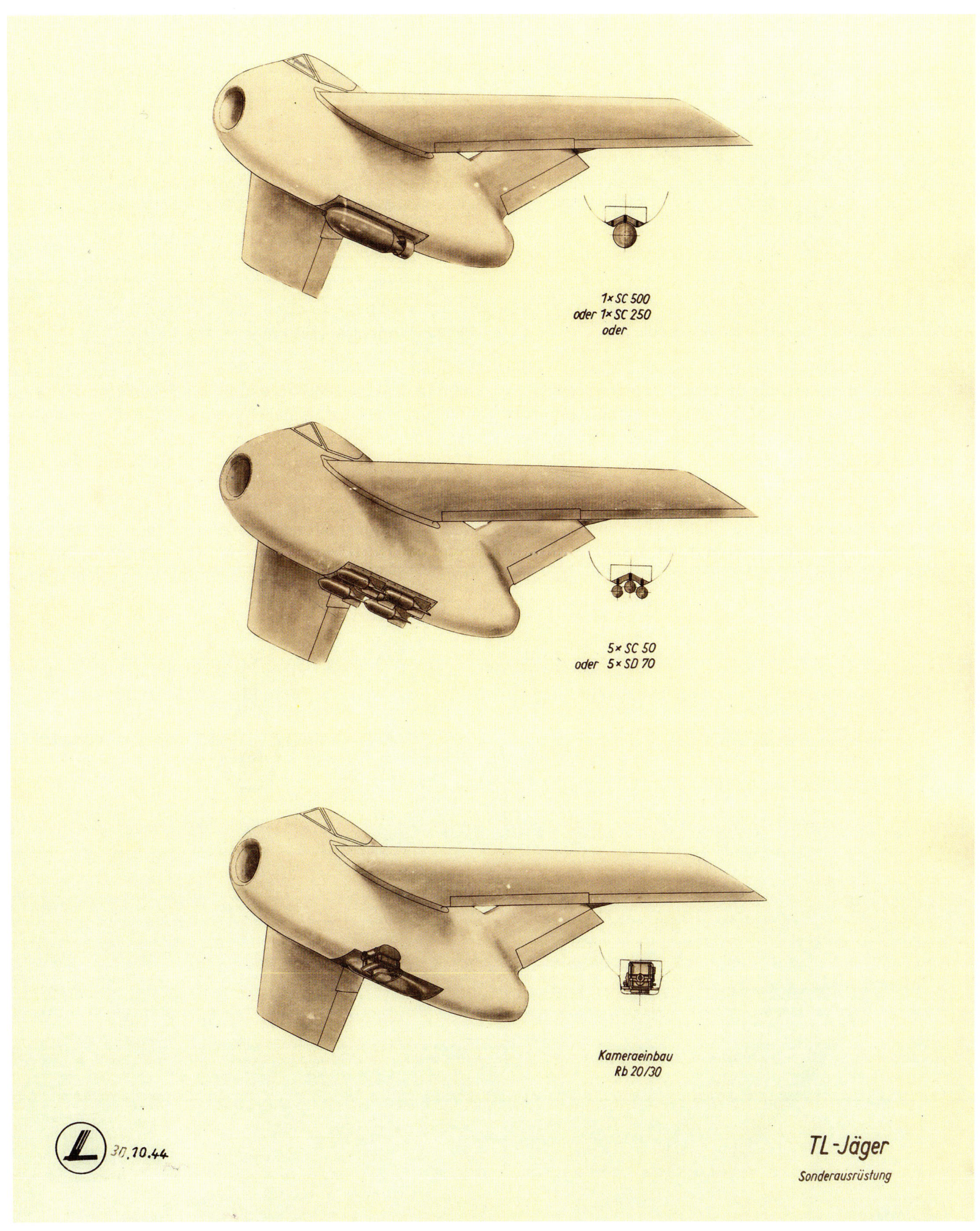

ABOVE: 'Huckebein' had a well-thought-out semi-recessed ventral bomb rack which doubled up as a reconnaissance camera bay.

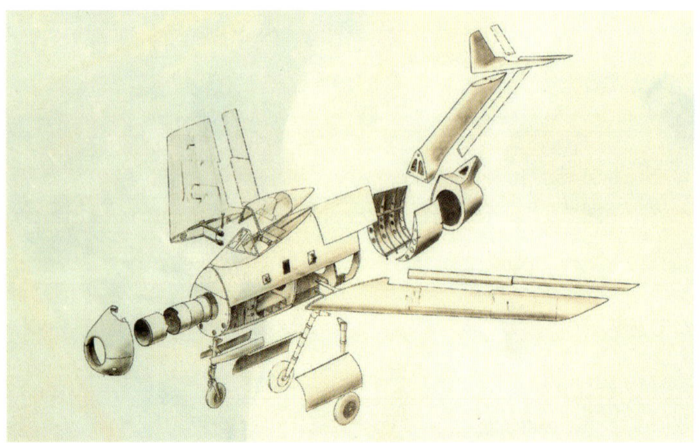

ABOVE: Component assembly for 'Huckebein'.

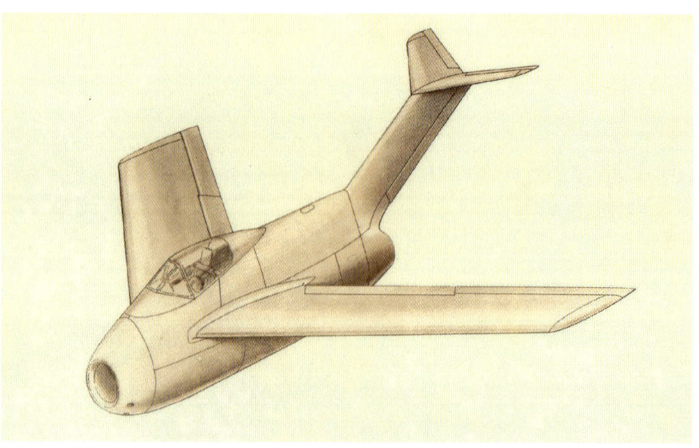

ABOVE: Perspective view of the Nr. 279 design.

conventional fuselage, swept wings and T-tail. Within the company this design was now being referred to as 'Huckebein' (after Hans Huckebein, a famously unruly but charismatic comic strip raven) but the formal project description, dated November 8, 1944,[13] was entitled 'Baubeschreibung Nr. 279 Einmotoriger TL-Jäger mit HeS 011' and began: "With the HeS 011 engine we have for the first time a jet engine available with which it seems possible to create a single-engine jet fighter, which in its level of performance can compete with the existing 2-engine jet fighters and with the best piston-engine fighters. Although it is possible to achieve relatively high airspeeds even with smaller engines, the climb rates are still rather unsatisfactory.

"For the design of the jet fighter, it is of utmost importance to decide the altitude where optimal flight performance is sought. Considering the bombers and fighters with great operational altitudes (B-29, Mosquito, Thunderbolt, Lightning) to be expected from the enemy in the next few years, we thought it was right to concentrate on

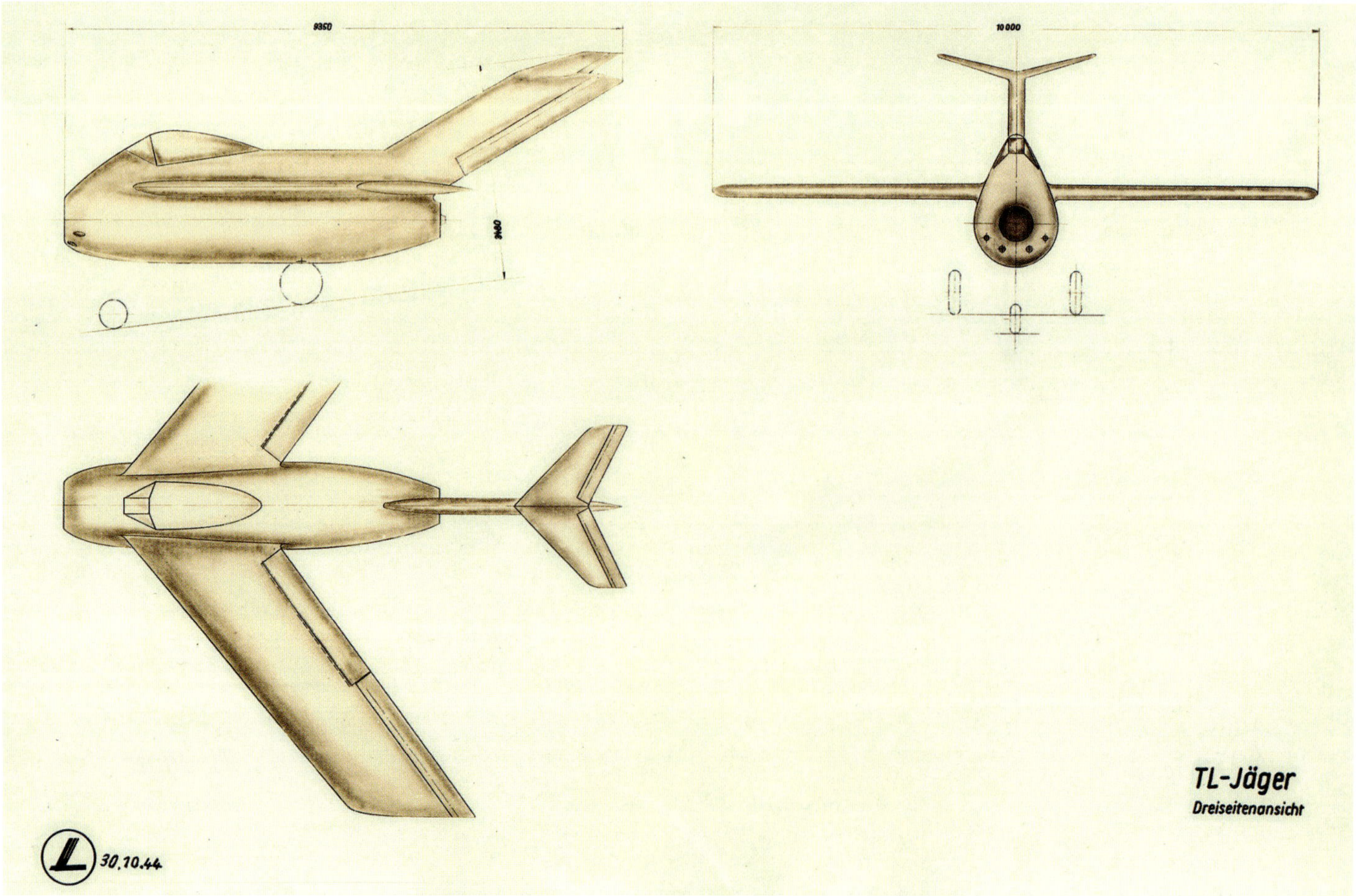

ABOVE: Focke-Wulf's Baubeschreibung Nr. 279 design as it appeared on October 30, 1944.

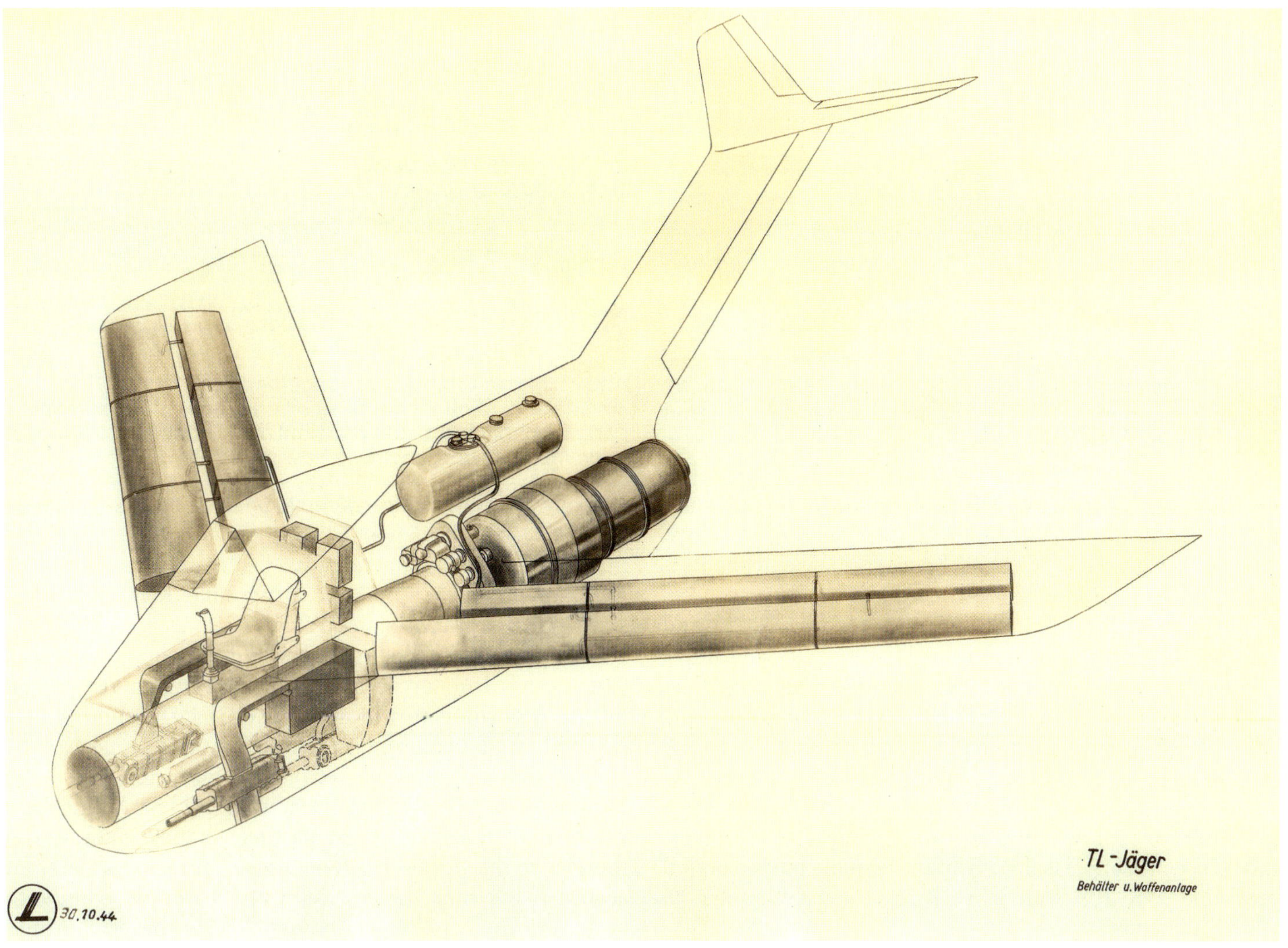

ABOVE: The weaponry, fuel tanks and engine of Focke-Wulf's 'Huckebein'.

good performance and handling properties in the range of 8 to 14km."

The report then went on to outline Focke-Wulf's attempts to lower drag at high Mach numbers through the clever use of aerodynamics – particularly with regard to the aircraft's thin sharply swept wings. It stated that for 'Objektschutz' or 'target defence' the aircraft could be equipped with a 1,000kg thrust rocket motor – allowing a climb to 10km altitude in just four minutes. The fuel for this would be carried in a pair of drop tanks which could be jettisoned when empty.

In addition: "For use as a fighter-bomber, a bomb load of 500kg can be accommodated in the lower fuselage area, whereby the bombs only protrude about halfway from the fuselage contour. The armament is limited to 2 MK 108s with 100 rounds of ammunition due to the high altitude performance." If maximum altitude of 14km did not need to be achieved, "it is easy to install 2 more MK 108s, each with 60 rounds of ammunition".

Maximum take-off weight was between 3,100kg and 4,300kg depending on loadout (4,900kg with rocket motor or 4,800kg with 500kg bomb), length was 9.2m, wingspan was 10m, wing area was 22.5m² and sweepback angle was 40°. The tricycle undercarriage utilised a 465 x 165mm nosewheel and 700 x 175mm main wheels with shock absorbers lifted directly from the Fw 190 production line.

MESSERSCHMITT P 1101 EXPERIMENTAL

A version of the P 1101 with a nose intake was drafted as drawing XVIII/125 on October 27, 1944, and in a meeting three days later Willy Messerschmitt announced to his staff that an experimental aircraft was to be built which would test the type's performance and flight characteristics.

On October 31, the Messerschmitt Probü staff held their own meeting to begin fleshing out the details of how such an aircraft could be built.[14] Present were the leader of the P 1101 project Hans Hornung, Karl Seifert, Stach, Faltermeier, Denk, Förster, Säfken and Stähle.

It was decided that the only usable space for the compass was at the far end of the fuselage and no steel could be used there for this reason. Antennae would be installed on the tailfin in the form of metal strips – which meant that the aircraft would need a wooden tail. There would be no de-icing system, hydraulics were to be avoided for undercarriage and flap actuation and "an attempt should be made to forego emergency landing gear actuation and instead to mount an unsprung skid for belly landing under the engine". A mock-up of the experimental aircraft was to be created immediately.

Two days later, a preliminary equipment plan was drawn up for a P 1101 series production model in six different variants:[15] two different fair-weather fighters, a bad-weather fighter, a night fighter, an interceptor and a close-reconnaissance version.

The first fair-weather fighter (Schönwetterjäger) was to be powered by a Jumo 004 with two MK 108s and Revi 16 B gunsight plus the option to fit a Schoss 503 bomb rack. The second, intended to replace the first, was to be powered by an HeS 011 and would have a pressure cabin, FuG 218, FuG 520 and FuG 206 radio equipment. There would be an EZ 42 gunsight for its two MK 108s and the option to equip it with two 21cm mortars, 12 SG 116s or eight Rohrblock 108s. Alternatively, Schloss 503 bomb racks would enable it to carry four X4 missiles or two Hs 298 missiles.

The bad weather fighter version added FuG 125 and 120 equipment. Instead of the various mortars and other weapons it could be fitted with oblique-firing guns. The option to carry four X4s or two Hs 298s remained.

The single-seat radar-less night fighter version, described as "Nachtjäger (einsitzig) 'Wilde Sau'" added the FuG 101 device to its equipment list and was only to carry two MK 108s plus four X4s or two Hs 298s. And finally, the close reconnaissance variant would have carried only one MK 108 with a Revi 16 B gunsight plus an RB 50/18 camera.

A full project description issued on November 10, 1944,[16] signed by both Hans Hornung and Woldemar Voigt, stated that the experimental aircraft was to be based on drawing XVIII/138. According to the introduction: "Task for the experimental aircraft: The experimental aircraft is designed as a test aircraft for the strongly swept wing (without engine nacelle influence on the wing) for take-off and landing as well as for maximum speeds and as a forerunner for a series. The spatial relationships between the fuselage and wing and the tail unit and the wing are therefore initially only to be roughly investigated and it is possible that the exact investigation or the testing will require changes when the transition to a series production model is made."

Drawings XVIII/131a and 131b gave details of the curves required for the aircraft's fuselage panelwork, particularly the air intake, and it was stated that "great importance must be attached to the exact adherence to the shapes at the air intake on the fuselage". The fuselage "consists of an upper and lower cylindrical section with an intermediate piece that connects the two sections". The design for the series production model was shown in drawing XVIII/130.

In designing the front of the fuselage for the experimental aircraft "the following points must be observed with regard to later prototypes: a nosewheel of 500 x 180 should be spatially possible ... the cockpit must be able to be made pressure tight; the canopy is constructed as a 'rennkabine'; in the case of the air duct, it must be ensured that the inner wall must be made as smooth as possible. Rivet heads and sheet metal seams are intolerable".

ABOVE: Messerschmitt was concerned that the P 1101's 3m intake duct might cause significant power losses under certain conditions and experimented by attaching a 3m duct to an Me 262's port engine in November 1944.

The upper part of the fuselage would contain the fuel tanks and retracted undercarriage mainwheels with the engine below. According to the design notes: "The following is planned for the experimental aircraft: Jumo engine 109-004 B. The HeS 011 must be retrofitted. Protected fuel tanks are not to be provided but the middle part of the fuselage can be riveted tightly so that 900 litres of fuel can be stored there. For the tightly riveted fuselage part (tank part) an access hatch must be provided at the top. The FuG 16 system of the Me 262 can be installed in the place of the first protected tank.

"Chassis space is to be provided for 740 x 210 wheels but 660 x 190 wheels will be installed for the time being. The larger wheels are decisive for the design. Make sure that the engine is easily accessible and easy to replace. Care must be taken to ensure the best possible sealing of the engine compartment from the fuel compartment. The fuselage end is shaped as a cone and contains the radio system, oxygen system, possibly a brake parachute, the course control and the compass. The tail unit is set on the fuselage end.

"Please note: It may happen that a flame may form at the end of the engine when it is started; the underside of the fuselage must be clad with sheet steel, at least near the engine. It is best to make the fuselage end entirely from sheet steel. The devices installed near the end of the engine are to be protected against heat radiation."

The description also stated: "In order to get the aircraft to the testing stage as quickly as possible, and to accelerate a later start of series production, the demand is made to use as many components and assemblies from the Me 262 series aircraft as possible without changes.

"The first prototypes are built as experimental aircraft, i.e. extensive changes to the wing, fuselage and tail unit can still be made during the tests. The experimental aircraft can become the basic pattern for the series, but it does not necessarily have to be".

Messerschmitt was concerned that the P 1101's long intake duct might result in a loss of engine power and carried out tests on November 15 by attaching a 3m tube to the intake of an Me 262 airframe 015's Jumo 004. A report was issued by engineers Caroli and Kaiser on November 23, 1944, outlining the results of these tests – a likely power loss of less than 3%.[17]

MESSERSCHMITT P 1106

With work under way on the P 1101 experimental aircraft, the Probü switched to a radical new single-jet design – the P 1106. It seems odd that the Messerschmitt team decided to pursue this new project with the P 1101 at such an advanced stage but nevertheless, the P 1106 was the design submitted for the December 15 meeting.[18]

The P 1106 variant put forward is known only from a drawing included with a DVL assessment of the December design published in January. The badly degraded drawing had a number in the XVIII series – probably /153, /156 or /157. The design featured a wide forward fuselage with the engine, undercarriage, weapons and fuel tanks all grouped together. The 40° swept $13m^2$ area wings also joined onto this section. The cockpit was suspended above the ground in the aircraft's tail, with the V-tailfins immediately behind it – echoing the configuration of the most radical P 1092 from mid-1943.

Messerschmitt's Probü was rapidly evolving the P 1106 design right up to the December 15 meeting; drawing XVIII/151 of December 11 showed a P 1106 with a T-tail behind its cockpit. This version was 8m long with a wingspan of 6.74m and the detailed view showed the mainwheels retracting into a very narrow space above the turbojet. Above them were two fuel tanks – one holding 200 litres and the other 1,100 litres. Armament was two MK 108s above the air intake in the aircraft's nose. Drawing XVIII/152a of the following day showed the same machine but without the internal detail. On December 14 another design appeared with the same configuration but extended to 8.65m long and with a wingspan of 7.37m appeared in drawing XVIII/154. This allowed for a larger fuel tank in the centre fuselage and

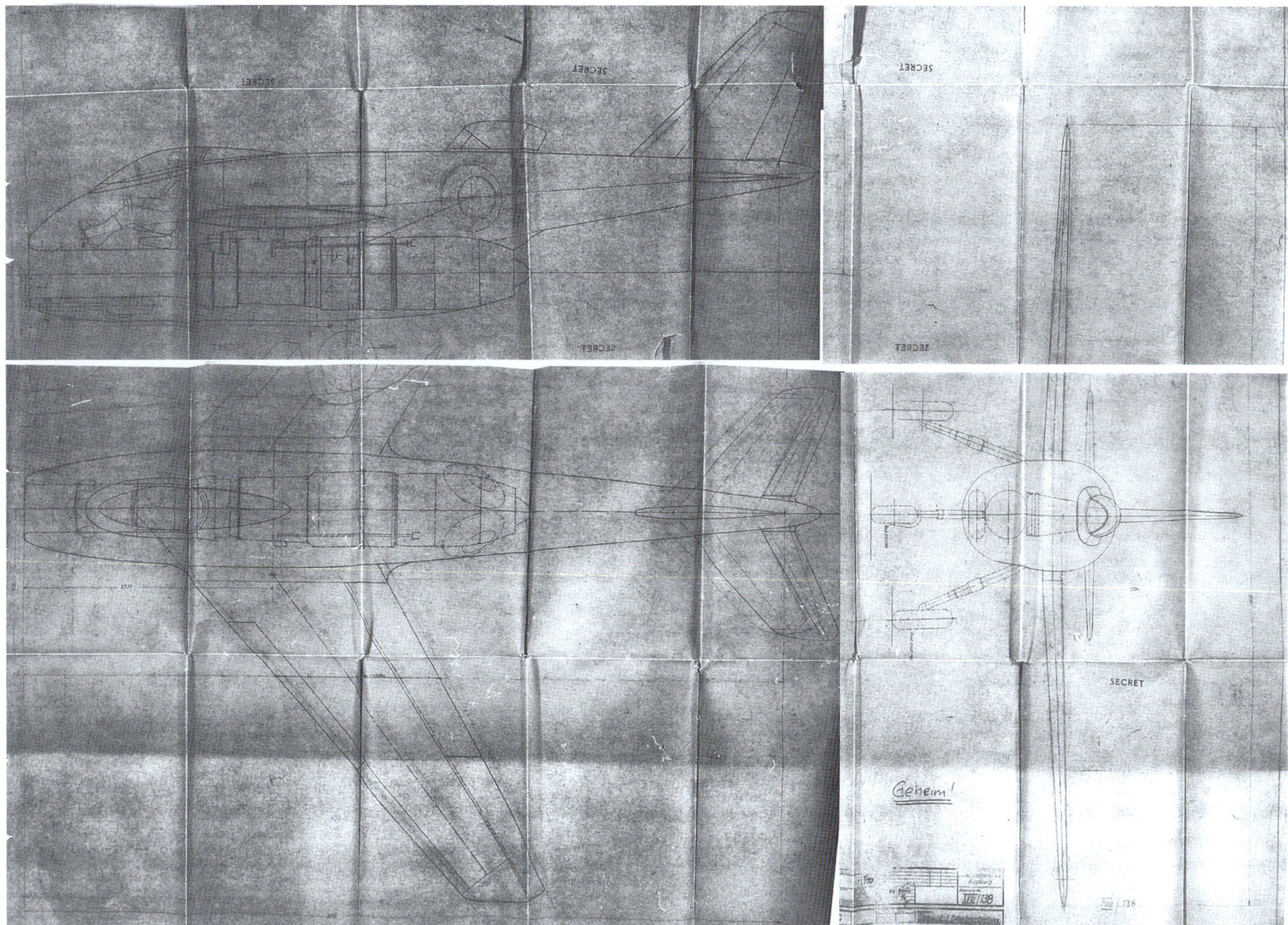

ABOVE: The experimental version of the Messerschmitt P 1101 was set out in XVIII/138, dated November 8, 1944. This is the only known surviving version of the drawing.

small additional tanks in the wing roots. Wing area was increased to 15m².

Yet another drawing in the sequence followed that same day, XVIII/155, which showed a modification of the V-tail design that would be submitted on December 15 – but with two 1,800kg thrust rocket motors in its lower fuselage instead of a turbojet. This highly unusual aircraft, labelled 'P 1106 als R-Flugzeug' was 8.42m long but with the wingspan of the earlier P 1106s: 6.74m.

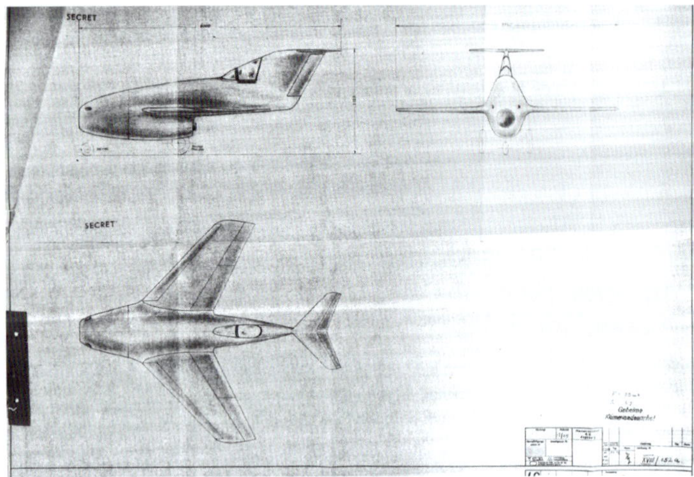

ABOVE: This version of the P 1106 appeared in drawing XVIII/152a of December 12, 1944.

JUNKERS EF 128

The precise origin of the EF 128, Junkers' December 15, 1944 entry, is unknown but it was certainly a contemporary of the Ju 248 detailed in Chapter 10 and work on it would have taken place under the supervision of Hans Gropler, head of the company's projects office. The efficient tailless design appears to have had no real precedent except in its resemblance to the Messerschmitt-derived 8-248. Each had a cigar-shaped fuselage and a similar undercarriage retraction system. The EF 128's cockpit was at the front of the aircraft, just behind the nosewheel bay, with the two MK 108s below and to either side of the pilot.

The HeS 011 turbojet in the rear fuselage was fed by a pair of intakes – one on either side of the aircraft beneath its 45° swept-back wooden wings. Control surfaces were set into the trailing edge of each wing, just inboard of the ailerons. In its earliest known form, from December 15, the EF 128 had a wing surface area of 15.3m². Other information about the design is lacking.

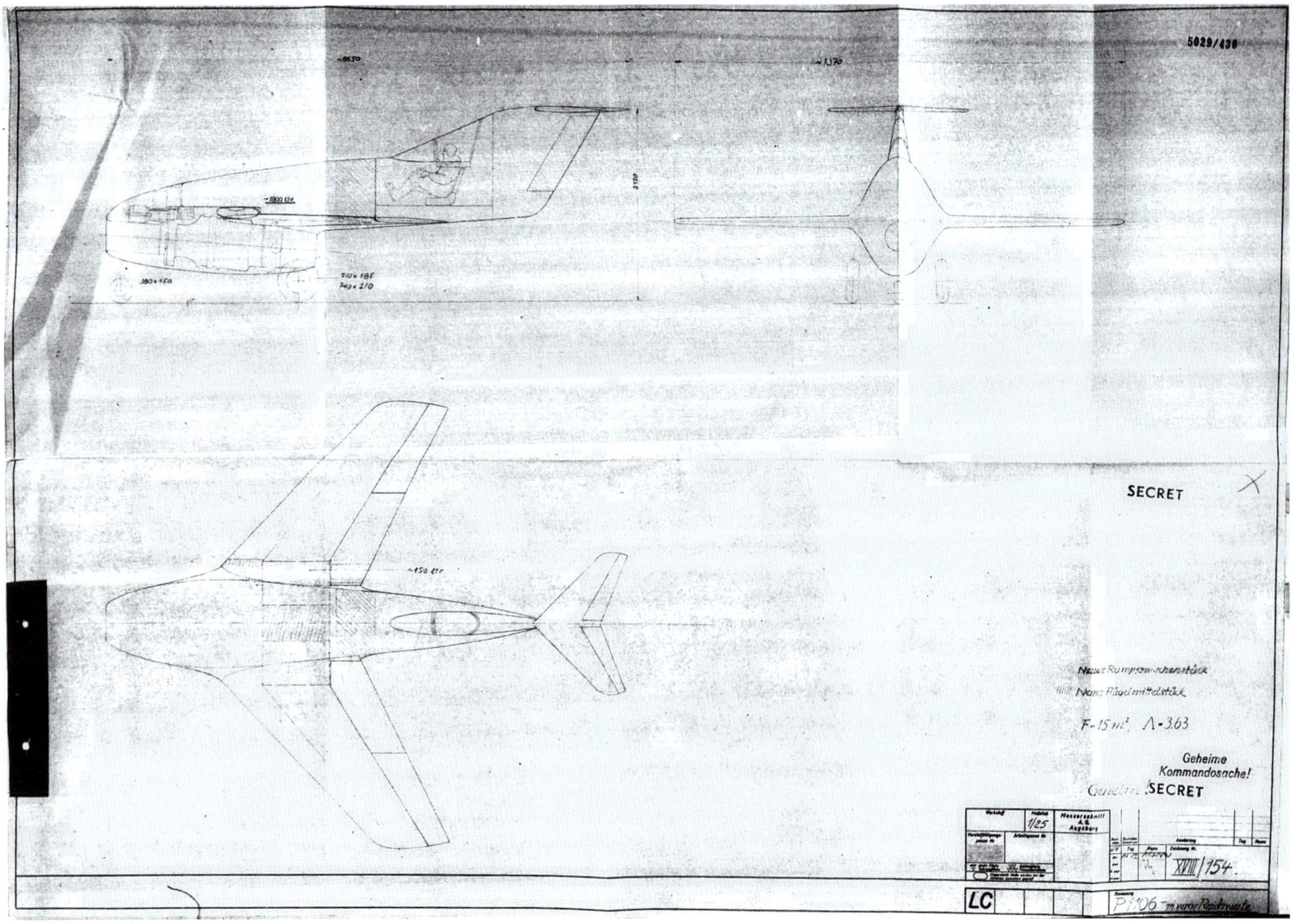

ABOVE: Messerschmitt showed an enlarged version of the P 1106 on December 14, 1944, in drawing XVIII/154.

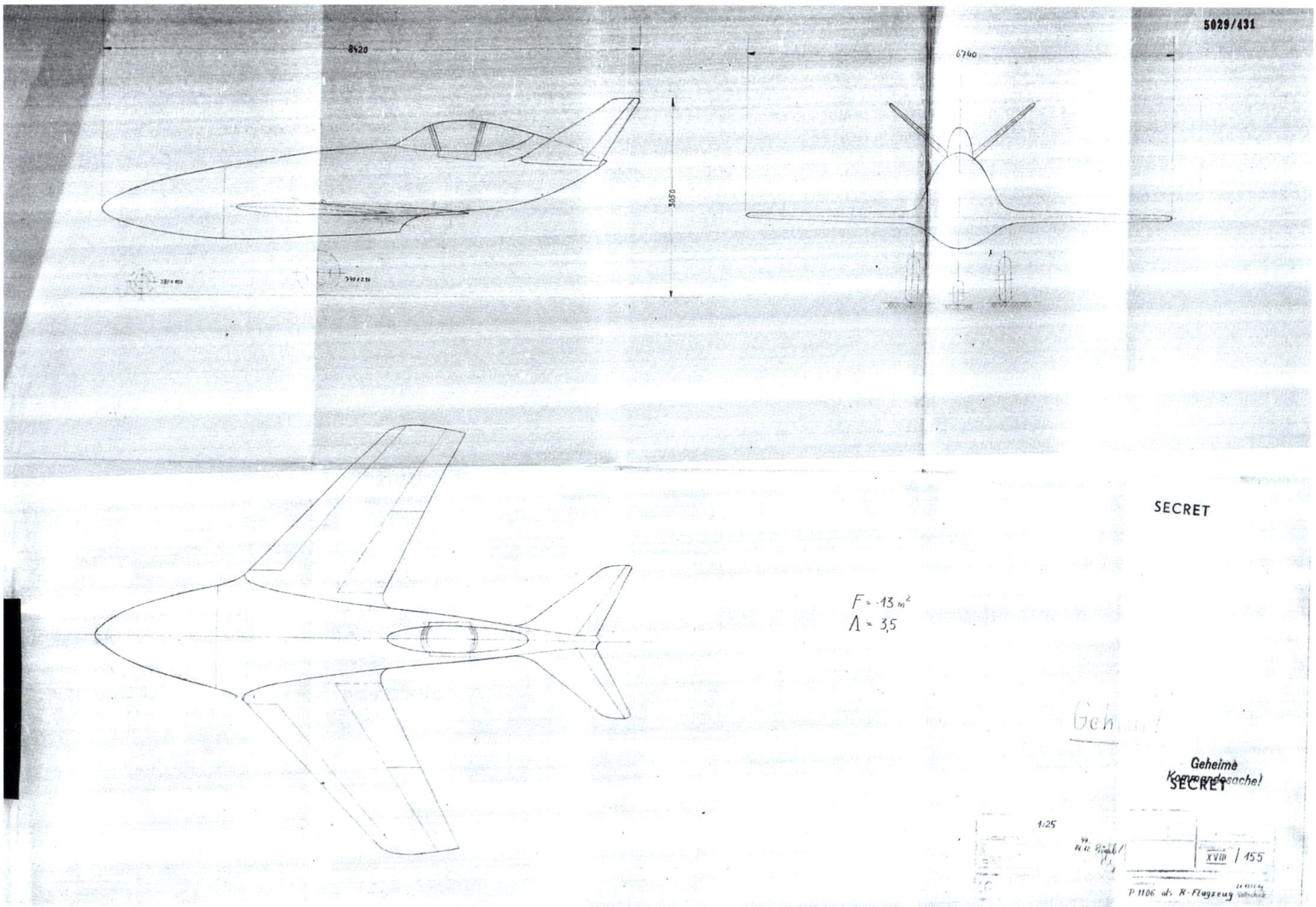

ABOVE: The oddest version of the Messerschmitt P 1106 appeared in drawing XVIII/155 of December 14, 1944 – an interceptor powered by a pair of 1,800kg thrust rocket motors.

ABOVE: This badly degraded drawing shows the version of Messerschmitt's P 1106 actually submitted for the single-jet competition on December 15, 1944.

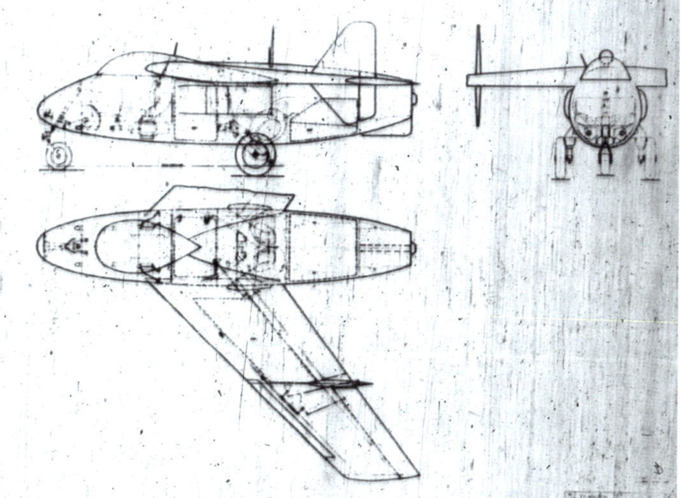

ABOVE: The earliest known version of the Junkers EF 128 from December 1944.

HEINKEL HE 162 DEVELOPMENT

Even less is known about Heinkel's entry for the December 15 design comparison meeting. It is described in the DVL assessment document as "Projekt He 162 Weiterentwicklung" or 'Project He 162 development'[19] but unlike the other submitted projects no drawing was appended and the charts showing calculated aerodynamic details of the other projects have only blank spaces where the Heinkel design's details should appear.

ABOVE: A model of the early **EF 128** being tested in Junkers' own wind tunnel.

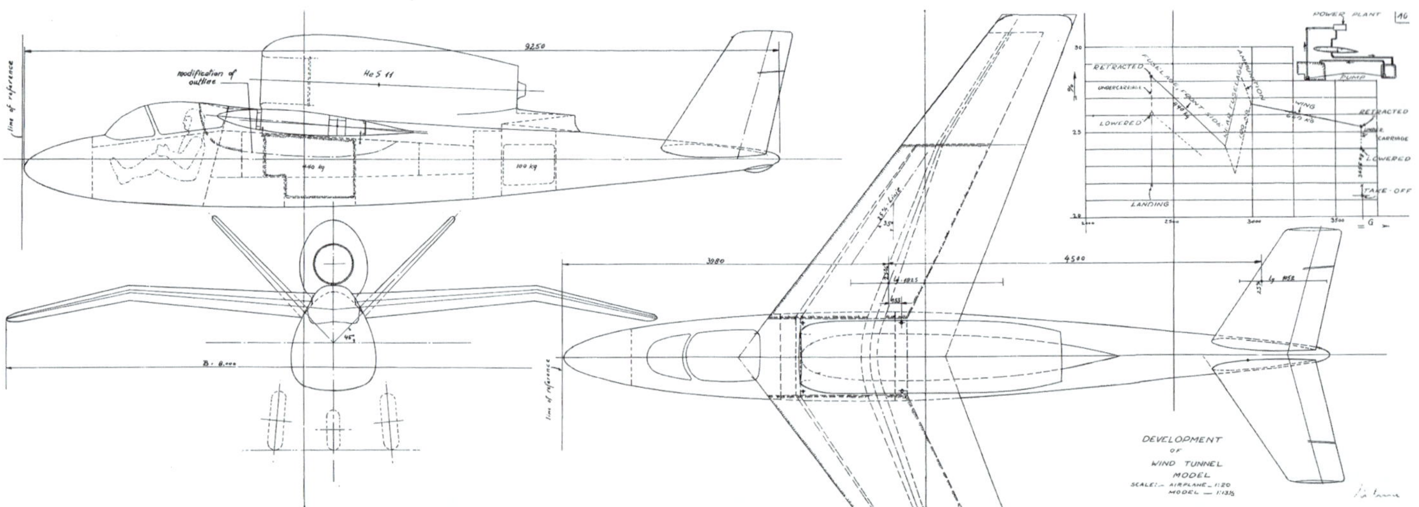

ABOVE: Exactly what Heinkel's **He 162** development looked like is unknown – however, the dimensions given for the design match those of the 'advanced' He 162 which Heinkel's engineers drew for the Americans immediately after the war.

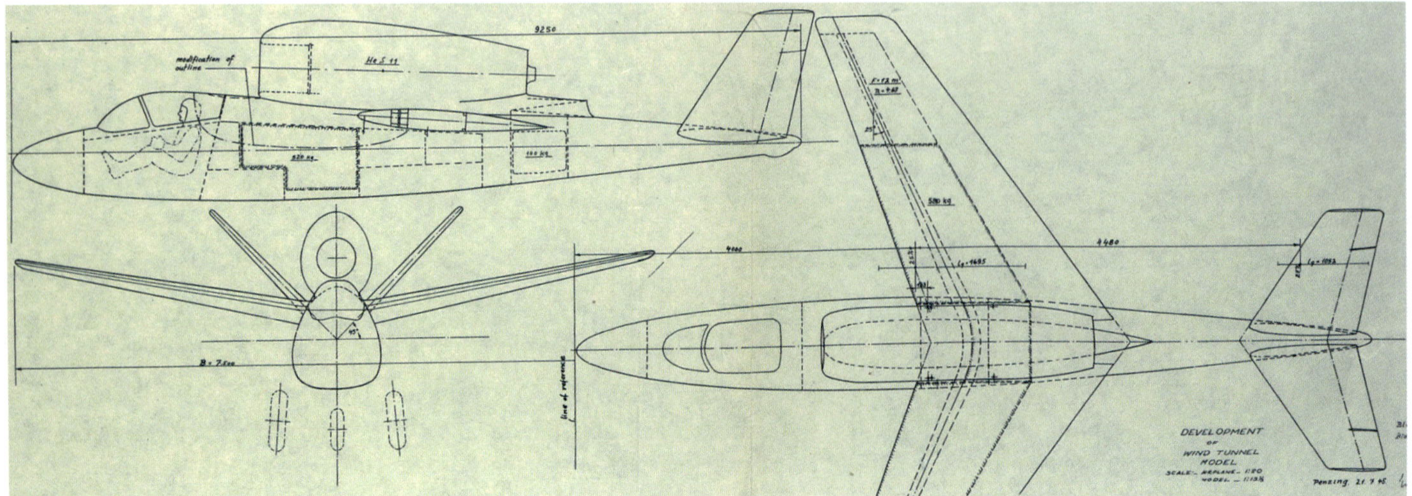

ABOVE: An alternative version of Heinkel's 'advanced' version of the **He 162** with forward-swept wings.

BLOHM & VOSS P 209

The first of Blohm & Voss's two entries was the radical P 209. The Hamburg-based company had initially designed the P 209 as a small tailless jet aircraft with swept wings and only minimal control surfaces attached to its wingtips on the end of short booms – a design which appeared as P 209.01-03. The earliest known Blohm & Voss design to use this innovative layout was the P 208.01-01, an aircraft powered by a single Jumo 222 F driving a pusher prop which appeared in a drawing dated August 30, 1944. This was an attempt to meet a requirement issued on July 21, 1944, for a new high-performance piston-engined fighter or 'Hochleistungsjäger'.

Blohm & Voss was no stranger to the use of unusual tail and control surface configurations – having pioneered several unusual arrangements – and company chief designer Richard Vogt is believed to have patented a layout where twin booms protruded from the trailing edge of an aircraft's wings with a complete fin and tailplanes on each during the mid-1930s. Yet the P 208/P 209.01 wing boom/angled control surface design appears to have had no precedent except for the early 1944 DVL sketches shown in Chapter 5.

The design submitted for consideration at the December 15 meeting, however, was dramatically different in its own right. The P 209.02-03 measured 9.2m long with a wingspan of 8.1m and a wing area of $14m^2$. It had a conventional fuselage with tailfin and swept tailplanes with a pronounced anhedral angle. A nose intake fed the turbojet mounted in the rear of the fuselage, exhausting under the tail, but the aircraft's most striking feature was its sharply forward-swept wings which, in contrast to the tailplanes, had a pronounced dihedral. They were to be built as a single-piece steel shell capable of holding 980 litres of fuel. Armament was three MK 108s – one on either side of the intake and one above it.

According to the company's project description:[20] "Our study on the Kleinstjäger, which we have largely worked through, that is our project P 211, showed such significant advantages in series production that it was obvious to apply the basic ideas of this project to the Schnellstjäger.

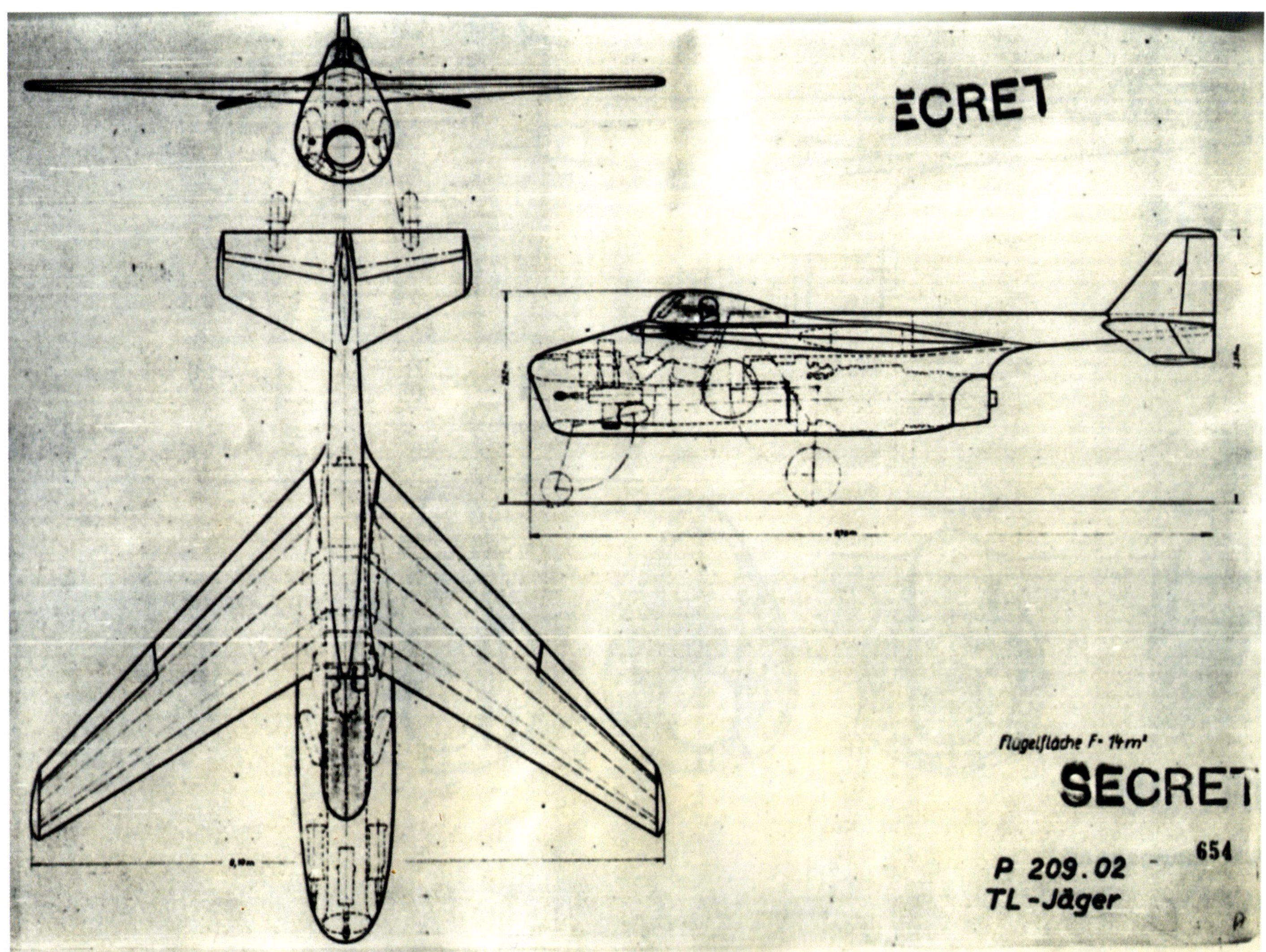

ABOVE: Blohm & Voss had high hopes for the marginally more conventional second version of the P 209 – the P 209.02 – which had forward-swept wings and downswept tailplanes.

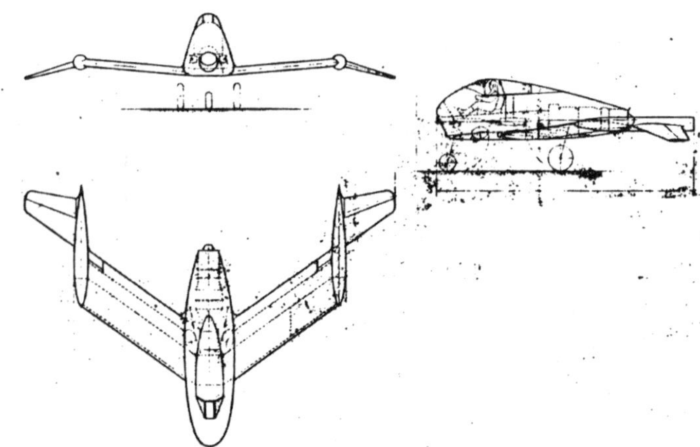

ABOVE: The short-lived tailless version of Blohm & Voss's P 209 single-jet fighter – this version being the P 209.01-03.

Of course, the aerodynamic questions associated with the much higher speed were given due consideration. This primarily concerns the sweeping of the wing that has become necessary. Although the design and manufacturing problems were deliberately placed in the foreground when working on the project – a basic requirement that is actually a matter of course these days – the results of the aerodynamic calculations show values that are within the scope of what was considered to be possible at the company meeting in Oberammergau at the time."

Blohm & Voss had ensured that there would be minimal thrust losses from the turbojet intake duct. In addition: "If the goal of aerodynamic equality with other projects is really achieved, then the advantages in terms of production technology become particularly valuable.

"The operational advantages of our solution also deserve the appropriate attention. The undercarriage attached to the fuselage makes transportation and assembly very easy and does not lead to any constructive complications.

"In the same way, the central intake duct not only has the pleasant manufacturing advantages associated with built-in installation, it also eliminates the complication in the wing support system associated with an opening on the wing and its structural disadvantages."

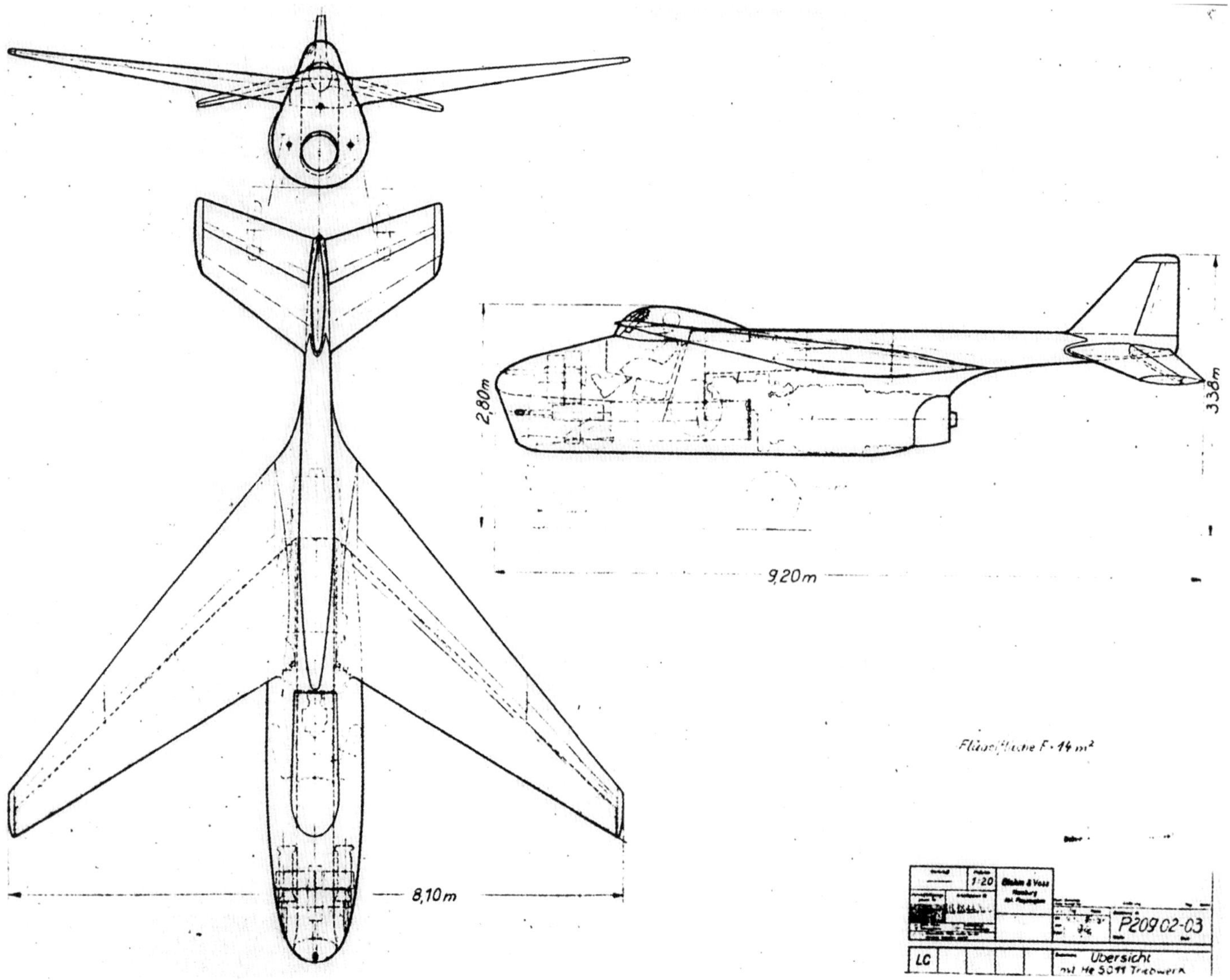

ABOVE: The last major evolution of the type P 209.02-03 featured a pronounced wing dihedral and swept tailplanes.

BLOHM & VOSS P 212.01 AND .02

The company's other entry was the P 212.02-01. While the P 212.01-01 had featured a cockpit and nose very similar to those of the P 209.02-03, the P 212.02-01 moved the pilot to the front of the aircraft, sitting just above the air intake. The swept wings were like those of the P 209.01-03 except the booms were shorter, making the control surfaces almost an extension of the wing itself. The design was outlined in a Blohm & Voss document dated November 1944 entitled 'Kurzbeschreibung – Schnellst-Jäger mit HeS 011 P 212 – Schwanzlos mit Randleitwerken' or 'Short description – fast fighter with HeS 011 P 212 – tailless with wingtip control surfaces'.[21]

The introduction said: "Compared to the P 209 project, the design described here shows a free outflow of engine gases that does not influence the stabiliser. This was achieved by the complete removal of the tail from the wing downwash.

"Incidentally, we had tried the same fuselage arrangement as that of the P 209 project. However, the performance of the new design resulted in such a considerable gain in speed – namely 35km on the ground and 45km in 6,000m altitude – that it was necessary to increase the sweep of the wing to 4°.

"This high sweep pushed the wing joint on the fuselage so far forward that a rearrangement of the cockpit and armament was required. The high sweep brings an increase in the flight weight but on the other hand it means that the tail booms can be reduced to insignificant stumps without affecting the effectiveness of the control surfaces.

"With regard to the production expense, everything that was said in the brief description of the P 209 project remains true. It is therefore advisable to consult them simultaneously. In summary it can be said that the technical benefits of the P 212 project against P 209 are both in the operational and performance areas.

"Operational – in order to avoid the consequences of an exhaust-fire – an unobstructed outlet for gases is always desirable. Performance-wise, the big speed gain is considered particularly important. With a 1,030km

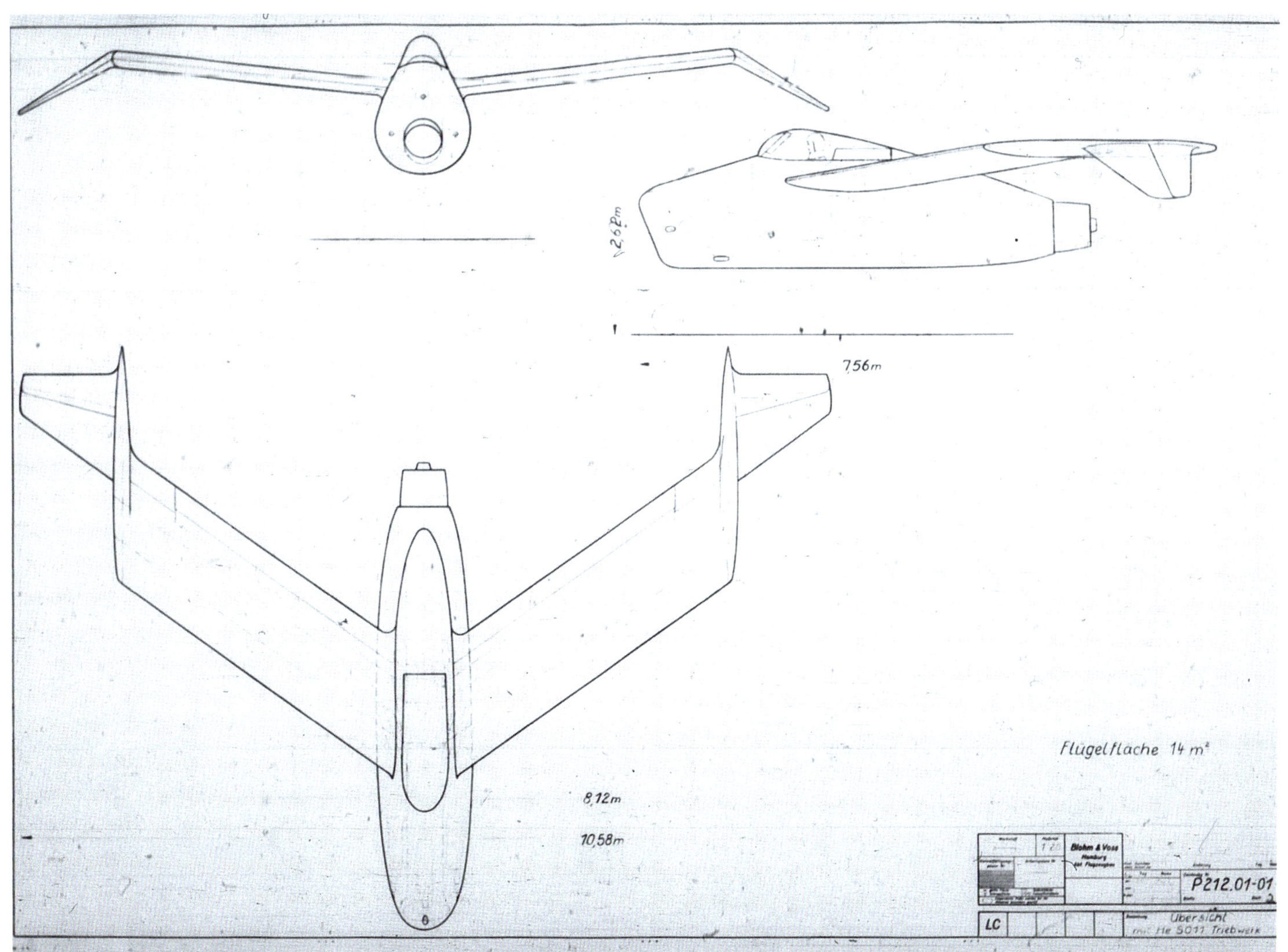

ABOVE: The original Blohm & Voss P 212 scheme was P 212.01-01. It had a nose intake similar to that of the P 209.02 but with the tailless layout of the P 209.01.

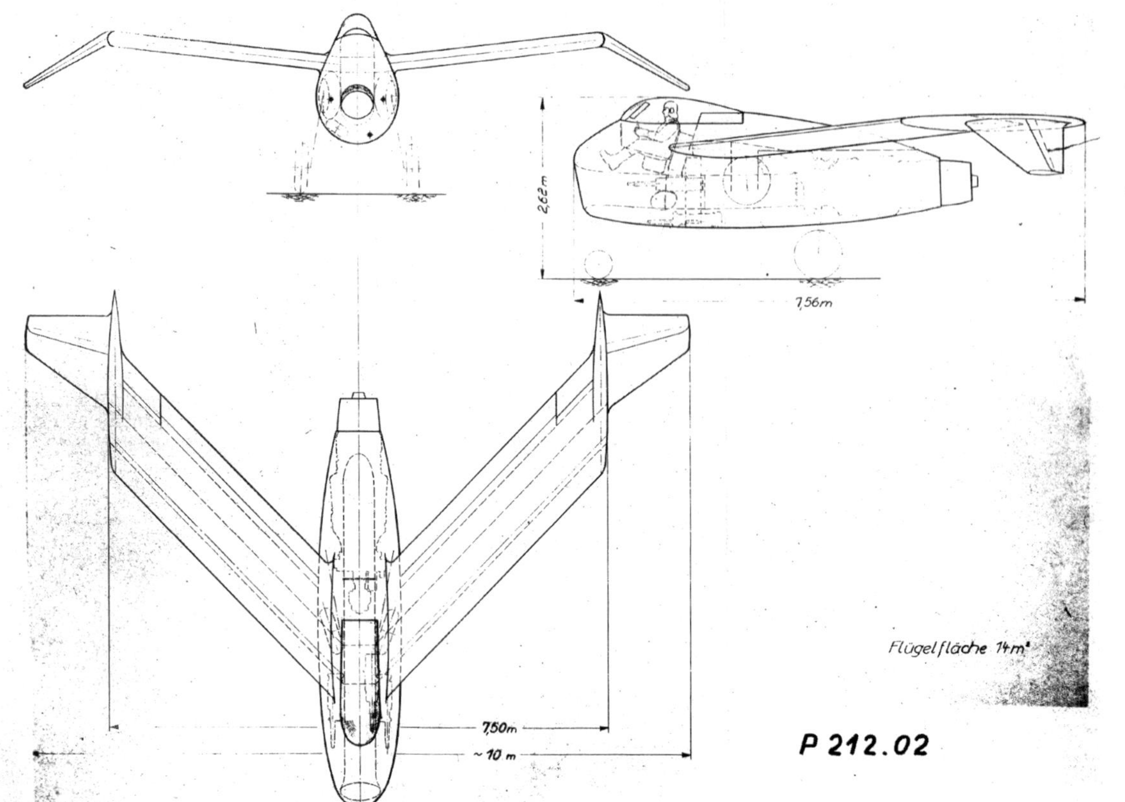

LEFT: The Blohm & Voss P 212.02 as it appeared in the Blohm & Voss P 212 project description produced by the company in November 1944.

BELOW: This intermediate design for the P 212.02 features a raised spine, a shorter fuselage and smaller wings than the P 212.01.

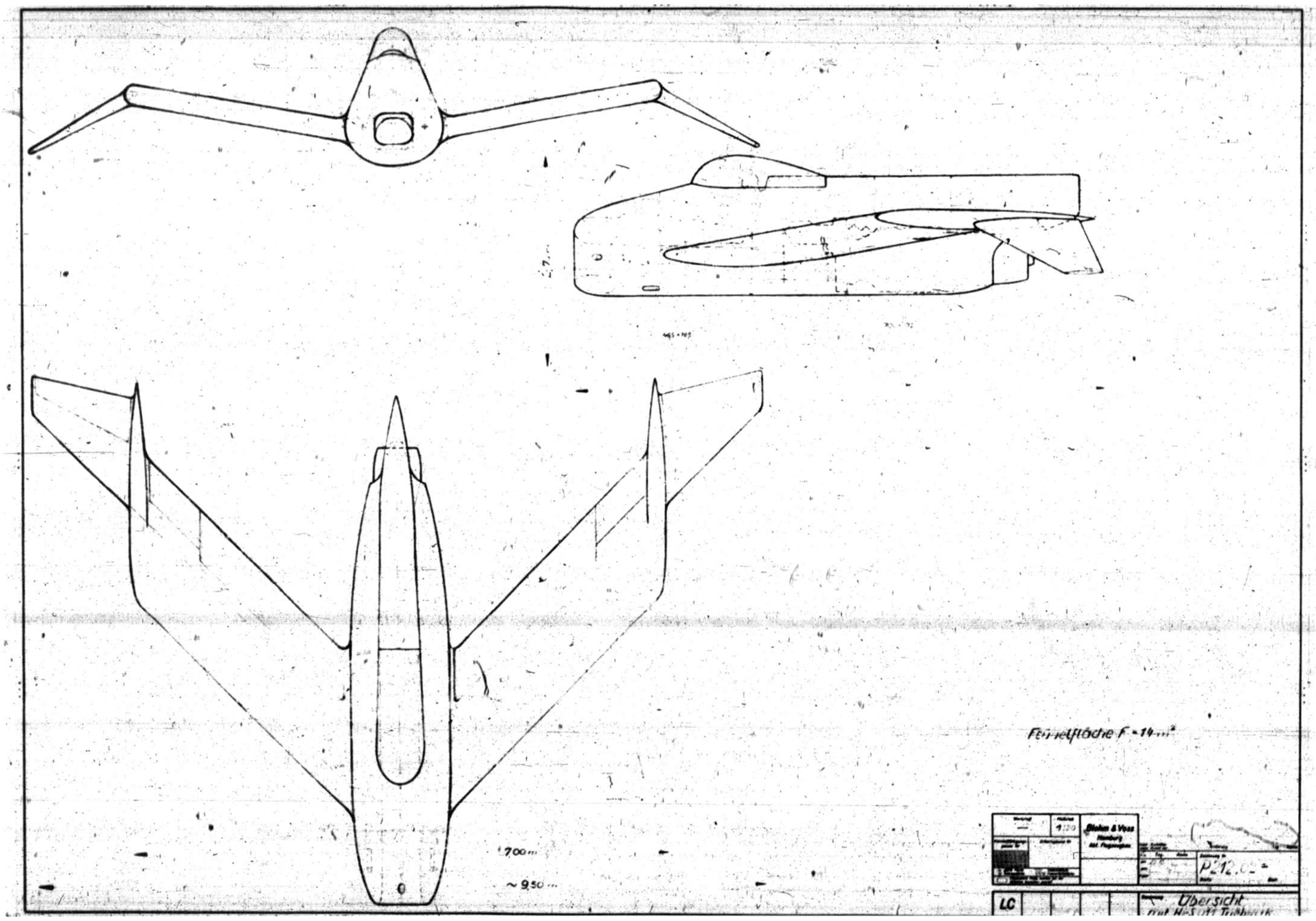

maximum speed the new project should be placed at the top of any comparison of projects with the same engine."

The P 212's wings, like those of the P 209, were to be "constructed as a steel shell" but could only hold 750 litres of fuel.

As with other Blohm & Voss designs of the period, the jet engine's intake pipe was to act as the main structural support for the whole aircraft. The brochure states: "Again, the TL inlet pipe is the inner support of the fuselage. On it all the control operations, the weapons, the instruments and the equipment are installed and so the installation of all these is completely accessible.

"Given the large size of the inlet pipe required for the engine, even without special bracing and with thin wall thickness, it is of sufficient strength."

Regarding the two main landing gear wheels: "The main landing gear is mounted on the fuselage boom and swings in forwards. The angle required for the retraction strut is favourable so only small pull forces or low hydraulic pressures are required.

"Extending the gear into the flight wind is done automatically, so it is particularly reliable. The nosewheel swings back at an angle to the fuselage. The bearing of the spring strut and retracting mechanism is carried on the fuselage pylon."

And regarding the engine: "The installation of the device is possible without a crane; the bodywork panels simply need to be folded away from it by the ground personnel."

Another 550 litres of fuel was to be contained in a fuselage tank – for an overall total of 1,300 litres. The wing tanks would be drained first, followed automatically by the fuselage tank. The aircraft's three MK 108 30mm cannon would be fitted around the jet intake, with the ammunition boxes, each containing 70 shells, positioned above them and behind the pilot.

DVL ASSESSMENT

Following this second comparison, during a further meeting of the EHK on December 19–20, responsibility for the 1-TL-Jäger designs was handed over to the Sonderkommission Tagjäger or 'special commission for day fighters', led

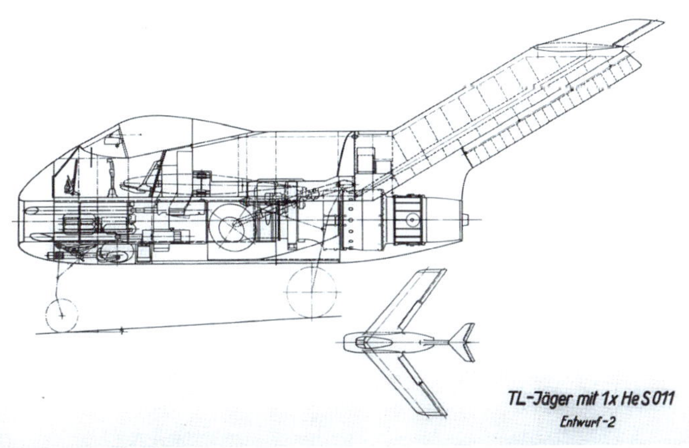

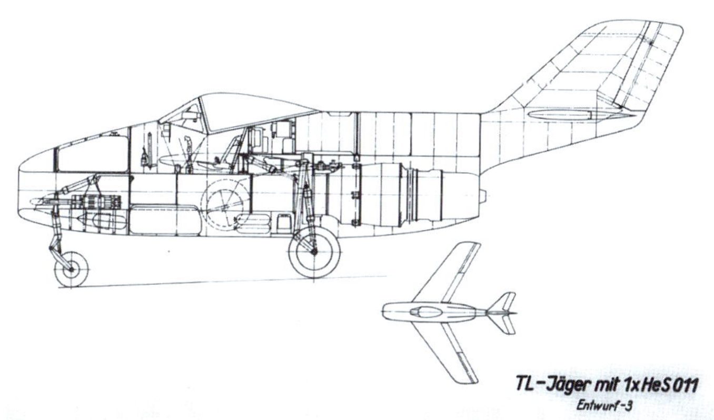

ABOVE: Focke-Wulf's 'Huckebein' as it appeared in January 1945 and the new 'Entwurf Mittelhuber which joined it as the company's second entry – a simpler design with a conventional tail. Since 'Flitzer' had been Focke-Wulf's 'Entwurf I', these designs were known as 'Entwurf II' and 'Entwurf III' respectively.

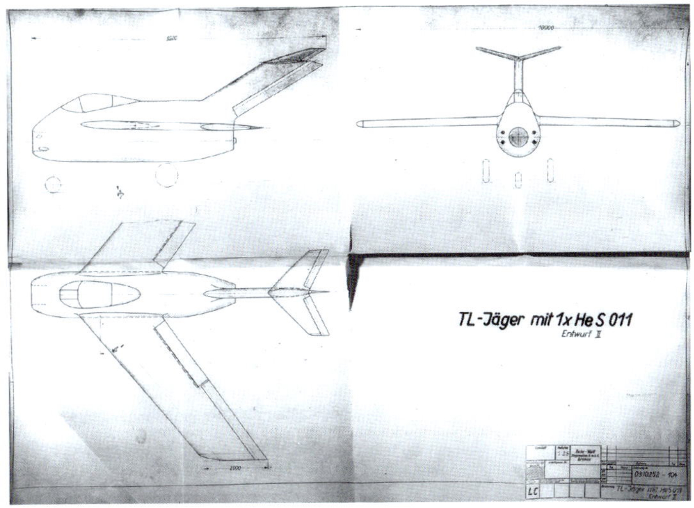

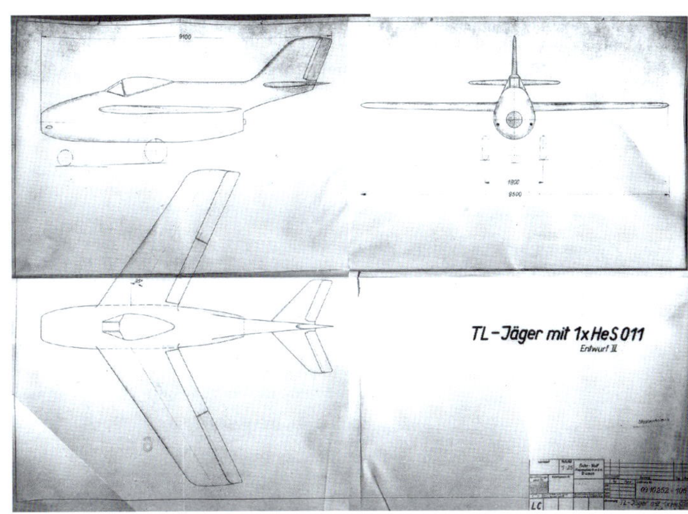

ABOVE: Three-view drawings of Focke-Wulf's Entwurf Multhopp and Entwurf Mittelhuber from January 1945. The former had been internally revised since it became the company's first choice at the end of October 1944.

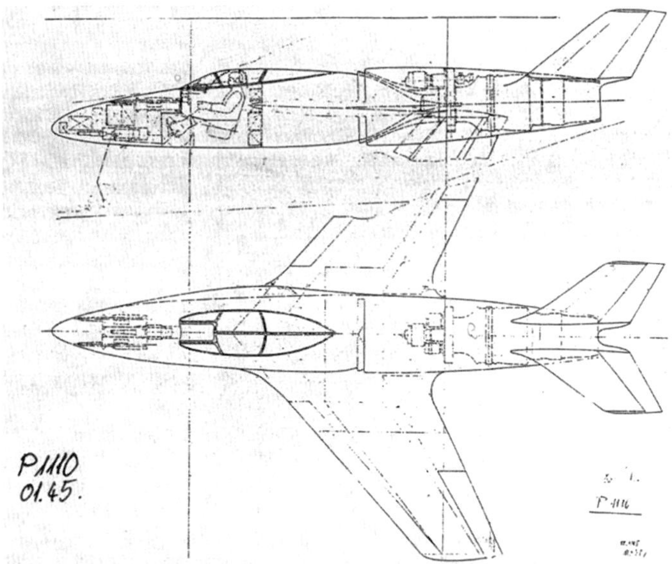

ABOVE: Another view of the P 1110 of January 1945 – this time showing the internal workings of the annular intake system.

by Willy Messerschmitt.[22] This was a sub-committee of the Entwicklungshauptkommission Flugzeuge (EHK) – a development committee established by Germany's minister for war production Albert Speer on September 15 to oversee work on new aircraft types.

Among the nine EHK sub-committee leaders were some of the German aircraft industry's biggest names: Messerschmitt himself, Kurt Tank, Heinrich Hertel, Günther Bock and Walter Blume. Each member was the head of a Sonderkommission – besides Messerschmitt, for example, Tank led the special commission on night fighters and Hertel the commission on bombers.

Messerschmitt began the meeting with a discussion described in the official minutes as "Otto and jet fighter or only jet fighter?" It was stated that "there is a clear, uniform view that only the jet fighter can be decisive in the performance race with the foreign country. It is known that the jet fighter development abroad is roughly on the same level as in Germany. Based on this knowledge, the jet fighter with the more powerful HeS 11 engine must be developed as quickly as possible, especially since the aircraft type 162 with the 003 engine is known not to be a best solution, but only a transition solution.

"Decision: The development of the HeS 11 engine must be accelerated with very special means. The Special Commission for Aircraft Engines (Schilo) has to create for this task preferred capacities, personnel etc. immediately. The Special Commission for Day Fighters (Messerschmitt) has to make the decision about the airframe project to be carried out at the beginning of January.

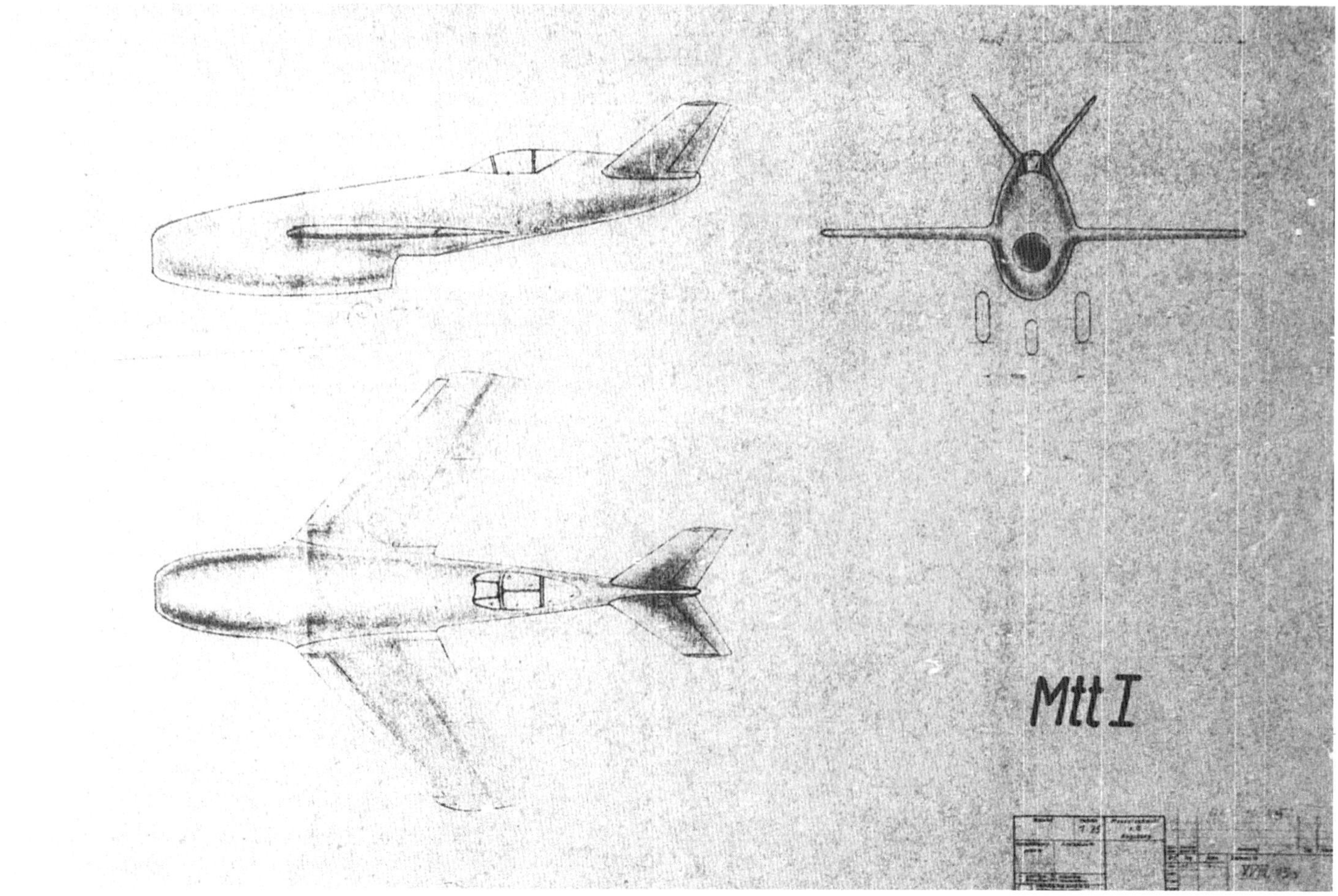

ABOVE: The last version of the P 1106 was submitted for the single-jet fighter competition in January 1945 as 'Mtt I'.

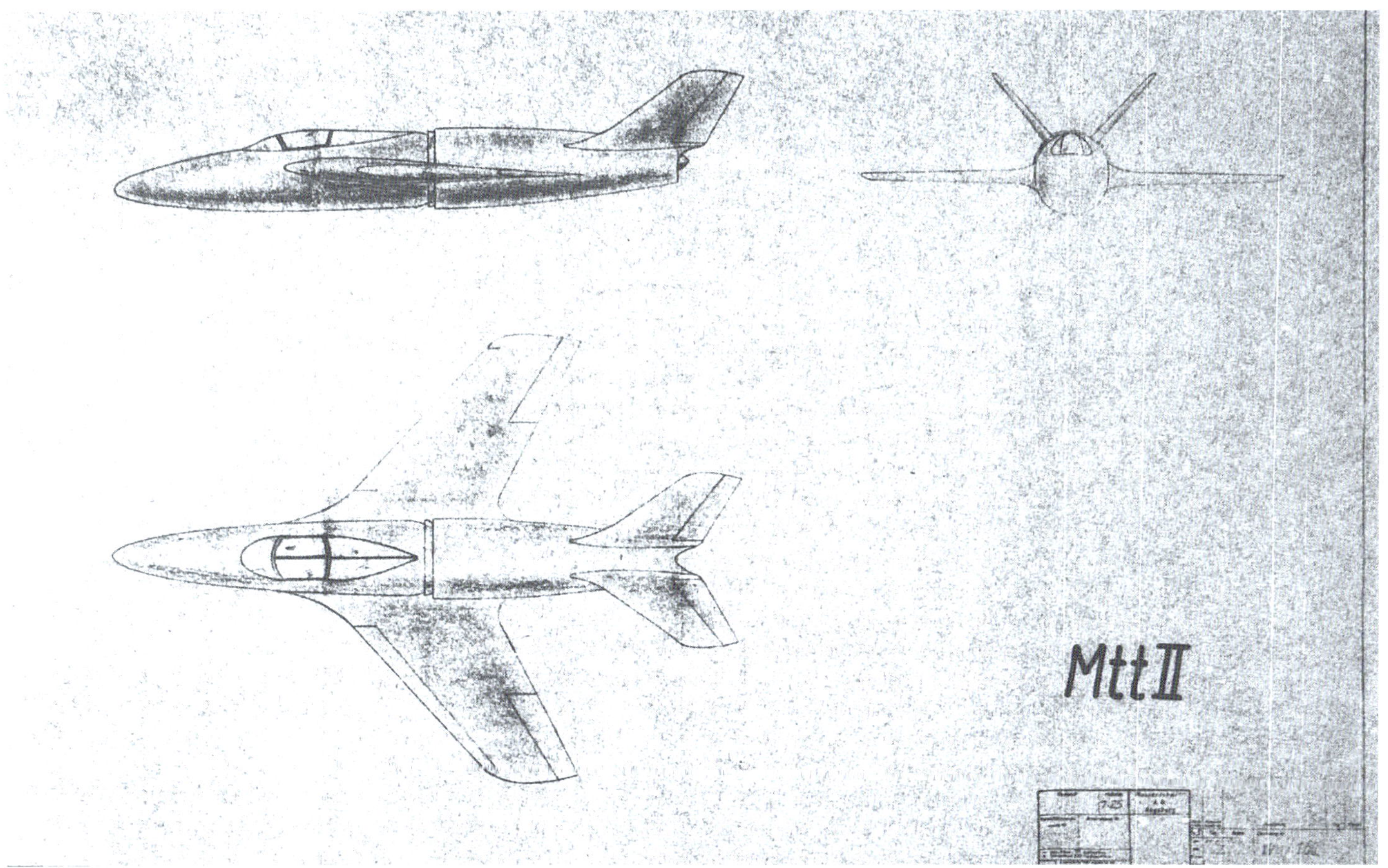

ABOVE: Messerschmitt's P 1110, submitted as 'Mtt II', was an attractive new design for January 1945 but featured a risky annular intake for its single HeS 011 engine.

"Since previous front line experience has not yet clearly shown whether the jet fighter is sufficient for all operational tasks due to its special properties compared to the Otto fighter, further development of the Otto fighter cannot yet be dispensed with.

"Decision: The further development of aircraft type [Ta] 152 must be continued on both the engine and the airframe side by all means in order to keep pace with the achievements of Otto fighters abroad in the near future. The Special Commission for Day Fighters (Messerschmitt) must also independently check at the beginning of January 1945 whether the new development of an Otto fighter still makes sense in terms of performance and effort compared to the 152 to be further developed. This requires a clear determination of how long and to what extent the performance of the 152 can still be increased compared to corresponding new Otto fighter designs."

During the meeting it was requested that the OKL – the Luftwaffe high command – formulate a clear list of what sort of fighters it wanted and what features they should have no later than the end of December.

The DVL assessment of the second-round single-jet designs was carried out by Professor August Quick and Dr P. Höhler between December 19 and December 21, 1944, and the results were published on January 9, 1945, as 'Untersuchungen und Mitteilungen Nr. 1448' ('Investigations and communications', the German term usually shortened to UM).[23] It was noted that the Huckebein's tail was likely to cause significant drag at its root during high-speed flight and rolling behaviour was "questionable". Doubts were also raised about the P 1106's rolling behaviour and its V-tail was likely to cause problems. The EF 128's intakes were expected to cause difficulties at high speed and although the assessors approved of its wing-mounted rudders these would need to be tested for effectiveness in a wind tunnel.

The He 162 development's V-tail with end plates was likely to cause disruption at high speed and the design had the same problems when it came to rolling characteristics as the Huckebein and the P 1106. For the P 212 "a vertical stabilizer appears to be necessary" and it appeared that the "large masses" at the ends of its wings would create the risk of flutter. Finally, the P 209 appeared to have the worst roll characteristics of all the competitors, with directional stability likely to be poor. It was also feared that the swept-forward wings would suffer from deformation at high speeds. The tail surfaces were too thick and the fin needed to be positioned further back.

1-TL-JÄGER THIRD COMPARISON

Just two days later, on January 11, the Luftwaffe high command finally issued its 'development request' for fighter aircraft – which the EHK had asked for by no later

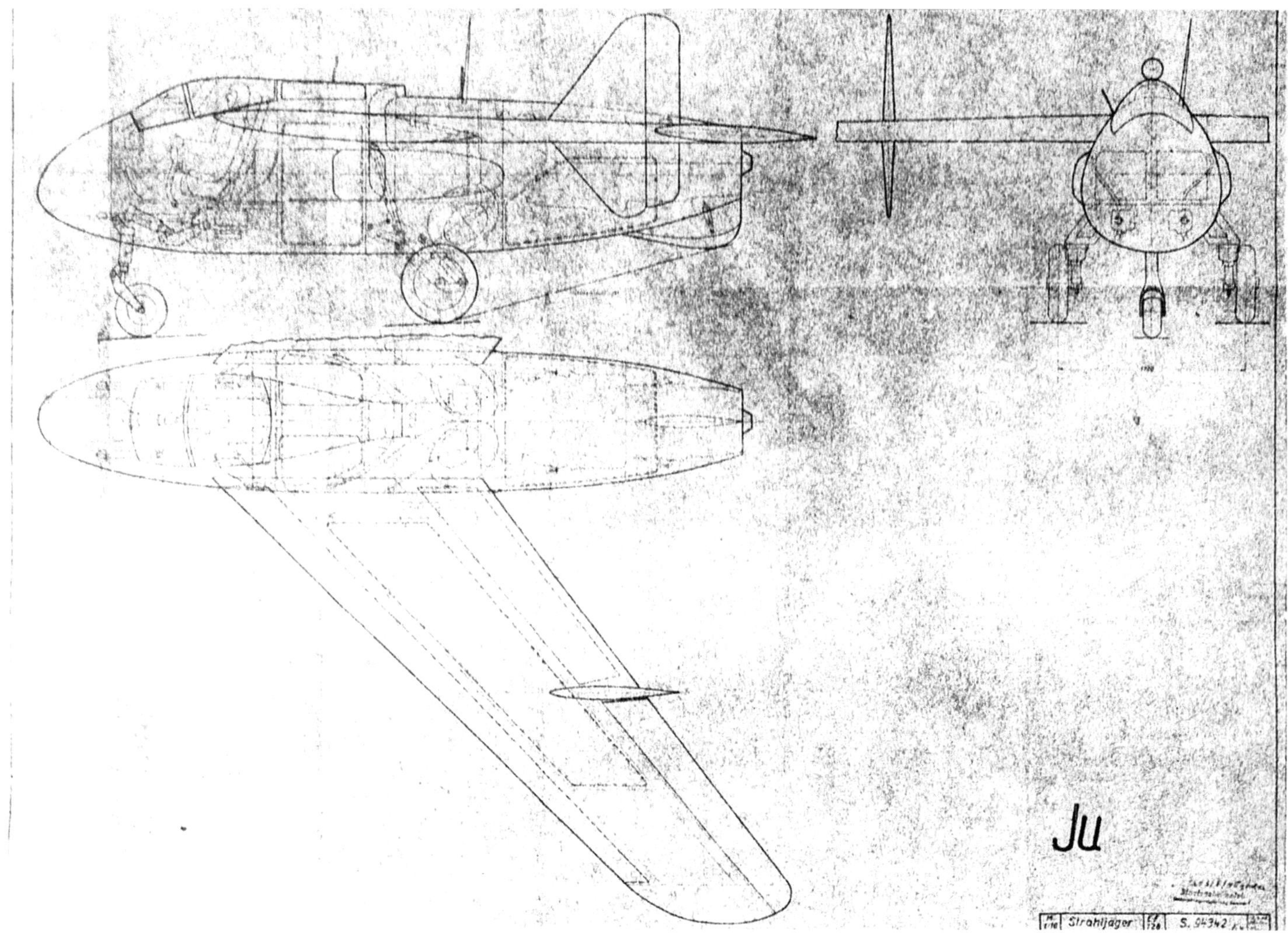

ABOVE: Junkers EF 128 in January 1945, with reprofiled nose, boundary layer suction and longer wings.

than the end of December.[24] It outlined specifications for three fighter aircraft – a "one-seat jet fighter with 1 x HeS 011 as successor for 8-162 with suitability for bad weather fighting and landing", a "one-seat propeller fighter with a performance-enhanced, air-cooled Otto engine if possible, as replacement for Ta 152 with suitability for bad weather fighting and extensive automatic mastery of bad weather landing" and finally "development of a three-seat night and bad weather fighter, equipped with a high-performance Otto engine in the fuselage driving the propeller and two jet engines in the wings".

The single-seat, single-jet aircraft had to be capable of 1,000km/h at an altitude of 8–10km. Flight time at that altitude, without drop tanks, had to be at least two hours at full throttle and armament was to be "4 x 3 cm weapons (MK 108 or MG 213/30) with 120–150 rounds per barrel".

It also had to carry standard fighter equipment such as a cabin heater, radio and oxygen gear plus an IFF system.

General requirements for all three aircraft were the ability to carry the 8-344 air-to-air missile, better known as the Ruhrstahl X-4, R4M rockets, Jägerfaust or high-calibre rockets full of incendiary shrapnel. The EZ 42 gunsight, or whatever its further development might be, was required too, as was course control, armour protection against 20mm rounds front and back, a pressure cabin, nosewheel undercarriage, ejection seat, blind-flying instrumentation, rough field take-off and landing capability and "sufficient strength of the undercarriage and tyres during overload starts".

The next day, January 12, a new two-day comparison meeting took place arranged by the Sonderkommission Tagjäger[25] – followed immediately by a further four days of design assessment by the DVL.[26]

A total of seven designs were put forward for the January meeting by Focke-Wulf, Messerschmitt, Junkers, Heinkel and Blohm & Voss. Focke-Wulf's Baubeschreibung Nr. 279 'Huckebein' was joined by a second entry from the company – known within the company as 'Entwurf Mittelhuber' after company project office leader Chefingenieur Ludwig Mittelhuber and described in Focke-Wulf Kurzbeschreibung Nr. 30,[27] dated January 10, 1944. The designs were known to the assessors as 'FW I' and 'FW II', though Focke-Wulf itself preferred 'Entwurf 2' and 'Entwurf 3', based on the Flitzer having been 'Entwurf 1'.[28]

The Mittelhuber design was essentially a less radical brother to Multhopp's Huckebein, featuring a conventional tail and a cockpit positioned more centrally in the aircraft's fuselage. It measured 9.1m long and had a wingspan of 9.5m, making it 10cm shorter than Huckebein and with a 50cm narrower wingspan of 9.5m.

Messerschmitt was also now offering two entries – the P 1106 and the P 1110, a new design. The former was an updated version shown in drawing XVIII/158 while the latter again continued the XVIII drawing sequence, appearing in XVIII/162. As with the Focke-Wulf designs, these were arbitrarily labelled 'Mtt I' and 'Mtt II' respectively for the competition. Both had a wingspan of 6.65m and a sweepback angle of 37.8° but here the similarities ended.

The P 1106 of XVIII/158 differed from earlier designs in having a more slender fuselage overall and an undercarriage which pushed the nose of the aircraft up, where the aircraft had previously been parallel to the ground. The cockpit canopy was wider and with more frames and the V-tail fins were more sharply swept.

MESSERSCHMITT P 1110

The new P 1110 was a much more handsome-looking aircraft. Its layout prefigured those of modern fighter jets, with a long aerodynamically formed nose and a turbojet mounted centrally within its rear fuselage. It retained a V-tail, however, and its turbojet was fed by a highly unusual annular intake – a narrow slot which ran around the circumference of its cylindrical central fuselage. Again, despite the work being carried out to construct a flying prototype at Oberammergau, the P 1101 was not submitted for consideration as the basis for the Luftwaffe's new single-jet fighter.

JUNKERS EF 128 REVISED

Junkers once again offered the EF 128. Since its last appearance on December 15, the design had undergone some revision – including a reshaped nose and cockpit canopy plus longer wings. It also now had an innovative boundary layer suction system missing from the earlier version. Its tailless fuselage was just 6.45m long and wingspan was 8.9m, with a sweepback angle of 43°.

HEINKEL P 1078

Heinkel's January design was entirely new – the P 1078.01 appears in two known drawings: 1078.01-03 and -04.[29] Both show the same aircraft but where the former is a close-up view of the fuselage, the latter shows the aircraft's wings and general layout. It was a small tailless design with shoulder-mounted wings which jutted straight from the short fuselage before the central section of each angled upwards while the outer section angled downwards.

The P 1078.01 had a tricycle undercarriage which barely had room to retract into a fuselage measuring just 5.2m long. The pilot sat right at the front above a slot-shaped intake and the turbojet was mounted to the rear.

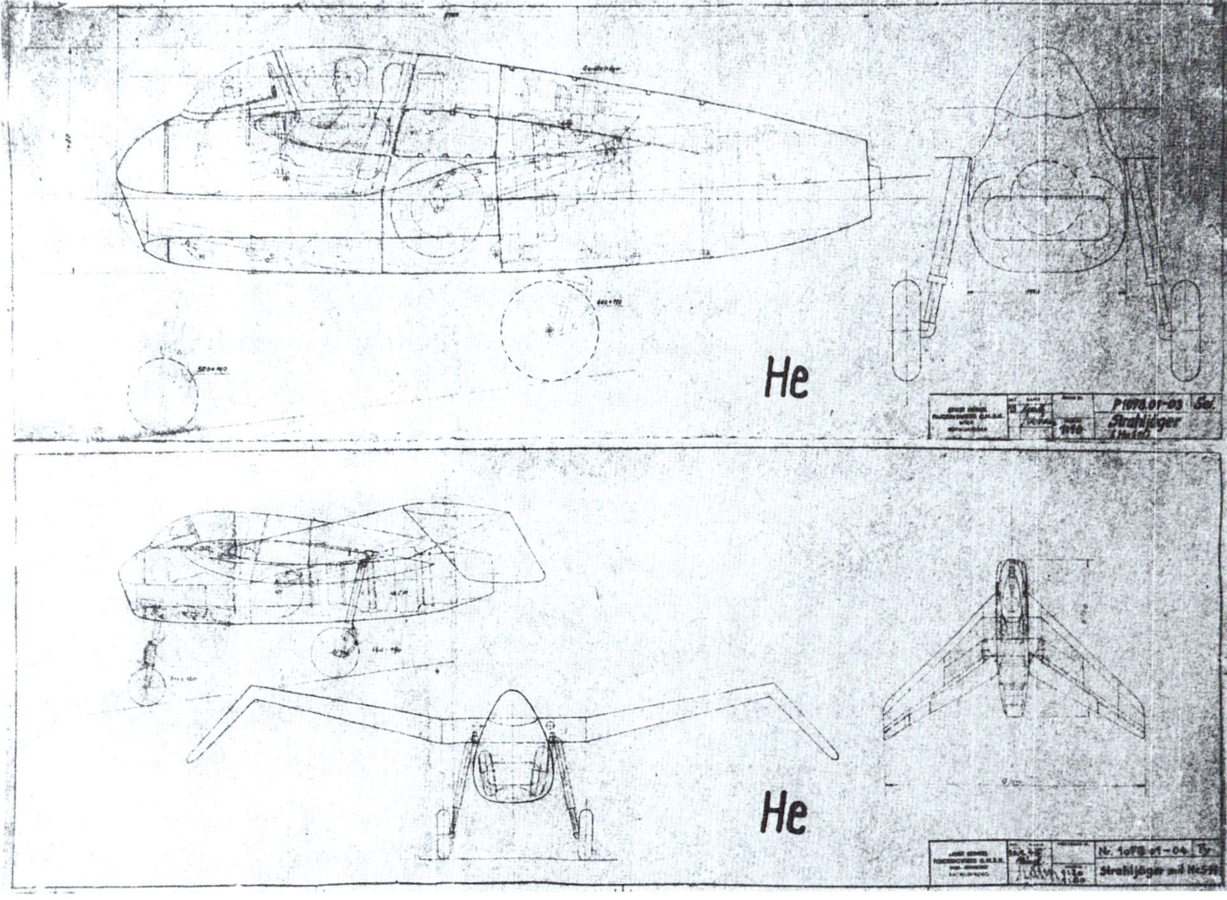

RIGHT: Heinkel's diminutive P 1078 as it appears in drawings P 1078.01-03 and -04 of January 10, 1945.

Wingspan was 9m with a sweepback angle of 43°. A report on the design was sent to the Sonderkommission Tagjäger by Heinkel on January 17, 1945.[30]

BLOHM & VOSS P 212.03

Blohm & Voss had, by now, withdrawn the P 209 from consideration and was concentrating instead on the P 212 alone. Like its competitors, the design had undergone some revisions since January, becoming the P 212.03. The cockpit had been pushed back from the nose, creating space for up to five MK 108s, and the whole fuselage had been reprofiled to become more slender. The wings had been shortened to give a wingspan of 7m, compared to 7.5m for the P 212.02, but chord had been increased to retain a wing area of 14m². A drawing of January 8, 1945, showed that Blohm & Voss had been considering fitting a fin to the rear of the aircraft's fuselage but the actual competition entry drawing shows a pair of fins instead – one on top of each prehensile wingtip boom.

DVL ASSESSMENT

Willy Messerschmitt gave a three-hour-long presentation on the 1-TL-Jäger designs during the second day of a two-day meeting of the EHK at Fährhaus Tornow, Potsdam, from January 23–24, starting at 2.30pm.[31]

The results of the DVL's second assessment, carried out by Quick, Höhler and Karl-Heinrich Doetsch, were published on February 26, 1945, as UM 1448 part 2.[32] According to the report, a detailed assessment of Heinkel's P 1078.01, Focke-Wulf's 'Entwurf Mittelhuber' and Messerschmitt's P 1110 could not be given "due to insufficient documentation".

Concerns regarding the Huckebein's tail were reiterated and the "large longitudinal extension of the fuel tanks should bring about difficulties in longitudinal stability". Also, "the pilot's exit requires the tail unit to be blown off or the like". Visibility was good but special attention needed to be paid to the behaviour of the tail with regard to stability and rudder effectiveness.

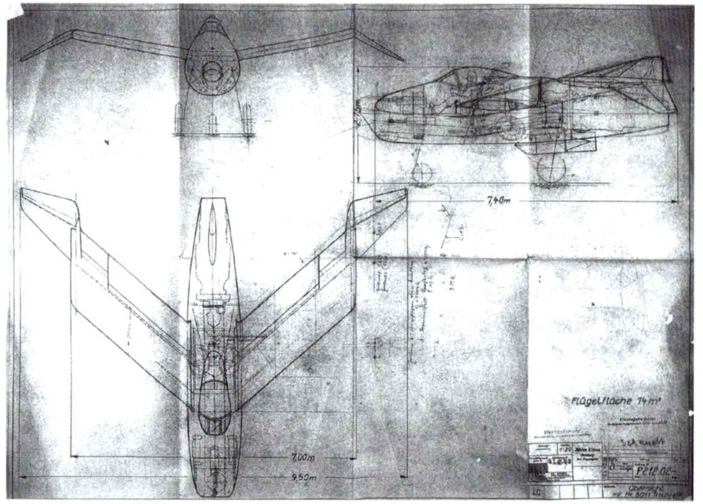

ABOVE: Despite being labelled P 212.02, this Blohm & Voss drawing dated January 8, 1945, appears to show the earliest version of the P 212.03 - but with a single fuselage fin.

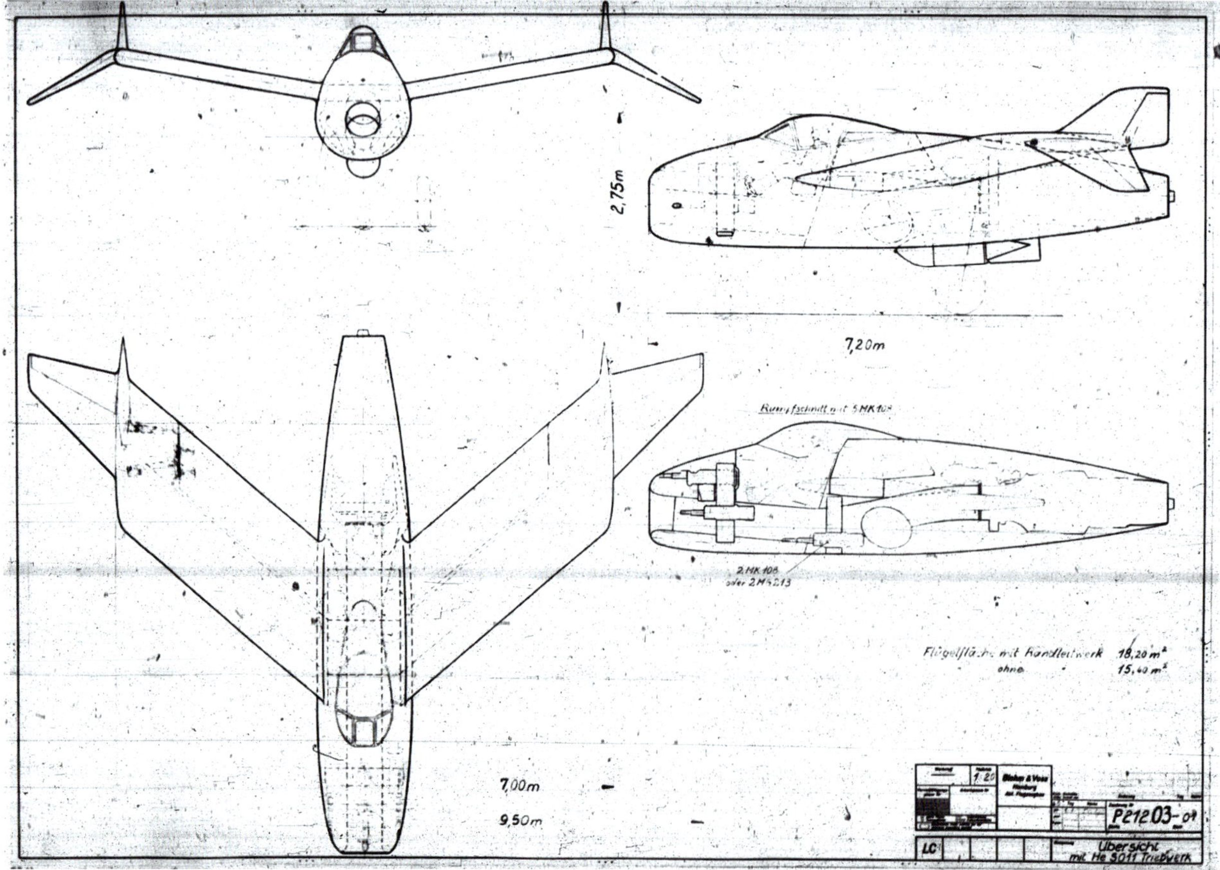

LEFT: The final version of Blohm & Voss's P 212 - P 212.03-01.

The P 1106's V-tail remained a likely area of problems, pilot visibility "does not appear to be satisfactory", "the elastic deformation of the fuselage due to the forces on the V-tail and its consequences on the flight characteristics must be examined in detail". The unusual layout of the aircraft in general brought about serious concerns about stability and wing loading was "extremely high".

Again, the EF 128's intakes were likely to be a problem – "it is expected that the first compression shocks will occur at this point". Rudder effectiveness could be reduced due to "the expected boundary layer runoff" on the wings and the "low down position of the weapons installation will make targeting difficult". Similar to the Huckebein, "the large longitudinal extension of the fuel tanks in the fuselage significantly affects the static longitudinal stability".

The only comments made about the P 212.03 concerned the flutter potential from the large masses at the ends of its wings and "stiffness difficulties".

1-TL-JÄGER FOURTH COMPARISON

On the same day that Quick, Höhler and Doetsch's report was published, another report was published by Messerschmitt. This document,[33] signed by Hornung and Voigt, was a pre-conference outline for the next two-day comparison meeting due to begin the following day, February 27, at Messerschmitt's Oberammergau facility. But where previous design comparison meetings had been held by Messerschmitt's Sonderkommission Tagjäger, the February meeting was a full meeting of the EHK where it was hoped that a final decision could be made on which designs, if any, would proceed to the prototype development stage and which would be eliminated.

The introduction to Hornung and Voigt's report stated: "A considerable interruption of the work necessary for this report was brought about by war conditions. On account of the bad traffic and communication facilities, it was not possible to obtain in written form the report of the DVL on general performance and criticism of the characteristics of the aircraft. For the same reasons it was not possible to compare in the general discussion with DVL the additional designs tendered (design from Heinkel and designs P 1101 and P 1111 from Messerschmitt)."

The report doesn't mention the fact that the DVL's report had only just been published on the same day.

Focke-Wulf's Baubeschreibung Nr. 279 aircraft (described as such) aka 'Entwurf Multhopp' aka Huckebein had been carried forward mostly unchanged from January but the Kurzbaubeschreibung Nr. 30 aircraft aka 'Entwurf Mittelhuber' had been altered somewhat. Where previously the design had featured a straight back it now had a curved spine. This had little impact internally but did serve to smooth out the transition from fuselage to tail externally.

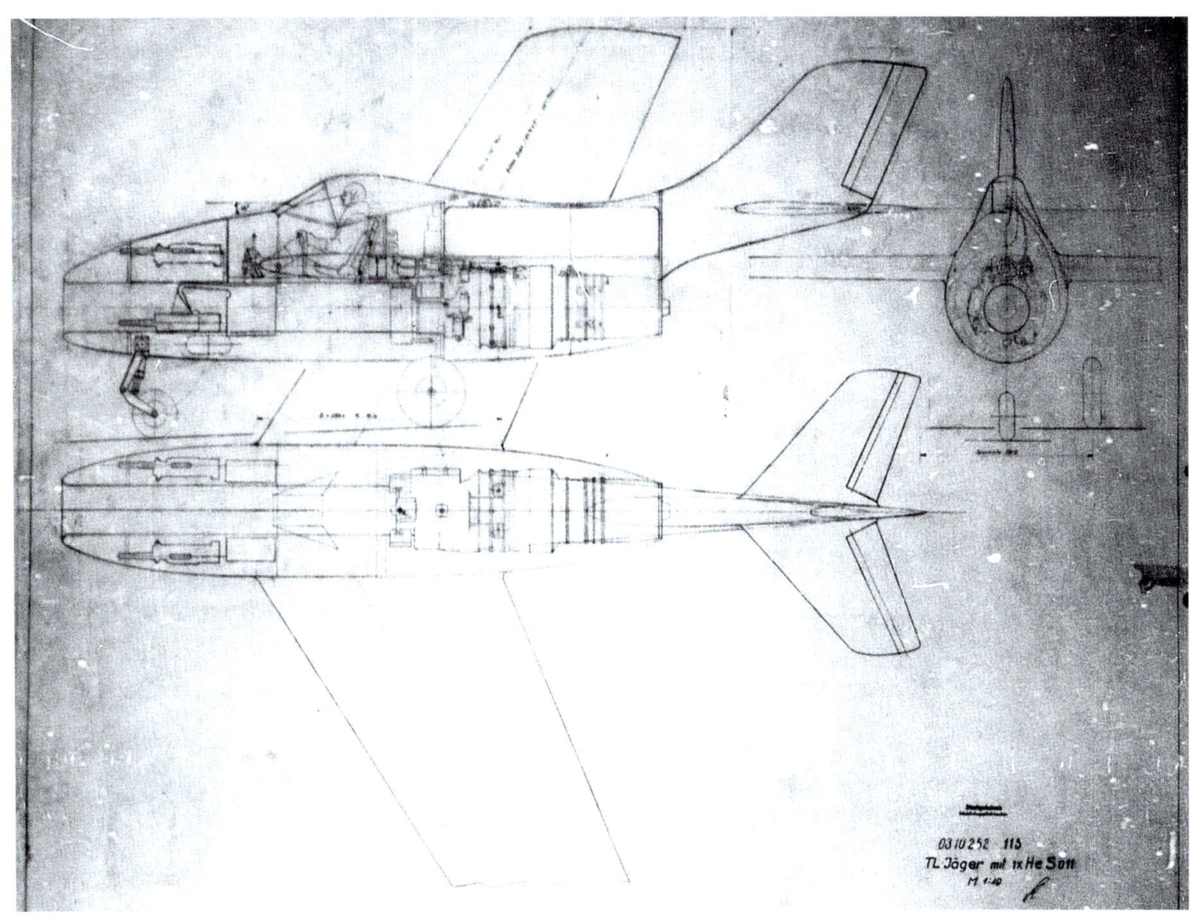

LEFT: The revised Focke-Wulf 'Entwurf Mittelhuber' with curved spine from February 1945 as it appears in drawing number 0310 252-115.

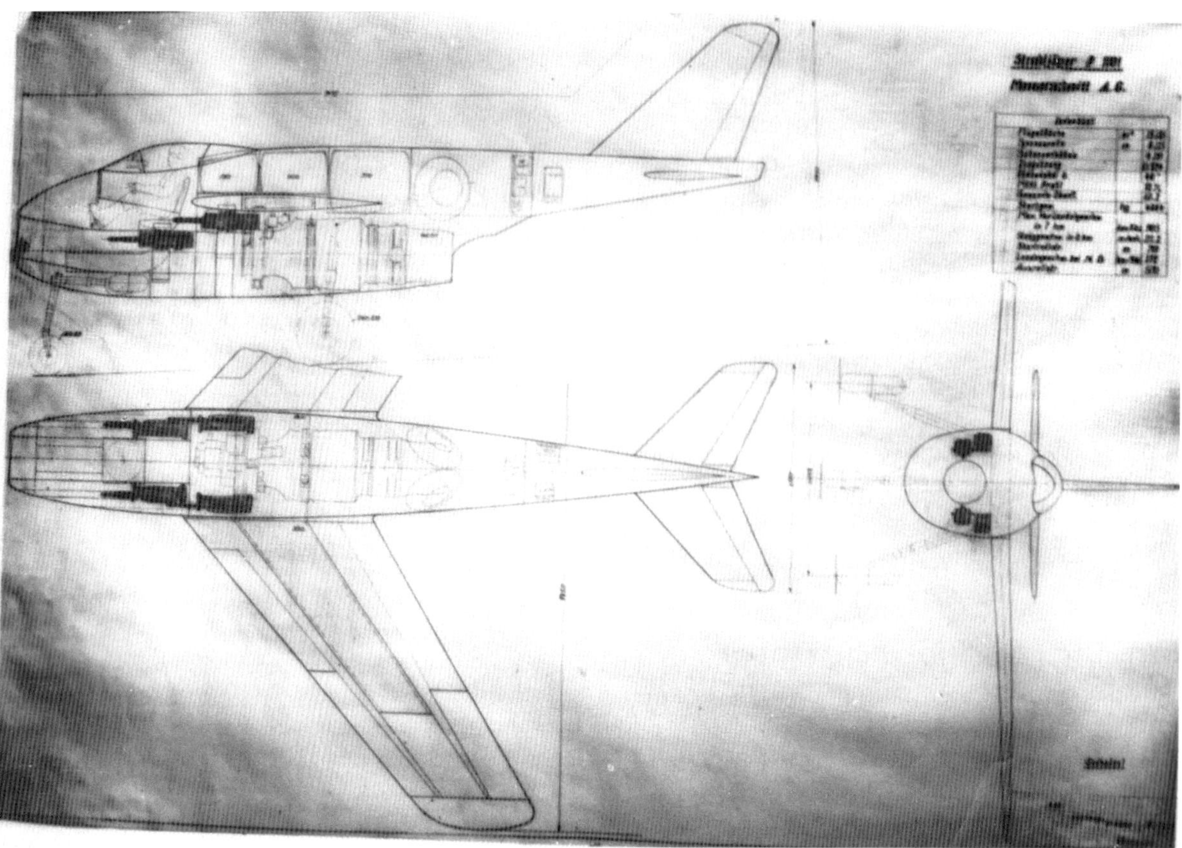

RIGHT: The Messerschmitt P 1101 of February 1945 had a reprofiled nose compared to the design of the previous November and therefore differed significantly from the experimental prototype already close to completion by this point.

LEFT: Messerschmitt regarded the P 1111 as inherently dangerous but presented it in February anyway – and it was chosen for development ahead of both of the other Messerschmitt designs and most of its competitors.

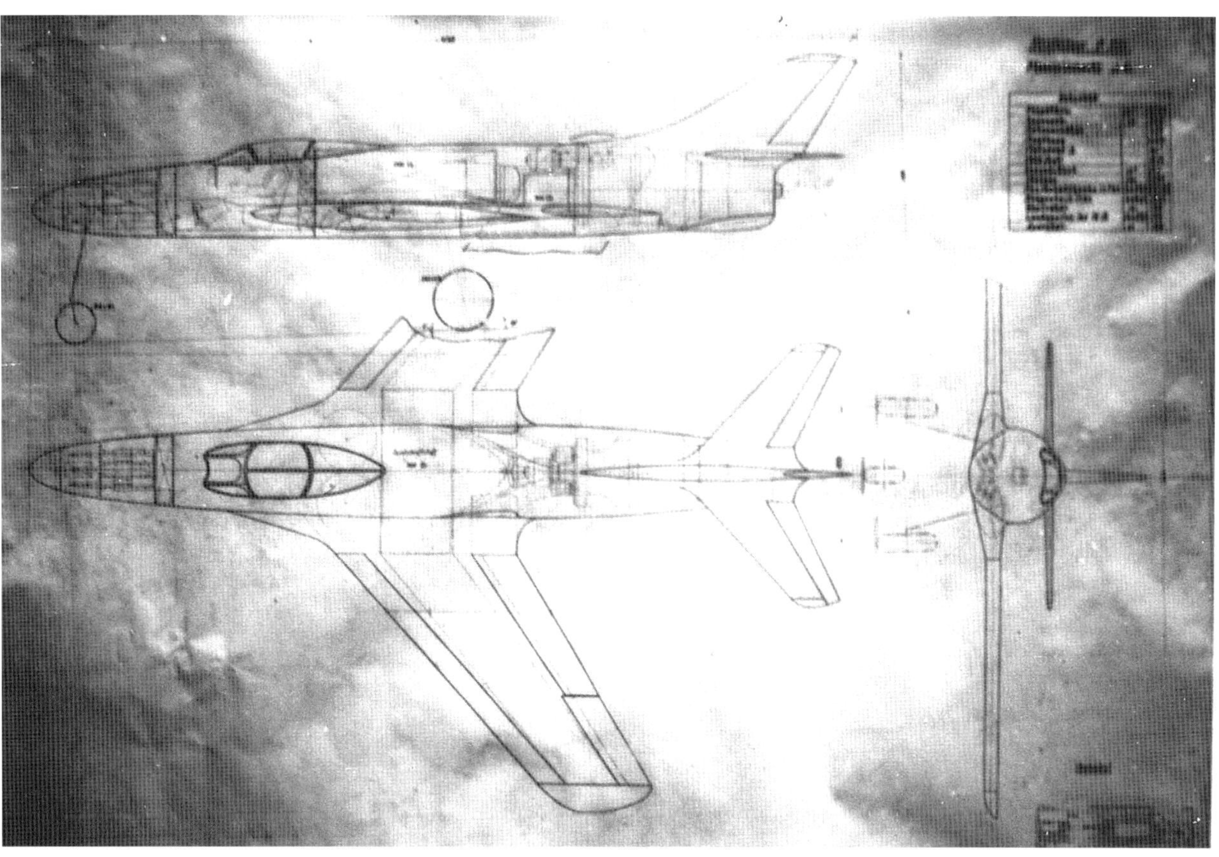

RIGHT: Messerschmitt's last P 1110 was, in hindsight, a remarkably modern-looking design. The unknown quantity of the January version's annular intake had been replaced with the slightly less unknown quantity of narrow intakes built into the sides of its fuselage.

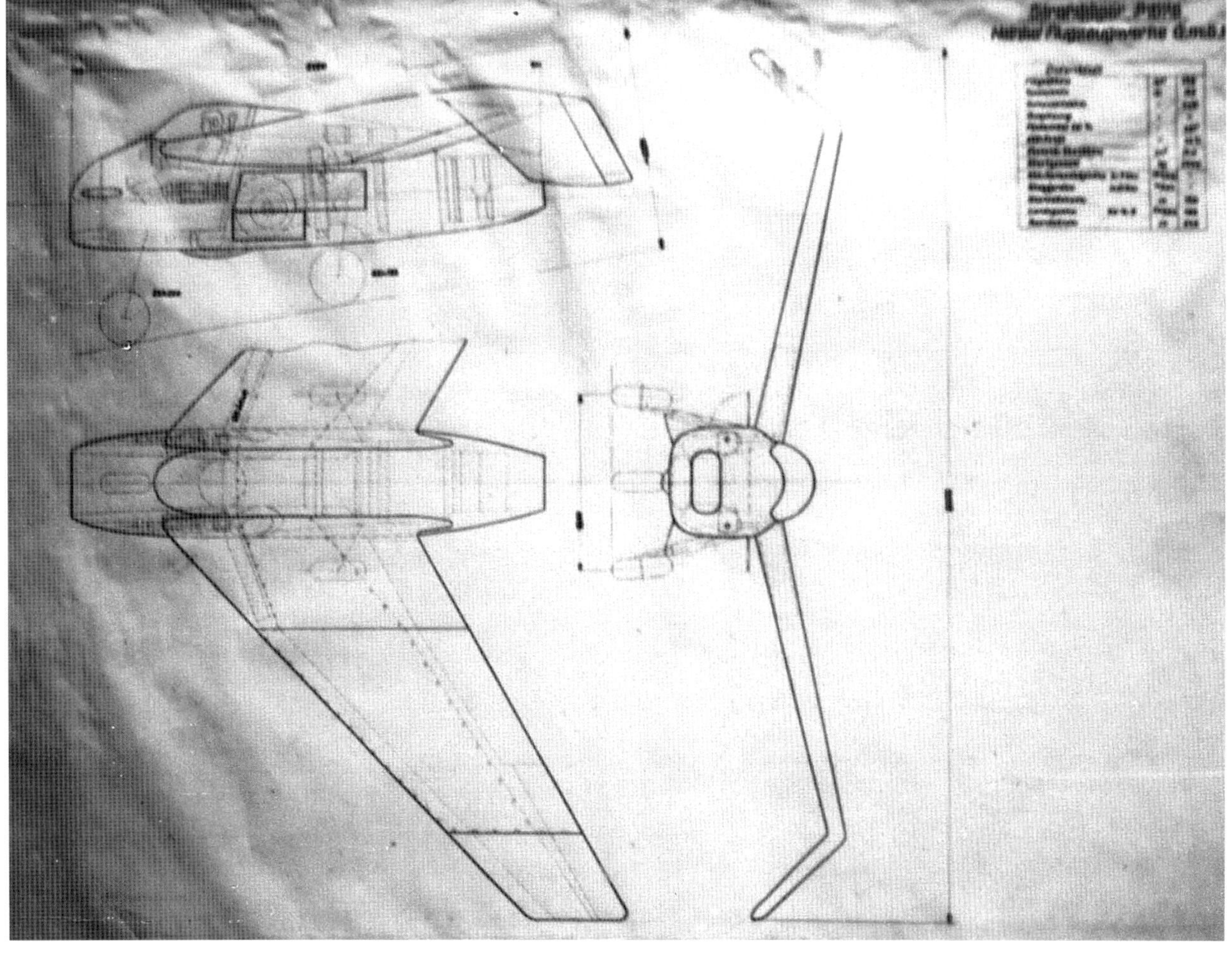

LEFT: The February 1945 revision of Heinkel's P 1078 had a slightly longer fuselage than the original, simplified wings and a broader cockpit canopy.

MESSERSCHMITT P 1110 AND P 1111

Messerschmitt's three designs were all different. The P 1106 had been withdrawn – never to be seen again. Now the P 1110 reappeared in modified form, with a wingspan of 8.25m compared to 6.65m previously, and a sweepback angle of 40° compared to 37.8° back in January. Wing area was 15.85m².

The P 1111 was a new tailless design with a wingspan of 9.16m, a sweepback angle of 45° and a wing area of 28m². Two MK 108s were mounted in the nose and two more were installed in the wing roots just outboard of large curved intakes. A large sharply swept fin overhung the rear of the fuselage and the wide-track undercarriage mainwheels retracted into the underside of the fuselage. The profile of the aircraft was extremely slender and streamlined, with the pilot's head almost touching the cockpit canopy and the bottom of his seat almost touching the bottom of the fuselage.

The new P 1101 presented had a wingspan of 8.25m, 17cm wider than the series version of the P 1101 designed in November 1944, a sweepback of 40° and a wing area of 15.85m². The front of the aircraft was also substantially altered, increasing its overall length.

HEINKEL P 1078 REVISION

The Junkers EF 128 remained the same, as did the Blohm & Voss P 212.03. The Heinkel P 1078 was slightly different, however. Compared to the January version, the nose had been lengthened and the short straight section of wing close to the fuselage had been eliminated, allowing the wings to angle upwards directly from the point where they extended from the fuselage. This meant that the undercarriage main legs now had to be attached to the fuselage sides rather than to the wing roots. The cockpit canopy was also reprofiled to become flatter and broader.

PROJECT ASSESSMENT

Several copies of Hornung and Voigt's pre-conference report for the February 27–28 meeting have survived and all are more or less identical except for Willy Messerschmitt's personal copy.[34] This had an additional 16-page section appended offering Hornung and Voigt's own detailed assessment of the competitors.

It began by comparing the performance characteristics of the designs, finding that "the Messerschmitt P 1111 project is undoubtedly at the top of the list for performance. Although, in terms of maximum speed, the P 1110 is

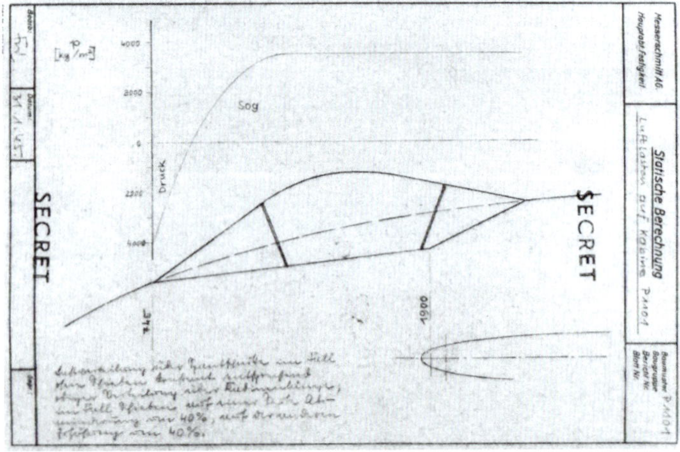

ABOVE: Drawing of January 31, 1945, showing the revised cockpit canopy for the experimental Messerschmitt P 1101 prototype - abandoning the low-profile canopy originally envisioned in favour of a more conventional design offering better visibility.

ABOVE: Rear view of the P 1101 prototype.

ABOVE: The incomplete experimental Messerschmitt P 1101 prototype as the Americans found it at Oberammergau. The advanced layout of the design has proven endlessly fascinating ever since - few realising that the P 1101 design had been rejected before the prototype could be completed.

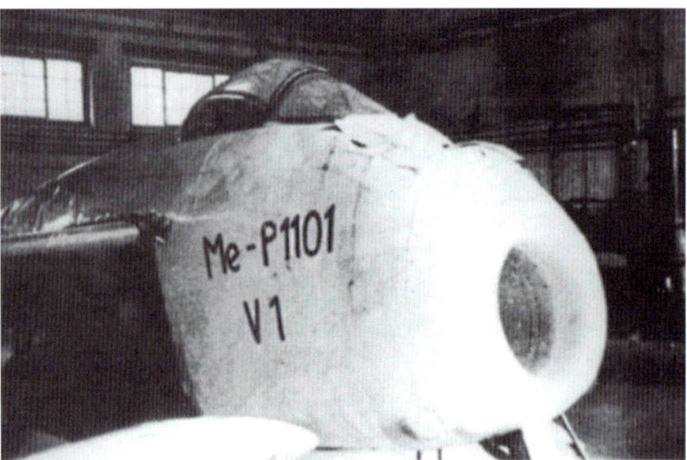

ABOVE: The front end of the experimental P 1101 with the type's designation written onto it.

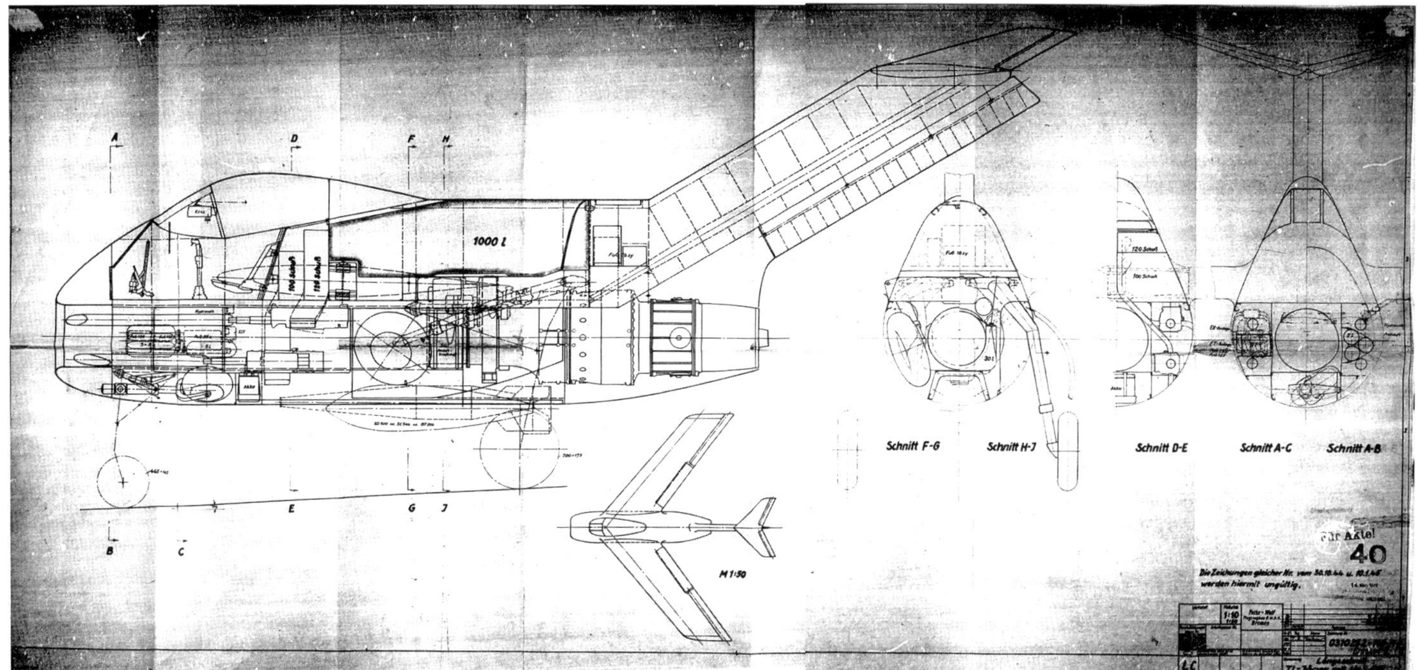

ABOVE: The latest known version of Focke-Wulf's Baubeschreibung Nr. 279/Entwurf Multhopp/Huckebein design from February 1945. The design finally received an official designation – Ta 183 A-1 – on or shortly before March 14, 1945.

slightly faster (difference 5km/h), it is at the top in all other calculated performances." The P 1111's rate of climb was 23.7m/sec compared to 'Entwurf Mittelhuber's' 23.3m/sec, take-off run was 600m compared to 650m for the Focke-Wulf design, landing speed was 155km/h compared to 164km/h for Focke-Wulf and landing distance was 450m compared to Focke-Wulf's 490m.

However, "Junkers' design is roughly equivalent to Messerschmitt's P 1110. In terms of top speed, it is inferior by 10km/h (it ranks third in terms of top speed), and it is somewhat better in the starting and climbing performance, slightly worse than the P 1110 in landing performance".

The Messerschmitt men clearly regarded the Junkers EF 128 as the main competition for their trio of designs, dismissing the Focke-Wulf, Heinkel and Blohm & Voss designs entirely, saying that "the other designs are so far behind in terms of speed performance that, in our opinion, they can be eliminated from the selection. Experience to date has shown that an inferior performance of around 30–50km/h compared to the fastest possible aircraft for a fighter aircraft would be an intolerable tactical disadvantage."

Both the P 1110 and EF 128 had risky air intake designs which relied on boundary layer removal but "although the arrangement chosen by Messerschmitt is more complex, it seems more reliable and probably associated with lower power losses". Furthermore, "the size of the performance uncertainty due to the boundary layer extraction is estimated to be quite considerable, especially since wind tunnel tests at low speeds cannot provide firm information; it is quite conceivable that at high Mach numbers a boundary layer thickening or detachment occurs as a result of a pressure surge ahead of the engine inlet, and that this will have a decisive influence on the conditions".

In terms of stability and controllability the conventional layout of the P 1110 and P 1101 made them superior to the EF 128 and P 1111. It was also likely that the P 1110 and P 1101 would have best spinning characteristics.

The P 1111 had the strongest armament at four MK 108s while the P 1110 could carry three. However, the P 1110 and EF 128 could carry up to 350kg or 400kg of weaponry of any sort whereas the P 1111 had more limited space available. The P 1101 was worst of the four because it could only carry four MK 108s – no other kind of weapon would fit.

In summary, Hornung and Voigt found that the P 1110 appeared to be the best all-round design but "on the other hand, the design P 1110, as mentioned above, poses the greatest risk because of technical failures (together with EF 128 from the arrangement of the intake openings)".

If the P 1110 was to be disqualified due to the risky nature of its engine intake, then "the P 1101 is rated as the best overall arrangement".

And "overall consideration of the individual designs gives the following, 1. The P 1110, P 1111 and EF 128, which are really satisfying at top speed, are associated with risks (engine intake) and disadvantages (unprotected tanks) which do not allow a decision in their favour now.

"2. All other designs are so inferior at high speed that a final decision in their favour also seems impossible because they would justify the use of the aircraft to be inferior to what is technically achievable (also to the opponent!). 3. The performance results of the P 1111 seem to indicate that a significantly better combination of top speed and take-off, climb and landing performance is possible compared to the normal aircraft and the tailless construction with a relatively large fuselage by development in the direction of the flying wing aircraft."

Finally, the pair made four recommendations: "1. An order with the immediate goal of series production cannot yet be placed. 2. The P 1101 study aircraft under construction is to be completed and tested as quickly as possible; the same can apply to other study aircraft that may be in progress. 3. New designs are to be created for the single-jet fighter to be procured in series, in which the performance of the best-performing designs to date are combined as far as possible and without substantial risks or disadvantages. A proposal for the technical requirements is available. 4. Based on the drafts submitted so far, only three companies receive the project order: this limitation is intended to avoid fragmentation of the design capacity due to parallel work by individual companies on too many tasks, as well as a concentration of German development on one task of the 1-TL-Jäger to the detriment of other militarily equally important tasks."

The proposal for the modified technical requirement was also appended to the report. Standard equipment was to be a pressure cabin and EZ 42 gunsight, two or if possible four MK 108s or four MG 213/20s and "what is desired is the possibility of installing heavier weapons, such as MK 103, MK 112 etc. Provide additional armament if possible".

The designs needed to be able to carry a 500kg external load and armour against 12.7mm ammunition from the front and 20mm from behind needed to be provided. Fuel load was at least 1,200kg and at least two thirds of it had to be "well protected". And "It would be desirable that a later enlargement of the wing and the available volume is possible for a larger amount of fuel, extended equipment or a stronger engine (up to 2,000kg thrust)."

Top speed at not less than 7km needed to be at least 900km/h, climb at 100% thrust had to be at least 20m/sec, take-off run fully loaded could not exceed 700m and landing speed with a third of a tank of fuel and full ammunition needed to be below 170km/h if possible. Landing distance could not exceed 600m.

TWO WINNERS OR FIVE?

EHK member Günther Bock noted in a letter dated March 12, 1945,[35] that at a meeting of the Entwicklungshauptkommission Flugzeuge on March 1, 1945, "the direction of the development work was discussed. In this case, the following picture emerged according to the intended use of the aircraft types: a) Day fighter. Development work for day fighters with the HeS 109.011 A jet engine, the companies Focke-Wulf and Messerschmitt are expected to have one development contract each. Whether the companies Junkers or Blohm & Voss are also activated for development is still undetermined".

Evidently the EHK had sat for an extra day in order to reach a decision and Focke-Wulf certainly came away from the meeting convinced that its Nr. 279 aircraft, 'Entwurf Multhopp', was about to receive a development contract. The company immediately began drawing up development schedules. The war diary of the Chef TLR for February 26 to March 4, 1945,[36] offers a slightly different perspective on the outcome: "1 TL project. In a final meeting at the EHK on February 27/28, 1945, it was agreed that the Focke-Wulf No. 279 project should be promoted as an immediate solution. The project is rejected by Fl-E due to technical and tactical considerations. Messerschmitt company should bring optimal solution in a tailless construction at a later time."

Since there was only one tailless Messerschmitt project pitched at the meeting so it would therefore appear that the company had won a development contract for its risky but high-performance P 1111 design – an 'optimal solution' that would be developed in parallel to Focke-Wulf's more 'immediate' design. The P 1101 and P 1110 appear to have fallen by the wayside at this point, with work on the almost completed P 1101 prototype being abandoned.

The war diary for March 5 to March 15, 1945,[37] goes on to say: "1 TL project. The following projects were requested from Fl-E: 1) Junkers, B&V and changed Focke-Wulf project. 2) Messerschmitt optimal solution. EHK has recently proposed, in addition Re: 2.), activate the Henschel firm as a second solution for the optimal solution (based on the work of Dr Zobel) and use Lippisch coupled with Heinkel appropriately for the Lorinjäger (solid fuels)."

It was now being proposed by the EHK that the Junkers, Blohm & Voss and Focke-Wulf designs should be considered for the 'immediate solution' while Messerschmitt's tailless high-performer was to be joined by a Henschel design as the longer-term 'optimal solution'. The 'Lorinjäger' is covered in Chapter 11.

Nevertheless, on March 14, 1945, Oberingenieur Ernst Lammel, head of Focke-Wulf's Technische Zentrale – the RLM liaison department – reported:[38] "The following series designation was chosen by the OKL for the TL-Jäger: 8-183 A-1. This number should be provided for the first aircraft. Accordingly, it is determined, for the flying mock-up: 183 V-1, 183 V-2, 183 V-3. For the prototype aircraft that correspond to the series version: 8-183 V-4 to V-14. For the two stress testing airframes: 183 Br-1 and 183 Br-2."

The company's Nr. 279 design (aka Projekt Multhopp, aka Entwurf 2, aka FW I, aka Huckebein) finally had an official RLM designation – Ta 183. A production schedule dated March 16 stated that the drawings for the "8-183 A1 (TL-Jäger) (Projekt Multhopp)" would be ready by April 3 and the V1 prototype would be ready to fly on June 15, 1945. This would be followed by the V2 on July 3 and the V3 on July 21. All three of these 'flying mock-ups' would be powered by the Jumo 004 B-1. The V4 to V13 would be powered by the HeS 011. Dates for the completion of these aircraft are also given, with the V13 due to be ready on January 10, 1946.

The next day, March 17, 1945, another document[39] was produced which stated that the 183 Ra 2 ('V1, V2, V3') would have a take-off weight of 3,100kg with 22.8m² area wings, 700 x 175 mainwheels and 465 x 165 nosewheels. But the 183 Ra 4 ('V4 / A Serie') aircraft would have their "wings pushed back 80mm on the fuselage compared to the Ra 2". Their mainwheels would be larger at 740 x 210, their wings would be slightly smaller at 22.6m² and their would each be equipped with "2 Waffen", although it doesn't elaborate on what the two weapons would be. Take-off weight was 4,000kg.

Work on the Ta 183 continued throughout the rest of March with numerous component blueprints being drafted but the last known example was dated March 29. Work on the design appears to have ceased at that point. Focke-Wulf's design offices at Bad Eilsen were overrun by British forces ten days later.

HENSCHEL P 135

Henschel had been working on single-jet fighter designs since at least July 1944. A BMW report of that month detailed work the company had been carrying out in partnership with Henschel to develop a single-jet aircraft powered by the BMW 003. The design was known as the P 108 and had its turbojet mounted in the rear fuselage. Two different intake layouts were tested by BMW – one where the intake was positioned above and behind the cockpit, like that of the Hs 132 or He 162, and another with intakes on either side and to the rear of the cockpit.

By January 1945, Henschel was working on a new single-jet design, the P 135, which had a nose intake instead. This was being developed in tandem with the P 130, a piston-engined pusher-prop design. Each had the same innovative cockpit design where the pilot lay in a supine position, almost on his back, with his head angled forwards in order to counter the effects of g-force during extreme combat manoeuvres. They also had the same armament: four MG 151s or rockets or a combination of both. In a postwar report,[40] Henschel chief designer Friedrich Nicolaus described the P 135 as follows: "Hs Proj 135: aircraft with one jet engine, initially BMW 003, later HeS 011. Jet engine centrally arranged in the fuselage tail, straight cylindrical air supply from fuselage tip with lowest friction losses. Pilot arrangement as before with the Hs P 130, armament the same. Nosewheel, gear only in the fuselage so that the wing smoothness was not disturbed.

"Wing with double kink, outer parts for the improvement of the rolling behaviour again kinked forward. The V-position of the outer parts should be adjustable in flight in the case of the prototype aircraft, in order to obtain the most favourable arrangement with respect to the design."

Top speed with the BMW 003 was expected to be 930km/h or 1,010km/h with the HeS 011. Maximum ceiling with both was 14-15km (46,000-50,000ft). Wind tunnel tests had been completed and "the construction contract was expected". The P 135 of the report had a 9.2m wingspan, 20.5m² wing area and take-off weight of 5,500kg. Whether the P 135 ever joined Messerschmitt's design as an 'optimal solution' is unknown.

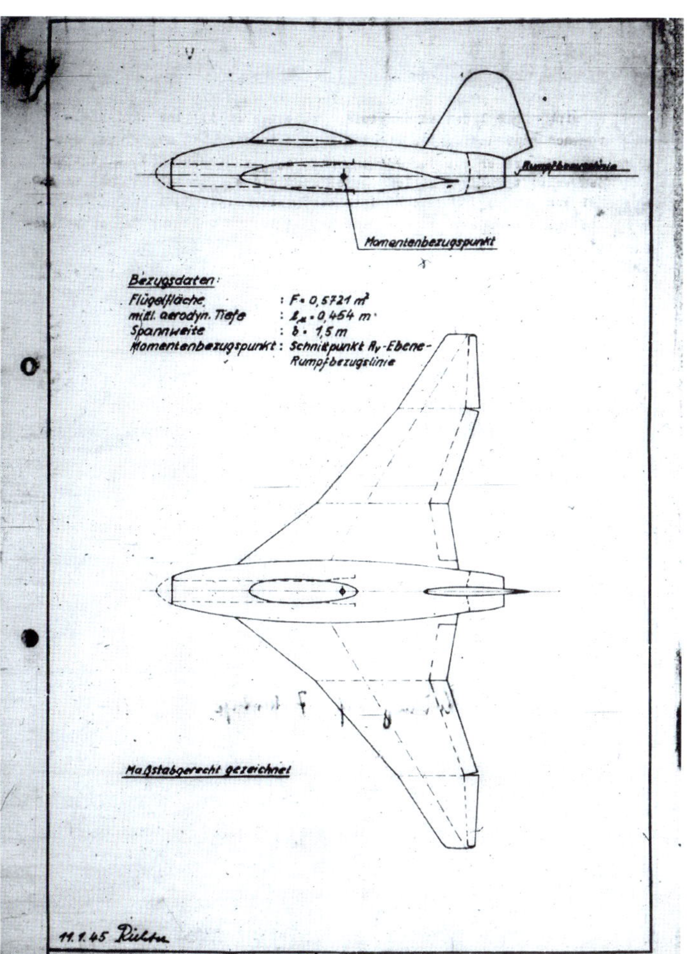

ABOVE: Henschel was an outsider when it came to single-jet aircraft design but its P 135, seen here as it appeared on January 11, 1945, was given serious consideration.

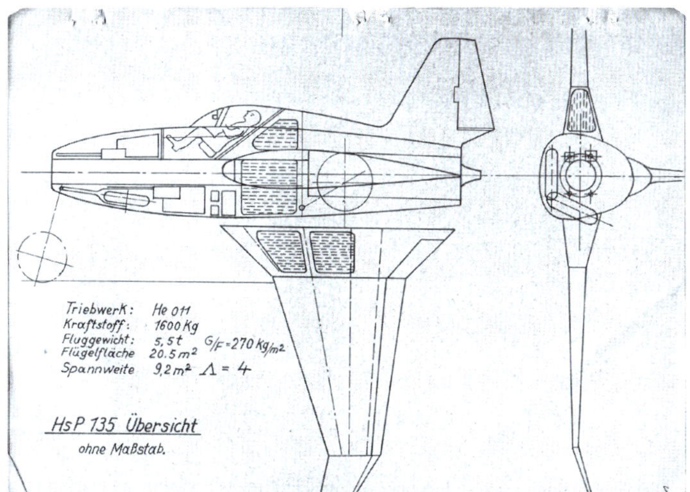

ABOVE: The final form of the Henschel P 135 as it appeared in a postwar summary of the company's projects compiled by its chief designer Friedrich Nicolaus.

MESSERSCHMITT P 1112

Rather than evolving its own tailless 'optimal solution' as the P 1111, Messerschmitt dropped this project number almost immediately after the February 27–28 meeting due to fundamental problems with the aircraft's layout and continued development work under the new number P 1112 in March 1945.[41] This would be the company's final project of the war. The British January 1946 report 'German Aircraft: New and Projected Types'[42] explains: "The P 1112 was designed to correct some of the faults which became apparent after study of the P 1111.

"The wing area was reduced to 236.5sq ft [22m² – compared to the P 1111's 28m²] since it was felt that the wing loading could be increased. The pilot's cockpit is situated at the extreme nose of the aircraft and 2 x MK 108 guns are fitted in the wing.

"The remainder of the design closely resembles the P 1111. Performance calculations were not completed."

Interviewed by American interrogators on June 8, 1945,[43] Woldemar Voigt gave a brief outline of the latest Messerschmitt "high speed model designs": "P 1101. Originally prototype fighter but changed to high-speed research plane. 3 different swept-back wing designs were to be tried. The construction of one plane was about 80% complete. One He 011 engine in fuselage was used which gave a top speed of 610mph estimated.

"P 1110. Design entered into last fighter competition in Germany. One He 011 engine, swept-back wing and tail surfaces and maximum speed estimated between 620 and 635mph. P 1111. A tail-less plane submitted as comparison with P 1110. Performance figures are given in U&M 1448. P 1112. Single engine jet tailless plane with smaller body than P 1111. Design study only begun."

Interviewed on September 7, 1945, by the British,[44] Voigt, said: "P 1101, P 1106, P 1110, P 1111, P 1112. A project investigation of the single-engined jet fighter was being carried out. The project drawings are known here, I assume.

"A final conclusion had not been drawn until the end of the war; the results secured by that time were: a speed of 1,000kph (620mph) is obtainable and had been guaranteed to the government.

"The tailless designs P 1111 and P 1112 showed the best performance out of the hitherto completed project series. They had the special advantage of combining best max speed with best landing speed. They seem to be dangerous at high Mach numbers (pitching moments). It seems to be possible to reach the same performance with more conventional designs (or even to exceed it) with less risk."

1-TL-JÄGER FIFTH COMPARISON

A mammoth five-day meeting of the EHK took place at Focke-Wulf's Bad Eilsen offices from March 20 to March 24. Various projects were discussed – including two days spent looking at single-jet fighters.

For the period between March 16 and April 4, 1945,[45] the Chef TLR diary stated: "1 TL-Jäger. EHK sitting on March 22 and March 23 in Bad Eilsen. No final decision on the proposals, since the chairman of the Sonderkommission Jagdflugzeug – Prof. Messerschmitt – was absent. Fl-E-Chef [Knemeyer] has, with the approval of the plenipotentiary for jet aircraft, SS-Obergruppenführer Kammler [Hans Kammler was appointed by Adolf Hitler to oversee jet and rocket development at around this time], commissioned the Junkers firm with the development of design EF 128."

It would appear as though, with Germany on the very brink of collapse, the tailless Junkers EF 128 was approved for development. With Germany now being invaded on both eastern and western fronts and defensive positions being hastily prepared around Berlin in anticipation of the

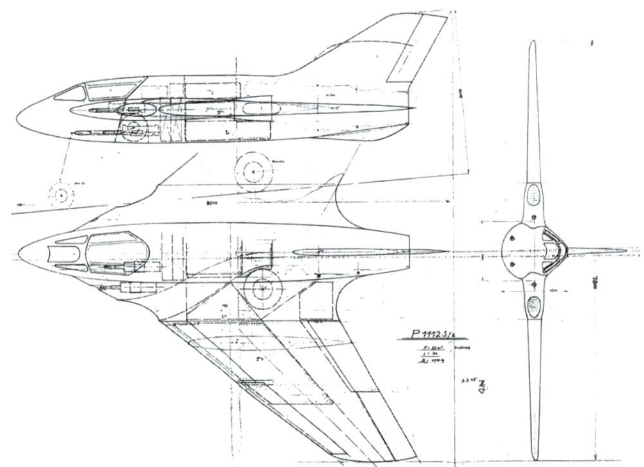

ABOVE: Just three days after the February 27-28 comparison meeting Messerschmitt had already ditched the P 1111, believing it to be fatally flawed, and replaced it with the new P 1112 which had its cockpit set further forward and a reduced wing area. This drawing is dated March 3, 1945.

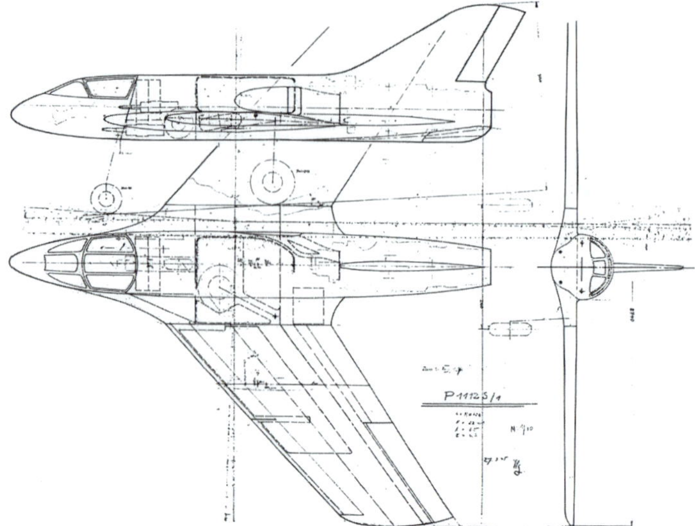

ABOVE: The Messerschmitt P 1112 S/1 of March 27, 1945, saw the design's wing area continue to shrink.

imminent arrival of Soviet forces, it seems unlikely that this development got very far. Junkers' Dessau headquarters was overrun by American forces on April 24, 1945.

ABOVE AND BELOW: Between March 1 and the end of the war, Messerschmitt's engineers at Oberammergau managed to construct a very basic mock-up of the P 1112's cockpit.

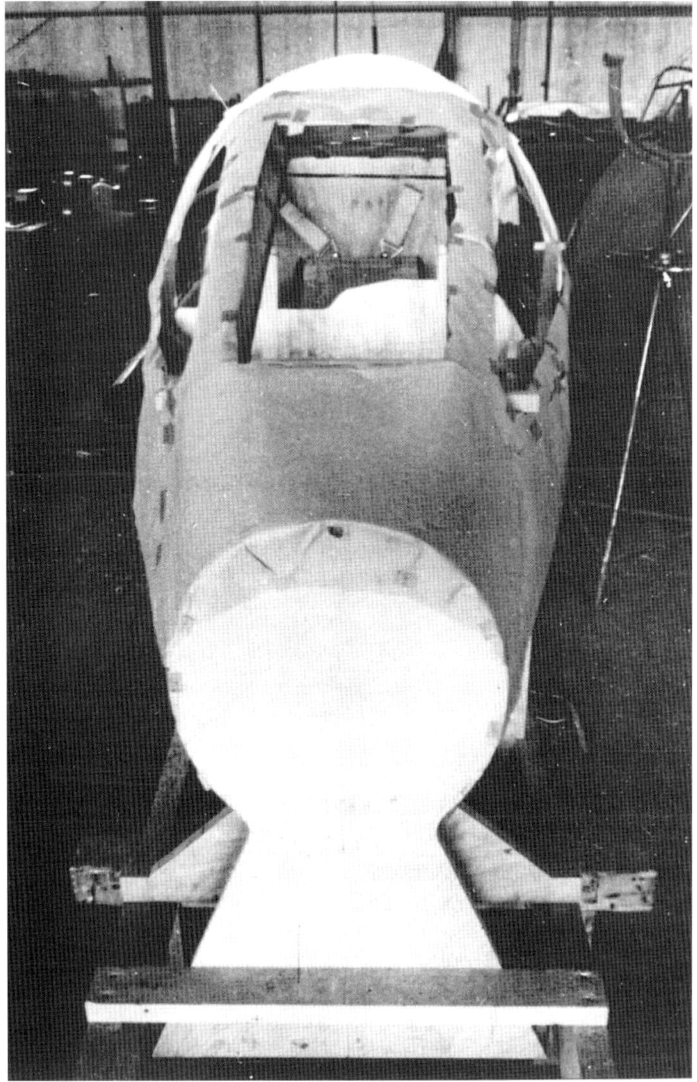

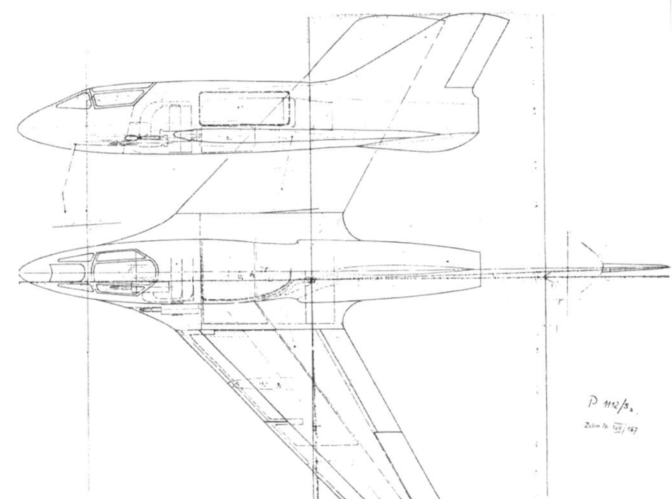

ABOVE: This undated drawing shows the Messerschmitt P 1112 S/2 – which had more sharply swept wings than those of the S/1.

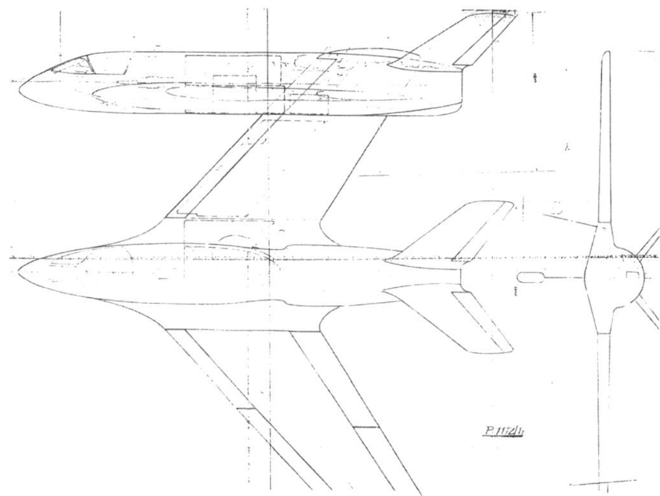

ABOVE: A further revision of the P 1112 design labelled 'P 1112/V1'. This design sees a surprising reintroduction of the V-tail right at the very end of the war.

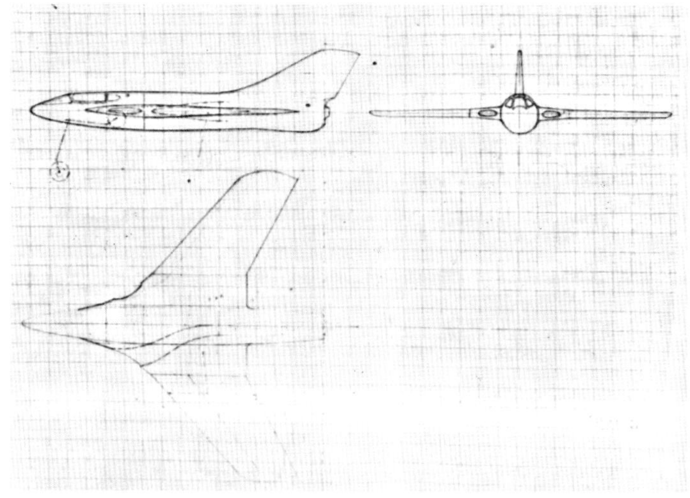

ABOVE: Undated Messerschmitt sketch showing an even more extreme version of the P 1112 – with a cylindrical fuselage, thin wings, wing-root intakes and enlarged tailfin.

HEINKEL P 1078 A AND B

Immediately after the war, Heinkel chief designer Siegfried Günter and a small remainder of the company's engineering team were employed by the Americans at Landsberg/Lech airfield, formerly Penzing Flugplatz, to recreate late-war Heinkel projects and research – evidently because all the original documents had been destroyed or captured by the Soviets.

Ernst Heinkel wrote in his 1956 autobiography:[46] "My Marienehe factory had been seized by the Russians and quickly dismantled. The Bleicherstrasse works in Rostock were blown up and the other factory in the same town had been taken over by the Russians. Oranienburg had been completely dismantled and taken piecemeal to Russia, as had the Waltersdorf branch of Zuffenhausen.

"The fate of the decentralised works during the last months of the war – those tunnels, mines and caves in Achensee, Stassfurt and Kochendorf, where we had worked to protect ourselves from bombing attacks, I knew only much later. They had all been destroyed, damaged or plundered.

"The repair shops which had been set up during the war behind the fronts in Poland, Denmark, France and even in Como, and which had been built and managed by Heinkel specialists, had been scattered to the four winds in the general collapse.

"The Jenbach factory was seized by the French occupation troops. Most of the machines were carried away and the rest of them handed over to the new Austrian government in trust in November 1948. Only the Zuffenhausen factory near Stuttgart remained in its damaged state on West German soil. But the American military authorities had laid hands on this, too, and given it over to one of those 'trustees' who like spotless blooms emerged in surprising numbers from the slime of defeat. I no longer dared to cross the threshold.

"As soon as I returned home on August 4, 1945, I wanted to see the position in Jenbach with my own eyes, and went to Lindau to try and get a pass from the French military government to visit their Austrian zone. It was a fruitless effort.

"I stopped in Landsberg and searched for Siegfried Günter who had withdrawn there in the middle of April

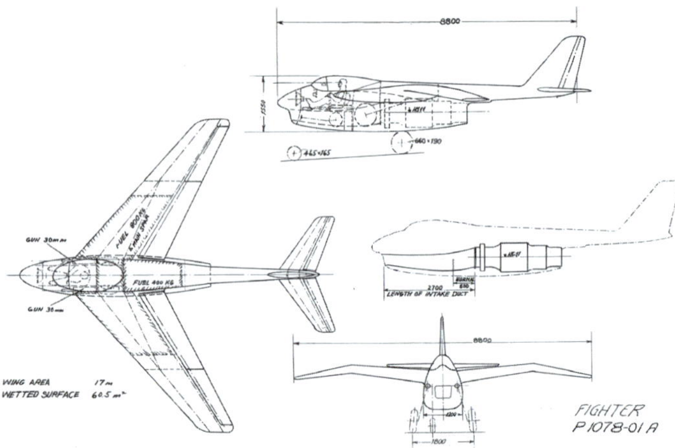

ABOVE: The Heinkel P 1078 A was evidently an alternative configuration for the P 1078 presented for the single-jet fighter competition in January 1945. Presumably at that time it was not known as the 'A' – a designation used in the postwar report produced on the design for the Americans.

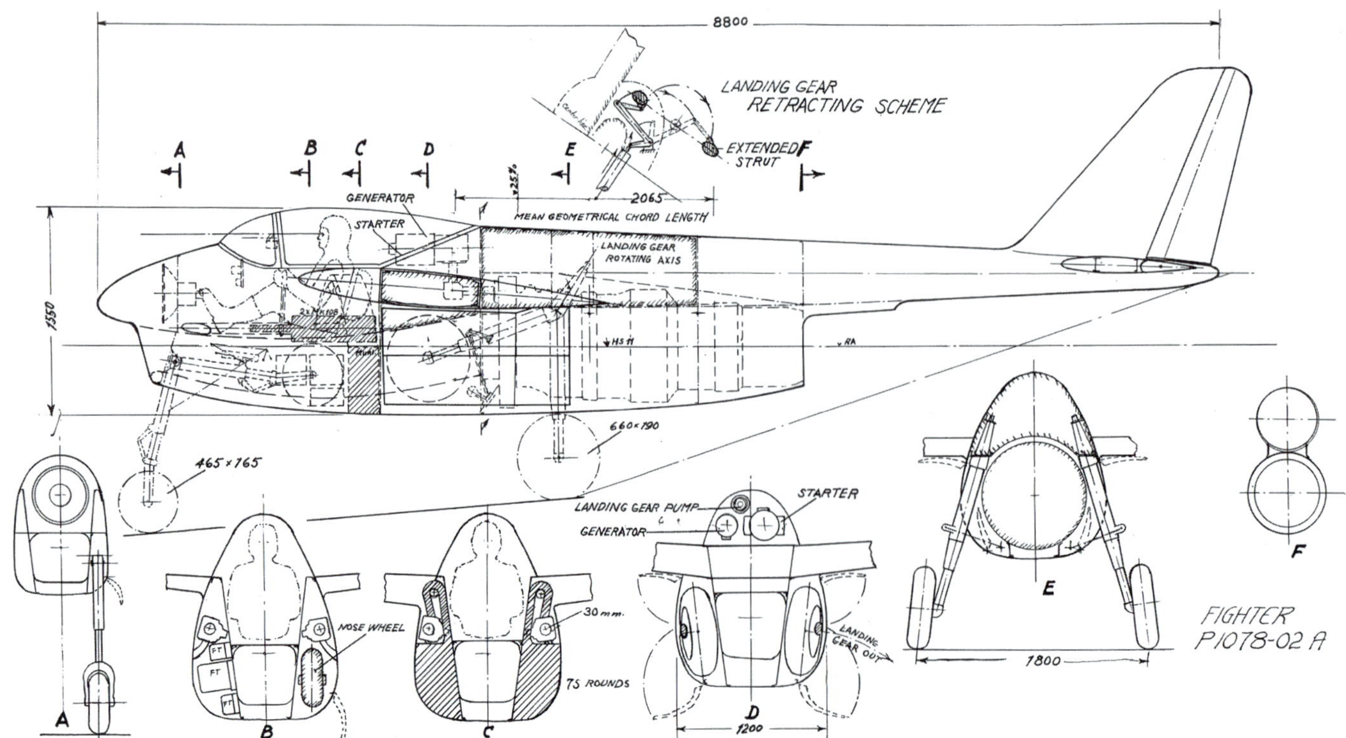

ABOVE: A postwar drawing showing the internal layout of the Heinkel P 1078 A.

1945, and had still managed to carry on a primitive office with 35 other employees of my design office. I found him – the most important expert on aircraft structures and aerodynamics in Europe at that time – living with his wife in a small room. He was working with 10 more of my people, including Töpfer, in a technical office which the Americans had established on Penzing airfield. Töpfer had tried in vain to draw the attention of the American military authorities to the presence and importance of Günter.

"Günter was as usual too ignorant of the ways of the world to push himself forward. Only when an American officer had been sent to Penzing from headquarters had he been able to go on working as a scientist. His work embraced everything that we had planned for the future in the way of fresh developments in jet propulsion, since the rejection of the He 280, and independently of the He 162. He was particularly engrossed, too, with new 'flying wing' types. I hoped on this visit to Landsberg that Günter would find a permanent outlet for his activities, either there or in America."

One of the reports produced by Günter's team during this time concerned the P 1078 – but rather than writing about the nose-intake tailless designs actually offered to the EHK in January and February, engineers Gerhard Eichner and Walter Hohbach chose very rather different single-jet layouts and called them 'P 1078 A' and 'P 1078 B'. Based on the text of their September 14, 1945, report[47] these appear to have been alternative wartime configurations – but it is also possible that they were not designed until after the war.

The P 1078 A was a relatively conventional swept wing fighter with a nose intake and with its HeS 011 engine built into its fuselage and exhausting beneath its tail boom. It was armed with two MK 108s, measured 8.8m long with a wingspan also of 8.8m and a wing area of 17m². It carried 400kg of fuel in its fuselage and a further 800kg in its wings.

The P 1078 B was a highly unconventional fighter similar to the original P 1078.01 but with two noses protruding from its central fuselage either side of the intake duct for its HeS 011. It measured 5.25m long with a wingspan of 9.4m and wing area of 20m². While one nose housed the pilot's cockpit, the other housed the two MK 108s. The aircraft's 1,200kg of fuel was housed in its wings.

The report summary said: "The extraordinary shortcoming of our oil production forced us to design lighter single-place fighters with reduced equipment. Armament: 2 x MK 108, 30mm calibre, the rate of fire would be increased to 900 rounds/min. per gun; also a radar aiming mirror. A construction with two turbo-jets (like the P 1079) would be the best regarding space for pilot and armament. But, in Germany, a new smaller turbo-jet with the same thrust/weight ratio and specific fuel consumption as the HeS-11 was not expected and the use of one HeS-11 had to be the basis for new fighter projects. The output of this turbo-jet gives the demanded increased flight performances with

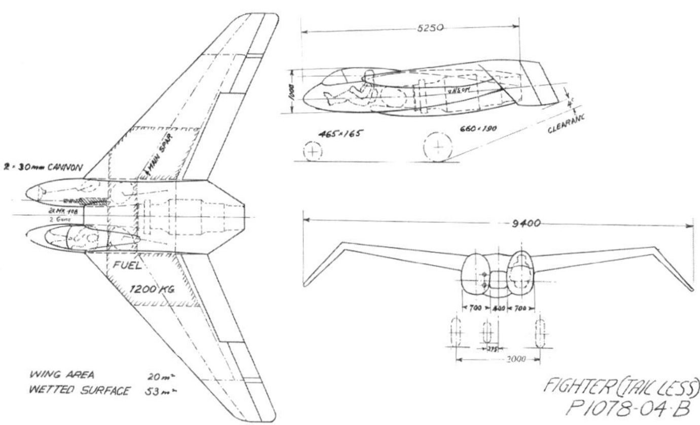

ABOVE: It is uncertain whether the radical Heinkel P 1078 B was a rejected alternative configuration from the wartime project or a postwar creation of the company's designers for the Americans. Its bizarre-looking twin-nose served to reduce intake duct length and separate the aircraft's armament from its pilot.

somewhat increased equipment and a 1 ton increase in gross weight as compared with the He 162.

"The fundamental question, whether with the same construction one big turbo jet has less weight and consumption than two small turbo-jets, each with half the thrust, was still not answered by our turbo-jet designers (a small turbo-jet for the V-1 was under construction by Dr Porsche; the data are not known.)

"This report contains two projects of single place fighters with one HeS-11; one project with tail, with turbo-jet in the lower part of the fuselage, the other is that of a tailless airplane, as far as possible designed as an all-wing airplane.

"The project with tail has a considerably smaller ratio of wetted area/wing area than the He 162. The idea for this new project was conceived before the construction of the 162 and it was also known that this design would give somewhat better performances. But the questioned production engineers of those firms which were engaged for the production of the 162 declared that the preparation for the series production of the He 162 would be considerably more rapid. Because of the situation at that time, this was the deciding factor."

In addition to its other features, the P 1078 A had detachable skin panels, an ejection seat and space in the nose for a radar dish. Of the tailless P 1078 B, the report says: "It was tried to construct this project as far as possible as an all-wing aeroplane, that is to make the ratio of wetted area/wing area as small as possible." The double nose sections, which each bent slightly inwards, were detachable and putting the armament on the 'inside wall' of the starboard nose kept "the moment about the normal axis due to firing recoil as small as possible".

After producing reports on the P 1078, P 1079 night fighter, P 1080 ramjet fighter, He 162 with pulsejets and other technical documents, the Heinkel team was dispersed, ending the company's projects output.

Chapter 9

Volksjäger

The people's fighter 1944–1945

Just as attempts to create a single-jet Me 262 replacement were getting into gear, a new single-jet fighter competition was launched. It was hoped that the result would be a 'National Fighter' or 'People's Fighter' manufactured without disrupting existing fighter production lines...

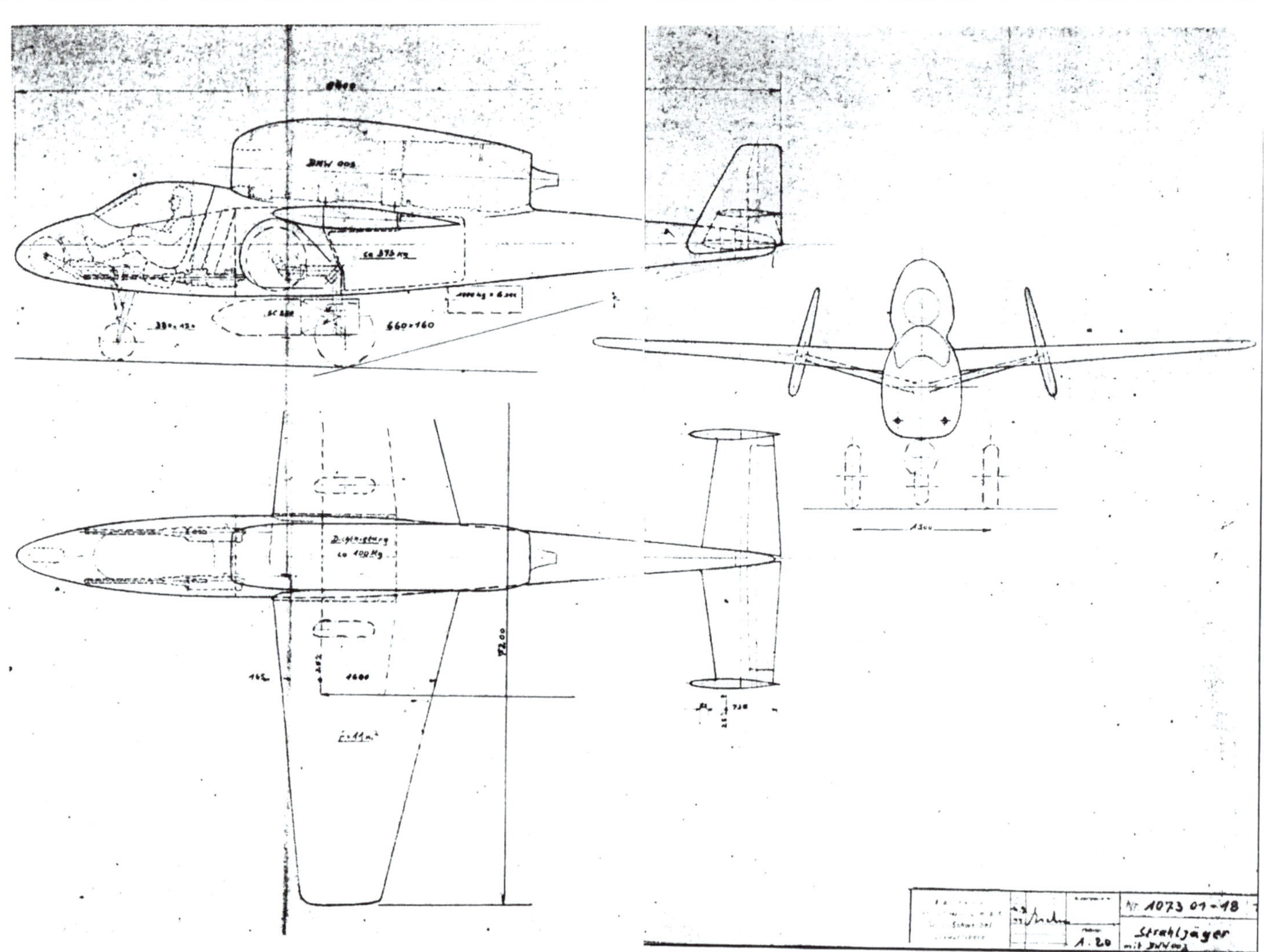

ABOVE: Heinkel's P 1073.01-18 design was a simplification of earlier iterations of the series – advanced features such as swept wings and a V-tail were replaced with safer options.

It was clear that the single-jet designs being worked on by the various manufacturers would take several years to reach full series production. And this was quite normal – even new conventional piston engine types using parts from existing models, such as the Ta 152, took at least 18 months to go from drawing board to active service.

But what if a relatively unambitious specification was put forward, one which would require performance only slightly better than that of existing piston engine types, and which would make use of an engine that was practically already in production? Such an aircraft could conceivably be ready in record time.

It was this line of thinking which resulted in the Volksjäger requirement. Exactly whose idea it was to launch this new downgraded single-jet fighter competition before the advanced single-jet presentations had even ended is uncertain – Heinkel general manager Karl Frydag, Heinkel director Carl Francke, Albert Speer's deputy Hauptdienstleiter Karl-Otto Saur and the Chef TLR's head of aircraft development Siegfried Knemeyer are good candidates – but clearly a rapid turnaround was demanded.

The single-jet fighter project comparison meeting begun two days earlier was just winding down on September 10, 1944, when, at around 11.40am, a telegram was sent to the offices of Arado, Blohm & Voss, Fieseler, Focke-Wulf, Heinkel, Junkers, Messerschmitt and Siebel[1] outlining a new requirement.[2] It called for a fighter "of the cheapest construction" powered by a single BMW 003 engine and made with "extensive use of wood and steel" that could reach a maximum speed of 750kph (466mph) and have an endurance of 30 minutes at full throttle. It also had to be able to operate from poor airfields, with a take-off roll of under 500m, and come equipped with two MK 108s or two MG 151s.

This was a much more achievable set of objectives compared to those of the HeS 011-powered Me 262 replacement but the companies were only given between three and five days to complete the design work.

Only three of the companies had something to present when the first design conference was called on September 14, 1944 – Heinkel, Blohm & Voss and Arado.

HEINKEL P 1073.01-18

The meeting took place at Berlin and was chaired by Heinkel technical director Carl Francke – who was also presenting the Heinkel design. He gave the delegates a detailed lecture which emphasised the fact that Heinkel had worked on its design longer than the others and that the original spec for armament, flight time and take-off distance could not be met so his company had only met a more modest brief – 20 minutes' flight time and smaller weapons.

This was a continuation of the earlier P 1073 series, the P 1073.01-18, with a wing area of 11m² and a wingspan of 7m. The Heinkel report outlining the design, dated September 11, 1944, was entitled 'Kleinst-Jäger P 1073 mit BMW 003'[3] and made it clear that the new fighter was a downgraded version of the fighter discussed from September 8-10.

It stated: "Simplifications compared to the design with HeS 11. 1) Due to the somewhat lower achievable speed, a swept wing can be dispensed with, consequently also leading edge slats. The result is simpler construction of the wings and less risk. The wing is made of wood. 2) For the same reason, the V-tail can be dispensed with and replaced with the proven vertical stabiliser with two vertical rudders.

"3) The undercarriage is drawn in towards the front, this results in a smaller fuselage surface, thus less material, and the rear part of the fuselage can also be made conical without any aerodynamic disadvantage. 4) The reduced fuel supply is accommodated in the wings and in a fuselage tank. The wing tanks empty into the fuselage tank by gravity. Pumps are only required in the fuselage tank.

"5) The equipment is more economical than usual with today's fighters, armour 50kg. It results with 2 x 151, 2 x 150 shots, 500kg load, with 2 x 108, 2 x 50 shots (less than half the shooting time 545kg load). 6) The speed remains somewhat higher than requested. A heavier TL-powered aircraft would result in insufficient climb speed."

BLOHM & VOSS P 210.01 AND P 211.01

Richard Vogt, representing Blohm & Voss was deeply unimpressed with what he saw – citing difficulties that would be caused by the design's dorsal engine position and the fact that it would be difficult to get the wings off for transportation by rail. Not only that, the P 1073 only used wood for its wing – the rest was of light alloy.

In a report entitled 'Aktenvermerk über Projektarbeiten für den Volksjäger und die geführten Besprechungen' or 'File notes on project work for the People's Fighter and the related meetings'[4] dated October 12, 1944, Vogt notes that when he pointed this out and said he had a design which suffered none of these problems, Francke "took the clock in his hand and called a time span of five minutes, which was all he could spend looking at my documents!"

Blohm & Voss presented two single-seat designs in a short brochure dated September 12, 1944, entitled 'Kurzbeschreibung BV P 210 und P 211 Jäger mit BMW 109-003'.[5] Heinkel had modified and existing design and it seems likely that Blohm & Voss did the same. Although its earliest P 209 designs had not been ready in time for the September 8–10 meeting, they were presumably far enough along to provide a basis for the P 210 and P 211.

The tailless P 210.01 had 15m² area swept wings, wingtip control surfaces, a short fuselage, tricycle undercarriage and one cannon on either side of its nose intake. The P 211.01 also had a tricycle undercarriage and swept wings of 15m² area but had a conventional fuselage with tail surfaces. The P 211 had a narrower wingspan at 8.4m compared to 11.4m for the P 210 due to the latter's slightly broader fuselage and wingtip control surfaces.

The P 210/211 brochure began: "When working on the design, structural considerations were in the foreground. Two projects have been worked out with the same wing arrangement. One can be described as 'risk-free', while the other is entering uncharted territory, i.e. requires development work to a greater extent.

"The internal structure is completely new. While with conventional aircraft the supporting structure of the fuselage lies in the outer skin and the entire equipment and installation has to be laborious and time-consuming to work on through openings, an attempt was made here to build from the inside out. The central air duct for the turbojet is also the inner structure – a kind of central frame.

"Steel brackets are welded onto this inner tube for everything that is available in terms of equipment, for example for weapon storage, the seat, the controls, the nose wheel, the turbojet etc. The outer casing is then the last component fitted in the form of plates similar to the engine cowling – fastened by quick connections when the maintenance of individual equipment groups requires it.

"It is easy to see that this construction process, in which the guts of the aircraft remain completely open and accessible during assembly, speeds up assembly considerably. With the central tube made of steel and all welded brackets, the proportion of light metal in the fuselage also drops for the first time. There is no reason not to expand this tendency even further and to manufacture the removable panels from wood and the engine cowling from steel, thereby avoiding dural altogether.

"In the P 210 project, the tail units were placed on the wing edges with the advantage of further savings in terms of construction weight and resistance. With P 211 the steel-welded suspension boom for the turbojet is extended as a tail boom. On the left and right there is a large access point for the turbojet and the fuselage end, so that the entire engine can be exposed in the simplest way.

"The wing – in both cases a swept wing of about 30° – is completely rectangular, so that only a single-rib template is required. In addition, aileron and landing flaps are interchangeable on the left and right. The construction of the wing is in line with our development tendency, which aims for a steel shell wing that is also used as a fuel container.

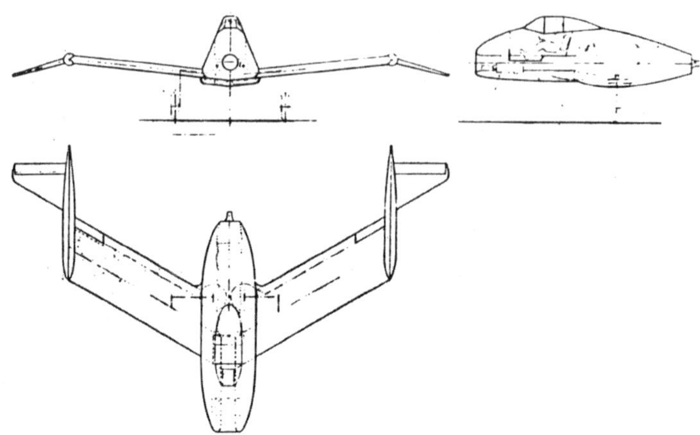

ABOVE: A second design pitched by Blohm & Voss but quickly discarded was the P 210. Like other designs from the company during this period, such as the P 208 and P 209, it was tailless with wingtip control surfaces.

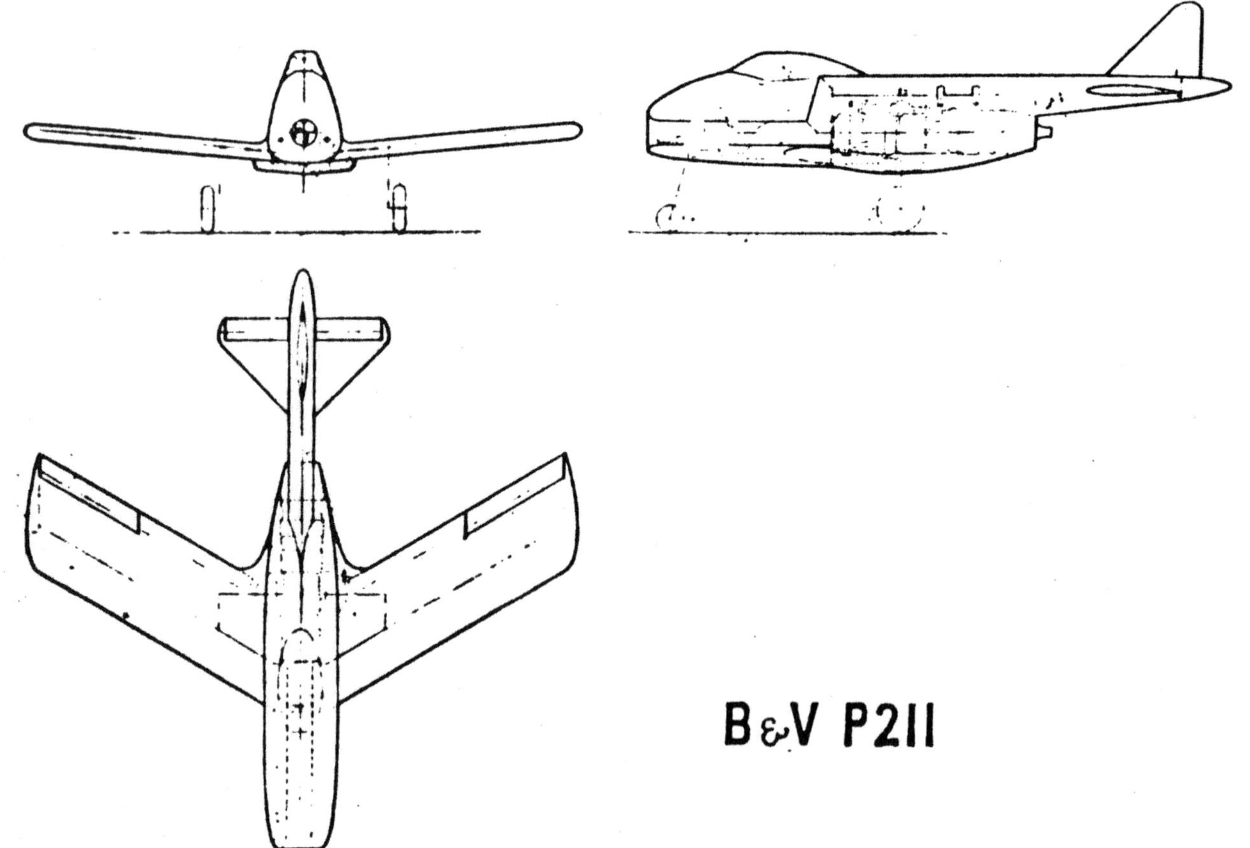

ABOVE: The Blohm & Voss P 211.01-01 was designed for maximum simplicity of construction - with a single internal structure that everything else was attached to. It also featured swept wings and a relatively long intake duct which made it less desirable for the purposes of the Volksjäger competition.

"For weight reasons, of course, only a cutout is built as a steel shell, while the nose and rear edge can be made of wood or dural. In any case, dural can also be avoided here. The protected rubber fuel containers are completely eliminated. For the undercarriage, a solution that works with large lever arms has been chosen.

"In summary, it can be said that the designs offered must lead to an aircraft that will finally relieve the light metal sector of our raw material production and at the same time bring savings in labour, and which, as far as maintenance is concerned, goes far beyond the usual. The last point in particular seems to be of crucial importance, because an increase in the operational readiness rate is equivalent to an increase in production, without the need for material and labour."

ARADO E 580

Arado's design was the E 580 – an 8m-long fighter, a 7.75m wingspan, its BMW 003 turbojet mounted on its back like that of the P 1073 and the rear part of the cockpit canopy slightly within the engine's intake. Heinkel had been able to produce a workable design in just one day because it had been forewarned about the requirement; Blohm & Voss (probably to the surprise of Heinkel) had only been able to do it because it could modify its almost complete P 209 designs to suit, but Arado was ready for a different reason.

According to Arado engineer Rüdiger Kosin, writing 39 years later:[6] "In mid-September 1944, without prior notice, a consultant from the Technical Office appeared in the development department, which had been relocated to Landeshut in Silesia, who wanted to have the project of a light fighter with a BMW 003 engine created within a few days. He seemed to know exactly what was going to come out. He did not leave the design office for two days and tried to steer the project in the direction he wanted.

"The company was fully occupied with the further development of the Ar 234, in addition to working on two other projects, and was therefore unable to devote itself seriously to the task. But after two and a half days, the consultant from the Technical Office was given a design with weight and performance estimates etc. which was similar in some respects to that of the Heinkel company."

The actual E 580 project description,[7] dated September 12, 1944, began: "Due to the short time available (2 days) the present project could only be worked out roughly. The following should be noted in detail: 1) Engine: The engine is arranged on the back of the fuselage in such a way that the cockpit canopy is located before the intake. To what extent this would disturb the air entry into the turbojet (loss of thrust, vibration) would still have to be clarified. Furthermore, it would have to be determined with BMW whether the existing connection points could be used, which requires a rotation of the device by 180°. In order to have sufficient flame protection behind the exhaust in the event of undesired afterburning, at least the top of the fuselage and possibly also the tail units must be made of sheet steel.

"2) Structure: The previous calculations have shown that the cheapest construction for the structure is wood as the material is a continuous spar with a torsion nose. The loads and dimensions are too unfavourable for a wooden shell construction, so that very thin shells prone to deformation would result. A simple spreading flap is suggested as a starting and landing aid.

"3) Tail: The engine position results in a double tail arrangement. It appears aerodynamically advantageous to give the vertical tail a V-position, so that a more favourable arrangement of the rudders is possible. A certain level of flame protection for the tail units appears necessary, whether a simple coating is sufficient or steel sheet must be used requires further investigation. 4) Fuselage: The fuselage tip is kept free for the installation of weapons. The accommodation of 2 x MK 108 with 60 rounds each is possible. Adjoining the weapon room is the cockpit, which is kept as low as possible in order to disturb the air intake for the turbojet as little as possible. The seating arrangement was chosen so that no special retraining to prone arrangement was necessary, although otherwise the prone arrangement would have great advantages for the project.

"The cockpit is armoured from the front against fire from 13mm ammunition. Armour against shelling from the rear was not provided due to the arrangement of the engine and the two fuel tanks. The construction of the fuselage requires the most extensive use of sheet steel for reasons of strength.

"5) Landing gear: The good rolling properties of the nosewheel arrangement meet the intended use of the aircraft with its difficult operating conditions. For the mainwheels 2 x 660 x 160 and for the nosewheel the braked wheel 500 x 180 were chosen. The nosewheel retracts backwards into the fuselage, with no further rotation of the wheel being necessary. The mainwheels fold inward about an axis in the direction of flight towards the centre of the fuselage. Operation hydraulic. In addition to the three brakable wheels, a brake screen is provided to dissipate the landing energy. Taking into account the threat to friendly airfields by the enemy air force, it should be considered to what extent these planes can be started using a primitive rail start method.

"6) Equipment: The equipment should be limited as much as possible. The installation should not go beyond an oxygen system, a FuG 15 and the most necessary flight and engine monitoring devices." Armament was to be either one or two MK 108s but "in our opinion, should the intended use of the aircraft be extended to attack enemy bomber formations, the possibility of installing a launcher system should be provided under all circumstances, which should probably also be more effective for attacks on ground targets than the intended extremely weak on-board armament."

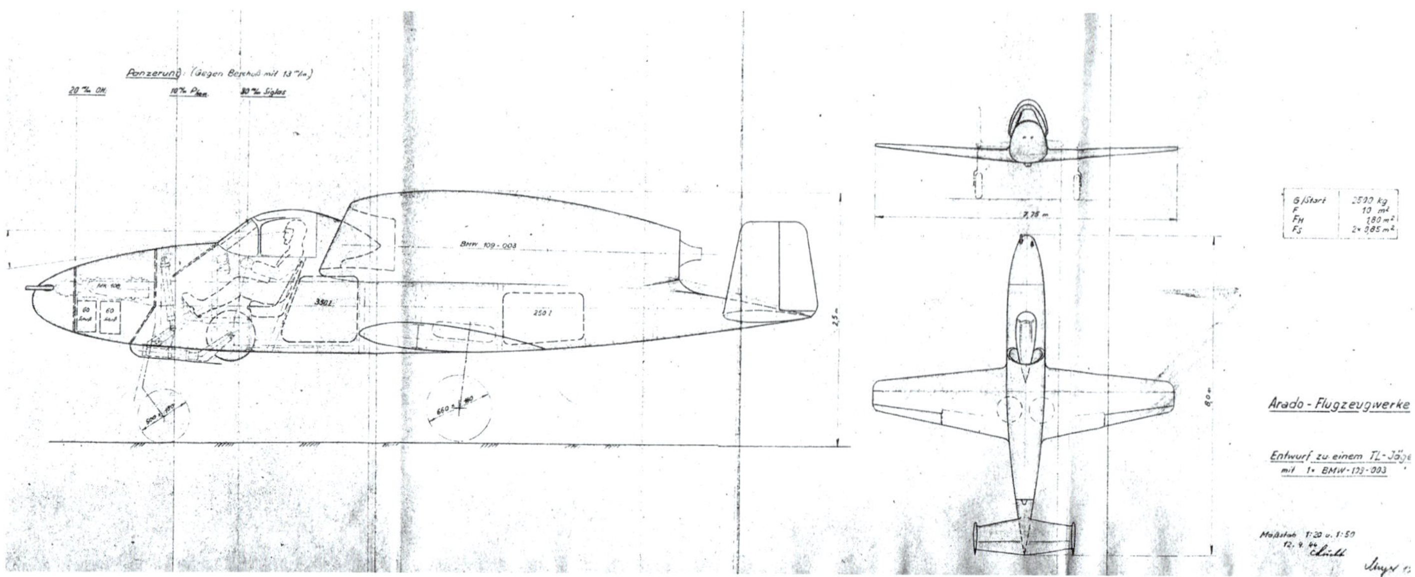

ABOVE: The Arado E 580 was designed very quickly under slightly mysterious circumstances. As was usual with Arado, the company did what was required without complaint and the result resembled the R-Jäger of nearly 18 months earlier.

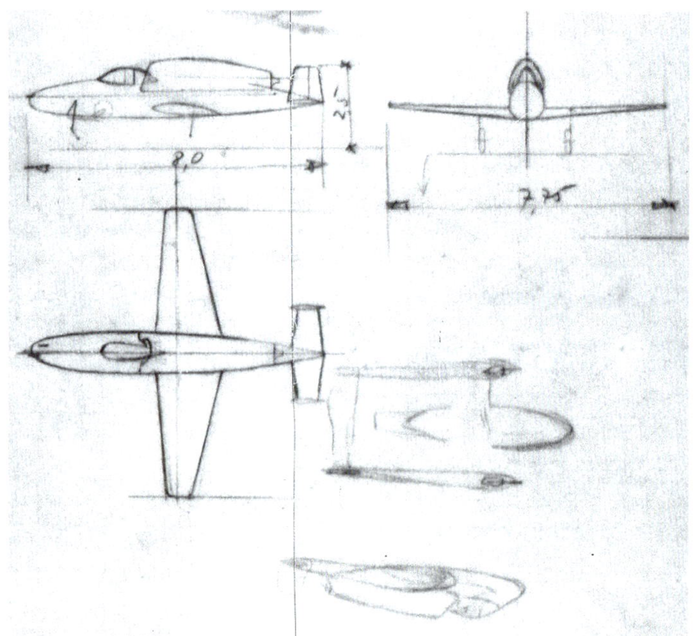

ABOVE: Contemporary Arado sketch showing what appears to be a potential alternative twin-boom layout for the E 580.

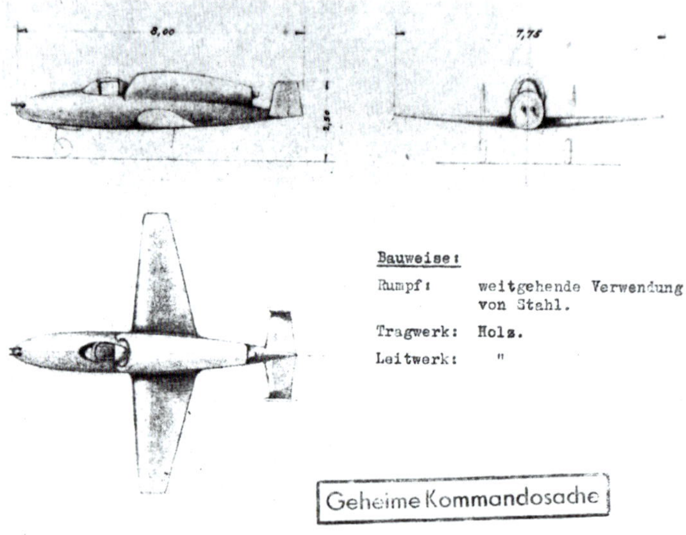

ABOVE: Another drawing showing the basic but not unattractive Arado E 580.

ASSESSMENT

It seems likely that compliant government-run Arado's E 580 was intended as little more than a fig leaf to make it seem as though Heinkel's pre-prepared design wasn't simply being rubber-stamped through uncontested. Both the RLM and Heinkel were apparently taken by surprise when Blohm & Voss offered such a well-thought-out proposal so quickly.

At the meeting on September 14, according to Vogt: "A project by the company Arado was also submitted, which was rejected by all involved.[8] The company Focke-Wulf participated in the meeting only informally, because the time available was not sufficient to complete a project. The company Messerschmitt was uninterested in the model under discussion."

No decision was made but "after leaving Mr Francke, we were asked to see Oberstleutnant Knemeyer". A second discussion on the Volksjäger now commenced.

The TLR minutes of the meeting[9] state: "Projects were submitted by the companies Arado, Blohm & Voss and Heinkel. According to the companies involved, the take-off requirement – rolling distance of 500m – is not achievable". Arado and Heinkel said they could meet the take-off distance but only by reducing flight time to 20 minutes. But "the project of the company Blohm & Voss is impressive in its construction and appears in this form extremely cheap and expedient for the intended construction. It also perfectly fulfils the demand for extensive use of wood and steel. Regardless of which company receives the final order,

ABOVE: Junkers offered this small fighter as its entry for Volksjäger – but was unsuccessful.

the aspects of Project B&V should be taken into account as much as possible.

"A production review of the B&V project will be carried out in the next few days. Everyone involved agreed at the final meeting with TLR Fl.-E-Chef [Knemeyer] that such an aircraft should be created without depriving the Me 262 of any capacity. The project is particularly justified if it is possible to use large numbers of measures in the shortest possible time through special means. The possible uses of such an aircraft are of course limited. The question of fuel quantity – 20 or 30 min. – is finally clarified between Chef-TLR [Diesing], TLR F.l-E-Chef [Knemeyer] and the General of Fighters [Adolf Galland]: 20 minutes' flight time is probably too low."

Incredibly, on Saturday, September 16, Vogt received a telex from Heinkel "in which the visit of its chief designer, Mr Schwärzler, is announced to me on Sunday! He was instructed to determine to what extent my suggestions could be worked into his own project."

The next Volksjäger meeting was on September 19, at which projects drawn up by Arado, Blohm & Voss, Focke-Wulf, Fieseler, Junkers and Siebel were presented. Today, the design of the Fieseler and Siebel projects is lost – though Vogt's account confirms that they were presented at the meeting and Focke-Wulf comparison documents show that the Fieseler design had a take-off weight of 2,900kg, a take-off run of 580m and a wing area of 13m².[10] The Junkers design is known only from photographs of a model rather than drawings.

FOCKE-WULF VOLKSFLUGZEUG

Focke-Wulf may have been slightly slower off the mark than Heinkel, Arado and Blohm & Voss but by September 15, 1944 the company was working on a 'Flitzer mit BMW 003-A1' with a reduced wing area of just 14m² compared to the normal HeS 011-powered Flitzer's 17m².[11] By the following day, the company was considering five different jet fighter designs[12] – two of them powered by a BMW 003 – each with a basic armament of two MK 108s and fuel load of 830kg.

The normal Flitzer and PTL 021-powered Peterle, each with an 8m wingspan and wing area of 17m², were shown alongside a 'V. F. m. zentr. Lw.' (Volks Flugzeug mit zentral Leitwerk – People's aircraft with central tail) and two different versions of a 'Kleiner Flitzer' (Smaller Flitzer), one with a BMW 003 and the other with an HeS 011.

The Volksflugzeug had a BMW 003 engine, a wingspan of just 7.5m and a wing area was 13.5m², while both Smaller Flitzers had the same wingspan as the normal Flitzer and Peterle but a wing area reduced to 14m².

The reduced power output of the 003 had a dramatic effect on performance, however. The normal Flitzer hit a top speed of 935km/h at 9km altitude whereas the Volksflugzeug could only manage 800km/h and the 003 Smaller Flitzer just 740km/h.

Take-off distance was even more badly affected. The normal Flitzer took 650m to become airborne, the Peterle a mere 360m (evidently reduced take-off distance was the Peterle's chief advantage over propellerless turbojet designs) – but the Volksflugzeug needed 1,000m and the 003 Smaller Flitzer an incredible 1,100m.

Precisely which of these designs was presented at the September 19 meeting is uncertain but on September 20 – the day after the meeting – the company issued a presentation document featuring a drawing of the Volksflugzeug,[13] which looked somewhat similar to the P 211 but with a T-tail and the option to fit either straight or swept wings. Also shown were the Peterle and the Smaller Flitzer design – now dubbed the Volksflitzer – which had a visibly reduced wing area compared to the normal Flitzer.

The document stated: "Volksflugzeug designs according to Arado, Blohm & Voss, Fieseler, Heinkel, Junkers, Siebel or Focke-Wulf will be inferior to enemy jet fighters when they appear on the front in large numbers (last third of 1945) or can only be used for a short time because 1) the BMW 003 is too weak; 2) the lack of streamlining due to manufacturing reasons results in insufficient performance and thus a strong inferiority in combat; 3) there is no possibility of increase in armament, armour and equipment. Runways of 800 to 900m are needed as a result of the weak BMW device – which rules out permanent use on field aerodromes.

"The immediate use of a final fuselage according to the above suggestion leads to the following consideration: 1) One-time series preparation; 2) The aircraft with 003, which has been overtaken shortly after its appearance, can be brought up to the performance level that can be achieved in the foreseeable future without any drop in output by installing the HeS 011 device and then represents a fighter superior for a longer period of time. (Installation of the rocket motor is provided, which gives climb times similar to Me 163); 3) The same aircraft is a replacement for the Me 262 and, with better flight performance, results in significantly lower fuel, labour and material expenses.

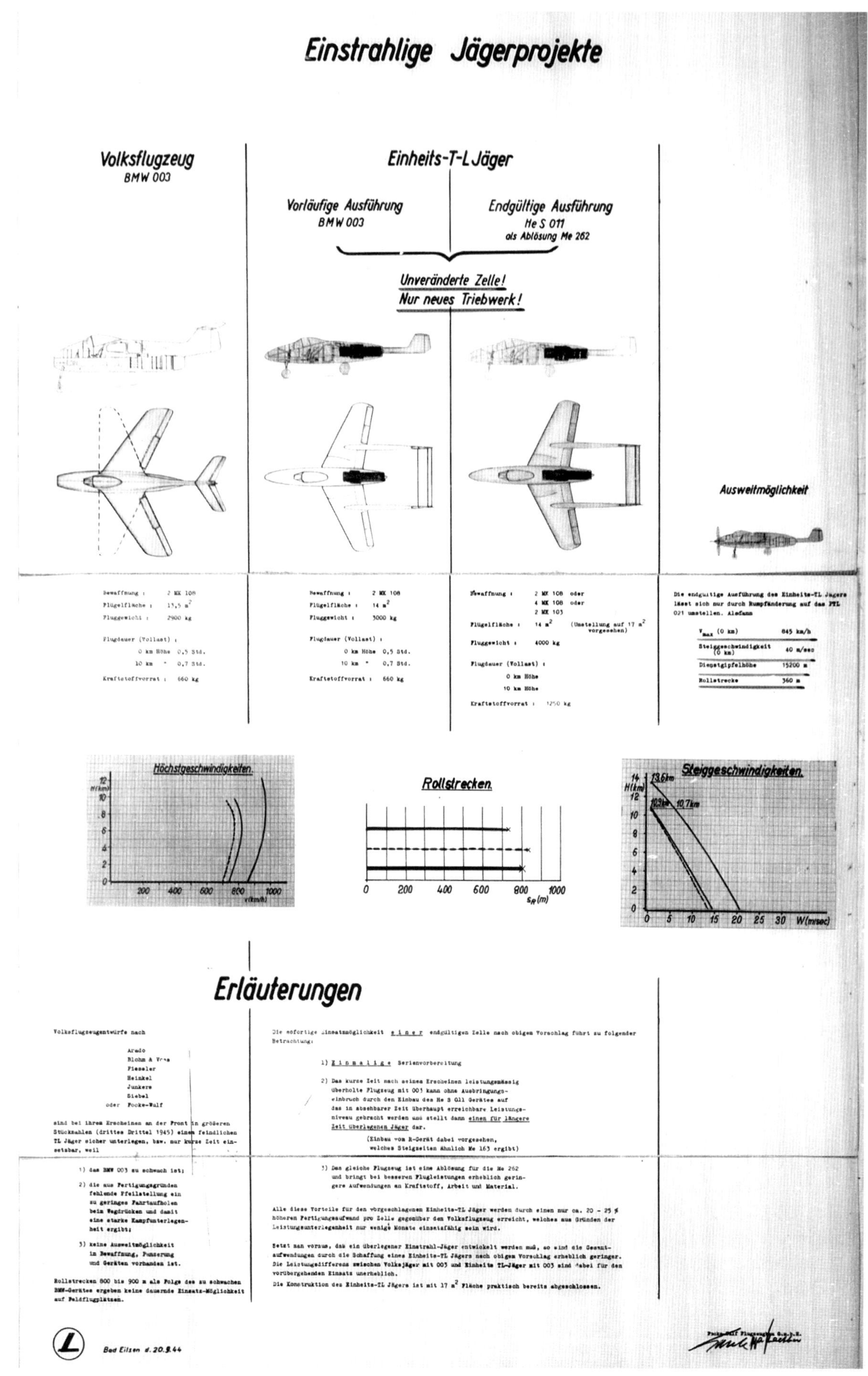

Einstrahlige Jägerprojekte

Volksflugzeug
BMW 003

Einheits-T-L Jäger

Vorläufige Ausführung
BMW 003

Endgültige Ausführung
He S 011
als Ablösung Me 262

Unveränderte Zelle!
Nur neues Triebwerk!

Ausweitmöglichkeit

Volksflugzeug (BMW 003)

Bewaffnung : 2 MK 108
Flügelfläche : 13,5 m^2
Fluggewicht : 2900 kg

Flugdauer (Vollast) :
0 km Höhe 0,5 Std.
10 km " 0,7 Std.

Kraftstoffvorrat : 660 kg

Vorläufige Ausführung (BMW 003)

Bewaffnung : 2 MK 108
Flügelfläche : 14 m^2
Fluggewicht : 3000 kg

Flugdauer (Vollast) :
0 km Höhe 0,5 Std.
10 km " 0,7 Std.

Kraftstoffvorrat : 660 kg

Endgültige Ausführung (He S 011)

Bewaffnung : 2 MK 108 oder
4 MK 108 oder
2 MK 103

Flügelfläche : 14 m^2 (Umstellung auf 17 m^2 vorgesehen)
Fluggewicht : 4000 kg

Flugdauer (Vollast) :
0 km Höhe
10 km Höhe

Kraftstoffvorrat : 1250 kg

Ausweitmöglichkeit

Die endgültige Ausführung des Einheits-TL Jägers lässt sich nur durch Rumpfänderung auf das PTL 021 umstellen. Alsdann

V_{max} (0 km)	845 km/h
Steiggeschwindigkeit (0 km)	40 m/sec
Dienstgipfelhöhe	15200 m
Rollstrecke	360 m

Erläuterungen

Volksflugzeugentwürfe nach

Arado
Blohm & Voss
Fieseler
Heinkel
Junkers
Siebel
oder Focke-Wulf

sind bei ihrem Erscheinen an der Front in größeren Stückzahlen (drittes Drittel 1945) einem feindlichen TL Jäger sicher unterlegen, bzw. nur kurze Zeit einsetzbar, weil

1) das BMW 003 zu schwach ist;

2) die aus Fertigungsgründen fehlende Pfeilstellung ein zu geringes Fahrtaufholen beim Wegdrücken und damit eine starke Kampfunterlegenheit ergibt;

3) keine Ausweitmöglichkeit in Bewaffnung, Panzerung und Geräten vorhanden ist.

Rollstrecken 800 bis 900 m als Folge des zu schwachen BMW-Gerätes ergeben keine dauernde Einsatz-Möglichkeit auf Feldflugplätzen.

Die sofortige Einsatzmöglichkeit e i n e r endgültigen Zelle nach obigem Vorschlag führt zu folgender Betrachtung:

1) E i n m a l i g e Serienvorbereitung

2) Das kurze Zeit nach seinem Erscheinen leistungsmässig überholte Flugzeug mit 003 kann ohne Ausbringungseinbruch durch den Einbau des He S 011 Gerätes auf das in absehbarer Zeit überhaupt erreichbare Leistungsniveau gebracht werden und stellt dann einen für längere Zeit überlegenen Jäger dar.
(Einbau vom R-Gerät dabei vorgesehen, welches Steigzeiten ähnlich Me 163 ergibt)

3) Das gleiche Flugzeug ist eine Ablösung für die Me 262 und bringt bei besseren Flugleistungen erheblich geringere Aufwendungen an Kraftstoff, Arbeit und Material.

Alle diese Vorteile für den vorgeschlagenen Einheits-TL Jäger werden durch einen nur ca. 20 - 25 % höheren Fertigungsaufwand pro Zelle gegenüber dem Volksflugzeug erreicht, welches aus Gründen der Leistungsunterlegenheit nur wenige Monate einsatzfähig sein wird.

Setzt man voraus, daß ein überlegener Einstrahl-Jäger entwickelt werden muß, so sind die Gesamtaufwendungen durch die Schaffung eines Einheits-TL Jägers nach obigem Vorschlag erheblich geringer. Die Leistungsdifferenz zwischen Volksjäger mit 003 und Einheits TL-Jäger mit 003 sind dabei für den vorübergehenden Einsatz unerheblich.
Die Konstruktion des Einheits-TL Jägers ist mit 17 m^2 Fläche praktisch bereits abgeschlossen.

Bad Eilsen d. 20.9.44

OPPOSITE: Focke-Wulf chart dated September 20, 1944, showing the company's two main Volksjäger entries – a new design somewhat resembling Blohm & Voss's P 211.01-01 and a slim-wing version of the Flitzer which could be upgraded to a more powerful engine when one became available. The Peterle DB 021 turboprop fighter was also offered as a 'wider possibility'.

"All these advantages for the proposed standard jet fighter are achieved by only about 20–25% higher manufacturing costs per airframe compared to the Volksflugzeug, which will only be operational for a few months due to inferiority in performance. Assuming that a superior single-jet fighter has to be developed, the total expenditure by creating a standard jet fighter according to the above proposal is considerably lower. The performance difference between Volksjäger with 003 and unit jet fighter with 003 are irrelevant for the temporary use. The construction of the standard jet fighter is practically complete with a 17m² wing area."

In other words, Focke-Wulf saw no future for the Volksjäger – it would very quickly be surpassed by superior Allied jet fighters. However, if the BMW 003 could be fitted to a standard Flitzer airframe, when the HeS 011 became available that same airframe could simply be upgraded to accommodate it. Eventually the design could be brought up to the full Me 262 replacement standard with the 17m² area wings always intended for the production-model Flitzer.

HEINKEL WINS

During the meeting on the 19th, chaired by Roluf Lucht, Vogt again raised his objections to the Heinkel design and later noted that "the way in which these important questions were dealt with at this commission meeting had a shattering effect on me. There was no question of an assessment of things, let alone a more serious examination. It remained with general discussion and remarks thrown over the table, which were for the most part untenable and worthless".

In return, Karl Frydag expressed doubt's about Vogt's design, "pointing out the difficulties involved in the production of removable skin panels. He did not say a word of appreciation for the unique assembly opportunity I could offer and not a word about the appropriate

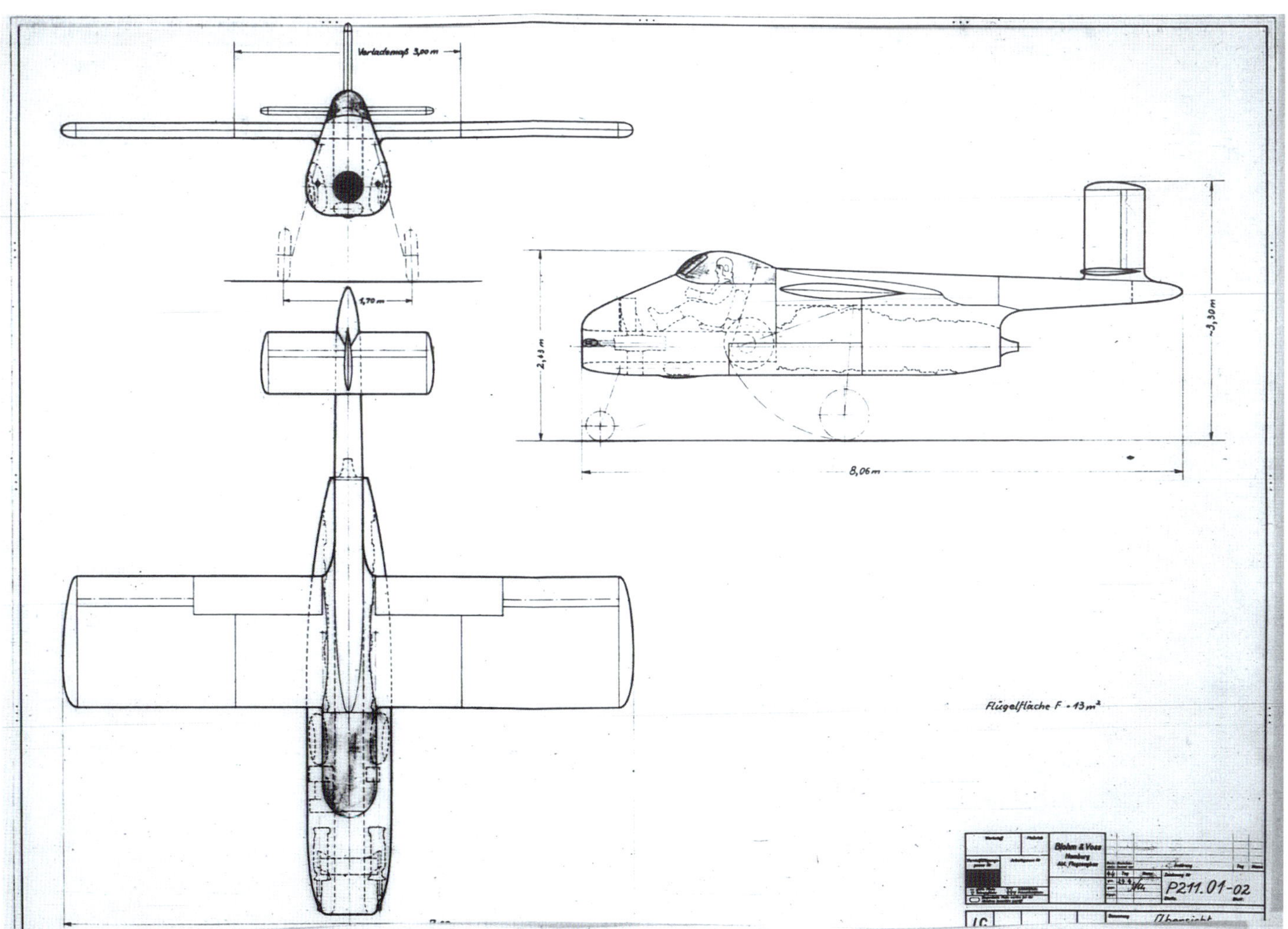

ABOVE: If Heinkel learned from Blohm & Voss, Blohm & Voss also learned from Heinkel. The second version of its design, the P 211.01-02, had straight wings, a straight tailfin and a straighter, easier-to-build nose section. This was how the design looked on September 29, 1944 – by which time it had already been defeated by the P 1073.

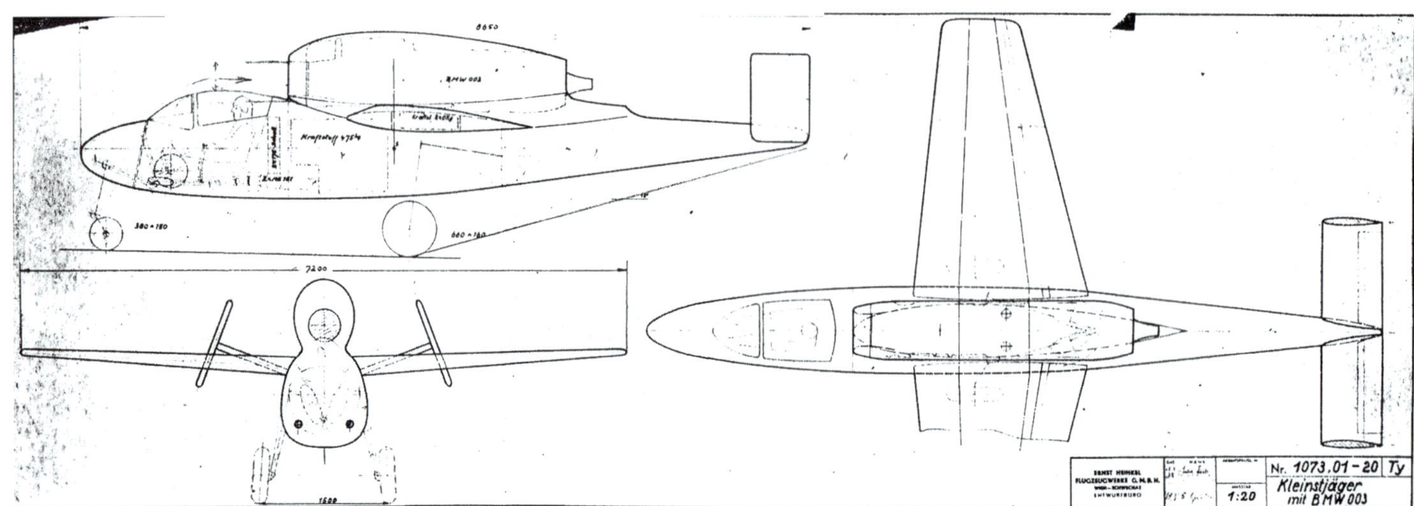

ABOVE: Heinkel's P 1073.01-20 appears to have been the last version of the design before it received the He 162 designation. Heinkel itself seems to have wanted to use 'He 500' but a single-engine 162 makes sense in the context of a twin-engine 262.

material types. No one contributed a single positive thought – only criticism".

Frydag was not only Heinkel's managing director but also held the post of Leiter des Hauptausschusses Flugzeuge or 'head of the main committee for aircraft' in Albert Speer's Reichsministerium für Rüstung und Kriegsproduktion, which made him one of the most powerful men in Germany's aviation industry as a whole.

During a meeting on September 23, 1944, the P 1073 was presented to Adolf Hitler for a decision.[14] The result was recorded by Saur: "The single-seat single-jet-engine fighter designed by Heinkel and proposed by the Entwicklungshauptkommission Flugzeuge, after detailed testing, is released for the immediate start of series production and ordered with a preliminary quantity of 1,000 per month.

"Completion of development, construction, testing and production are carried out in a forceful action as a collaborative effort between the departments involved and

ABOVE: A wind tunnel model of the He 162. The type's characteristic downturned wingtips were entirely absent during its development, only being introduced during prototype testing.

industrial plants with the utmost concentration. There is also clarity that with the method of deciding on the series consolidation from the first draft, the risk of a possible failure that has not yet been overlooked must be accepted."

Knemeyer announced that a decision had been made and that Heinkel's P 1073 would be built.[15] Vogt, convinced that this was a terrible error, went back to the RLM on October 2 and gave a convincing presentation to Knemeyer,[16] demonstrating that his design was superior to that of Heinkel. All those present, including Heinrich Beauvais the head of fighter testing at Rechlin, agreed that the P 211 ought to have won the contest.

BLOHM & VOSS P 211.01-02

The P 211 presented at this point, the P 211.01-02 from a new brochure dated September 29, 1944,[17] differed significantly in appearance from the original design, the P 211.01-01.

The wings were straight and their surface area was reduced from 15m^2 to 13m^2. Wingspan was also reduced from 8.4m to 7.55m and a high- rather than low-wing configuration was adopted. The tailfin and tailplanes had become rectangular where they had been triangular, the cockpit canopy was shortened and the undercarriage mainwheel track was narrowed. Rather than folding up into the underside of the fuselage, the mainwheels now retracted into the sides of the fuselage.

The new Blohm & Voss brochure offered a detailed comparison between the proposed P 211 and the Heinkel P 1073, inevitably suggesting that the former offered superior performance and economy over the latter.

But Knemeyer "did not want to argue the decision [to choose the P 1073] and defended it with psychological reasons". He promised Vogt to put his design back up for discussion. In the afternoon, Vogt attended a meeting with the RLM's production specialists, who were also convinced that the P 211 would be better than the P 1073. Vogt rang Lucht and "informed him of my exasperation that he

ABOVE: The He 162 M6 prototype during flight testing in early 1945 – perhaps less than five months after the type was approved for development.

ABOVE: A captured He 162 A-2 being flown over Farnborough in Britain during the summer of 1945 – a year earlier the aircraft existed only as concept drawings.

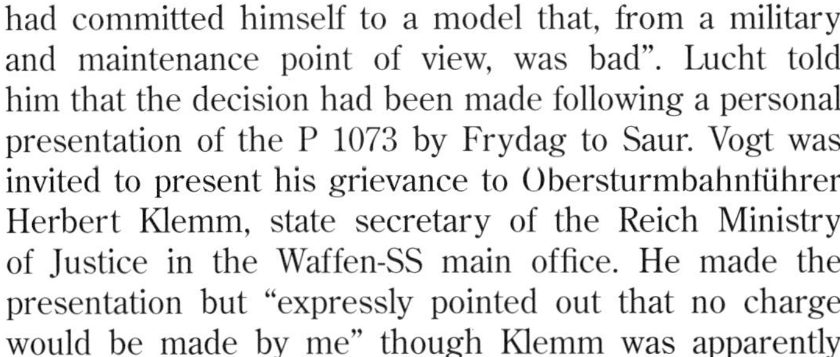

had committed himself to a model that, from a military and maintenance point of view, was bad". Lucht told him that the decision had been made following a personal presentation of the P 1073 by Frydag to Saur. Vogt was invited to present his grievance to Obersturmbahnführer Herbert Klemm, state secretary of the Reich Ministry of Justice in the Waffen-SS main office. He made the presentation but "expressly pointed out that no charge would be made by me" though Klemm was apparently also convinced that the wrong decision had been made.

Vogt was told that Saur would be asked to consider whether the wrong type had been approved for construction but it then transpired that Frydag had raised doubts about the P 211's intake duct. At this, Vogt enlisted the AVA to assess his design and after wind tunnel tests they found that there was no problem with it.

By now, however, it was far too late and the P 1073 went on to be built as the He 162.

ABOVE: Partially assembled He 162s in a former salt mine at Tarthun, Saxony-Anhalt, Germany in mid-1945. The speed with which He 162 mass production began – even as development of the type continued – is staggering.

Chapter 10

Objektschutzjäger

Target defence fighter 1943–1945

The only role that the Me 163 could fulfil was that of point defence - and it wasn't particularly good at it. As the bombing of Germany intensified, the Objektschutzjäger or 'target defence fighter' competition was launched to create a new rocket-powered interceptor.

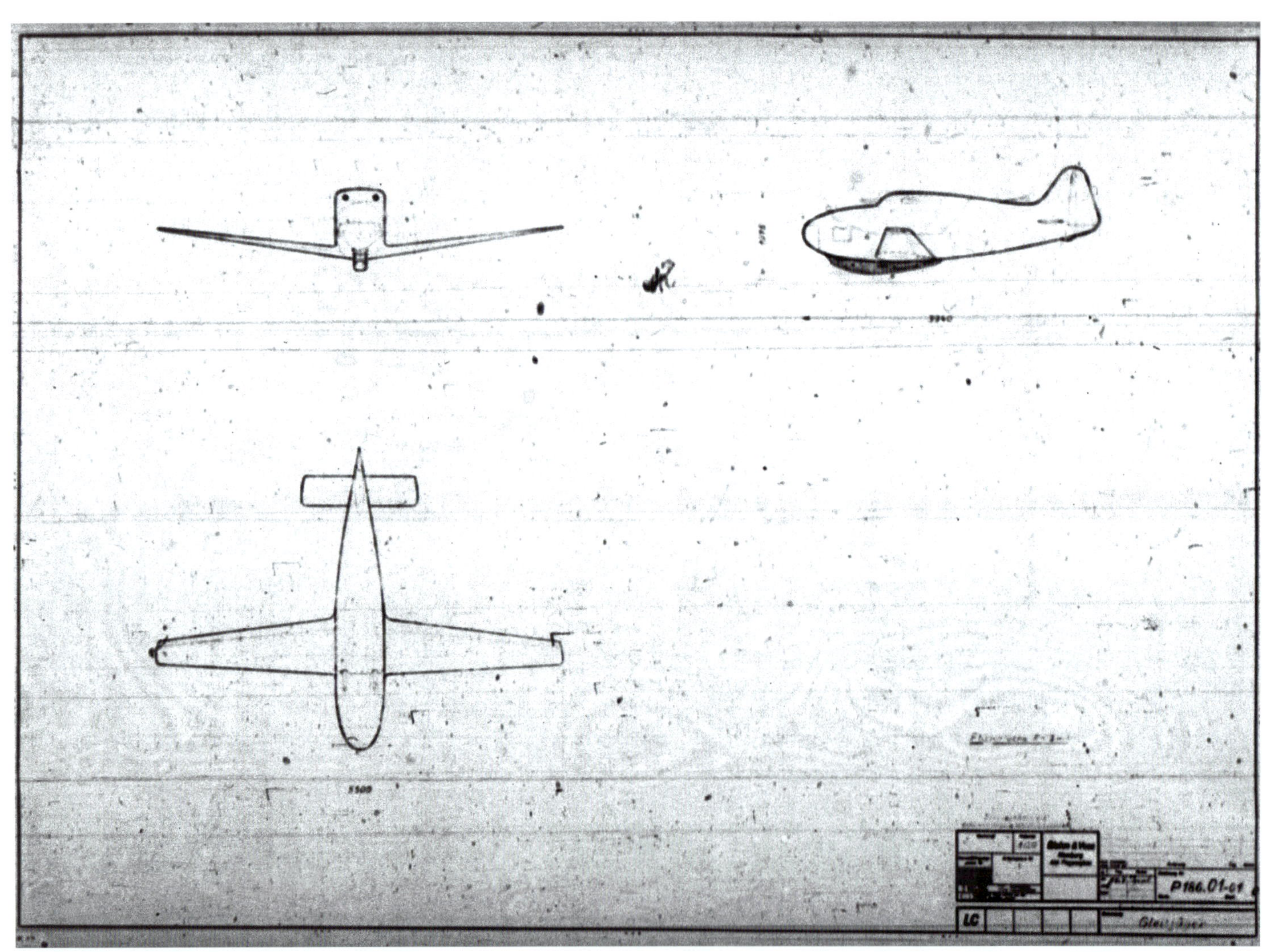

ABOVE: The earliest known version of Blohm & Voss's P 186.01-01 Gleitjäger - originally conceived as a rammer. The drawing is too degraded to make out much detail but the aircraft had a landing skid and two cannon in the upper fuselage.

BLOHM & VOSS P 186/BV 40

The bombing of Germany intensified during the summer of 1943 and the whole nation was shocked when Hamburg was subjected to six devastating incendiary attacks between July 24 and August 3. Some 41,800 people were killed and around 37,400 more were injured.

The perceived inability of the Luftwaffe to defend against these raids using conventional aircraft gave rise to serious discussions about less orthodox methods of attack.

The design staff of Hamburg-based Blohm & Voss, led by Dr Richard Vogt, experienced the horrific bombing first hand and within a fortnight had come up with a new heavily armoured Gleitjäger or 'glide-fighter' for use against enemy bombers. The unpowered P 186 would be towed aloft behind a Messerschmitt Bf 109 G then released to make a "ramming stroke against the rudder of the opponent".[1] A single 30mm MK 108 cannon was included in the design for keeping enemy gunners' heads down during the approach.

On August 19, 1943, the RLM responded to the company's proposal with an appeal for details of how quickly the P 186 could be designed and put into service – also requesting that development work should be progressed quickly.

Blohm & Voss continued development work and the RLM's head of aircraft development (GL/C-E) Major Siegfried Knemeyer wrote to General of Fighters Adolf Galland on November 2, 1943 to explain the project to him. Galland's office responded on November 13 to say: "The General of Fighters asks that Blohm & Voss continue to develop a so-called ramming aircraft. If the losses in climbing performance due to attachments to fighter and destroyer aircraft are within acceptable limits, the ramming aircraft in conjunction with rope-cutting devices could be used as a valuable support for home defence. The General of Fighters asks for documents about the Blohm & Voss glider fighter and, after inspection, will make appropriate proposals for the development."[2]

However, in the cold light of day Blohm & Voss evidently realised that the pilot of a 'glide-fighter' which crashed into its target would have little chance of escaping his own wrecked aircraft and in a report of November 29, 1943, stated that "the modification of the fighter as a ramming aircraft is not currently being investigated".[3] Nevertheless, the project was given the official RLM designation BV 40 on December 15, 1943, and 12 prototypes were ordered.[4]

Work on these progressed quickly but official interest in the project soon waned. Armed with only a single MK 108 the BV 40 was unlikely to be much use if it couldn't ram its target. On March 16, 1944, Vogt wrote to Otto Malz at the RLM[5] with a proposal to fit the BV 40 with the same rocket motor as the Hs 293 guided missile – the 109-507 – which could extend the aircraft's attack time by almost two minutes. A "later device" could provide four minutes of additional power.

A proposal to fill a BV 40 with explosives and carry it aloft on a Heinkel He 177 bomber followed in April but Blohm & Voss's Berlin representative reported back to Vogt on April 20 that "the General of Fighters is no longer interested".[6] Even so, at least five flying prototypes (V1, V2, V4, V5 and V6 – V3 was used for static tests) were completed and the company was still trying to get the type built by subcontractor Siebel in November 1944.

Although the P 186/BV 40 glide-fighter was a flawed concept, the simple idea of a small cheap rocket-propelled

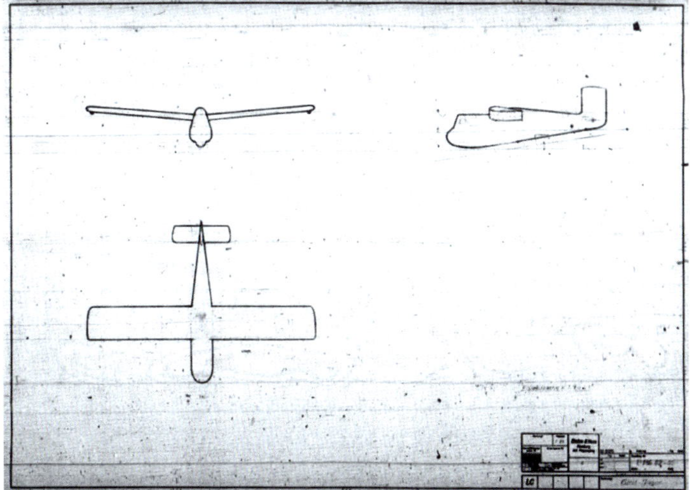

ABOVE: The second version of Blohm & Voss's towed 'glide-fighter' – the P 186.02-01. The design now features only a single cannon in the upper fuselage.

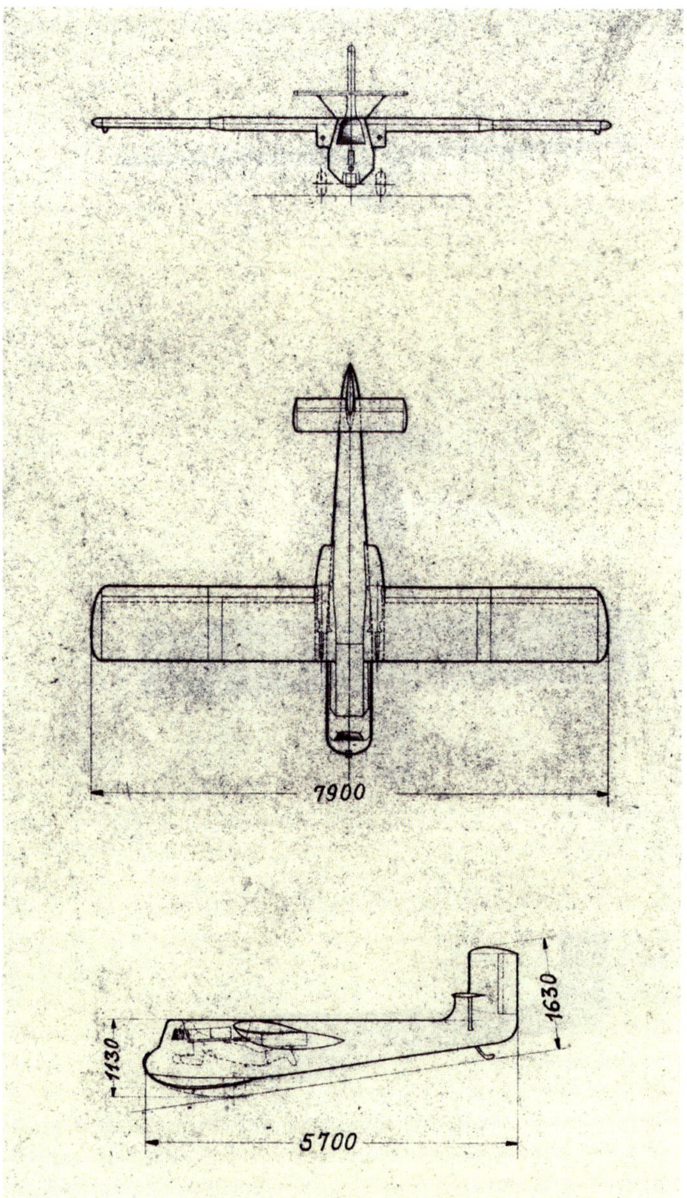

ABOVE: The Blohm & Voss Gleitjäger in its final form with the official designation BV 40. This drawing comes from a description report dated January 1944.

ABOVE: During the early war years the DFS experimented with a wide range of glider-towing combinations – such as this Heinkel He 111 towing a DFS 230. Blohm & Voss planned to employ this method for launching its 'glide-fighter'.

ABOVE: Another DFS experimental towing combination – a Heinkel He 177 bomber towing a Gotha Go 242 glider.

fighter or rammer that could be towed or carried aloft by a more conventional fighter would prove attractive to several other designers in 1944.

LIPPISCH'S 'ARROWS OF DEATH'

An even more radical approach to destroying enemy bombers was proposed by physicist Dr Paul Karlson, author of *Du und Die Natur* (published in English as *You and the Universe: Modern Physics for Everybody* in 1938), in a report dated September 9, 1943, based on an idea conceived by his friend Alexander Lippisch.

It was entitled 'Ramm-Raketen nach Dr Alexander Lippisch, Wien' or 'Ram Rockets According to Dr Alexander Lippisch, Vienna' and it discussed the idea of building thousands of small manned scaffold-launched rockets designed to destroy enemy aircraft by ramming them.

Karlson seems to have rapidly typed out two versions of this proposal,[7] the first version being short and rather vague, the second being more colourful and detailed. What follows is from the second version: "Who does not know the unnerving feeling of impotent rage when one is helplessly stuck on the ground during an enemy air raid, knowing that the ominous birds are above our heads, their grinding engine noise, their shadowy outline, snatched by the arms of the searchlights from the protective darkness of the night sky, for seconds, yes, for minutes gliding over the sky, or even flying over our land in bright daylight? Every shot of the flak is accompanied by secret wishes; you clench your fists and wish to go up to the enemy plane, to be able to throttle its crew; you want to control the flight of the projectile, give the little corrections that would bring it to the target – but you can't.

"Does it have to stay that way? Sure, night fighters and flak do their utmost – but the anti-aircraft shell is, once fired, blind and uncontrolled; the night fighter, on the other hand – what an overpowered technical marvel is his machine, how finely wrought and complicated is the direction-finding mechanism, and how high the level of flying skill, laboriously acquired over a long period of time, needed to master the bird – and all of this only to finally get within a few hundred metres of the enemy and give him a few bursts of fire, a few kilograms of iron and explosives, gone in seconds.

"Today's technical war demands the highest level of engineering performance, and as a result it tends to become inevitably more complicated, refined, and overbred. The machines are more susceptible to problems, with every new improvement, every new device that is already being added to dozens; the crews are highly trained flying engineers and, despite the best training, will soon find themselves unable to cope with the stress. What is necessary is a reflection on the real task and radical exploration of the simple.

"In war, it is important to smash your opponent and his material. Most combat actions can be reduced to destructive forces striking rapidly moving masses; it doesn't matter whether this force is applied by the hand axe of the Stone Age man or by the jagged steel splinters of a modern explosive bullet fired from an on-board cannon. The fighter is basically just an extremely complicated and powerful gun carriage, and the gun, of whatever kind, is only the extended arm of the warrior. The most straightforward way to strike an enemy plane is to get up close and deliver the largest possible blow with the largest possible mass. In other words: ram it!

"That sounds insane, like the worst kind of outburst from an armchair warrior. However, the suggestion does not come from a fanatic, but from the ranks of our aviators themselves – that is, from men who know what they are talking about and who are always ready to try it out themselves. And really, if you forget the unfamiliarity of the idea, if you familiarise yourself with the thought and assume it as a given, you will immediately see a number of decisive advantages, with almost no disadvantages.

"An aircraft weighing 500kg and travelling at 250m per sec has a force of [blank space in document]; this corresponds to the force of a fully loaded goods train falling from a height of 50m onto the ground. With this energy you can cut the fuselage of a bomber at the tailplane as smoothly as a knife passes through a loaf of bread. The millimetre-thin sheet metal cladding, the light metal components buckle and shatter under the elemental force of this blow, without the pilot of the ramming plane feeling more than a moderately hard impact; his flying projectile has no time to slow down. A bomber without a tail unit is still capable of flying, and so it is possible to cut a wing. However, the rammer pilot has to watch out for the heavy masses of the engines and the armoured cockpit.

"A prerequisite for this way of fighting is a technically worked-out new device. It has to combine high speed with stable construction. It has to be straightforward, to reach the enemy as fast as possible and – literally! – fly through him. But otherwise no armament, no equipment, and none of the flying properties you would expect from an aircraft. The manned projectile, more precisely, the manned missile, ideally fulfils all these requirements.

"A tubular fuselage is provided at the front with a solid steel ram tip and at the rear with three strongly swept wings and driven by a rocket engine. The best thing is the tried and tested starting aid rocket with a thrust of 1,500kg. The pilot lies prone, ignites the engine and flies like a rocket vertically upwards. No flying skill is necessary because the projectile is inherently stable; it goes in a straight line as long as the fuel lasts.

"Climbing speeds of 250m per second are easy to reach and easy to endure. With a simple wheel control modelled on that of the motor vehicle, the pilot operates the tail surfaces and can therefore fly moderate curves. There is no need for any aviation skill. He targets his opponent before the start, keeps an eye on him, making the necessary corrections – until it crashes! Heil Hitler! Only small deviations and movements are to be made using the crosshairs. The flight takes about 40 seconds from take-off to ramming at 6,000–8,000m. At night the aircraft can fly along the searchlight beam. With additional launch rockets you can accelerate the start, shorten the flight time, increase speed and range or climb altitude.

"At the peak of its climb, the projectile slows down to zero. The pilot then releases a parachute on which the rocket slowly floats to the ground. Sooner or later he will also parachute out himself. There is also the option of turning back around, picking up speed again in a steep gliding flight and starting a second ramming attack. In the future one could think – varying the process – of firing a single strong explosive charge from the tip of the fuselage shortly before hitting the target, 50–100m away."

Disadvantages were the need to find pilots psychologically capable of ramming their target, the need to hit the target on the first pass, lack of defensive weapons, the risk of parachuting after every launch and frequent damage to the rammer itself. But Karlson believed finding pilots would be easy, normal fighters were often only able to make one pass, defensive weapons were unnecessary and the physical strain on the pilot would be "no greater than endured by paratroopers or dive-bomber crew".

And the advantages were great: a hit would almost certainly destroy the target and "the effect of this weapon on morale must be a terrible one: the enemy crew sees the lethal, controlled projectile race towards them from the ground without the slightest possibility of defending themselves. From the bomber you can see how the razor-sharp bullet is approaching, effortlessly cutting through a comrade's aircraft and shooting away from it in a fraction of a second, with no loss of speed or change of direction. Whole swarms of such arrows of death, close to each other, race towards the squadron on parallel tracks; and in a single second the work of destruction is started and ended.

"This weapon would only realise its true potential if it was used in large numbers, not in hundreds but in thousands. This is easily possible. The engines – take-off booster rockets – are available in large quantities and are easy to manufacture. The equipment of the machine is almost zero. It is limited to a visor and a lever for the parachute release."

The pilot wouldn't even need to be a pilot as such: "Today's aviator must be an artist, and a flying engineer. But the ramming plane flies itself – the man inside it doesn't need to be able to fly. The most suitable crew is not aviators, but paratroopers. Anyone who can control an automobile, a bobsleigh, or even a scooter, can also bring the rammer, the rudders of which are easily operated, to the enemy. Training would be completed in a month: test flights to get used to the acceleration and to become familiar with the device. Then final flights against towed light cloth discs that are pierced smoothly. That's all."

Karlson proposed a programme of model tests using powder rockets followed by tests using an unmanned full-scale model. The finished vehicle would be a 5m steel tube with three equally spaced swept tailfins and a 'wingspan' of 2m. The tube would be tipped with a steel ram, behind which would be the pilot's armoured pressure cabin. The engine would be a 1,500kg RATO booster supplemented by three Borsig rockets with 1,000kg thrust each. It would take off vertically from a pipework scaffold. Take-off weight was 800–1,000kg, top speed 250–300m/sec, climb to 10,000m 40 seconds and 60 seconds overall flight time.

Following on from the Blohm & Voss P 186/BV 40 towed or carried take-off idea, the Karlson/Lippisch concept added a second theme to discussions about target defence in Germany – the vertical take-off rocket-propelled fighter or rammer.

MESSERSCHMITT ME 163 C

A third theme – the horizontal take-off rocket-propelled fighter – continued a much earlier thread of development. Following the dissolution of Abteilung L and the departure of Alexander Lippisch in April 1943, Messerschmitt continued preparing the troublesome and dangerous Me 163 for front line service but did little to further develop the design.

A model designated Me 163 C had been undergoing tests at the AVA during February 1943 but this was not Lippisch's extended fuselage design with a second rocket motor for cruising – it was an Me 163 B with modified wings and under-wing weapons gondolas.[8]

Another model which did have an extended fuselage was being tested in parallel however[9] under the simple description 'Me 163 B with longer fuselage'. Now both projects were put on hold until July when Messerschmitt decided that the Me 163 C designation should be revived[10] – although exactly what form this revived 'C' took is unclear.

Certainly by September 1943 it is clear that the Me 163 C would be powered by the new BMW P 3390 C – a rocket motor with two combustion chambers. The larger of these was for take-off and climb, while the smaller chamber was for more fuel-efficient cruising at altitude.[11] Progress on both the engine and airframe would prove to be painfully slow, however.

Hauptmann Wolfgang Späte, commander of Me 163 test unit Erprobungskommando 16, reported on January 13, 1944:[12] "At the beginning of December last year, Messerschmitt promised to deliver at least 23 operational Me 163 Bs by the end of 1943. In fact, only six aircraft were delivered between December 24 and January 13. These delays cannot be blamed on more technical problems encountered in flight testing. They are simply the result of negative comments made about the Me 163 in the last few days by Messerschmitt's directors.

"Messerschmitt AG promised Dr Lippisch on his departure from the company last year that it would use all possible means to develop and produce the Me 163 for front-line operations. In practice, almost all work in the design offices and factories immediately began to stagnate ... EKdo 16 has already pointed out, in the progress report submitted in June last year, that the Me 163 was of 'no

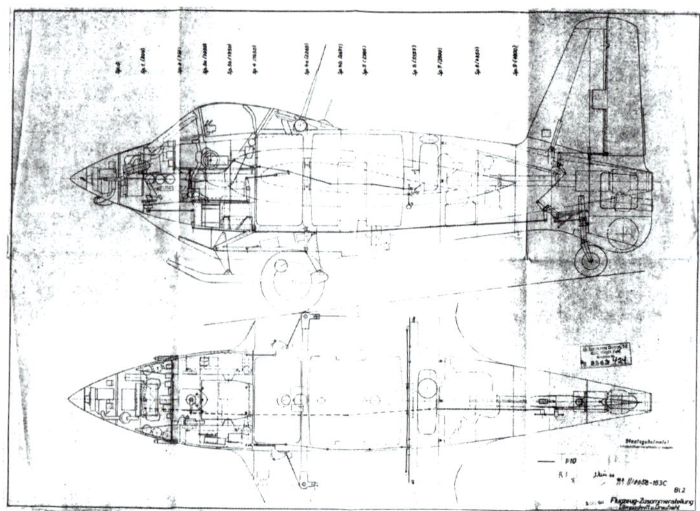

ABOVE: The last known version of the Me 163 C, with long tailfin, backwards retracting tailwheel and short skid.

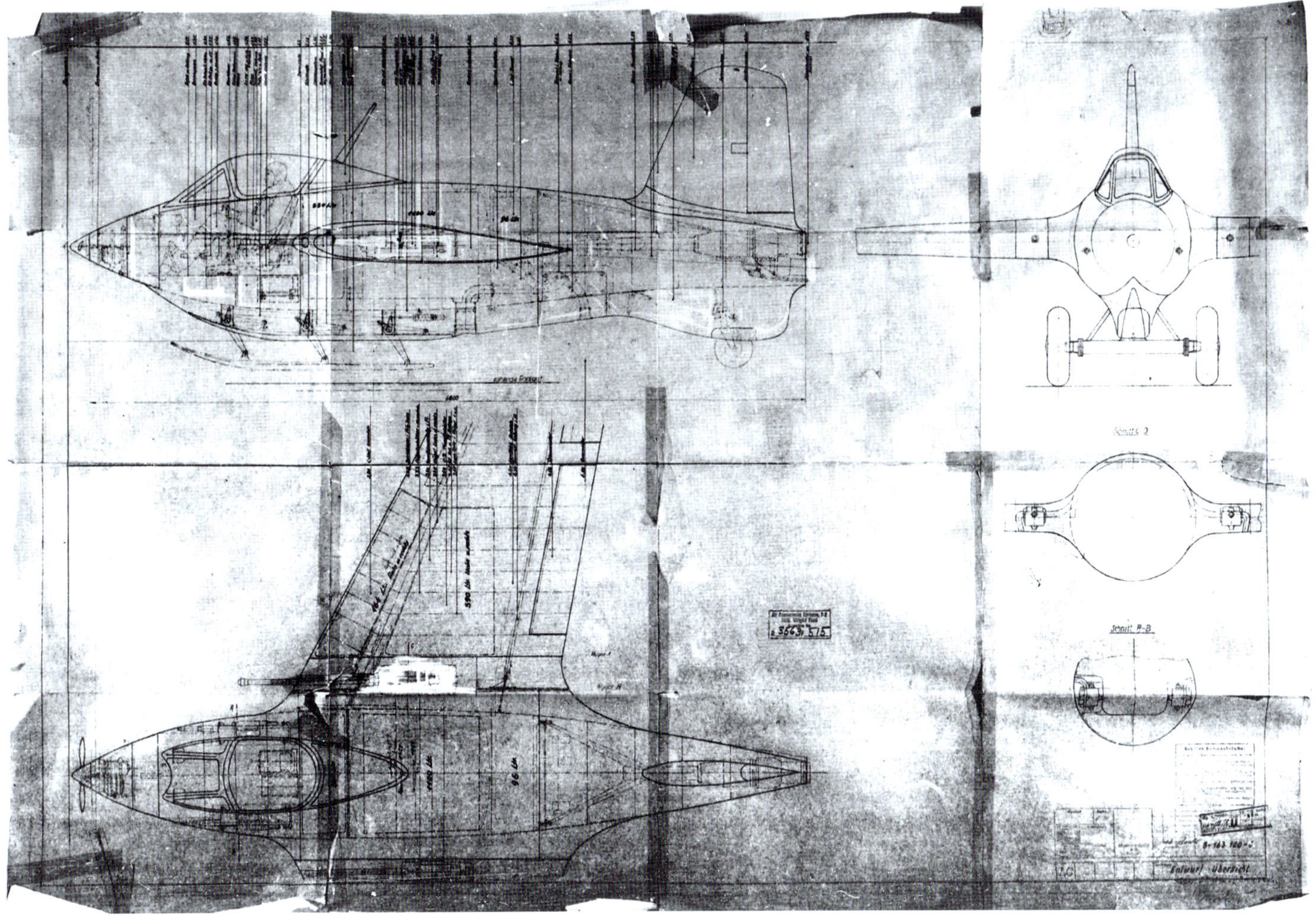

ABOVE: An earlier version of the Messerschmitt Me 163 C with short tailfin, vertically retracting tailwheel and long skid.

interest' to Messerschmitt's technical directorate. It was discovered later that this statement had been made by Herr Prof. Messerschmitt himself.

"Recently (at the end of December 1943), his lack of interest again became apparent as Flg.-Stabsing. Antz asked him why he (Antz) had not heard anything more about the Me 163 C since May 1943, almost nine months ago. Comments have been made in a similar vein by members of the design and project offices (among them Dr Wurster), who believe there is no point in working on the Me 163 since it will be 'killed off' anyway."

By May 1944, the Me 163 C and its rocket motor were still being listed by the RLM as developments in progress[13] but evidently little work had been done.

BACHEM NATTER

Back in December 1941, while he was still working for Fieseler, engineer Erich Bachem produced a report on two versions of a rocket-powered high-altitude fighter based on an original concept created by Wernher von Braun two years earlier but rejected at the time by the RLM.

The Fieseler Fi 166 designation covered both a turbojet-powered fighter launched vertically on top of a rocket and a pure rocket fighter. The former was to have a flight time of 45 minutes thanks to its assisted take-off, although it could not be guaranteed that the launch rocket, once detached and hopefully floating back to earth on a built-in parachute, wouldn't land on somebody's house or on high-voltage power lines. The latter would have a flight time of only five minutes and would therefore only be able to make one attack before having to break off and land on a belly skid.

Neither concept went any further but Bachem remained interested in the potential of a vertical-launch rocket-propelled interceptor. On July 16, 1944, Bachem sketched a tube-shaped manned rocket remarkably similar to the Karlson/Lippisch concept of 10 months earlier. The pilot lay prone within the fuselage with a parachute on his back and the nose of the rocket was designed for ramming. Bachem later recalled that von Braun may have sketched out the idea for him initially – once again providing him with the concept.[14]

Oberst Siegfried Knemeyer, head of aircraft development at the new Chef der Technischen Luftrüstung department, replacing the old Technisches Amt at the RLM, was evidently impressed with Bachem's design and on August 1, 1944, commissioned further development of the project which was dubbed BP 20 or 'Barak I'.[15]

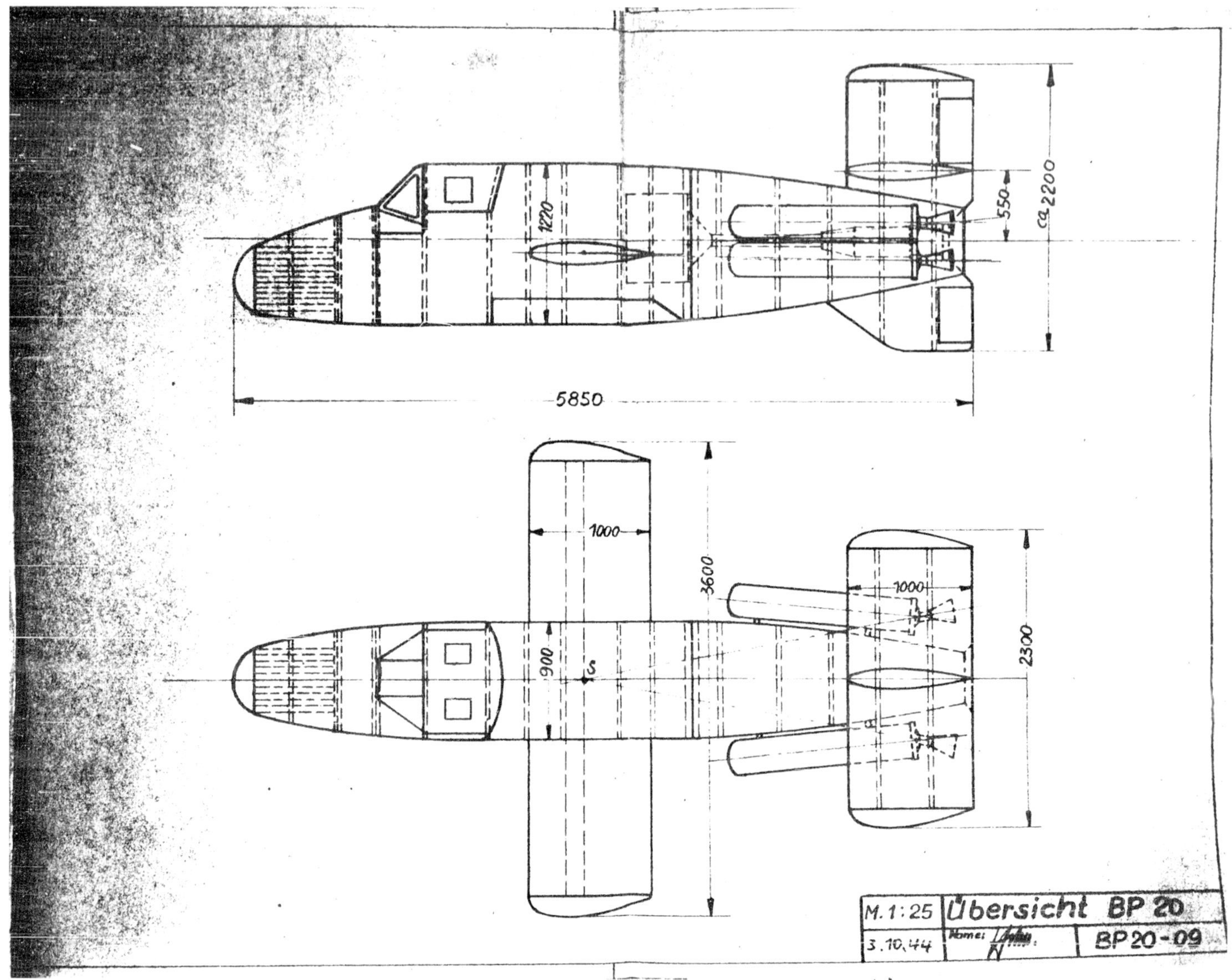

ABOVE: The Natter was known as the BP 20 within Bachem. This drawing shows the ninth scheme for the interceptor, dated October 3, 1944.

The earliest design produced by Bachem-Werk was the BP 20-01. This was a 6.5m-long tubular manned rocket similar to the one shown in Bachem's earlier sketch. It had two wings level with the prone pilot's position and a wingspan of 2.6m. An outline report on the design, headed 'Projekt Natter', was produced on August 9.[16]

This stated the project's purpose and requirements: "Annihilation of enemy aircraft especially bombers by bringing up a gunner within the immediate vicinity of the enemy and discharging rocket projectiles at him with the smallest possible amount of manoeuvring and propellant.

"No self-destruction of the pilot, on the contrary, armoured protection of him. Smallest possible production cost, maximum use of wooden parts, reduction of iron. No burden on standard aircraft industry. Exploitation of the large, partly free, timber resource. Repeated use of the most critical airframe and propulsion unit parts by parachute recovery.

"Little flying requirement of the pilot, due to the omission of a steered take-off as well as omission of a normal landing. Little ground input. Little transport cost. Easy transferability. Good camouflage potential."

The design went through rapid changes as it was worked on. Wing and tail shapes were altered, the pilot's position was shifted upwards to allow him a better view and a more detailed arrangement was worked out for the positioning of fuel tanks within the fuselage.

During September 1944 the design was changed so that the pilot could be seated rather than prone, and the single rocket engine in the rear of the fuselage was augmented by a pair of disposable Schmidding SG 34 solid fuel boosters attached onto each side of the aircraft. As the design evolved, four SG 34s would be specified.

The launch sequence was to involve the pilot bracing himself against padded head and back rests and gripping handles on either side of the rear nose frame. On the left

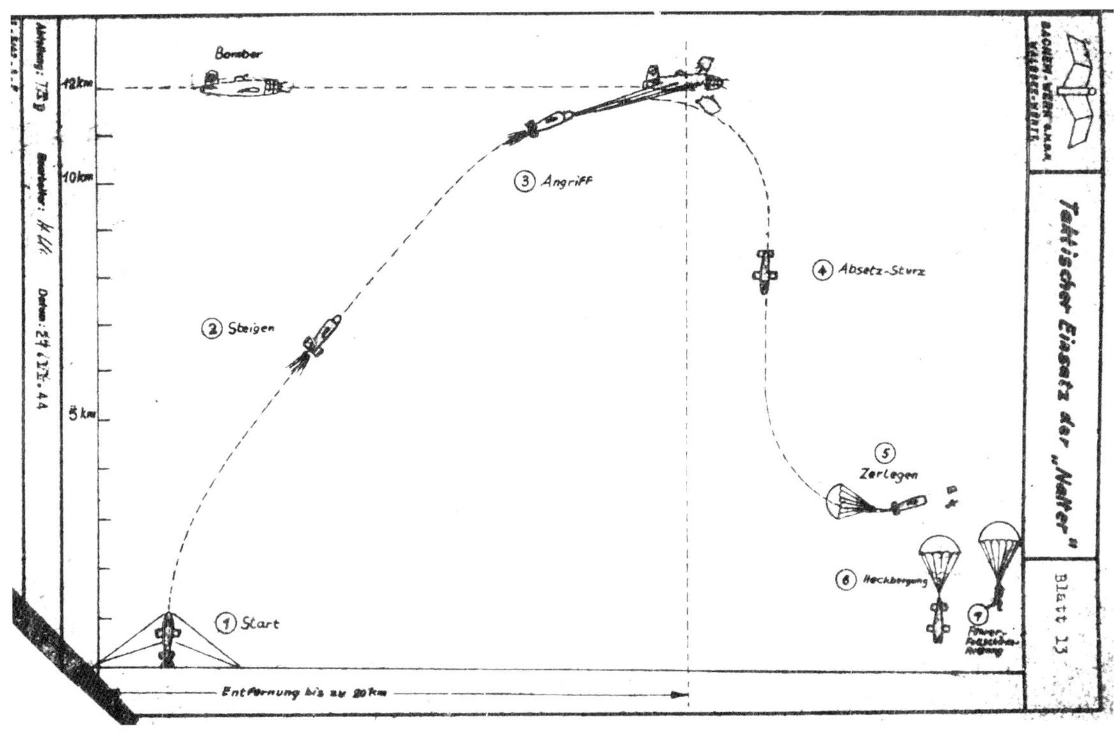

LEFT: This Bachem drawing from November 27, 1944, headed 'Tactical Operations for the 'Natter' shows the different stage's of the Natter's attack from launch to climb, the attack itself, descent, separation and finally recovery.

BELOW: A carriage/launcher for the Natter, shown in a drawing dated January 4, 1945.

BELOW LEFT: Plan from a Bachem document showing how the Natter could be launched using three ropes to hold it upright.

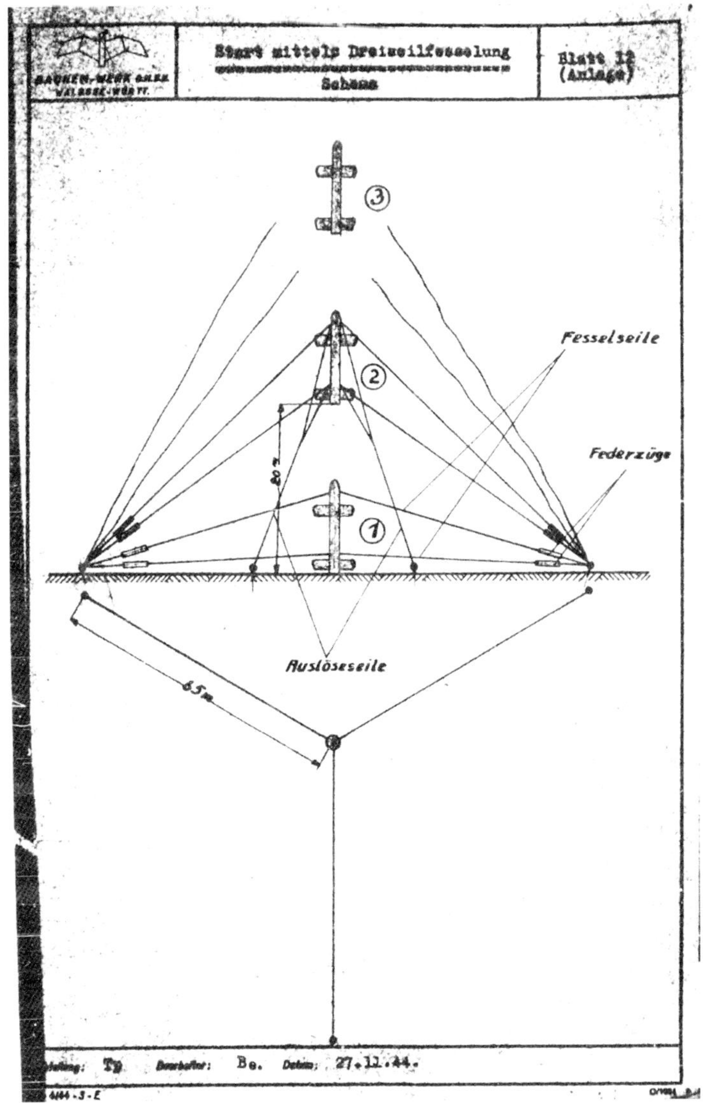

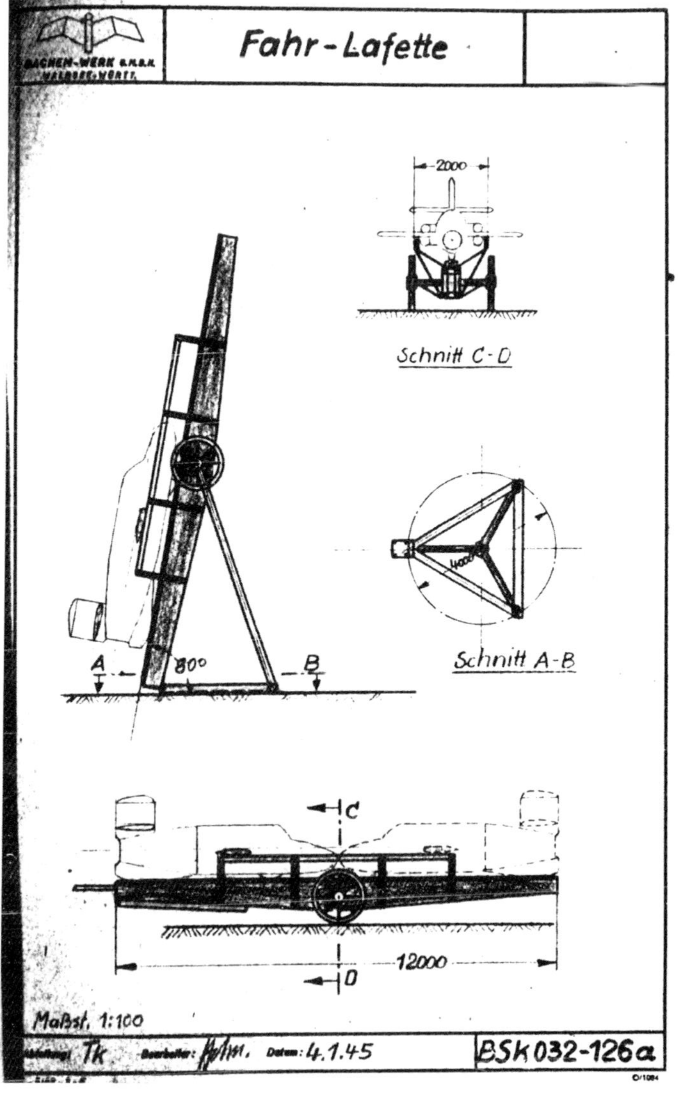

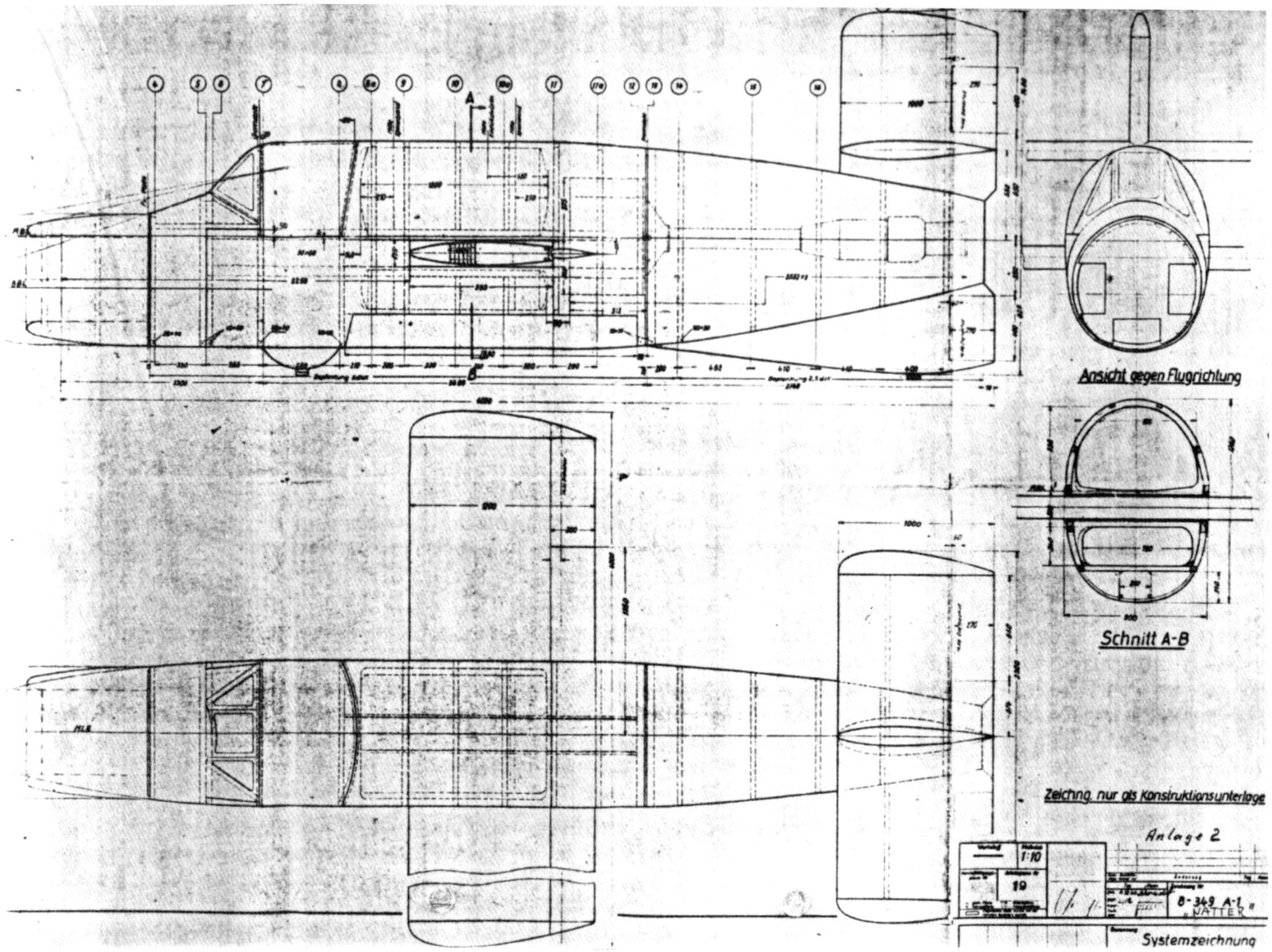

Ansicht gegen Flugrichtung

Schnitt A-B

Zeichng. nur als Konstruktionsunterlage

Anlage 2

1:10

B-349 A-1 „NATTER"

Systemzeichnung

ABOVE: The Bachem Ba 349 A-1 Natter, as it appeared on December 8, 1944.

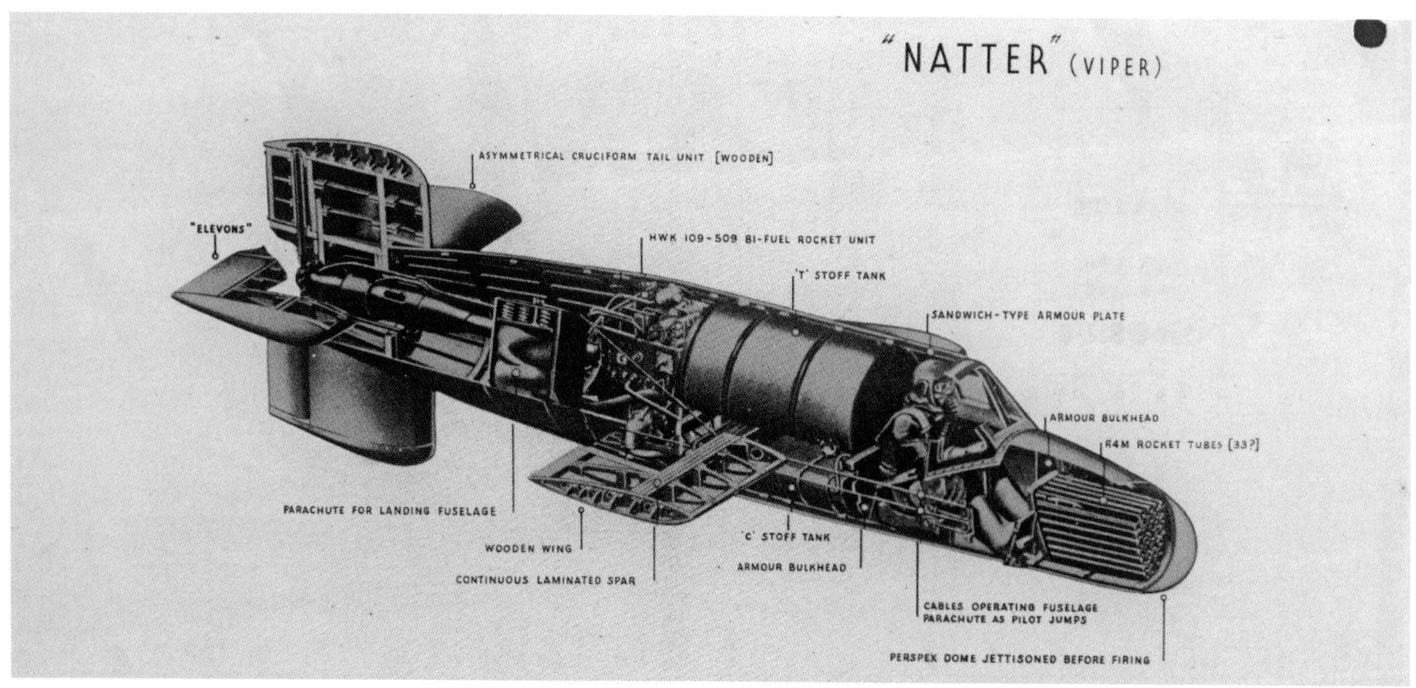

ABOVE: A schematic of the Natter from an original Bachem document given English captions for a British intelligence report.

handle was the button to fire the boosters. Once this was pressed, the Natter would rise under the control of a basic autopilot and after 10 seconds the exhausted boosters would automatically detach and the Walter engine would take over.

Within another 50 seconds, the Natter would reach the same altitude as the bomber formations. The pilot would then turn off the autopilot, grab the control column and bring the aircraft around in a shallow curve to line up a target.

At a distance of 300m he would fire the Natter's weapons at the bomber. The Walter engine would then run out of fuel and the pilot would put it into a dive. Pulling out at low level, the pilot would activate a mechanism to detach the aircraft's nose, which would fall away, and he would then be catapulted out by the sudden deceleration – activating his parachute.

The Natter's fuselage, minus the nose, would then descend to earth on its own parachute, ready to be refurbished and reused.

An updated 'Projekt Natter' report was drawn up in September[17] which described the vehicle as a hybrid of rocket and aircraft. Although the text of the report was dated September 20, 1944, it came with drawings dated October 3 and October 5, suggesting a slightly later publication date. Length was now 5.72m and wingspan was 3.2m.

MESSERSCHMITT P 1104

At Messerschmitt, with the Me 163 C stalled and the Me 163 B just beginning to enter service, some thought was being given to a cheaper alternative for target defence based on the P 186/BV 40 towed take-off model but powered by a rocket motor during its attack phase. The P 1101 had been followed by the P 1102 jet bomber and the P 1103 rocket-propelled parasite fighter in July 1944 so the Me 163 replacement was given the project number P 1104.

Messerschmitt drawing XVIII/125 of August 10, 1944, is the earliest known depiction of the design which was evidently drawn up by Hans Hornung. It had a short fuselage with two large tanks in the centre for its rocket fuel. The pilot sat up front in the nose with a single MK 108 underneath him. The mid-wing aircraft would have been towed aloft by an Me 262 with a detachable wheeled dolly provided for the initial ground run. A skid was provided for landing.

By September 25 the design had been switched from mid-wing to shoulder wing and the cockpit was moved back from the nose. Rather than a tow, take-off would now involve two 1,000kg thrust expendable rocket motors. The only known text description of this version[18] stated that it had a wing area of just 6.5m² (although project notes from mid-September show that a wing area of 5.55m² was initially worked on), a wingspan of 6.2m and an overall length of 5.48m. It was powered by a single HWK 109-509 A-2 rocket motor and armament was one MK 108 with 100 rounds of ammunition. Take-off weight was 2,563kg.

During early October, Willy Messerschmitt evidently came to hear about Bachem's Natter and sent Erich Bachem a personal letter on October 6, 1944,[19] saying: "Dear Mr Bachem! You will be surprised that I turn to you again, since we haven't had any contact points for a long time. However, since in these times it is necessity to avoid, if possible, any duplication of work, I come to you with a suggestion to modify the special aircraft you have designed according to a design that Mr Hornung of my project office made for the purpose of making the aircraft cheaper and to make it more versatile.

"The Horning design would fulfil all of your bird's tasks, including the vertical take-off, but can still start and land on a skid on wet grass or wet wooden rails without any aids (undercarriage). The aircraft is not always lost, as with you, but can often be used.

"The Hornung design also has the advantage that the flight duration would be sufficient for several approaches, and the amount of ammunition for the armament is also designed accordingly. Nevertheless, according to rough calculations, the aircraft will hardly need more than 500–600 working hours for a production of 500–1,000 per month, plus about 50 working hours for parts that are from V1 (Fi 103), which is known to run in bulk, so that the procurement of which should be relatively easy.

"In the interest of saving on construction and workshop time etc. I would be grateful if you would examine the matter and if you might be willing to combine your proposal with that of Mr Hornung. This aircraft would then probably make the considerably expensive 163 obsolete. I would be grateful for a very early communication of your opinion." He appended a brief description of the project entitled 'Kurzbaubeschreibung P 1104 II. Ausf. v. 25.9.44'.

Bachem presumably ignored this, having already made significant progress towards finalising the design of his Natter. A third design worked on in August 1944 was a vertical launch interceptor given the codename 'Julia' by its creator – Heinkel's Wilhelm Benz. Apparently the name derived from Benz's Viennese girlfriend of the time.[20] Little information has survived about the genesis of this design but the earliest known drawing of August 16, 1944, shows an aircraft 6.98m long with a wingspan of 4.6m, a prone pilot position and a twin-rudder tail. Armament appears to have been two MK 103s or two MG 151s in under-wing pods and propulsion was provided by an HWK 109-509 C. A single skid was provided for landing and four solid fuel boosters were attached for launch.

By now the Natter, P 1104 and Julia fell within the decision-making remit of the new Entwicklungshauptkommission Flugzeug and their development was to be overseen by Robert Lusser, designer of the Fieseler Fi 103, who had been appointed as head of the Sonderkommission Sonderflugzeuge or Special Commission for Special Aircraft which covered both flying bombs and rocket-propelled target defence aircraft.

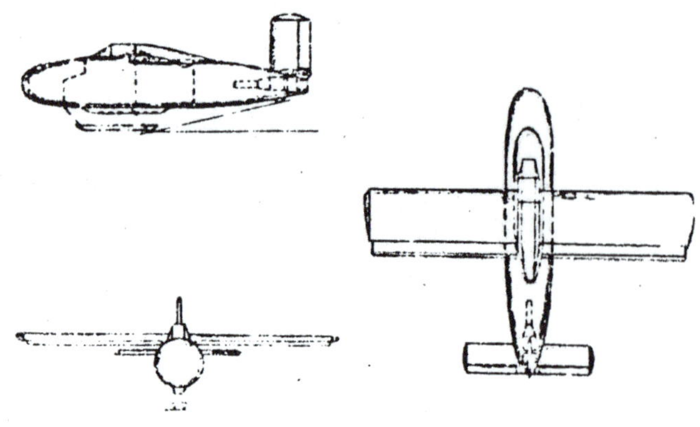

ABOVE: The Messerschmitt P 1104 was designed with towed take-off in mind but the last version was intended to take off on its own using booster rockets.

Over the next month, possibly at Lusser's instigation, several new target defence projects would appear in direct competition with the Natter, P 1104 and Julia.

JUNKERS JU 248/8-263

With Messerschmitt eager to wash its hands of the Me 163 B, responsibility for producing and developing the aircraft was handed to Junkers with effect from September 1, 1944. The government-run company immediately began implementing minor changes to the design intended to speed up production and lower costs – such as revising all interface drawings to ensure that major assemblies were interchangeable between airframes built in different factories.[21]

Having inherited the Me 163 C as well, Junkers also set about creating its own version of the design – intended to resolve all of the Me 163's inherent design flaws. A description comparing this new design, already known as the Ju 248, to the old Me 163 C was issued on October 11, 1944.

The company was awarded a contract[22] to develop the Ju 248 on October 21 and Junkers began by developing an interim design known as the Me 163 D – essentially the

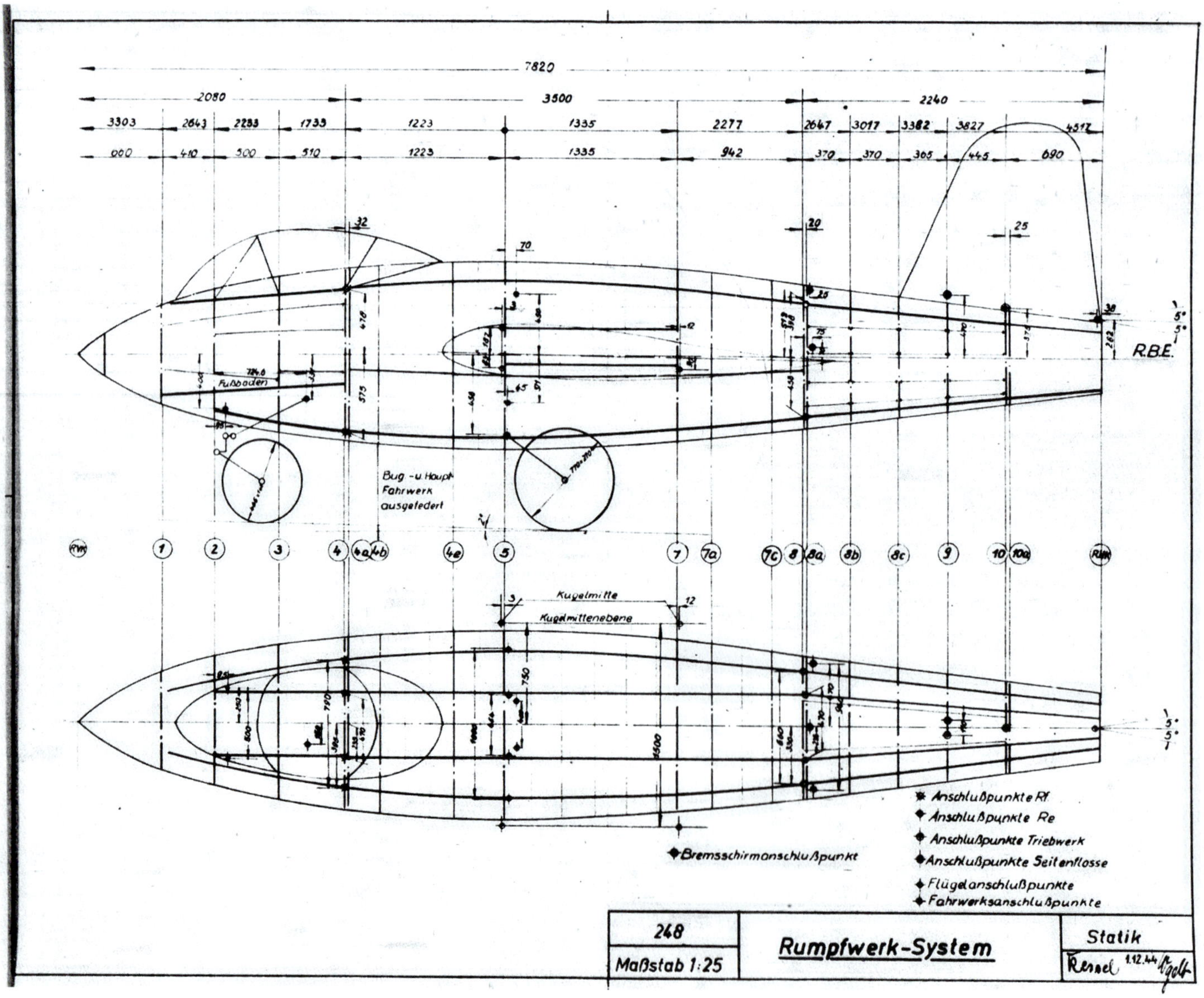

ABOVE: Fuselage dimensions of the Junkers Ju 248, dated December 1, 1944.

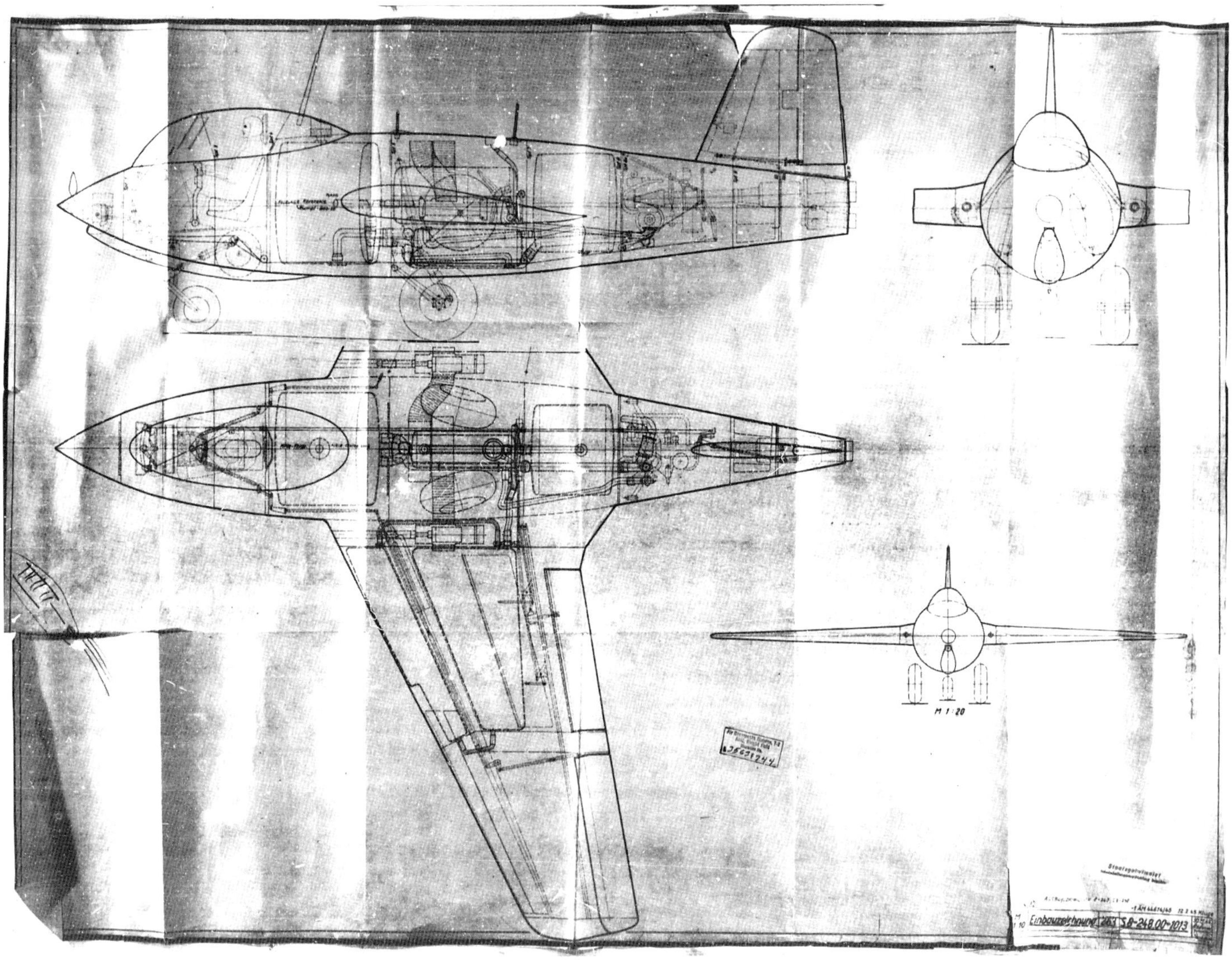

ABOVE: A detailed plan of the Ju 248, dated December 30, 1944. The number '263' now appears in the info panel.

Me 163 B V13 and V18 rebuilt as gliders with lengthened fuselages and fixed tricycle undercarriages.

A new description was issued on November 3, 1944,[23] which said: "The 248 is a further development of the 163 B. It arose from the endeavour to create a new series in which the weaknesses identified in the 163 B were eliminated, but the components currently in production were largely adopted without modification.

"The main defects of the 163 B are considered: 1) Flight time too short and therefore too little possibility of enemy contact. 2) Unsatisfactory solution to the starting problem with a jettisoned undercarriage. 3) Unsatisfactory landing

ABOVE: Wind tunnel models of the Ju 248 with undercarriage in the retracted and lowered positions.

ABOVE: A mock-up of the Ju 248 showing the cockpit and a view over the 'shoulder' looking towards the tail end.

characteristics of the skid (risk of overturning to the side, ground angles too small). 4) Inadequate manoeuvrability and thus operational hindrance due to skid solution. 5) Not taking advantage of the possible maximum lift when landing due to unsuitable centres of gravity.

"The 248 has the following key features: The outer wings of the 163 B are carried over. Only a few minor modifications are necessary due to the enlargement of the wing tanks. Due to a widening of the fuselage (see below) the wingspan increases by 0.2m. This results in wing area 17.8m^2 and wingspan 9.5m [fuselage length was 7.85m].

"The later possibility of a new wing, profiled for high speed, is taken into account in the airframe construction.

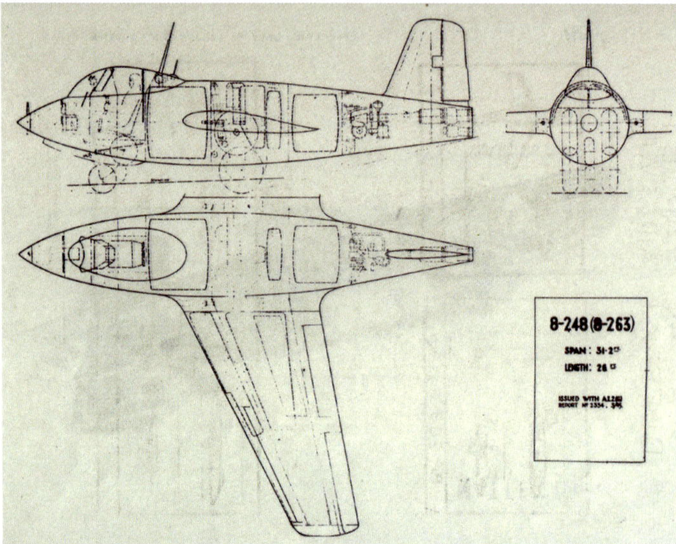

ABOVE: The Me 263 as it appears in a British intelligence report of May 1945. The drawing seems to have been copied from an original German document and is unusual for showing a narrow undercarriage which retracts directly upwards.

In order to eliminate the shortcomings of a skid, an undercarriage is installed in a nosewheel arrangement."

The undercarriage was designed to absorb particularly heavy impacts without injury to the pilot. In addition, "the need to accommodate the wheels in the retracted state and the larger fuel supply in the fuselage given the spatial conditions of the B-wing used made a completely new division of space and thus a new construction of the fuselage necessary."

The fuselage was to be constructed in three assemblies: "The cockpit, which is designed as a pressure chamber, the middle fuselage and the end of the fuselage. The pressure chamber is connected to the middle part of the fuselage by means of bolts, the end of the fuselage can be removed as a shell, as with 163 B, so that the necessary access for the engine is guaranteed. In order to facilitate manufacturing, the cockpit and the centre fuselage section are designed to be rotationally symmetrical. Double panes with silica gel drying are provided for the pressure chamber. Entry and emergency exit are through the detachable canopy. Controls are largely carried over from the 163 B, as far as the hull construction allows.

"The vertical tail of the 163 B is taken over unchanged. The engine will be taken over in the version planned for 163 C with a chamber for cruising flight [HWK 109-509 C]. The support tube is shortened by about 250mm. The fuel system consists of the four C-Stoff containers, which are stored in the wing and are larger than 163 B, with a capacity of 2 x 245 litres and 2 x 100 litres. Another C-Stoff container with a capacity of 152 litres is provided in the fuselage , the rear part of which is sealed off with a capacity of 62 litres.

"The T-Stoff is housed in three fuselage tanks. Behind the cockpit bulkhead is a large container with a total volume of 980 litres, the back part of which with 175 litres is sealed

ABOVE: The only completed Ju 248/Me 263 prototype. The aircraft was completed in January 1945 and test flown as a glider the following month.

off. In the space between the wing connection frames there is a tank with a capacity of 100 litres and behind the connection of the second wing carrier there is another tank with a capacity of 520 litres."

Radio equipment consisted of FuG 16 ZY and FuG 25a although "a later exchange of FuG 16 ZY and FuG 16 ZVG for FuG 15 is taken into account in the space requirement, also a FuG 24 can be installed instead of the FuG 16".

Armament was a single MK 108 in each wing root with 75 rounds apiece. The pilot had armour protection against 20mm rounds from both front and rear. The 2000W generator and Knorr air compressor were driven by a Seppeler propeller on the aircraft's nose with a maximum output of 12hp. A brake parachute was built into the rear fuselage.

Another description of the aircraft issued on December 15, 1944,[24] differed primarily in being entitled 'Baubeschreibung 263' rather than 'Kurzbaubeschreibung 248' – apparently Willy Messerschmitt had insisted that the type should revive this old designation from 1941 because the type represented a direct evolution of his company's 163.

The new description also elaborated on the thrust provided by the different chambers of the HWK 109-509 C – 2,000kg thrust from the primary combustion chamber and 400kg from the smaller 'cruising' chamber. In addition, 80 rounds were now provided for each gun but armour had been reduced to protecting against 12.7mm rounds coming from the front. Detailed diagrams were included showing how the fuel system would operate.

ABOVE: The remains of the Ju 248/Me 263 V1 after the war surrounded by parts of both American and German aircraft.

HEINKEL P 1077 JULIA

After the August 16 design, the only three known original drawings of the Julia all date from October 15, 1944, and accompany Heinkel report Nr. 111/44 of November 16, 1944, simply entitled 'Julia',[25] written by Benz. They show the aircraft slightly reduced in length at 6.8m but with the same wingspan of 4.6m and a wing area of 7.2m^2.

Two of the drawings – a side view and a three-view – show the prone pilot with a single MK 108 now on either side of his tiny cockpit rather than under the aircraft's wings. The third drawing shows an alternative nose – with the pilot in a more conventional seated position. The four expendable solid fuel take-off rockets are not shown in the drawings.

The twin-tail arrangement is the same but the wings now have pronounced 'ears' at their tips similar to those of the He 162. Two landing skids are now fitted – one beneath the pilot and the other below the fuel tanks.

The report began: "In order to be able to defend ourselves more effectively from the current mass of enemy bombers better than anti-aircraft guns and current fighters, the construction of a device is proposed which represents an intermediate solution between a manned anti-aircraft rocket and a fast, cheap miniature fighter with rocket drive.

"In order to achieve good flight characteristics in the shortest development and testing time, the device has been planned with the normal tail design. The size and construction of the aircraft is chosen so that there is sufficient flexibility and adaptability in terms of engine, armament and possible uses to be able to respond appropriately to the opponent's defence tactics with other weapons and other methods of attack.

"The following design principles have been used: speed – the achievable speed should be greater than that of the fastest fighter, so that a successful attack and a safe withdrawal from the opponent is ensured. Because of the lower combat strength and the smaller range, the greatest possible superiority in speed is of particular importance.

"Range – because of the low range of the rocket aircraft, it is advisable to provide an amount of fuel so large that two to three approaches are still possible after the enemy has

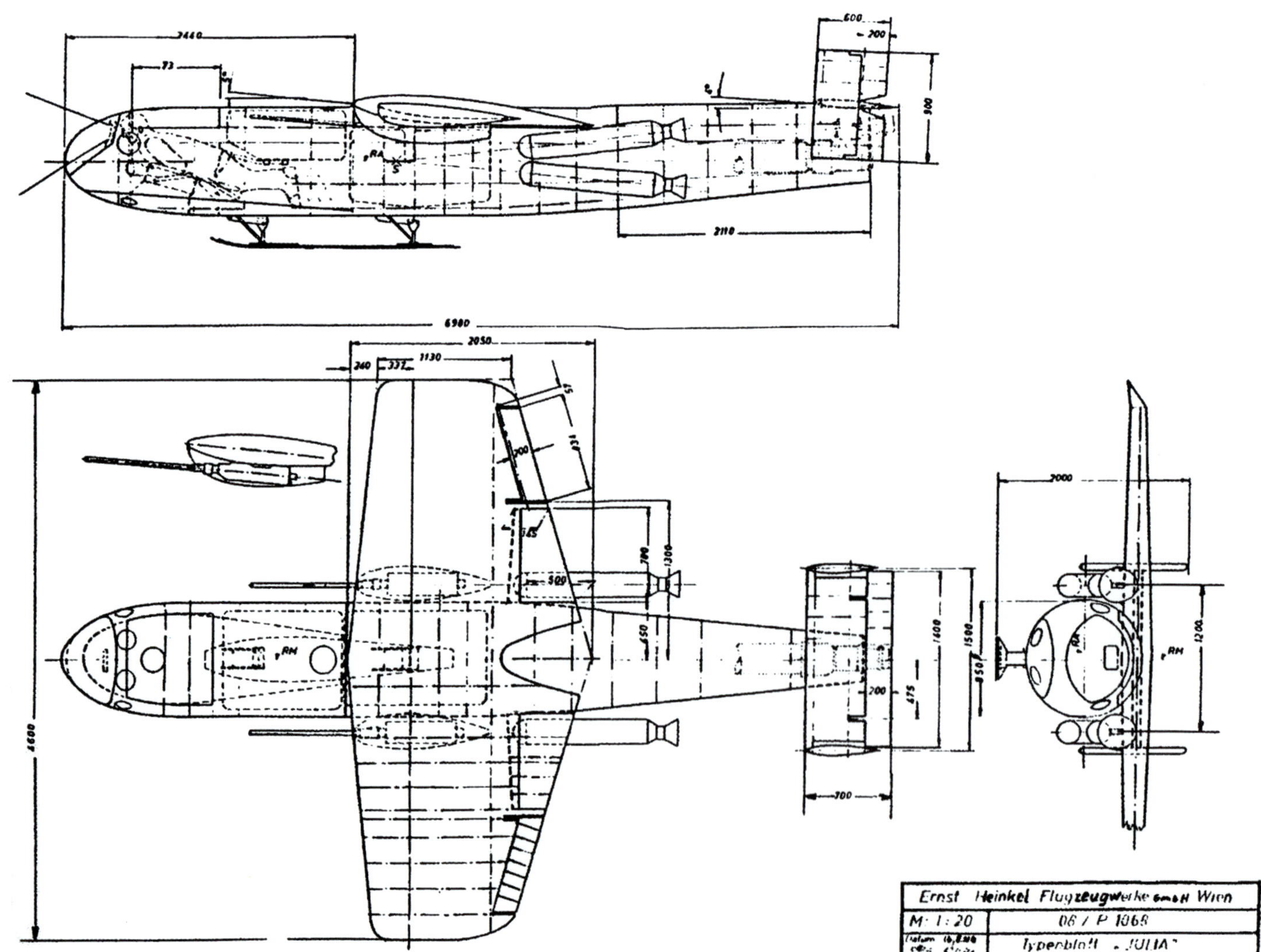

ABOVE: The earliest known version of the Heinkel Julia vertical take-off interceptor from August 16, 1944. The fact that the drawing refers to the design as 'P 1068' is inexplicable.

been found. In addition, there are still sufficient reserves to find the opponent during evasive movements and to safely withdraw from the pursuer.

"The possibility of being able to make more than one approach per start is necessary for fuel-saving reasons, since then about 40% fuel is saved per contact with the enemy. Given the current lack of fuels, this is an important design consideration. Fuel reserves for more than two to three approaches would unnecessarily make the device larger and more expensive. In addition, the overall structure and construction are such that, in addition to simple manufacture, the greatest possible weight savings and thus fuel savings are achieved.

"Usage – the device is mainly used to combat bomber groups during the day. It should be used for target defence at home and near the front. A horizontal start is intended as an immediate solution. However, in order to ensure the most effective and successful use possible and to fully exploit all possibilities of a manned rocket-device, the following demands should be made: a) No airfields. Airfields should be avoided considering the effects of the enemy, distance, independence of terrain, ground mobility. b) Mass attack. A simultaneous mass attack requires a mass start. Time-consuming gathering in the air may not be possible for reasons of range and because of the high flight speeds.

"c) Straight line direct enemy approach. It must be possible to fly to any point of the combat area on sight without complicated fighter guidance procedures in the shortest possible time. However, all of the above requirements can only be met by the vertical or steep start and steep climb.

"With this starting method, it was required that the launch frame should consist of a cheap, simple, easily transportable device. A number of successful vertical start model tests have already been carried out.

"Production time: Size, construction and equipment are kept small and simple to the limit of practicality, so that the device only takes a fraction of the development or production times and the material costs of a current propeller or jet fighter under development or in series.

"The individual construction parts were designed so clearly and simply that they can be manufactured in the smallest workshop and at home with the simplest devices. In general, the aim should be that the adjustment of this miniature fighter can be carried out with construction capacity that has so far been inadequate for aircraft construction. Parts that require a long development or trial period should not be used.

"Additional claims: a) Safe escape of the pilot can be ensured even at high speeds. b) Departure procedure must be simple in terms of flight and equipment. c) Training requirements should remain as low as possible. d) Skid landing.

"With a pure 'consumable device' [the word used by the report is 'Verbrauchsgerät' – presumably a reference to the Natter] there is a risk that the opponent will shoot down the pilots hanging on the parachute. For this reason and for reasons of material conservation, a skid has been provided which generally allows an outside landing. The basic principle was to arrive with a minimum of weight and effort for landing.

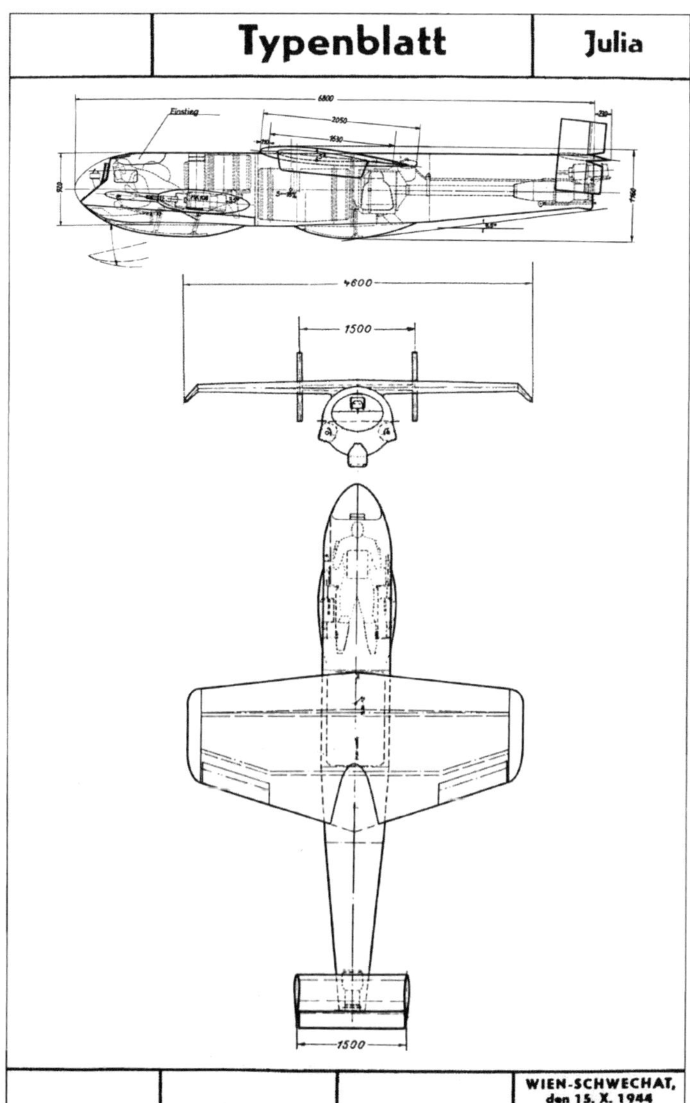

ABOVE: Type sheet for Heinkel's Julia from October 15, 1944. The design was included in a full report the following month.

"Armament and attack method: The device was designed in such a way that all types of new armament can be provided without major changes or loss of performance. Visibility and armour conditions, arrangement of the pilot as well as flight performance and strength allow a wide variety of attack methods."

Eight days later, a Heinkel 'Allgemeine Geräteliste (AG Liste)' or 'general equipment list' for the M1 and M2 (also mentioned as V1 and V2) prototypes of a rocket-propelled interceptor referred to as '5035' was produced.[26] On the list were a trip meter, altimeter, variometer, 'armbandkompass' for the cockpit – the latter presumably being a wrist compass. An RH 28 parachute was provided for the pilot but the M1 and M2 had no armament. Further alterations to the '5035' equipment list were made on December 9 including the addition of a pressure gauge and by December 29, equipment lists were being produced for the '5035' M3 and M4 (also given as V3 and V4) which included a master compass. Two propulsion options were added when the lists were updated on January 16, 1945 – two 500kg 109-502 rockets or two 1,000kg 109-503 rockets.

Another AG Liste for the '5035' was produced for the M3 and M4 in March 1945 (the document is date-stamped March 21, 1945, but the printed date is too faded to read) which now included a 109-509 A-1 rocket engine for the fuselage and four Schmidding 109-533 A-1 rockets for launch, each with 1,200kg thrust. Armament was two MK 108s with 40 rounds each.

It seems likely that the '5035' was in fact the Julia.

MESSERSCHMITT ME 262 WITH ROCKETS

Messerschmitt's other option, besides the P 1104, was a variant of the Me 262 augmented by rocket propulsion – either in the form of BMW 003 R combination turbojet/rocket engines or with a rocket motor installed in the rear fuselage. Both concepts had been under examination since 1943 and in August 1944 the company began to

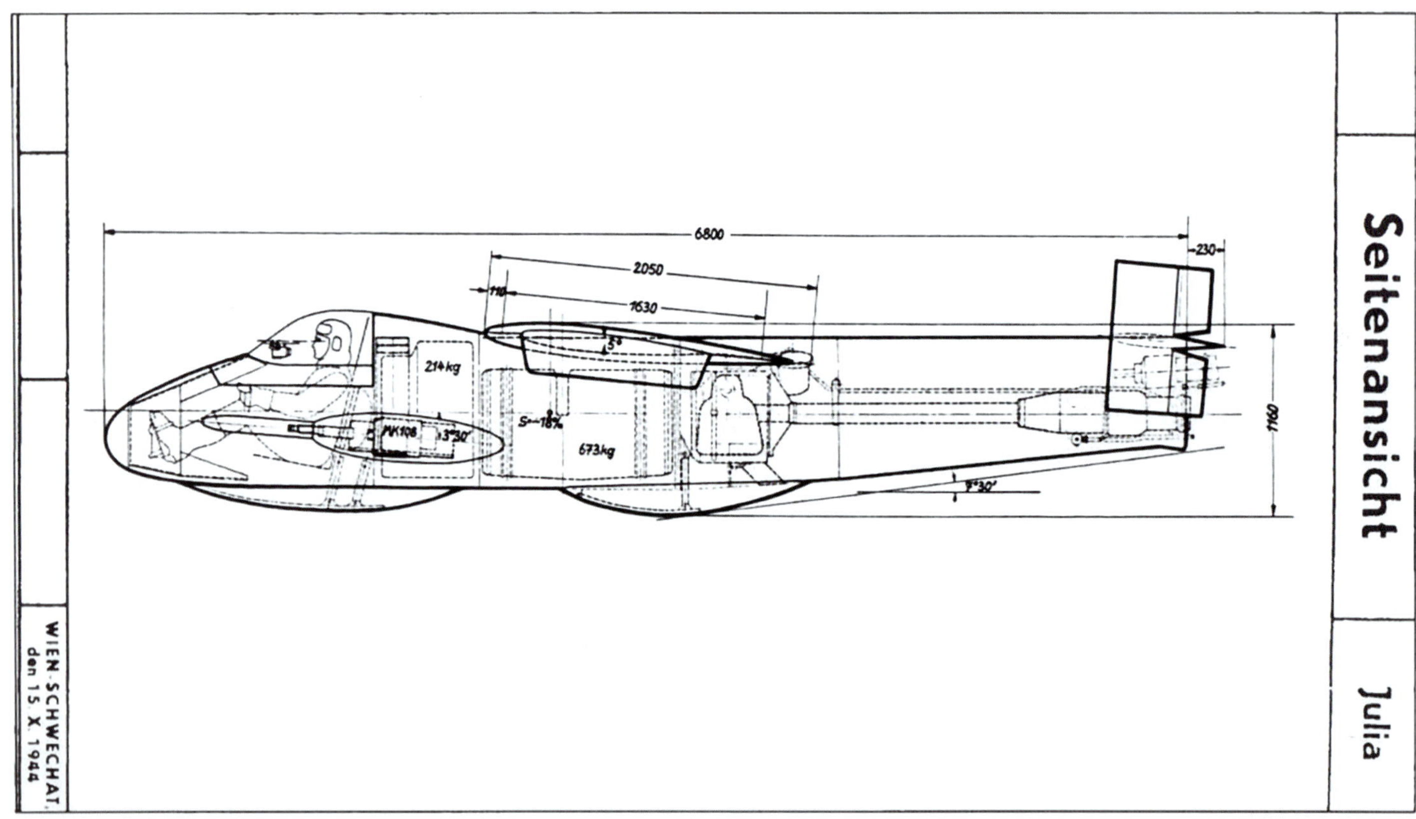

ABOVE: A seated cockpit version of Julia was designed but wasn't mentioned in the report that the drawing accompanied.

ABOVE: The prototype Me 262 C-2b, Me 262 A-1a W.Nr. 170074 is ground-tested in spectacular fashion with BMW 003 R turbojet/rocket engines. For months, fitting the Me 262 with additional rocket propulsion was the best solution to Germany's problems in finding a suitable target defence interceptor.

ABOVE: Production model Me 262 A-1, W.Nr. 130186, was fitted with a Walter HWK 109-509 rocket engine and became V186, the prototype Me 262 C-1a. After numerous delays and problems with its fuel system, the aircraft finally flew powered by both turbojets and its rocket engine on February 27, 1945.

develop prototypes under the name Heimatschützer (Home Defender).

A production model Me 262 A-1, W.Nr. 130186, was taken off the assembly line and fitted with a modified tail unit with fittings for a Walter HWK 109-509 rocket engine. A part of the rear fuselage and the lower part of the rudder were also cut out to provide an exhaust for the engine.

A T-Stoff fuel tank was installed in place of the aircraft's usual 900-litre forward fuel tank and the rear tank was used for C-Stoff. This left one other 900-litre tank and a 170-litre tank for the aircraft's ordinary jet fuel.

The aircraft's weapons load of four MK 108s was left unchanged and it was given the Messerschmitt designation Me 262 C-1. Its first flight, under jet power only, was on September 2, 1944, during transfer from Leipheim to Lechfeld – where it was damaged during a USAAF bombing raid on September 12. After repairs, a second flight followed on October 18.

This was to test the vital system for dumping the aircraft's volatile T-Stoff fuel in an emergency. Coloured liquid was used instead of the fuel itself and the test showed that dumping the fuel might well make matters worse – since it would then coat the underside of the aircraft, particularly clinging around the undercarriage doors.

The aircraft's rocket engine was ground tested on October 25, 1944. This revealed a host of problems that needed to be corrected – from incorrectly welded seams to fumes leaking into the rest of the fuselage – and it was November 23, 1944, before further ground tests could take place.

A document dated November 21-22, 1944 presumably the minutes of an EHK meeting,[27] indicates that the various target-defence fighter or 'Objektschutzjäger' designs were now in competition with one another. It stated: "Target-defence Aircraft: 4. The importance of target-defence was emphasised and consideration was narrowed down to the 8-248 (8-263), a development of the Me 163 B; the Heinkel 'Julia'; Bachem 'Natter'; and the Me 262 interceptor with supplementary rocket propulsion.

"It was decided that since these developments were in an advanced state it was not expedient to abandon any of them. A proposal by the Special Commission for Jet Aircraft and Special Aircraft to defer or reject the 8-263 in favour of the He 162 was opposed on the ground that further development and series production of the 263 could be based on the work already undertaken in connection with the 163.

"The four types of target-defence aircraft already enumerated were to be developed in the following priority: (i) Me 262 with supplementary rocket propulsion. (ii) Heinkel 'Julia'. (iii) 8-263. (iv) Nachem 'Natter'. The development of the BMW rocket 109-708, using nitric acid, was to proceed on a high priority as this unit was intended for the three last-named developments."

Presumably the reference to 'narrowing down' the options for target defence amounted to dismissing the P 1104, since Messerschmitt's Me 262-based interceptors offered significantly better performance and used mostly existing components.

THE CONTEST

Where Bachem's Natter and Messerschmitt's P 1104 had previously been developed in isolation from one another, with the advent of the Junkers Ju 248/8-263 and Heinkel Julia, a competition came into being – even though the Natter was already much closer to becoming a reality than the others.

DFS EBER

Meanwhile, yet another target defence aircraft based on the towed launch concept was now being worked on by the DFS under the codename 'Eber' or 'Boar'. The earliest known evidence of Eber comes from a telegram sent to Professor Paul Ruden at the DFS in Ainring on November 9, 1944, by Karl Leist of the Forschungsführung des Reichsluftfahrtministers und Oberbefehlshabers der Luftwaffe – the Luftwaffe's aviation research command, known as FoFü for short.

Leist says: "According to telex Dr Jennissen, on behalf of Mr Temme I expect to come to Ainring for a week on Monday, November 13, as I am working on the similar Project Bembo. I will bring documents with me and make them available. You will be asked to arrange accommodation for a week."

Joseph Jennissen was a government supervisor for aviation research and Heinrich Temme was another member of the FoFü who had worked on the Fi 103 flying bomb. Jennissen appears to be putting Leist and Ruden together because they were both working on the same sort of project – although Project Bembo is entirely unknown today. Presumably Ruden hadn't been working on the project for very long at this stage and could therefore benefit from Leist's input.

The following day, November 10, Versuchsingenieur or 'test engineer' Arno Schieferdecker of the DFS sent a telex to the RLM in Berlin requesting "procurement of drawing and installation documents of the machine cannon MK 108 and the rocket R4M in 28x tube battery".

Three days after that, Rheinmetall-Borsig wrote to the DFS with drawings showing the MK 108 and two different projects for the installation of R4M rocket tubes, "a drawing showing the arrangement of seven by seven blocks and a second showing a battery of 32. This latter is intended only as a trial execution. For the installation, only the seven by seven block would be eligible".

It goes on to query the request: "We ask, for the sake of order, that you tell us for which installation you need these documents, since we are allowed to issue them basically only by official requirements. It is also appropriate, after the existence of the first project to discuss the internals with the firearm installation subgroup. You should treat the documents of the tubular battery in a particularly confidential manner."

The earliest known mention of the name 'Eber' comes from a drawing dated November 23, 1944 – drawing number 5026-0000. This shows a tiny aircraft just 5.16m long and with a wingspan of 4.16m. Its wing area is 5m² and it has an almost entirely rectangular fin. The pilot sits within an angular box made of armour plates clearly designed to separate it from the rest of the aircraft down a dividing line just behind the pilot's seat.

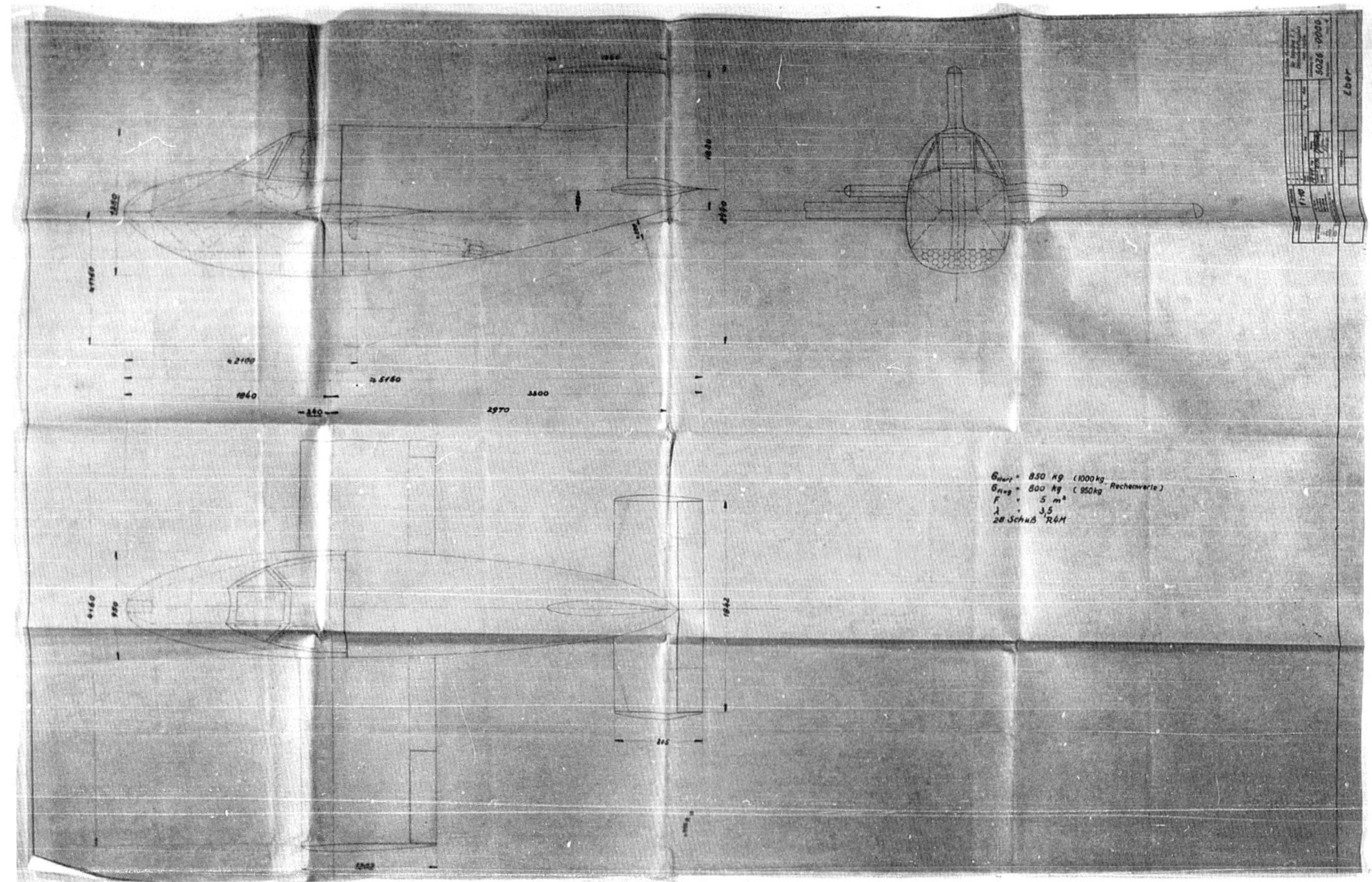

ABOVE: The first DFS Eber drawing dated November 23, 1944. The forward view shows the aircraft's honeycomb-shaped R4M rocket launcher and the angular armour plates protecting the pilot. The side view shows the line where the disposable aircraft would split in two, allowing the pilot to escape after making his two attacks.

Beneath the pilot is a box containing a honeycomb of 28 R4M missiles and behind him is the aircraft's rocket motor. Just visible below the aircraft is a wheel intended to represent a trolley or other detachable undercarriage for take-off. The aircraft's starting weight is just 850kg.

Also dated November 23 is a drawing – number '5026' without the zeroes – showing Eber being towed by a Messerschmitt Me 262. The rigid steel pole linking the two is 5m long and rigidly fixed to Eber, only the Me 262 end having flexibility.

Another drawing dated November 28, 1944, number 5026A-0000, shows the aircraft 1cm shorter at 5.15m with the same wings and starting weight. This time, however, the aircraft carries 44 R4M rockets and the fuselage split has been replaced with a more conventional canopy.

A set of handwritten notes and calculations on DFS graph paper headed 'Eber I', dated November 30, 1944, shows that the anticipated performance of the Eber vehicle was based on data supplied by Henschel relating to the rocket-propelled Hs 295 guided missile.

A meeting was held to discuss Project Eber on December 3 and the summary states: "After a detailed discussion of the project, the following points were noted for the further work: 1. The DFS pursues above all the Eber Project for the shooting approach from behind and above. The aircraft pilot is seated.

"The weapon is R4M device, installed in the direction of a straight line and designed for two attacks. The required number of shots and the probability of hit as a function of distance and scattering are recalculated by the LFA ... as a view finder, a simple reflex sight is provided. The FoFü provides the DFS with documents about the necessary armour of the cabin against 12.7mm projectiles.

"The protection of the aircraft pilot against explosion of the R4M rockets shall be provided by the form of the cabin. At the DVL, the Institute for Aerial Medicine is to provide an estimate of the necessary reaction time for the target, the attack and the flight of the aircraft. The information should be accurate to a few tenths of a second. As an additional drive, the RI 502 device is to be used. Protection of the aircraft against fuel explosions is not provided.

"The landing of the Eber is not planned. The aircraft is to be designed as a disposable device. In order to rescue the pilot by a parachute, the DFS designs the cockpit to break off or installs an ejection seat. For the parachute selection FGZ is to be consulted.

"2. Regardless of the examination of the firing range, the work for the rammer project continues. DFS continues the started model measurements on the destruction work of shell constructions. DVL will begin large-scale trials with a V1 slingshot in Adlershof as soon as the requested equipment and operator's attendant have arrived. The design for DFS completed rammer is built at thc DVL. In the DVL flight medical institute, medical officer Dr Henschke will provide the measurements on tolerable acceleration."

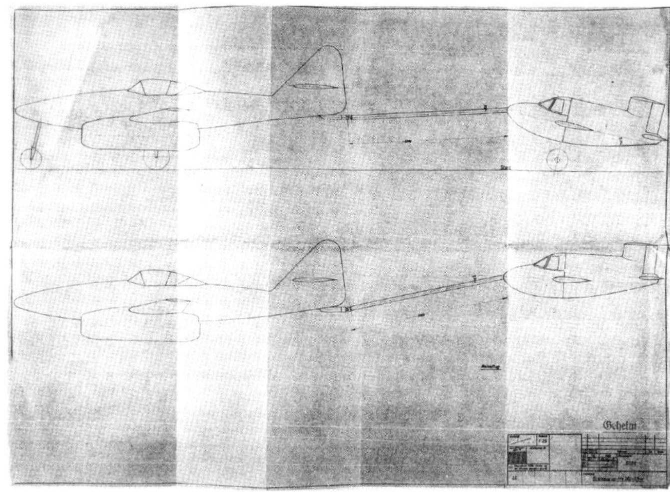

ABOVE: Diagrams showing how Eber would take off, fixed to a 5m pole behind a Messerschmitt Me 262.

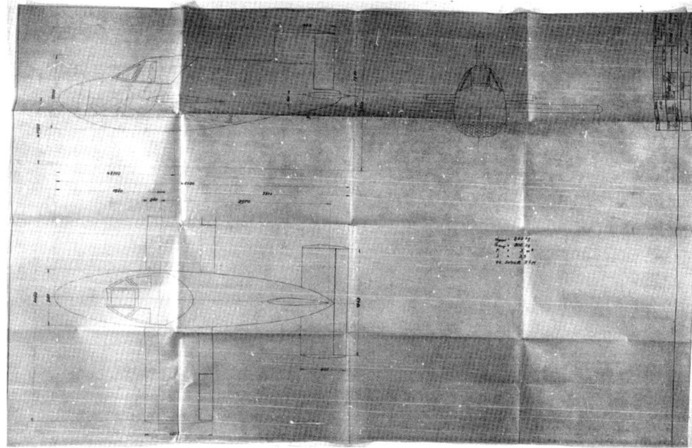

ABOVE: The second Eber from a drawing dated November 28, 1944. The split fuselage concept has been replaced by a primitive drag-chute ejection seat and the number of rockets carried has been nearly doubled.

Eber was to have an armoured cockpit with a conventional seat for the pilot. His mission would be to make two attacks on enemy aircraft using rocket projectiles before escaping from the vehicle and parachuting to safety. At this point it seems that although ramming was being considered it wasn't yet a definite part of the design.

A new version of Eber appears in a drawing dated December 13, 1944. The earlier square fin has been replaced with a twin-fin arrangement though the aircraft's length and wingspan remain unchanged from the previous design. Now the rocket weapon contains '2 x 20' R4Ms and the original single-rocket motor has been replaced with a pair of motors installed at different angles. Overall weight has increased to 870kg and the aircraft is depicted with two large landing gear wheels and legs which presumably would have been jettisoned at take-off. Another drawing from the same date, number '5026B', shows this version of Eber on tow – the arrangement unchanged from before.

The last known drawing of Eber was made the following day, December 14, 1944, and shows that the canopy of the

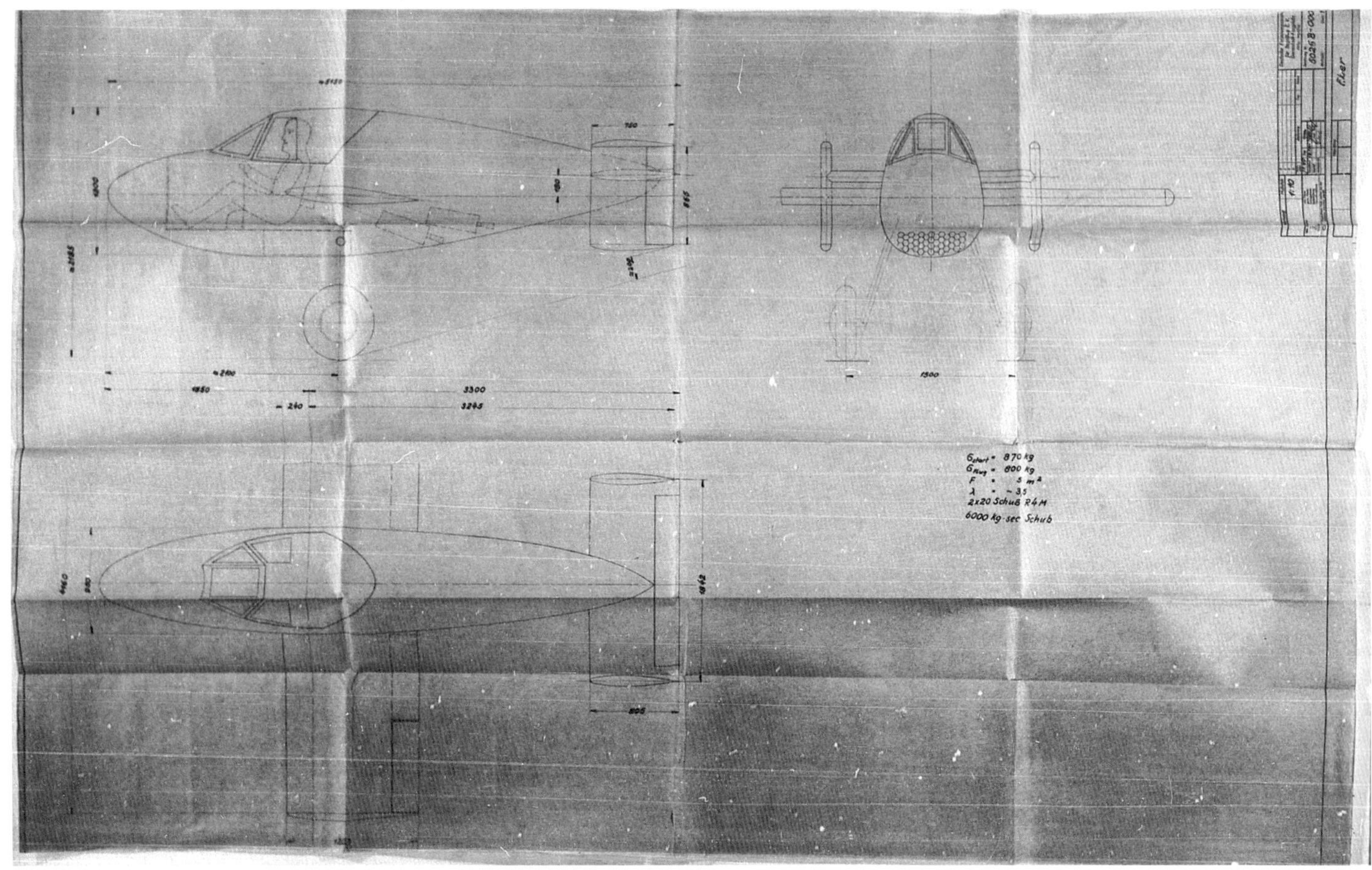

ABOVE: The last known version of the Eber, in a drawing dated December 13, 1944, has a twin-fin tail and two rocket motors.

two later versions was intended to facilitate the inclusion of an ejection seat. This design has no drawing number.

Two weeks later, on December 28, a full project description was issued. This states: "Required was a disposable device that allows two attacks after being towed by an Me 262 or Fw 190 to the appropriate distance (about 2,000m) from the enemy unit. The first attack should normally be made from about 300m above. The attack position for the second approach must be ensured by rocket-units. The thrust of the rocket-drive should be such that it is possible to attack from an under-elevation of about 700–1,000m.

"The original demands went on a ramming flight and a shooting approach. A closer examination of the ramming flight resulted in the following: provided that the aircraft's armour possesses sufficient strength a successful ramming at an approach speed of 150–200m/sec is always carried out. The speed decrease when piercing a fuselage is 6 to 10m/sec. If the ram-fighter performs the impact longitudinally, the accelerations that occur here have the order of magnitude 100g. It was assumed that the pilot could withstand a maximum acceleration of no more than 16g. The resilient or damped seat assembly is able to reduce the pilot's acceleration load from 100g to 16g, but with a construction that is difficult to accommodate.

"The construction cost is not insignificant, especially when the shock stress is not guaranteed safe except in the longitudinal direction. It must always be expected that even in the direction of high and transverse axes significant accelerations must be recorded, against which effect on the pilot is yet to be estimated. (The values of the shock delay and the acceleration ability of the pilot have not been determined correctly yet.)

"Apart from the fact that the mechanical process after the shock is completely unpredictable, the effort associated with the impact process has led to the view that it would be better to forego the ramming altogether and to make two shooting passes instead.

"For the armament was the choice of the device R4M, the MK 108 and the battery MK 108. Due to its low production costs the R4M device is evidently the most suitable for a disposable aircraft. It also has the advantage that its installation can be recoil-free. The pulse of the MK 108 battery would result in a decrease in speed of 6–8m/sec at launch with a weight of 800–1000kg. High shock sensitivity counts against the R4M however.

"For this reason, the R4M battery was positioned far below the aircraft's armour, so that the chances of them meeting from the front are low. For centre of gravity reasons, this situation has proven to be expedient. It was assumed that 20 rounds would suffice for a successful shooting approach. The armour is intended to cover the pilot essentially against fire from the front.

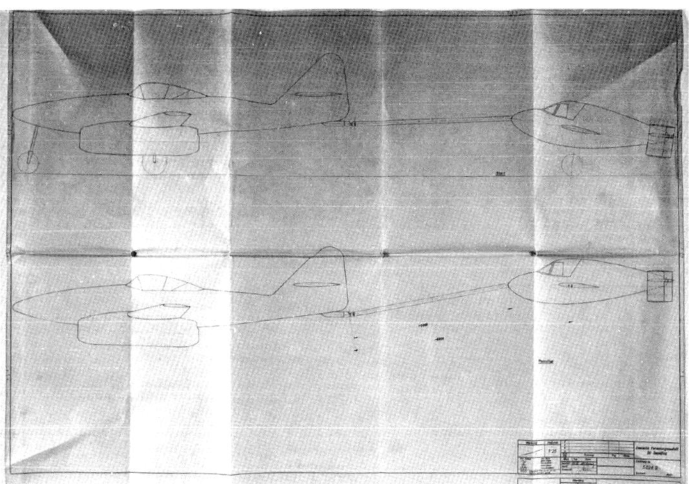
ABOVE: There appears to be no change in the towing procedure for the final version of the Eber in this drawing dated December 13, 1944.

"The fuselage itself was kept as simple as possible for disposability. For this reason, both the wing and the tail unit use the unbroken rectangular shape. Sweepback was deliberately omitted. The critical Mach number must be postponed by high-quality profiling. The choice of the double fin seemed to be expedient, because it was possible to avoid rolling moments at rudder deflection. A certain caution in this regard seemed necessary, since the moment of inertia around the longitudinal axis must be extraordinarily small. The fuselage is designed to be made from wood.

"The Eber should be linked to its towing aircraft with a drawbar. In our experience, between the drawbar and the aircraft only two degrees of freedom should be allowed, namely the rotation to release the transverse and vertical axis, while the towed vehicle in the longitudinal axis is rigidly connected to the towing vehicle. It has been shown that the trailer can be towed unmanned, so during the launch and during the climb no requirements are made on the pilot of the trailer.

"When firing its engines on approach the Eber is expediently set to the correct curve. This includes a well-functioning ram sight, which must be given to the tow plane. To comply with the straight ramming curve requires the Eber only to have a simple mirror visor. However, there is still the question as to how big the chosen angle can be. That depends on the initial speed of the R4M weapon on launch.

"After a shooting attack the pilot gets out of Eber. The landing was waived, 1) because the aviation requirements of the pilot must be very high, 2) because with unprepared terrain the percentage chance of a safe landing is low and 3) because the transport difficulties of retrieving the aircraft are too great.

"In order to make it easy for the pilot to get out, the following path was taken: a small parachute rips the seat with the pilot out of the plane and gives it a descent speed of about 40m/sec. As the aircraft continues to fly at its own speed, a separation of both should be ensured. The pilot then loosens himself from the seat, which is now braked to about 25–30m/sec. After getting away from this, the pilot opens his own parachute."

Shortly after the end of the war, Professor Ruden described his work for the Allies in CIOS XXXII-66: "The project 'Eber' was worked out from the institute of the undersigned [Ruden] under commission of the research management. The 'Eber' is a short-range fighter similar to the Natter with the difference of the Natter where the self start from the ground is abandoned.

"More consideration was given to start and raising the apparatus in pole-tow with help of the Me 262 or the Fw 190. At the altitude of the bomber squadron the 'Eber' was to be released at a distance of 5–10km behind the squadron and speeded up with rocket apparatus. At a greater height than the bomber squadron of approximately 300–400m. The run was calculated as being sufficient in order to reach and attack the bomber squadron which was assumed with a closing speed of at least 100m/sec. A second attack was possible through the resumed power of the installed rocket apparatus. The equipment of the 'Eber' consisted of a salvo-gun and heavy armament of the pilot.

"At first it was considered to carry out one attack as a ram attack. Closer examination persuaded us to drop the ram attack and perform two attacks. The project was not completed owing to the lack of production capacity and also to the limited capacity for use. The consideration originated from the fact that the towing fighter and 'Eber' at comparably small performance loads experience a loss of speed of approximately 100km/h.

"With strong air superiority it must be reckoned with that the towing and towed pair will be easy prey for the enemy if the latter attacks from greater height."

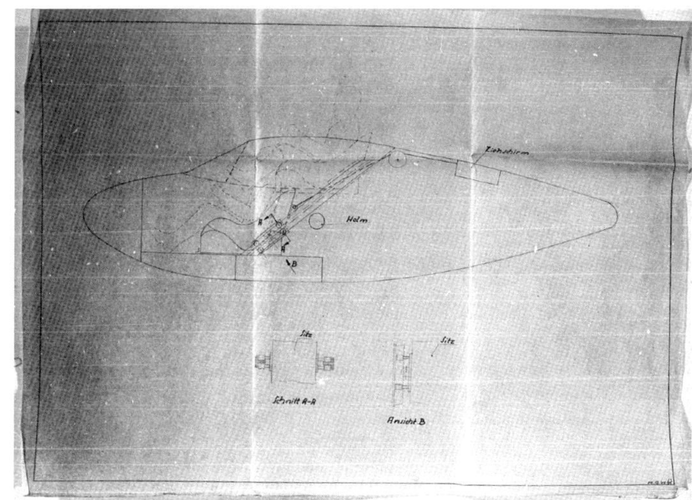
ABOVE: A very basic ejection seat was proposed for the final Eber – a parachute would 'rip' the pilot's seat from the fuselage before slowing its descent enough for the pilot to unstrap himself and open his own 'chute.

In short, Eber was originally intended as a rammer but ended up as a two-attacks-only disposable rocket fighter. The project was abandoned because no production capacity was available for it and because it was worked out that the fighter towing Eber would suffer a huge performance penalty, making it extremely vulnerable to Allied fighters roaming unchallenged through German airspace.

JUNKERS EF 127 WALLI

Nearly a month after the November meeting, on December 16, 1944, Junkers produced an outline description of another rocket-propelled interceptor – the EF 127, codenamed 'Walli'. This aircraft had a wing area of 8.9m^2 and a wingspan of 6.65m.[28] It had a conventional tailfin arrangement, straight wings and a skid undercarriage – in stark contrast to the tailless Ju 248/8-263 with its Me 163-derived wings and wheeled undercarriage.

The description included graphs showing Walli's performance against those of the Natter, Julia and 248 when intercepting a B-17, Mosquito or B-29, and while the 248 tended to be way out in front in performance terms, Walli was significantly cheaper to build – utilising the wings and tailplanes of an Fi 103 flying bomb – and could carry more fuel than the Natter or Julia. It also shared most of its components with another aircraft design proposed on the same day: the EF 126 'Elli' pulsejet ground-attack aircraft.

The next discussion of the Objektschützer designs by the EHK took place at Arado's Brandenburg offices on December 19–20, 1944. According to notes of the meeting signed by Arado chief executive Walter Blume on December 22, the chairman of the EHK, Roluf Lucht "discusses Natter, Julia, Walli; rejects Natter". He also "rejects prone pilot arrangement, since during acceleration head pressed on chin rest and no longer rotatable".

General discussion followed, during which it was stated that both range and flight duration were too short. Then Willy Messerschmitt "suddenly says that he had considered that it would be possible to quickly bring the jet fighter up to scratch with cheap means and to throw in additional material, so that the task would already be

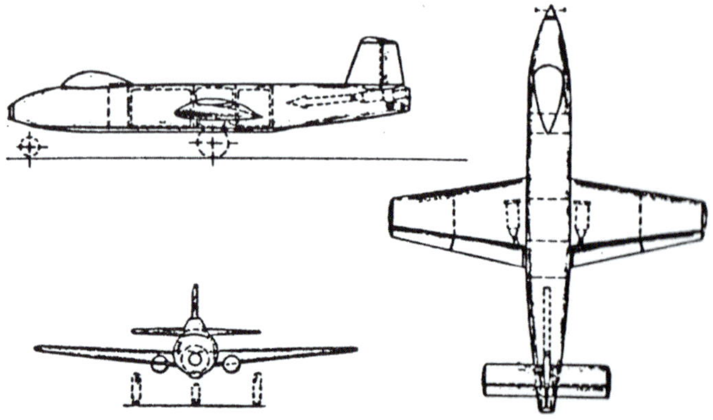

ABOVE: Contemporary drawings exist showing the rocket-powered Junkers EF 127 'Walli' with both wheeled and skid undercarriage arrangements. Which came first is unclear but the very similar pulsejet-powered EF 126 began with wheels before switching to a skid.

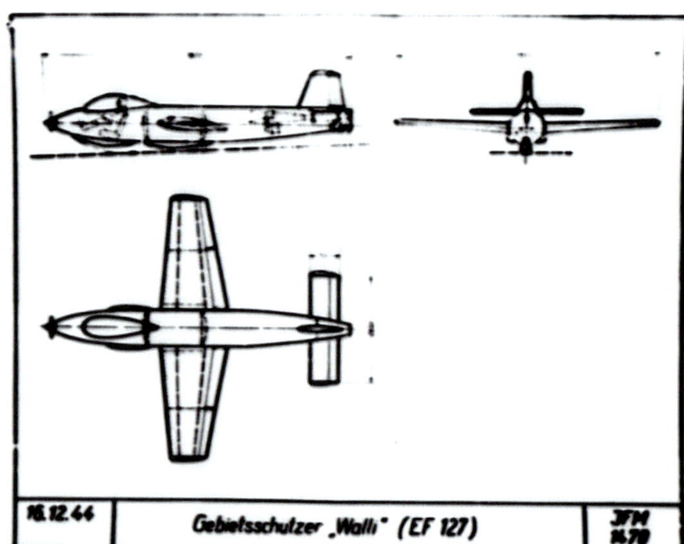

ABOVE: Slide showing Junkers' EF 127 'Walli' with skid undercarriage from a presentation on December 16, 1944. The design differs significantly from the wheeled version in having a reprofiled forward fuselage and wings that are set further forward.

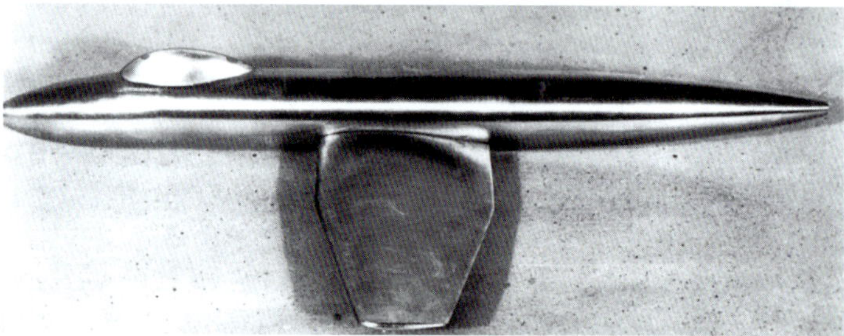

ABOVE: Wind tunnel model of the EF 127 in wheeled configuration.

ABOVE: Photo showing a mock-up of the wheeled EF 127. The same arrangement was also planned for the original EF 126 ground-attack aircraft.

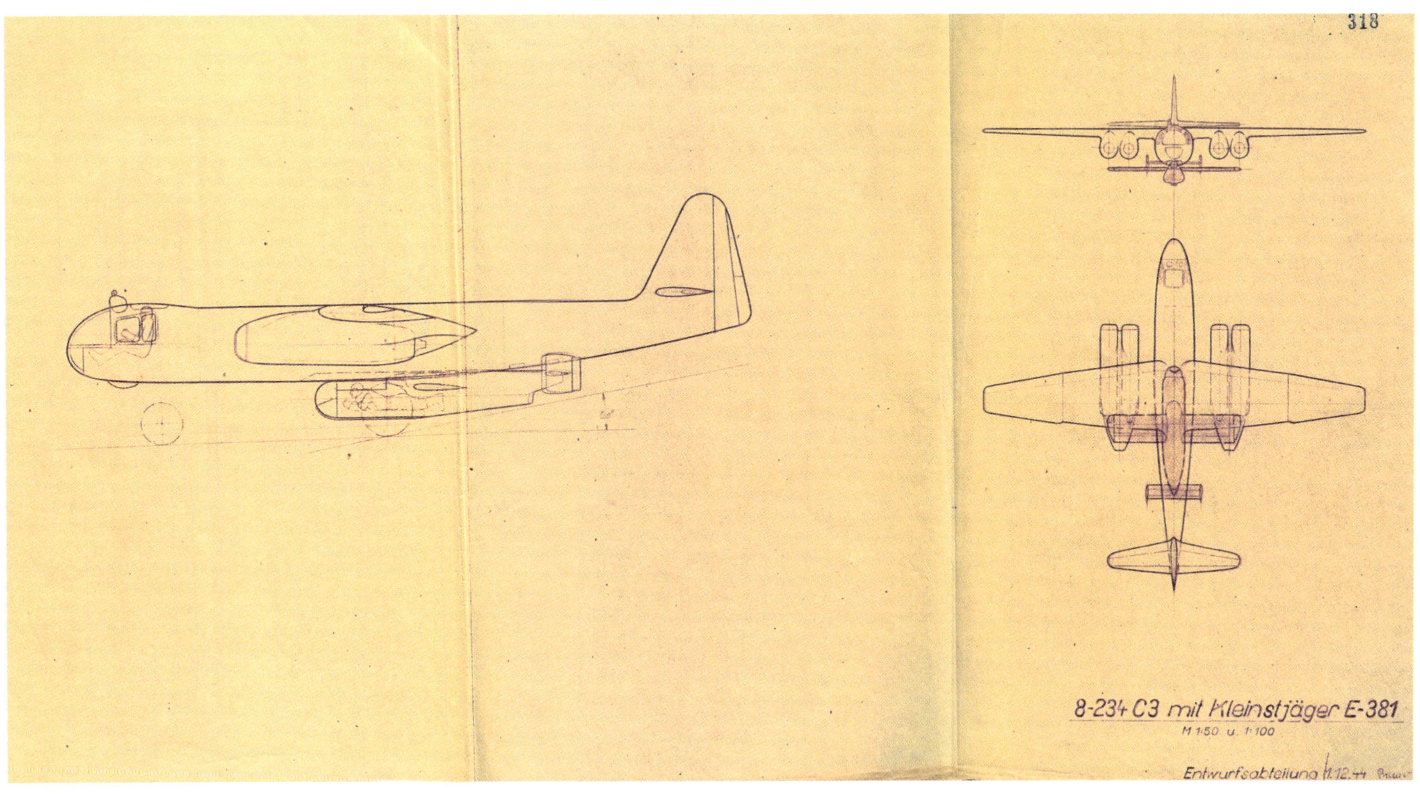

ABOVE: The Arado E 381 Kleinstjäger would have been carried into battle slung beneath a four-engined Ar 234 C-3. The E 381's pilot could communicate with the Ar 234 via telephone and warm air was circulated to the tiny fighter during transit and before separation.

solved". General Ulrich Diesing said he "welcomes this and expects further details".

ARADO E 381

Blume then showed the meeting an "anhängejäger" or 'attached fighter' and its "range made an appropriate impression". The ratio of the "number necessary" compared to Walli was 500 to 5,000. However, "return organisation is strongly criticised".

It would appear that what Blume showed his fellow commission members was the Arado E 381 Kleinstjäger 'midget fighter'. This was a small rocket-powered 'parasite' fighter with a prone pilot position designed to be taken aloft attached to the belly of an Ar 234 C and the official Arado report describing it is dated December 5, 1944. The E 381 had been worked on since at least early November 1944 and bore a strong similarity to Heinkel's Julia, albeit somewhat shorter in length at 4.95m but with a longer wingspan of 5m. It was armed with a single MK 108 with just 45 rounds.

The report states: "The parent aircraft flies to within sight of the hostile formation, as far as possible above it, in order to increase the effective range of the midget as far as possible. Upon recognition of the adversary, the midget slips the connection with the carrier, goes into a glide and switches on the rocket unit, which, when developing its mean thrust, gives the fighter a speed about 200km/h in excess of the speed of the enemy formation.

"It is so protected by armour, that it has every chance of penetrating the enemy fire barrage without serious damage, and opening fire at closest range from its MK 108 guns."

However, regarding landing and recovery ('return organisation') – it is little wonder that the other members of the EHK raised concerns. The Arado report goes on: "When the ammunition is expended, the rocket unit is stopped, to have an available reserve of fuel on landing, and the fighter glides down. A landing parachute brake is provided to enable a belly landing on a relatively confined space, a carefully designed shock-absorber skid and springing of the pilot's support mitigating the landing shock.

"To facilitate towing-off after landing, the midget should be brought down near some main road, airfield, or the like, which should easily be possible in view of the considerable gliding range of the aircraft. Meanwhile, the parent aircraft will be in a position to attack the enemy in the capacity of an escort fighter. If the operational situation permits, it will endeavour to spot the landing place of its midget before returning to base, where one or more further midgets should be available for operational use.

"Collection and return of the landed midgets is a matter for the ground services. The ideal organisation is approximately as follows: each two parent aircraft are served by one special motor truck, on which two midgets can be accommodated. These trucks are directed to the landing points, either determined by the parent aircraft,

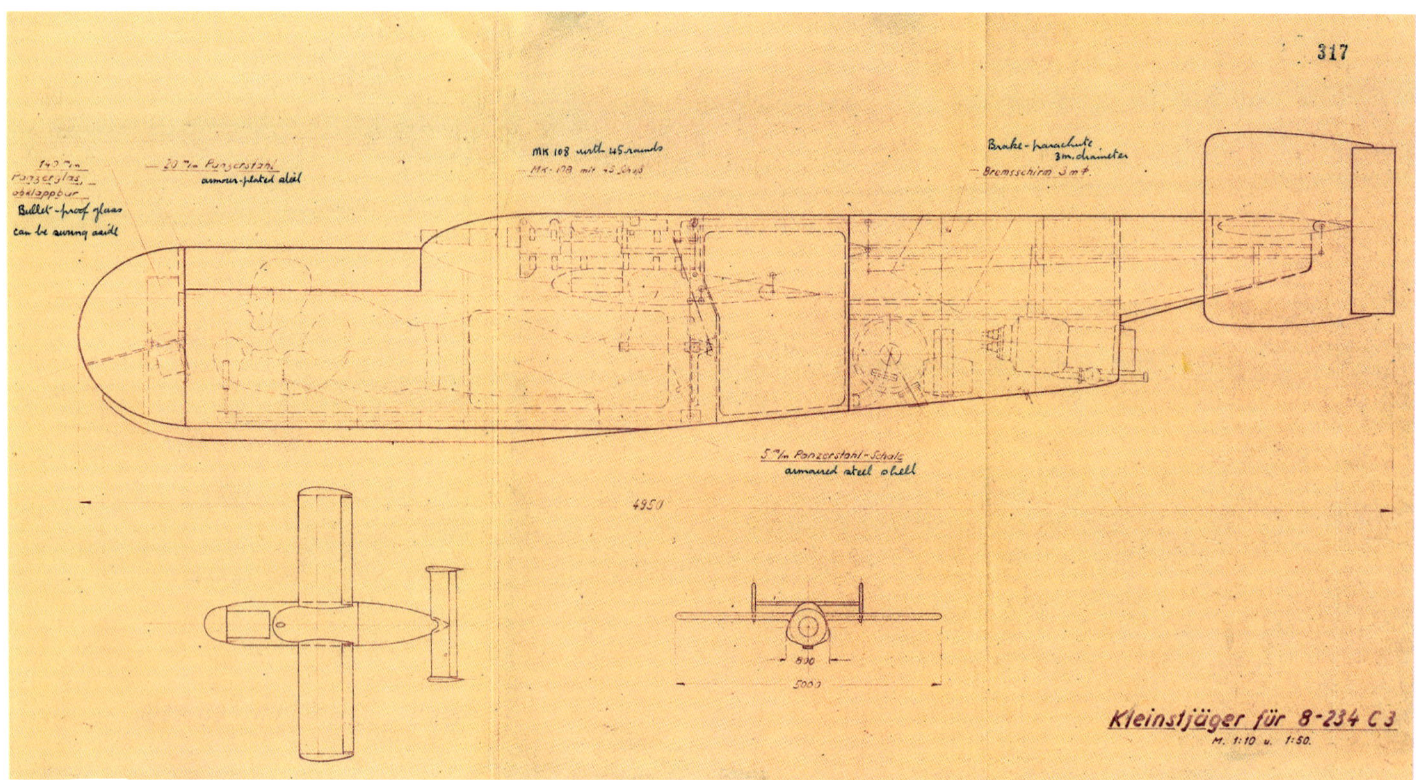

ABOVE: The compact dimensions of the Kleinstjäger are illustrated in this Arado drawing, with handwritten English annotations added after its capture by the Allies.

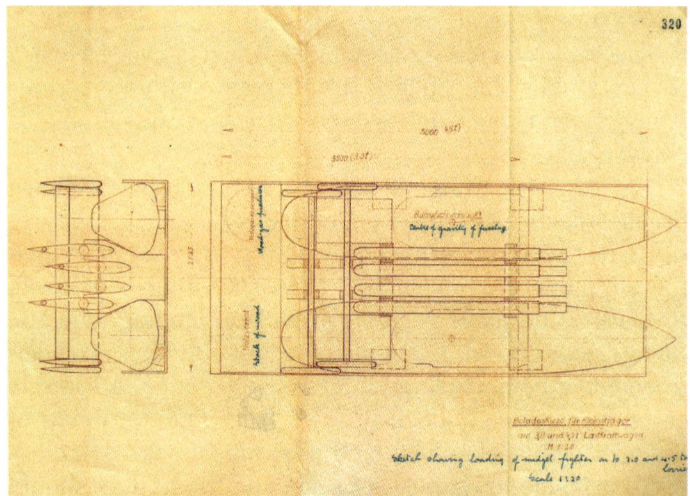

ABOVE: Once released, it was unlikely that the E 381 would be able to return home under its own power. It was proposed, therefore, that trucks would be made ready to collect it from wherever it had landed. This drawing shows how two aircraft could be neatly fitted onto a three-ton or five-ton lorry with the detached wings and tail sections positioned vertically between and above the fuselages.

or reported by telephone. At the landing point the midget is disassembled into wing unit, fuselage, tail unit. These components are so designed that their individual weight can be handled manually without difficulty."

In other words, having landed on a road in the middle of nowhere, the midget's pilot would have to extract himself from the aircraft before waiting for a recovery lorry. When

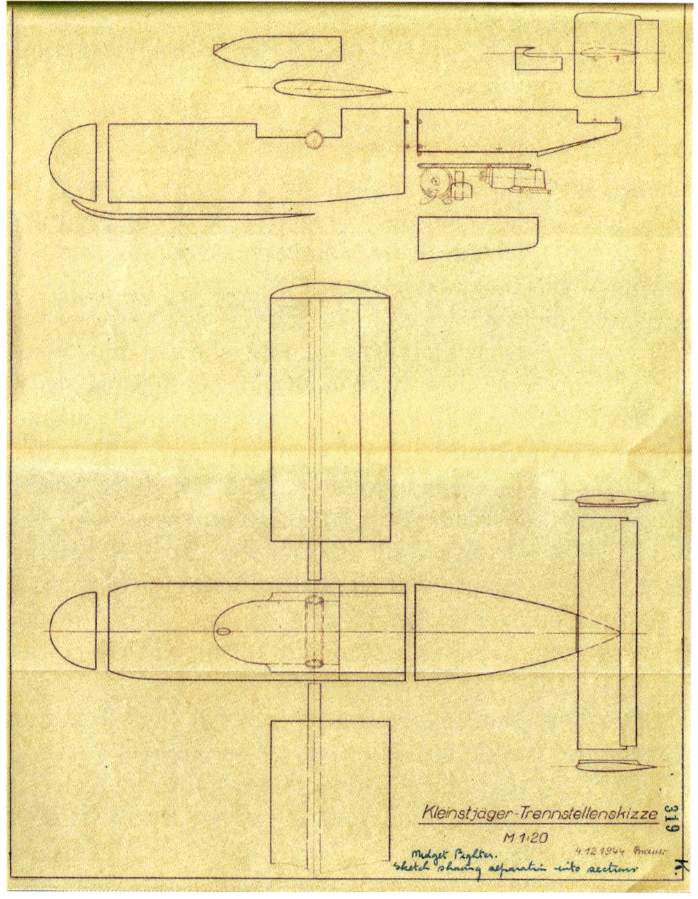

ABOVE: Another key feature of the E 381 Kleinstjäger was the ease with which it could be taken to pieces for transport.

this arrived, the ground crew would have to completely disassemble the midget, load it up, then attempt to find a second midget to make the trip worthwhile before taking that to bits too and finally hauling both back to base.

The only further note on the E 381 was "Arado to continue work on attached projects" but permission to build a mock-up was not yet granted.

According to these notes, the decision of the EHK meeting was "to put Julia and Walli back, in order to make a final statement on the basis of testing Ju 248. Natter: Series of 100 examples under construction. Lock out". The latter comment appears to suggest that there was nothing that could be done about the 100 Natters already under construction at this point.

OFFICIAL DECISION

The official minutes of the meeting – as opposed to the notes signed by Blume – were signed by Lucht[29] and were somewhat clearer: "Objektschutzjäger. The EHK came to the unequivocal opinion that projects with insufficient flight duration cannot be justified in operational terms, whereby the 163 B flight duration, which is hardly sufficient today, is assumed. From this it follows that the use of the patterns 'Julia' (Heinkel) and 'Natter' (Bachem) will not be promising.

"Decision: a) The development of the 263 as a further development of the 163 must be accelerated by all means. b) Likewise, the development of the 262 with rocket-drive should be continued, because if the result is good, this

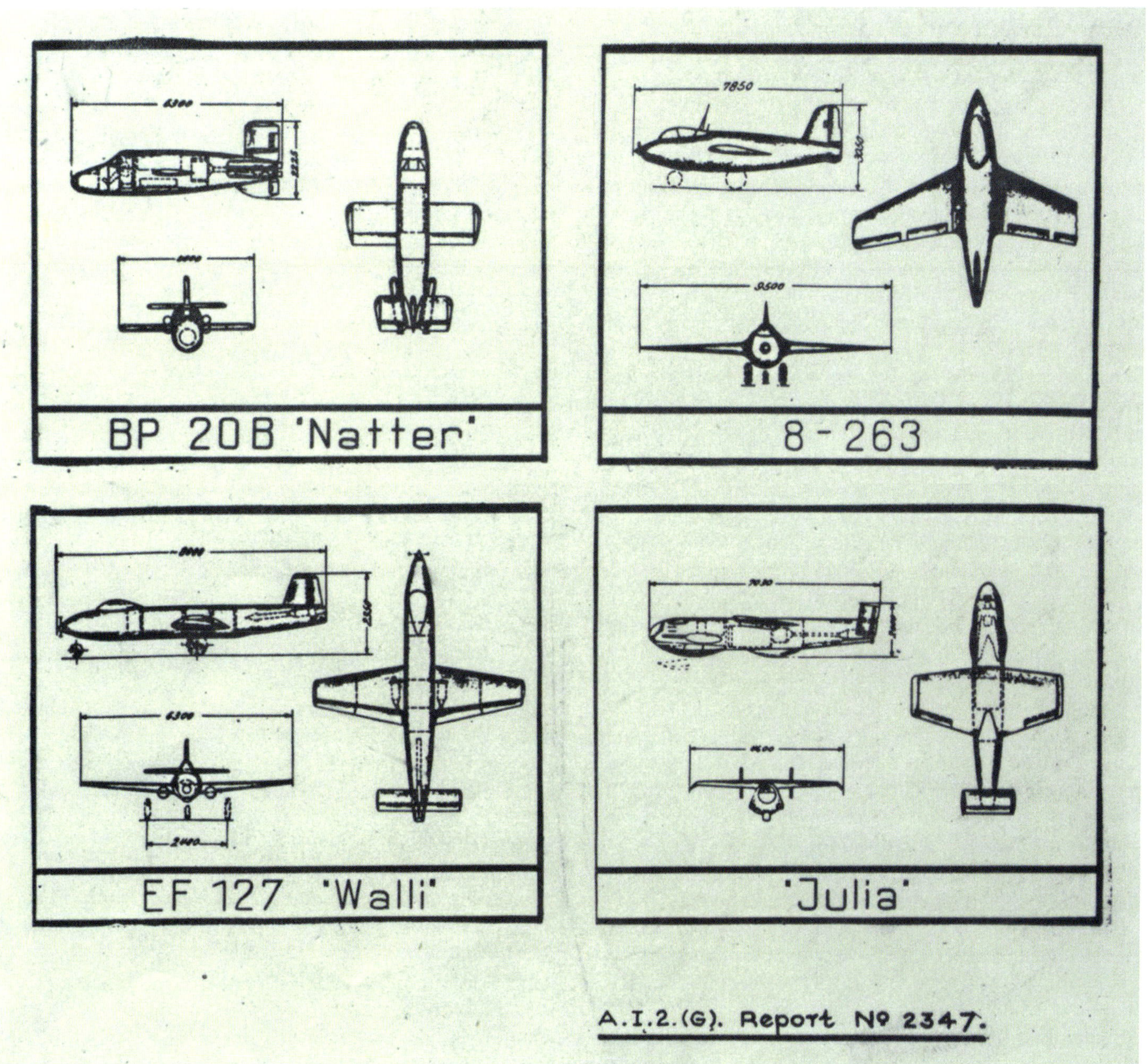

ABOVE: Drawings from a British intelligence report showing Objektschutzjäger competitors.

development may make all other target defence fighters superfluous. c) The development of the 'Julia' will be discontinued with immediate effect for reasons of range. d) The development of the 'Walli' (Junkers) is stopped immediately. A continuation of the development has to be made dependent on the result of the further use of the 163, 263 and 262.

"e) Although the EHK is fundamentally opposed to the project 'Natter' for technical and tactical reasons, the development is approved up to and including the attempts to launch it, since these are to take place in January. Any series preparations are absolutely to be rejected for the above reason. f) Messerschmitt's proposal to add rocket motors with external special containers under existing fighter planes (piston engine or jet) should be examined in detail by Messerschmitt company by the beginning of January, since a positive outcome of this examination is the simplest solution for target defence (lecture Lusser)."

The Ju 248/8-263 was to continue in development, as were the two Messerschmitt Me 262 Heimatschützers. In fact, work on the latter would continue until the end of the war. Heinkel's Julia was cancelled. Junkers' Walli was cancelled. Natter's development could continue but no series production would be authorised; as far as the EHK was concerned it was cancelled. Messerschmitt's proposal was a rocket pack that could be fitted to any standard Me 262 A-1 – instantly converting the jet fighter into a rocket interceptor.

A third perspective on the same set of projects comes from a management report produced by the Chef TLR's Fl-E 2 department[30] on December 21, 1944. The Natter entry states: "Delays due to personnel difficulties mainly caused by the Lusser Special Commission. Triggering the rear parachute, dropping the nose cap, falling out of the parachute mannequin and descent of the rear part on the brake parachute after previous attempts in order. Unmanned free flight attempts with cockpit dropping and operation of the rescue parachutes, meanwhile went well."

Things had not been going well for the Julia either: "Target defence, ready to take off and land, mock-up burned during attack on Vienna. Work not far beyond project status."

The Walli entry says: "Project 'Wally' [with a 'y', although Junkers spelled it with an 'i'] (= EF 126 with Walter engine). Preliminary economic considerations between 8-163 and 248 and Wally were in favour of Wally. However, sample [the Ju 248/8-263] is already so far advanced (V1 ready to fly at the beginning of the year 1945) that it is not easily justifiable to stop it in favour of Wally."

Finally, under "new development tasks (orders)", the report states: "Project 'Eber' (Mauch, Lusser and Research Leadership) becomes a loss device – so contrary to the opinion previously propagated by Lusser."

Evidently the TLR was unhappy with the way Bachem's Natter project was being treated by Lusser and his commission – and found it ironic that Project Eber, apparently backed by Lusser as a reusable system – was being changed to become a 'loss device'.

FINAL OUTCOME

Production of the Me 163 was formally cancelled by Reichsmarschall Hermann Göring on January 5, 1945.[31] During his interrogation by the Allies on June 7, 1945, HDL Saur was asked why the Me 163 had been cancelled[32] and he replied: "1. Lack of fuel. 2. This model was much too short in the flight duration. Of the 163, 217 machines were produced in 1944, many of which failed when flying, etc. About 30 machines were used, which could only destroy five [enemy] machines."

Just three factories produced all the C-Stoff and T-Stoff required for Walter's 109-509 series motors – Chemische Werke Transehe at Gersthofen near Augsburg (C-Stoff), Electrochemische Werke, Hollriegelskreuth at Munich (T-Stoff) and the Bad Lauterberg plant in the Harz mountains (T-Stoff). Gersthofen relied on erratic supplies of hydrazine from elsewhere, which ensured that its output remained low throughout the war. The Munich T-Stoff plant was completely bombed out in June 1944 and the Bad Lauterberg plant had to stop producing T-Stoff by January 1945 due to a lack of coal and electricity. According to Hauptmann Späte,[33] the only C-Stoff and T-Stoff available going into 1945 was whatever could be found in storage at airfields and supply depots.

Junkers was now hard at work building the Ju 248/8-263. The 8-263 V1 was completed and flown at least 13 times before the war's end, starting on February 8, 1945, with Karl Wendt at the controls.[34]

The Natter survived cancellation by the EHK in December thanks to direct support from the SS and a single manned launch was carried out with Natter prototype M23 on March 1, 1945. This resulted in the death of its pilot, 22-year-old Lothar Sieber. The aircraft had taken off vertically from its launch tower but at an altitude of about 100m it curved backwards by about 30°. After climbing another 1,500m, the Natter's rocket motor cut out and it turned over and nosedived into the ground.

Fourteen production-model Natters were completed under the designation Ba 349 A-1, four of them for tests, though none of them flew. Junkers and Arado, both government-run, abided by the EHK's decision without question. Walli appears to have been quickly abandoned as does the Arado E 381. However, Heinkel persisted with Julia almost until the end of the war.

In an end-of-year projects summary under the heading 'Julia',[35] Heinkel technical director Carl Francke wrote on December 29, 1944: "Me 163 output was reduced by half to an output of 100 per month, as no more fuel was available. From this and for reasons of the leadership (range), Julia, Natter and Walli were stopped by the Entwicklungshauptkommission in favour of the 263. In my opinion, the work started should definitely be brought to

ABOVE: Natter airframes under construction within the cramped confines of the Bachem-Werk factory.

a conclusion of development (2 flying prototypes). If the idea of object protection by means of an interceptor is to be started correctly beforehand, then the aircraft available for this should be as cheap as possible."

Essentially, Francke disagreed with the EHK's decision – because the Julia was much cheaper to build than the 8-263 – and decided to simply ignore it and press ahead with construction of two flying Julia prototypes.

However, this strategy threatened to come unstuck less than a month later when it transpired that the production manager for wooden construction working on the prototypes had discovered that the type had been cancelled and it was feared that he might communicate with Berlin about the ongoing work.[36] Francke wrote to Lusser and Heinkel managing director Karl Frydag on January 27, 1945, asking "for a short message on whether the four Julia samples (two glider, two powered) should be finished according to our conversation a few days ago".

Francke's fears were confirmed on February 5 when Lucht wrote to him: "Dear Francke! In December the EHK Flugzeuge decided at the meeting in Potsdam to stop working on the Julia with immediate effect. As I now find out, despite this decision and explicit instructions to Heinkel, the work has continued "behind our backs", so to speak. If this message is really correct, such a view and attitude of the Heinkel company to the EHK Flugzeuge is completely incomprehensible and I would be grateful for an investigation and ask you to arrange that the work should be stopped immediately despite improvements that may have been made to the project in the meantime.

"Unfortunately, I have to ask for confirmation from you Mr Francke, because otherwise I no longer see any guarantee that the decisions of the EHK aircraft will actually be brought to bear."

On February 15, Francke wrote to his fellow director Huffziger to say he had just spoken to Hans-Martin Antz at the RLM who had relayed the message that Frydag "has decided that we should not stop Julia in the current situation, but let what is going on run now, so that at least the people are busy. Frydag wants to speak to Lucht tomorrow".

ABOVE: The unmanned M17 test vehicle was brightly painted for visibility, with asymmetrical stripes painted on its wings so that its orientation could be better discerned from photographs of the launch.

Frydag's level of authority, thanks to his other position within Speer's ministry of war production, appears to have been slightly above that of Lucht – which seems to have put Heinkel in the unique position among Germany's privately owned aircraft companies at the end of the war of being able to defy the established administrative structures.

The following day, February 16, Francke wrote to his subordinates at Heinkel, Jost and Benz,[37] to say: "EHK (Lucht) expressed his surprise at the fact that, despite the fact that the Julia had been stopped, work continued. After discussion with Frydag the four prototypes (two with, two without engines) continue as before, so that people do not stand around. Under no circumstances may further attempts and other work that I have not approved be started or continued. That includes vertical test starts. I ask Mr Benz not to go on business trips without my permission."

The day after that, Benz reported[38] to Francke that Flugbaumeister Otto Malz of Fl-E 2 "wants to have the vertical start tests with the Julia test vehicles. The vehicles are ready. We have stopped trial on EHK order".

Francke wrote to Jost on February 21,[39] saying: "After I received the message from General Staff Engineer Lucht that Julia should now finally be stopped immediately, I informed Huffziger. He immediately stopped working on Julia. To make sure, I spoke to Frydag, who ordered Julia to keep running so people didn't stand around.

"I am giving you a letter from the OKL stating that the mock-up room in Neuhaus should be closed immediately. I ask you to do the following: set up the Neuhaus mock-up room as an emergency joinery as soon as possible. Try to get as many carpenters as possible, but not here at our Vienna plant, which we are looking for here; you have to try to get the carpenters there somehow. With this carpentry you can do mock-up work for 162 glider training aircraft, for the small wing or Julia you want to build and similar things. The main purpose of this joinery, however, is to be able to absorb the 162 0-series and changes."

On the same day, Francke also wrote to Huffziger[40] to say that "on February 16, 1945, it was said that the

ABOVE: Natter M23 fuelled and ready to make the first vertical launch of a manned rocketplane.

ABOVE: Test pilot Lothar Sieber, in the foreground wearing overalls and a flying helmet, shortly before his first and only flight in BP-20 M23.

ABOVE: With nothing between him and a plunge to the ground, test pilot Lothar Sieber clambers into the Natter's cockpit legs first, assisted by ground crew.

work [on Julia] should be stopped due to the decision of the EHK. Gen. Dir. Frydag decided that the work on Julia should continue as before, so that the work capacity remains secured. I also gave this in writing in my communication of February 15, 1945 – No. 66/45. I think it is right to let the Julia work be done where workers are free to do it. I believe that there is free working capacity at the table warehouses in Krems, St Pölten and Melk [all in Austria]."

Francke wrote to the RLM on March 3, 1945, to request that Lucht or Georg Manigold (seconded to the EHK from Focke-Wulf) issue a written order for further work on Julia be continued.[41] A meeting was then held on March 5, 1945, to further discuss the Julia.[42] Francke noted that he had telephoned Frydag that day and had been assured that development of the aircraft should continue. He said that since work on the Julia at the Geppert company in Krems had been stopped, and that Geppert was unlikely to be able to continue because it was too busy with the He 162, "it should be examined whether the company Schaffer – Linz can take over the task Julia and on what dates. The previous order to Schaffer – Linz was two gliders, [Julia] version M 2, and two powered aircraft, version M 4.

"Since the aircraft have been dismantled due to the stop command, it is examined whether it is easier to finish building the two M 2 aircraft, which are already 90% complete, or to go ahead and build all other aircraft according to model M 4 with the first two of them engineless and flying in as gliders.

"Under the deliberate limitation of variants, work on Julia should be carried out rapidly so that a useful development result can be expected shortly. The Julia task is divided into the following questions: a) Execution and completion of some flight mechanical calculations. b) Wind tunnel model measurements, especially in the fast tunnel. c) Execution of fire horse trial in Waldee. Mr Jost seconded a man to take care of this in agreement with Schrenk and Benz. d) Construction of the model aircraft. e) Carrying out flight testing. Mr Sorge is dismissed as the liaison officer. 4) Jost takes over as the model manager for Julia. 5) All Julia works are to be relocated to Neuhaus for type management."

The last known communication on Julia was a note sent by Francke to Ernst Heinkel himself on March 28, 1945,[43] which said: "Lucht has not yet approved Julia. I brought the case to him and suggested that I let it go now. I am afraid that he himself would not like to return from his decision to stop because he made it in front of a large auditorium and is afraid that it could be said that industry does not do what he wants. He wanted to speak to Frydag and give me written notice."

Julia had been kept alive, it would appear, simply to give Heinkel's subcontractors something to do. Quite why Heinkel would go to such lengths to preserve this production capacity – the alternative no doubt being the carpenters being drafted to fight with the Volkssturm against the Russians – is unclear. Nevertheless, both Francke and Frydag were willing to disobey the EHK's ruling on their project and even put pressure on its chairman for several weeks in order to secure it.

ABOVE: The business end of the Bachem Ba 349 Natter's Föhn rocket launcher. This aircraft was among several Natters abandoned at a DFS facility at Sankt Leonhard im Pitztal, western Austria in May 1945.

Exactly how complete the Julia prototypes were and what form they took when the war ended is equally unclear. The only known drawings date from October 1944 and even assuming that the design had remained largely the same by March 1945, there were clearly differences between the M 2 glider version and the powered M 4 which went beyond the simple lack or inclusion of an engine. It seems likely, too, that these were unmanned test articles rather than fully equipped manned aircraft. Like so many of Heinkel's projects, however, the surviving documentation is insufficient to draw any firm conclusions – the company's projects documents having been either destroyed or captured by the Soviets.

Chapter 11

Jäger mit Lorinantrieb

Ramjet projects 1942–1945

Austrian rocket scientist Eugen Sänger had worked on ramjet or 'Lorin' powerplants on and off since 1938 and during 1943 proposed a fighter powered by one. This work would lead to a host of innovative and unusual projects during 1944 and 1945.

ABOVE AND BELOW: A Dornier Do 217E-2 was used to carry Eugen Sänger's 19,700bhp ramjet for flight tests. These determined that the design produced so much drag that the huge thrust was mostly negated.

DFS (SÄNGER) STRAHLJÄGER

First described as a concept by Frenchman René Lorin in 1908, the ramjet was nothing new even at the beginning of the Second World War. Just as the Germans called the piston engine 'Otto' after the man they saw as its inventor, so they called the ramjet 'Lorin'. The British tended to refer to it instead as the athodyd – a contraction of Aero THermoDYnamic Duct.

A ramjet engine needs to be moving forwards before it can produce any thrust – with the air being 'rammed' into its intake by this motion acting like the compressor of a jet engine. Lorin couldn't get his engine moving quickly enough to work, nor did he have the necessary materials to build it. However, these problems were more easily overcome in Germany during the late 1930s and engineer Hellmuth Walter set to work on a ramjet in 1936 before switching to rocket propulsion instead.

Studying Lorin's invention while working at the LFA in 1938,[1] Sänger had been pessimistic about the design – arguing that it would be very difficult to make one work efficiently in practice. But he also noted that if the numerous difficulties associated with the ramjet could be overcome, the engine's performance potential would be significant.

Interviewed by the Allies immediately after the war,[2] Sänger explained that he had returned to the ramjet three years later: "In view of the war situation in 1941, I suggested making use of the thrust concentration in a modified Lorin engine with a high firing temperature, at the expense of rapid climbing and employing this engine in the supersonic flying region in place of rocket propulsion.

"On account of the fact that it consumed about five times less fuel it would have the following advantages: extreme climbing capacity e.g. two minutes' climbing time to an altitude of 12,000m, as actual rocket fighters; considerable horizontal flying time e.g. one hour at great altitudes similar to the jet turbine fighter; extraordinarily low cost of construction as a result of the simple engine and thus suitable for the cheapest mass production; very suitable for the fuel programme of cheap petrols over medium and heavy oils, alcohols, furfur oils up to liquid coals and pure coals.

Sänger began practical experiments on ramjets at the DFS during 1942, including carrying out tests of his engines fixed to the backs of lorries and later to Dornier Do 17 Z and Do 217 E-2 aircraft. In October 1943, with his frequent collaborator (and future wife) Irene Bredt, he produced a report entitled 'Über einen Lorinantrieb für Strahljäger' or 'On a Lorin Engine for Fighters'.[3] This proposed the creation of a flying "special structure" for the testing of ramjets which could be towed to an appropriate altitude before the engine was ignited and the "structure" put into a dive so that it could build up sufficient speed for the ramjet to produce maximum thrust. The report noted that work on this project was "under way".

Elsewhere in the report, he discusses the potential uses of a ramjet-powered fighter or fighter-bomber particularly for low-level operations – the ramjet would burn too much fuel if it was required to climb steeply for any length of time and would be increasingly ineffective running on thin high-altitude air.

The long tube of the ramjet itself is suggested as forming the fuselage of the fighter, with cockpit, fuel tank and tail all attached to the top of it. Landing gear would be attached to the lower surface of the tube and the wings could be either low- or mid-set.

He wrote that "the whole aeroplane has no moving parts, but represents a pure shell design, when discounting the few auxiliaries for landing gear operation, cabin air system, fuel injection, nozzle and flap setting and so forth".

Take-off would be made with the help of two conventional rockets and once airborne the ramjet fighter would have a range of more than 100km at speeds ranging between 110km/h and 720km/h carrying 1,000kg of bombs or other weaponry.

Sänger produced several designs for his Strahljäger. The first one measured an incredible 16.39m long with a wingspan of 11.9m – making it slightly longer than a Do 17 but with a wingspan only slightly greater than that of an Fw 190. It was effectively a set of aircraft components wrapped around the tube of an enormous ramjet. Undercarriage consisted of a relatively normal set of mainwheels and a tailwheel positioned well away from the actual rear of the aircraft.

Two smaller versions of the same design were drawn up but the last known configuration, the one included with the report, was labelled 'Strahl-Jäger Variante 2a' and was much

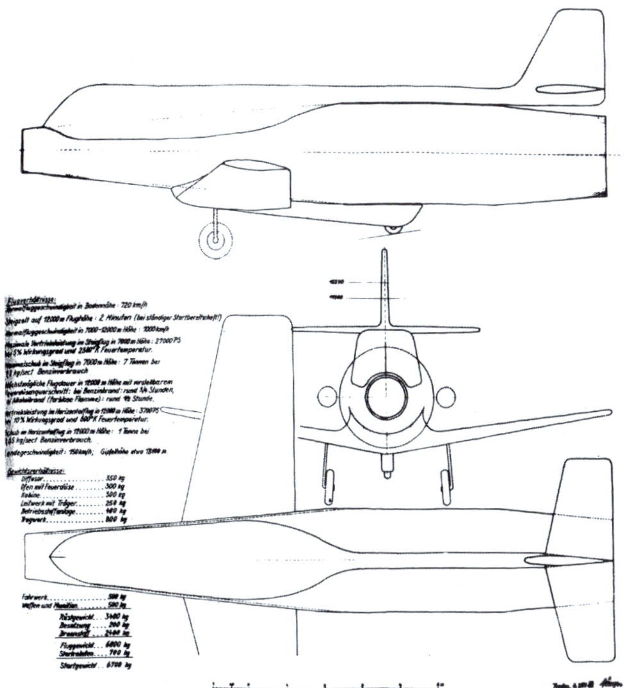

ABOVE: Eugen Sänger's first design for a ramjet-powered fighter. It is essentially a single large ramjet tube with aircraft parts attached to it.

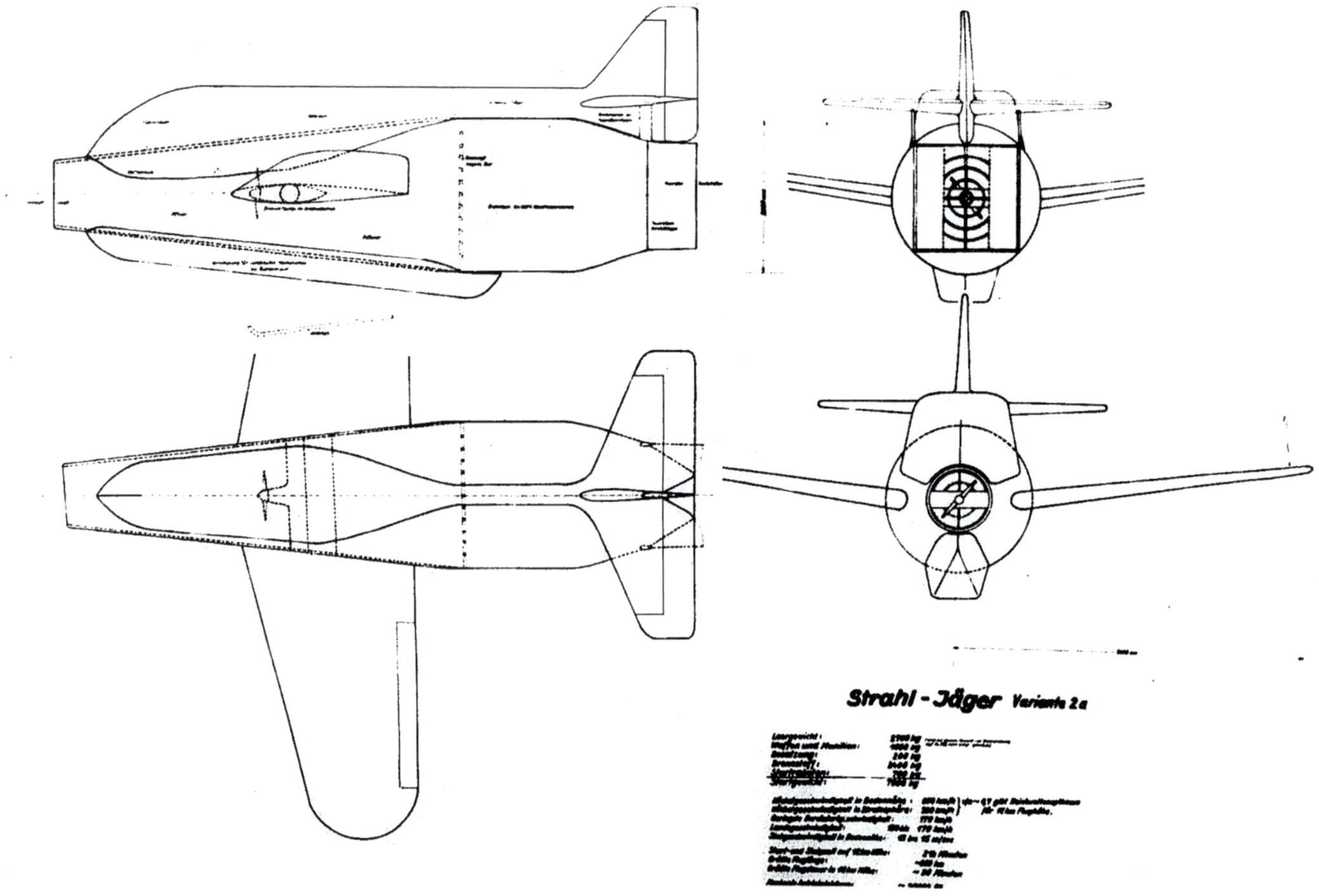

Strahl - Jäger Variante 2a

ABOVE: The ramjet fighter design Sänger included in his 1943 DFS report 'Über einen Lorinantrieb für Strahljäger' – labelled as 'Strahl-Jäger Variante 2a'. Work on designing it in more detail would not take place until around two years after its publication.

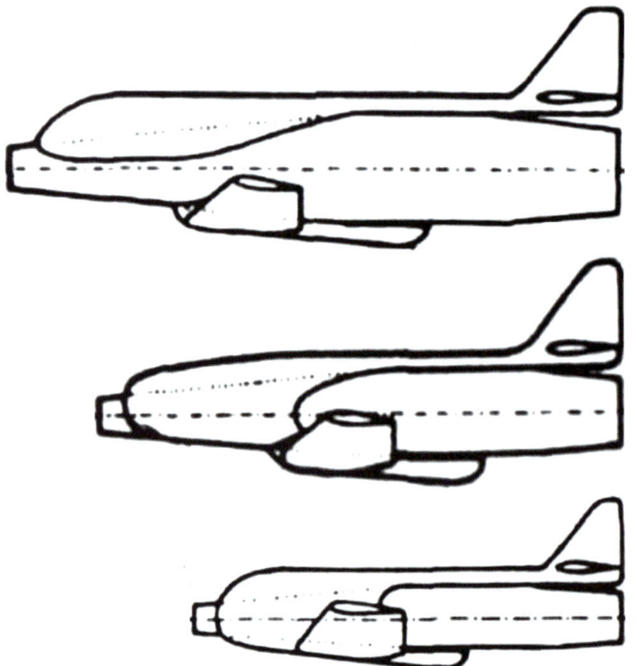

ABOVE: Three differently sized ramjet fighters were outlined by Sänger: the uppermost one was to be 15.59m long (by way of comparison, a Messerschmitt Bf 109 G-6 was 8.95m long), the one in the middle was 13.27m long and the one at the bottom was 10.95m long.

shorter at 10.9m long with a slightly greater 12m wingspan. Annotations on the drawing showed the pilot's position in the upper part of the nose with space for weapons wrapped around the forward section of the ramjet tube. This design had a skid landing gear rather than wheels.

DFS JABO

Five months after Sänger's ramjet fighters report, in March 1944, the DFS produced a second report on ramjet propulsion for military purposes, this time offering the propulsion system for either a fighter or fighter-bomber in a report[4] entitled 'Kurzbeschreibung des vorgeschlagenen Jabo bzw. Jägers mit Lorinantrieb' or 'Brief description of the proposed Jabo or fighter with Lorin drive'.

In its surviving form, there is no author's name on the document and it is undated. However, it is accompanied by a DFS drawing dated March 15, 1944, which is referred to in the text, thereby offering a good idea of when the report itself was produced. All other drawings and diagrams mentioned in the report text are missing. It is likely that Sänger also produced or was involved in producing this report too.

The one drawing that is included shows an extension of Sänger's earlier concept and one of the more bizarre-looking German 'secret projects'. It is a 1m diameter tubular ramjet with the pilot lying prone in an armoured capsule over the intake at the front. A narrow vestigial tail emerges from the top of the ramjet tube and two different wing/tailplane arrangements are depicted – one set straight and the other swept. Both side and forward views show a 1,000kg bomb suspended from the tube. There is little else to the design except for a skid landing gear.

The report itself begins by discussing how this highly unusual creation was intended to be operated as a fighter: "The launch takes place by means of rocket propulsion by horizontal acceleration to 800km/h, after which the climb is made using the fuel carried along. The low production costs and the low demands on starting area (runway of 800m length in the main wind direction) allow an increase of the total number of places where it can be used.

"Fig. 1 shows the climb and flight times for different heights. For arming and ammunition 500kg are used. You can see that the fighter reaches an altitude of 7,000m after 157 seconds; during this time the target can not have covered 20km at 400km/h. Assuming that the estimated 500kg weapon load consists of launcher ammunition, so could even a single fighter, stationed as local protection of a target at a distance of 30km, break up the enemy formation before it reaches the goal. The actual strength of the device is however in mass use."

This makes it clear that the DFS ramjet fighter was intended to tackle incoming bomber formations in force. It would be cheaply mass-produced then launched in huge swarms to defend vulnerable targets.

The mission profile was somewhat different when the aircraft was used as a fighter-bomber, however: "From the point of view of favourable range, the following use is considered: Mistel start up to 12,000m; launch from the carrier and transition to the dive until the operating speed of 800km/h is reached, then straight flight at 10,000m altitude to the goal. Target approach and attack at speeds around 900km/h; climb with full performance up to 11,000m and return at this altitude.

"With careful consideration of the experimental results available in certain speed and altitude ranges, it seems penetration depths of more than 200km will be achievable ... During the combat (approach, dive, climb and return), the speed is so high that attack by enemy fighters or heavy flak seems impossible.

"During the final part of the attack near the ground, frontal attack by machine guns is likely. Against this, extremely strong armour is provided, which covers a large part of the fuselage cross-section in addition to the aircraft pilot. Here, too, the high speed in conjunction with the smallness of the entire vehicle results in an additional measure of security. The landing speed is around 180km/h."

As a 'Jabo', the DFS ramjet aircraft would be launched at altitude from a carrier aircraft – perhaps a Dornier Do 217 or Ju 88, though no particular type is specified in the report. It would then fly to its target, dive to attack, then climb back to altitude. Close to the ground, the pilot would be protected from defensive machine-gun fire by a heavily armoured shroud over the cockpit.

A third use was also envisioned for the ramjet aircraft: "When used as a course-controlled, remote-controlled or television-controlled bomb, there are ranges of over 500km, since no altitude change takes place and there is no return flight. The extremely low production costs make the device particularly suitable for this purpose. By eliminating crew and fuel tank armour, the payload can be increased to about 1,400kg = 46% of the flight weight.

"The basic idea of the project is the approach to the limits of technical and operational feasibility. It is propelled by the favoured means for ease and simplicity in manufacturing and engine monitoring, and by focusing on attacking armament, which is made possible by the high speed, so you can do without weapons of defence."

The extreme simplicity of the DFS aircraft would have made it attractive from a production perspective – but getting it to work correctly and in the manner envisioned would most likely have been rather difficult and made it significantly less attractive.

The report next describes the aircraft's structure: "The hull, together with the engine and fuel tank are of sheet metal construction, forming a closed unit. Here are three main frames, which serve to complete the cockpit, wings, load, and tail.

"Frame 1 is the cockpit connection point (also with wing connection), frame 2 as wing and load connection and for the front of the tail boom, frame 3 as burner chamber, spur and tail support. The intermediate space can readily serve as a fuel container, which is restricted only by a smooth tube required as a control channel and plus a recess for the load and skid connection point.

"The cockpit is heavily armoured frontally for attacking bombers so that projectiles striking in the direction of flight or at a slight angle to this encounter a wall thickness of 6cm, measured in the longitudinal direction, so they either bounce or fail to penetrate. From this one can derive a safe protection of the pilot and a large part of the fuselage cross-section. Hits at right angles to the surface are hard to expect. In the fighter version, the armour can be changed depending on the expected calibre. (Today aircraft guns are at 2cm.)

"In order to avoid viewing slits in the case of the bomber attack, a two-mirror arrangement is provided, whereby only the outer mirror is exposed to the bombardment and can be replaced automatically or by hand from a mirror magazine ... As entry and exit for the pilot is a large, hinged ejectable hatch, which is incorporated into the armour.

"The prone control offers no difficulties. The pilot's couch is designed in such a way that the pilot has the necessary freedom of movement in a comfortable posture and is well-protected from forces during acceleration. The skid can

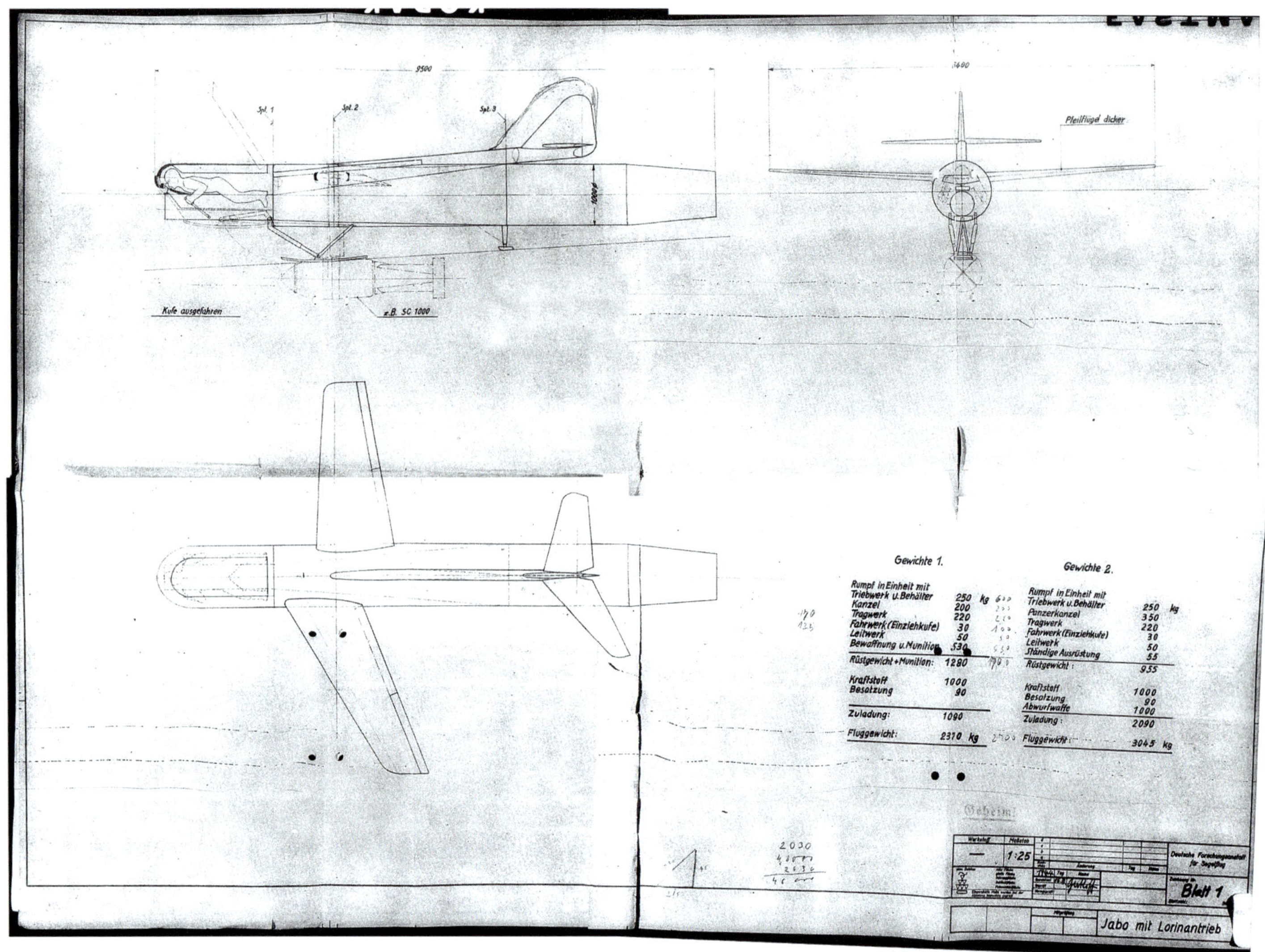

ABOVE: The only known drawing of the DFS Jabo mit Lorinantrieb or 'fighter-bomber with ramjet'. It depicts a tiny aircraft that is little more than a ramjet tube with wings and a small tail attached. It measures 9.3m long with a wingspan of 6.4m.

be arranged in different ways ... In the front view the centre of gravity seems to be very high due to the tall skid. However, this applies only to the moment of touchdown, whereupon the struts spring back into the aircraft to a third of their height, absorbing the impact, and do not then change from this point.

"The pilot's instruments are limited to altimeter, airspeed indicator, fuel gauge (rough) and compass. The choice and accommodation of the latter must still be clarified because of the steel masses present at the cockpit. Because of the space restrictions, mirror reading of the instruments is thought necessary, so they would be positioned to the side of the pilot with the mirror showing them in front of him.

"The equipment should also be limited to the minimum of what is operationally required. Radio receivers and signal cartridges can meet the requirements. In addition, there is oxygen equipment and heating, for their simple design good opportunities exist."

The DFS Jabo mit Lorinantrieb is a singularly focused design, produced when Germany still had the resources to put it into production if desired. However, it came at the wrong time. The Jägerstab was being formed, or had already been formed, by March 1, 1944 in order to streamline Germany's aviation industry in favour of existing types that were deemed critical to the war effort. Experimental designs such as this were unlikely to find favour. Whether this was the fate that befell the DFS design is unknown.

LIPPISCH DELTA VI WITH RAMJET

A month after the DFS Jabo mit Lorinantrieb report, on April 18, 1944, another ramjet design was produced by the DFS, this time verifiably by Sänger since it has his signature on the only surviving drawing[5] – a variant of Alexander Lippisch's Delta VI powered by a built-in ramjet.

The only note on the drawing says: "Vorschlag zur Kombination eines 20000-PS-Lorin rohres mit der Lippisch-Delta VI-Zelle" or 'Proposal to combine a 20000 PS Lorin tube with the Lippisch Delta VI airframe'. While this design seems to have gone nowhere, Lippisch himself was now involved in designing ramjet-powered aircraft.

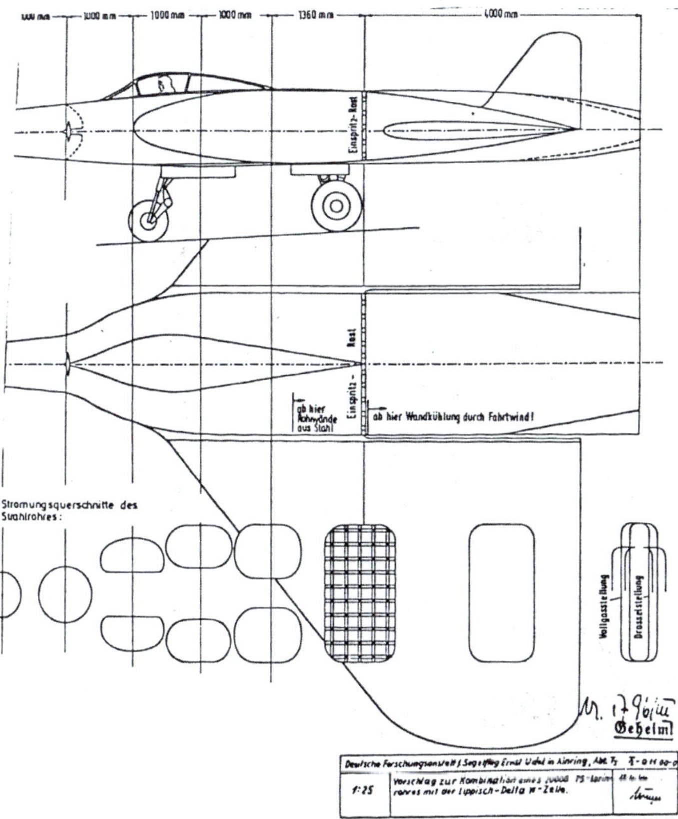

ABOVE: Drawing signed by Eugen Sänger and dated April 18, 1944, showing a Lippisch Delta VI airframe fitted with a 20,000PS ramjet.

LIPPISCH P 12

Lippisch, still working at the Luftfahrtforschungsanstalt Wien (LFW), mentioned his involvement with Sänger and his own early work on ramjets in his memoir, *Erinnerungen*:[6] "Together with some competent thermodynamicists at the Vienna Institute, I succeeded in starting some basic work on the ramjet engine and also to carry out some fundamental and promising experiments. The engine, already invented by Lorin in 1912, but which can only operate economically at very high velocities, was moved into the interior of a thicker delta wing of small span, the jet emerging along the trailing edge being simultaneously used by appropriate deflection to control it.

"Such a combination is called a 'power wing'. The tests for the corresponding burners, which were originally to be operated with liquid fuel, later with solid fuel (carbon dust or paraffin-impregnated brown coal), were carried out by Dr Schwabl in Vienna and by Dr Sänger at the DFS.

"In March 1944, a corresponding design was created for a piloted aircraft, P 12, which would be brought to the required initial speed using a Mistel combination (on the back of another aircraft) or with the aid of a launching sled with rocket propulsion. After finding a more favourable form of the air intake, we had successfully let fly a corresponding model of the engine in May 1944 at the Spitzerberg in Vienna.

"This led to an improved draft, P 13, which was very much promoted by the Ministry of Defence."

Work on the P 12 appears to have lasted for around three months. A drawing labelled 'P 12 I. Entwurf' dated March 11, 1944, and signed by Lippisch himself[7] seems to be the first iteration of the design.

The aircraft shown bears a marked resemblance to the Me 163 B, with a similar swept-wing planform, one-piece bubble canopy and landing skid. It differs from the Me 163 B, however, in having a circular intake directly in front of the pilot's legs rather than a rounded snub nose and in its sheer size. Where the Me 163 B measured 5.7m from end to end, the P 12 I. Entwurf was just 5m. Where the Me 163 B had a relatively generous wingspan of 9.3m, the P 12 I. Entwurf's was just 7m.

Already visible in the P 12 I. Entwurf's plan view are the air channels leading from the nose intake to the burners just behind and to either side of the pilot. It would appear that the type would also have had fuel tanks in each wing as well as another in the fuselage just behind the pilot's head.

This drawing appears to have been later labelled P 14-001 but why this should have been so is unknown, since Lippisch himself would later explain that the P 14 was a later turbojet-powered delta-wing design. A second drawing, labelled 'P 12 II. Entwurf'[8] is less ambiguous but shows a very different design. All resemblance to the Me 163 is gone – replaced with a curved delta planform including twin wingtip fins instead of a single central fin. The aircraft is slightly longer than the I. Entwurf at 5.5m but it only has just over half the wingspan at 3.58m.

As with the I. Entwurf, air for the ramjet motor passes around a sealed-off central section containing the pilot but this time the space is also used to house not only the single

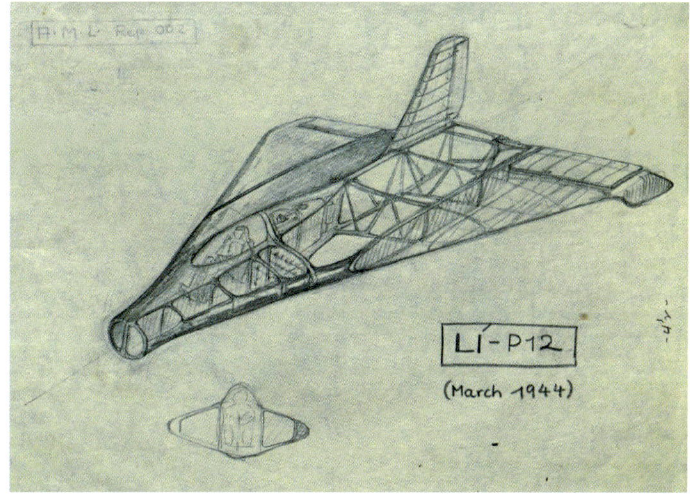

ABOVE: The final form of the P 12 as sketched by its designer Alexander Lippisch for his Allied interrogators immediately after the Second World War. The drawing appears to show the aircraft looking flatter and more aerodynamically clean than wartime models and drawings show it to have been.
IOWA STATE UNIVERSITY LIBRARY SPECIAL COLLECTIONS AND UNIVERSITY ARCHIVE

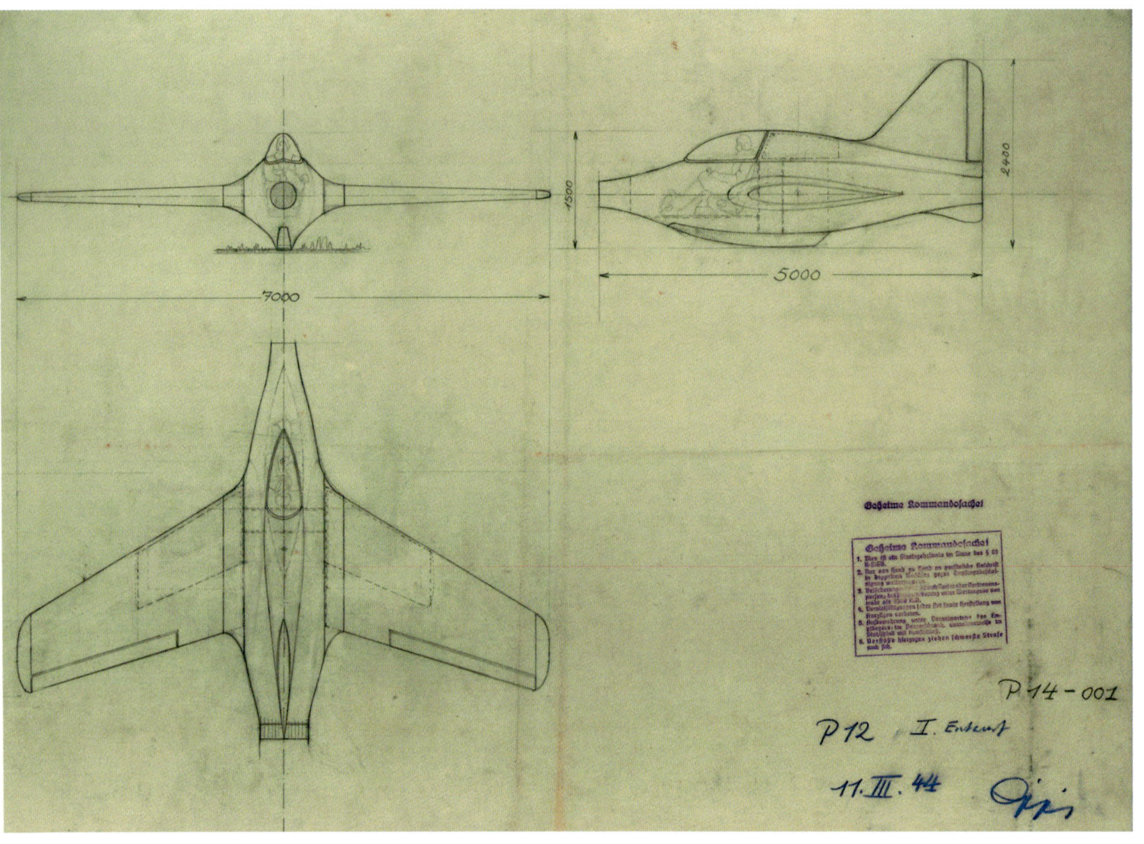

LEFT: Lippisch's first attempt at designing a ramjet fighter – the P 12 I. Entwurf, dated March 11, 1944 – appears to have been based on the rough outline of the Me 163. Note the apparently circular exhaust nozzle. *IOWA STATE UNIVERSITY LIBRARY SPECIAL COLLECTIONS AND UNIVERSITY ARCHIVE*

BELOW: The P 12 II. Entwurf marks a radical departure in design from the I. Entwurf. However, it still has a split intake for its ramjet motor and the pilot, landing gear and armament are all crammed in together between the intake tubes. *IOWA STATE UNIVERSITY LIBRARY SPECIAL COLLECTIONS AND UNIVERSITY ARCHIVE*

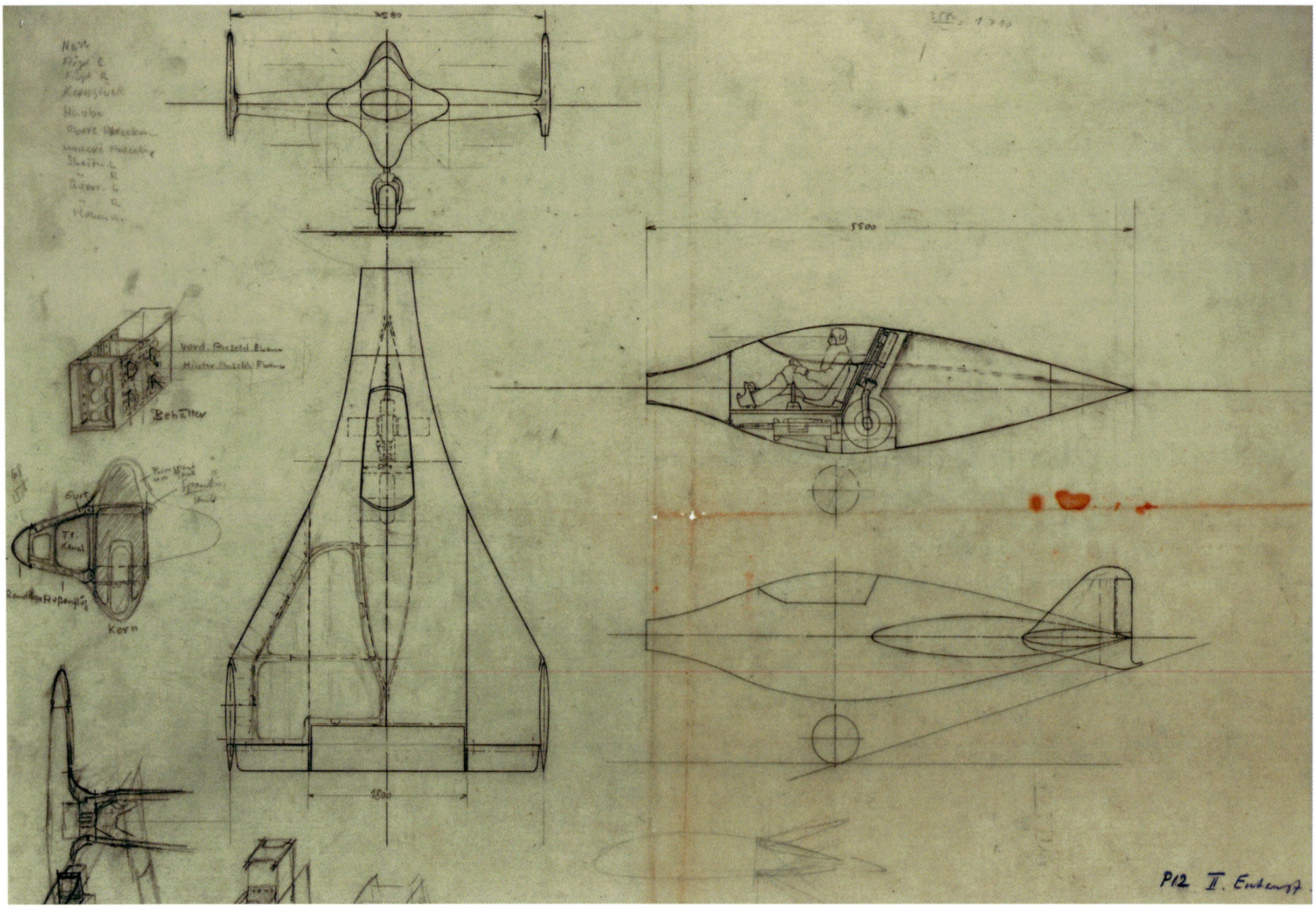

MK 108 armament but also an unusual monowheel landing gear complete with shock absorber directly behind the pilot. It appears from the cutaway diagram to the left of the top view that the centre section is also intended to house a saddlebag-type fuel tank wrapped around the area housing the landing gear.

Another notable feature of the I. Entwurf is the ramjet's highly unusual exhaust. At first it appears as though the motor has no exhaust – however, on closer inspection there is a faint sketch at the bottom centre of the sheet showing what appears to be a variable geometry exhaust. Presumably the ramjet exhausted through the 1.8m-wide section to the rear of the aircraft with the pilot able to control the upper and lower surfaces of the exhaust to effect manoeuvres.

The final P 12 design[9] combines elements of both the I. Entwurf – the central fin and belly skid – and the II. Entwurf with its wingtip surfaces, delta planform and slot exhaust with controllable flap. However, it also adds new design features such as the bifurcated nose intake and the reduction of the wingtip surfaces to what would later become known as 'Lippisch ears'.

The P 12 offers an interesting insight into Lippisch's design process. Despite having left Messerschmitt almost a year earlier and having spent the intervening time working on twin-jet fast bombers, when it came to designing a ramjet fighter Lippisch went back to what he knew – the familiar form of the Me 163, albeit with a few necessary adjustments. The further development of the P 12 was then necessitated by the needs of its propulsion system – with all of the aircraft's other necessary parts being crammed together in the only apparent space available: sandwiched between the spit intakes.

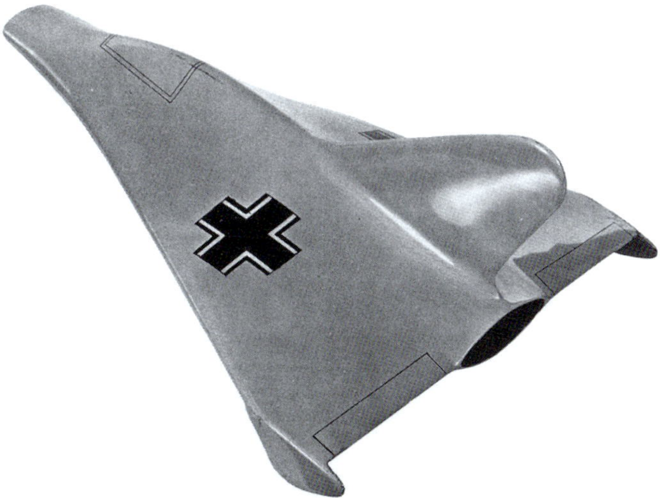

ABOVE AND BELOW: Views of a wartime model of the P 12 in its final form. These images appeared in numerous Allied reports immediately after the war.

LIPPISCH P 13

The P 12 had been designed to carry weaponry but Lippisch's next project, which followed on directly from it, was not. Work on the P 13 began in May 1944 and while it utilised the ramjet technology of the P 12 its role harked back to the 'arrows of death' conceived back in September 1943 – it was a rammer.

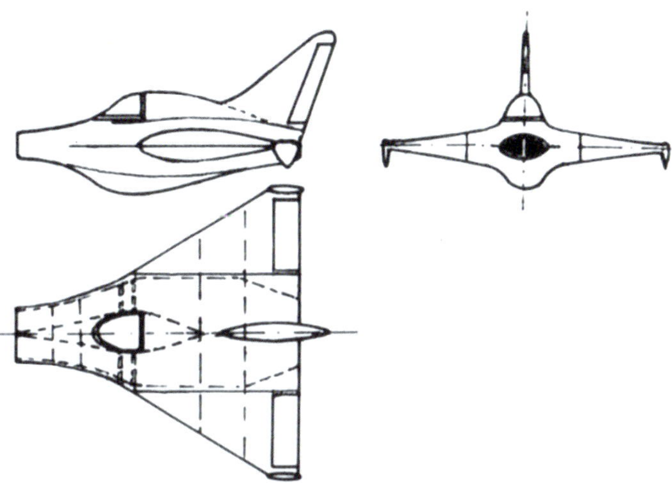

ABOVE: A postwar drawing showing the P 12's final form. Having been chosen by Lippisch himself to represent the P 12 in his own books, and corresponding to the shape of the wartime model, it is presumed that this image is a fairly accurate representation of the wartime design.

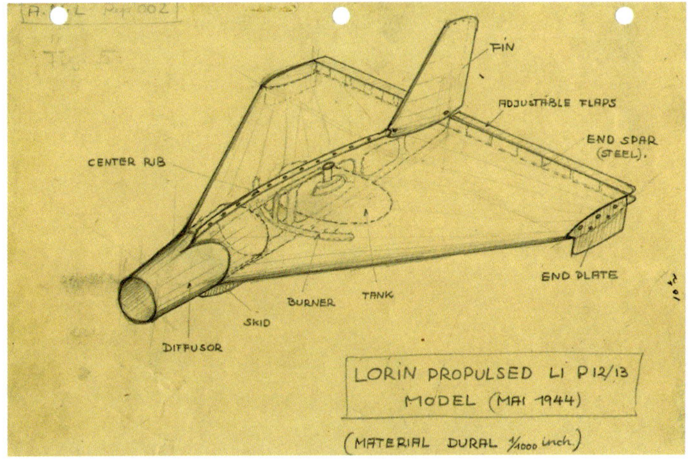

ABOVE: By May 1944 Lippisch was already considering how to free up more space for his ramjet's combustion chamber, as shown by this postwar sketch Lippisch made of a proposed flight test model.
IOWA STATE UNIVERSITY LIBRARY SPECIAL COLLECTIONS

It also differed from the P 12, in layout. Where the P 12's pilot sat low down in the fuselage, with the air ducts feeding the ramjet wrapped around his narrow cockpit, the P 13's ramjet intake was completely unobstructed – with air entering the intake almost immediately entering the large combustion chamber. This was made possible by shifting the cockpit upwards to create a large combined fin/cockpit sitting on top of the combustion chamber.

A third difference was fuel – the P 12 carried tanks of liquid fuel for its ramjet but the P 13 would be fitted with a wire mesh grate containing a quantity of slow-burning coal or other solid fuel. Under interrogation by US Army officers[10] in July 1945, Lippisch explained his decision to switch fuels: "He felt that at high velocities, solid fuels were more desirable than liquid fuels, the reason being that the solid fuel did not flow out into the velocity stream and, therefore, its burning took place at a predetermined place.

"For a solid fuel to be acceptable it must have good 'gas producing qualities', such as has bituminous coal. The most intense heating among the solid fuels was provided by German normal pine wood 'cooked' in oil or paraffin in pieces 1 x 1 x 1cm."

Neither of Lippisch's 1976 books[11] makes any mention of how the P 13 was to be used on active operations. When the British asked him whether the P 13 had been designed as a rammer,[12] it was stated that: "The possibilities of using the P 13 as a ramming aircraft had been considered but Dr Lippisch did not think that athodyd propulsion was very suitable for this purpose owing to the risk of pieces of the rammed aircraft entering the intake. This would be avoided with a rocket-propelled rammer."

However, the LFW's undated P 13 Baubeschreibung or construction description of 1944, probably produced in October or early November,[13] states: "The aircraft presented here is a jet with the task of fighter. The propulsion is by a Lorin engine, not using liquid fuel (gasoline, gas oil, J2 etc.) but with solid fuels (coal). Studies have been carried out on two forms – bulky pieces in a real grate or secondly with pressed coal in plates and hollow cylindrical shapes. Either can be used.

ABOVE: Artwork from the original P 13 project brochure. No armament was ever planned for the P 13 despite dozens of blueprints being drawn up – it was designed to ram Allied bombers.

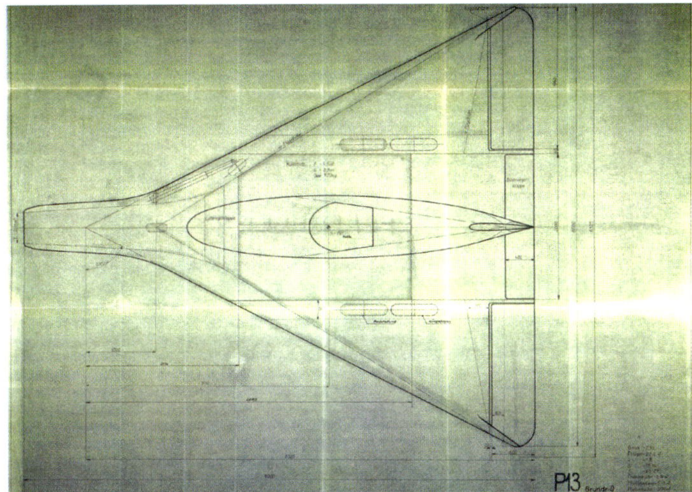

ABOVE: Plan view of the P 13 with wheeled undercarriage and straight trailing edge, dated October 4, 1944. The main wheels fit into the spaces between the main body section and the outer wings. *IOWA STATE UNIVERSITY LIBRARY SPECIAL COLLECTIONS AND UNIVERSITY ARCHIVE*

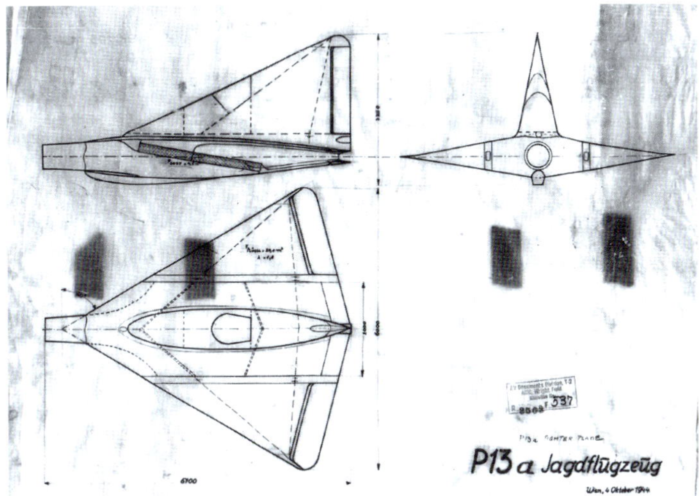

ABOVE: Lippisch seems to have worked on both simple delta- and diamond-wing configurations, and wheeled and landing skid arrangements, for the P 13 simultaneously.

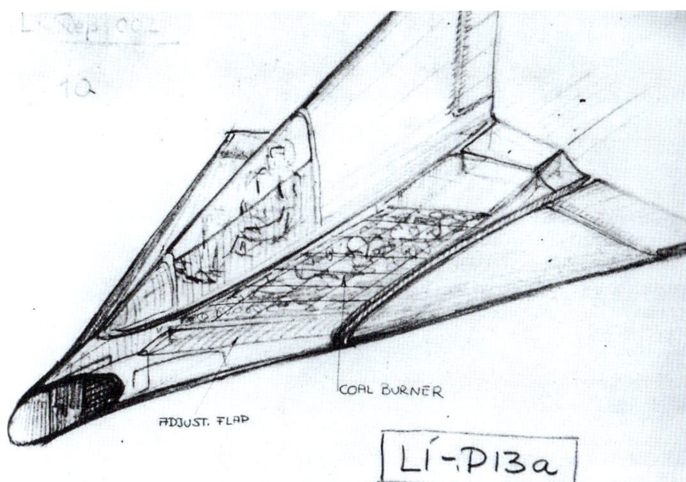

ABOVE: Sketch produced by Lippisch shortly after the war to illustrate the layout of his P 13 design.

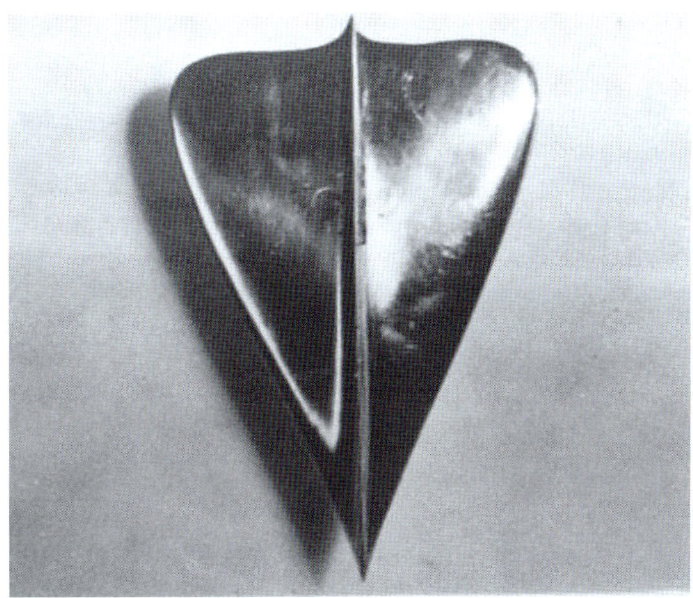

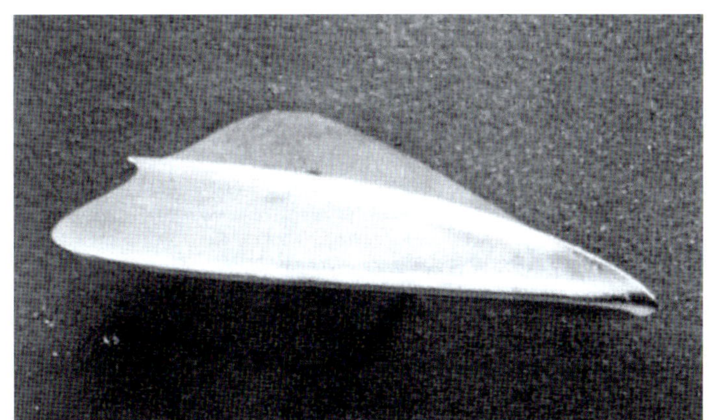

ABOVE AND BELOW: The Lippisch P 12/13 project began with aerodynamic forms devised by Alexander Lippisch during 1942 while he was working on the Messerschmitt Me 163. These arrowhead shapes were the model for the basic P 12/13 platform.

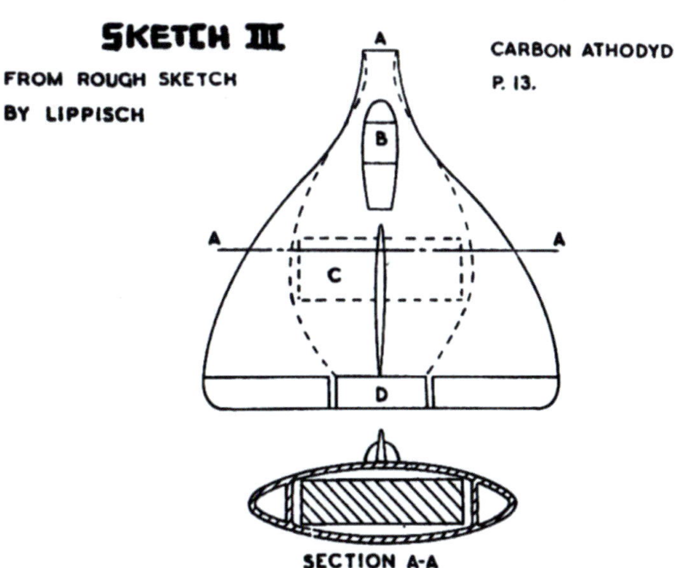

ABOVE: Another sketch produced by Lippisch to show the ramjet arrangement of his P 13.

"The engine used allows, for the time being, no self-launch. The launch itself must be carried out with (powder) rockets, a Madelung catapult or similar. Due to tactical considerations, among other things, the speed difference of fighters and bombers, attack from behind is preferable, though thought was given to the installation of brakes (brake parachute, retractable screens similar to dive brakes etc.), and although ample room for weaponry is present, the task of ram fighter has been taken into account – so that the ramming attack will not lead to the loss of the aircraft, thanks to its shape and static structure."

The small rammer aircraft would be easily transported: "When the outer wings are folded, transport of the entire aircraft on a small open car is possible on the Deutsche Reichsbahn."

The fuselage section of the description gives the length of the aircraft as 6m, saying: "The wing attachment fittings (outer wing to wing centre section) are formed so that the

ABOVE: The cover of *Projekt baubeschreibung P 13*, the brochure which outlines in detail how the type was to be operated in practice.

ABOVE: Illustration from the P 13 brochure showing how its wings would fold up for transportation on a rail car. The hinged wings also allowed easy access to the coal grate so that the aircraft could be refuelled between missions.

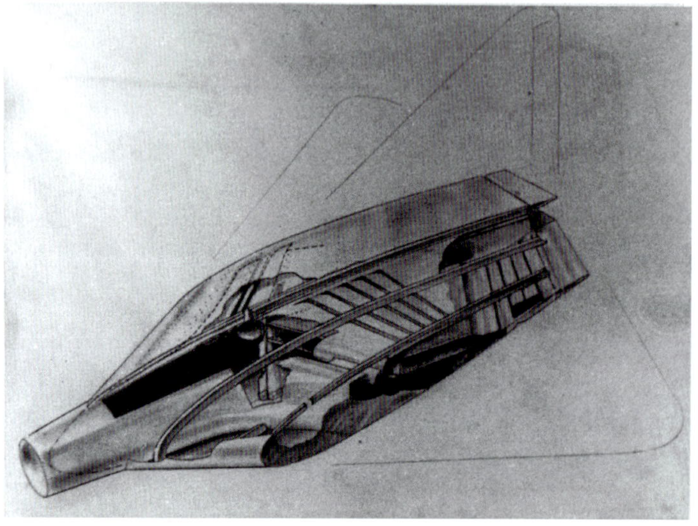

ABOVE: The refined form of the P12/13 combined ramjet/wing centre section. This was the basis upon and within which a variety of different fin and cockpit combinations were mounted.

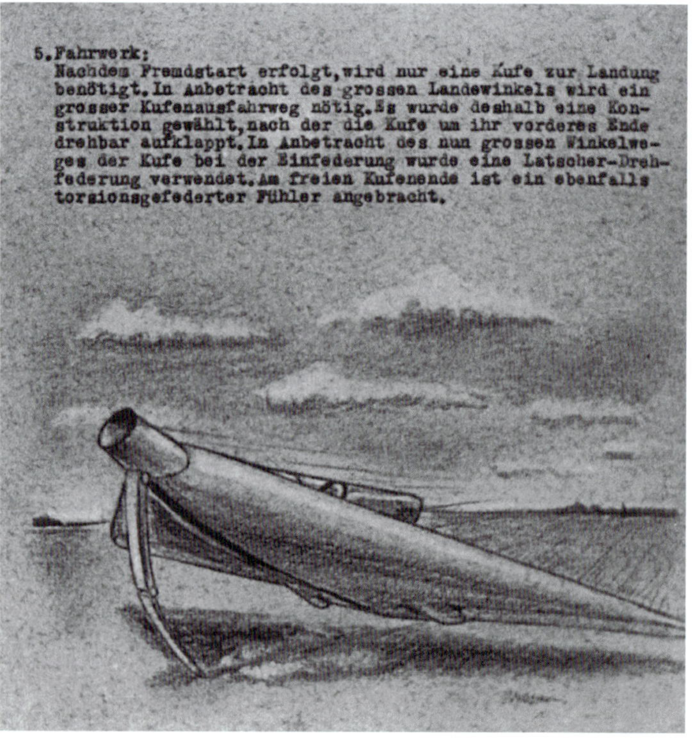

5. Fahrwerk:
Nachdem Fremdstart erfolgt, wird nur eine Kufe zur Landung benötigt. In Anbetracht des grossen Landewinkels wird ein grosser Kufenausfahrweg nötig. Es wurde deshalb eine Konstruktion gewählt, nach der die Kufe um ihr vorderes Ende drehbar aufklappt. In Anbetracht des nun grossen Winkelweges der Kufe bei der Einfederung wurde eine Latscher-Drehfederung verwendet. Am freien Kufenende ist ein ebenfalls torsionsgefederter Fühler angebracht.

ABOVE: Another page from the brochure, showing the multi-jointed torsion-sprung landing skid proposed by Lippisch. This also graphically illustrates the high nose-up attitude pilots would have been obliged to use in bringing it in to land.

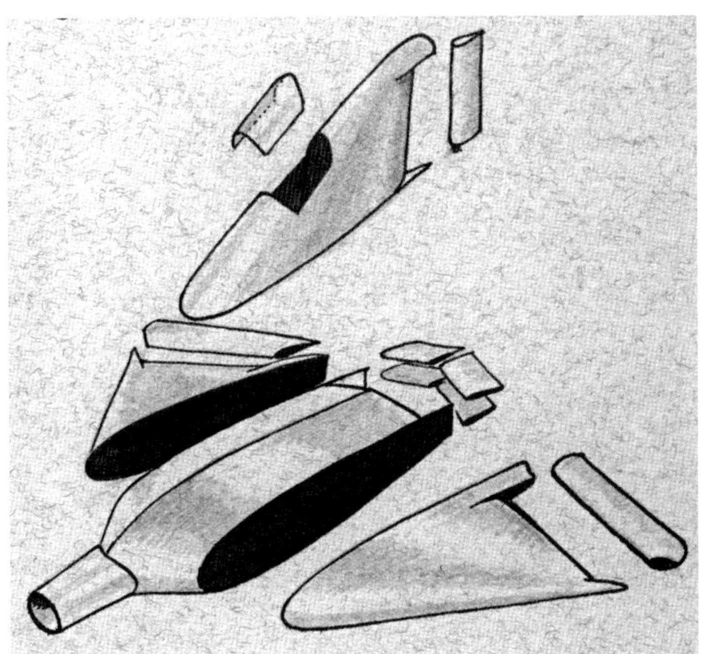

ABOVE: A breakdown of how the P 13 was to be constructed, again from the brochure.

ABOVE: Lippisch's brochure for the P 13 even included details of how the cockpit controls were to be laid out.

outer wings can be folded upwards. This is necessary for two reasons. The 'up' position allows access to the engine through the wing centre section, and secondly for loading and offloading onto vehicles during transport."

In other words, folding up the wings would allow ground crew to simply reach in and remove the exhausted coal grate before sliding a new one in – completing the type's refuelling in a matter of minutes.

The P 13 would cut through Allied aircraft using existing technology: "The edge cap is constructed as a deflector to avoid rudder damage if possible during ramming. The entire wing leading edge is reinforced with a knife (similar to the Kutonase)."

The Kutonase was originally developed to cut the steel cables anchoring barrage balloons. It was a hard steel blade that was fitted right the way along the leading edge of an aircraft's wing beneath an aerodynamic fairing of thin flexible metal. This concealed the Kutonase and made the wing fitted with it appear normal.

The fin was "designed just like the wing and built along similar lines" – presumably meaning that it too housed an aircraft-slicing Kutonase along its length. It was 2.28m high, with this measurement being determined by the need to provide the pilot with good visibility, and was attached to the wing by four fittings, two of them on rockers to allow for expansion of the wing's surface due to intense heating by the ramjet.

The point of connection between the centre section of the aircraft and the outer wings served "as a cooling passage" and when the inner side changed in length due to heating, this was offset by a rocker.

The cockpit itself was "equipped with front and rear walls and is planned to be extremely spacious" and the glazing

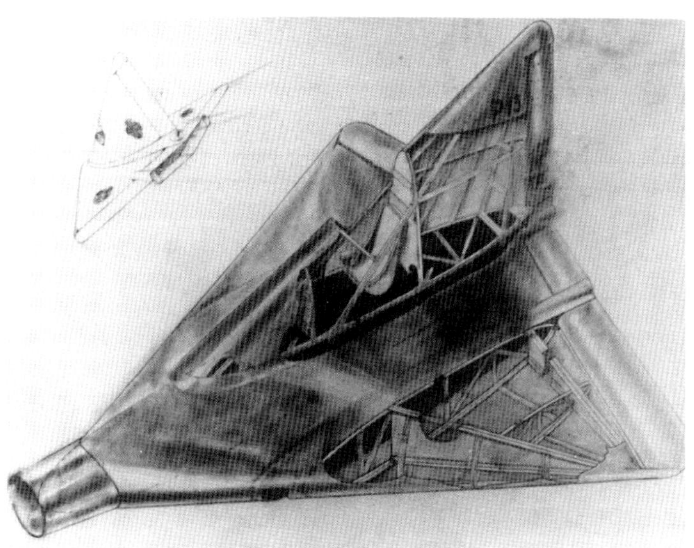

ABOVE: A grainy cut-through of the P 13 from the brochure shows the wing apertures for cooling.

was designed to be simple to manufacture and distortion-free. The seat and joystick were to be "mounted on the detachable cabin floor" for "production reasons" and the portion of the fin to the front of the cabin was designed to be easily removable for ease of access to the equipment housed beneath it.

Landing the P 13 would have been tricky given the nose-up attitude required. To this end, "a design was chosen where a blade flips open that is rotatable around its front end". This blade-skid with multiple torsion springs was only attached to the fuselage at one end, the other was intended to contact the ground and allow the aircraft to settle using the springs to cushion the impact.

Equipment fitted was to be "in accordance with its task" – horizon, artificial horizon, trip and altimeter, variometer and AFN2, compass, clock, FuG 16 ZY and FuG 25a. There was also an oxygen system installed but no pressure cabin.

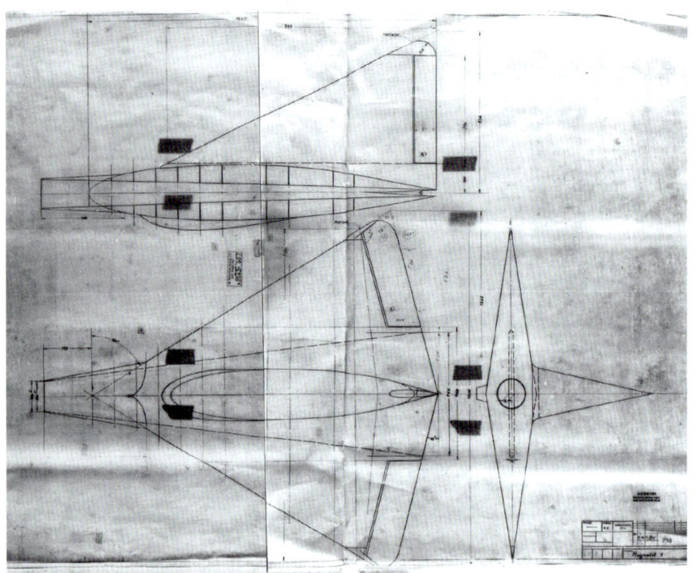

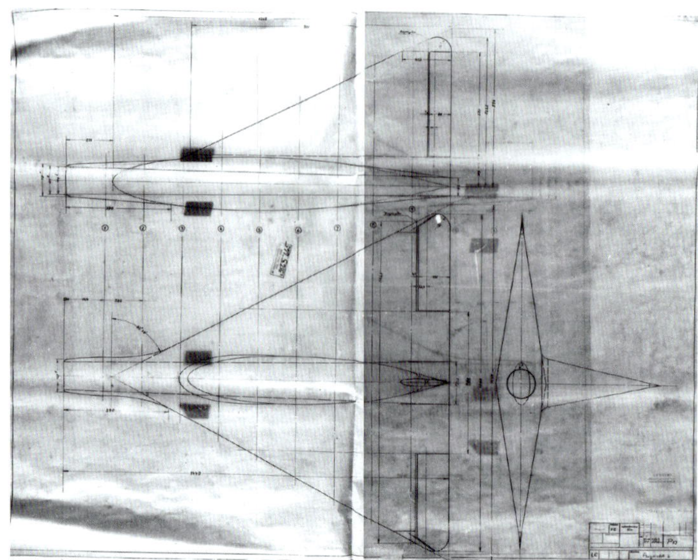

ABOVE: Drawings of diamond and simple delta planform P 13 models dated October 12 and October 25, 1944, respectively.

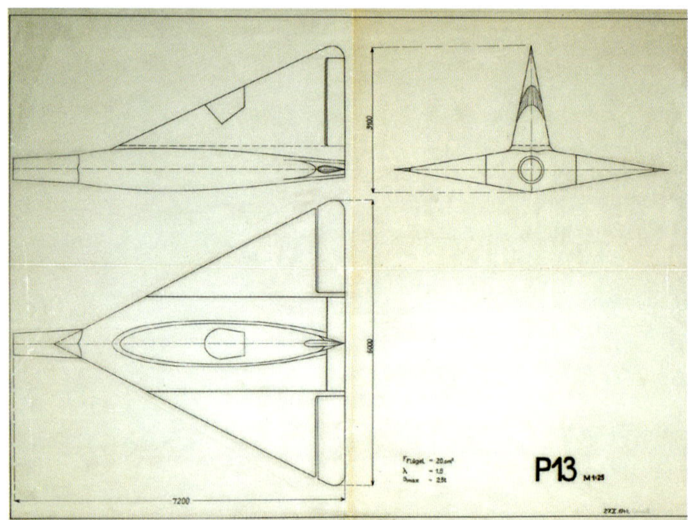

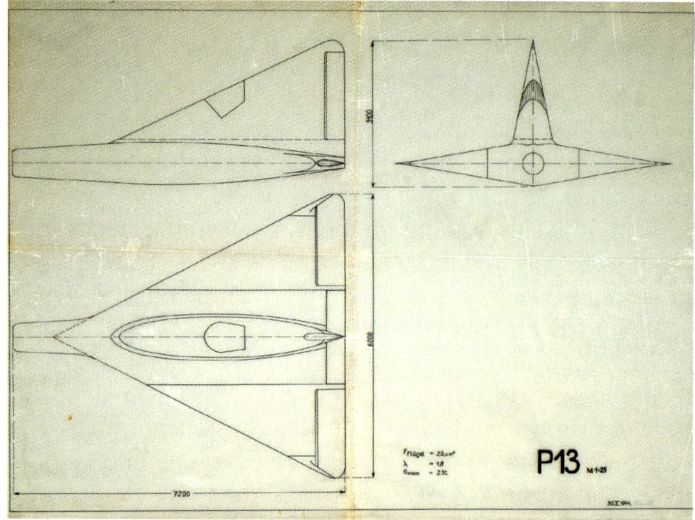

ABOVE: Two drawings of the P 13 simple delta, dated October 27 and October 30, 1944, showing minor detail refinements to trailing edge flaps and combustion chamber. *IOWA STATE UNIVERSITY LIBRARY SPECIAL COLLECTIONS AND UNIVERSITY ARCHIVE*

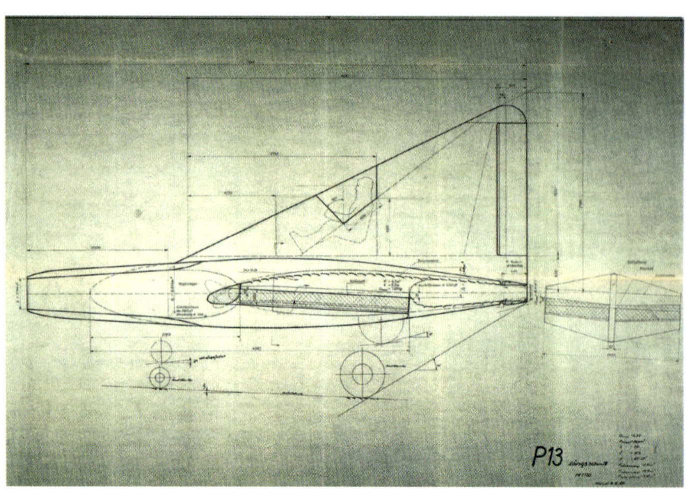

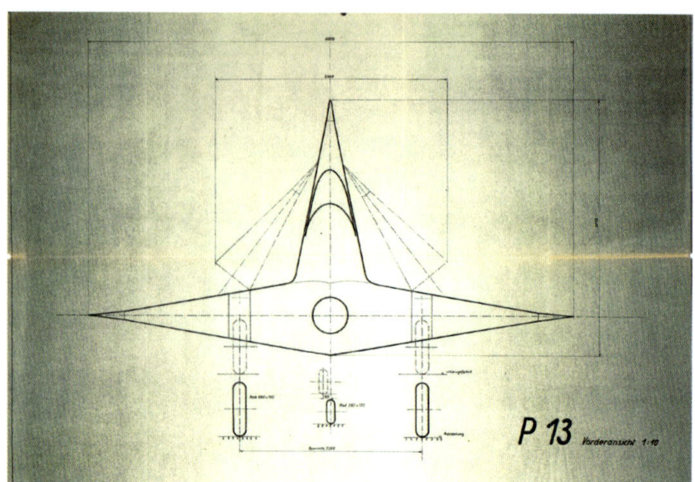

ABOVE: Side and front views of the wheeled undercarriage P 13, dated November 8, 1944 – over a month after the plan view. *IOWA STATE UNIVERSITY LIBRARY SPECIAL COLLECTIONS AND UNIVERSITY ARCHIVE*

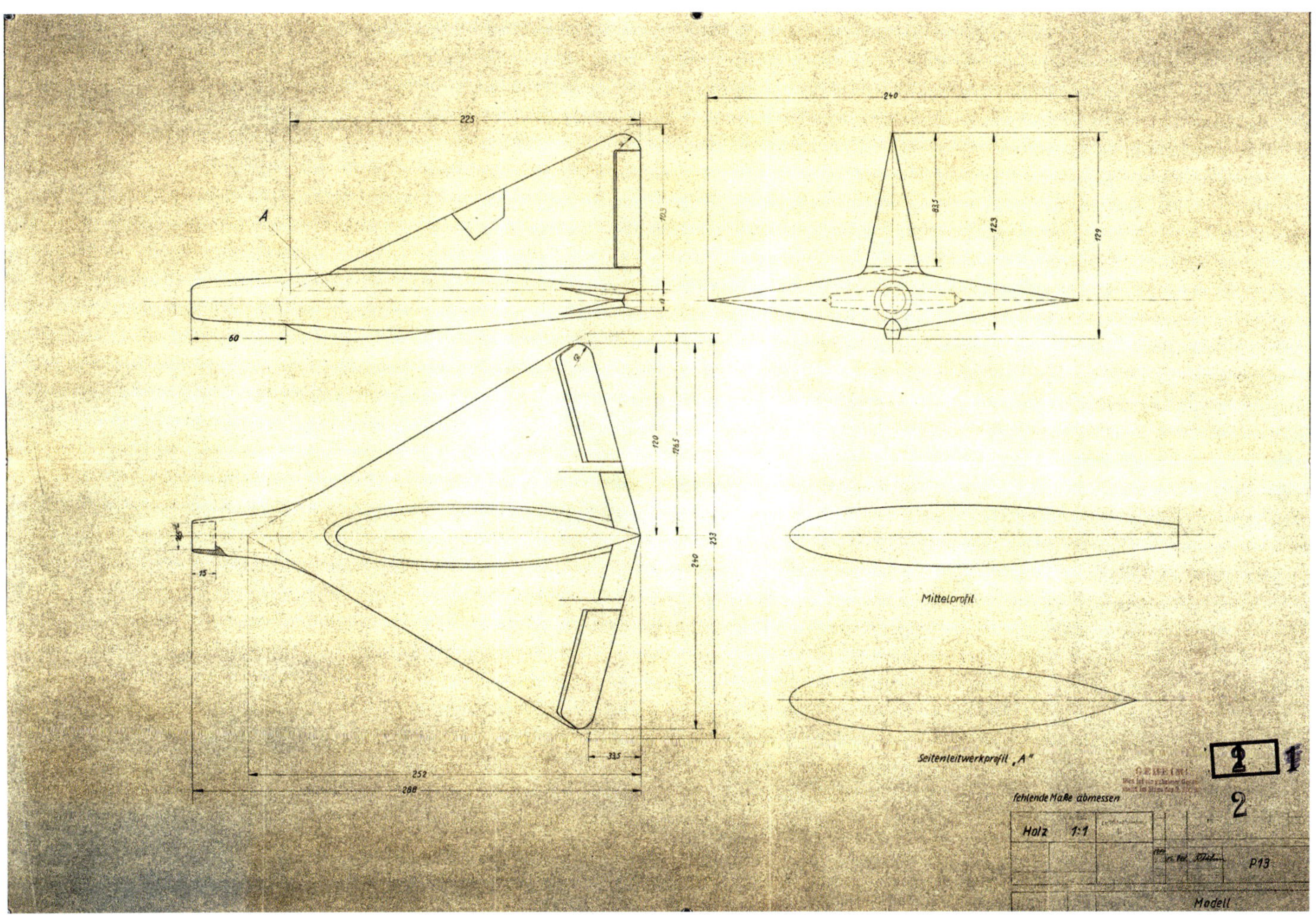

ABOVE: A drawing showing a model of the P 13a dated December 15, 1944. The external appearance of the design has changed little in two months.

Under the heading 'research aircraft' the translated minutes of a meeting on November 21–22, 1944, probably of the EHK,[14] stated: "Emphasis was placed on the athodyd propulsion system of the Lippisch P 13" and Lippisch says in *Ein Dreieck Fliegt*[15] that "for a short time, series production of this type was planned".

A Chef TLR management report[16] of December 21, 1944, said of the P 13: "Test aircraft, flying wing with coal-Lorin drive; V1 as a glider is supposed to fly at the end of February, but a date for the aircraft with engine has not yet been determined (depends on the behaviour of the engines)."

By this time Lippisch had arranged for the construction of an unpowered prototype of the P 13 (see below). No further mention of the P 13 appears to have been made in official Chef TLR records however and it did not appear in the Entwicklungs-Notprogramm formulated by the EHK at the end of January 1945 – but this was not the end of the P 13 story.

LIPPISCH P 13b

Further experimentation with the P 13's coal grate arrangement seems to have resulted in the original P 13 being dropped (and henceforth being referred to as the P 13a by Lippisch) and replaced by a new design which combined a circular rotating coal grate design with an airframe derived from that of the P 11/Delta VI – the P 13b. According to a British intelligence report:[17] "It was originally proposed that the solid fuel in the form of small pieces of brown coal should be carried in a wire mesh container set in the duct at a small angle to the air stream.

"The free flow of air through the lower portion of the duct was thus obstructed and it was hoped to obtain a progressive reaction with the oxygen passing through the fuel burning to CO which in turn would combine with oxygen in the air passing through the upper, unobstructed, portion of the duct to form CO^2. This arrangement proved inefficient and was abandoned.

"A later design called for a circular basket of oval axial section supported within the duct and positively rotated about its vertical axis at some 60rpm. Combustion is initiated by a gas flame and liquid fuel may be employed to facilitate starting up. Alternatively a more easily combustible material in the form of granules made from coal dust and an oxygen carrier may be distributed around the outside of the charge.

This was the P 13b. The first sketches[18] for this seem to have been made on November 25–26, 1944. Although these show the form of the fighter emerging, they do not show a circular basket for the ramjet's fuel. This does not appear, even in outline, until January 7, 1945. Then the fine detail of the basket and its construction is shown in drawings dating from early February.[19]

No more work on the P 13b is evident after February 1945 but a P 13 V1 prototype, based on the original arrowhead-shaped P 13, was ongoing by this point.

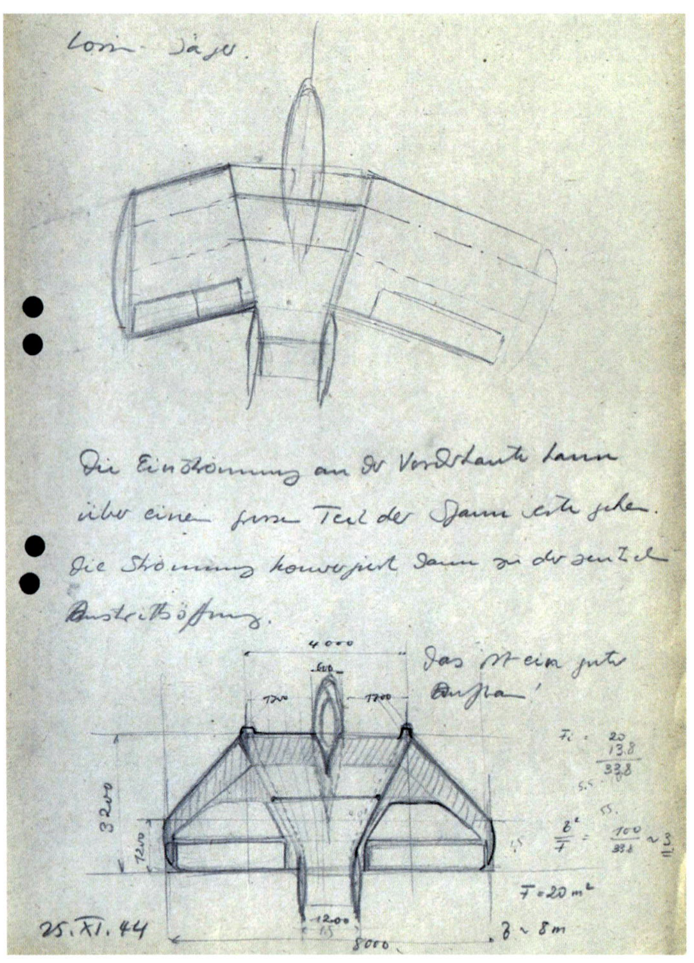

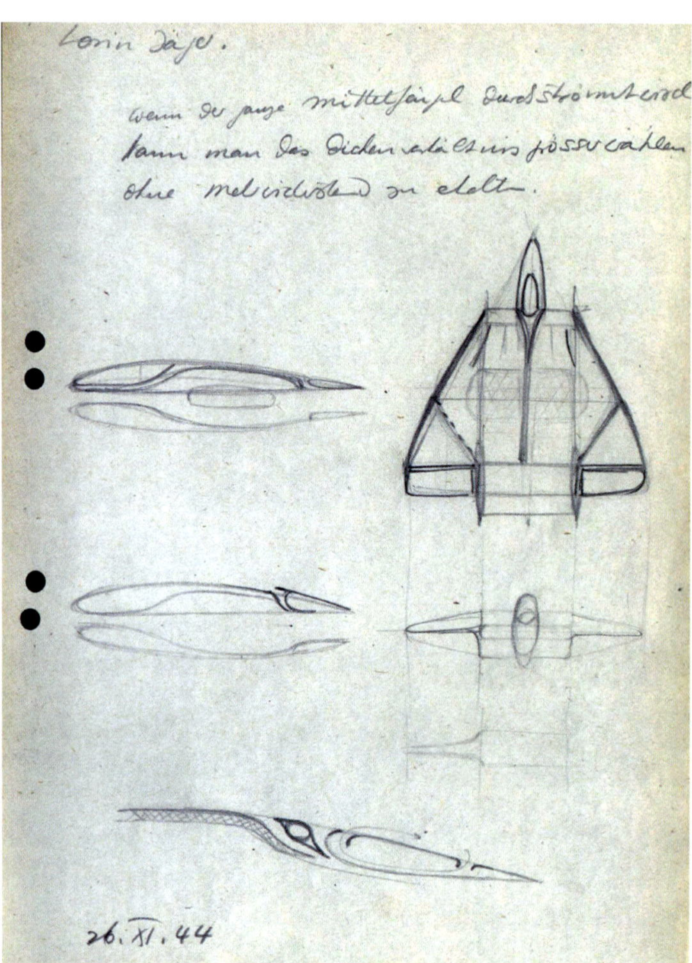

ABOVE: Lippisch's initial sketches for what would become the P 13b, dated November 25–26, 1944. *IOWA STATE UNIVERSITY LIBRARY SPECIAL COLLECTIONS AND UNIVERSITY ARCHIVE*

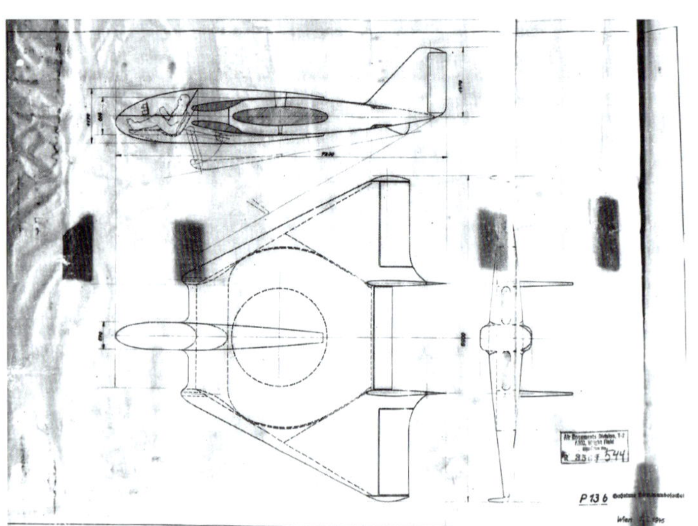

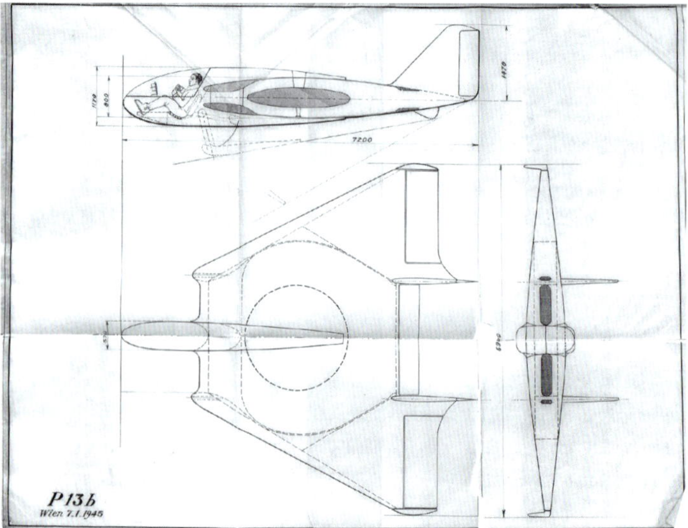

ABOVE: Two different versions of the P 13b three-view drawing, both dated January 7, 1945. The latter, from Lippisch's personal papers, appears to show a younger version of Lippisch himself at the controls! *IOWA STATE UNIVERSITY LIBRARY SPECIAL COLLECTIONS AND UNIVERSITY ARCHIVE*

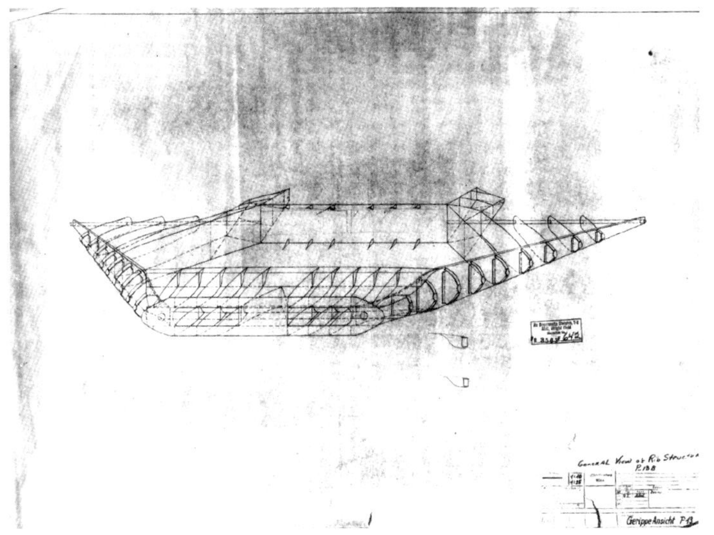

ABOVE: The proposed internal structure of the P 13b, showing the vents that would allow fast-flowing air to reach its ramjet's rotating circular coal grate.

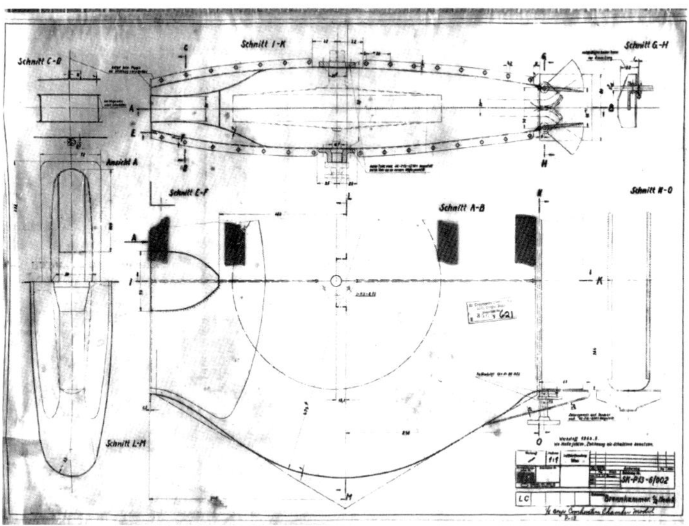

ABOVE AND BELOW: Drawings showing a model designed to demonstrate the workings of the P 13b's rotating coal grate and of the coal grate itself.

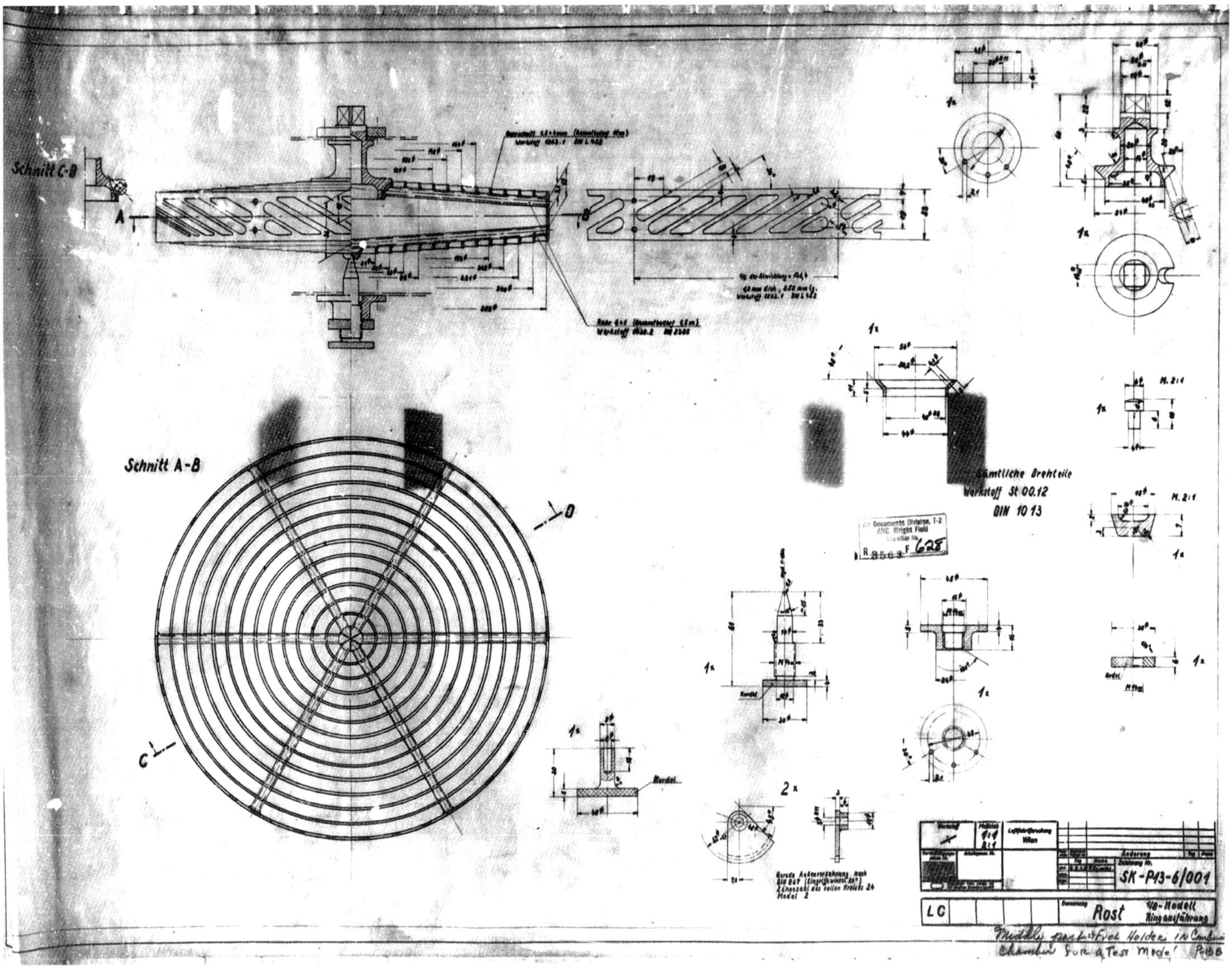

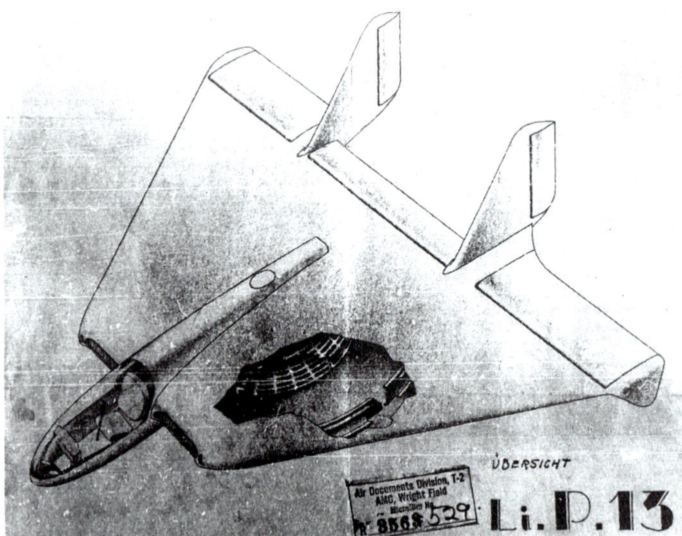

ABOVE: Contemporary artwork showing how the P 13b was intended to look, with cutaway showing its ramjet's combustion chamber.

LIPPISCH P 13a V1 (DM-1)

Lippisch had always been a firm believer in testing prototypes of new designs rather than simply relying on a combination of wind tunnel testing and mathematical formulas to determine what their flight characteristics might be.

As mentioned above, when the German armaments ministry showed an interest in his P 13 (later P 13a) he decided that practical flight tests were needed to work out how it was likely to handle in the air. However, since every aircraft company was already operating at maximum capacity, he needed an alternative if his P 13 V1 was to be constructed. So he enlisted the aid of some students.

In his book, *Ein Dreieck Fliegt*,[20] Lippisch writes of the P 13: "Admittedly this project of a delta wing of very low aspect ratio and with the location of the cockpit on top of a wing that had been converted into a combustion chamber was rather unusual and there loomed the possibility of flight attitudes which might have to be mastered by special procedures.

"However, wind tunnel and free-flight model tests did not reveal any especially problematical characteristics. We therefore deemed it practical to start construction of an experimental, manned powered craft of this design (scale 1:1).

"The difficult military situation demanded the drafting of many of the remaining civilian employees for service in the Volkssturm, which was regarded by the powers-that-be as the last line of defence. This meant that the aeronautical engineering students also had to join.

"Only work of highest priority rating could protect these students from being drafted. Aeronautical students of the Institutes of Technology of Darmstadt and Munich approached me in the hope of obtaining work in connection with the P 13. Since the end of the war obviously was imminent anyway, I created the project of a wooden flying-glider of the P 13, which the students were to build under the direction of my assistant, Heinemann, in a hangar of the small airfield in Prien on the Chiemsee.

"The students designated this project D 33 (Darmstadt 33) which was later changed to DM-1. We succeeded in obtaining deferment for the students. The DM-1 was almost completed when the war ended and the Prien airfield was occupied by American forces.

"In response to the suggestion of Professor Theodor von Karman the project was completed under US direction."

So the DM-1 was only built to help some students dodge the draft. But in his memoir *Erinnerungen,*[21] Lippisch's account is slightly different: "After finding a more favourable form of the air intake [than that of the P 12], we had successfully let fly a corresponding model of the engine in May 1944 at the Spitzerberg in Vienna. This led to an improved draft 'P 13', which was very much supported by the Rüstungsministerium.

"Since the design of a thick delta wing with a low aspect ratio and a cockpit on top was novel and the Academic Aviation Groups Darmstadt and Munich had approached me with a high degree of urgency, I handed over the construction of a flyable wooden model in a 1:1 scale. They built this P 13a V1 in Prien-am-Chiemsee under the designation 'DM-1'.

"In Prien the students there, under the direction of my assistant Heinemann, handed over the almost completed DM-1 to the Americans as their own development, without mentioning that it was actually my P 13a V1. At the instigation of Theodor von Karman, it was completed and brought to the USA for survey."

While the two versions are not incompatible, there are two facts presented in *Erinnerungen* which are omitted from *Ein Dreieck Fliegt* – firstly the reference to strong support for the P 13 from Albert Speer's Rüstungsministerium and secondly the fact that Lippisch very much regarded what was being constructed at Prien as the P 13a V1, whatever the students chose to call it.

The Rüstungsministerium's support was based on the model tests Lippisch mentions plus wind tunnel tests of the P 13 in model form carried out by the AVA up to August 1944. Wolfgang Heinemann, who Lippisch mentions as his assistant, had been studying at Darmstadt but a bombing raid on September 11–12 had resulted in him and his fellow students moving to Prien am Chiemsee in Bavaria.

Evidently as part of his studies Heinemann had been given a work placement at the LFW in Vienna, where he met Lippisch, and in November 1944 Heinemann persuaded Lippisch to let him build the P 13a V1 with the help of his fellow students back at Prien. It was initially called 'D 33' as the 33rd Darmstadt project, then 'DM-1' for 'Darmstadt and Munich' because the Darmstadt students were using the Munich students' workshop and enlisted their help during the build.

In fact, Lippisch was wrong about Heinemann – he did tell the Americans exactly what the DM-1 really was. An American intelligence report based on Heinemann's interrogation[22] states: "Target: Flugtechnisches Fachgruppe. Location: Prien Chien See. 1. Interviewed Heinemann. This firm evacuated to Prien from Darmstadt and Munich last September. Previously, they had only made gliders, but were now working on a prototype glider of the P 13 (DM-1) type (see report on Dr Lippisch).

"One interesting fact quoted by him was that on the P 13, the speed was too great to use a gun in attacks on bombers, and it was hoped to use a kind of sword suspended below the plane by a pendulum suspension which would cut through the fuselage of the bomber as the P 13 flew over.

"The Flugtechnische Fachgruppe Prien – before 1933 known as Academische Fliegergruppen (Akafliegs) – were independent associations of students of aircraft construction and similar faculties, who in addition to their studies voluntarily and without any payment designed, built and flew aeroplanes in order to complete their knowledge and instruction.

"Present aim of the Akaflieg Prien. During the war, the rooms of the Akaflieg Darmstadt and Munich were destroyed by air bombardment, and they therefore moved to Prien. In November 1944, they began in cooperation with Dr Lippisch, Vienna, the development of a completely new aeroplane type in order to conquer the stratosphere and attain supersonic speed, both necessary for aeroplane traffic over long ranges and 'to other stars'. In consequence of low air density at high altitudes, load capacity of the aeroplane could only be utilised if necessary lift was attained by high flying speed.

"In order to increase the speed of an aeroplane of standard construction, a long and difficult development would be

ABOVE: The P 13a V1 aka Akaflieg Darmstadt/Akaflieg München DM-1 as it appeared from the front when American forces occupied Prien am Chiemsee on May 3, 1945.

ABOVE: Viewed from the side, the P 13a V1/DM-1's true nature is readily apparent. Even though the glider's pilot was to sit at the base of the fin, in the nose, it is easy to imagine the P 13's pilot seated entirely within the fin above the wing, looking out through a canopy about halfway up the slope.

necessary. Velocity of sound could not be surmounted because of resistance, rigidity and flying qualities. It was, therefore, very uneconomical to continue working in this direction.

"Though not difficult to develop an aeroplane for supersonic speed only, it was very difficult to design a plane which could fly as well in subsonic as in supersonic. The present state of flight research and flight testing in the sphere of high speed promised that the form chosen would correspond to both demands. Relation of rigidity to weight and of surface to useful inner room as well as the change of the flying qualities in passing the velocity of sound would be much better than in planes of standard construction. Only one tenth of performance would be needed.

"In order to be able to fly slowly as well a compromise was necessary, which made aeroplanes of standard construction superior to the DM-1. The main point however was to design a plane which could fly in both cases. At present, the principal problem was to find out the necessary shape of aeroplane. Afterwards, the development of an appropriate power plant might be started."

Heinemann told his interrogators that it would be impossible for a propeller-driven aircraft to reach supersonic speeds and that it "was uncertain whether it would be possible by a turbine though at least higher speed could be attained. But it could positively be stated that the supersonic speed would be attained by means of rocket propulsion of the same kind as in the Messerschmitt 163. This rocket propulsion however began to be economical only in flying speeds of more than 2,000km/h".

ABOVE: A rear view of the P 13a V1/DM-1. The shape of the glider suggests that Lippisch favoured the diamond planform over the simple delta for his P 13.

ABOVE LEFT: The P 13a V1/DM-1 was taken apart piece by piece at the Langley Memorial Aeronautical Laboratory at Langley Field, Virginia, during 1946 and reassembled in this radically altered form. Its wings received sharp leading edges, the fin was removed, a new smaller fin was fitted and a canopy borrowed from a P-80 was installed. The goal was to test the likely effect of these innovations on lift and handling performance of a sharply swept delta form.
ABOVE RIGHT: Tests on the P 13a V1/DM-1 in its original form began at NACA Langley in April 1946.

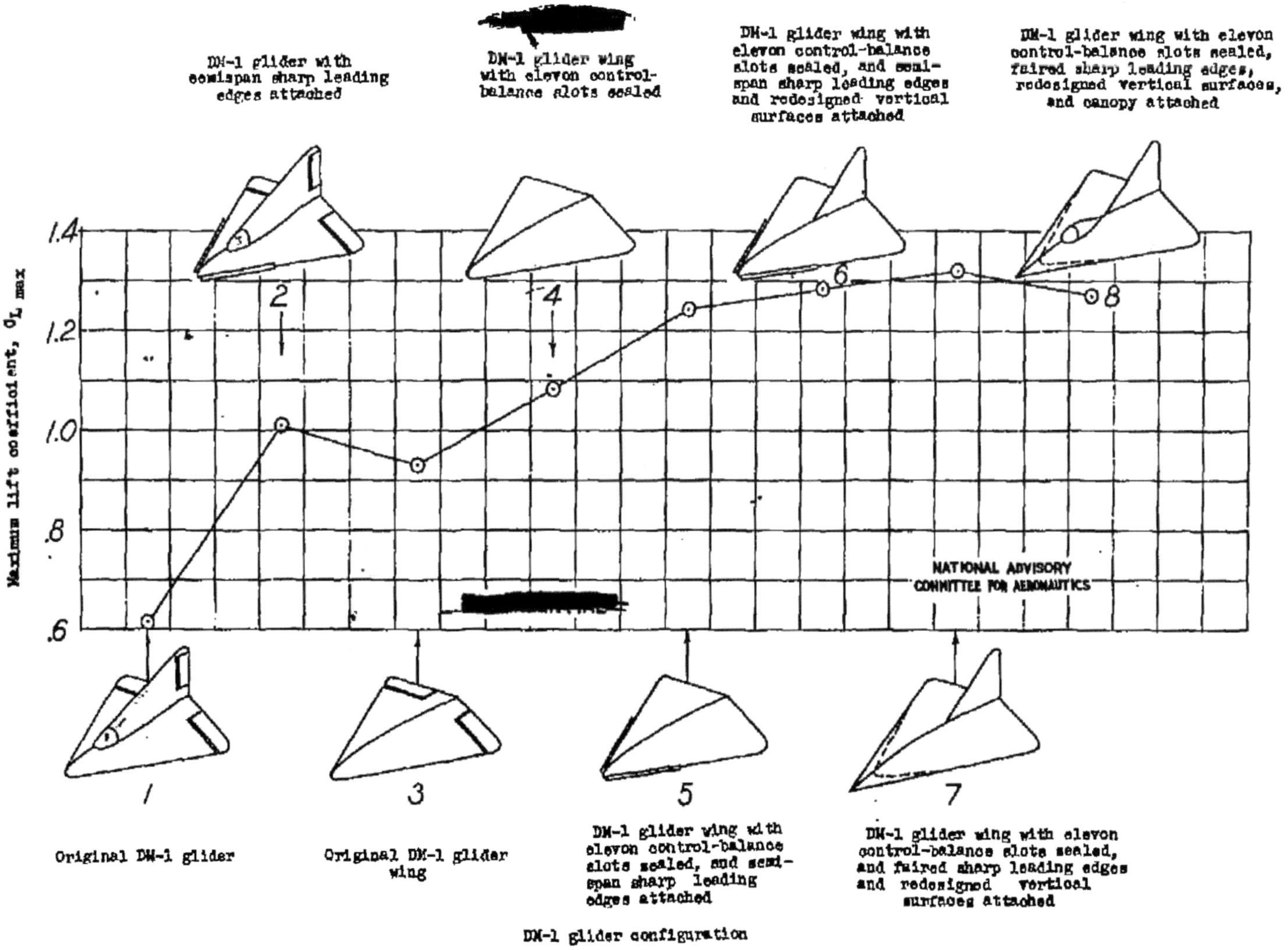

ABOVE: Diagram showing the eight different stages of modification carried out on the glider.

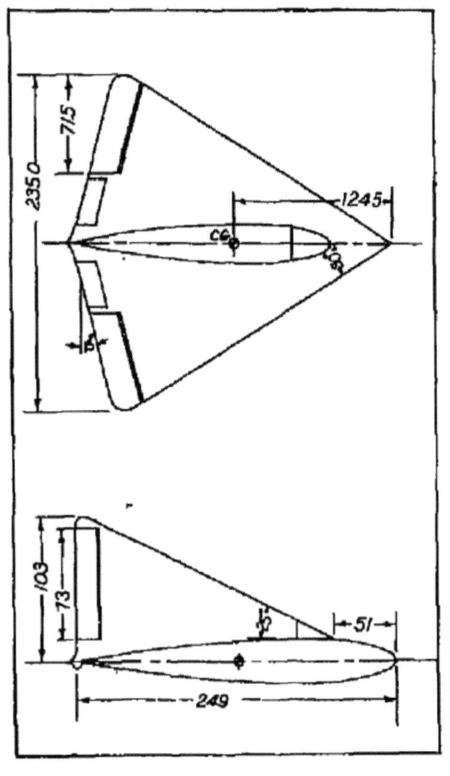

(a) Principal dimensions of glider configuration 1 (original DM-1 glider).

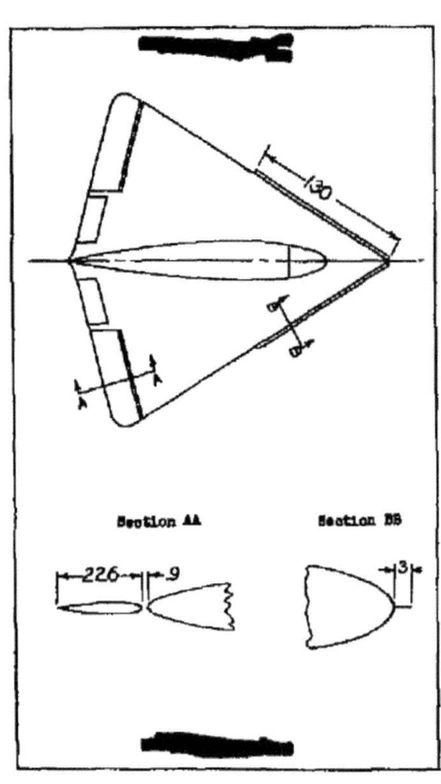

(b) Dimensions of the semispan sharp leading edges and of the elevon control-balance slots.

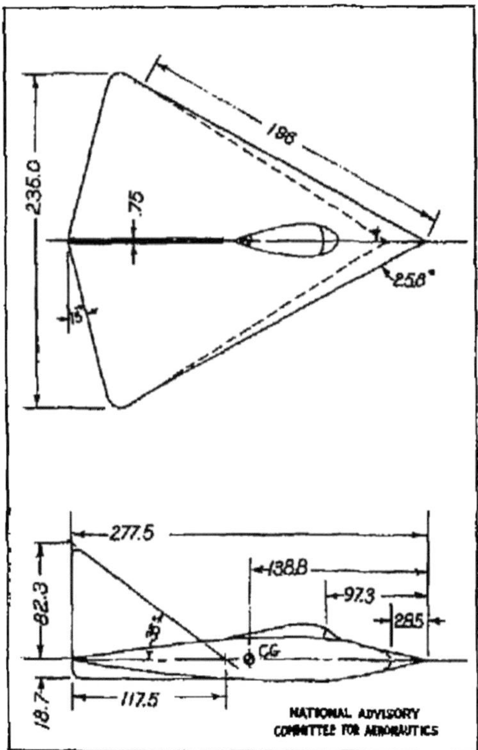

(c) Principal dimensions of glider configuration 8.

ABOVE: Dimensions of the glider in unaltered and altered forms as shown in a NACA report.

Evidently Lippisch had mapped out a three-stage testing process for the P 13: "The first model would have no power plant and would only show whether the chosen shape would be corresponding to the demands of slow-speed flight though assimilated to the demands of high-speed flight, especially whether it could start and land. This aeroplane would be towed in 'Huckepack' by a normal plane to 8,000–10,000 yards, when released and in glide would test the qualities of slow speed flight, in diving those of high-speed flight (perhaps a rocket would be built in) until about 800km/h (500mph).

"The second model would have a jet power similar to that of the Messerschmitt 262 or P 38 [presumably Heinemann meant either P 59 or P 80 here]. With it the critical range of speed of 800–1,200km/h (500–750mph) would be tested.

"The third model would have a rocket propulsion (as for the Me 163) and with it the plane would attain a maximum speed of 2,000km/h (1,250mph). The newly developed shape of aeroplane could also be used for testing other possible drives for high speed."

Some technical details of the DM-1 followed, primarily concerning its wing form. On its structure, the report states: "The plane is built in wood construction. Wings and tail are developable surfaces and cantilever shells without spar. As the DM has no power plant, the pilot's seat is not completely put into the tail, but a little in front on account of the centre of gravity. On account of the low aspect ratio unusual angles of attack can be obtained, in theory up to 35°. Therefore, there are two hoods for the pilot, a normal one at height for his head for high-speed flight, the second is set at the bottom of the fuselage for landing.

"Also the undercarriage is a new construction. It is a tricycle undercarriage with a stroke of 60cm and permits a sinking speed of 6m/sec in landing. Stressed steel tubes will be the material for it. Therewith a good method of retracting will be obtained. The holes for accommodating the retractable undercarriage have the size of the lowest wheel section."

The glider also had two water tanks on board – one in the nose and one in the tail – connected via a hand pump so the pilot could move 35 litres of water back and forth to alter the centre of gravity as required. It would seem that the solid shock absorber-less undercarriage was of the students' own invention and may have only been intended for moving the glider around the airfield, since landing on it would probably have damaged the DM-1's delicate wooden structure.

Take-off would have been on the back of a powered aircraft, such as a Siebel Si 204 or Focke-Wulf Fw 58, with the glider's pilot detaching once a suitable speed was achieved. A test pilot had been recruited to fly the DM-1, Hans Zacher from the DFS, but by the time American troops arrived at the airfield on May 3, 1945, the glider was still incomplete.

While he clearly outlined every other aspect of the P 13's design process and its purpose, Heinemann does seem to have entirely omitted the P 13's intended ramjet powerplant from his account. Why he should have left this out is unclear, unless Lippisch had kept it from him. The Americans were interested in the DM-1 and shipped it back to the US for study. The team at NACA – the forerunner of NASA – took it apart and rebuilt it in a variety of different arrangements, even to the extent of adding new parts. Six major revisions were attempted: installation of sharp leading edges for its wings, removal of the vertical fin, sealing the elevon control-balance slots, installing a new smaller thinner fin, installing faired sharp leading edges and finally installing a conventional cockpit canopy borrowed from a P-80 Shooting Star jet fighter.

American historians have recently re-evaluated the significance of the DM-1 glider and credited it as the inspiration for Convair's successful series of delta-winged Cold War interceptors beginning in 1948 with the XF-92 and progressing on to types such as the F-102 Delta Dagger in 1953 and Delta Dart in 1959, which themselves went on to inspire further delta-wing designs – a lasting legacy for Alexander Lippisch's years of work on tailless deltas.

FOCKE-WULF STRAHLJÄGER

Just as Lippisch had apparently been inspired to work on ramjets by Sänger, so too were engineers at Focke-Wulf – leading to yet more radical fighter designs during 1944. The company appears to have worked on two ramjet fighters simultaneously, one with fixed wings known as the Strahlrohrjäger and another with rotating wings referred to as the Triebflügeljäger.

The earlier history of Focke-Wulf's involvement in ramjets is revealed in a US report[23] on the interrogation of company aerodynamicist Dr Otto Ernst Pabst on May 21, 1945: "Pabst stated that his last job was on the ramjet

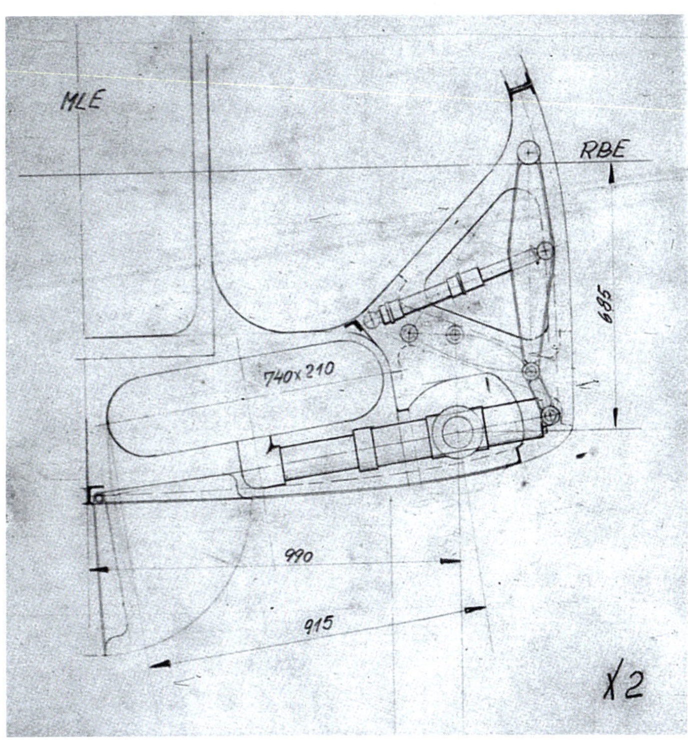

ABOVE, BELOW LEFT AND BELOW: Fuselage and landing gear sketches for Focke-Wulf's X2 Strahlrohrjäger.

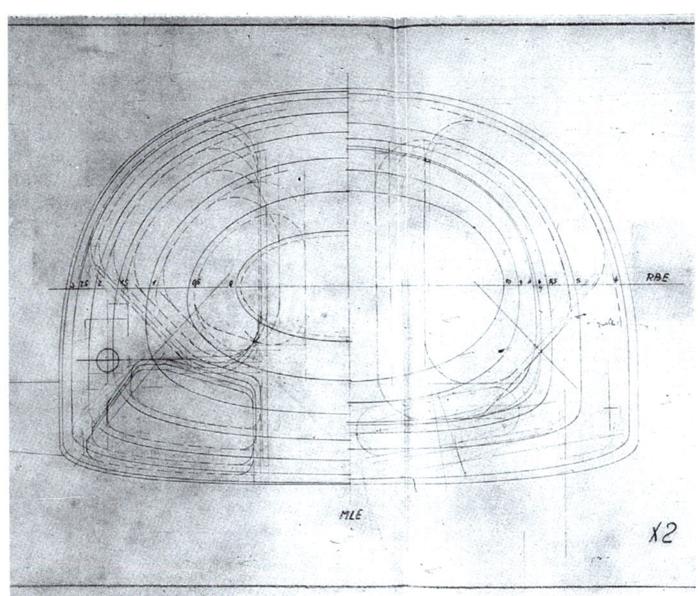

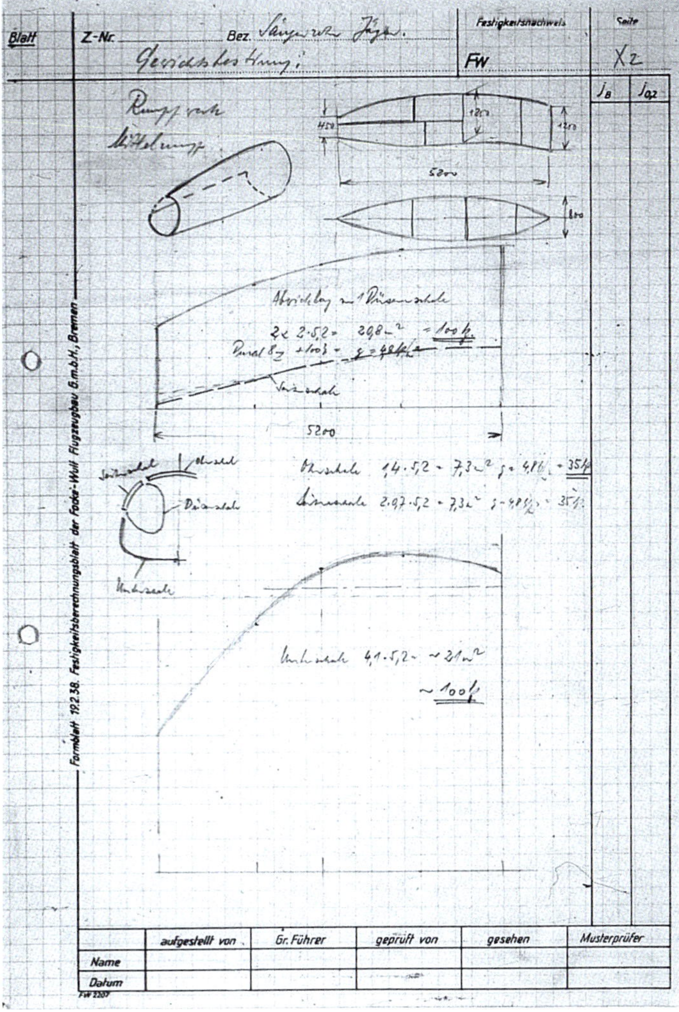

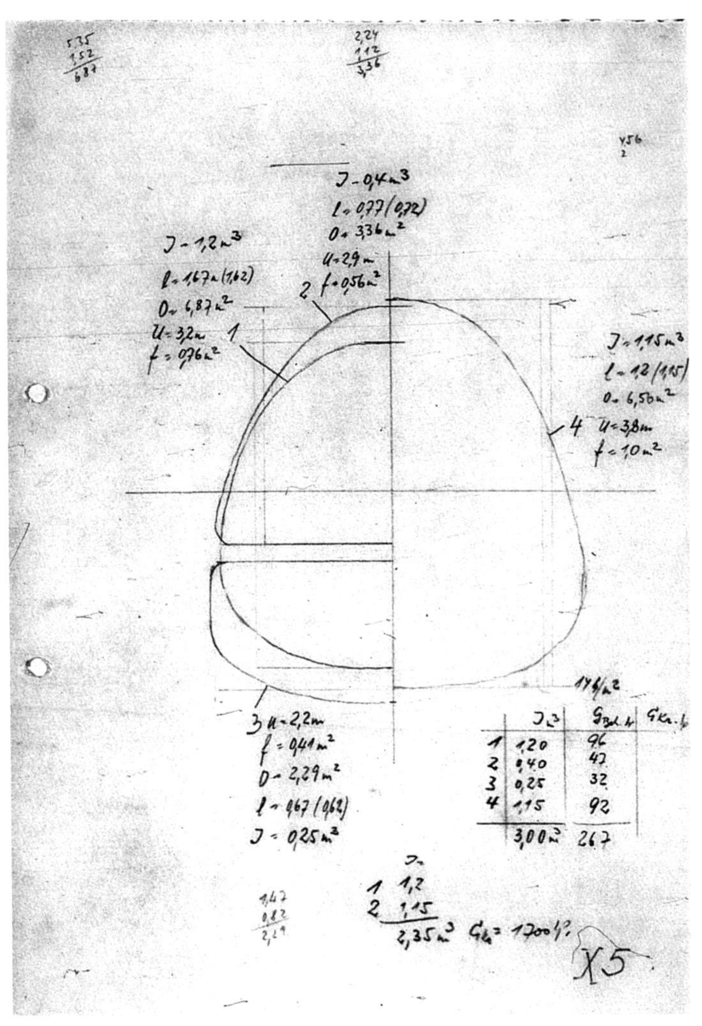

ABOVE: Sketch showing a cross section of the X5's fuselage.

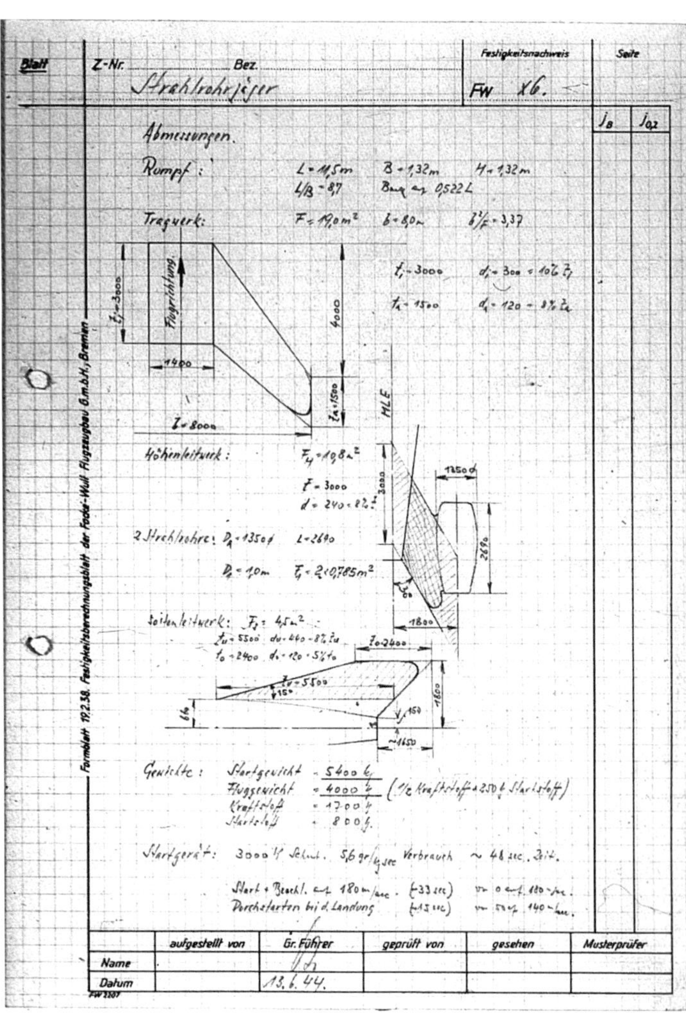

ABOVE: Designs for the wings, ramjet booms and tailfin for the X6.

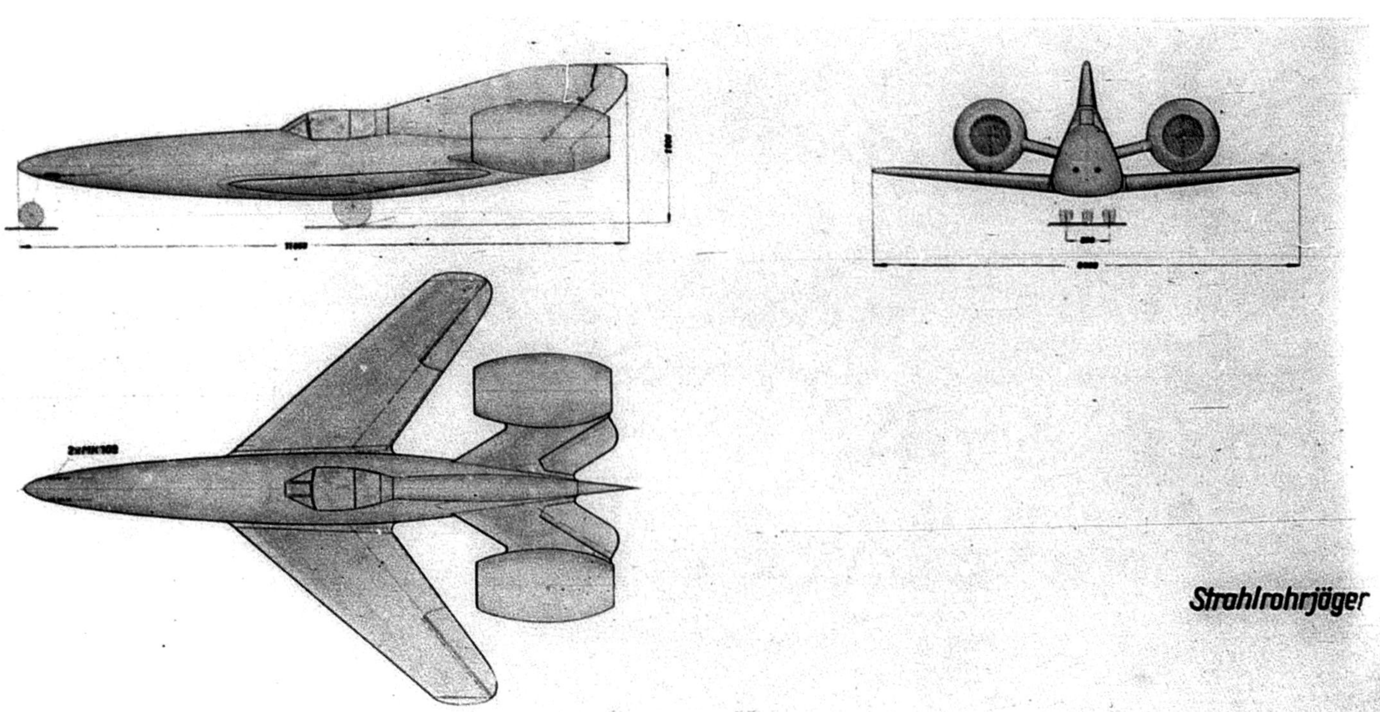

motor. This project originated from a visit of Sänger to Tank. Sänger had built and flown a ramjet motor, but it was no good because it was too long and had high drag and high skin friction.

"In fact the skin friction alone was so high that there was no net thrust. Pabst suggested to Tank that he could do a better job, i.e. get lower drag."

According to a British report,[24] Pabst stated that he had begun work on ramjets in 1941, Focke-Wulf having already been made aware of the technology, and simply kept a close eye on reports of Sänger's work as it progressed. A research station under Pabst was set up at Kirchhorsten in 1943, just to the north of Focke-Wulf's design and research headquarters at Bad Eilsen, and work was begun on combustion problems with the goal of making the ramjet a viable means of propulsion.

The earliest known mention of a '2 Strahlrohr Jäger' or 'two ramjet fighter' in surviving Focke-Wulf company documents is on a number of handwritten papers[25] dated February 11, 1944. No drawing seems to have survived for the type,[26] which was designated X1 and had a wingspan of just 7m with a wing area of 10m². It was to be a single-seater powered by a pair of ramjets and had a take-off weight of 5,000kg. This compares to a take-off weight of 4,900kg for the Fw 190 A-8.

The X2 had followed[27] by April 24, 1944. This had 45° swept wings which, despite retaining a wingspan of 7m now had a much larger wing area of 17.5m². The fuselage appears to have been quite broad and flat with very short undercarriage legs, judging by an existing fuselage cross-section drawing. It was to have been armed with three MK 103 cannon – which appear in a table of weights for the aircraft's components.

By May 16, the sequence had reached the X5 with nothing of the X3 or X4 appearing to have survived.[28] The X5 had a more oval-shaped fuselage, only two MK 103s and a single-rocket motor with 3,000kg thrust for starting from the runway, since ramjets cannot work at a standstill when no air is being 'rammed' into them. Its nosewheel was to measure 380 x 150 and its two mainwheels 740 x 210. Its maximum take-off weight was 5,072kg.

The X6[29] had taken over by June 6, 1944. This had a wingspan of 8m and a wing area of 16.8m², later amended to 19m². Again, it had two MK 103s and a tricycle undercarriage arrangement but now the nosewheel was 475mm x 173mm although the mainwheels remained 740mm x 210mm. A table of weights cites drawing number 0310 246/10-01 but this may no longer exist. The maximum take-off weight is 5,400kg.

By June 13, 1944, the X6 was becoming a much more fleshed-out aircraft design, with details being given for how long the 3,000kg thrust rocket motor would fire for on take-off – 48 seconds – and the shape of both the aircraft's ramjet motor nacelles and its tailfin. It was calculated that it would take 2.4 minutes for the X6 to reach an altitude of 11km (36,000ft), while overall flight time was 50 minutes and range was 800km. It would measure 11.85m long and 2.9m high with a wing sweep of 45°.

On July 5, 1944, the X6 was the subject of a full Focke-Wulf baubeschreibung or construction description –

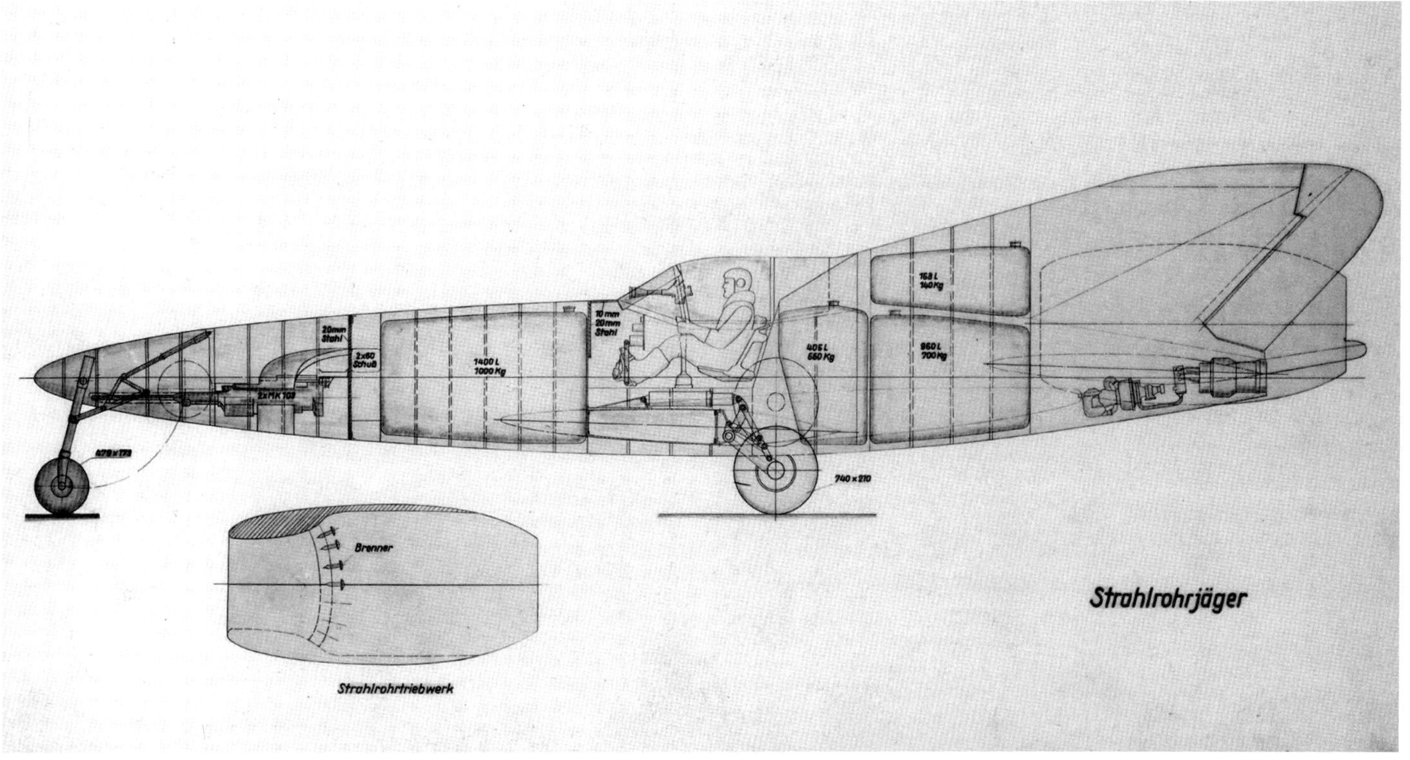

ABOVE AND LEFT: X6 – the finished design, now known simply as the Strahlrohrjäger and described in Baubeschreibung Nr. 283.

ABOVE: Weights of the X6's different component parts.

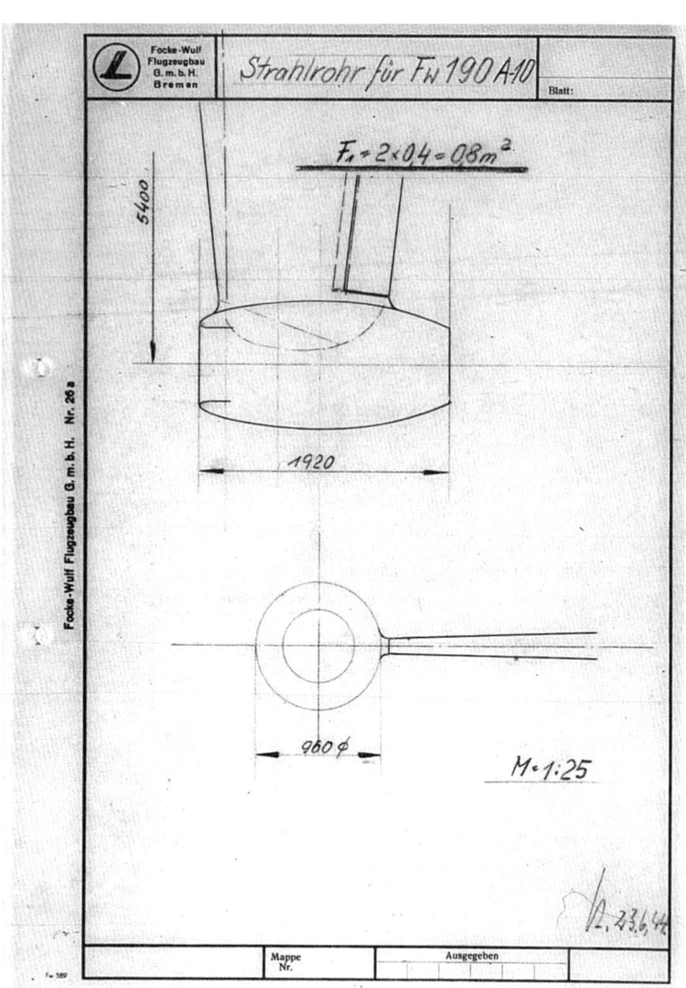

ABOVE: This drawing showing ramjets attached to the wingtips of an Fw 190 A-10 comes from the same file of notes and sketches as the X-series.

Nr. 283 Strahlrohrjäger.[30] The company also produced Baubeschreibung Nr. 290 Ta 152 C and Baubeschreibung Nr. 292 Ta 152 H, indicating that the construction description number was not the project's 'Fw' or 'Ta' number. So the X6 Strahlrohrjäger was never the 'Ta 283' – this postwar designation was the result of a simple misunderstanding about the nature of Focke-Wulf's project numbering system. The company knew the design formally as the Strahlrohrjäger or Baubeschreibung Nr. 283 aircraft and internally as the Strahlrohrjäger or X6 but not the 'Ta 283'. An undated, unsigned Focke-Wulf report on ramjet propulsion projects appears to suggest that the Strahlrohrjäger/Baubeschreibung Nr. 283 aircraft was given the codename 'Diogenes'[31] but no other documents appear to use this name.

Using the name 'Diogenes' in connection with the Strahlrohrjäger might be suggestive of Focke-Wulf's pessimism about the design, since the ancient Greek philosopher Diogenes was known as Diogenes the Cynic.

The introduction of Baubeschreibung Nr. 283 states: "The ramjet fighter is a particularly fast aircraft, with a high ceiling and good climbing power. The use of such fighters calls without doubt for new tactics, which must be developed accordingly. Their high rate of climb and horizontal speed allow their use against bombers with 700km/h average speed, which fighters with Otto engines cannot fight successfully. Their speed superiority of about 500km/h, compared with the normal bomber formations makes pursuit possible to a greater distance than with Otto engine fighters.

"In combat itself, an approach directly from behind is often the most favourable, for it offers the longest time of fire. There is possibility for the installation of two MK 103, four MK 108 or four MG 213. Considering the short duration of attack, density of fire must be preferred to an extended flight path with possible greater range, so that the installation of four MG 213 appears to answer the purpose best.

"Nor must it be forgotten that a synchronising of the speed of the ramjet-propelled fighters with the bomber speed can hardly be possible, because of the limited controllability of their engines."

Shortly before work on drafting a presentable version of this document commenced however, on June 23, 1944, the ramjet designers examined what a Pabst-type ramjet

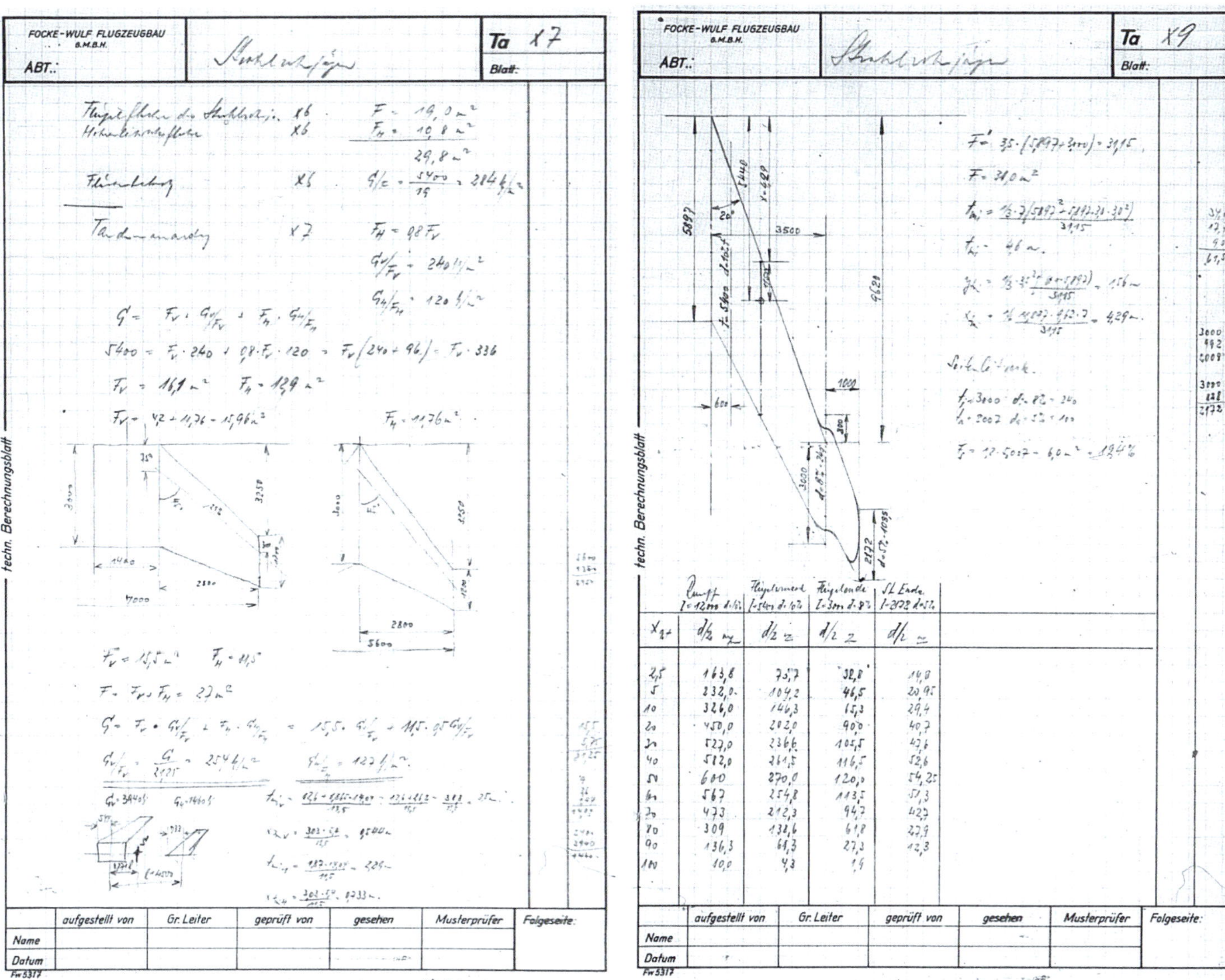

ABOVE LEFT: A progression of the Strahlrohrjäger beyond the X6 to the X7. ABOVE RIGHT: Sharply swept-back wings would have been a feature of the X9 – the last known design in the series.

would look like attached to the wing of a Focke-Wulf Fw 190 A-10.[32] Although Baubeschreibung Nr. 283 was completed on August 4, 1944, an X7 Strahlrohrjäger was then designed which seems to have gone back to the smaller earlier designs of the series, with a wingspan of 7m and a wing area of 15m². It had a maximum take-off weight of 3,600kg and was armed with two MK 103s. Nothing is known of the X8, if there was one, but finally there was the X9 Strahlrohrjäger with a wing area of 31m² and a wingspan of at least 10m. Incredibly, its wings had an 80° sweepback and while a drawing of the wing survives, there is no further visual representation.

FOCKE-WULF TRIEBFLÜGELJÄGER

Work on the fantastic-looking Focke-Wulf Triebflügeljäger (power-wing fighter) tail-sitter with vertical take-off appears to have begun at around the same time as work on the Strahlrohrjäger, if not earlier.

A design for a vertical take-off tail-sitter was first patented on September 10, 1938, by Otto Muck[33] – a prolific Austrian-born inventor – but Focke-Wulf's Triebflügeljäger married Pabst's work on ramjets to a concept invented by Professor Erich von Holst, with input from the AVA's Göttingen-based jet intakes specialist Dr Dietrich Küchemann.

Von Holst, who worked at the Zoological Institute at Göttingen, had entered two model ornithopters for the Second Reichswettbewerb für Saalflugmodelle or Reichs Contest for Indoor Flying Models in Breslau on November 20 and December 1, 1940, and won second prize. While one model was intended to recreate the effect of a bird's two wings flapping, the other was based on the movements of a dragonfly's four wings.

Research around this latter movement formed the basis of a paper published in the *Jahrbuch des Deutsches Luftfahrtforschung 1942* by von Holst, Küchemann and

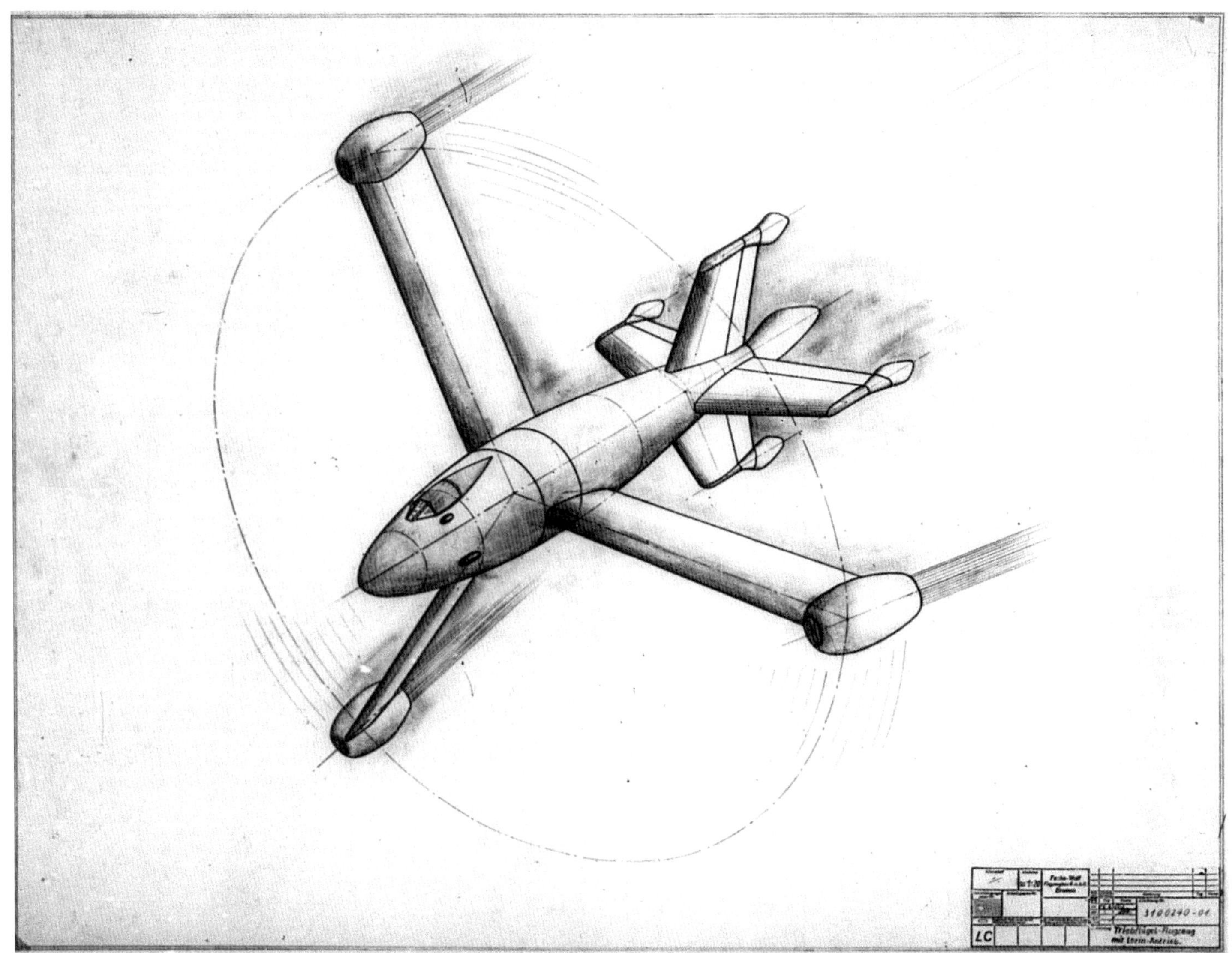

ABOVE: A design quite unlike any other – Focke-Wulf's ramjet-propelled Triebflügeljäger as it appears in drawing 310 0240-01.

K. Solf[34] which looked at how von Holst's models could be scaled up to a full-size aircraft – effectively replacing the flapping dragonfly wings with two counter-rotating sets of helicopter blades.

Pabst published a set of calculations[35] detailing how his ramjet engines might perform when attached to the tips of the von Holst/Küchemann aircraft's helicopter blades in March 1944 and during May Focke-Wulf engineer von Halem worked on determining the best layout for the aircraft. His efforts focused on a cigar-shaped fuselage with three spinning wings attached either at the nose tip, with the pilot positioned behind them; part way back from the nose with the pilot still positioned behind them; or nearly halfway along the fuselage with the pilot positioned in the nose.

The single-seater aircraft's basic form was established by September 15, 1944, and a short preliminary description was produced.[36] The aircraft was designed to sit vertically on the ground, resting on four 380 x 150mm outrigger wheels on the ends of its cruciform tailfins and a single 780 x 260mm wheel housed within a fairing on the tip of the fuselage. Maximum fuselage length, with the mainwheel retracted into its housing during flight, was 9.15m.

Each of the three wings was tipped with a combination ramjet and 300kg thrust solid fuel Walter starter-rocket. Once the rockets had got the wings spinning fast enough the ramjets could be activated and would then take over. Maximum wing diameter was 11.50m and total wing area was 16.5m^2 (3 x 5.50m^2). Fuel for the ramjets – either gas or liquid – totalling 1,500kg was housed within the fuselage. Armour protection was given as 'usual fighter armour' and armament consisted of two MK 103s with 100 rounds each and two MG 151s with 250 rounds each in the aircraft's nose – one of each on either side of the pressure cabin cockpit.

A second description[37] was produced on October 12, 1944, which altered the aircraft's dimensions slightly – fuselage length increased to 9.35m while wing diameter was reduced to 11.40m but wing area remained 16.5m^2.

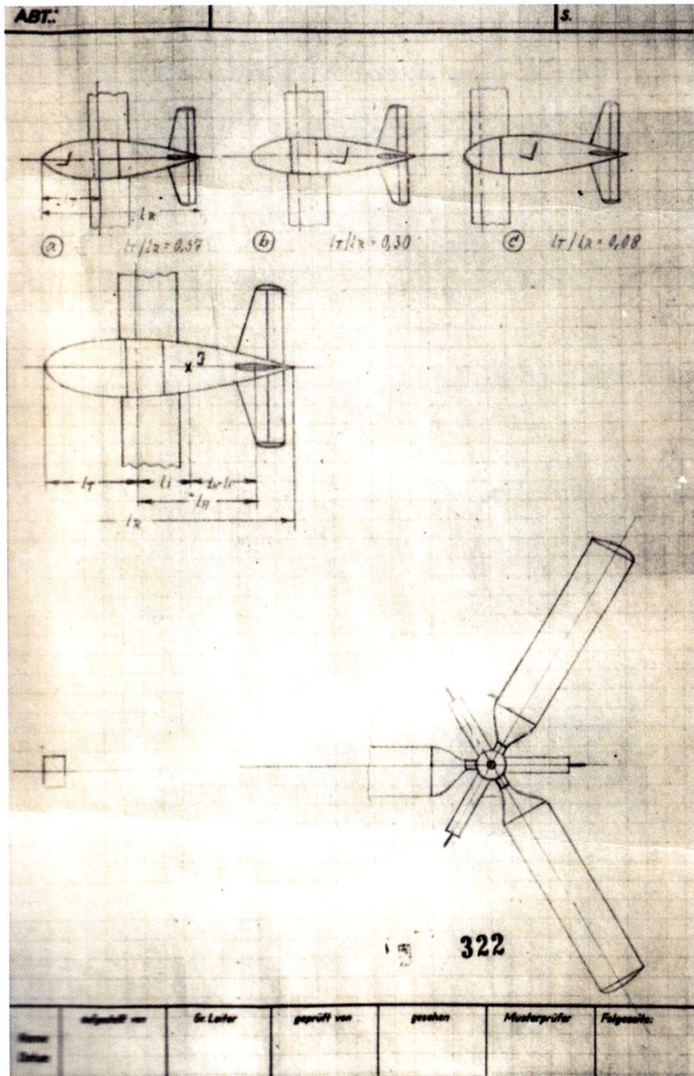

ABOVE: Sketches showing the three different layouts considered for the Triebflügeljäger, with the pilot positioned either in the nose, in the rear behind the wings or in the centre behind the wings.

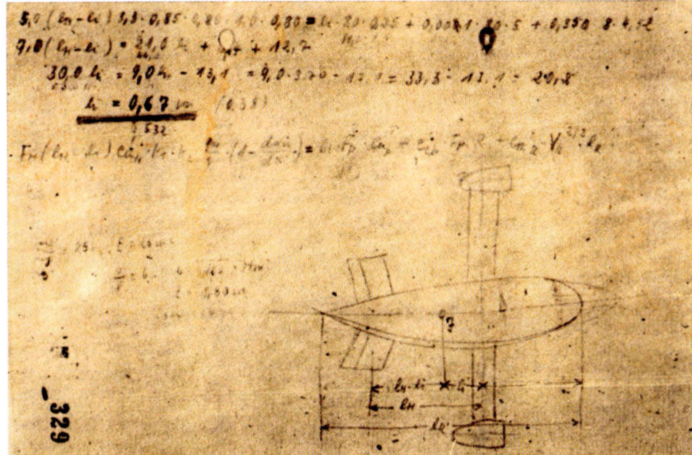

ABOVE: Early design sketch showing the Triebflügeljäger's podded ramjets on the ends of its wings.

RIGHT: A neatened-up version of the three layout sketches from a file of notes written by Focke-Wulf engineer Heinz von Halem.

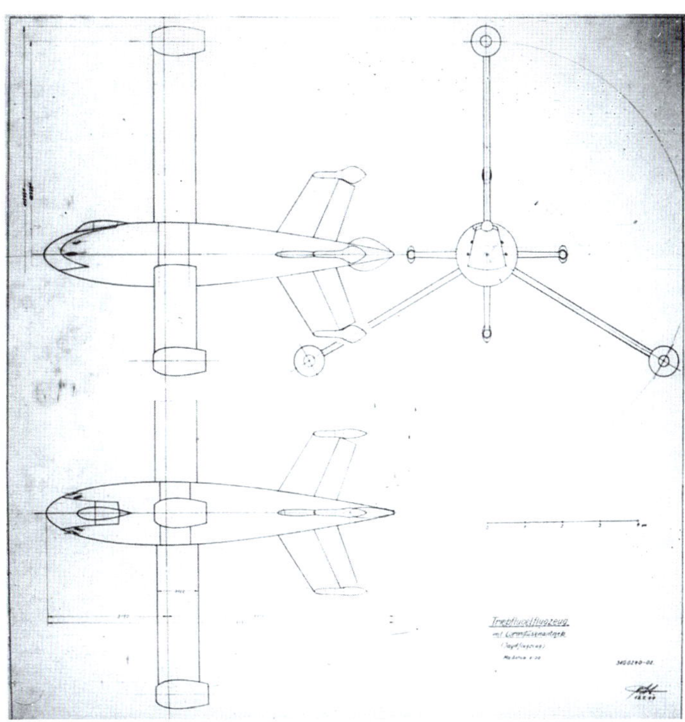

ABOVE: Focke-Wulf drawing 310 0240-02 of July 1944 shows a three-view of the Triebflügeljäger with its distinctive cruciform landing gear.

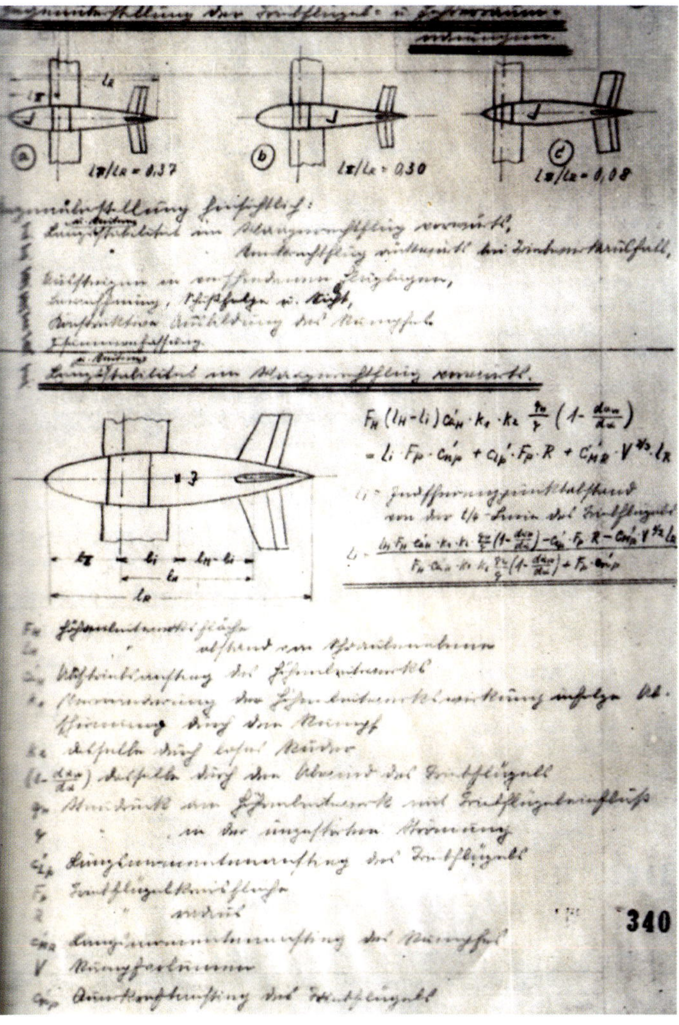

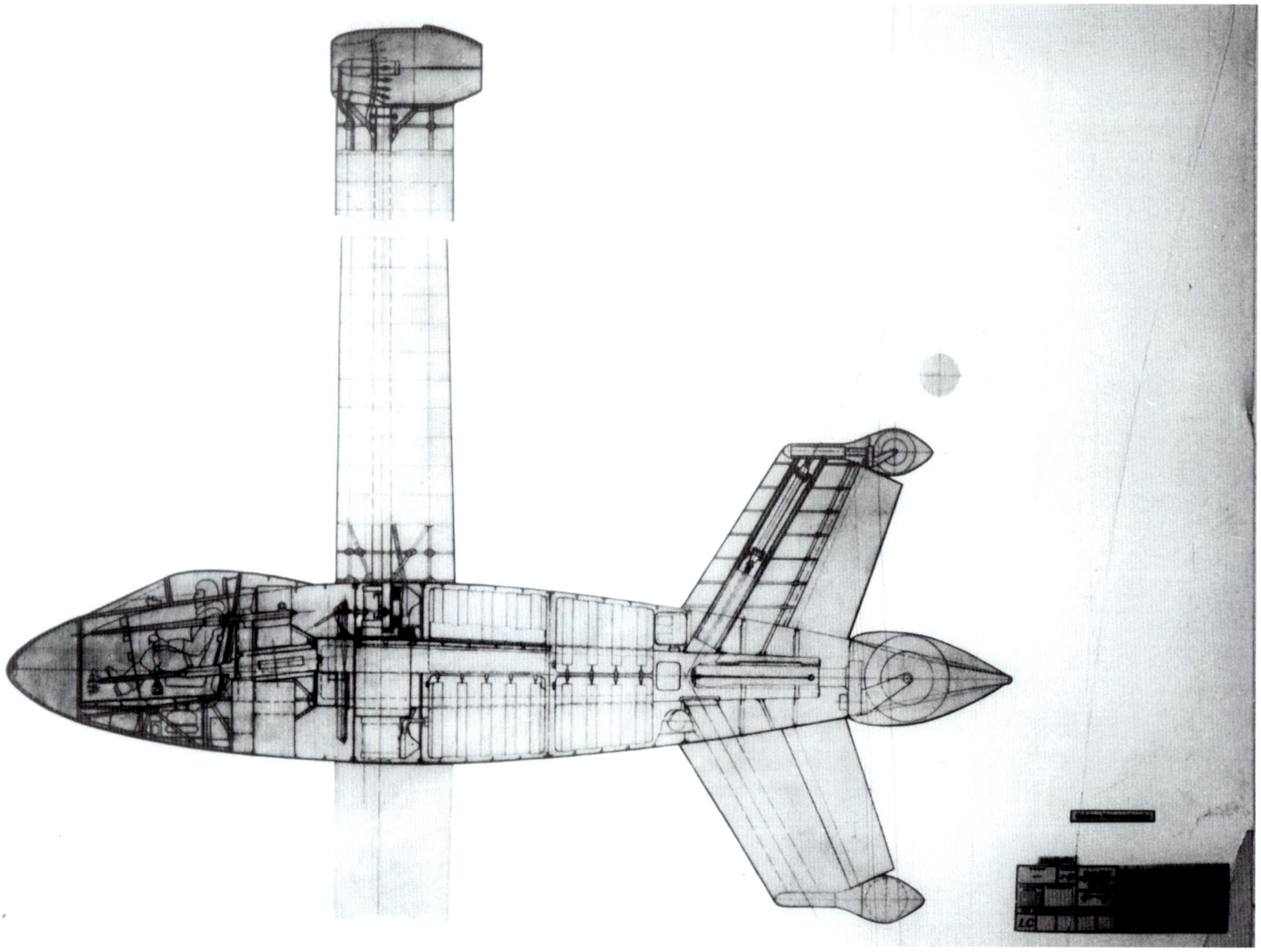

ABOVE: Another drawing from the 310 0240 sequence, presumably 310 0240-03, this time showing the internal structure of the Triebflügeljäger, complete with angled cannon positioned around the pilot.

According to a description in a British intelligence report.[38] "In this aircraft the normal wing is replaced by three rotating 'wings' or vanes each with an athodyd (Lorin duct) at the tip. The 'pitch' of the 'wings' can be adjusted by the pilot and the maximum peripheral speed which is normally used for climb is 670ft/sec (455mph). It will be appreciated that with such an arrangement the air speed of the athodyds may be considerably greater than that of the fuselage.

"In this way efficient operation of the athodyds can be achieved when the aircraft speed is relatively low and one of the main objections to this form of propulsion is thus overcome. The higher the forward speed of the fuselage the slower the rotation of the 'wings' and the smaller the difference between athodyd speed and fuselage speed.

"For take-off the aircraft stands vertically on its tail (in which the wheels are housed) and initial rotation is imparted to the 'wing' assembly by three auxiliary Walter rockets incorporated in the athodyds. As soon as the peripheral speed appropriate to athodyd operation has been attained the main units are started up. During this stage the 'wings'

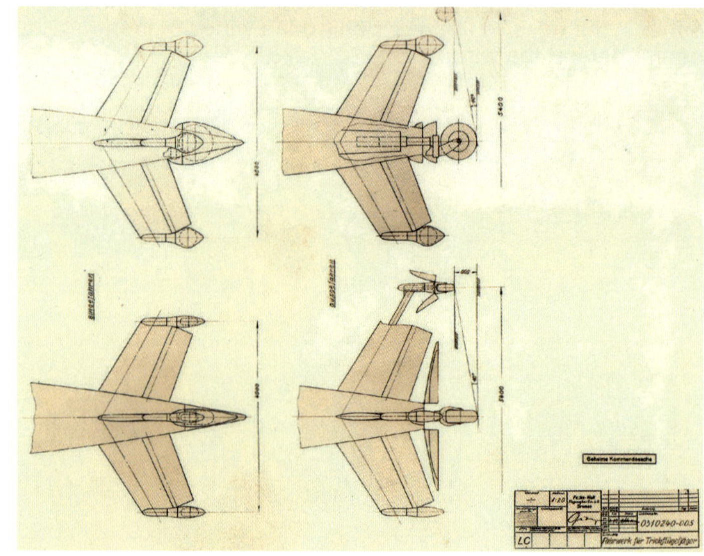

ABOVE: The last known drawing in the Triebflügeljäger series - 0310 240-005, dated October 4, 1944. It shows the type's distinctive undercarriage with one large central wheel in its own streamlined housing and four smaller wheels on extending legs.

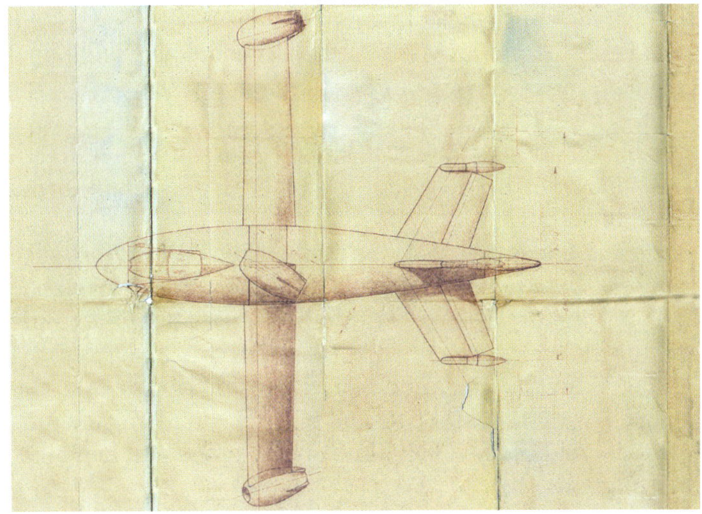

ABOVE: Top view of the Triebflügeljäger from **0310 240-004.**

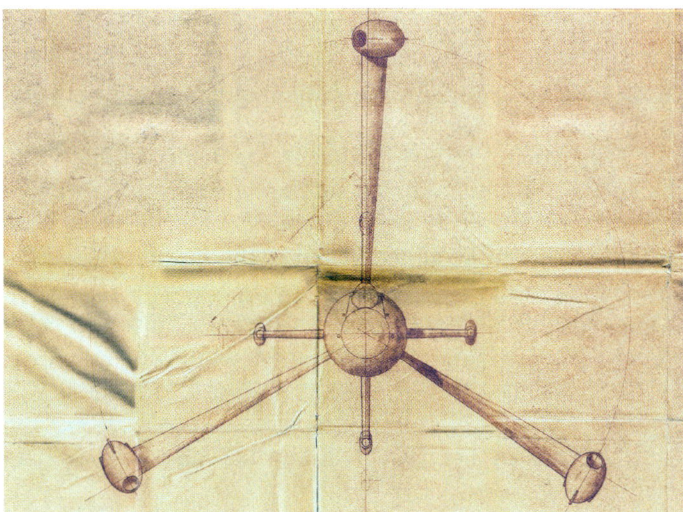

ABOVE: Forward view of the Triebflügeljäger.

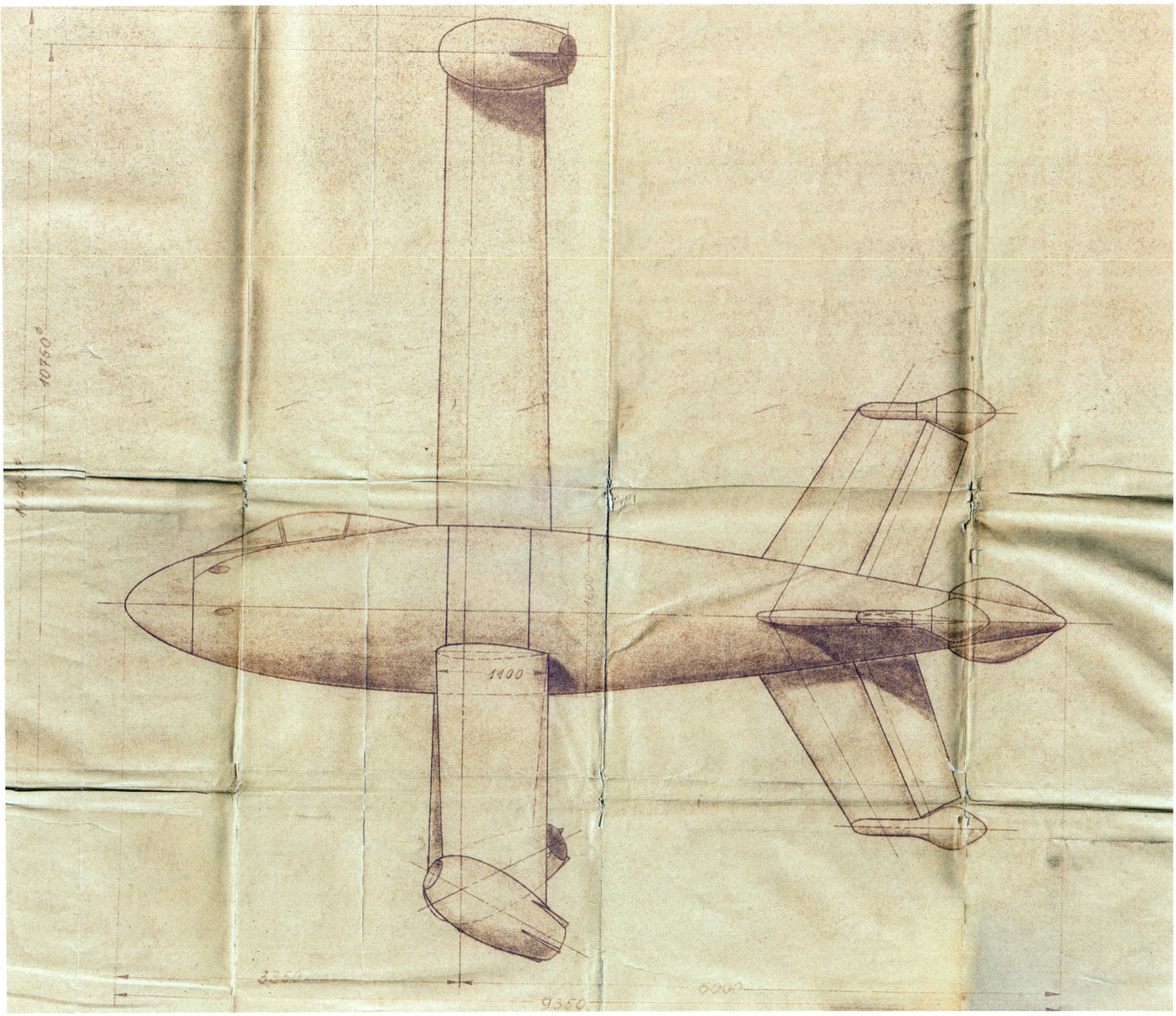

ABOVE: Detail from Focke-Wulf drawing **0310 240-004**, dated **September 30, 1944** – another three-view of the Triebflügeljäger.

are in neutral pitch. When they are moved into fine pitch lift is imparted to the aircraft partly due to the rotation of the wings and partly due to the components of athodyd thrust parallel to the fuselage axis.

"After levelling out, the pitch is further coarsened and the speed of rotation correspondingly reduced in order to maintain a constant Mach number of 0.9 at the wing tips. At the maximum designed speed of the aircraft the wings rotate at 220rpm."

The Triebflügeljäger was claimed to have six key advantages: "1) High efficiency and low fuel consumption. 2) High ceiling. 3) No runway required for take-off or landing. 4) Low weight. 5) Simplicity. 6) Any combustible medium (gas or liquid) which can be vaporised may be used as fuel."

The project appears to have gone no further after October 1944.

RAMJET-ASSISTED ME 262

At the end of 1944, the RLM asked Sänger to resume his ramjet work but to take it in a different direction. The DFS was, according to Sänger's report[39] of January 31, 1945, given a contract to "draw up designs and performance calculations on a possible increase in performance of the Me 262 with the help of additional Lorin engines".

He wrote: "It is our opinion that the Lorin engine is very suitable as an attachment to finished airframes, although these may not be designed to perform well close to the speed of sound.

"Nevertheless, the task was undertaken as a compromise solution to giving the Me 262 better climb performance and a higher ceiling with the least possible time and effort."

Sänger states that much of the work detailed in the report – which mostly consists of graphs and calculations – was carried out by Dipl. Math. W. Peterson and Dipl. Ing. W. Lungstras.

The conclusion of the report was that the Me 262 could be made to go faster but at a cost of dramatically increased fuel consumption.

SKODA P14

When Lippisch and Focke-Wulf got to work on their ramjet fighter designs, Sänger's original designs had been left on the shelf. Then, as Sänger was working on the ramjet-

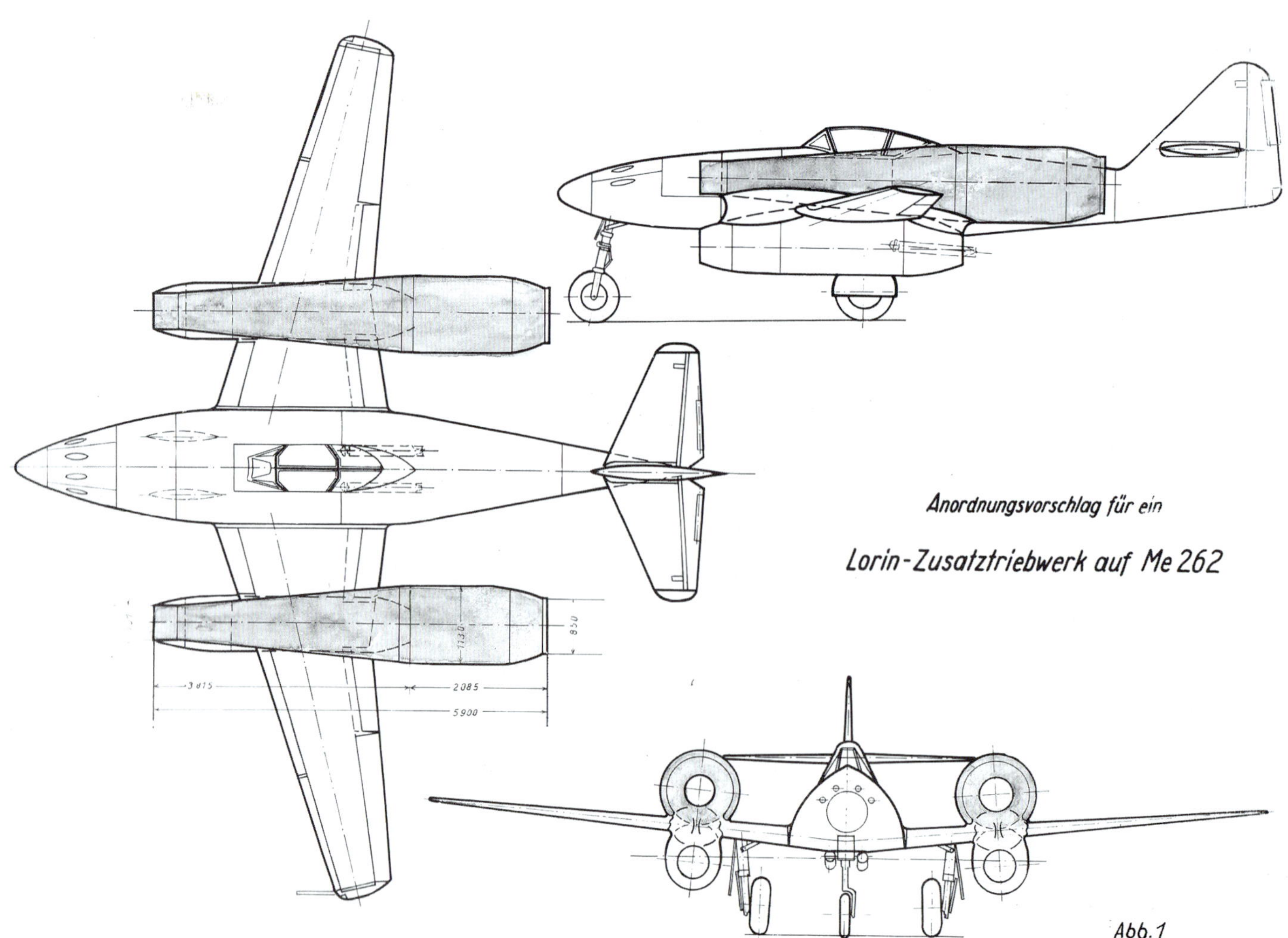

ABOVE: Messerschmitt Me 262 with ramjet tubes as it appeared in Sänger's report of January 31, 1945.

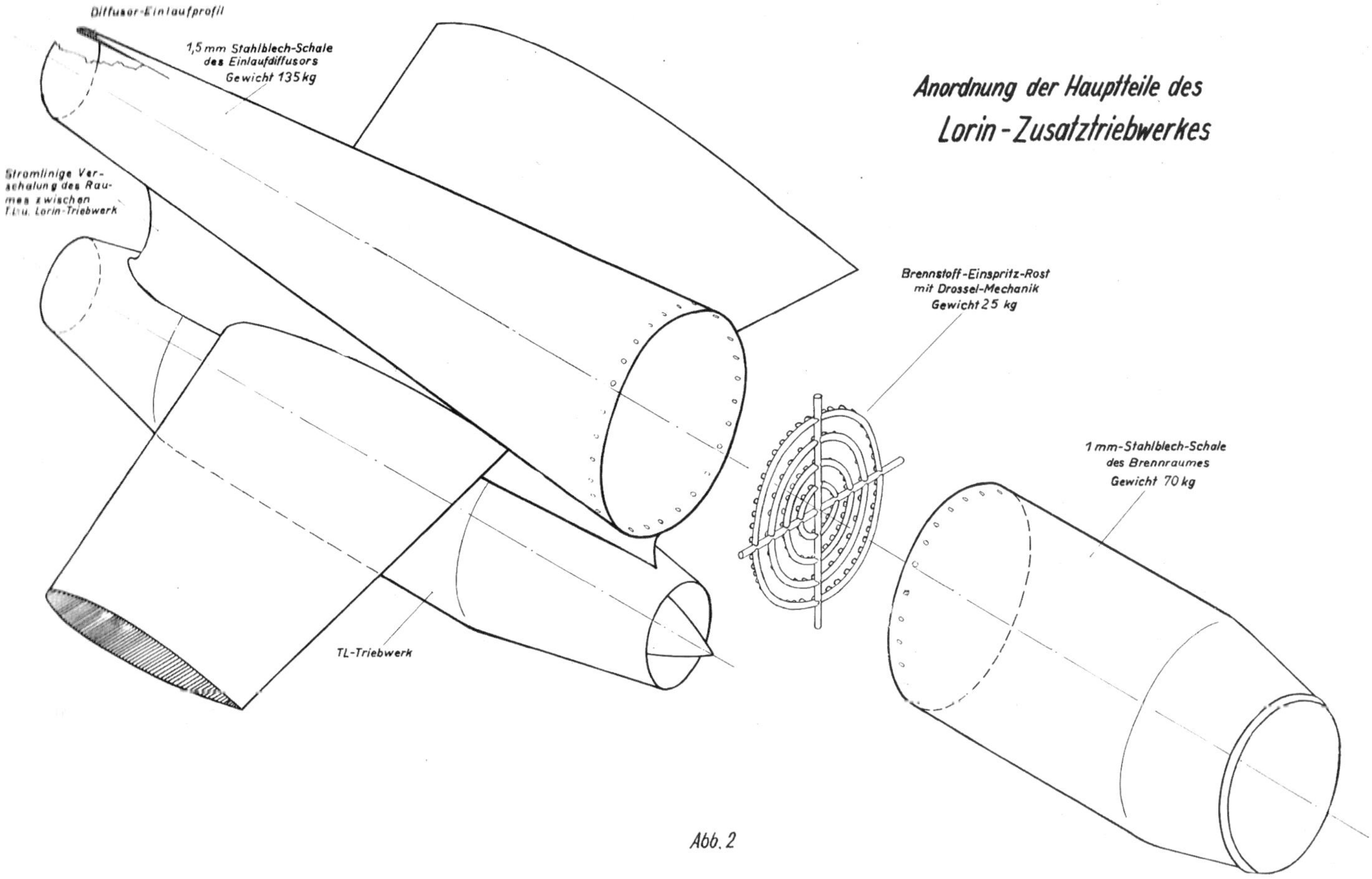

Diffusor-Einlaufprofil

1,5 mm Stahlblech-Schale
des Einlaufdiffusors
Gewicht 135 kg

Anordnung der Hauptteile des
Lorin-Zusatztriebwerkes

Stromlinige Verschalung des Raumes zwischen T.L.u. Lorin-Triebwerk

Brennstoff-Einspritz-Rost
mit Drossel-Mechanik
Gewicht 25 kg

1 mm-Stahlblech-Schale
des Brennraumes
Gewicht 70 kg

TL-Triebwerk

Abb. 2

ABOVE: Diagram showing how a Sänger ramjet could be fitted above the wing of an Me 262.

boosted Me 262 in January 1945, the RLM commissioned both Skoda-Flugzeugbau and Heinkel to each design a new fighter based on Sänger's work. Sänger reportedly gave each company a different layout to work on.

According to an American intelligence report of June 21, 1945, on the DFS at Ainring:[40] "Dr Sänger ... was head of Triebwerkes Abteilung T-2, and was directly responsible to Professor Georgii. Triebwerkes Abteilung T-2 was concerned with development of athodyd propulsion units known in Germany as the Lorin Triebwerk. His department consisted of a total of 20 scientists, technicians and mechanics."

The report outlines the type and size of ramjets Sänger and his team were working on then states: "At the direction of RLM, Dr Sänger's group was in the process of designing a new large athodyd, previously referred to, to be utilised in conjunction with a high performance fighter aircraft.

"The performance requirements demanded the transporting of 1,000kg of armaments equipment from sea level to 12,000m in two minutes. It was further desired that the level flight endurance at altitude be of appreciable duration.

"Based on Dr Sänger's calculations, two airplanes were laid out, both of 5,000 to 6,000kg gross weight, carrying the aforementioned armament loads plus 2,400kg of fuel. Of this fuel weight 600kg was required for climb to 12,000m and the balance would give the airplane one hour endurance at 12,000m when flying at a Mach no. between 0.7 and 0.8. Of the two airplanes laid out, the one airplane was designed by Skoda in Prague, and the other by the Heinkel organisation.

"The Skoda design was in a more advanced stage than Heinkel, although both designs were originated in 1945 and could hardly have progressed beyond the preliminary

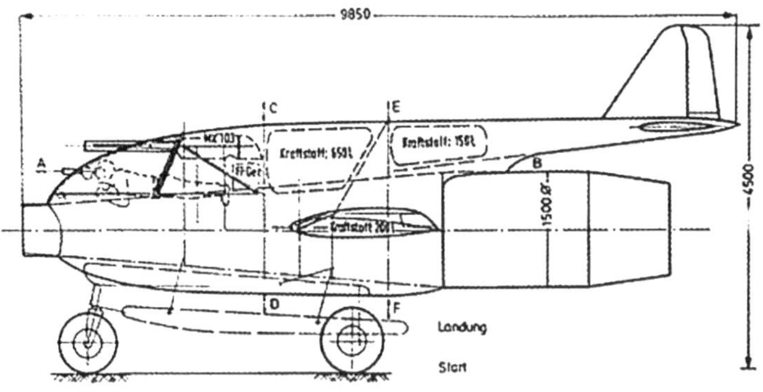

ABOVE: Skoda P14 design with separated tail boom.

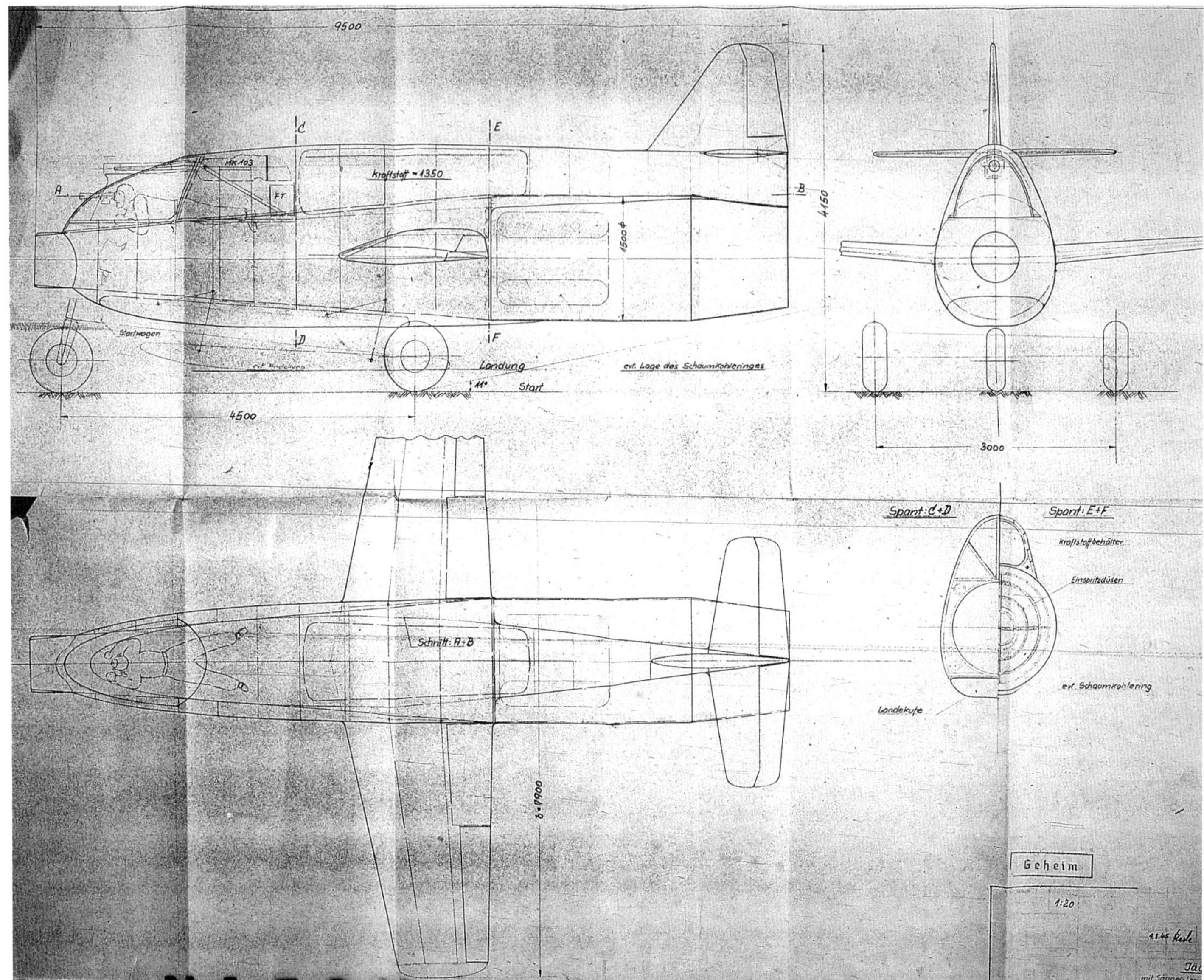

ABOVE: The Skoda P14-02, shown in a drawing dated March 9, 1945, very closely followed the design of Sänger's Strahljäger.

design stage. Discussions on the Heinkel airplane took place in Jenbach, Tyrol works with Dr Günter, but no evidence has been found to indicate that the actual design work was under way there."

Sänger evidently told his interrogators that both aircraft would have to overcome some problems specific to ramjet-powered aircraft: "For starting and take-off of the airplane, the use of solid fuel assisted take-off rockets of conventional design was projected. An interesting point brought out by Dr Sänger was that an athodyd is no longer capable of providing sufficient thrust to accelerate or maintain constant forward velocity of the airplane at low forward speeds, as for example 170km/h on the proposed aircraft under design.

"The designed landing speed for these aircraft was 120km/h. The landing characteristics of the airplane would accordingly have been very poor. The institute was ordered to develop a coal-burning version of the athodyd, and the Skoda airplane was being designed on this premise.

"While work was going ahead on this basis, Dr Sänger's group considered the arrangement very inferior to liquid fuel, and such developments in this direction as were carried on were apparently quite half-hearted in nature."

Skoda-Flugzeugbau's project was outlined in a report[41] produced on March 23, 1945. The aircraft itself, shown in drawing SK P14-02, was remarkably similar in layout to Sänger's original design – being essentially a 1.5m-diameter ramjet tube with bits of aircraft attached to it. The upper portion housed the prone pilot in the nose with a single MK 103 firing right above his head. In the centre of the 'fuselage' was a fuel tank containing 1,350 litres and parallel with the ramjet exhaust was a conventional tail. Overall length was 9.5m and wingspan was 7.9m with a wing area of 12.5m^2. The aircraft used a tricycle trolley fitted with rockets to take off and a belly skid for landing.

According to the project summary: "The aircraft design Skoda-P14 is a fast fighter with Sänger-Lorin jet engine, which is particularly characterised by short ascent times of a few minutes to operational altitude and level flight times of over half an hour at altitude. Due to the thrust of the engine ... the aircraft has particularly favourable top speed performance and ... the mentioned good climbing performance and high ceiling ... in conjunction with a pressurized cabin makes it appear particularly suitable as a high-altitude fighter at 10 to 15km altitude.

"The simplest design and particularly low costs for the airframe and engine components that have been developed into a single unit are further special features of the design. In addition, engine tests have already started to replace half of the fuel consumption of 1,200kg B4 or I2 for the three-quarter-hour flight with Schmitt system [i.e. replacing half the fuel with powdered coal] so that the need for liquid fuels per flight can be assumed to be around 600kg medium oil.

"The aircraft is a middle-decker design with a continuous box spar. In order to get by with the smallest possible fuselage cross-section compared to the main frame of the combustion chamber given by the selected engine, the prone pilot arrangement was chosen, which seems justified, since the aircraft has only a total flight time of 1 hour max.

"The actual load-bearing structure of the fuselage is the diffuser of the jet engine (pipe carrier). Above the front part of the diffuser is the cockpit with the weapon storage, then the fuel tank, separated by a fire wall. The removable cylindrical combustion chamber of the engine is flanged to the inlet diffuser. Above the combustion chamber and outlet nozzle, the tail boom supports extend in the symmetrical profile placed on the engine, at the end of which is the tail in the usual arrangement.

"The spring-loaded landing skid of the aircraft, which can be extended by means of dynamic pressure, lies under the diffuser. The start is made by launch rockets, aircraft towing or similar on a three-wheeled starting car (nose wheel landing gear), from which the machine detaches itself when it takes off."

The following day a covering letter[42] was written for the P14 description which compared the design's projected performance against that of the ramjet-boosted Me 262 based on calculations made on March 13 and signed off by an engineer called Fleck. This gave the P14's service ceiling as 18.5km compared to 11.5km for the normal Me 262 and 15km for the ramjet-boosted Me 262. Fuel consumption to reach that altitude was just 875kg for the P14 compared to 1,000kg for the Me 262 and an incredible 2,100kg for the

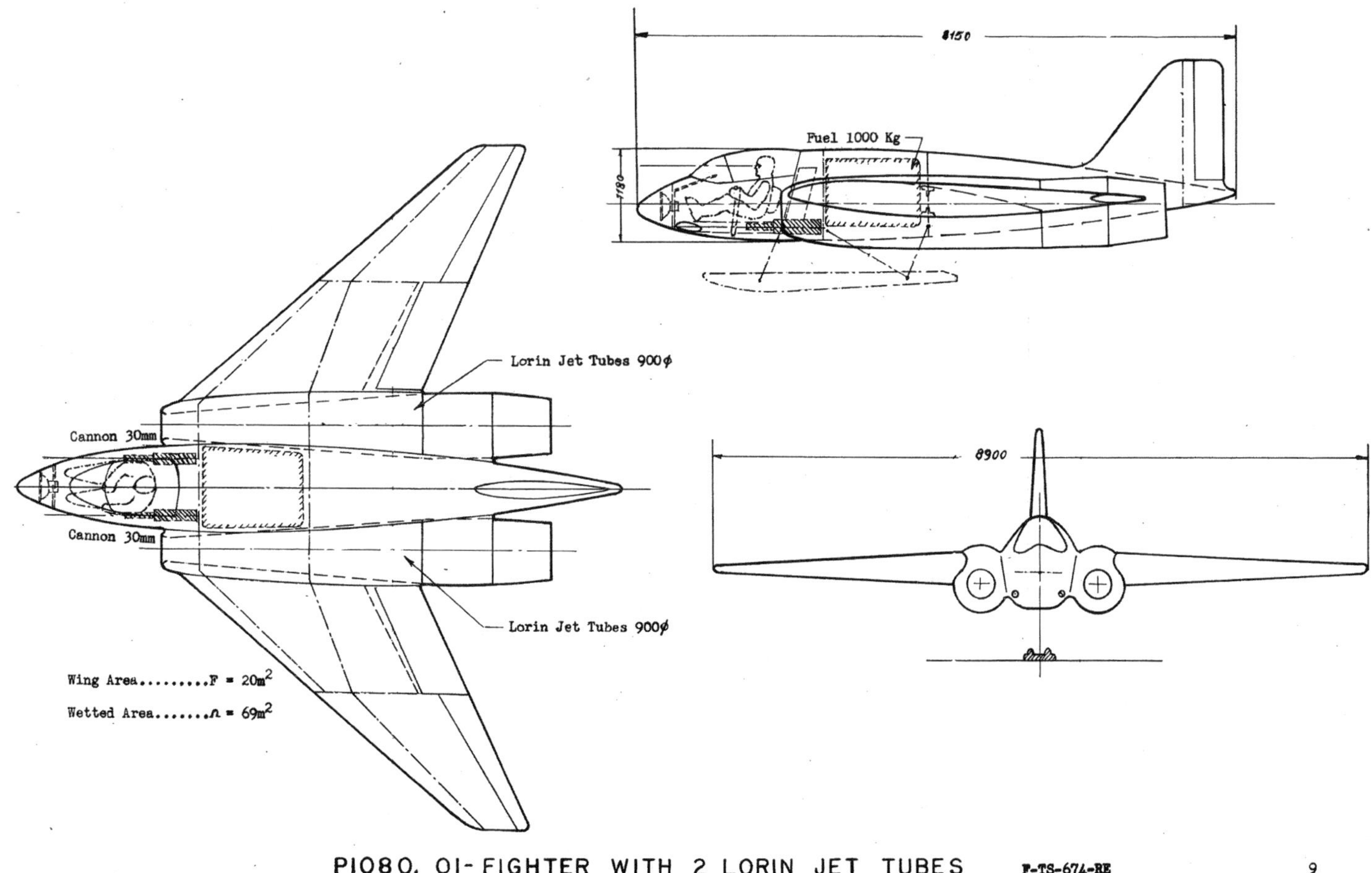

ABOVE: Heinkel's final project of the war – the tailless P 1080. It was a compact design based on two small ramjets rather than one large one.

ramjet 262. And time to climb to 10km was three minutes, 26 minutes and 6.3 minutes respectively. The report concluded that the P14 would have 43 minutes of flying time remaining at 10km altitude on this basis, compared to 38 minutes for the normal Me 262. The ramjet 262 would by now have used all its fuel.

Prague held out longer than any other German-held territory at the end of the Second World War – the Soviets launching their Prague offensive on May 6, 1945, and finally mopping up the last pockets of resistance on May 11. It is unclear exactly how far Skoda got with its work on the P14 but it is unlikely that work on a prototype was begun.

HEINKEL P 1080

Similarly, exactly how far Heinkel got with its ramjet fighter design, designated P 1080, before the war ended is uncertain – though the end came for Heinkel's designers much sooner than it came for Skoda's. What is known about the P 1080 – Heinkel's last wartime project – comes from a report[43] written in German by company chief designer Siegfried Günter while interned with a handful of his remaining staff at Penzing airfield long after the war ended.

According to the report: "Shortly before the end of the war the project design office of the Heinkel firm received data concerning the Lorin jet developed by Dr Sänger from the RLM. The RLM ordered the design of a single seat fighter with this powerplant."

He then outlines the same take-off and landing problems mentioned by Sänger during his interrogation, stating that the P 1080 would use four 500kg solid fuel rockets providing 1,000kg thrust each for 12 seconds attached to a take-off trolley to get airborne, then the same number of rockets attached to the aircraft itself would enable it to climb up to the ramjet's operating altitude and speed.

The next section of the report is headed 'comparison of Lorin-jet with turbojet' and begins: "In order to obtain a judgement of the possibilities of the new propulsion power plant, a comparison with the turbojet engine is computed. Basis for the computation: as an example for a turbojet airplane the tailless P 1079 B [although this appears to be an error – the aircraft referred to throughout is in fact the P 1078 B – a design believed to have been produced after the war ended] with HeS 011 (static thrust = 1,300kg) is used. The weight of structure, military load, mean flight weight, wing area and drags were likewise taken from the tailless airplane P 1078 B. The computation was accomplished for several very different modifications of the Lorin and the turbojet power plants.

"As a base for these power plants, the Lorin tube suggested by Dr Sänger and the HeS 011 were chosen.

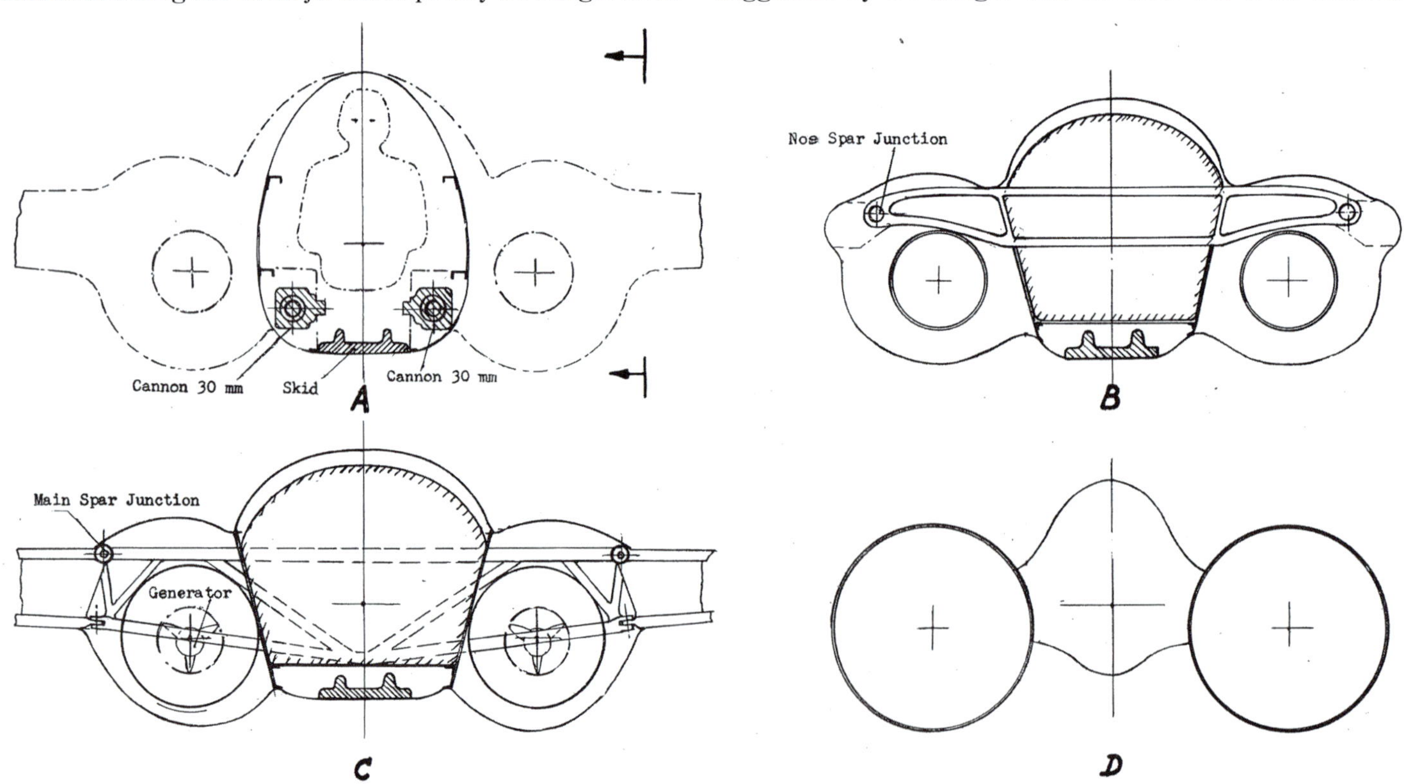

ABOVE: The diminutive size of the **P 1080** is illustrated by this sectional view – showing the pilot, cannon and ramjets all positioned within a short distance of one another.

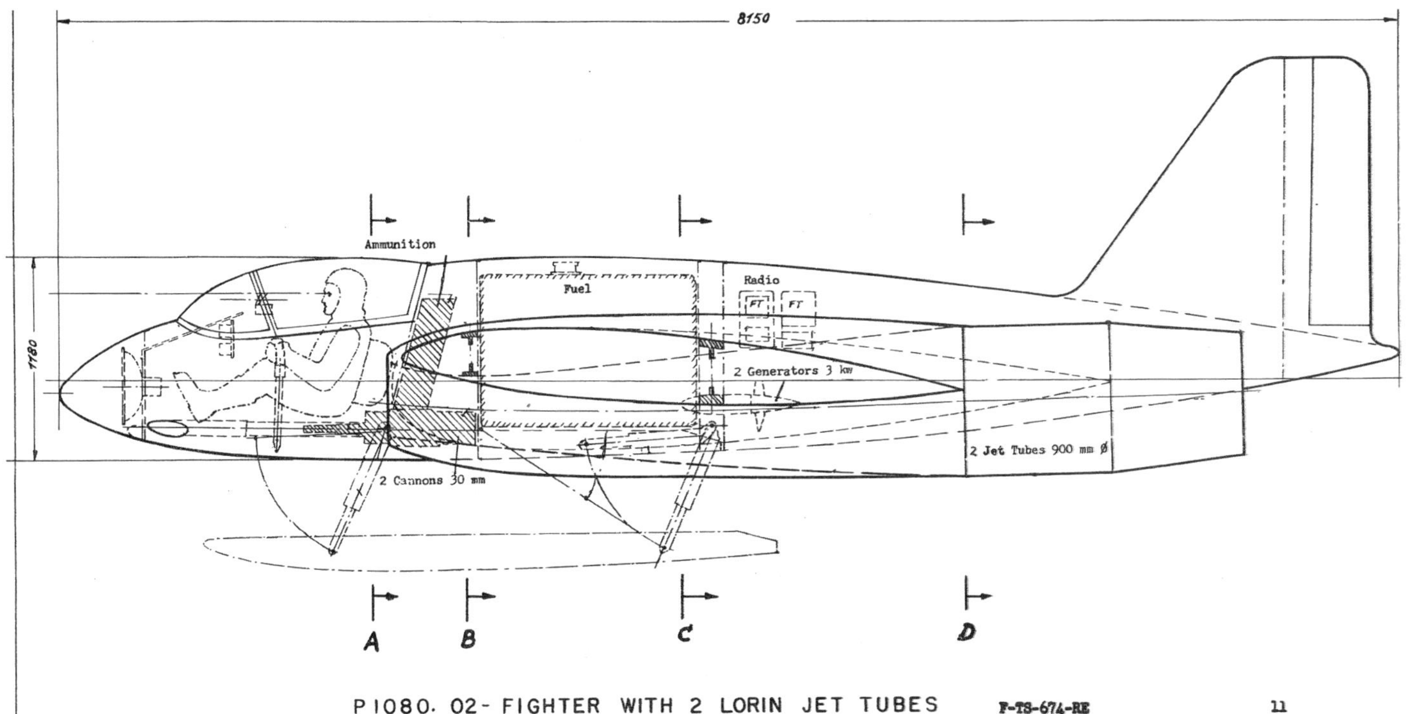

ABOVE: The P 1080 would have been a relatively simple aircraft, utilising what appears to be an He 162 canopy and cockpit but with a different nose.

The weight per thrust unit of these power plants was retained for the modifications with changed thrust. By these reasons the different thrusts of power plants result in different fuel weights."

A full comparison of performance is then made between the tailless P 1078 B and the tailless P 1080 – presumably to eliminate any question of performance being affected by the type's layout and avoiding any discussion concerning the merits of a tailless design.

Günter concludes: "The airplanes with turbojet engines attain essentially smaller climb performances. By enlarging the turbojet engine, the climb performances are only a little improved while endurance is considerably decreased.

"By reducing the turbojet engine there is a great gain of endurance with a great decrease of speed. The latter airplane also needs rockets for take-off, but essentially smaller ones. The loss of endurance because of the fuel consumed for take-off is unimportant for turbojet airplanes.

"The comparison shows that airplanes with Lorin-jet propulsion are eminently suited for flying in the stratosphere. However, it must be kept in mind that the data of the Lorin-jet are not yet sufficiently tested."

In the following section, headed 'remarks regarding the design', he says: "There are two important advantages of the design fitted with two Lorin-tubes as compared to the design with one tube. First, the favourable installation of the cockpit and the better vision. Secondly, two small tubes result in smaller unstable moments about the normal axis.

"The tubes are placed at the most aft possible position in order to keep the hot part of the tube free of structure details and cowling, and to get effective cooling. A Lorin-jet-powered aeroplane is not able to accelerate again in the case that the landing cannot be completed (because of landing errors or emergency). Thus, decreased landing speed is urgently necessary.

"Power required for pumps, compressor, armament and radio is furnished by two generators installed inside the Lorin-tubes. The considerable decrease of air speed in the diffuser of the Lorin-jet improves the efficiency of this device.

"When starting, the skid is retracted. The springs of the retracted skid legs are compressed. It is then possible to extend the skid without additional power."

This was where the very brief text of the report ended. No mention was made of armament, but the drawings provided showed the P 1080 fitted with a pair of 30mm cannon. The P 1080's wingspan was shown as 8.9m, length was 8.15m and wing area was 20m². Its ramjets were each 90cm in diameter.

It is tempting to suggest that the design worked on by Heinkel was superior to that worked on by Skoda if only due to the deceptive simplicity of its tailless layout – particularly the clever way in which it enables the inclusion of a seated rather than prone cockpit by using two smaller ramjets rather than one large one. It was also capable of carrying heavier armament while remaining relatively compact in size thanks to its departure from the strictures of Sänger's 1943 configuration.

In the end, none of the ramjet projects of 1944 and 1945 had any chance of becoming even a working prototype – the time remaining was far too brief and the technology far too new and untested.

Chapter 12

Night and bad-weather fighters

2-TL-Jäger 1944–1945

Efforts to provide the Luftwaffe with effective new night fighters began in earnest during the late summer of 1944. Requirements initially encompassed a new high-performance piston-engine type and interim jet fighters before attention switched to a next-generation twin-jet design shortly before the war's end.

Germany's night fighter force going into 1944 largely consisted of repurposed heavy fighters and bombers such as the Messerschmitt Bf 110, Dornier Do 217 and Junkers Ju 88. The new Heinkel He 219 was not a sufficient improvement over the Ju 88 G to be worth manufacturing in large numbers and the Focke-Wulf Ta 154 was suffering from developmental issues.

The earliest known evidence that jet aircraft were being considered for night-time operations comes from a meeting about the Arado Ar 234 held at E-Stelle Rechlin[1] on December 13, 1943, between four engineers from the test centre and Arado engineer Hans Rebeski. Right at the end of the meeting, after various aspects of the design had been discussed, it was noted that according to a letter from the General der Kampfflieger Generalmajor Dietrich Peltz, the Ar 234 B and C were being considered for night operations.[2]

A discussion about Ar 234 project work took place during another meeting,[3] this time at Arado's Landeshut facility on June 3, 1944, and it was stated that the RLM's Technisches Amt had required changes to the Ar 234 C's pressure cabin design in order to make it easier to use for blind flying – e.g. operations at night and during bad weather.

The same meeting heard that two fighter versions and two heavy fighter versions of the aircraft were being considered but none of them for operations at night.

The earliest known solid evidence of plans to build a jet night fighter appears to come from a different company – Messerschmitt. On July 17, 1944, Herr Jung of the firm's Abteilung Flugerprobung Gruppe Leistungen or 'Flight test department – performance' conducted tests to determine how visible the engines of an Me 262 were likely to be a night.[4]

He recorded the occasion for the tests as "At various requests, an assessment of the engines should be carried out at night, for eventual night operations."

The experiments were carried out with the aircraft (the W.Nr. is given as '262 015' – presumably 130015 aka. V015) stationary on the ground under visual observation. Tests began at 11.45pm on a "starry night without moonlight".

The engine speed at which "the strongest light is seen from behind, was found to be full throttle = 8,800rpm. At this speed, a faint glimmer can be seen from behind from 900m. Close up you can see a ~2m-long building light. In addition, strong flying sparks can be observed. The thruster inside glows bright red. The needle dark red.

"The partially incandescent nozzle is only visible at speeds below 8,000rpm (needle on). When observing from a distance of 50m, the brightest fire cannot be seen directly from behind; in this case you can only see the glowing needle, but outside at an angle of 10° in the longitudinal direction; from there to ~ 30° you can see the bright red thruster inside. A firelight can be seen 20° forwards in the transverse direction from the next distance. No light is visible from the front. There is no reflection on the cabin window from the glowing needle."

The destruction of the factory that made the glue used to hold the wooden Ta 154 together meant that by the beginning of August it was scheduled for cancellation but on August 4 Focke-Wulf recorded preliminary technical specifications for a new bad-weather day and night fighter.[5]

The required top speed was 800km/h "with additional turbojet about 60 to 80km higher speed desired". Minimum ceiling was 13km but 15km with exhaust gas turbocharger was desirable. Endurance was four and a half hours at full throttle.

By comparison, top speeds for the Ta 154 and He 219 were 635km/h and 670km/h respectively.

The new night fighter was to have a crew of three – like the Ju 88 G – a pilot, radio operator and gunner. Armament was three MK 103s or MK 108s with Lichtenstein SN-2 target intercept radar. The aircraft would additionally have two MG 213s or two MG 151s in a remotely controlled turret.

General requirements were: pressure cabin, wing de-icing, nosewheel undercarriage arrangement, "as smooth a wing as possible without interference from the engine nacelles (laminar profile, wheels in the fuselage etc.). The possibility of taking 2 x 500kg bombs on the wing must be available". Finally, the spec sheet stated: "Assessment: Flight characteristics come first, speed performance second."

Three and a half weeks later, on August 28, the Chef TLR's FL-E 2 department formally sent out a new 'Zerstörer' (heavy fighter – literally 'destroyer') requirement to Blohm & Voss, Dornier, Focke-Wulf, Heinkel and Messerschmitt[6] – although it was clear that in this instance the desired aircraft was a bad-weather and night fighter, rather than another daytime destroyer in the Bf 110 mould.

The requirement began: "The development of a new destroyer to combat enemy fighter jets in bad weather conditions during the day and at night is urgently required ... Purpose: bad-weather destroyer for day and night with Lichtenstein complete blind-flying ability including take-off and landing with the best possible flight performance including flight duration."

Engine options available were the Jumo 222 E/F, As 413, DB 603 L or DB 613 although their "number and arrangement are not specified to ensure high flight performance. Combination piston engine/turbojet is to be tested".

Again, a crew of three was specified – pilot, radio operator and navigator/defensive gunner. Armament was four forward-firing MK 108s with 100 rounds per gun plus two MG 213s with 250 rounds each. These weapons were to be used in conjunction with the Lichtenstein system but the installation of a semi-fixed weapons system with Bremen radar had to be possible.

In addition, the aircraft was to carry a pair of MK 108s in its fuselage pointing upwards at a 70° angle plus two MG 151s with 250 rounds each in an FHL 151 turret – for a grand total of 10 guns. It also had to be possible to carry two 500kg bombs externally as an overload.

The requirement lists the following electronic equipment for the aircraft: FuG 15, Fu Hl 3 F, Peil G 6, FuG 139, Gerät Bremen, FuG 130, FuG 226 (or initially 25 A) and FuG 101 A.

The heated and pressurised cockpit had to be armoured against 20mm rounds from the sides and 13mm from below, while both the wings and tail needed to have a de-icing system.

Performance characteristics were the same as those included in the August 4 preliminary specification although it was now also specified that landing speed should not be above 180km/h. "Tactical braking and landing brake" were to be investigated in detail and "the construction cost is to be kept as small as possible by constructive measures. Wood and plastic as materials are to be used as far as possible".

FOCKE-WULF NIGHT FIGHTERS

Apparently the only company that would even attempt to meet such an ambitious requirement was Focke-Wulf. The company had been working on designs for a new piston-engined single-seat fighter requirement[7] known as Hochleistungsjäger – literally 'high-performance fighter' issued on July 21, 1944. This had resulted in a swept-wing

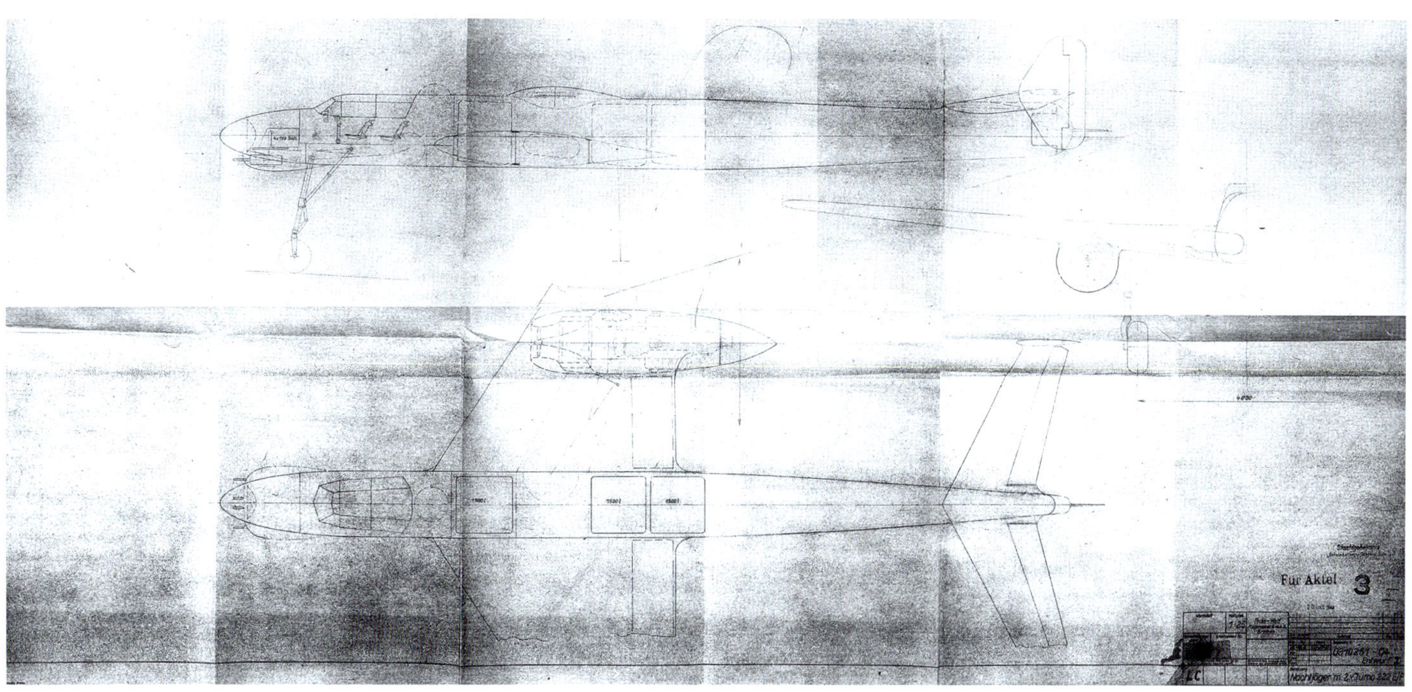

ABOVE: Focke-Wulf drawing 0310 251-04 of September 12, 1944, shows the company's 'Entwurf I' – its first attempt to meet the night fighter requirement of August 28, 1944. This large aircraft was to be powered by a pair of Jumo 222 E/Fs.

pusher-prop design of August 30 which had wing-root intakes for its Jumo 222 E/F engine and was in competition against designs from Blohm & Voss and Dornier.

An alternative version was designed to accommodate Argus's 24-cylinder H-block As 413 engine and featured an enormous nose intake for its radiator – the project description stating that the design, though produced to meet the July 21 requirement, could be adapted for bad weather operations with the addition of a second crewman.[8]

In seeking to meet the August 28 requirement, the company created a new design then compared it against scaled-up versions of the two Hochleistungsjägers to see which would best accommodate the additional fuel, weaponry and crewmen specified.

The bespoke design appeared in drawing 0310 251-04 of September 12, 1944, and was described as 'Entwurf I – Nachtjäger m. 2 x Jumo 222 E/F'. This large twin-tail swept-wing aircraft had a tricycle undercarriage and its engines were positioned beneath its wings but in a pusher-prop configuration. The crew sat together, all facing forwards,

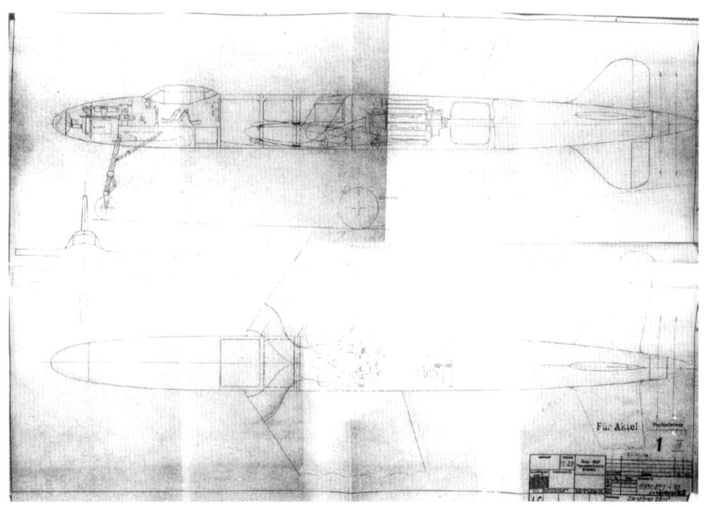

ABOVE: This pusher-prop Focke-Wulf night fighter, 'Entwurf III', had a single As 413 mounted in its fuselage plus a pair of BMW 003 turbojets under its wings. It appears in drawing 0310 251-10 of October 12, 1944. This layout was chosen by the company to provide the basis for several further night fighter and heavy fighter designs.

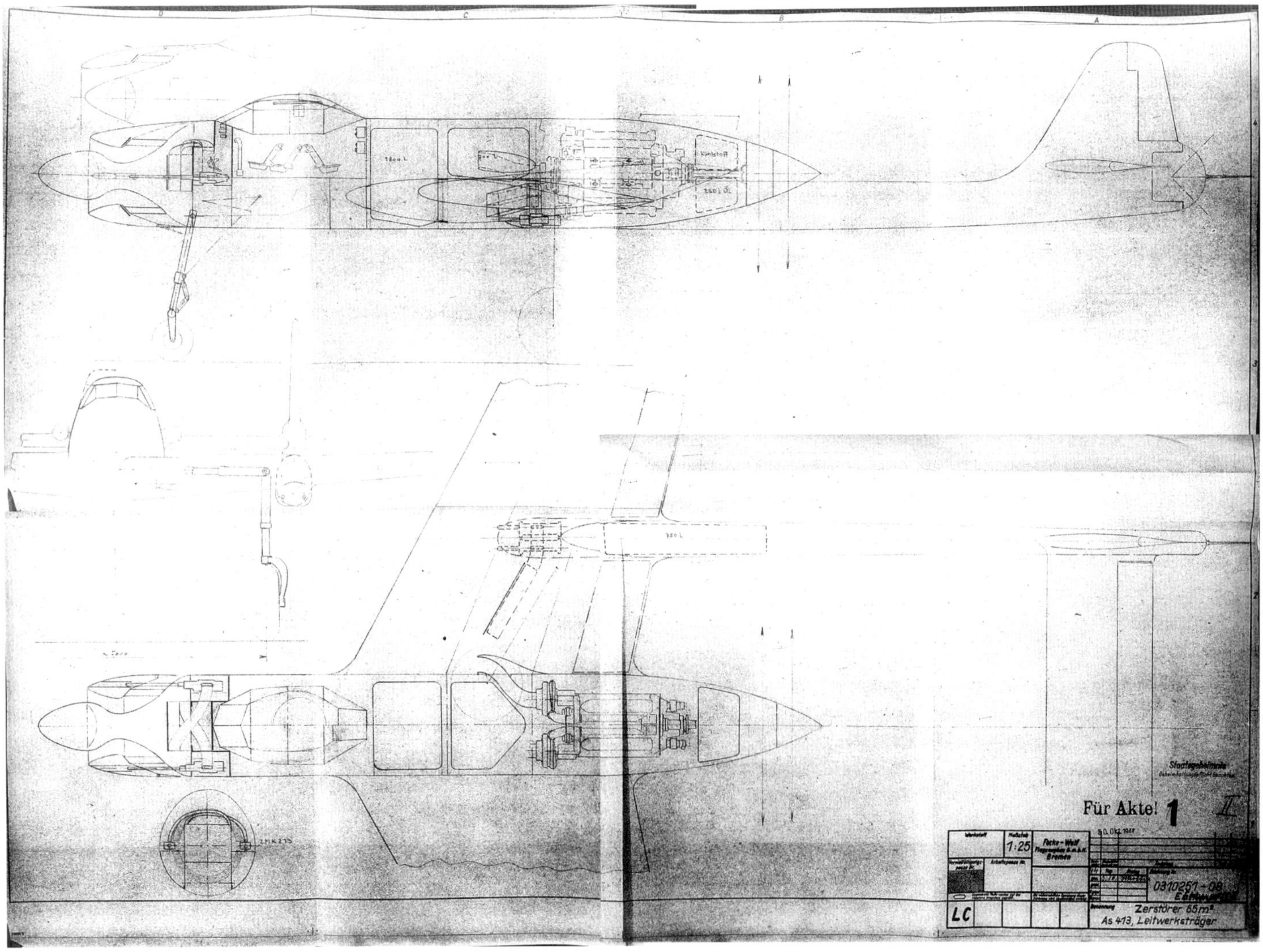

ABOVE: The twin-boom night fighter 'Entwurf II', powered by a single As 413, appears in Focke-Wulf drawing 0310 251-08 of October 3, 1944.

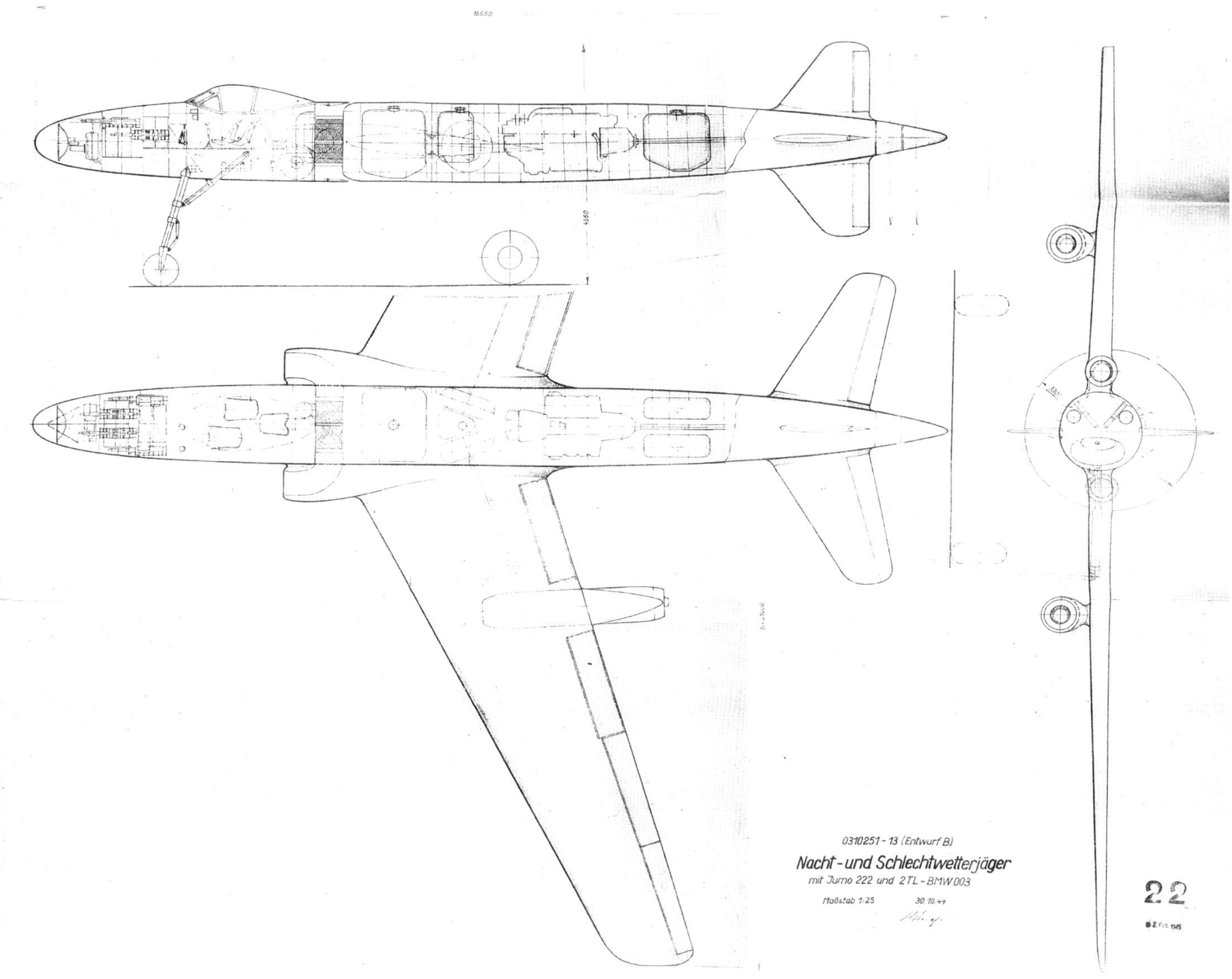

0310251 - 13 (Entwurf B)
Nacht- und Schlechtwetterjäger
mit Jumo 222 und 2TL - BMW 003
Maßstab 1:25 30.10.44

22

ABOVE: Focke-Wulf produced Kurzbeschreibung Nr. 21 on November 23 outlining a design derived from the earlier night fighter 'Entwurf III'. It appears in drawing 0310 251-13 of October 30, 1944.

and the aircraft's radar dish was in its nose. Armament consisted of four chin-mounted MK 108s and a tail turret. It was about 19m long – precise measurements are not given – with a wingspan of about 20m, making it slightly larger than a Dornier Do 217.

The design based on the As 413-powered Hochleistungsjäger appeared in drawing number 0310251-08 of October 3, 1944, and was the 'Entwurf II – Zerstörer 65m² As 413, Leitwerksträger'. The original design had had an unusual forward-swept tail form known as 'Viktoria-Leitwerk' but although this had mathematically been shown to offer less drag and weigh less than a conventional tail the risk and the amount of testing it would require were deemed to be too great.[9]

Therefore, the As 413-powered heavy fighter was given a tail boom configuration broadly similar to that of the Fw 189– except the single engine was mounted to the rear of the central fuselage driving contra-rotating propellers.

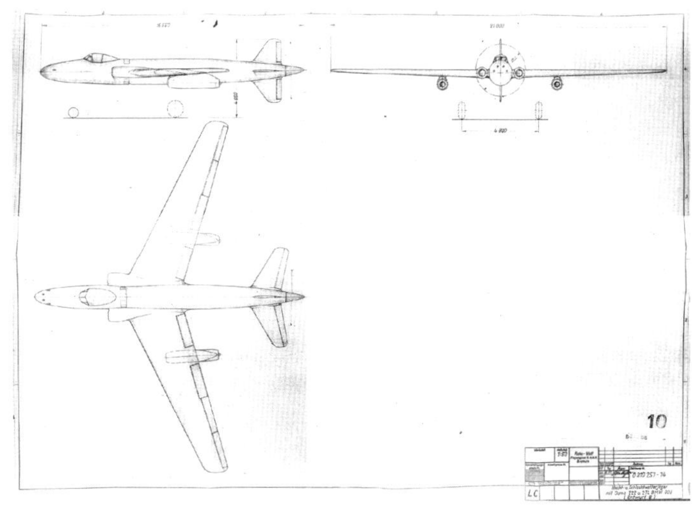

ABOVE: Another view of the Kurzbeschreibung Nr. 21 design with a Jumo 222 E/F and two BMW 003s shown in drawing 0310 251-14 of November 6, 1944.

The crew of three sat behind an enormous nose intake, which housed the radar dish within a central 'bullet', with the navigator now facing to the rear.

Armament consisted of a forward-firing MG 213 on either side of the intake, two forward-firing MK 108s in the nose of each boom and a tail turret on the tip of each boom for a total of eight guns. Again, the design was about 19m long but the only known drawing does not appear to show the full extent of its wingspan.

The third design, based on the layout of the Jumo 222 E/F-powered Hochleistungsjäger, was shown in drawing number 0310 251-10 of October 12, 1944, described as 'Entwurf III – Zerstörer 66m² As 413 und 2 x BMW 003'.

The single As 413 was mounted within a central fuselage, driving contra-rotating pusher propellers at the rear of the aircraft. The cruciform tail was similar to that of the Do 335 and there was a slot intake in the root of each swept wing. In addition, a BMW 003 turbojet was suspended in a simple cylindrical nacelle outboard of the undercarriage wells under each wing.

The crew sat in the same positions they had been given for the Entwurf II design and in front of them, in the nose, were the aircraft's radar dish, two MK 108s and two MG 213s.

Evidently Focke-Wulf's engineers found the concept of a piston engine backed up by turbojets appealing and dropped the first two designs – concentrating on the third. The result was a project description produced on November 23, 1944, entitled 'Kurzbeschreibung Nr. 21 Nachtjäger und Schlechtwetterzerstörer Jumo 222 E/F und 2 x BMW 003'.[10]

This stated: "When the German fighter defence grows stronger, the enemy will carry out his attacks against the Reich territory mainly at night and in bad weather during the day, using fast fighter planes in an ever-increasing range. It is therefore a task of the utmost urgency to develop a bad weather fighter aircraft to combat these fast combat aircraft, which is considerably superior to the enemy machines, above all in terms of speed.

"Since at night and in bad weather the use of enemy jet bombers is not to be expected for the time being, the anticipated speed of the enemy aircraft can initially be assumed to be around 700km/h in 8–10km altitude according to the performance of the Mosquito version with 36.6 litre Griffon motors. However, it should not be overlooked that the opponent is able, e.g. to achieve even higher speeds with an airframe similar to Do 335.

"The requirements for a bad weather fighter in terms of flight duration, equipment, number of crew members and armament result in an aircraft, when equipped with Otto engines, the dimensions of which do not differ significantly from those of the fast combat aircraft. It is therefore impossible to build an aircraft with piston engines that can be used against enemy fast fighter jets with the prospect of success.

"Bad weather fighters with a pure jet drive have such a large fuel consumption due to the long flight duration required that they will only be used when the necessary speed superiority can no longer be achieved in any other way. It is therefore obvious to look for a solution in which the relatively low consumption of the piston engine and the high performance of the jet engine are used simultaneously."

The piston engine would be used to bring the aircraft up to the point of entering combat, whereupon the turbojets would be fired up to provide superior performance. The report continued: "The implementation of all of Fl E2 III's preliminary development guidelines for a 'new destroyer to combat enemy combat aircraft in bad weather conditions' (28.8.44) results in an aircraft with a take-off weight of at least 16 tons and a wing area of about 70m². This effort is hardly bearable for a fighter, which has to be used in considerable numbers, both in terms of construction costs and in terms of the large fuel consumption. In addition, such a solution will not be sufficiently superior in speed to the fast combat aircraft.

"However, if you restrict yourself to the equipment and armament that are absolutely necessary for the task at hand, then it is possible to use a petrol engine of the performance class Jumo 222 to provide the power required for flight with inactive jet engines. An aircraft equipped with this engine will perform best in terms of speed with the least possible construction effort and fuel consumption.

"With a flight weight of 12 tons and a wing area of 55m², a Jumo 222 E/F and two BMW 003 jets results in a maximum speed of 880km/h at an altitude of 9.5km. The flight duration is 4.1 hours when the jet engines are off."

The design described in the report appeared in drawing number 0310 251-13 of October 30, 1944. Clearly based on the earlier 'Entwurf III' design, it was 16.55m long with a wingspan of 21m. It's centrally mounted Jumo 222 E/F, its radiator fed by two circular wingroot intakes, drove a single pusher prop and beneath each wing was a BMW 003.

The three crew sat in a pressure cabin behind a nose containing the radar dish and armament of four MK 108s. Further weapons options not shown in the drawing were two MK 103s and two MK 213s or one MK 112 and two MK 108s. Two further MK 108s, angled upwards, could accompany any of the three options but exactly where they would be positioned is unclear. Equipment was FuG 15, Fu Bl 3F, FuG 130, FuG 101a, Peil G 6, FuG 244 Bremen, FuG 226 (25a), FuG 139 and FuG 280 Kiel.

The aircraft's wings had a 25° sweepback and "with the small wing dimensions, it is not possible to accommodate the main chassis completely in the wing. The main wheels (1015 x 380 or 1140 x 410) are drawn into the fuselage so that only the struts remain in the wing. The aim is to use the undercarriage of the Ju 287. The nose chassis (wheel 630 x 220 or 685 x 250) is in the retracted position under the cabin floor".

Top speed with the Jumo 222 alone was expected to be 580km/h at 10km altitude at full throttle or 490km/h at 10km at reduced throttle. Endurance was 4.1 hours and

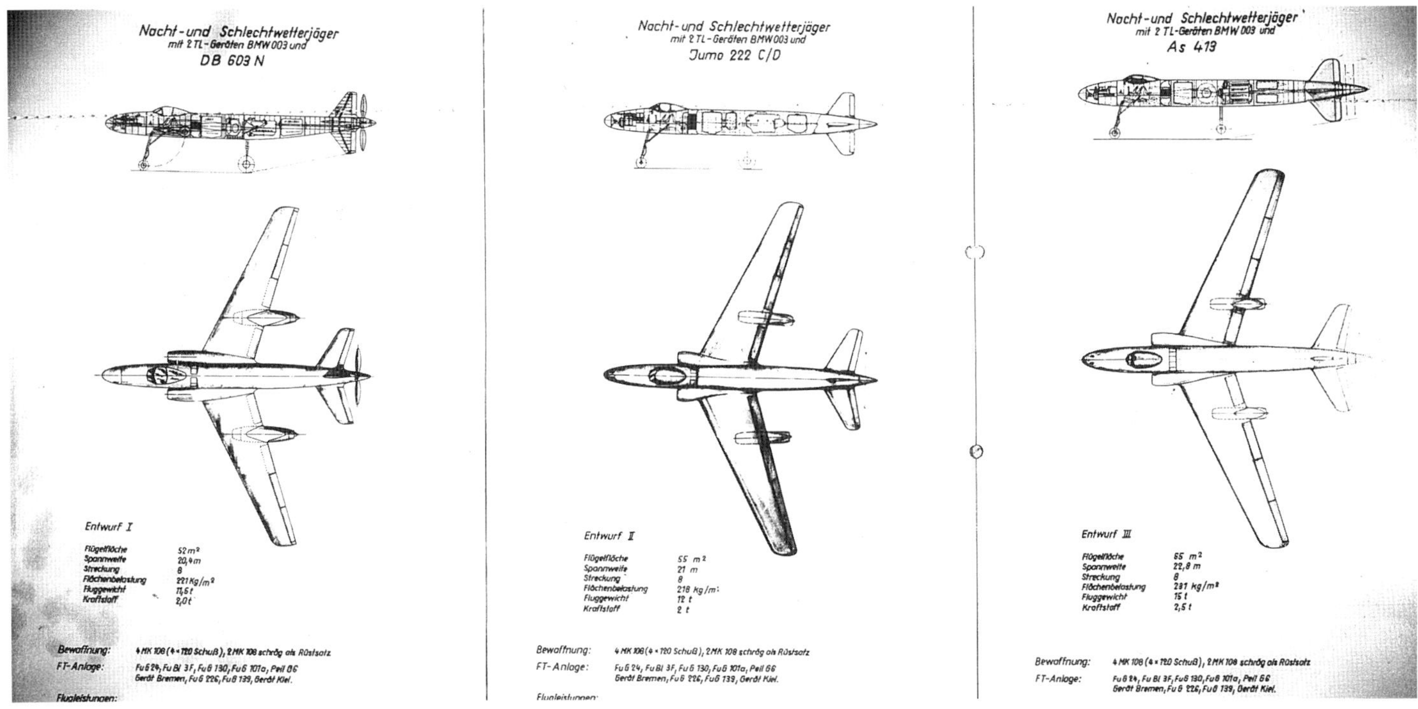

ABOVE: An overview of the three progressively larger variants of the Kurzbeschreibung Nr. 21 type design – the first powered by a DB 603 N, the second by a Jumo 222 C/D and the third by an Argus As 413.

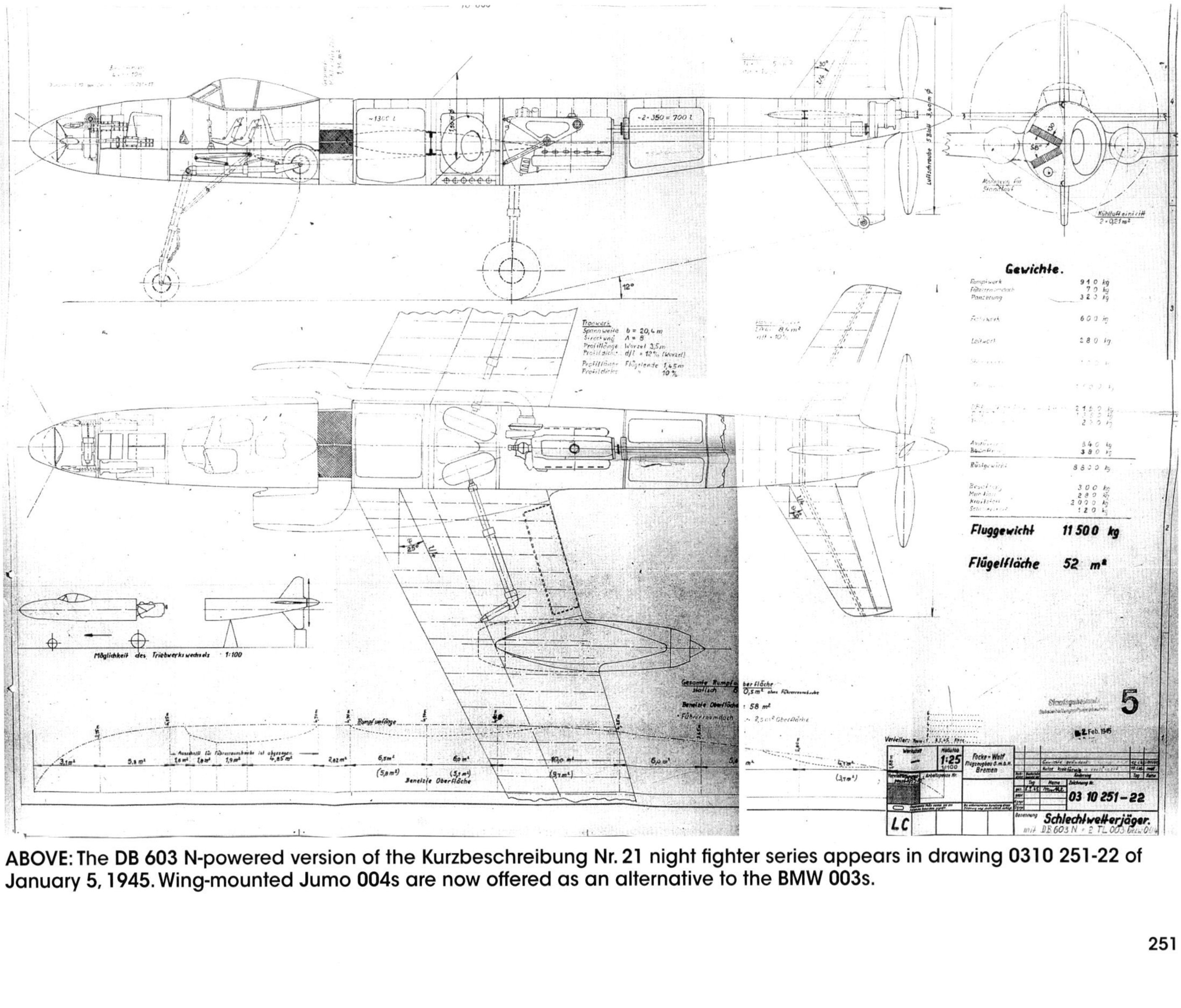

ABOVE: The DB 603 N-powered version of the Kurzbeschreibung Nr. 21 night fighter series appears in drawing 0310 251-22 of January 5, 1945. Wing-mounted Jumo 004s are now offered as an alternative to the BMW 003s.

5 hours respectively. With both the turbojets and the Jumo 222 at full power, top speed rose to 800km/h at 10km with endurance shrinking to just 1.6 hours.

While the report outlined a 55m² wing area design fitted with a Jumo 222, two further designs – one powered by a Daimler-Benz DB 603 N with 52m² wings and another with an Argus As 413 and 65m² wings. The latter was particularly unusual in having its engine fitted slightly at an angle within the fuselage, which meant that the contra-rotating propellers at its tail end were slightly tilted forwards. Work on these designs would continue into 1945 with the last known drawing in the series, 0310 251-23 of January 13, being a side view of the As 413 version.

On the same day that Kurzbeschreibung Nr. 21 came out, November 23, 1944, Focke-Wulf also produced a second report – 'Kurzbeschreibung Nr. 23 Zweimotoriges TL-Jagdflugzeug mit HeS 109-011'.[11] This was a follow-up to work already carried out earlier in the year on a high-altitude fighter powered by a single BMW 018 turbojet.[12] The report said: "It is possible to use two jet engines of the type HeS 011 with the least possible development risk to create a twin jet fighter aircraft, which combines better flight performance and stronger

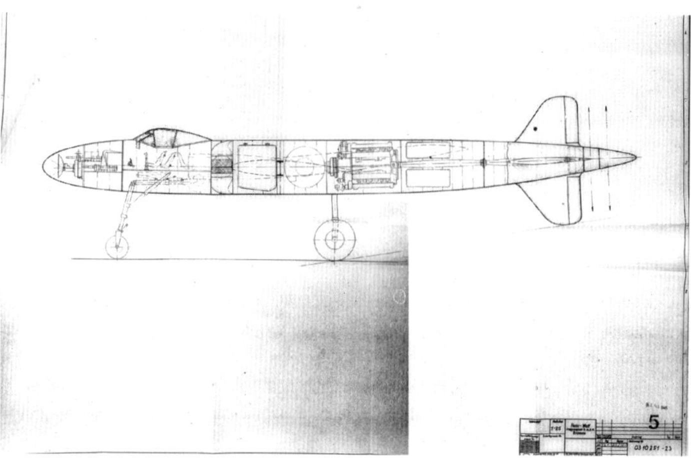

ABOVE: The last known drawing in the Kurzbeschreibung Nr. 21 series is 0310 251-23 of January 13, 1945, showing only a side view of the As 413-powered version.

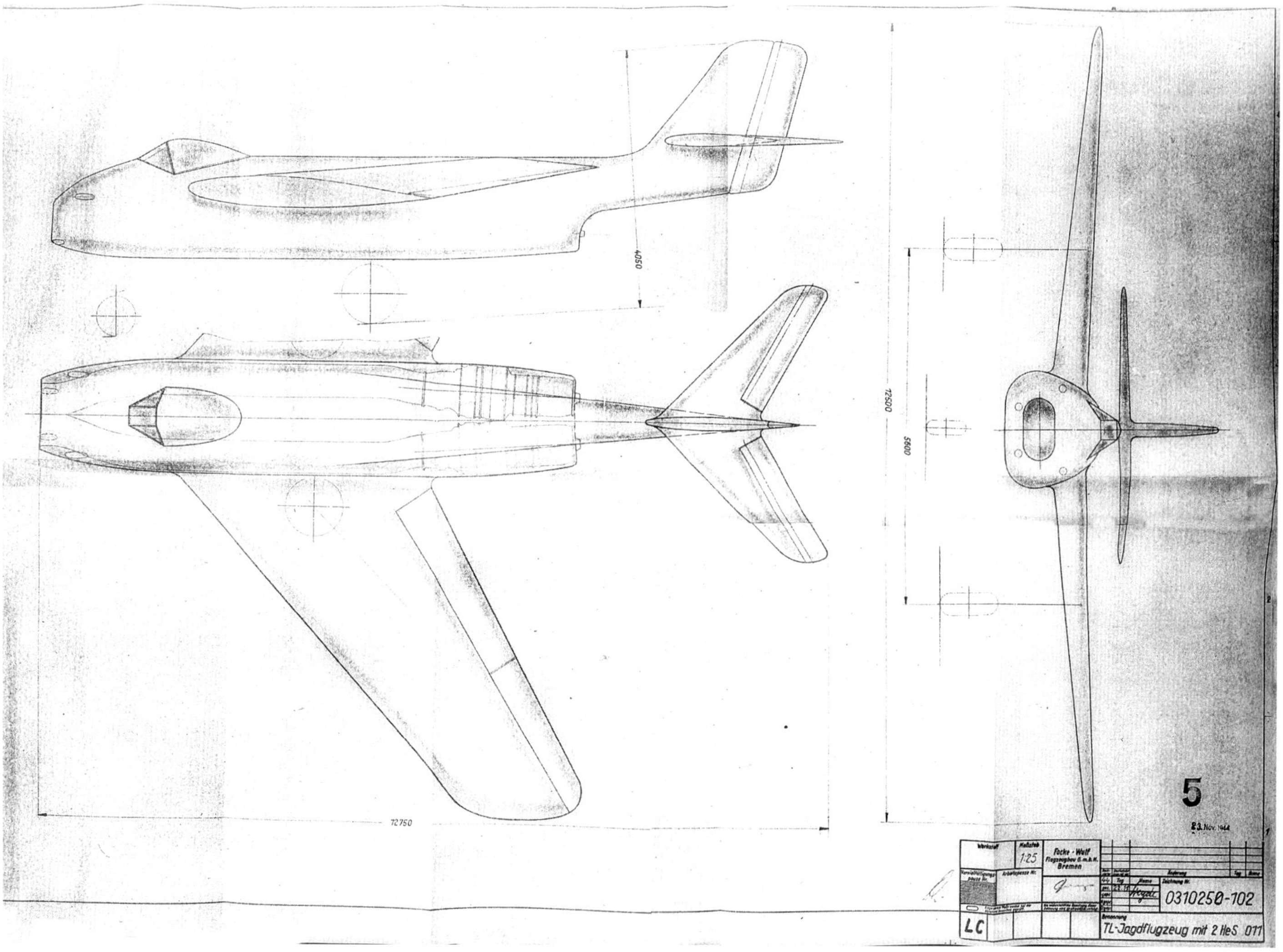

ABOVE: Focke-Wulf twin-jet fighter design of Kurzbeschreibung Nr. 23 shown in drawing 0310 252-102 (at first glance this appears to be 0310 250-102 but closer inspection reveals that the third '0' has been overwritten with a '2') dated November 22, 1944.

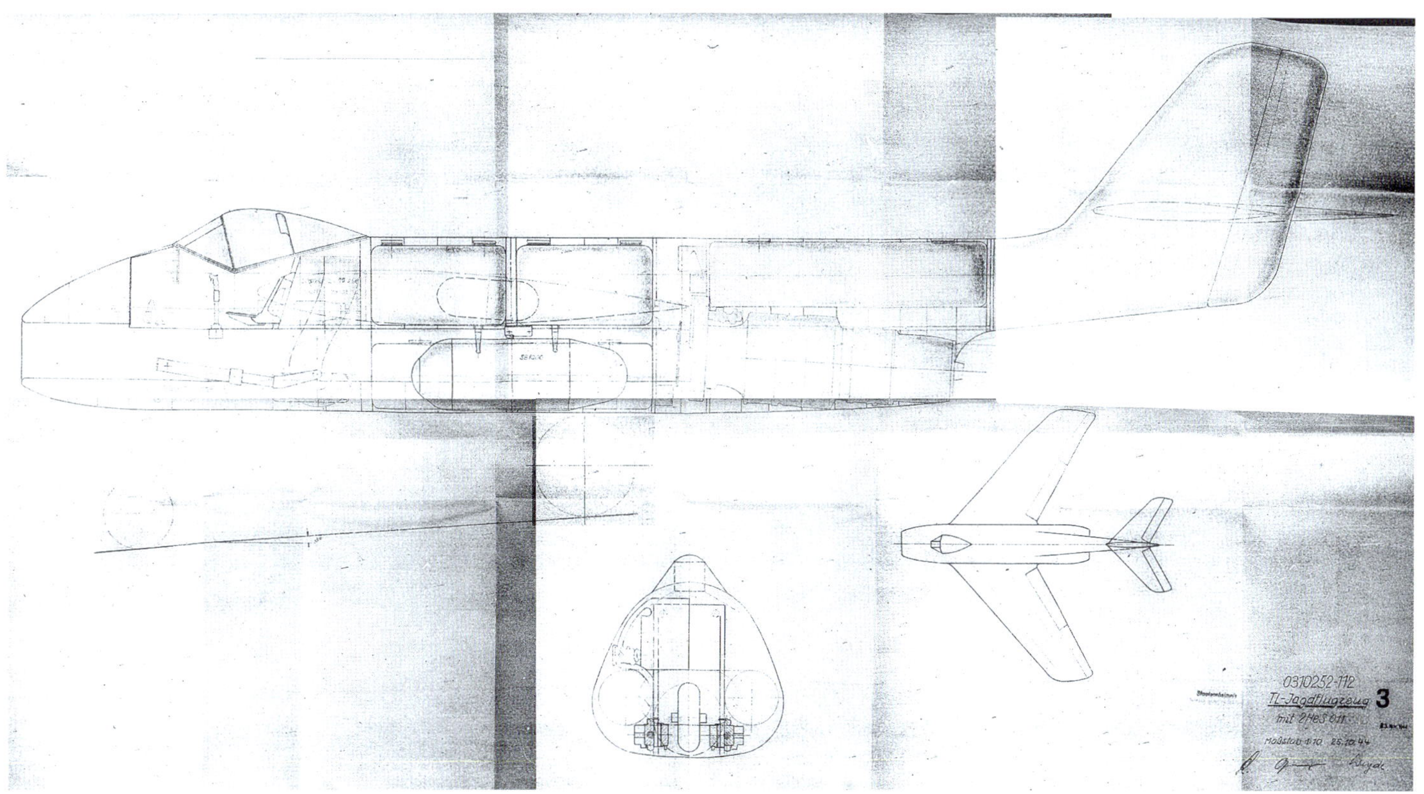

ABOVE: The second drawing included in Kurzbeschreibung Nr. 23 was 0310 252-112 of October 25, 1944. It shows the internal structure of Focke-Wulf's twin HeS 011-powered fighter.

throttle capability with the possibility, compared to the single-engine jet fighter, of carrying a bomb of up to 1,000kg in the fuselage.

"The greater throttle capability results from the division of the propulsion into two jet devices arranged close to each other in the fuselage, one of which can be switched off during cruise flight without the aircraft being forced to make noticeable rudder deflections and thus being forced to shift angles. Since in the present design the intake ducts are led along the fuselage sides to the tip of the fuselage, where they unite to form a common intake opening, the nose landing gear and the bomb load, or an additional fuel tank, can be arranged between the intake ducts."

At a take-off weight of 7,400kg the single-seater aircraft could carry 1,620kg of fuel, giving a flight time of 45 minutes at 14,000m altitude. A total of 3,000kg of fuel could be carried – all of it in the fuselage – giving a maximum range of 2,440km at 865km/h and 11km altitude. Flying at the same height at 610km/h would increase endurance to a respectable 3.74 hours.

According to the report: "As can be seen from a comparison of the enclosed performance sheets with the corresponding performance specifications for the Focke-Wulf draft 'High-altitude fighter with BMW 109-018', there is a speed superiority of the twin-engine design in the altitude range from 4 to 13km by about 20km/h, the remaining performance, as climbing speed, climbing time and ceiling still show satisfactory increases. The top speed of the aircraft is 1,075km/h at 7,000m."

Two drawings of the design were appended, 0310 252-102 of November 22, 1944 and 0310 252-112 of October 25, 1944. These showed an aircraft not dissimilar to the single-jet 'Entwurf Multhopp' with swept wings and a nose intake. The aircraft was 12.75m long with a wingspan of 12.5m and wing area of 39m². It was armed with two MG 213s and two MK 103s, all firing through the nose, and there was indeed space between the engine intake ducts for a 1,000kg load – an SB1000 aerial mine is shown in one of the drawings. Weapons alternatives given in the report were four MG 213s or four MK 108s and normal day-fighter equipment was fitted, including an EZ 42 gunsight.

While the latter was not a night fighter, it would soon provide the basis for one.

MESSERSCHMITT ME 262 B-1A/U1

While plans were being laid for a new piston engine or possibly combination turbojet/piston engine night fighter – and while Focke-Wulf was working on its designs for just such a new aircraft – discussions were taking place about whether Germany's existing jet aircraft might be converted into night fighters.

According to a Messerschmitt document dated October 5, 1944, entitled 'Projektübergabe Me 262 Nachtjäger (Einbau Schulflugzeug Stufe I):[13] "With the protocol Me 262 No. 40 from September 1, 1944 Messerschmitt AG is commissioned with the development of a two-seater night fighter based on the training aircraft without dual controls. In a meeting on

September 14, 1944 in Staken [Berlin-Staaken – home to workshops operated by Deutsche Lufthansa (DLH)], it was determined that not only the pattern installation, but also the construction would be carried out by the DLH."

A two-seater trainer version of the Me 262 had been planned back in 1943 and Blohm & Voss was subcontracted to begin creating the first Me 262 B-1a prototype[14] from a fully built Me 262 A-1a in April or May 1944. This involved reducing the size of the original aircraft's fuel tanks to create space for a second crewman behind the pilot and full dual controls. A longer cockpit canopy was also required and the reduction in fuel capacity was remedied through the use of two 300-litre drop tanks. A delivery schedule was drawn up on August 30, 1944, showing DLH joining in conversion work in November and the first full production model examples were completed by Blohm & Voss during September.

Night fighters required staying power so the Me 262 B-1a's lack of fuel tank capacity was a serious drawback. Therefore, the conversion process of turning a completed Me 262 B-1a into a night fighter focused heavily on restoring some of the original aircraft's tank capacity. Removing the dual controls gained some additional space and allowed the back-seater's position to be reduced in size – moving their seat closer to the pilot and allowing larger tanks to be installed behind them and on either side. Consideration was given to the addition of external fuel tanks and the possibility of a towed fuel tank.

The wall separating the front and rear seats was altered to incorporate cut-outs for the back-seater's feet and further modified to accommodate the frame of the FuG 16 wireless set.

Particular attention was paid to getting the back-seater's footrests right and to ensuring freedom of movement for his legs. New banks of equipment were to be fitted on either side of him. The second man would operate the FuG 350 ZC Naxos radar direction finder and other equipment.

The resulting machine was given the designation B-1a/U1[15] and was described by Messerschmitt as a 'Behelfsnachtjäger' or 'Makeshift night fighter'. It is believed that around six examples, fitted with FuG 218 Neptun V airborne interception radar, were completed by DLH before the end of the war.

ARADO AR 234 B-1 BEHELFSNACHTJÄGER

The first proposal for Arado's interim night fighter is dated September 12, 1944. It was decided that 30 Ar 234 airframes should be converted under the designation

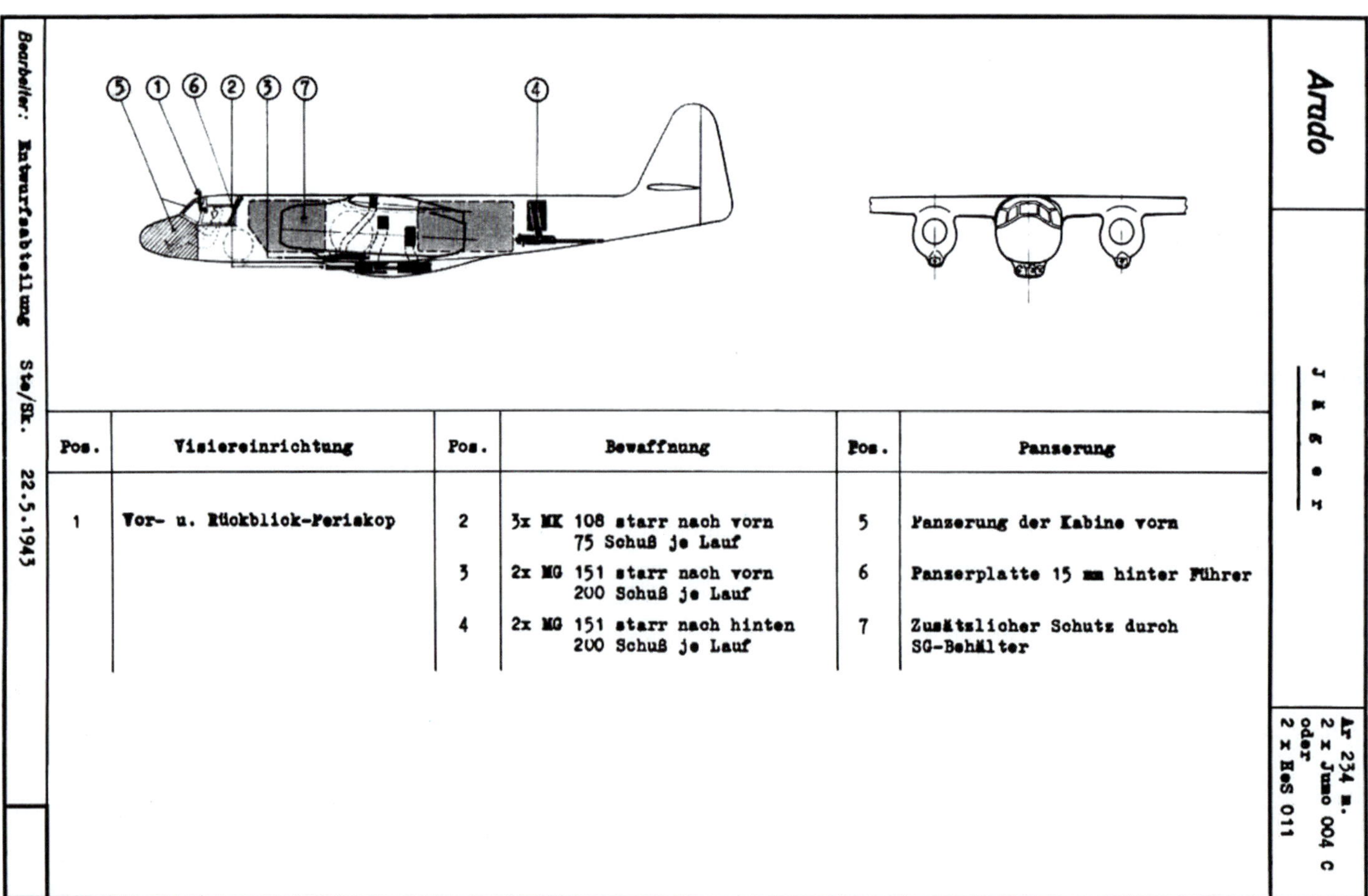

ABOVE: Arado had proposed this fighter version of the Ar 234 in May 1943. It was to be armed with three ventrally mounted MK 108s, an MG 151 under each engine nacelle and two more MG 151s pointing to the rear.

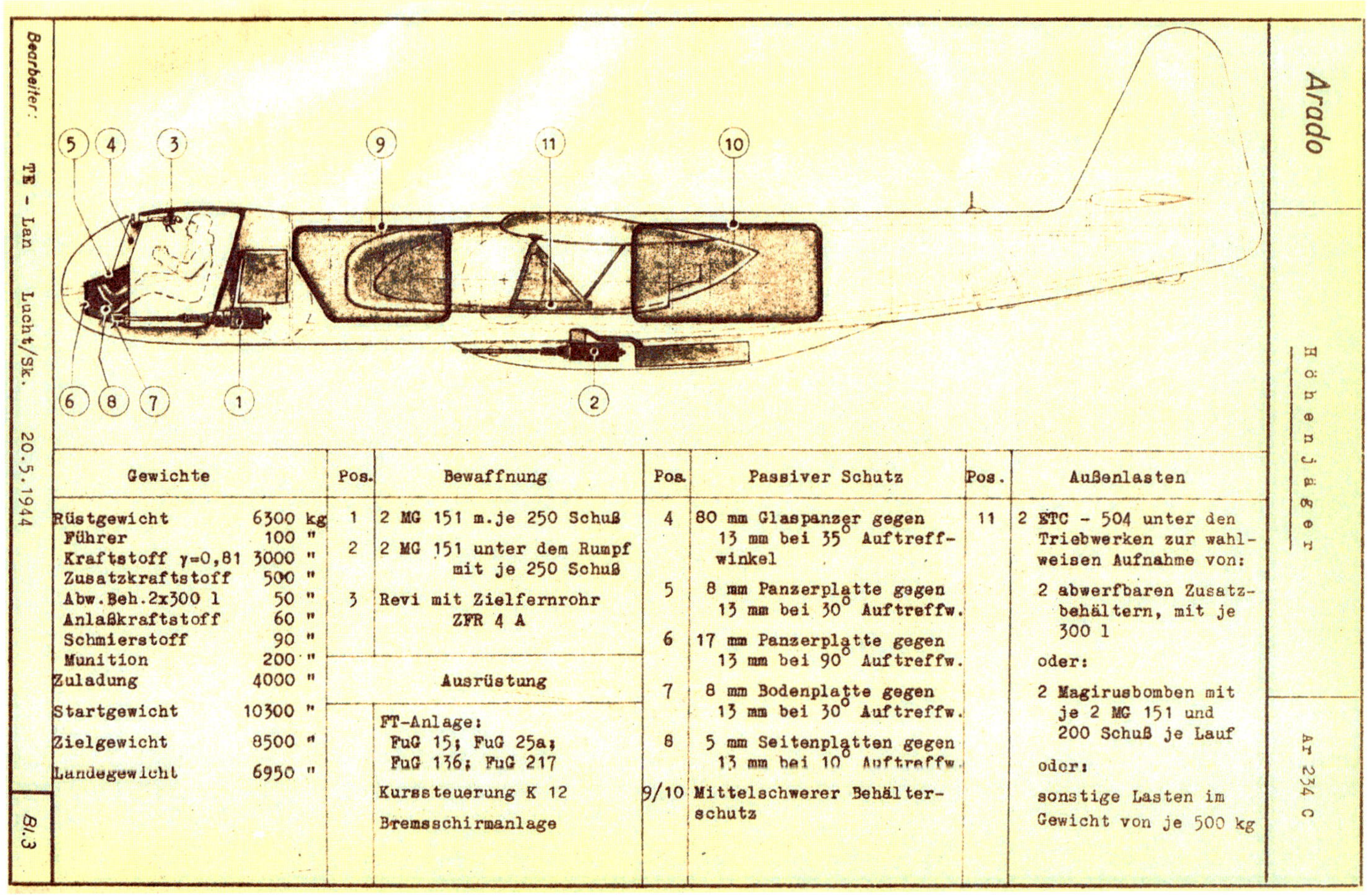

Bearbeiter: TE - Lan Lucht/Sk. 20.5.1944

Arado

Höhenjäger

Gewichte	Pos.	Bewaffnung	Pos.	Passiver Schutz	Pos.	Außenlasten
Rüstgewicht 6300 kg Führer 100 " Kraftstoff γ=0,81 3000 " Zusatzkraftstoff 500 " Abw.Beh.2x300 l 50 " Anlaßkraftstoff 60 " Schmierstoff 90 " Munition 200 " Zuladung 4000 " Startgewicht 10300 " Zielgewicht 8500 " Landegewicht 6950 "	1 2 3	2 MG 151 m.je 250 Schuß 2 MG 151 unter dem Rumpf mit je 250 Schuß Revi mit Zielfernrohr ZFR 4 A	4 5 6 7 8 9/10	80 mm Glaspanzer gegen 13 mm bei 35° Auftreffwinkel 8 mm Panzerplatte gegen 13 mm bei 30° Auftreffw. 17 mm Panzerplatte gegen 13 mm bei 90° Auftreffw. 8 mm Bodenplatte gegen 13 mm bei 30° Auftreffw. 5 mm Seitenplatten gegen 13 mm bei 10° Auftreffw. Mittelschwerer Behälterschutz	11	2 ETC - 504 unter den Triebwerken zur wahlweisen Aufnahme von: 2 abwerfbaren Zusatzbehältern, mit je 300 l oder: 2 Magirusbomben mit je 2 MG 151 und 200 Schuß je Lauf oder: sonstige Lasten im Gewicht von je 500 kg
		Ausrüstung				
		FT-Anlage: FuG 15; FuG 25a; FuG 136; FuG 217 Kurssteuerung K 12 Bremsschirmanlage				

Bl.3 Ar 234 C

ABOVE: A high-altitude fighter version of the Ar 234 C was put forward in May 1944. It was to be armed with four forward-firing MG 151s.

Ar 234 B-1 Behelfsnachtjäger. Like the Me 262 B-1a/U1, the Arado aircraft was fitted with a compartment for a second crew member who operated a FuG 218 Neptun V radar. Unlike the Messerschmitt design, the Arado's radar man had to sit in a very cramped compartment built into the rear fuselage with a small window above his head. Armament was two MG 151 cannon in a self-contained pod slung beneath the fuselage known as a Magirus-Bombe.

The first converted airframe, Ar 234 B-2 W.Nr. 140145, made its first flight on November 8 but work on further examples was suspended in mid-November because all available Ar 234s were required for operational use as bombers.

It was noted in the Chef TLR war diary for the week of January 22 to 28 that a meeting of the EHK had been held concerning night fighters. It had heard that: "The advantage of the 234 with four BMW 003s compared to 262 with two Jumo 004s of only 40km/h more speed stands as a disadvantage of greater TL effort. Cockpit vulnerability, reduced flight time. It has to be decided whether the development of new night fighter is justified. Number requirement for 262 night fighters has not been increased (three per month)."

A further Ar 234 night fighter had been built by March 1, 1945 and a third example had been completed by the beginning of April.

MESSERSCHMITT ME 262 B-2

Once work on converting the Me 262 B-1 trainer into a night fighter was under way, Messerschmitt began making preparations for a purpose-built 262-based night fighter: the Me 262 B2.[16]

A plan[17] for the weights and rough positioning of electronic equipment in the "Nachtjäger Me 262-Serie" was produced on October 20, 1944, showing the aircraft fitted with FuG 16 ZY, FuG 25a, FuG 125a, FuG 120a, FuG 218 and FuG 353 Zc. A full description of the aircraft,[18] under the designation Me 262 B2 was produced on January 18, 1945.

This stated: "The night fighter described here is a replacement for the makeshift night fighter currently under construction at Deutsche Lufthansa, which has been converted from the Me 262 training aircraft, with its short ranges and limited space.

"The night fighter Me 262, type designation B2, was developed from the basic model Me 262 fighter A1 by extending the middle part of the fuselage. In this way it was possible to accommodate the second man needed for navigation tasks and to increase the amount of fuel carried in the fuselage compared to the day fighter by about 500 litres.

"With the protected fuselage tanks the night fighter thus has a better flight time of 2¼ hours at a 6km altitude, which can be increased to approx. 2¾ hours by using 2 x 300 litre external tanks suspended beneath the fuselage and in a further development step can be increased to approx. 4 hours with larger external tanks and 'rider' tanks built onto the wing. In addition, a further increase in flight time can be achieved by using towed tanks.

"With Jumo 004 engine, the take-off run can be kept within normal limits by means of an additional thrust and,

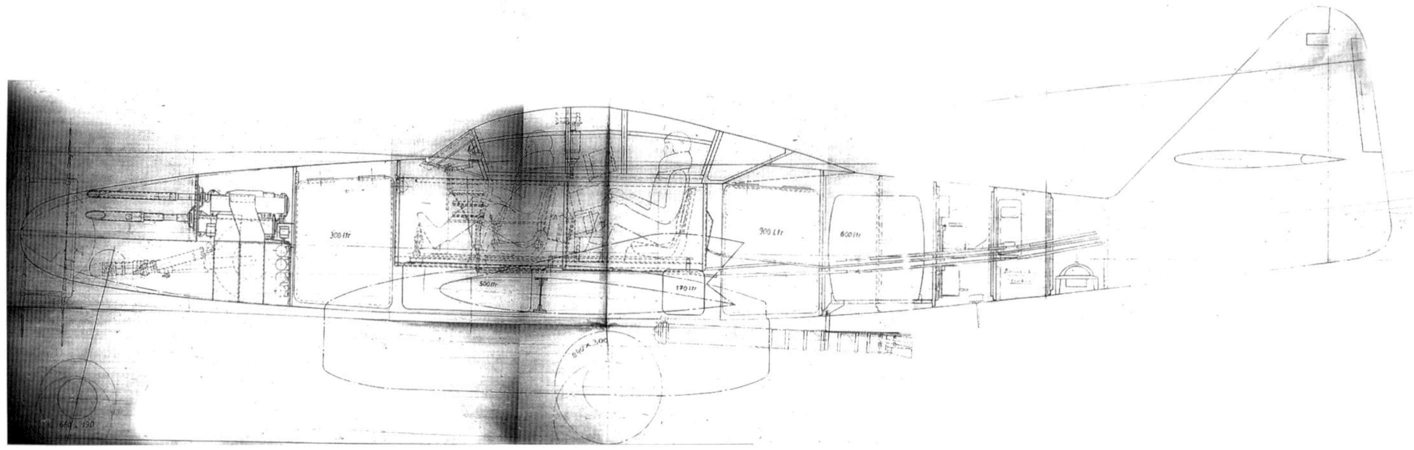

ABOVE: The Messerschmitt Me 262 B2 night fighter as it appeared on October 4, 1944.

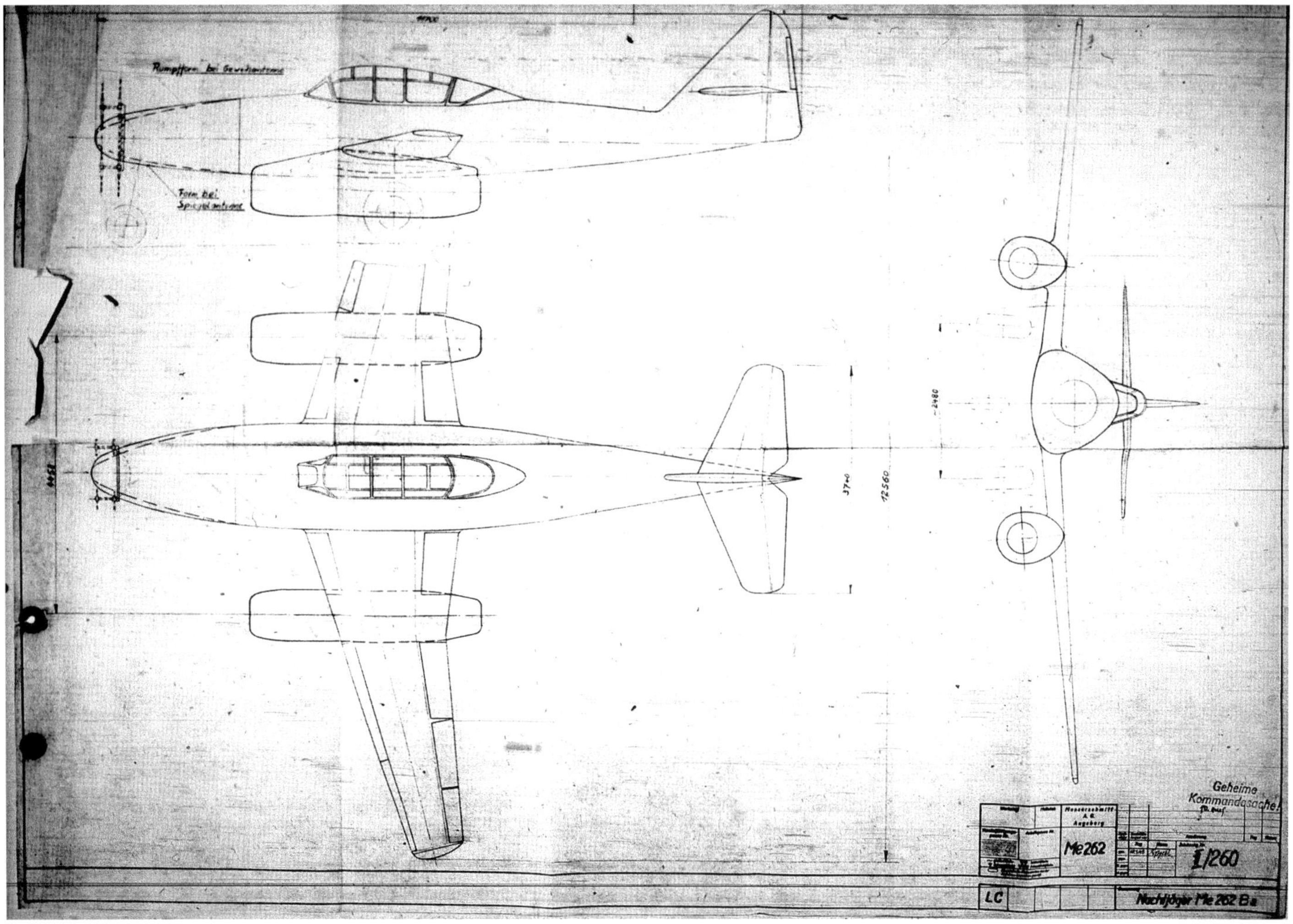

ABOVE: The first full project description of the Me 262 B2, dated January 18, 1945, included this drawing dated January 17.

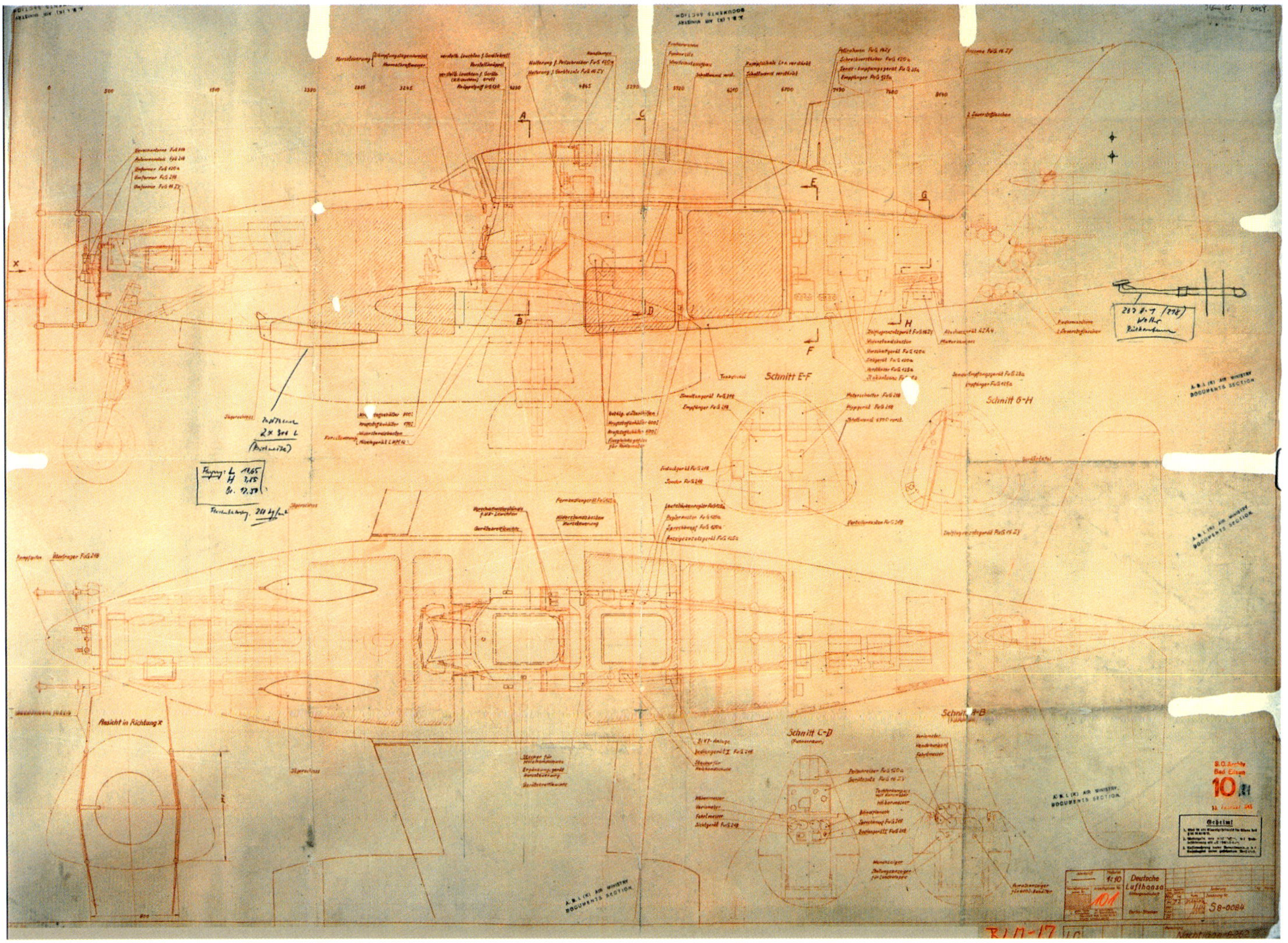

ABOVE: This drawing from Deutsche Lufthansa, dated February 7, 1945, shows the detailed design of the Me 262 B2. DLH was already converting two-seater Me 262 B-1 trainers into Me 262 B-1a/U1 night fighters by this time.

when the Heinkel HeS 109-011 engine is installed, remain in the same order of magnitude without additional thrust.

"The normal armament 4 x MK 108 gives the night fighter considerable firepower, which can be increased by installing another two cannon or multi-barrel weapons. Since the attack by flying underneath the target is likely to be the most successful attack tactic, this fact is taken into account by the preparation for the installation of an angled armament in addition to a reduced horizontal armament.

"Due to the unchanged takeover of the fuselage front part of the day fighter, the night fighter also receives normal fighter armour against shelling from the front, which, even if armouring should not be necessary in the near future, protects against the great disintegration effect when the enemy machines are fired on with 3cm explosive ammunition.

"The aim is to quickly install the concave dish antenna currently under development for the search devices, which practically does not cost any speed because it can be accommodated in the fuselage. First, however, the planes will have to be equipped with the Siemens antler antenna, which in its current version entails a speed loss of approx. 70km/h. An attempt is made to reduce this speed loss to approx. 20km/h by suitable profiling of the previously round bars.

"The possibility of taking measures to increase the critical speed (high-speed development), such as swept wings with a centrally installed engine, will be taken into account later."

The fuselage extension mentioned took the aircraft's overall length from 10.6m to 11.7m and allowed an increase in internal fuel capacity from 2,570 litres to 3,100. And the two crewmen would be able to communicate with each other using the EiV 7 system.

ARADO AR 234 C AND P NIGHT FIGHTERS

While Messerschmitt was working on the B2, Arado had drawn up plans to build a host of Ar 234 night fighter variants. Work initially focused on the C-series, which was already being prepared for production. The Ar 234 C was intended to be extremely flexible with the option to fit four BMW 003 engines, two HeS 011s or two Jumo 004 Cs depending on what might be available. It would also

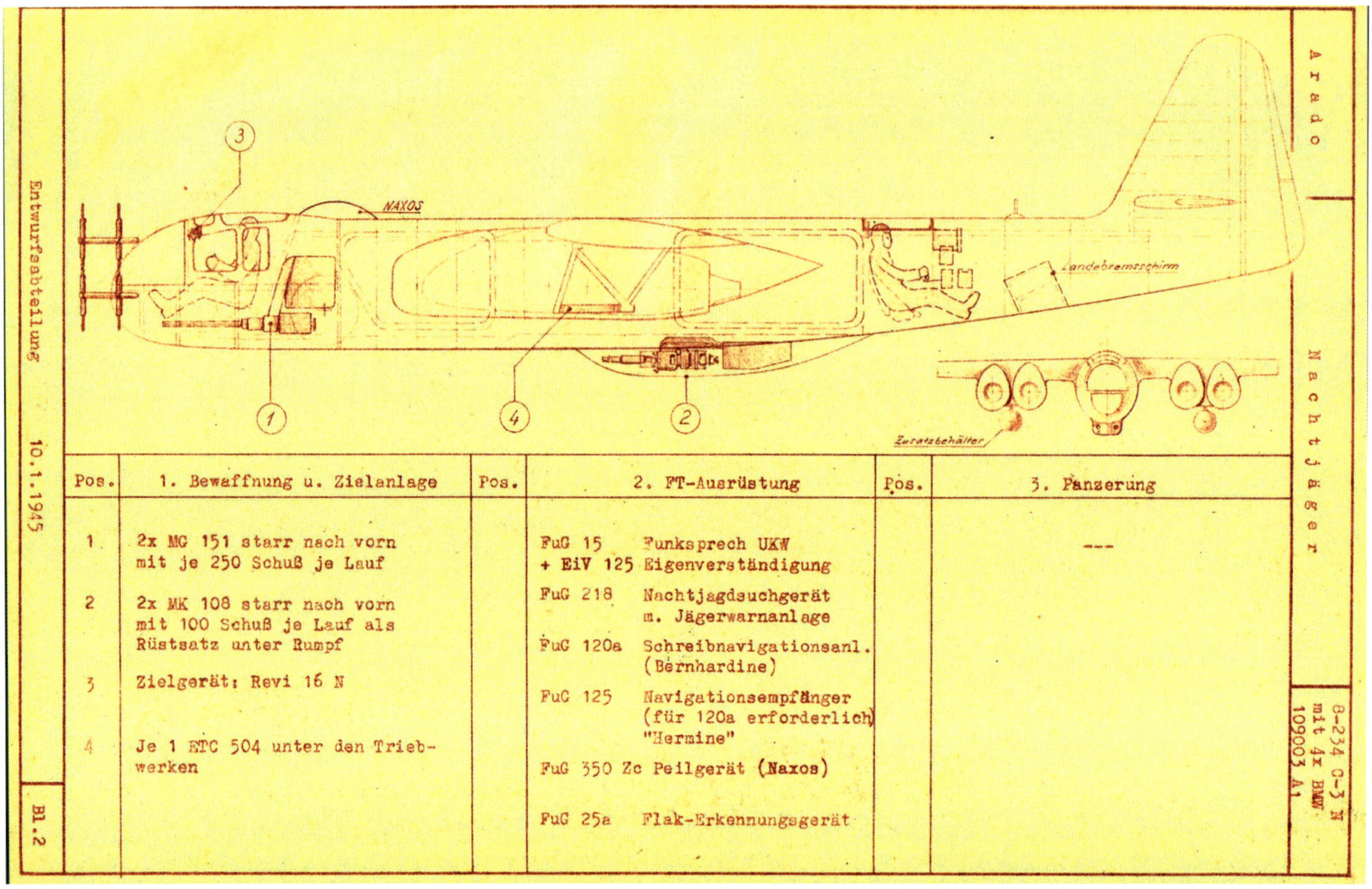

Arado

Nachtjäger

8-234 C-3 N mit 4x BMW 109003 A1

Entwurfsabteilung 10.1.1945 Bl.2

NAXOS

Landebremsschirm

Zusatzbehälter

Pos.	1. Bewaffnung u. Zielanlage	Pos.	2. FT-Ausrüstung	Pos.	3. Panzerung
1	2x MG 151 starr nach vorn mit je 250 Schuß je Lauf		FuG 15 Funksprech UKW + EiV 125 Eigenverständigung		---
2	2x MK 108 starr nach vorn mit 100 Schuß je Lauf als Rüstsatz unter Rumpf		FuG 218 Nachtjagdsuchgerät m. Jägerwarnanlage		
3	Zielgerät: Revi 16 N		FuG 120a Schreibnavigationsanl. (Bernhardine)		
			FuG 125 Navigationsempfänger (für 120a erforderlich) "Hermine"		
4	Je 1 ETC 504 unter den Triebwerken		FuG 350 Zc Peilgerät (Naxos)		
			FuG 25a Flak-Erkennungsgerät		

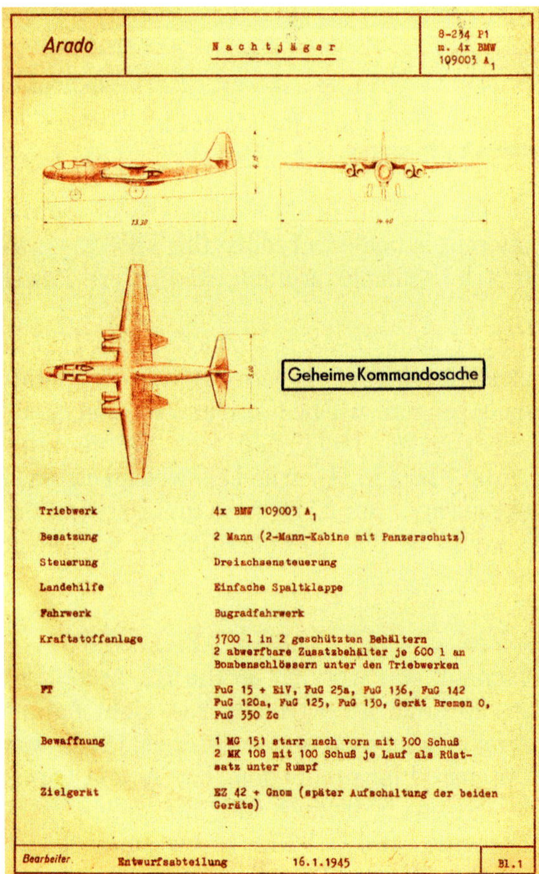

Arado | Nachtjäger | 8-234 P1 m. 4x BMW 109003 A1

Geheime Kommandosache

Triebwerk	4x BMW 109003 A1
Besatzung	2 Mann (2-Mann-Kabine mit Panzerschutz)
Steuerung	Dreiachsensteuerung
Landehilfe	Einfache Spaltklappe
Fahrwerk	Bugradfahrwerk
Kraftstoffanlage	3700 l in 2 geschützten Behältern 2 abwerfbare Zusatzbehälter je 600 l an Bombenschlössern unter den Triebwerken
FT	FuG 15 + EiV, FuG 25a, FuG 136, FuG 142 FuG 120a, FuG 125, FuG 130, Gerät Bremen 0, FuG 350 Zc
Bewaffnung	1 MG 151 starr nach vorn mit 300 Schuß 2 MK 108 mit 100 Schuß je Lauf als Rüstsatz unter Rumpf
Zielgerät	EZ 42 + Gnom (später Aufschaltung der beiden Geräte)

Bearbeiter. Entwurfsabteilung 16.1.1945 Bl.1

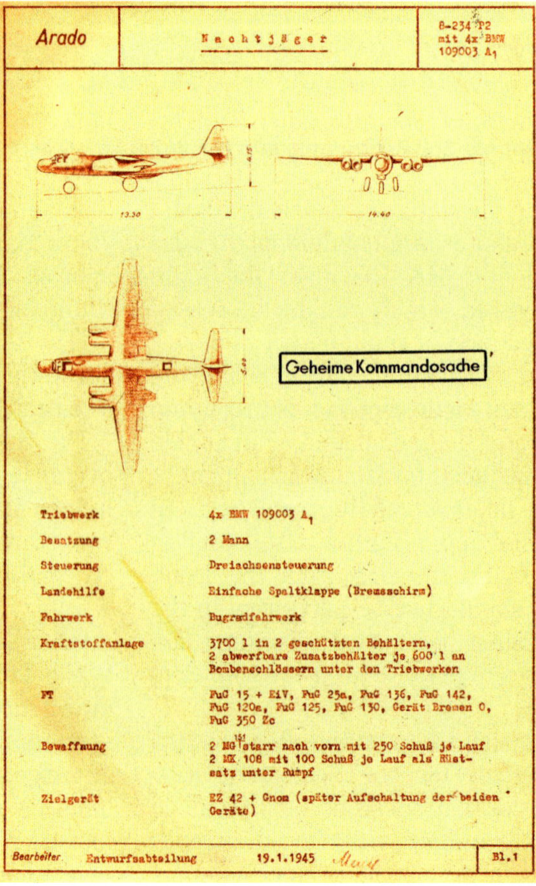

Arado | Nachtjäger | 8-234 P2 mit 4x BMW 109003 A1

Geheime Kommandosache

Triebwerk	4x BMW 109003 A1
Besatzung	2 Mann
Steuerung	Dreiachsensteuerung
Landehilfe	Einfache Spaltklappe (Bremsschirm)
Fahrwerk	Bugradfahrwerk
Kraftstoffanlage	3700 l in 2 geschützten Behältern, 2 abwerfbare Zusatzbehälter je 600 l an Bombenschlössern unter den Triebwerken
FT	FuG 15 + EiV, FuG 25a, FuG 136, FuG 142, FuG 120a, FuG 125, FuG 130, Gerät Bremen 0, FuG 350 Zc
Bewaffnung	2 MG 151 starr nach vorn mit 250 Schuß je Lauf 2 MK 108 mit 100 Schuß je Lauf als Rüstsatz unter Rumpf
Zielgerät	EZ 42 + Gnom (später Aufschaltung der beiden Geräte)

Bearbeiter. Entwurfsabteilung 19.1.1945 Bl.1

ABOVE: Arado Ar 234 C-3 N of January 10, 1945. Its layout closely followed that of the interim night fighter Ar 234, including the vulnerable glass cockpit and cramped radar operator position in the tail.

FAR LEFT: The Arado Ar 234 P1 featured an armoured cabin where its two-man crew could sit side by side. Its bulbous nose concealed its radar dish and a single MG 151 was installed beneath the cockpit. Two MK 108s were carried in a ventral pack. The design is dated January 16, 1945.

LEFT: Unlike the P1, only the pilot could sit in the armoured cockpit of Arado's Ar 234 P2 with the radar operator positioned inside the aircraft's tail. Two MG 151s were fitted beneath the cockpit and two MK 108s were carried beneath the fuselage. It was powered by four BMW 003s.

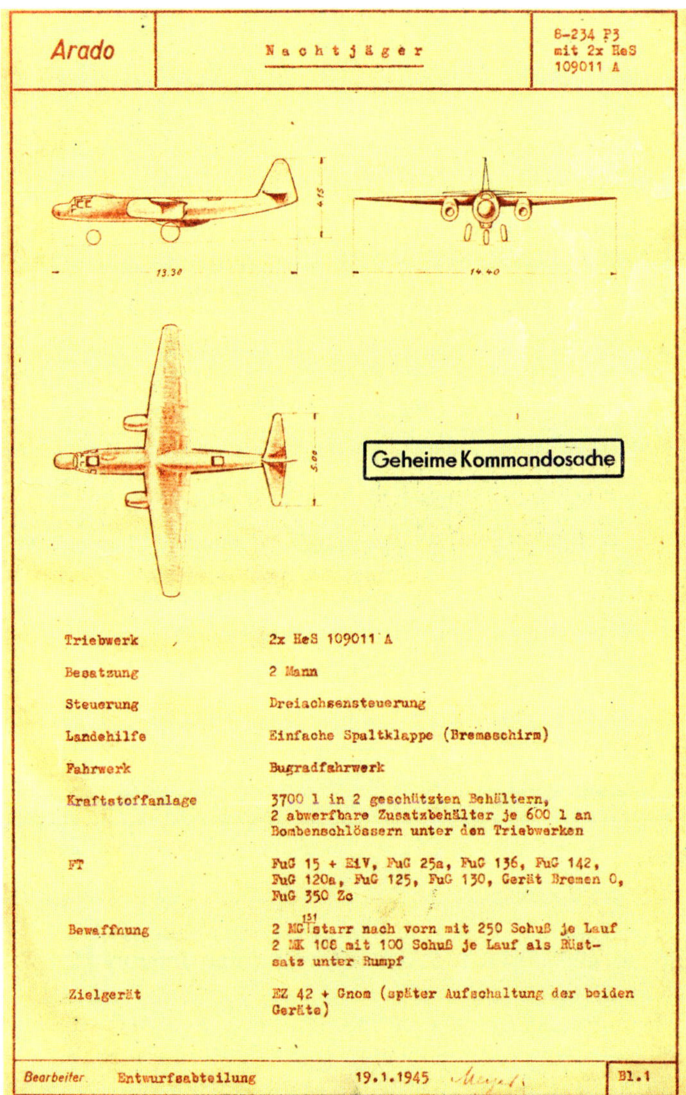

Arado | Nachtjäger | 8-234 P3 mit 2x HeS 109011 A

Geheime Kommandosache

Triebwerk	2x HeS 109011 A
Besatzung	2 Mann
Steuerung	Dreiachsensteuerung
Landehilfe	Einfache Spaltklappe (Bremsschirm)
Fahrwerk	Bugradfahrwerk
Kraftstoffanlage	3700 l in 2 geschützten Behältern, 2 abwerfbare Zusatzbehälter je 600 l an Bombenschlössern unter den Triebwerken
FT	FuG 15 + EiV, FuG 25a, FuG 136, FuG 142, FuG 120a, FuG 125, FuG 130, Gerät Bremen O, FuG 350 Zc
Bewaffnung	2 MG 151 starr nach vorn mit 250 Schuß je Lauf 2 MK 108 mit 100 Schuß je Lauf als Rüstsatz unter Rumpf
Zielgerät	EZ 42 + Gnom (später Aufschaltung der beiden Geräte)

Bearbeiter Entwurfsabteilung 19.1.1945 Bl.1

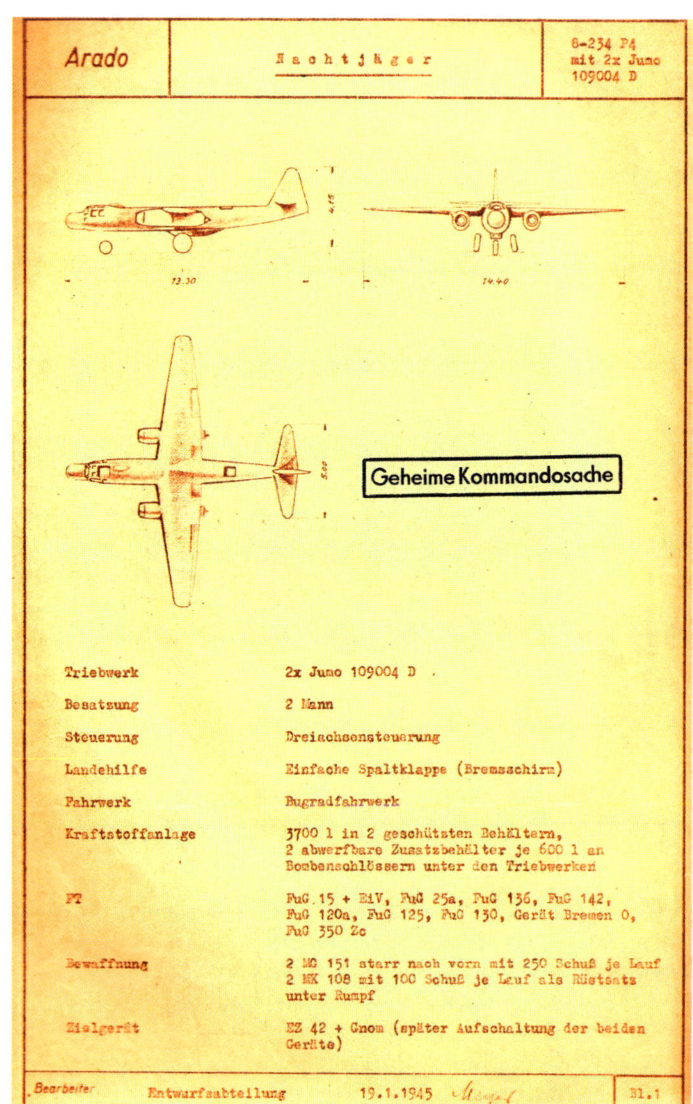

Arado | Nachtjäger | 8-234 P4 mit 2x Jumo 109004 D

Geheime Kommandosache

Triebwerk	2x Jumo 109004 D
Besatzung	2 Mann
Steuerung	Dreiachsensteuerung
Landehilfe	Einfache Spaltklappe (Bremsschirm)
Fahrwerk	Bugradfahrwerk
Kraftstoffanlage	3700 l in 2 geschützten Behältern, 2 abwerfbare Zusatzbehälter je 600 l an Bombenschlössern unter den Triebwerken
FT	FuG 15 + EiV, FuG 25a, FuG 136, FuG 142, FuG 120a, FuG 125, FuG 130, Gerät Bremen O, FuG 350 Zc
Bewaffnung	2 MG 151 starr nach vorn mit 250 Schuß je Lauf 2 MK 108 mit 100 Schuß je Lauf als Rüstsatz unter Rumpf
Zielgerät	EZ 42 + Gnom (später Aufschaltung der beiden Geräte)

Bearbeiter Entwurfsabteilung 19.1.1945 Bl.1

ABOVE LEFT: The Ar 234 P3 was largely the same as the P2 except for its engines – a pair of HeS 011s rather than four BMW 003s. ABOVE RIGHT: Another engine swap resulted in another version of the Ar 234 P. This time the four BMW 003s were swapped for a pair of Jumo 004s to create the Ar 234 P4.

incorporate a new cockpit and a wide range of possibilities for weapons installation.

Arado chose the Ar 234 C-3, originally designed as a bomber, to be the basis for a new night fighter designated C-3 N. This would have the two-seater pressure cabin which Arado had been working on, with the pilot and FuG 244 Bremen radar operator seated side by side. Weaponry would consist of two MG 151s built into its cockpit and a pair of MK 108s in an under-fuselage pack, aimed using a Revi 16 N gunsight. It would, however, have no armour protection.

The second night fighter in the C-series was the C-7. This appears to have been a very late or short-lived addition to the C family and differed from the C-3 N primarily in having only a single cockpit-mounted MG 151. Presumably the weight saving allowed more fuel to be carried or armour protection but this is unclear.

In addition to adapting the Ar 234 C as a night fighter, Arado also prepared a new dedicated night fighter series – the Ar 234 P. The P1 (Arado seems to have dropped the hyphen for the P-series, as Messerschmitt had done with the Me 262 B2) featured the new two-seater pressure cabin with a blister on its nose to house the Bremen radar dish. A single MG 151 was installed in the lower port side of the cockpit and a pair of MK 108s were fitted in the same under-fuselage pack as that used on the C-3 N. But where the C-3 N's crew were unprotected, the P1 carried 250kg of armour. It was also fitted with the advanced Revi EZ 42 lead computing gunsight.

The Ar 234 P2, P3 and P4 each appear to have had single-seat cockpits, with a consequently reduced armour load of 100kg, and with their second crewman housed in the rear of the fuselage. The main difference between them seems to have been their engines – with the P2 having four BMW 003s, the P3 two HeS 011s and the P4 a pair of Jumo 004 Ds. Each was armed with two MG 151s and two MK 108s.

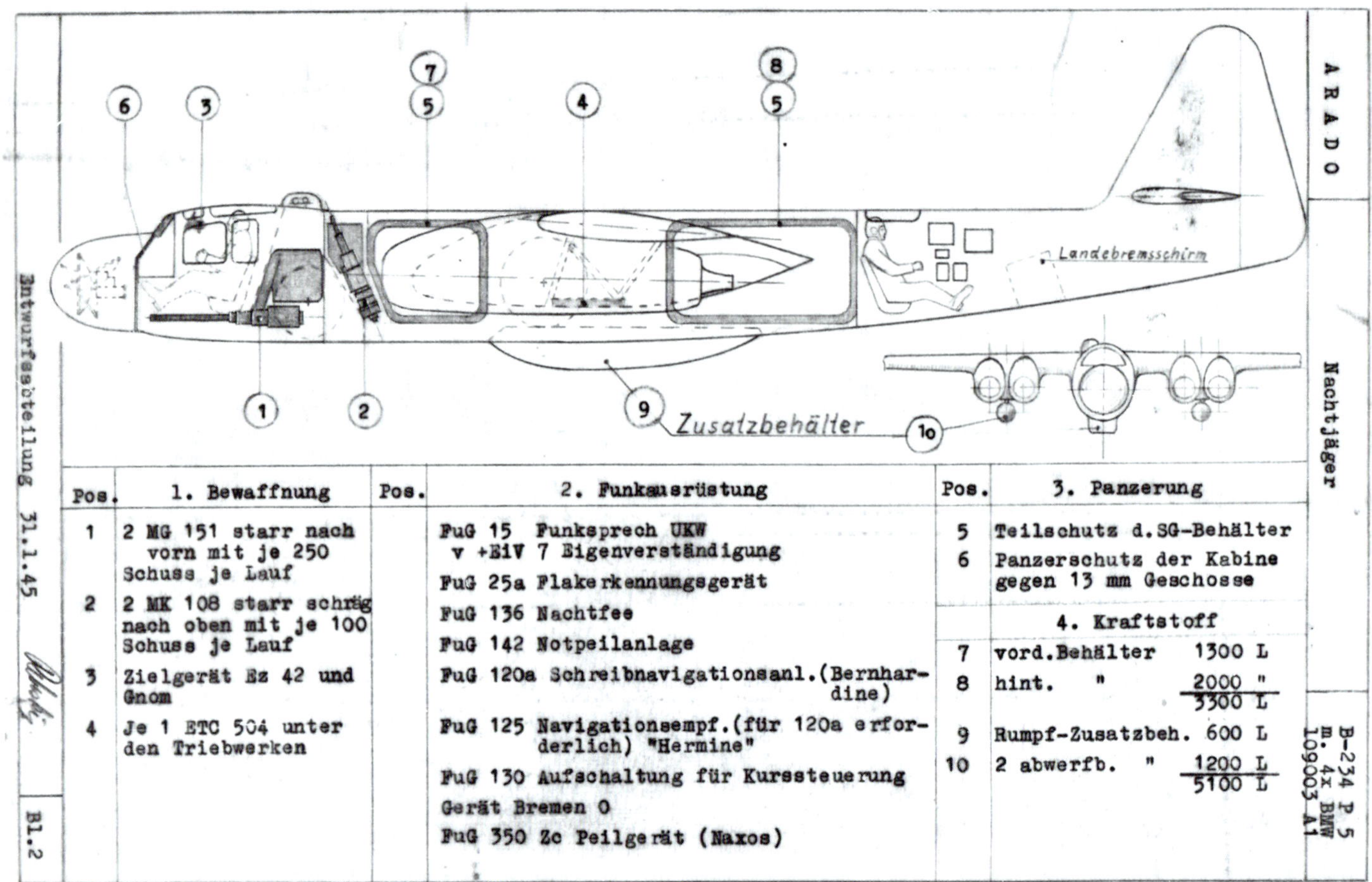

ABOVE: The Ar 234 P5 was similar to the P2 but with its MK 108s positioned behind the cockpit and pointing upwards at an oblique angle. The vacated ventral pack could now house an additional 600 litres of fuel. The design is dated January 31, 1945.

The final design in the series, the P5, was a late addition. The other four were all in existence by January 19, 1945,[19] but the P5 does not appear to have been drafted until later that month.

It had another new pressure cabin design – also a two-seater – with a third crewman in the rear fuselage. It was powered by two HeS 011s or four BMW 003s and armament was two MG 151s in the cockpit, two MK 108s on its underside and a further two MK 108s firing upwards at an oblique angle from a position just behind the cockpit. Equipment included FuG 15, EiV 7, FUG 25a, FuG 136 Nachtfee, FuG 142, FuG 120a, FuG 125, FuG 130, Bremen 0 radar and FuG 350 Zc Naxos.

INTERIM NIGHT FIGHTER COMPETITION

Following the formation of the EHK in September 1944, Focke-Wulf betriebsführer Kurt Tank had been appointed as head of the Sonderkommission Nachtjagd-Flugzeuge with responsibility for the future development of night fighters. During an EHK meeting on December 19–20, 1944, he briefly outlined Germany's position, stating: "a) There is agreement that the operational capability of today's bad weather fighters is not yet fully satisfactory for organisational and management reasons. b) In addition, there is clarity that a bad weather fighter plane for combat without visibility is still not technically solved. c) The achievements of the German night fighters are at present and in the near future in no way sufficient for the fight against the fast night fighter aircraft (Mosquito) of the opponent. d) The current developments of the 234 and 335 in the night fighter version are makeshift solutions and are unsatisfactory for the required night use with a long flight duration and sufficient navigation.

"Decision: The development of a superior night fighter with a long flight duration (5 to 8 hours?) and sufficient navigation capability (three man crew) is particularly urgent. In January the Sonderkommission Nachtjäger (Tank)[20] has to make clear decisions about a possible further development of the 335 using an additional jet engine as well as a fundamentally new development of the night fighter. The EHK must at short notice, in other words before the start of January 1945 at the latest, receive the complete tactical demands from the Luftwaffe."

As mentioned previously, on January 11 – 11 days later than hoped – the Luftwaffe high command issued a "rough guide" to its development requirements[21] for three new aircraft – a replacement for the He 162, a replacement for the Ta 152 and a new night fighter. The night fighter was to be a three-seater powered by a single-piston engine in the fuselage and a jet engine in each wing.

Automatic blind landing had to be a feature and it had to be possible for the aircraft to fly and carry out attacks on four-engined enemy bombers on piston-engine power alone – the turbojets being there to ensure superiority over enemy night fighters.

Top speed with the piston engine alone needed to be at least 550km/h between ground level and 10km altitude and 900km/h at 8–10km using the turbojets. Endurance had to be at least five hours at full throttle without the use of drop tanks using the piston engine and one and a half hours using the jets.

The forward-firing armament requirement was "four 3cm weapons, four MG 213/30 if possible. Oblique armament two MG 213/20". If only fixed rearward firing weaponry could be installed, this could be dropped entirely but for the remaining weapons "amount of ammunition has to be adjusted to the possible consumption during the long flight time".

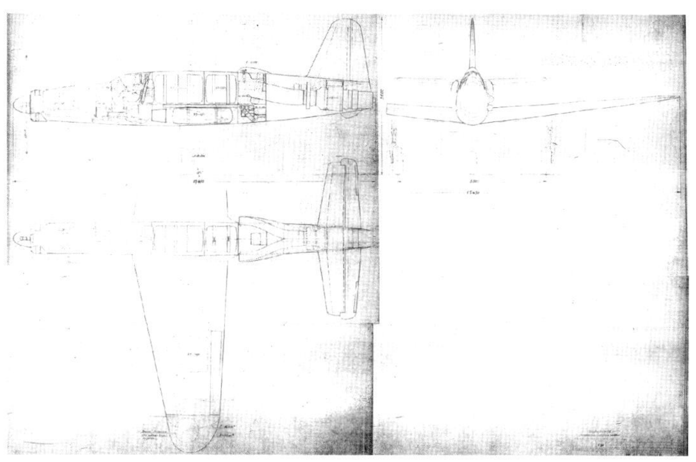

ABOVE: Dornier's combination piston engine/turbojet Do P 254/1 night fighter as it appeared on January 16, 1945, in the company's Baubeschreibung Nr. 1605. The pilot has the Do 335's usual ejection seat, unlike the crewman in the cramped rear fuselage cabin.

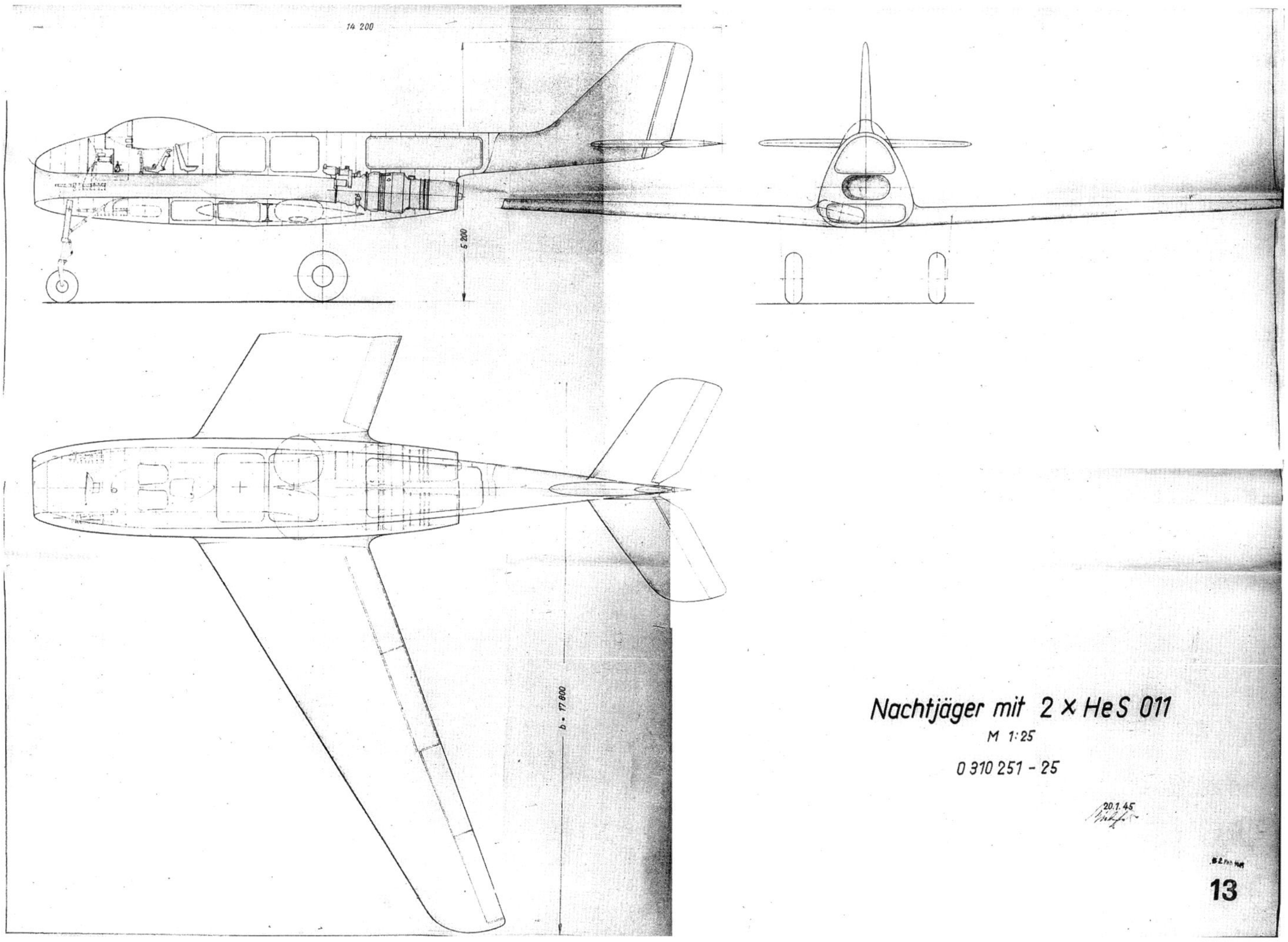

Equipment needed to include de-icing for wings, tail unit, propeller and canopy, "short and long wave on-board radio system for touch, speech and direction finding", a navigation device, airborne intercept radar with a range of at least 10km, airborne early warning radar system, device to track Rotterdam and Meddo (Allied airborne

ABOVE: Focke-Wulf's remarkably modern-looking Kurzbeschreibung Nr. 26 twin-jet night fighter design as it appears in drawing number 0310 251-25 of January 20, 1945. This would be this first in a series of pure jet night fighters designed by the company during the final months of the war.

radar mapping systems) and a device to track enemy using jammers.

The night fighter also had to be capable of carrying the 8-344 (aka Ruhrstahl X-4 air-to-air missile), R4M rockets, Jägerfaust or "on-board rockets of larger calibre with incendiary shrapnel filling". It needed a gunsight suitable for use at night, armour protection against 20mm ammunition from the front and behind, a pressure cabin, nosewheel undercarriage, ejection seat, the option to carry an overload, and rough field take-off and landing capability.

Tank appears to have requested descriptions of all night fighter designs currently in progress and by January 19, 1945 had received them for Messerschmitt's Me 262 B2, Arado's Ar 234 C-3 N and Ar 234 P1, and Dornier's Do 335 A-6 and Do 335 B-6. The latter two projects were both purely piston-engine designs, the A-6 being described in Baubeschreibung Nr. 1600 of November 20, 1944.[22] It was a modified version of the Do 335 heavy fighter with a radar operator squeezed into a space behind the pilot beneath a small window. The Do 335 B-6 was largely the same as the A-6 but based on the strengthened B-series airframe which also featured a new windscreen and nosewheel.

Over the next few days further project descriptions trickled in to Tank's offices at Bad Eilsen including Dornier's Do P 254/1, outlined in Baubeschreibung Nr. 1605 of January 16, 1945 – received on January 24. The Do P 254/1 took the forward fuselage, wings of tailplanes of a standard Do 335 airframe and added a completely new rear fuselage and fin. This section incorporated an absolutely tiny space for a radar operator and behind him, beneath the fin, an HeS 011 turbojet. The intakes for this wrapped around the radar operator's cramped cabin and protruded from the sides of the aircraft.

Focke-Wulf's engineers, meanwhile, appear to have had a revelation. They had dropped the idea of a combination turbojet/piston-engine night fighter and had instead been concentrating on turning their twin HeS 011-powered single-seat day fighter of Kurzbeschreibung Nr. 23 into a night fighter. The result was outlined in Kurzbeschreibung Nr. 26 of January 27, 1945.[23] This really was a short description, with only the bare essentials outlined – an overall length of 14.2m, a wingspan of 17.3m (although the appended drawing shows 17.8m) and a wing area of 50m². The three crew sat together in a pressure cabin with a full-vision canopy. The radar dish was in the nose, directly in front of the crew and above the intake. Armament was kept to the bare essentials – either two MK 108s firing forward and two upwards at an 80° angle or four MK 108s firing forward.

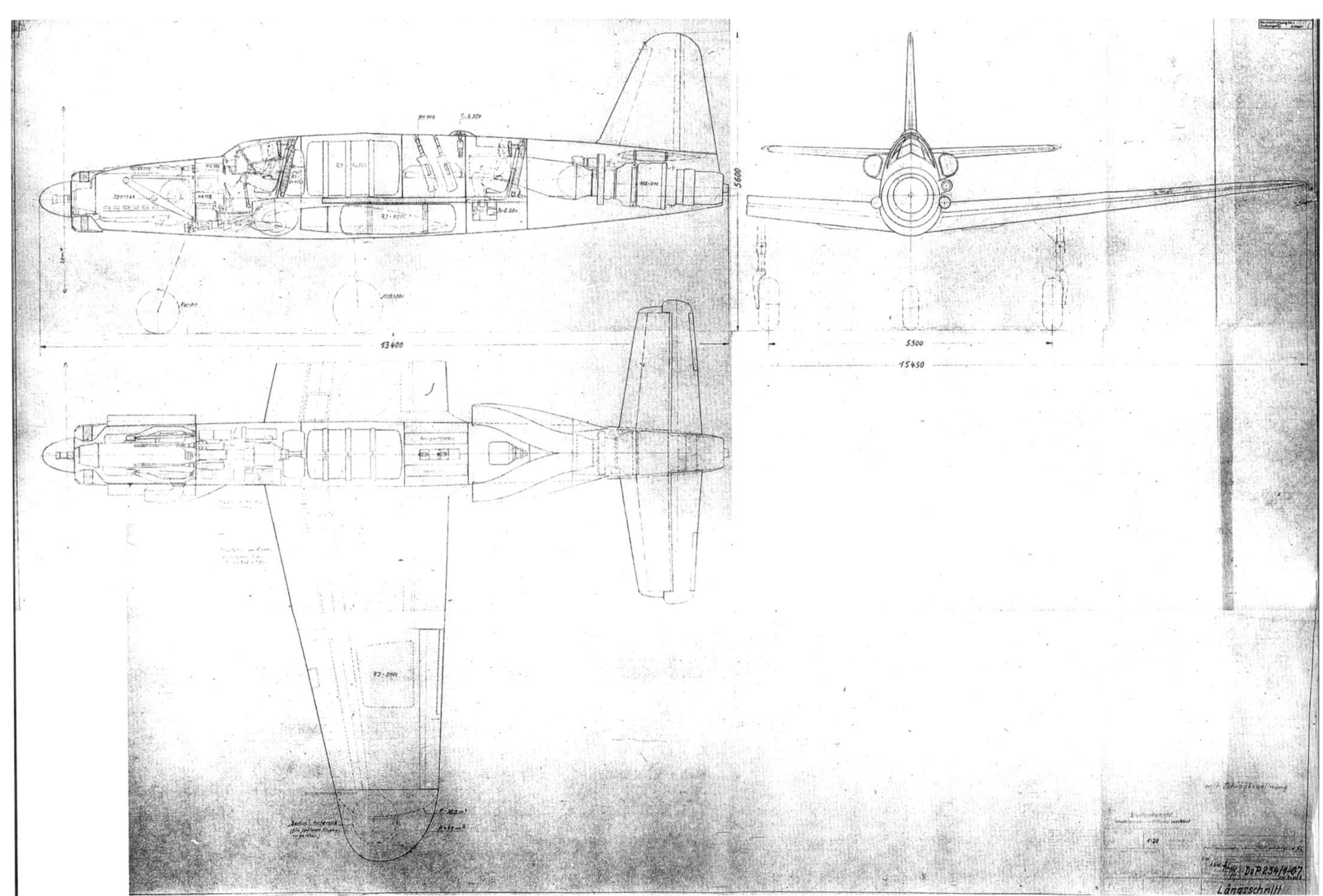

ABOVE: Dornier's revised Do P 254/1 of February 5, 1945 – a fuselage fuel tank has been replaced with a pair of angled MK 108s and the rear crewman now has an ejection seat.

Equipment included FuG 15 (24), Fu Bl 3F, FuG 130, FuG 101a, Peil G 6, FuG 244 Bremen, FuG 25a (226), FuG 139, FuG 280 Kiel and FuG 350 (Naxos). Take-off weight was 11 tons and top speed at 10km altitude was 920km/h.

A meeting of the EHK was held on January 24 and the night fighter designs were discussed – though not Focke-Wulf's because it was too late – and the only decision reached was that none of the designs really met the Luftwaffe's requirements. Consequently, it was decided that another new specification should be issued to the manufacturers. This was duly drawn up and sent out on January 27.[24]

ADVANCED NIGHT FIGHTER COMPETITION

The specification included a crew of three – pilot, radio operator and navigator who had to be "as close together as possible without separating components". The aircraft had to be purely jet propelled since "to reach the required maximum speed of 900km/h only turbojet propulsion is possible". But the number and arrangement of the engines – BMW 003s, Jumo 004s or HeS 011s – was optional. Standard armament was four forward-firing MK 108s or MG 213/30s, each with 120 rounds. Targeting would be via an EZ 42 or Bremen. A further two MK 108s or MG 213/30s would need to be fitted firing upwards at a 70° angle with 100 rounds apiece. Two remote-controlled MG 151s were to be provided for defence and it had to be possible to carry two 500kg bombs externally as an overload.

Electronic equipment would consist of FuG 24 SE with ZVG 24, FuG 25 A, FuG 139, Fu Bl 3 with AWG 1, Bremen, FuG 280, FuG 218 R, FuG 101 A, Naxos and an EiV system. There had to be armour protection against 20mm rounds from the front and the fuel tanks needed to be protected. Required safety features were a pressure cabin, ejection seats, de-icing for wings, tail and windows, cabin heating and a fire extinguishing system.

The top speed mentioned had to be achievable at an altitude of 8–10km and endurance needed to be four hours including climb at full throttle, without drop tanks. But "the

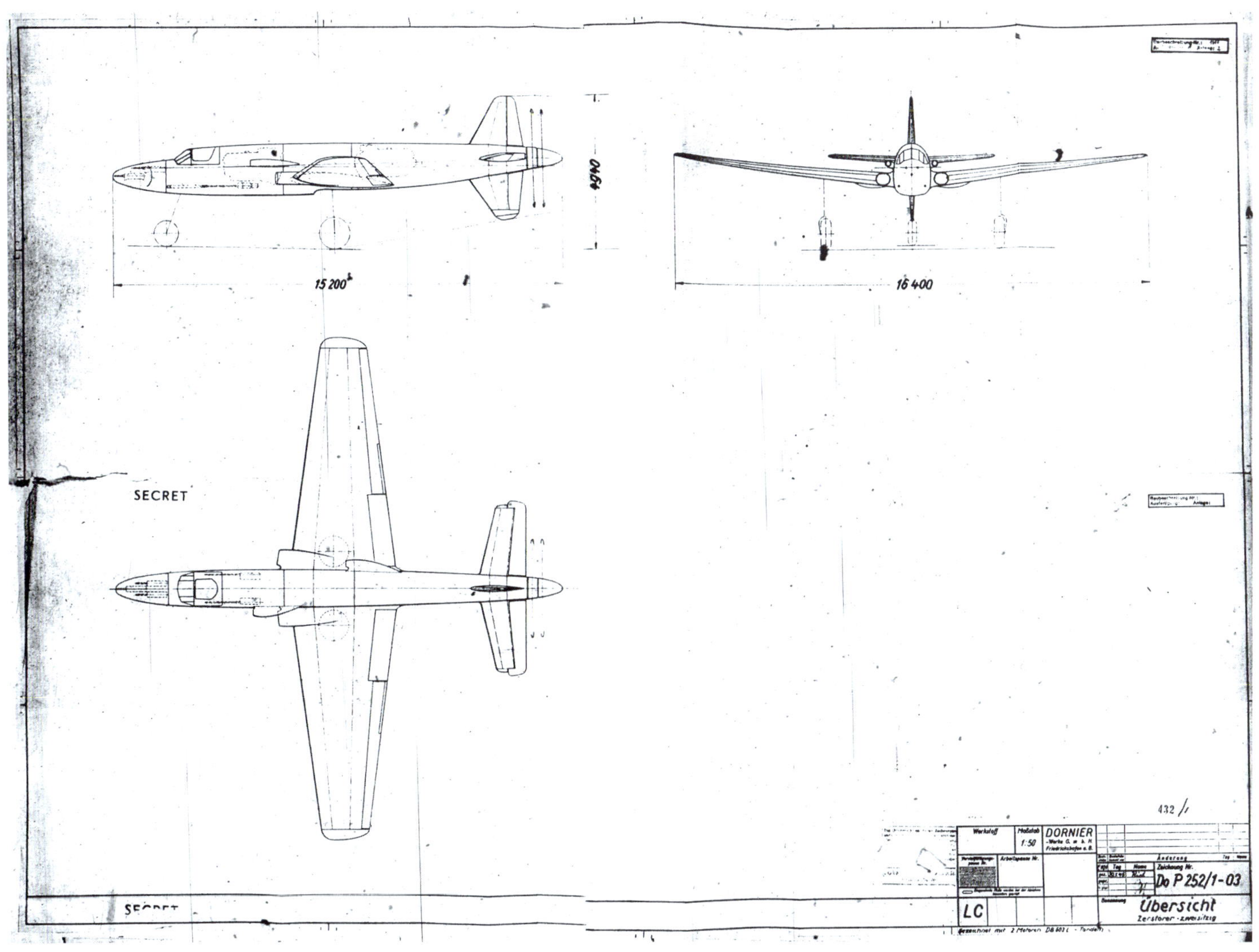

ABOVE: Arriving just at the point when piston-engined night fighter projects were unlikely to be considered was Dornier's Do P 252 design. The aircraft was to be powered by a pair of DB 603 Ls fitted in tandem and armament was three MK 108s plus two MG 151s. The drawing is dated January 20, 1945, and it was included in a description dated January 27.

possibility of accommodating more internal fuel for the increased performance engines must be ensured". Service ceiling needed to at least 13km and landing speed had to be below 180km/h. Rockets could be used to assist take-off if necessary and when it came to building the fighter "the construction effort is to be kept as low as possible by constructive measures. Wood is to be used as a material to a large extent".

This three-seater jet-only specification sent out to Arado, Blohm & Voss, Dornier, Heinkel, Junkers and Messerschmitt (Focke-Wulf is listed on the document as 'Sonderkommission Schlechtwetter und Nachtjäger' rather than by its own name) by Generalingenieur Herrmann, resulted in a transition phase since there were already numerous combination piston engine and turbojet projects, most of them two-seaters, already in existence and under consideration and these were not immediately dropped.

Arado was continuing to work on its Ar 234 designs; Blohm & Voss began working on a scaled-up twin-jet version of its P 212 single-jet fighter under the project number P 215; Dornier was in the unfortunate position of having just completed a new two-seater twin-piston-engine night fighter – the Do P 252 – outlined in Baubeschreibung Nr. 1608 of January 27 and wasn't quite ready to give up on it yet. It also continued to refine the Do P 254/1; Heinkel does not appear to have worked on any night fighter designs at this point, nor does Junkers, and Messerschmitt kept on with its Me 262-derived series of designs.

BLOHM & VOSS P 215.01-01

Work on the P 215 seems to have commenced in mid-February 1945 and its earliest known appearance is in a drawing dated February 21, 1945. This showed an aircraft similar in overall shape to the P 212.03-01 but with a wing area of 50m² compared to 15.4m². Wingspan was 17.6m compared to 9.5m and it stood 5m tall on its undercarriage, compared to the single-jet fighter's 2.75m height. Within its upper nose it housed four MK 108s with two more shown below but firing up at an angle. Another pair of MK 108s were positioned within the rear fuselage pointing upwards at a 70° angle.

It is unclear from the drawing how many crewmen sat inside the pressure cabin cockpit – the side view shows only two, back to back, but a list of weights[25] produced on February 22 stated that the aircraft had three crew. Presumably, as with the later version of the design, the pilot and radio man sat side by side at the front while the navigator sat behind them. In addition, the list of weights only provides for a total of four MK 108s even though the drawing clearly shows six. The reason for this is unclear.

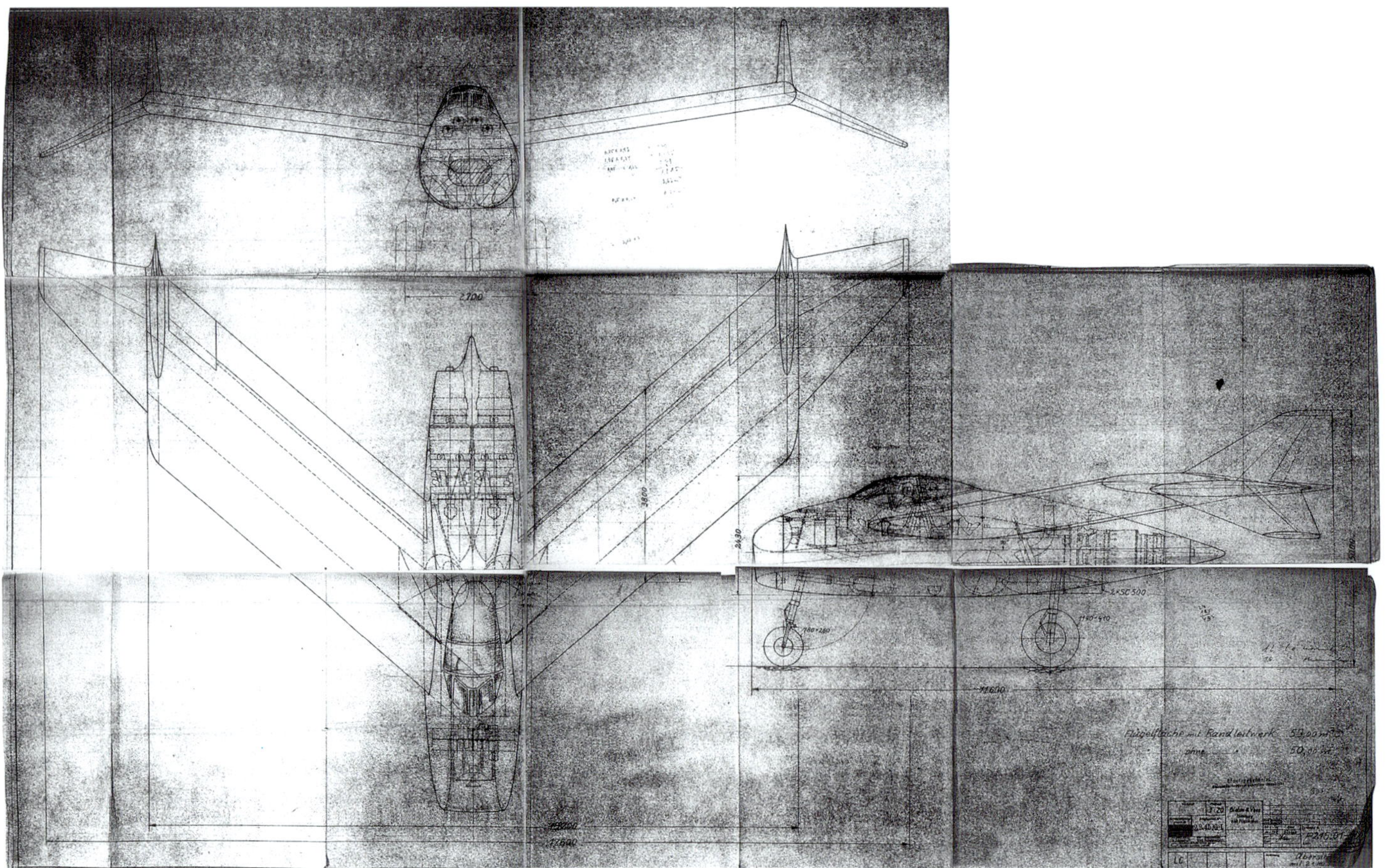

ABOVE: Only two copies of Blohm & Voss's P 215.01-01 drawing are known to exist – one is very faint and the other, shown here, is somewhat fragmentary. The aircraft itself was essentially a scaled-up P 212 but it nevertheless came close to meeting the latest night fighter requirements.

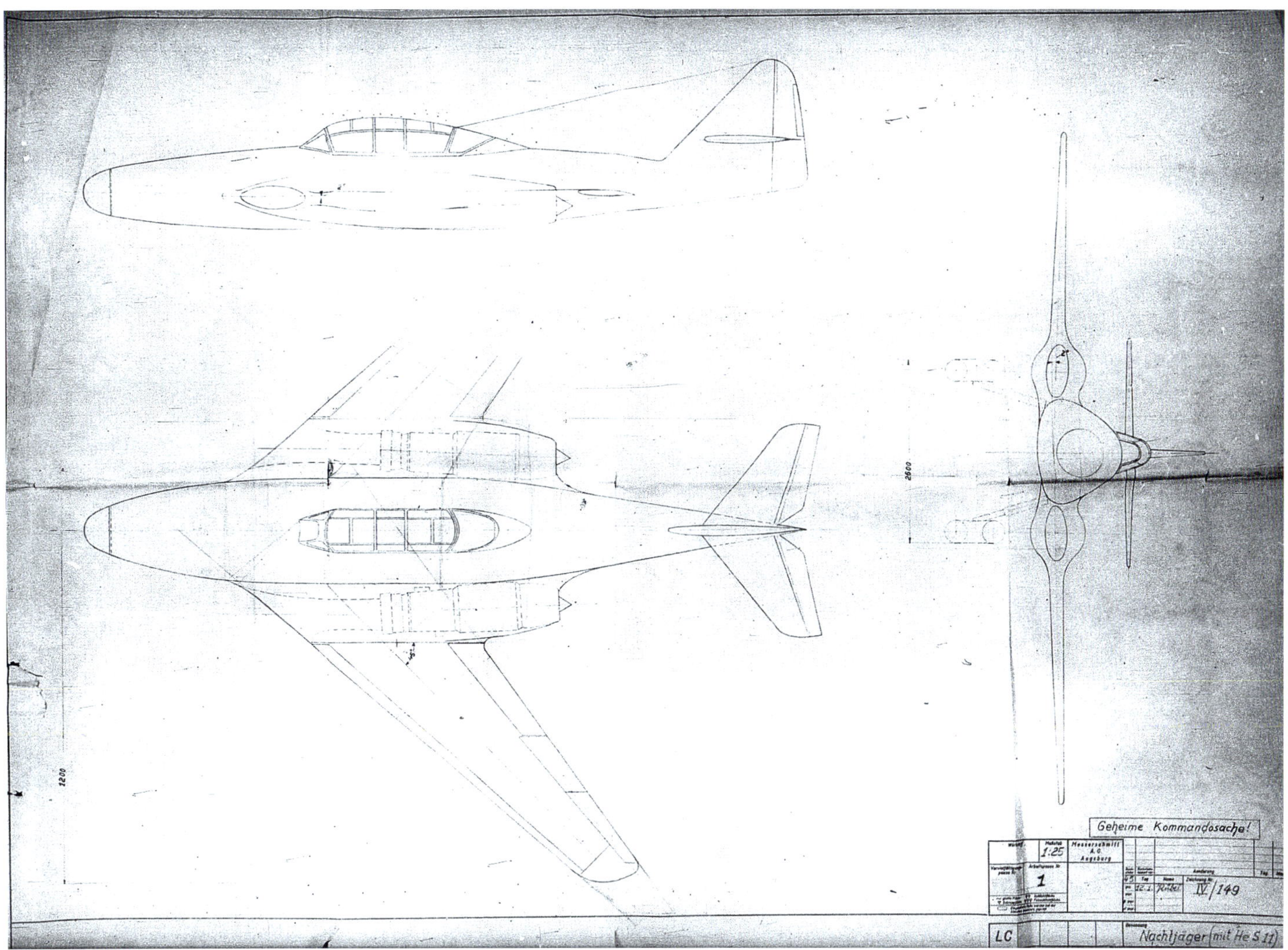

ABOVE: The Messerschmitt Me 262 B2 as it appeared in a description document of February 12, 1945 – utilising the layout intended for the HG III fighter. The drawing itself, number IV/149, is dated a month earlier.

ME 262 NACHTJÄGER MIT HES 011

The penultimate variant of the Me 262 produced by Messerschmitt appeared in a project description document[26] dated February 12, 1945. The introduction stated: "To increase combat value and performance, as a replacement for the current night fighter Me 262 B2 under construction with 2 x Jumo 004 B engines, a night fighter has been designed with HeS 011 engines, which largely meets the technical guidelines from January 27, 1945.

"In addition to armament, equipment, protection and strength, the night fighter B2 already met the demand for speed after switching to the HeS 011 A engines. To further increase performance, as is also planned for the day fighter, a sweeping of the wings with a drag-reducing arrangement of the engines in the wing root was provided. The fuselage corresponds to the Me 262 B2 version. This means that the amount of fuel stored in the fuselage is the same – approx. 3,100 litres. The flight duration with full throttle is approx. 2.5 hours at 6km altitude. The required flight duration of four hours can be achieved at an altitude of 12 km with 2,600 litres of external tanks.

"In both cases, i.e. the night fighter version B2 with HeS 011 engines and the night fighter with swept wing, the demand for the third man can be realised by lengthening the cylindrical fuselage centre section without reducing the fuel content. A further increase in the fuel supply accommodated in the fuselage is possible at the expense of the armament or the number of the crew. However, it seems advisable to use this space for an expansion of the FT system, the development of which is constantly in flux, and to accommodate the fuel in detachable external tanks or in low-drag fixed external tanks."

This new Me 262 had four MK 108s in its nose and another two in its fuselage pointing upwards at an angle of 70°. Its swept wings and wingroot engine positions were based on what Messerschmitt had already learned from its work on the HG III and it was based, albeit loosely by now, on the tried and tested Me 262 A-1 fighter. However, it was only able to meet the requirement's stringent endurance specification through the use of external fuel tanks. No figure was given for fuselage length or wing area but wingspan was 12m.

FOCKE-WULF ENTWURF I AND II

With the course of the night fighter competition having been steered towards their preferred twin-jet configuration, the Focke-Wulf team set about refining their design. Kurzbeschreibung Nr. 27 Nachtjäger mit 2 x HeS 011 of February 7, 1945,[27] outlined a design similar to that shown in Kurzebschreibung Nr. 26 of 11 days earlier. Appearing in drawing 0310 251-36 of February 6, it was 13.8m long – 40cm shorter than the preceding design – but with the same wingspan of 17.3m and wing area of $50m^2$.

The shape of the fuselage had become cylindrical where before it had had a triangular cross section and the wings were moved forwards and given a stronger sweepback of 35°. The aircraft now sat lower on its tricycle undercarriage too, with its tail pointing towards the ground. Take-off weight had risen to 11.3 tons and armament options remained the same.

However, as the date set for the next EHK meeting, February 27–28, 1945, approached, the Focke-Wulf team continued to rework and refine their design. Shortly before the meeting they produced what they dubbed 'Nachtjäger mit 2 x HeS 011 Entwurf II', with the evolution-in-progress designs of Kurzbeschreibung Nr. 26 and 27 having been 'Entwurf I'. No paperwork was available in time for this new design, however, so Kurzbeschreibung Nr. 27 had to be submitted along with a decidedly rough-looking drawing, 0310 251-38, showing the new Entwurf II.[28]

Wingspan was now reduced to 16m, though wing area remained $50m^2$ and no value was given for overall length. Entwurf II differed from the earlier designs primarily in having a much shorter intake for its two turbojets. This meant that they exhausted under a very long section of rear fuselage. And consequently where the undercarriage had previously retracted slightly forwards and up into the lower sides of the fuselage next to the intake duct, it now had to pivot through a huge arc to enter the sides of the fuselage higher than the aircraft's wings.

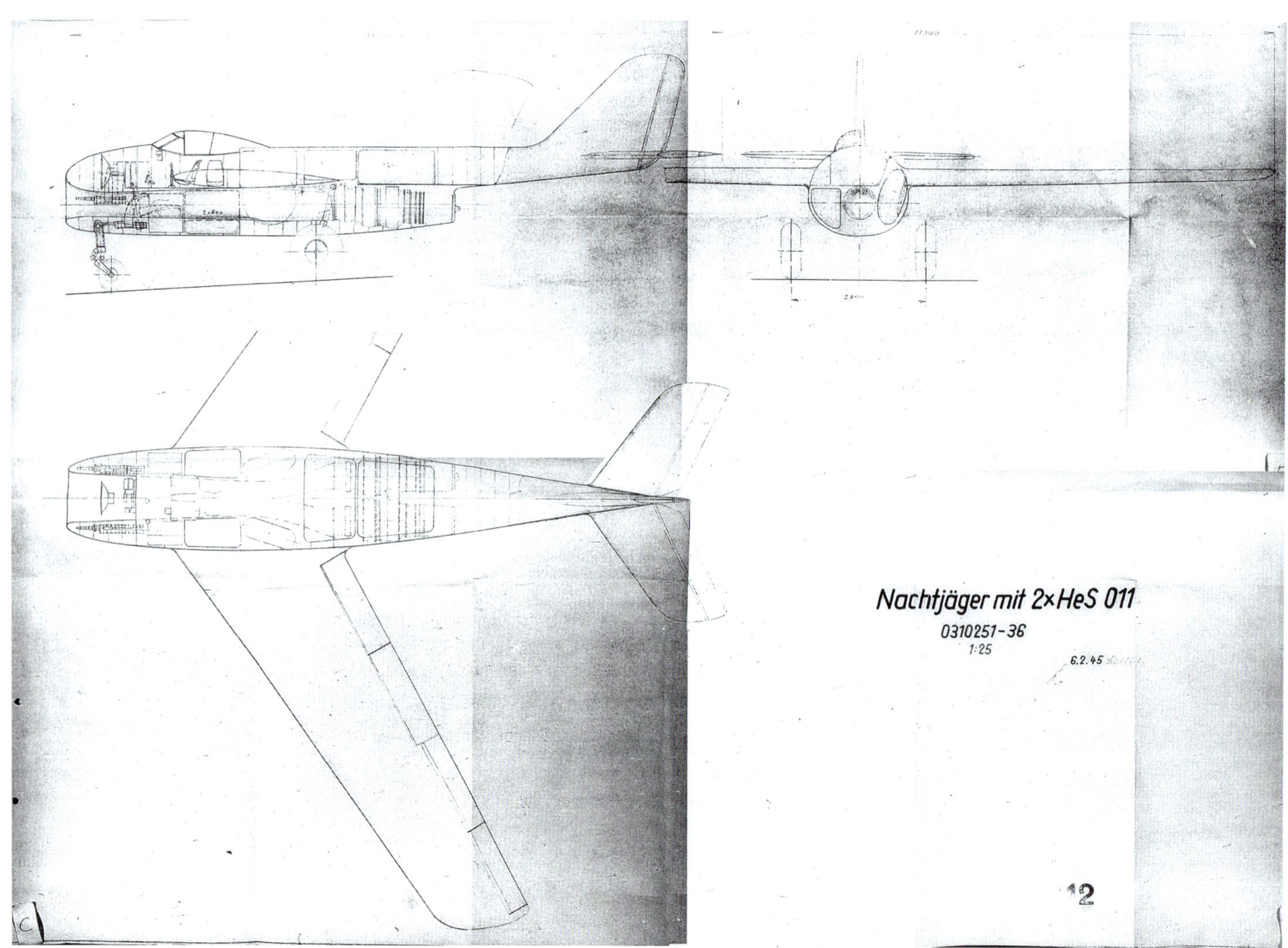

ABOVE: Produced just 11 days after the Kurzbeschreibung Nr. 26 design, this new Focke-Wulf twin-jet night fighter appeared in Kurzbeschreibung Nr. 27 of February 7, 1945 – the drawing itself, 0310 251-36, being dated February 6. It appears more practical and less conceptual than the previous design, with a simpler cylindrical fuselage, lower undercarriage and multi-pane canopy. Focke-Wulf regarded it as 'Entwurf I'.

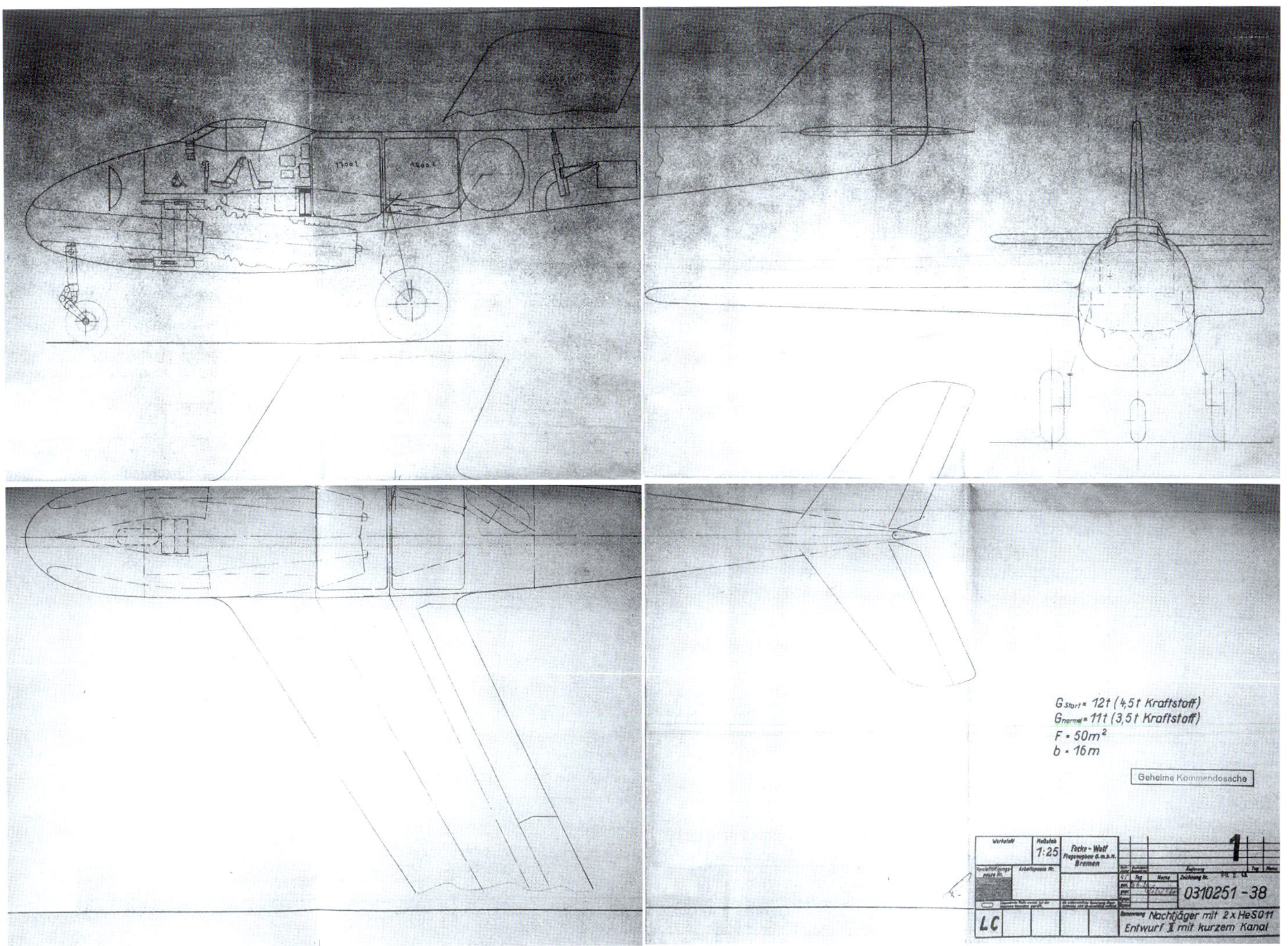

ABOVE: Like Messerschmitt, Focke-Wulf had serious concerns about turbojet intake duct length. This drawing, 0310 251-38, shows a new night fighter design dubbed 'Entwurf II' with a shortened intake – resulting in a shorter lower fuselage and therefore an unusual undercarriage retraction system.

PROJECT COMPARISON I

The engineers of Focke-Wulf's Arbeitsgruppe Flugmechanik worked feverishly to compare the various night fighter projects and by February 26 had narrowed the field to five designs[29] – Blohm & Voss's P 215.01, Dornier's Do P 252/1 and Do P 254/1, Messerschmitt's Me 262 night fighter project and Focke-Wulf's own entry, described as Nachtjäger mit 2 HeS 011 Entwurf II.

The Do P 252/1 was 15.2m long with a wingspan of 16.4m and a wing area of 43m^2. Assessment highlighted a number of shortcomings – a crew of only two clearly meant that "the project offered does not meet the requirements of the tender in this point". Dornier had failed to supply a detailed list of weights, it was noted, so it was difficult to assess that aspect of the design but based on a comparison with the other designs the Do P 252/1 appeared to be relatively lightweight. It was also unclear "how and where the extensive FT equipment is to be accommodated".

Another problem with the Do P 252/1 was the cooling system for the two piston engines fitted in tandem within its fuselage – either two DB 603 LAs or two Jumo 213 Js.[30] Standing still, the aircraft's radiators would be ineffective so Dornier had been forced to provide "50 litres of water for evaporation per engine for free-standing and starting. In view of the uncontrollable water losses, this system is strongly discouraged. Even if the planned water flow would be able to remove the relevant quantities of heat, such large quantities of water can be carried away by the escaping steam that there is a risk of the system failing".

In addition, the aircraft's upwards-firing guns were in its nose – a position likely to create "strong moments around the transverse axis" when firing, causing a deterioration in accuracy. Armour protection was good but in summary the Focke-Wulf engineers found that building the Do P 252/1 would take a considerable amount of effort and further development of the straight-winged design to accommodate swept wings "should always be taken into account, since when the wings are swept practically a new aircraft is created".

The Do P 254/1 (as a night fighter – it was also offered as a heavy fighter with smaller wings) was 13.4m long with a wingspan of 15.45m and a wing area of 41m². The Focke-Wulf team were equally unimpressed by it however, again stating that its two-seater configuration failed to meet the requirements of the tender, noting a lack of information about weights and stating that Dornier's drawings offered no clues as to where all the radio and radar equipment would go. Engine-wise, they were concerned that gases from the nose-mounted piston engine would end up being fed into the turbojet, resulting in a drop in its performance.

The extensive fuel tanks in the Do P 254/1's wings were unprotected. In summary, the Focke-Wulf team concluded that "the Do P 254/1 project is not very suitable for fighting at night or in bad weather" although "the construction effort for this aircraft is slightly less than for the Do P 252/1 project".

Messerschmitt's claims that a third man could be accommodated in its Me 262 project of February 12, 1945 were dismissed due to a lack of drawings showing the arrangement and the design as presented, with two crew, failed to meet the tender requirement. As with the Dornier projects, a detailed breakdown of weights was lacking and the Focke-Wulf men were worried that putting the aircraft's engines in its wingroots would result in a loss of rigidity in the wings and furthermore "the specified outline of the turbojet housing does not correspond to the recently enlarged dimensions of the turbojet device".

Unfortunate positioning of the aircraft's upwards-firing weapons was also a major drawback: "In the Messerschmitt project, the oblique weapons are arranged on both sides of the rear man, in such a way that the muzzle is located close to the radio operator's head in the immediate vicinity of the glazing. A circumstance that will most likely lead to disturbances and complaints". And "the visor angle of 12° necessary for the full use of the EZ 42 sighting device does not seem to have been observed".

The summary focused on dismissing what might have been seen as the Messerschmitt project's main strength – the opportunity to build it from existing components: "The present project can only fall back on existing assemblies of the previous Me 262 in very subordinate components. The overall structure of the fuselage, the new main landing gear, the strong sweepback, the novel positioning of the turbojets with the intake located in the wing leading edge, the swept tail, can be expected to give a completely different flight behaviour than the previous Me 262 version. In particular, difficulties regarding stability about the transverse axis are to be expected, caused by the arrangement of the turbojets close to the fuselage."

Surprisingly, given its radical design, the Focke-Wulf team had little to say about the Blohm & Voss P 215.01-01. It had three crew – meeting the requirement – and "the weights given correspond to current existing experience and can probably be adhered to". The 6.3m-long intake duct for the two turbojets would likely result in a thrust loss of around 4% and "the crew members and equipment arranged in the fuselage can be easily protected" but "the entire fuel supply is housed in the wings and is unprotected".

In summary, "a decision on the BV project can only be made based on experience with the flight characteristics of tailless aircraft at high Mach numbers, as can probably be obtained from the Horten aircraft". The only concern was stability.

Focke-Wulf did not subject its own design to analysis but it certainly adhered more closely to the requirements than its rivals.

NEW SPECIFICATION

The five designs were presented when the EHK met on February 27–28, 1945, but the new General der Jagdflieger Gordon Gollob – who had officially replaced Adolf Galland in the role on January 31 – was apparently not satisfied by any of them and "expressed new demands".[31] Therefore it was decided that a new specification should be drawn up and issued to the companies, with a requirement that their revised projects should be ready by March 20 when the EHK was scheduled to begin its next five-day meeting.

As it turned out, Gollob's amended requirement, issued to Arado, Blohm & Voss, Dornier, Heinkel, Junkers, Messerschmitt and Tank's Sonderkommission Schlechtwetter und Nachtjäger on March 1, 1945,[32] though dated February 27, was mostly the same as that of January 27 but with a few key amendments. The section relating to crew had a new line added: "Arrangement: radio operator at the front, navigator at the rear. The latter must have good visibility to monitor the rear airspace".

The angle of the aircraft's oblique weaponry was changed from 70° to 80° but it was to be checked whether the defensive weaponry and the oblique weaponry could be combined – saving weight by eliminating one set of guns. There was also a new requirement to fit R4M rockets if necessary.

Where the original spec had called for "ejection seat", it was now spelled out that there should be "ejection seat for each crew member, if it is possible to arrange the openings without impairing the functional safety of the pressurised cabin".

Greater endurance was also specified. It had been "four hours including climb" but now the requirement said: "Climb to 10km with full throttle, then two hours' flight at full throttle, then two hours' flight at 60% throttle".

There was more detail, too, in the construction specification: "The construction effort is to be kept as low as possible through constructive measures. Wood is largely used as a material. Good maintenance and repairs by troops are largely to be considered. It must be possible to park the aircraft outdoors (earth connections, etc.) The use of fuels of the poorest quality ... has to be considered constructively. (Heating devices for fuel in tanks, insulated pipes, heater, etc.)"

This new set of guidelines, along with the tight deadline, seems to have galvanised Arado, Blohm & Voss, Dornier, Focke-Wulf, Heinkel and Messerschmitt into action – though the designs produced by the latter two would not be included when the night fighter projects were presented to the EHK during its meeting of March 20–24. In fact a seventh company, Gothaer Waggonfabrik, would also submit a design which was presented to the EHK, although exactly how the company came to be admitted to the competition is unclear.

ARADO I AND II

Having abandoned any attempt to make an Ar 234 variant fit the spec, Arado started work on an entirely new series of designs, the first of which, labelled 1. Entwurf zum Nacht- und Schlechtwetterjäger, was drawn on March 7. This was a tailless design with no vertical control surfaces and the distinctive kinked wing that the company had developed during an aerodynamics project designation E 560. It had an armament of six forward-firing MK 108s, two more firing upwards and a small rear turret set between its turbojet exhausts mounting two unspecified remote-control weapons.

The 2. Entwurf zum Nacht- und Schlechtwetterjäger (mit verkleinertem Rumpf), dated March 8, 1945, was another tailless design which had, as the name suggests, a shortened fuselage. Again it had six forward-firing MK 108s but no rear turret and its swept wings had a dead straight leading edge. 3. Entwurf zum Nacht- und Schlechtwetterjäger dated March 9 was also tailless but switched back to the longer fuselage. Now, however, it also featured a vertical fin towards the end of each wing. The forward-firing weaponry stayed the same but a rear turret mounted what appeared to be a pair of MK 103s on a pivot allowing them to track left and right. This was the first design to feature any measurements – having a wing area of 75.6m² and

ABOVE: Arado's tailless 1. Entwurf night fighter design from March 7, 1945. The drawing seems to be incomplete since it offers no clues as to where vertical control surfaces might go – without which the aircraft would have lacked stability.

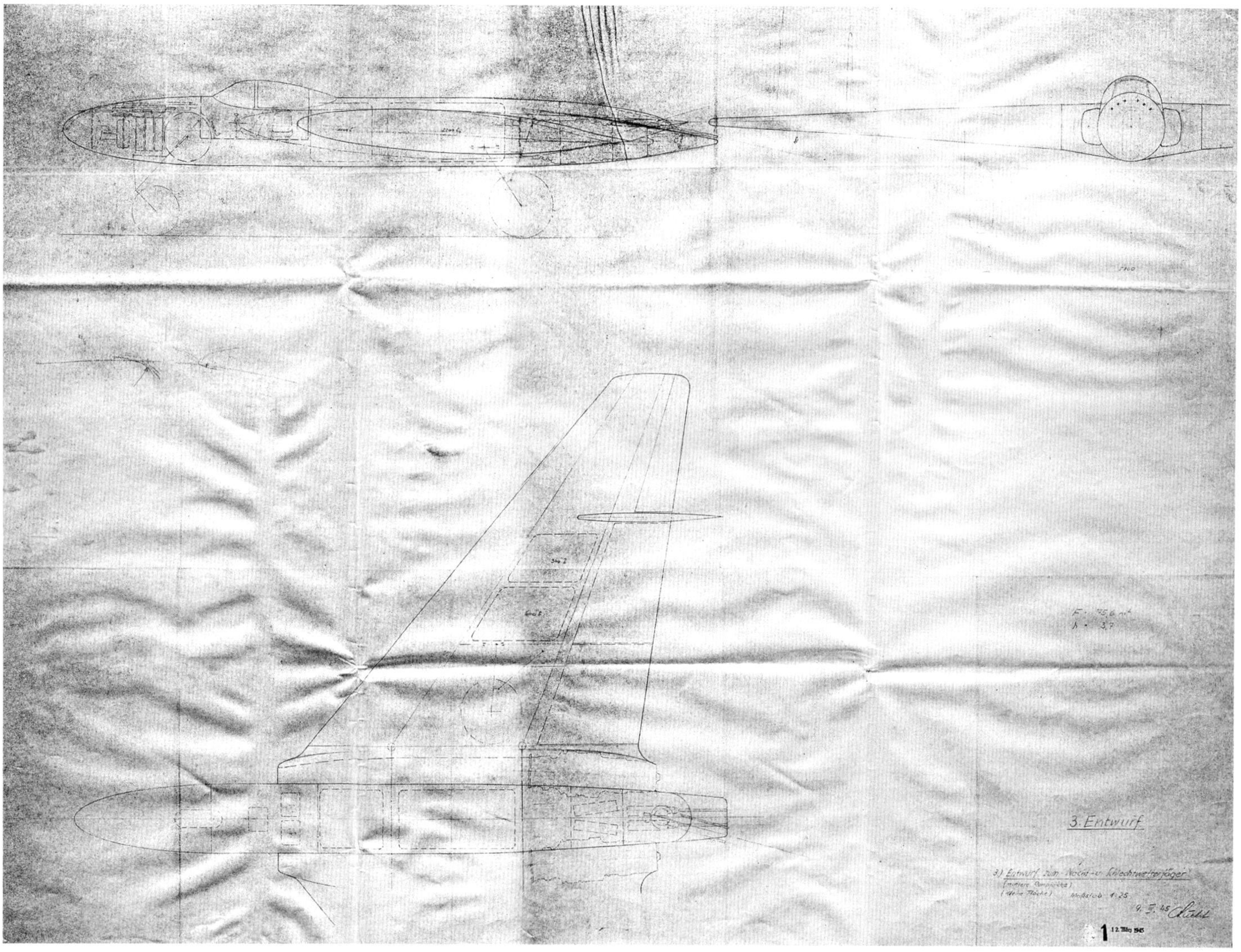

ABOVE: Arado reshaped the intakes for the engines of its twin-jet night fighter again for the 3. Entwurf of March 9, 1945. This design had smaller wings than its predecessors, a pair of vertical fins for stability and a seating arrangement where only the pilot faced forwards – his fellow crewmembers positioned back to back behind him.

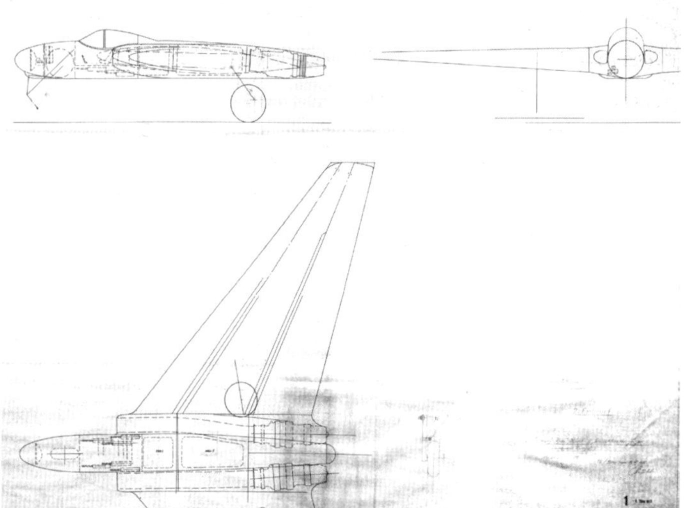

ABOVE: The 2. Entwurf Arado night fighter was another tailless design, dated March 8, 1945. It featured a narrower and shorter fuselage than the 1. Entwurf with reshaped intakes.

5.8m between the centre of each landing gear mainwheel in the forward view.

The fourth design is lost but the 5. Entwurf zum Nacht- und Schlechtwetterjäger, dated March 13, was a substantial refinement on what had gone before. Forward armament was the same but now two oblique MK 108s made an appearance and the MK 103 rear turret could track up and down rather than left and right. Wingspan was 16.8m and wing area was again 75.6m². The 6. Entwurf zum Nacht- und Schlechtwetterjäger was similar to the fifth design but with a wingspan of 18.4m and a wing area of 75m². Fuselage length was given for the first time too: 12.95m.

Curiously, the 6. Entwurf is dated March 14, whereas the 7. Entwurf is dated March 13. This design is the first in the sequence to feature a conventional fuselage – albeit an extremely long one – with a huge swept tailfin and engine nacelles integrated into its wings. There is no forward armament depicted – only oblique-firing cannon and a tail turret. Wingspan was 16.8m and wing area was 75m².

The 8. Entwurf is again missing and the 9. Entwurf zum Nacht- und Schlechtwetterjäger, dated March 17, 1945 – just three days before the conference – is another conventional layout design. Shorter than the 7. Entwurf, everything about it seems designed to be straightforward. The wings are swept but not dramatically so, the engines are fitted into underwing pods making them easy to access for maintenance, the three crew sit together under a blister canopy offering good visibility over the not-overlong nose and the tricycle undercarriage appears not dissimilar to the tried and tested undercarriage of the Ar 234 B-2. A V-tail was presented as an option. Length is given as 17.25m, wingspan 15m and wing area 50m².

From these nine designs, Arado chose the 6. Entwurf to offer as 'Arado I', albeit with a slightly modified undercarriage retraction arrangement, and the 9. Entwurf to offer unaltered as 'Arado II'. These two were sent to Tank's special commission for night fighters.

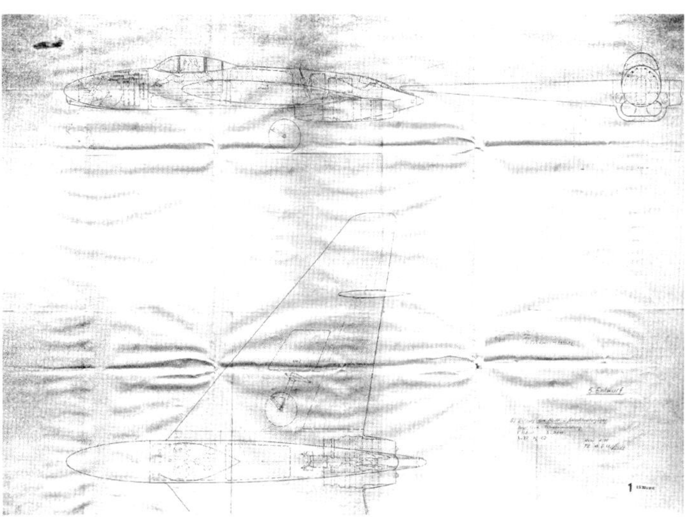

ABOVE: For the 5. Entwurf of March 13, 1945, Arado repositioned the night fighter's turbojets to eliminate the lengthy intake ducts of the previous designs.

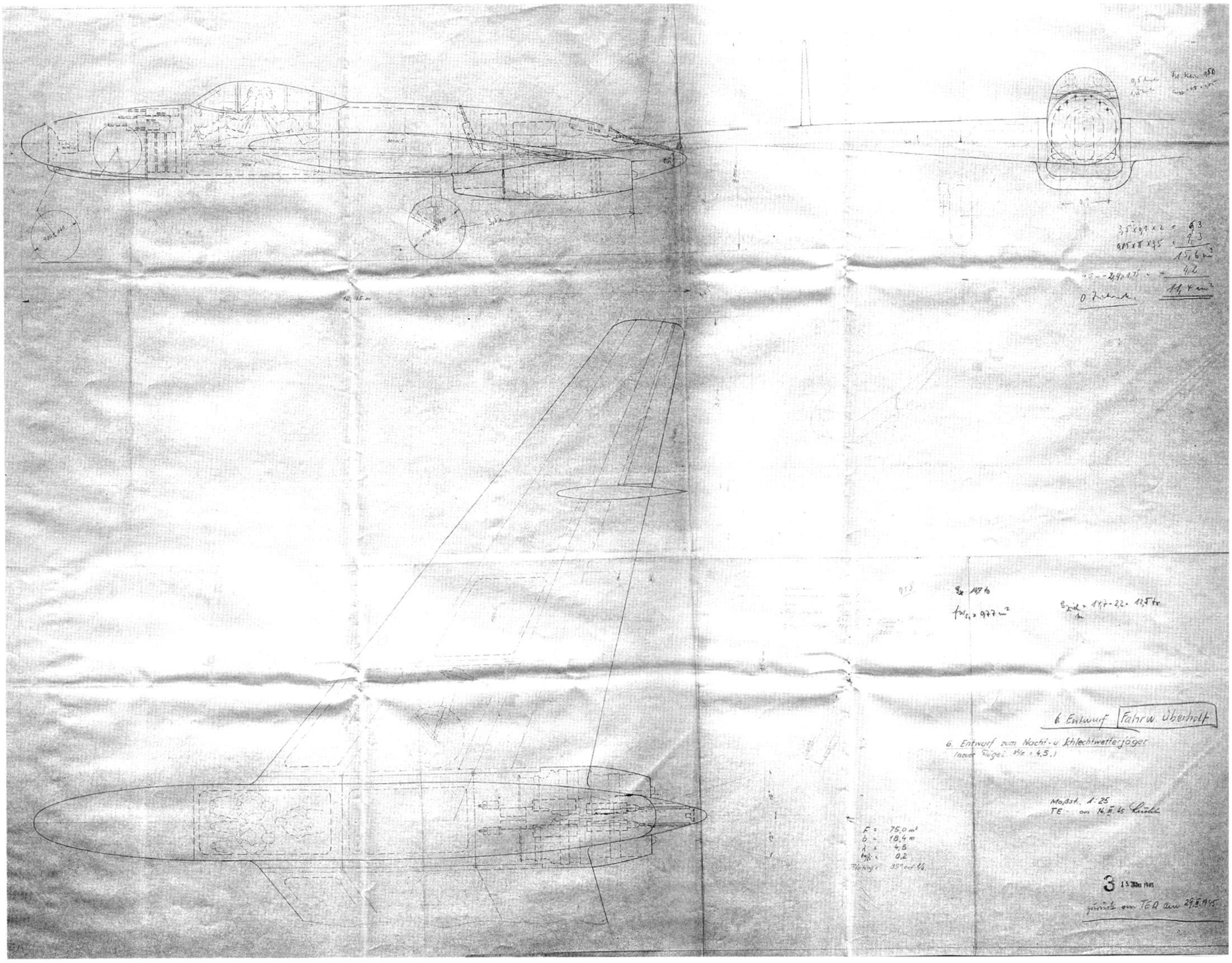

ABOVE: Having removed any concerns about intake duct length, Arado next widened the rear section of its night fighter design's fuselage - providing extra space for fuel. The 6. Entwurf design was dated March 14, 1945 and would be one of the company's two entries for the last round of the night fighter competition.

ABOVE: The Arado 7. Entwurf night fighter re-imagined the design with a conventional fuselage and wing-mounted turbojets. The result was a very large design, dated March 13, 1945.

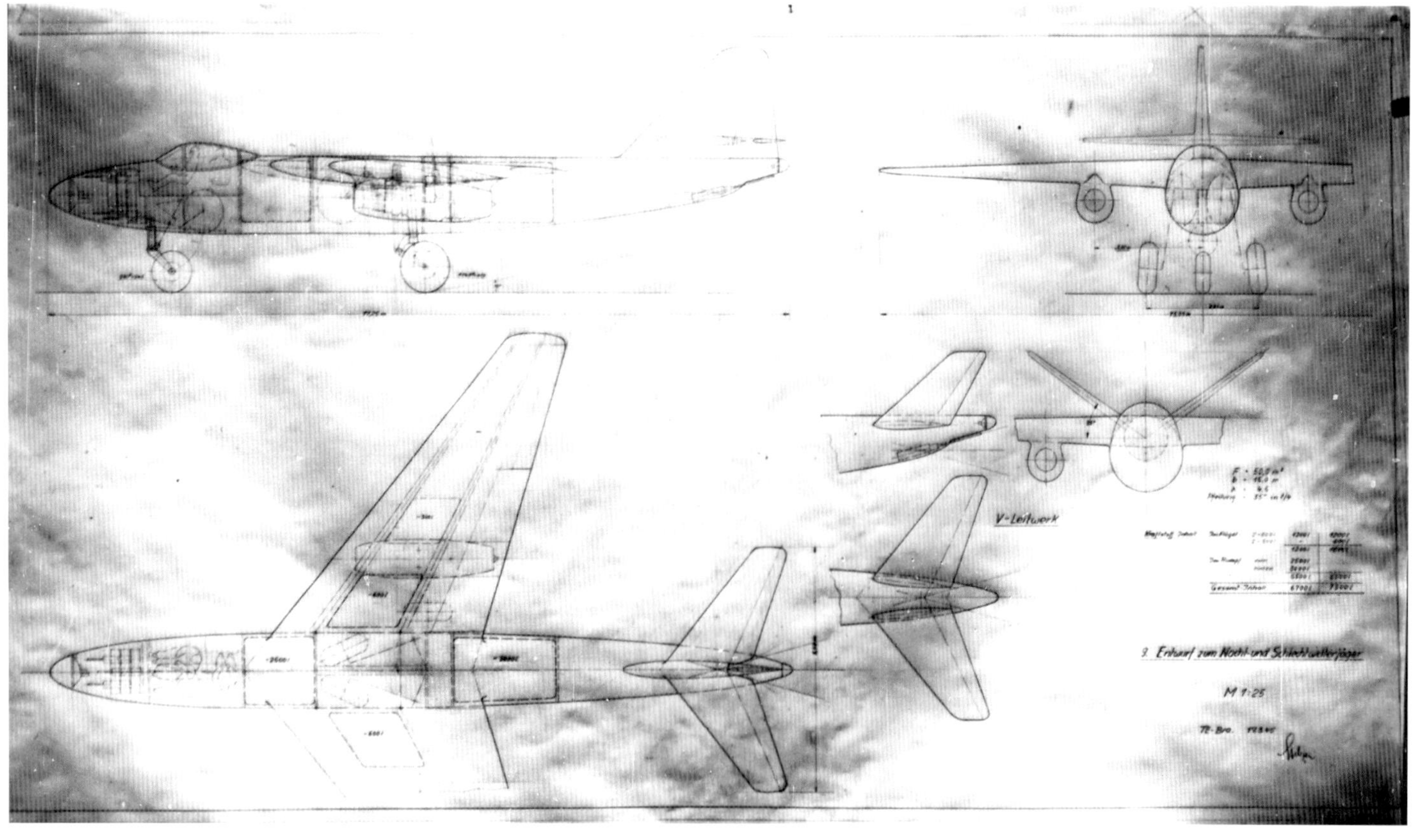

ABOVE: The last known Arado night fighter design was the 9. Entwurf of March 17, 1945. It was clearly intended to be a 'safe' option with a normal cylindrical fuselage and tailfin – though a V-tail option was also offered. This was Arado's second entry for the night fighter competition.

BLOHM & VOSS P 215.02-01

The Blohm & Voss P 215.02 appears almost identical to the P 215.01 at first glance but in fact it is a different, larger, aircraft in almost every respect. The wingspan, including control surfaces, is 18.8m, compared to 17.6m; wing area is 63m² compared to 59²m and fuselage width is 2.2m compared to under 2m. The only dimension which remains the same is overall length at 11.6m.

According to the introduction to the P 215.02 brochure, dated March 1945: "The 215 project grew out of the P 212 design, a day fighter with only one turbojet."

The brochure quickly acknowledges the P 215.02's sheer size as it continues: "With a three-man crew, with reinforced armament, with increased flight time and with two turbojets the dimensions of the aircraft are correspondingly large. However, the entire arrangement is largely similar to that of the day-fighters.

"The two turbojets are built into the rear fuselage and get fed from a front fuselage nose intake via a common channel. The armament is gathered in a weapons section, which is directly behind the antenna devices and in front of the crew cabin. Nose and main landing gear are both hinged to the fuselage. The aircraft is therefore transportable after removing the wings. This brings advantages in the final assembly, because the transport scaffolding can be omitted."

The landing gear consisted of an He 219 nosewheel, according to the brochure, and large mainwheels which tucked up neatly inside the fuselage.

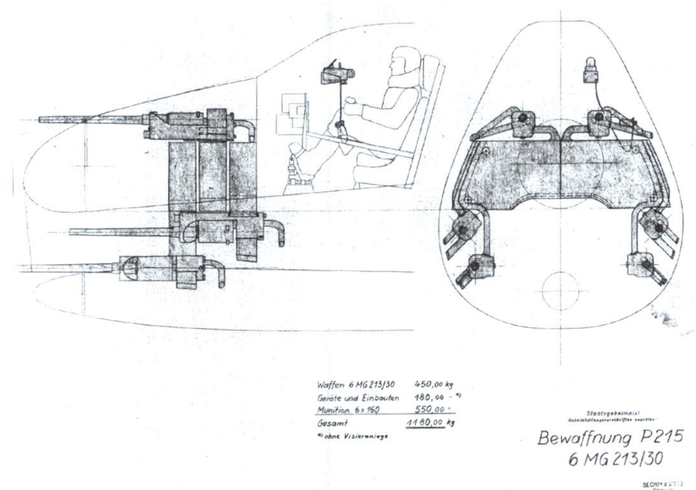

ABOVE: A straightforward armament of six MG 213/30 cannon would have seen the P 215.02's nose crammed with ammunition containers.

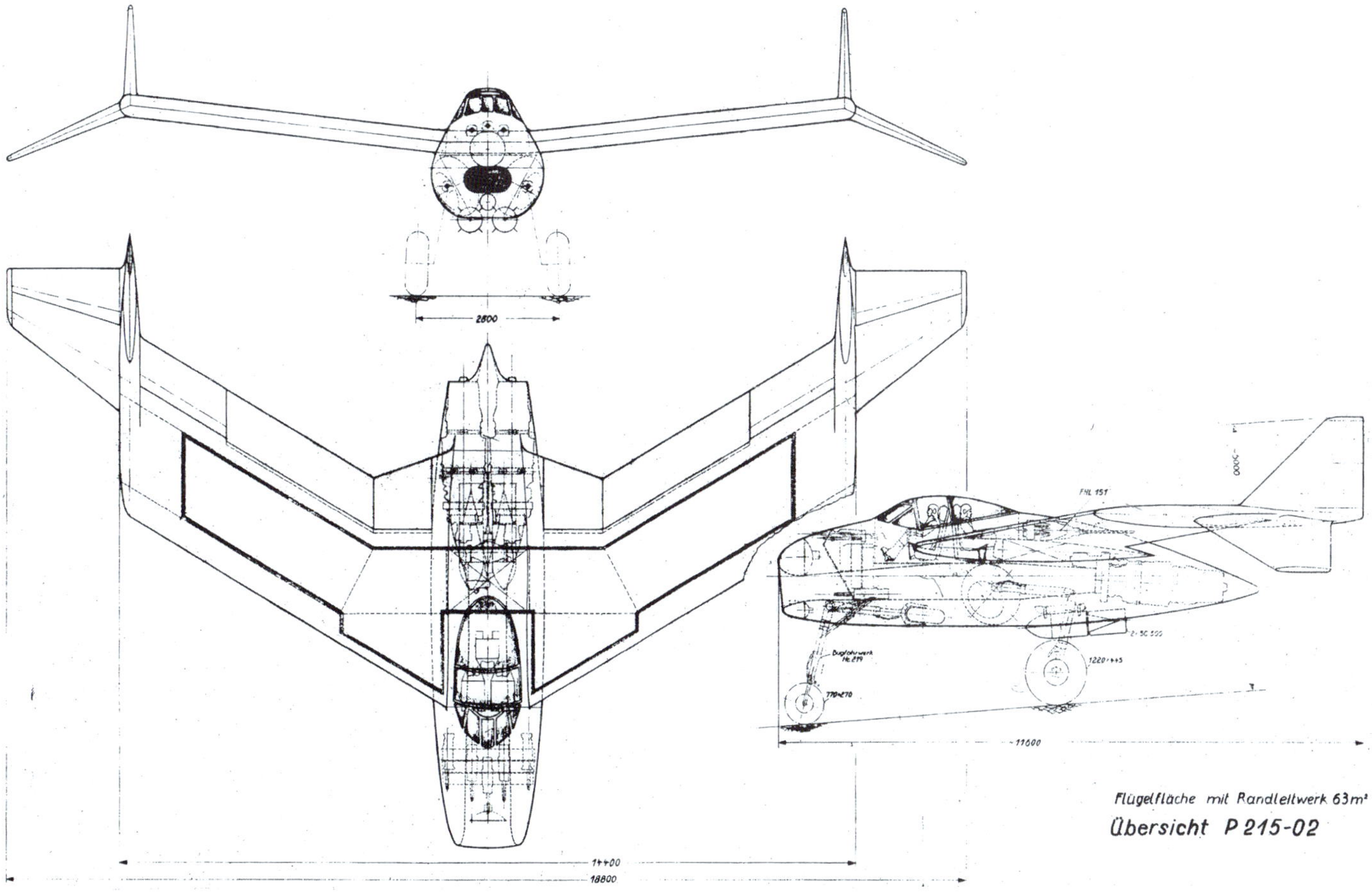

ABOVE: The Blohm & Voss P 215.02-01 looked superficially similar to the P 215.01-01 but was actually a completely different design. It was larger with differently shaped wings, different fins and a new remote-controlled defensive turret.

Schnitt A-B
Schnitt G-H
Schnitt C-D
Schnitt E-F
Längsschnitt P 215

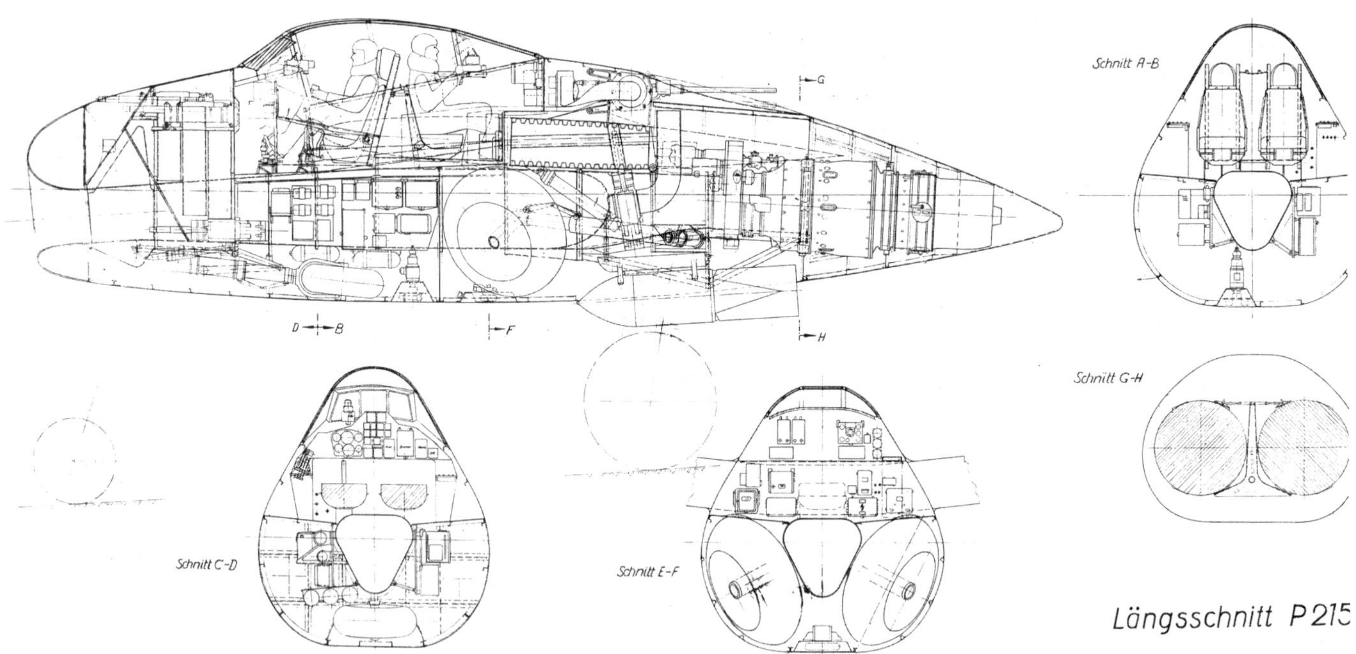

ABOVE: Detailed side view of the **P 215.02** from Blohm & Voss's description report showing a full-vision canopy and with the navigator positioned higher up than in other views.

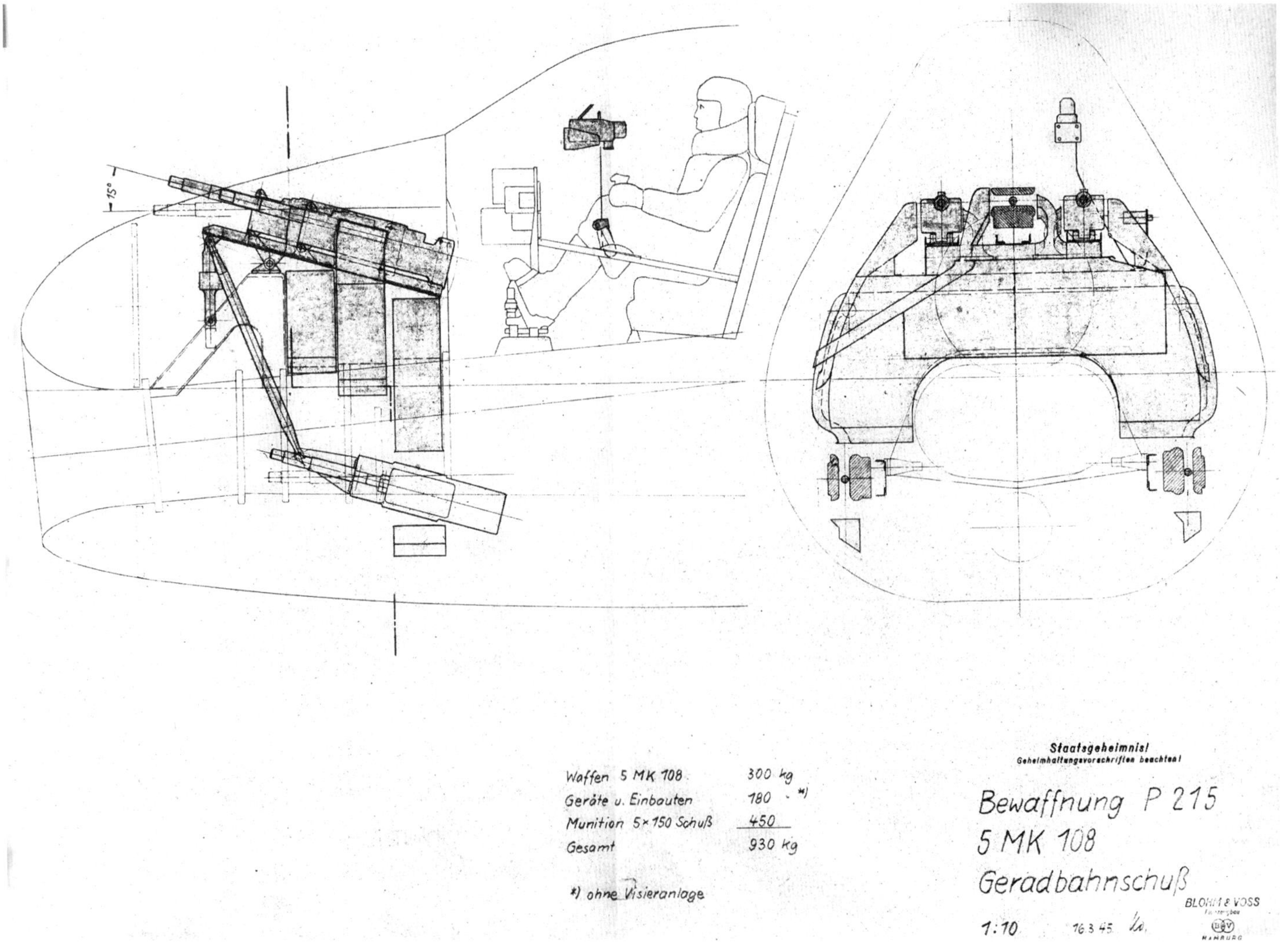

15°
Staatsgeheimnis!
Geheimhaltungsvorschriften beachten!
Waffen 5 MK 108 300 kg
Geräte u. Einbauten 180 *)
Munition 5×150 Schuß 450
Gesamt 930 kg
*) ohne Visieranlage
Bewaffnung P 215
5 MK 108
Geradbahnschuß
BLOHM & VOSS
HAMBURG
1:10
16.3.45

ABOVE: In this configuration a novel mechanism would have allowed the **P 215.02**'s pilot to tilt its five **MK 108** cannon upwards for attacking enemy bombers from behind and below.

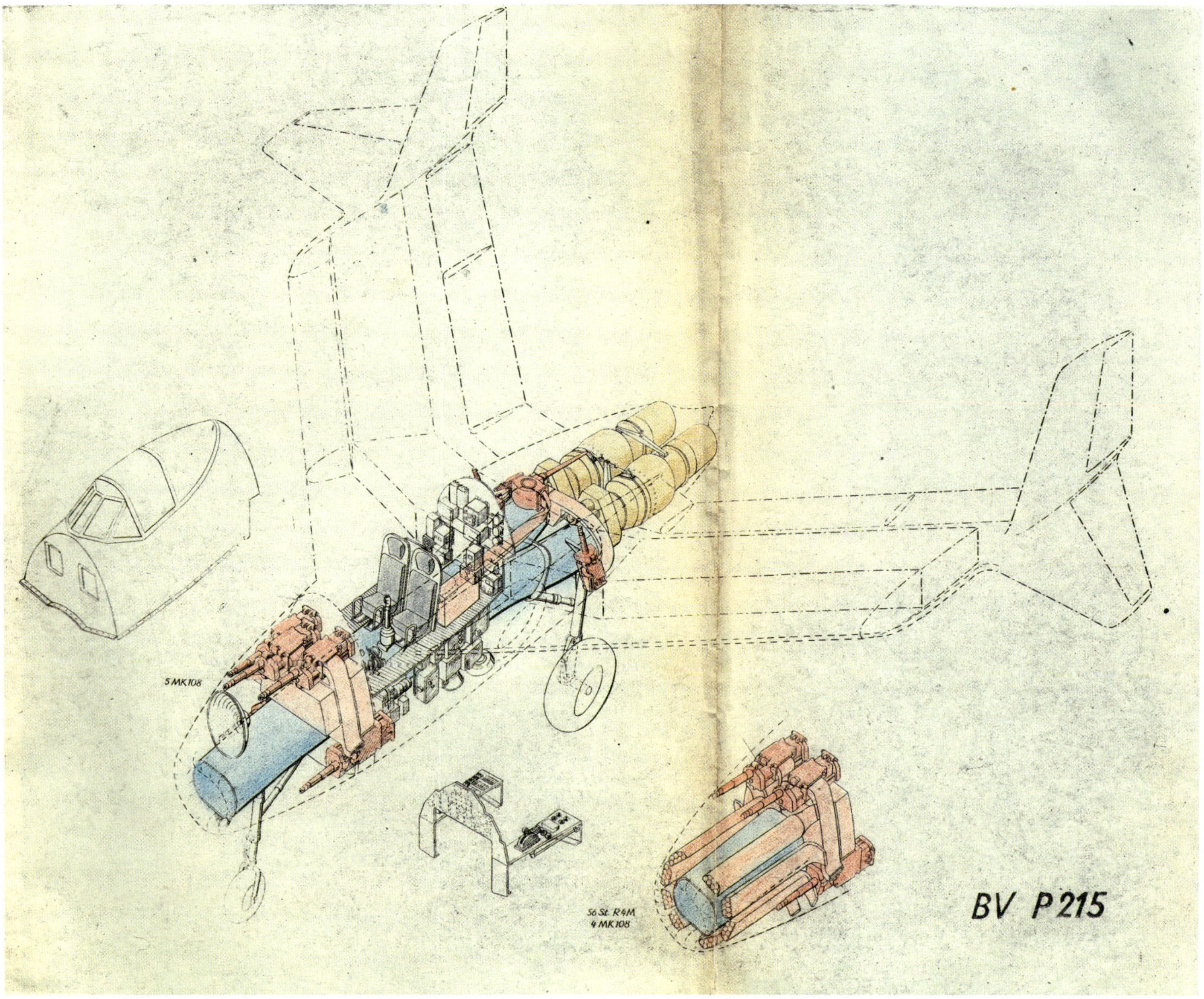

ABOVE: Colour artwork from the original Blohm & Voss P 215 night fighter project description document. Towards the bottom centre of the drawing is the type's alternative armament of 56 R4M rockets and four MK 108 cannon in red around the blue jet intake tube.

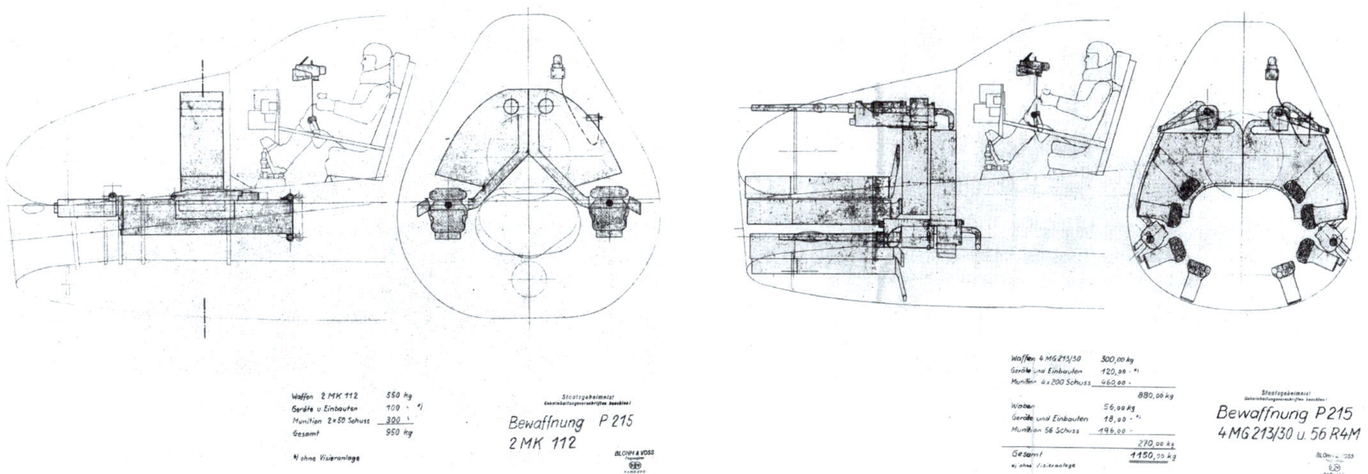

LEFT: The P 215.02 fitted with a pair of 55mm MK 112 cannon – a scaled-up variant of the 30mm MK 108. **RIGHT:** Possibly the deadliest potential armament for the P 215.02 was a quartet of MG 213/30 revolver cannon combined with 56 R4M rockets arranged in eight pods of seven within the nose of the aircraft.

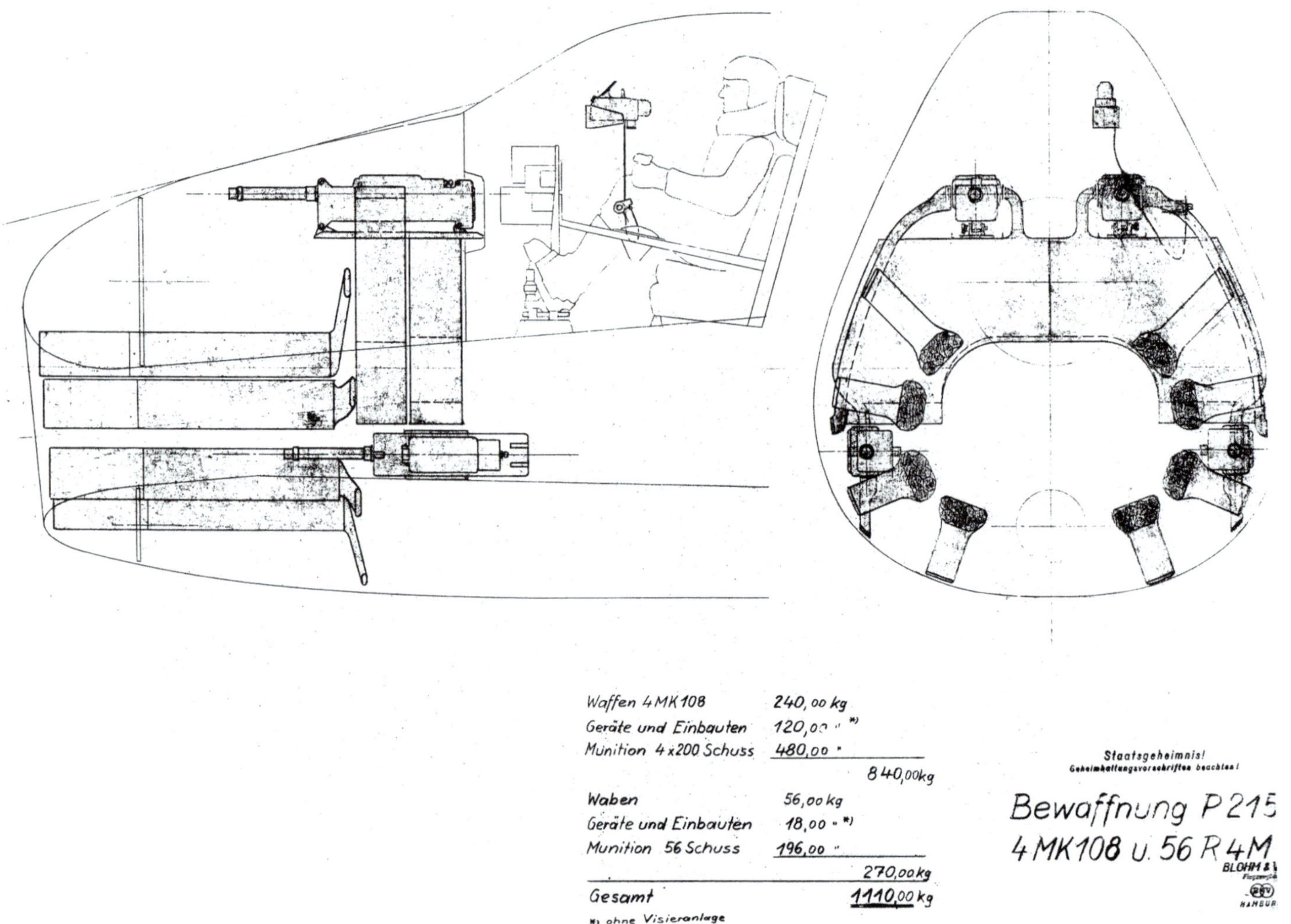

ABOVE: Yet another P 215.02 armament option was four MK 108s and 56 R4M rockets.

Normal armament (option A) was five MK 108 cannon with 200 rounds each. This could be exchanged for (option B) five MK 108s with 150 rounds each that could be tilted upwards at an angle of up to 15°.

Option C was four MG 213/30s with 200 rounds each and option D was two massive MK 112 cannon with 50 rounds each. Finally, option E allowed for 56 R4M unguided rockets to be installed in the P 215's nose – alongside either four MK 108s, four MG 213/30s or two MK 212/214s.

Like the P 215.01, the P 215.02 could also be fitted with a pair of MK 108s aimed to fire upwards. There was also an option to carry a pair of 500kg bombs semi-recessed within the fuselage. Where the P 215.02 again differed significantly from its predecessor was the inclusion of a rear-facing FHL 151 remote-controlled turret armed with a single MG 151/20 cannon. This was to be aimed and fired by the navigator.

DORNIER DO P 256/1

Unlike Arado, Dornier was still attempting to base its night fighter on existing designs, at least partially. The company's entry for this latest round of the advanced night fighter competition was the Do P 256/1 described in Baubeschreibung Nr. 1615 of March 15, 1945.[33] Time was clearly short when the project was being drawn up because the description is decidedly threadbare.

The introduction amounts to fewer than 60 words: "The present design is based on the proven design of the Do 335. Instead of the piston engines in the fuselage, however, two 109-011 jet engines are provided under the wings on the left and right. As a result, the proposed solution combines the aerodynamically favourable conditions of the Do 335 with the advantages of jet propulsion."

The Do P 256/1 effectively used bits and pieces of the original Do 335 with a new nose and new rear fuselage – the turbojets simply being slung beneath the Do 335's straight wings in nacelles. The pilot and radio operator sat side by side at the front but the navigator was forced to sit in a cramped position on his own in the rear fuselage.

The nose housed the radar dish and four MK 108s. Length was 13.8m and wingspan was 15.45m with a wing area of

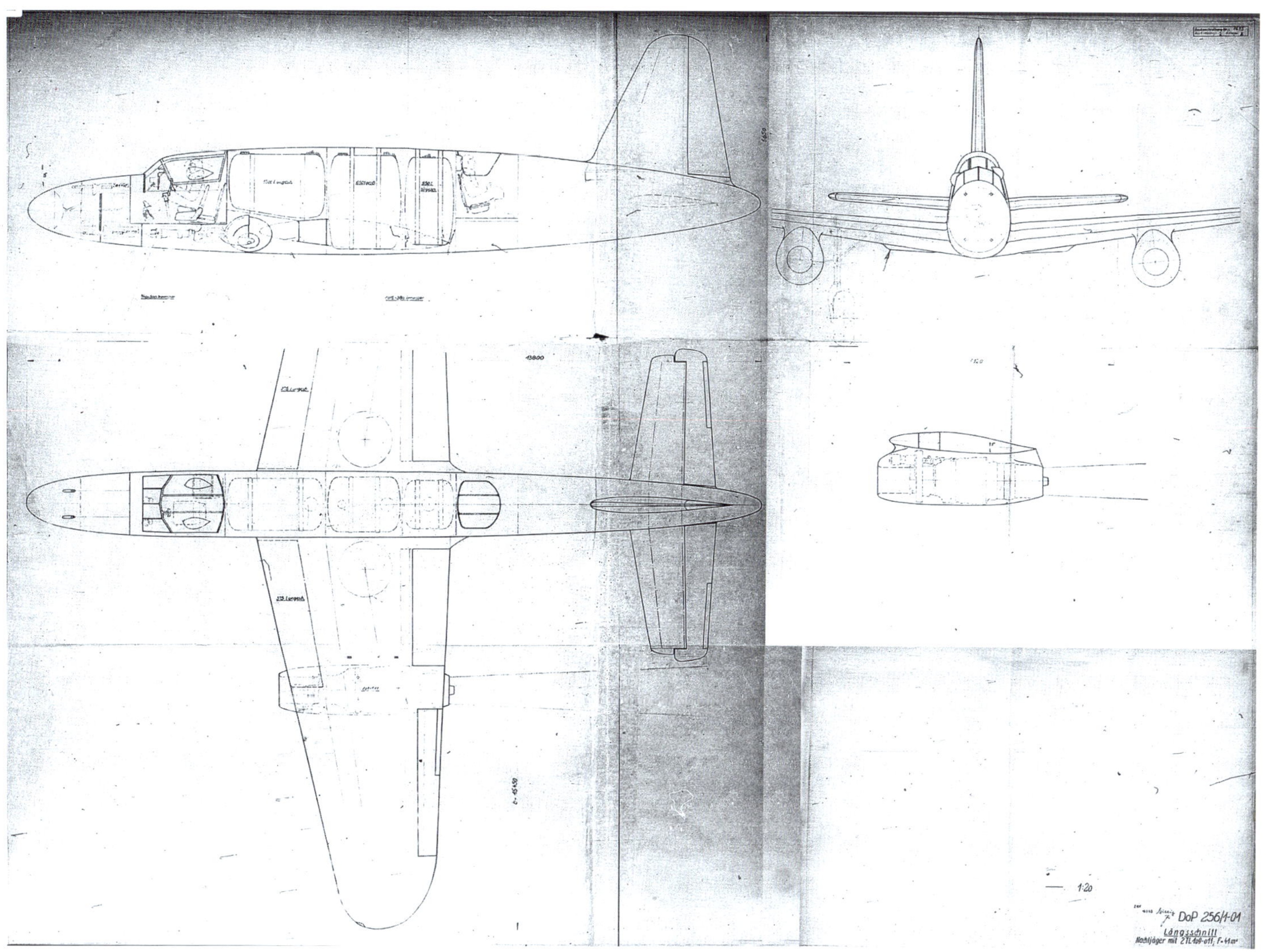

ABOVE: Dornier's Do P 256/1 night fighter was the very last of many designs intended to utilise existing components from the Do 335 heavy fighter – in this case the central fuselage and most of the wings. The drawing is dated March 10, 1945.

41m². Equipment was given as FuG 24 SE with ZVG 24, FuG 29, EiV, FuG 25 A oder C, Fu Bl 3, Kurskoppler, FuG 101A, Peil 6 with APZ 6, FuG 280, FuG 350 and Bremen.

FOCKE-WULF II AND III

Having just barely completed the design of their night fighter Entwurf II in time for the February EHK meeting, Focke-Wulf now worked to refine it. In the process they created four further night fighter designs – Entwurf III, IV, V and VI.

The Entwurf II–V were bundled together in a single dossier which lacked any description number and was dated March 19, 1945.[34] Details concerning Entwurf VI are somewhat sketchy, Focke-Wulf having only produced a few hand-written data sheets on it[35] on March 20. It was powered by two HeS 011s, carried 5,400 litres of fuel and had an all-up weight of 11.4 tons. Wing area was 74.5m² with a wingspan of 17m and a sweepback of 38°. Armament was four MK 108s. Most surprisingly of all, Entwurf VI was a flying wing.

Entwurf II was presented in drawings 0310 251-50 and -51 of March 7 as having a wingspan of 16m (although the text description said 15.8m) with a wing area of 50m² and sweepback of 30°. Equipment and other details remained the same as before. All–up weight was 12 tons including 4,500 litres of fuel.

The similar-looking but somewhat smaller Entwurf III was shown in drawing 0310 251-52 of March 10. It was 13.52m long and had wings of 40m² area with a wingspan of 14.1m, though sweepback was the same at 30°. Weight was reduced to 10.5 tons including 3,750 litres of fuel.

Entwurf IV had three turbojets rather than two. Only a single engine was carried in the fuselage, fed by a small circular intake, and the other two were beneath its wings. This allowed for a more slender cigar-shaped fuselage. Wing area was 75m², it was 19.1m long and had a wingspan of 19.4m with the same 30° sweepback. A total of 9,000 litres of fuel could be carried for an overall weight of 19.6 tons.

Lastly there was another three-jet design: Entwurf V of drawing 0310 251-54. This returned to having two

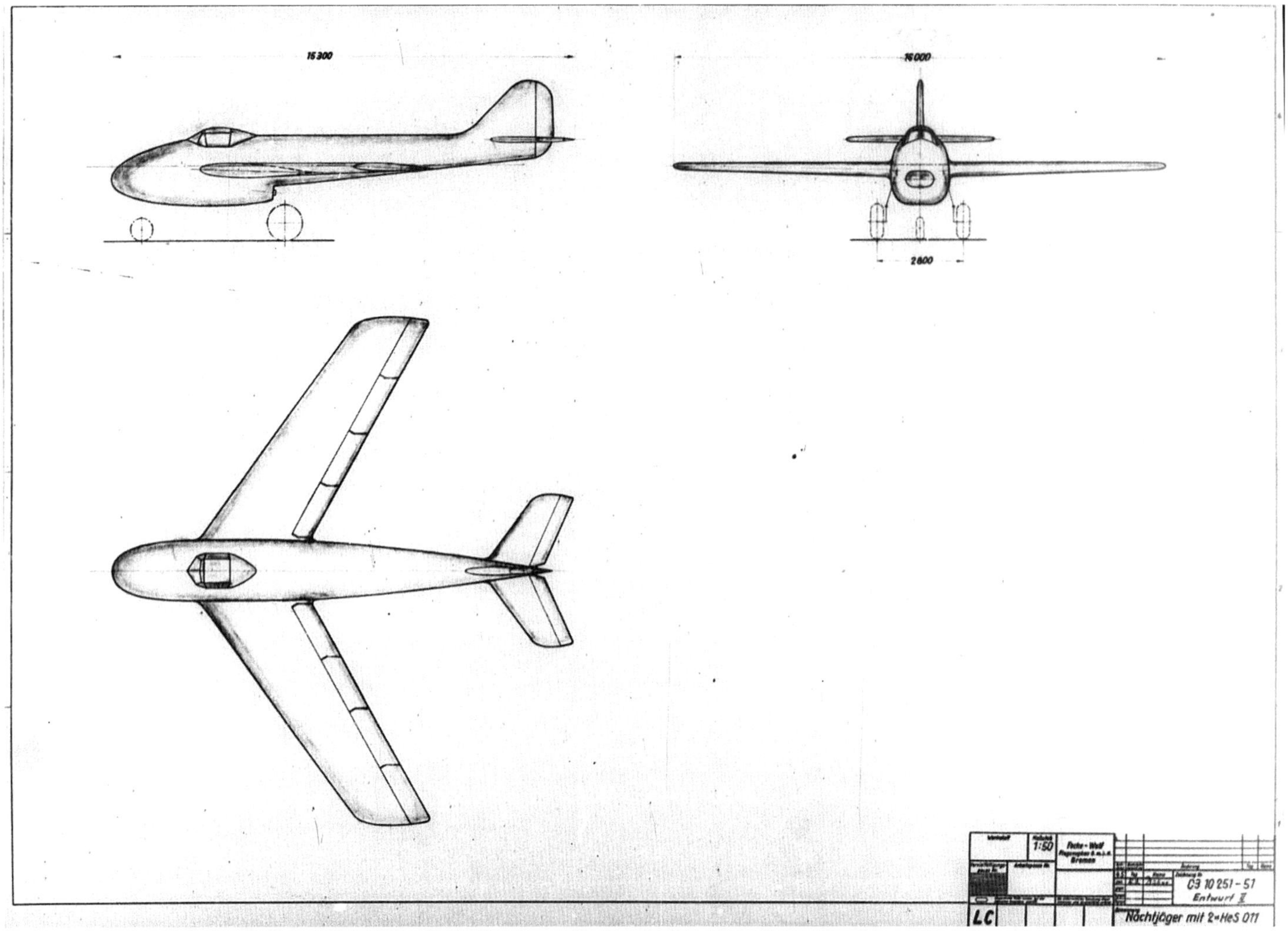

ABOVE: The tidied up Focke-Wulf night fighter Entwurf II appears in drawing 0310 251-51 of March 6, 1945. The design is clean and workmanlike but looks slightly odd due to its very short intake duct.

turbojets in the forward fuselage but added a third at the tail end, fed by side intakes on the fuselage. It was 17.8m long with a wing area of 74.8m² and a wingspan of 19.3m. Sweepback was reduced to 25°. A total of 10,800 litres of fuel could be carried and overall weight was 19.4 tons.

Only Entwurf II and III were submitted for comparison with the other companies' projects.

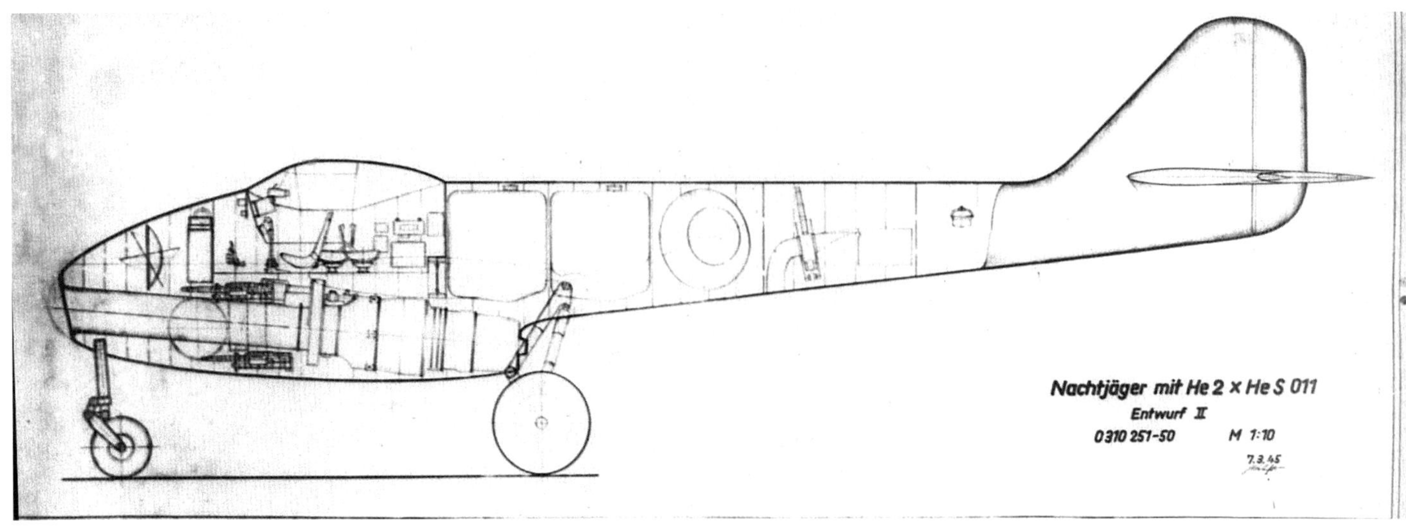

ABOVE: This side view of the Entwurf II night fighter is shown in 0310 251-50 of March 7, 1945 – a day later than the three-view despite the lower drawing number.

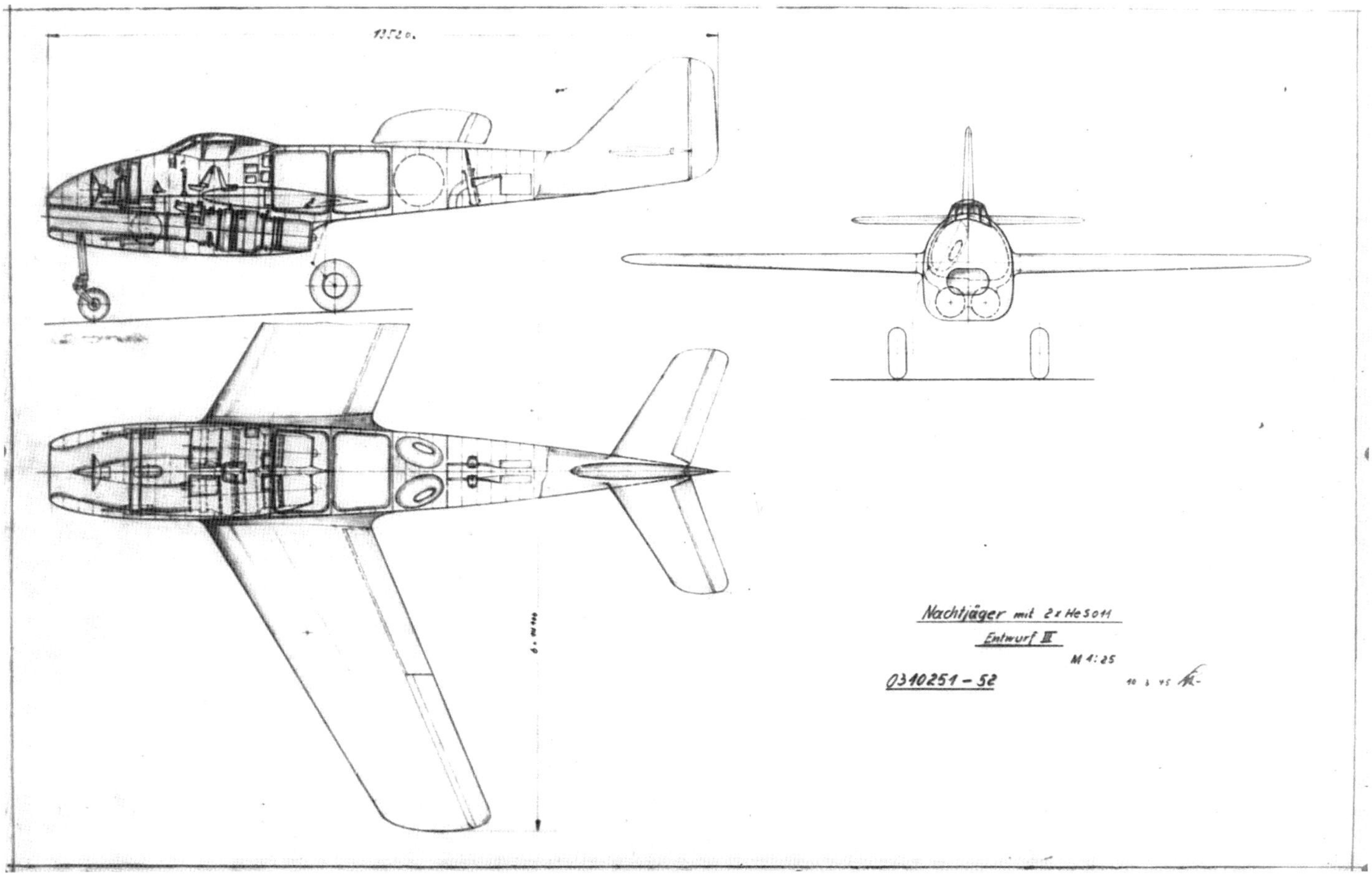

ABOVE: Focke-Wulf's Entwurf III design of drawing 0310 251-52, dated March 10, 1945, appears almost identical to Entwurf II but was actually a more compact design, being shorter in both length and wingspan.

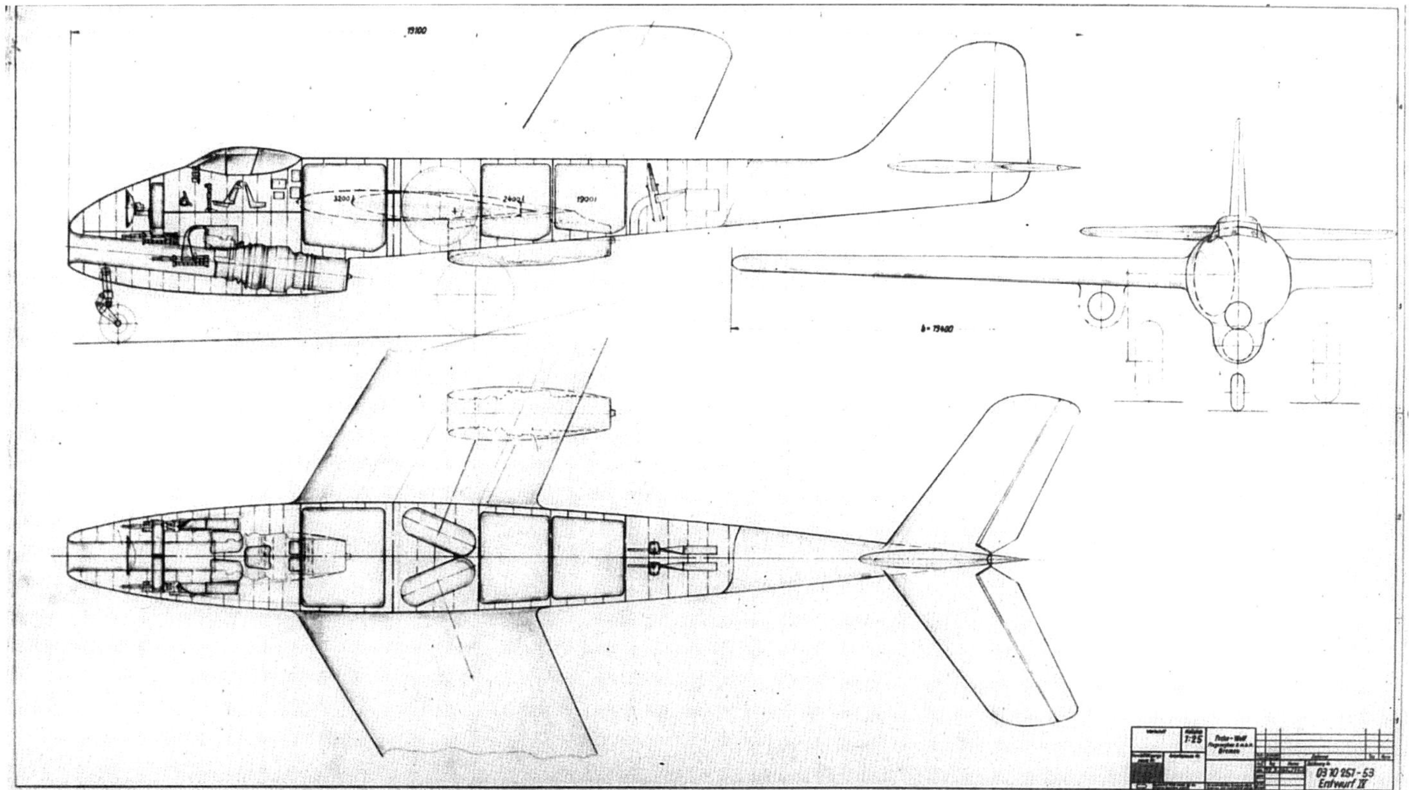

ABOVE: The Entwurf IV night fighter from Focke-Wulf had three turbojets rather than just two – one in its fuselage and the other two beneath its wings. It appears in drawing 0310 251-53 of March 12, 1945.

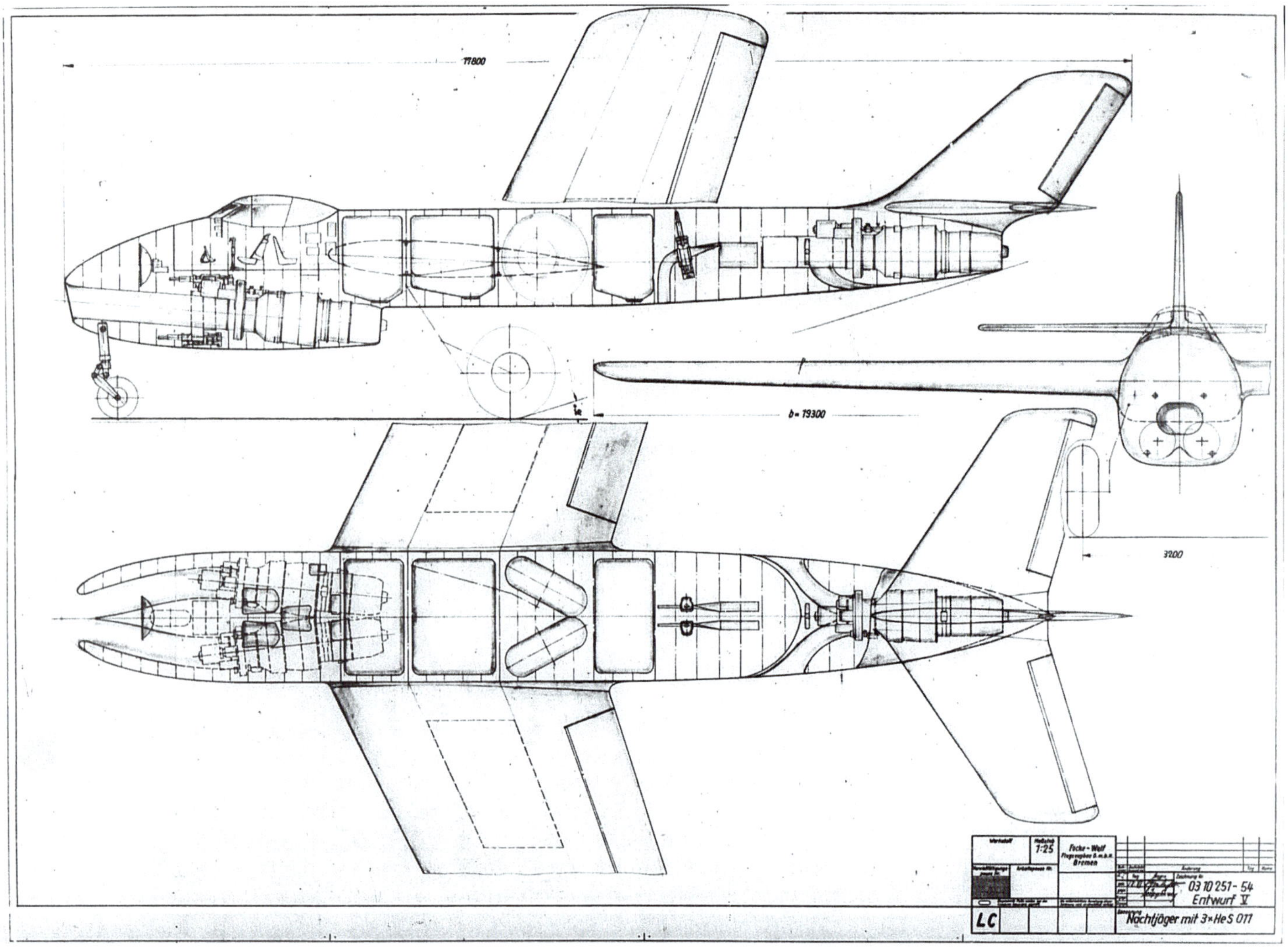

ABOVE: Another triple-jet design, Focke-Wulf's Entwurf V of drawing number 0310 251-54, dated March 12, 1945, was the largest of the five designs at 17.8m long and with a 19.3m wingspan. This made it slightly larger overall than a Dornier Do 217.

GOTHA P-60 C

Just nine days before the conference, Gothaer Waggonfabrik's P-60 C had been a two-seater with the crew sitting in tandem. The report produced by the company on March 11, 1945, outlining the P-60 series of designs[36] said that the C-variant had been specifically designed as a night fighter: "With the exception of the wing centre section, P-60 C corresponds to the project P-60 B. P-60 C was designed especially with regard to the use of night fighting.

"The prescribed FuG 240 ['Berlin' airborne intercept radar – later renamed FuG 244 Bremen, which was the type actually specified in the March 1 requirement] made it necessary to provide a protuberance on the middle part of the wing. This measure brought about the seated arrangement of the two crew members and also made the use of normal vertical stabilisers necessary."

The original P-60 C drawing included with the report, dated March 10, 1945, showed the tailless aircraft just 10.9m long with a wingspan of 13.5m. Like the P-60 A and B it had one HeS 011 turbojet positioned on its back and the other on its underside, though the conventional seated cockpit's canopy was now partially in the way of the upper engine's intake.

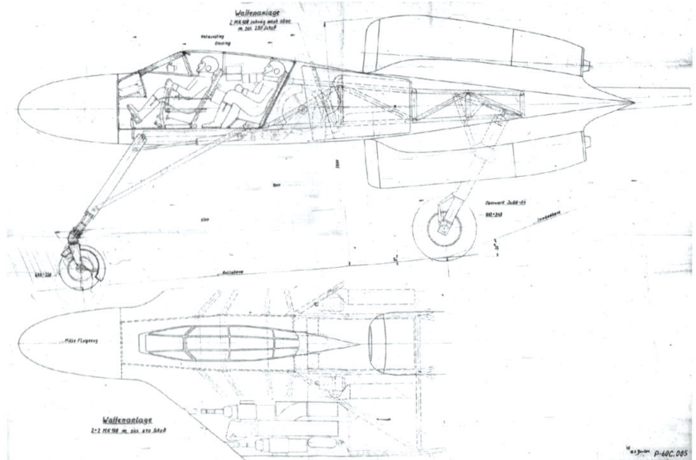

ABOVE: Side view of the Gotha P-60 C two-seater from March 10, 1945. The cramped crew accommodation is clear, as is the proximity of the upwards-firing MK 108s to the radio operator's head.

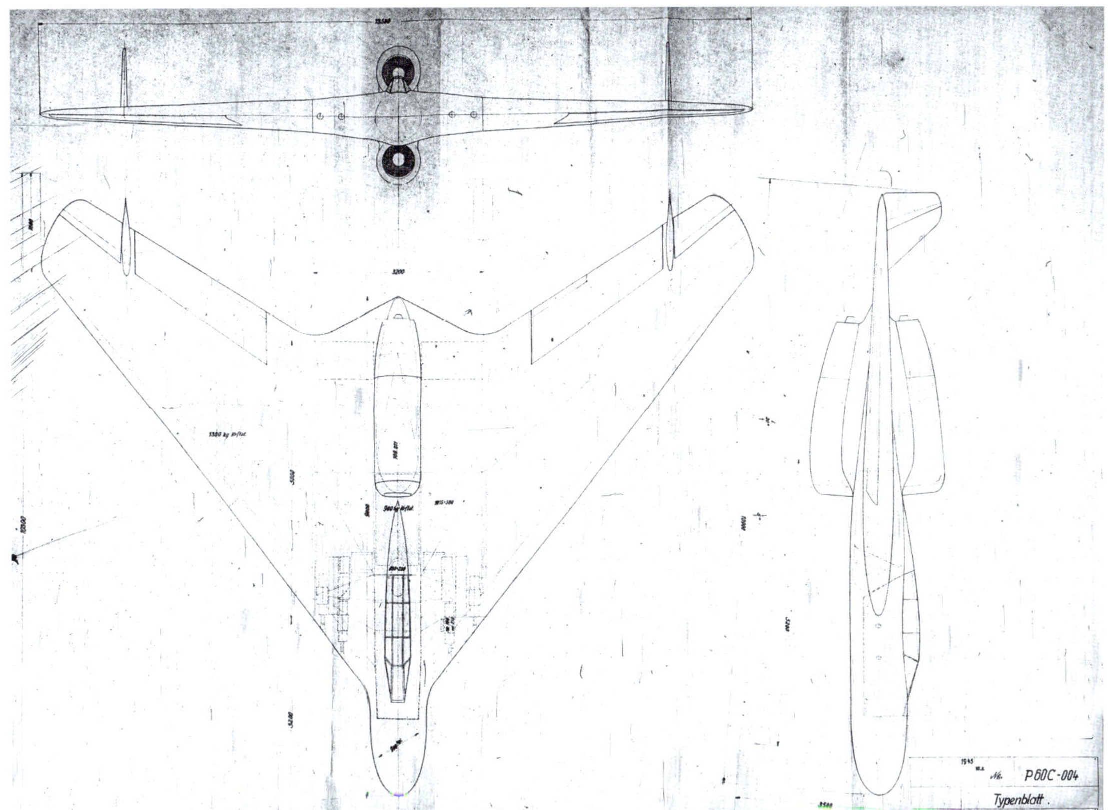

LEFT: Gotha's **P-60 C** tailless night fighter design as it appeared on March 10, 1945 – with only two crew seated in tandem.

BELOW: The Gotha **P-60 C** redesigned as a three-seater with only the pilot housed within the fuselage. The radio operator and navigator each had a reclined seat within one of the wings beneath a large Perspex cover. The drawing is dated March 21, 1945.

Furthermore, it had been necessary, as the report mentioned, to put a vertical fin on each wing towards the tip for stability. Four of the aircraft's MK 108s fired through the leading edge of its wings – two on either side – while the remaining two fired upwards from behind the radio operator's position.

However, Gotha had not been copied in on the relevant specifications and only at the last minute realised that three crew were regarded as essential. Within the space of 10 days, therefore, the company came up with a novel solution: leave the pilot where he was but put a seat on either side of him within the aircraft's wings. There is no mention of it in the documentation but the drawing shows these two positions each covered by a large curving sheet of Perspex – giving their occupants a reasonable degree of visibility. Armament remained the same at four fixed-forward MK 108s plus two more angled upwards although the latter were very close to the reclined head positioned of the wing seats.

Dimensions remained the same too but remarkably fuel capacity had actually increased from 3,500kg for the two-seater to 4,000kg for the three-seater because additional fuel could now be positioned centrally in the fuselage where the rear cockpit had been.

MESSERSCHMITT 3-SEATER NIGHT FIGHTER

Messerschnitt produced two separate night fighter descriptions on March 17, 1945. The first was for a two-seater Me 262 B2 powered by HeS 011 engines[37] and the second was for an Me 262-based three-seater.[38]

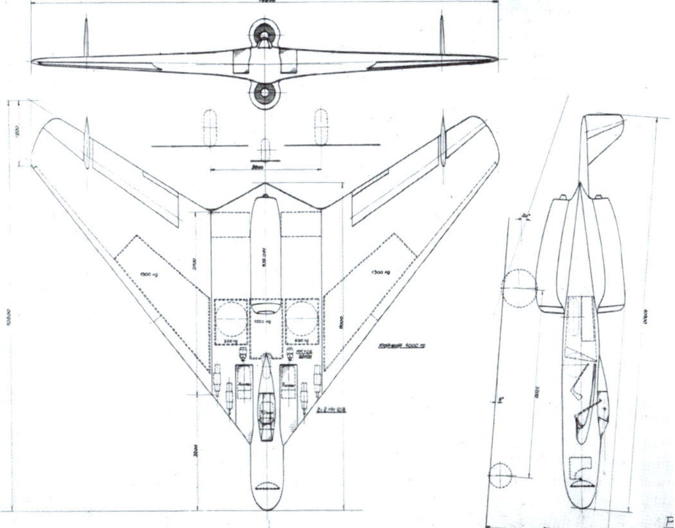

While the former was intended to extend the longevity of the nearly-in-production Me 262 B2 by swapping its Jumo 004 Bs for more powerful HeS 011s – plus a stronger undercarriage so that heavier weapons and/or more fuel could be carried – the latter was an attempt to meet the latest night fighter requirement. The introduction to its description said: "Based on the technical guidelines for a bad-weather and night fighter from February 27, 1945, Messerschmitt AG suggests a further development of the previous night fighter. The advantage of this design is the use of large components of the Me 262, which means the least effort in testing and device capacity.

"The tip of the fuselage with nosewheel is taken over by the night fighter, the end of the fuselage, the tail boom, the outer wings, the two main fuel tanks, the 600-litre additional tank and the pressure trough with equipment for the pilot correspond to the Me 262 A1.

"To achieve the required maximum speed, HeS 011 engines were provided in conjunction with a swept wing with a 3° sweep. The wing area was enlarged to 28m² using a new middle section and the 262 outer wings.

"The engines are installed centrally in the wing. Because of the simple structure and because test results with HG II are available in a short time, the engines can be installed under the wing. The accommodation of the third man is made possible by a renewed fuselage

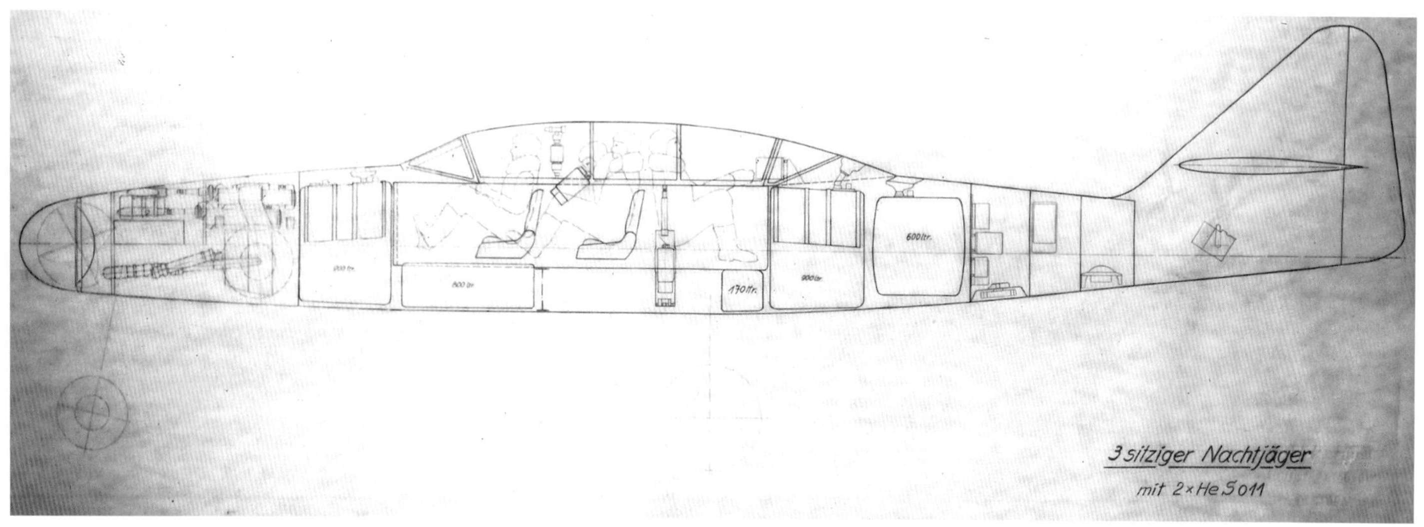

ABOVE: Messerschmitt's three-seater night fighter version of the Me 262, dated March 17, 1945, as viewed from the side. Note the proximity of the upwards-firing MK 108s to the rear crewman.

ABOVE: Perspective view of the Me 262 three-seater with wingroot turbojets.

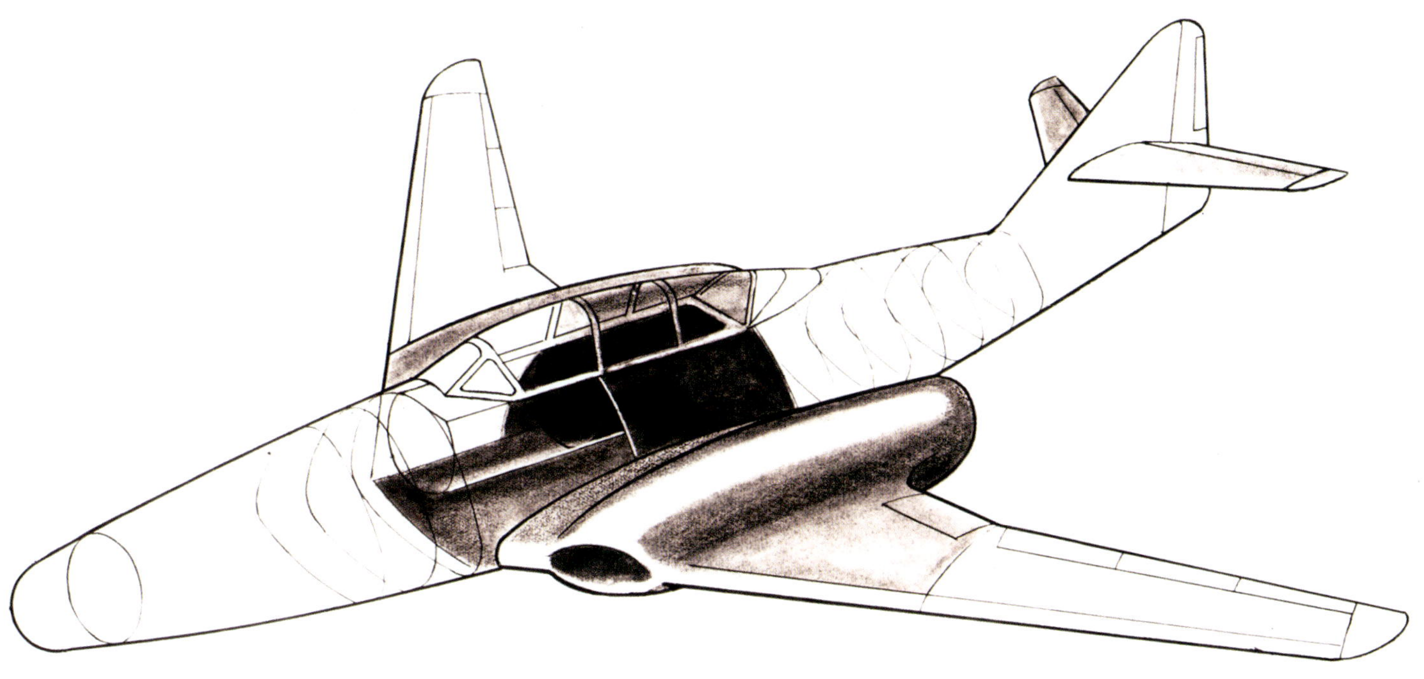

ABOVE: The shaded areas on this perspective view of the Me 262 three-seater night fighter show the new assemblies required for the design.

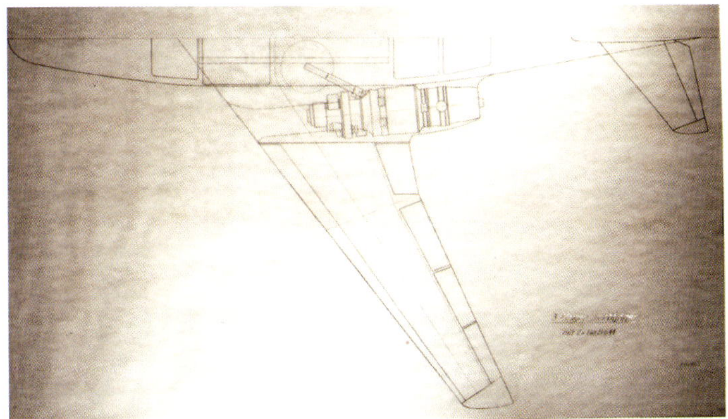

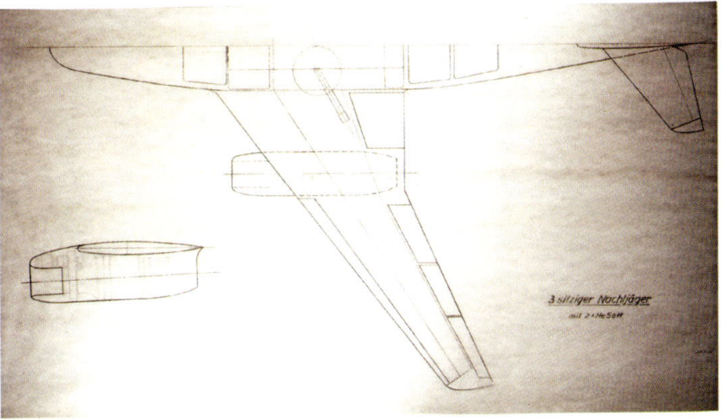

ABOVE: Drawings from the March 17 Me 262 three-seater report showing the two different options for installation of its engines – in the wingroots or in underwing nacelles.

extension (total 2m) compared to the normal fighter. The fuel supply of 3,200 litres accommodated in the airframe corresponds to a flight duration of 2.5 hours at a height of 10km.

"With 2 x 600-litre fixed external tanks, the flight time can be increased to 3.4 hours at a height of 10km. The requirements regarding armament can be regarded as fulfilled by using the previously developed Me 262 armaments, insofar as they can be used with the aiming device, and by two MK 108 as inclined armament. A rear armament was initially dispensed with. However, this can be installed in principle, but at the expense of fuel.

"The electronic equipment complies with the guidelines which correspond to the equipment previously intended for the two-seat night fighter Me 262."

Drawings included with the report partially showed two variants of the three-seater – one with the engines buried in the wingroots and the other with them in the more familiar under-wing position – but there was no full three-view nor were precise dimensions for the design included with the description. As such, it would appear to have been rejected before the final selection phase of the competition and ahead of the EHK meeting.

HEINKEL P 1079

Among Focke-Wulf's night fighter papers prepared ahead of the March 20–24 meeting is a single sheet headed 'Vergleichsgewicht Heinkel P 1079'.[39] This indicates that Heinkel did offer a project for the competition but like

Messerschmitt's three-seater Me 262 it was not considered alongside the other designs – possibly because it only included provision for a crew of two, warranting automatic disqualification.

A report on the project, dated August 11, 1945, was drafted by Heinkel chief designer Siegfried Günter assisted by engineers Gerhard Eichner and Walter Hohbach at Landsberg-Lech air base, formerly known as Flugplatz Penzing, then under US occupation.

A translation of this report, 'German Report G-77',[40] was produced by American intelligence in November 1945, entitled simply 'Night Fighter and Destroyer Fitted With Two HeS-011 Power Plants P 1079' and offers no real context or preamble before commencing a straight forward technical description. It begins: "The subject of the following report is a night-fighter airplane fitted with two HeS 011 turbojet engines (static thrust = 2 x 1,300kg). By variation of weights and dimensions the determination of the best compromise with regard to characteristics of take-off, climb, range and speed will be facilitated."

In other words, several different shapes and sizes of aircraft had been looked at in trying to determine which combination of features was likely to offer the best performance.

Some details were common to all, however. The report goes on: "The basic fixed data used for all designs are: powerplant – 2 x HeS 011; armament – 4 guns type MK 108 (30mm) forward firing, 2 guns type MG 151 (20mm) rearward firing; armour plate – 220kg, radio equipment – set for long wave and short wave bands, direction finding set, blind landing equipment, set for distance measurement and aiming (with parabola registering mirror) provided fore and aft; crew – 2."

Four different layouts were assessed: three with 35° swept-back wings and a V-tail, but with wing areas of 30m², 35m² and 45m², and a 30m² area design with a sharper 45° sweepback. It states: "High flight Mach numbers require the use of sweptback wings. In order to reduce the wetted surface of the airplane and to avoid interference effects to obtain a minimum drag, the power plants are mounted into the wingroots. Because of the sweptback wings the V-tail is also sweptback.

"The maximum speed of the three layouts with a 35° sweepback having respective wing areas of 30, 35 and 42m², lies between 960km/h (30m²) and 915km/h (45m²). Calculations based on a 45° sweepback for the 30m² layout gave an increase of speed from 960km/h to 980km/h."

The crew were seated back to back and armament was four MK 108s firing forward and "rearward defence is accomplished by two fixed 20mm cannon [MG 151s] mounted between the fuselage and engine nacelles".

The landing gear used would have been that designed for the long-since-aborted He 343 bomber. A supplement added to a later re-translation of the report showed a tailless version of the design and incorporated drawing numbers P 1079-06B, P 1079-07B and P 1079-08B. This presumably postwar design had a wingspan of 13m and was 9.15m long but it still only had two crew.

PROJECT COMPARISON II

While the EHK meeting lasted from March 20 to March 24, it appears that the section dealing with night fighters took place right at the end – giving Focke-Wulf time to assess all submitted designs deemed eligible for comparison.

The assessment report produced on March 23 stated bluntly: "The requirements set out in the invitation to tender by the Chef der TLR on February 27, 1945 are not met in their entirety by any of the drafts. Every attempt to meet the required long flight duration with two jet devices had to lead to insufficient take-off performance due to the high take-off weights.

"Designs with three jet devices, which are not considered further here, result in flight weights of approx. 19t, which

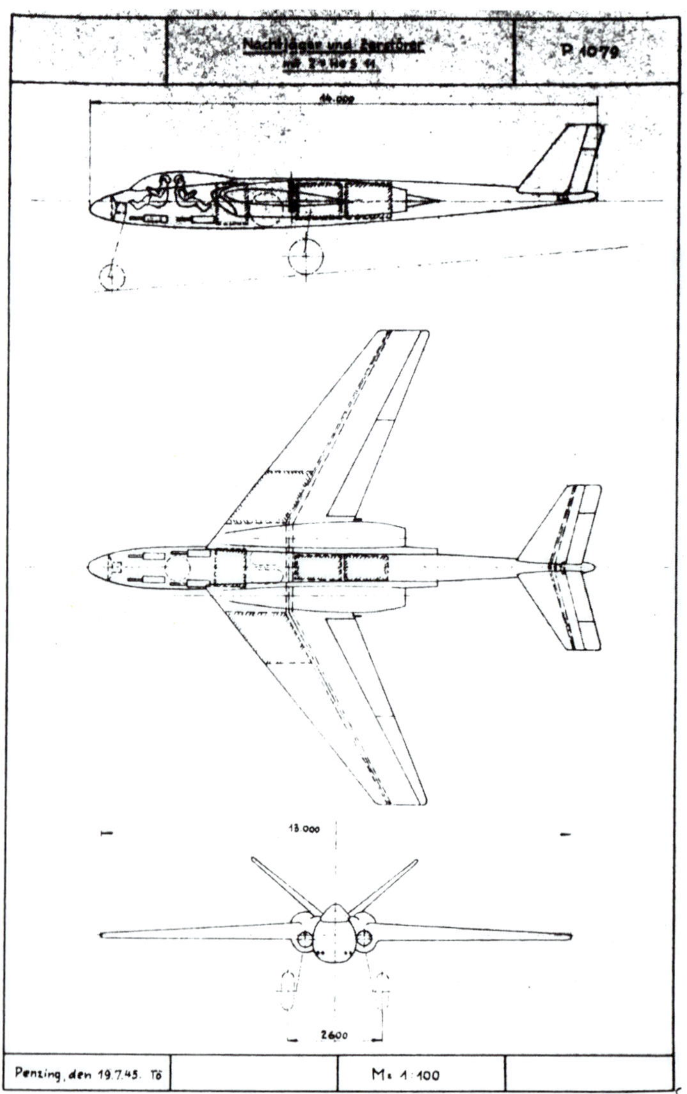

ABOVE: Three-view of the conventional Heinkel P 1079, dated July 19, 1945, showing its rear-firing MG 151s in its wingroots. This aspect of the design appears to have been changed later.

Geheime Kommandosache!

S.O. Archiv
Bad Eilsen

3. Ausfertigungen
3. Ausfertigung

Vergleichsgewicht Heinkel P 1079.

Rumpf	700 kg	
Rumpfeinrichtung	50 "	
Rumpfverkleidung	40 "	
Panzerung	300 "	
Rumpfwerk	1090 kg	
Hauptfahrgestell	410 kg	Rad 1015 x 380
Bugfahrgestell	150 "	Rad 770 x 270
Fahrwerk	560 kg	
Schwanzleitwerk	250 kg	
Flügelleitwerk	100 "	
Leitwerk	350 kg	
Steuerwerk	90 kg	
Tragwerk	1160 kg	
Triebwerksverkleidung	80 "	
Tragwerk	1240 kg	
Flugwerk		3330 kg
Triebwerk	1790 kg	
Behälter u.s.w.	400 "	
Triebwerk	2190 kg	2190 kg
Ausrüstung	780 kg	780 kg
Bewaffnung	577 kg	577 kg
Rüstgewicht	6877 kg	6877 kg
Besatzung	300 kg	
Kraftstoff	4000 kg	
Anlasskraftstoff	50 kg	
Munition	517 kg	
Zuladung	4867 kg	4867 kg
Fluggewicht		11744
		~ 11750 kg

Bei Verwendung der von Heinkel vorgesehen Räder 1140 x 410 bezw. 840 x 300 erhöht sich das Fluggewicht um ~ 100 kg.

ABOVE: A page from a report on the P 1079 that was among Focke-Wulf's papers captured at the end of the Second World War. This indicates that Heinkel's P 1079 was a genuine wartime design.

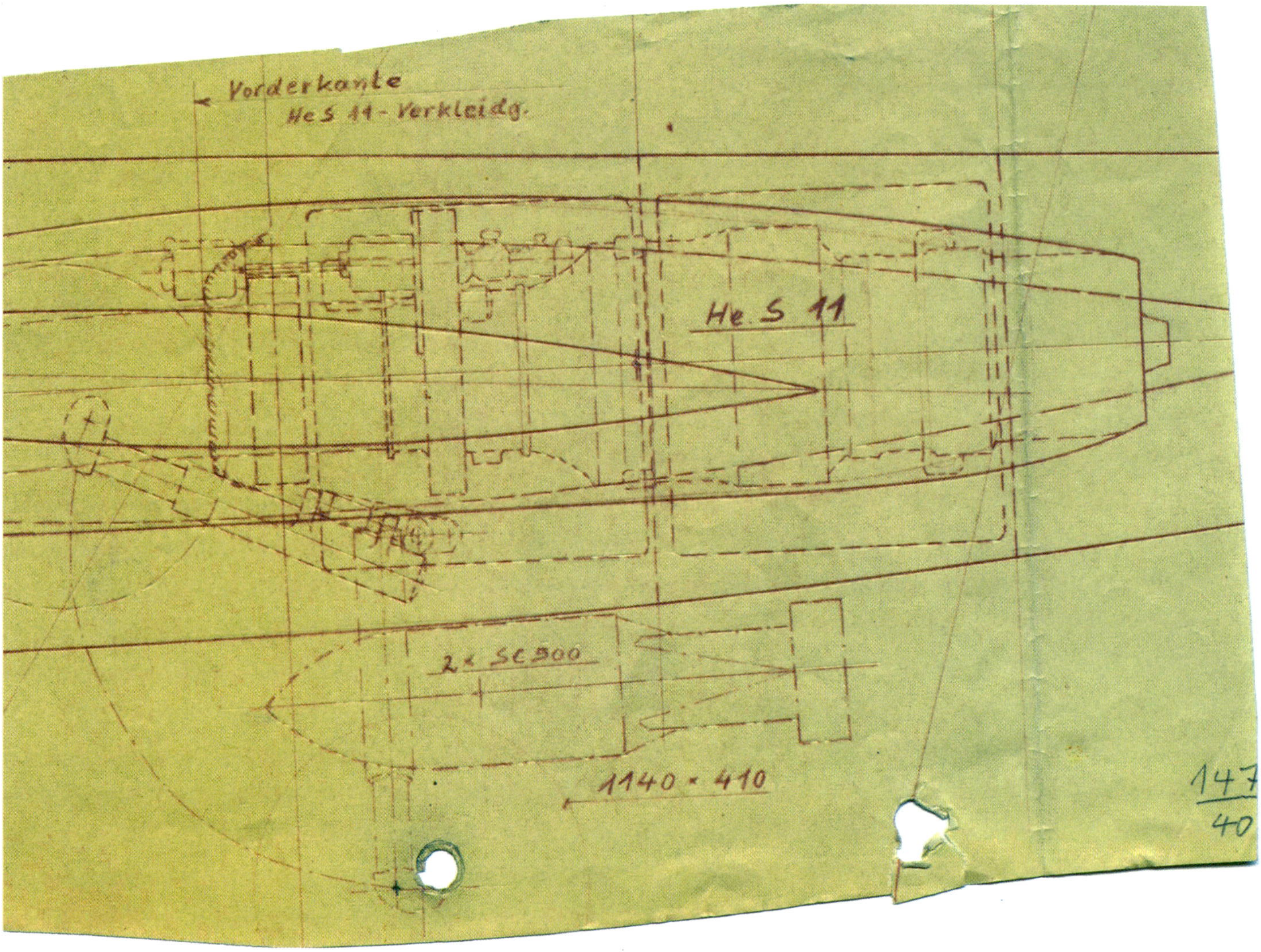

ABOVE AND BELOW: These partial drawings from a file of miscellaneous Heinkel documents appear to show sections of a P 1079 design with its rear-firing MG 151s in its wingroots. These may represent the wartime version of the P 1079.

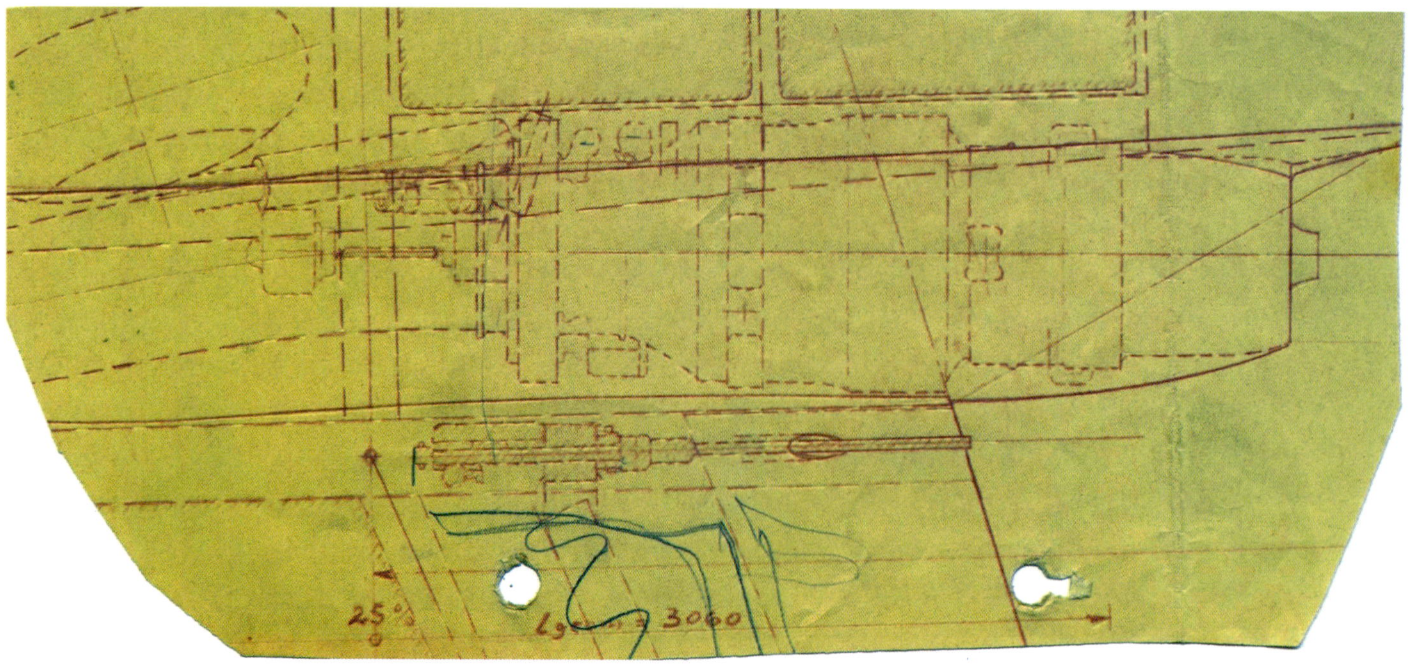

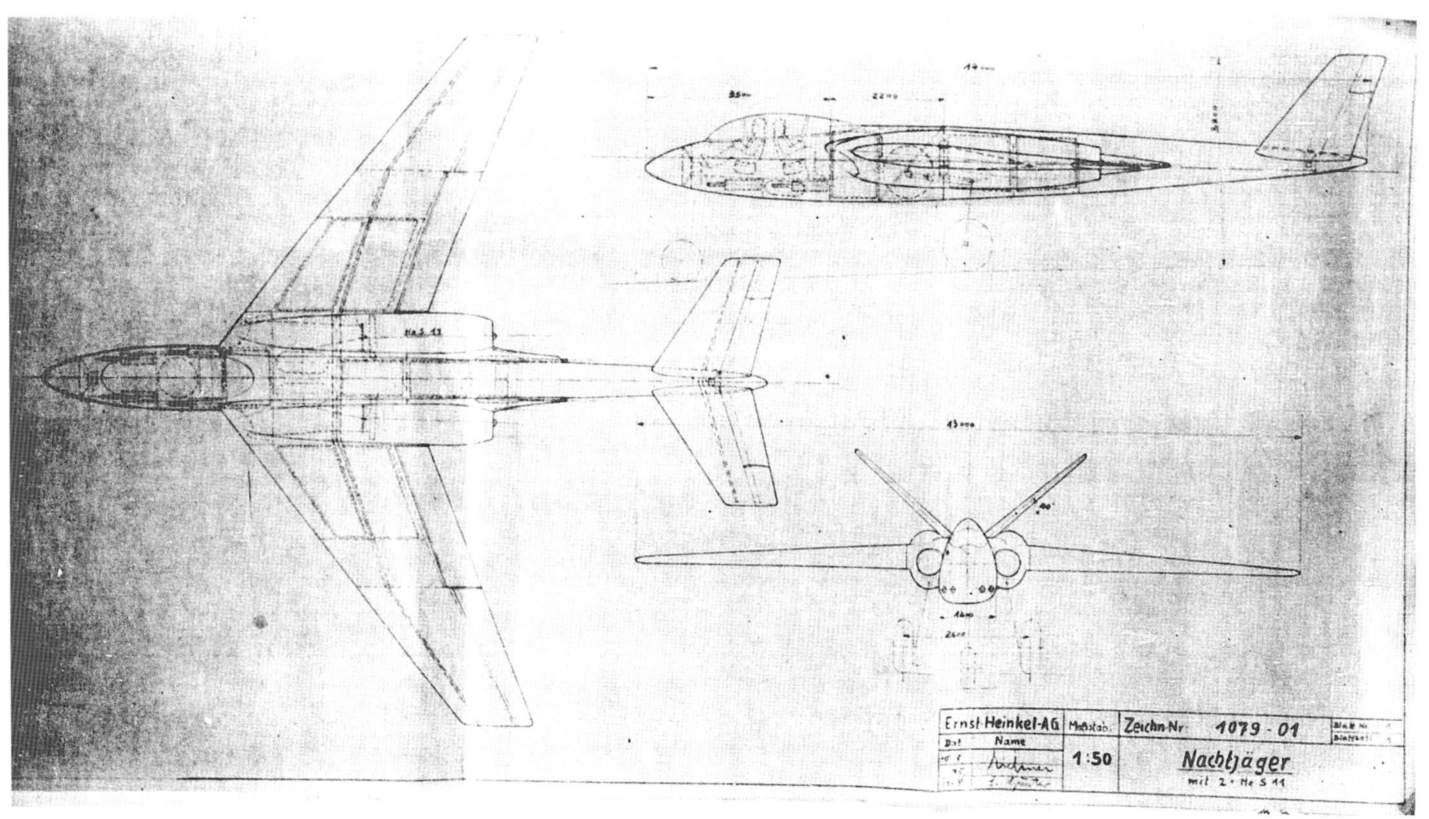

ABOVE: Three-view drawing of the conventional layout Heinkel P 1079 night fighter dated August 1945 – with the rear-firing MG 151s in aerodynamic housing on the rear fuselage sides.

are not only to be rejected because of the high hourly consumption, but also result in bulky aircraft for use as night and bad-weather fighters. Experience with the Ju 88 has shown that the manoeuvrability of this aircraft, especially around the longitudinal axis, is perceived as insufficient. Therefore, the payload had to be limited compared to the tender and the aircraft had to be kept as small as possible.

"Investigations into the possible use and flight duration indicate that a fuel quantity of around 4,000kg seems sufficient for the two-engine jet night fighter with HeS 011; this results in a wing area of approximately 45 to 55m^2 and a flight weight of 11.5 to 13 tons.

"When rear defence is demanded, the flight weight increases by another 750kg and the wing area by about

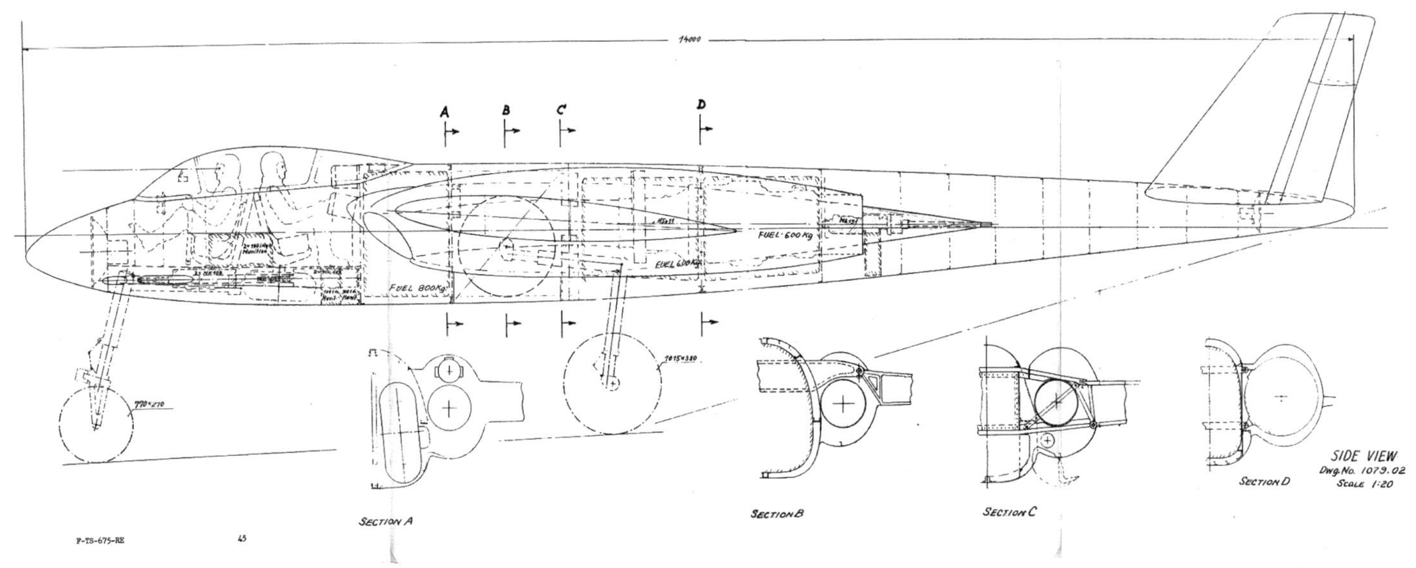

ABOVE: Side and sectional view of the P 1079 from a report dated November 1945, based on a report produced in August 1945.

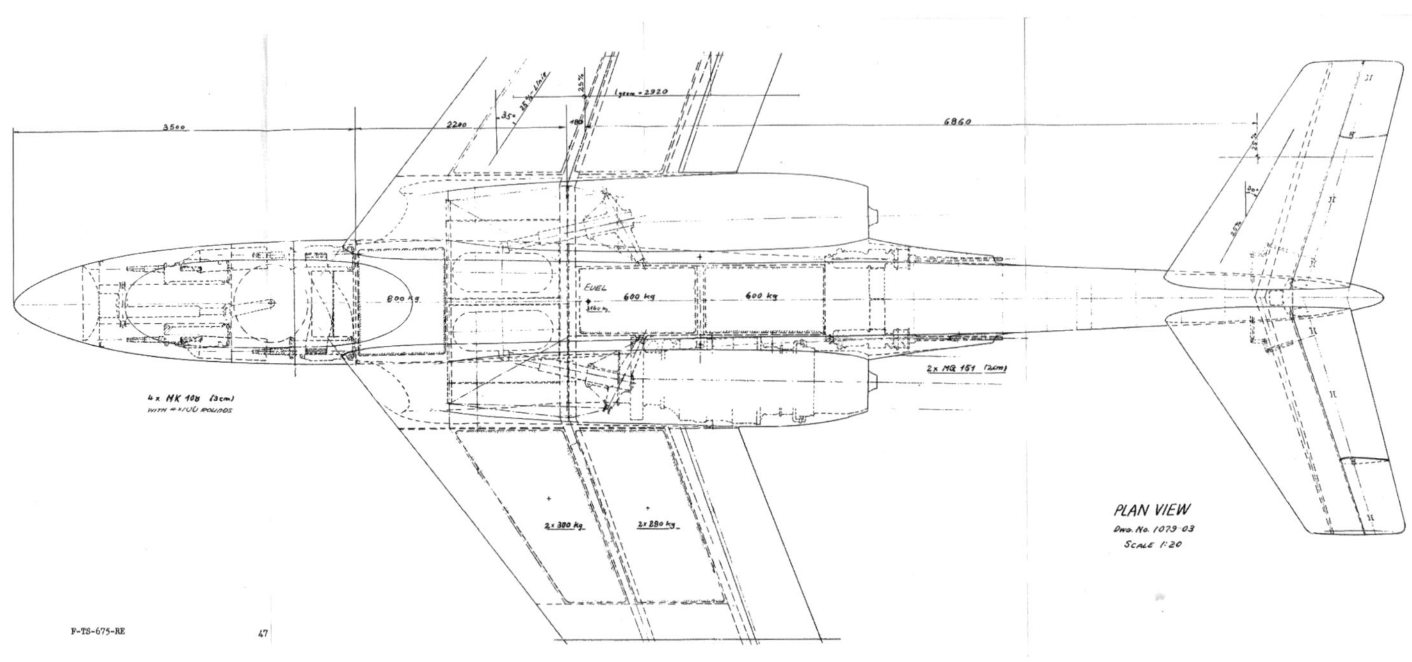

ABOVE: Plan view of the P 1079.

3m². As a result, the flight performance and flight characteristics are correspondingly worse, at which the top speed is of particular importance, a renewed review of this requirement seems necessary. A final statement by the inspector of the night fighters will still be given.

"Most of the participating aircraft companies tried to meet the requirements of the tender regarding flight duration as far as possible. For this reason, projects were submitted that are not satisfactory in terms of performance. An improvement in performance can only be achieved by reducing the payload. This also results in smaller airframe weights, which are taken into account in the following lists. It is not taken into account that the absolute size of the aircraft can also be reduced in accordance with the weight saving,

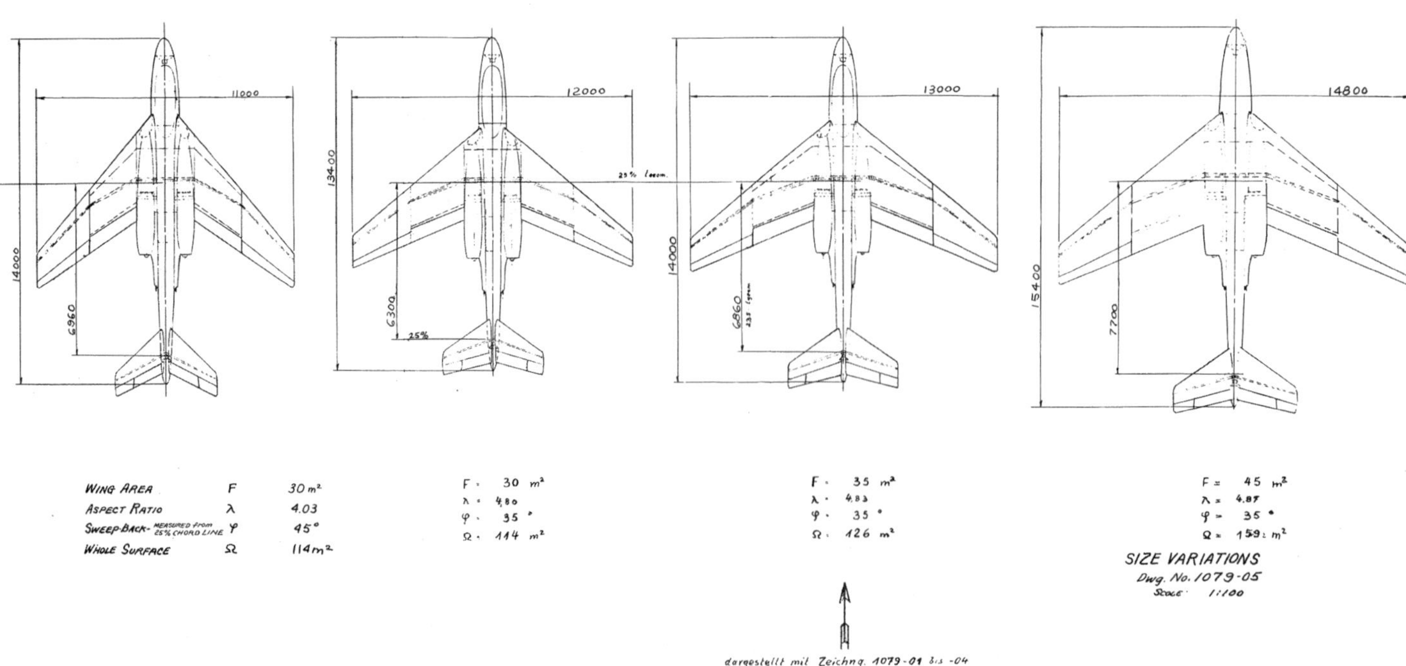

ABOVE: The P 1079 design was assessed with various different wing sweepbacks and wing areas. The Heinkel report focused primarily on the 35m² wing area design.

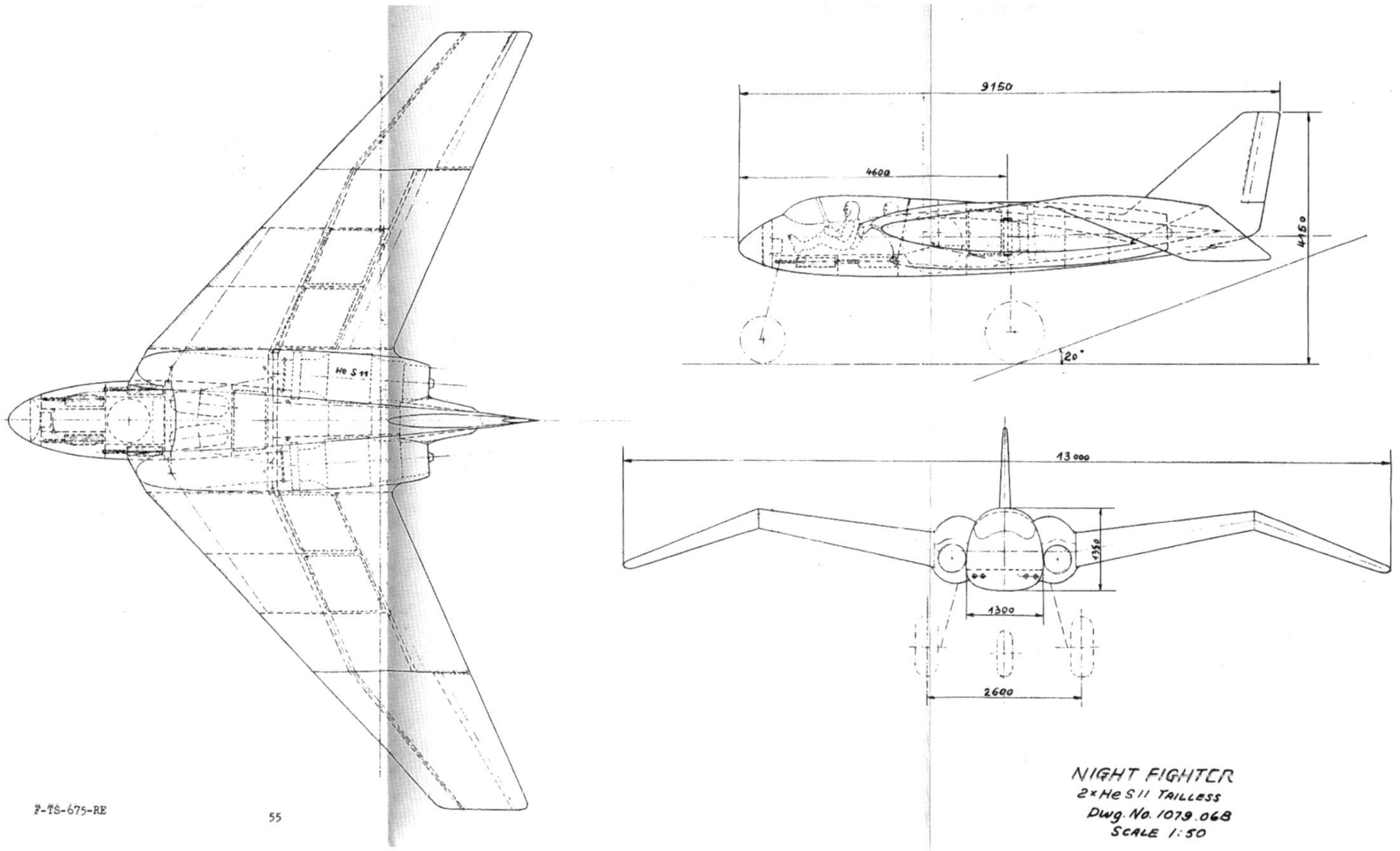

NIGHT FIGHTER
2×He S11 TAILLESS
Dwg. No. 1079.068
SCALE 1:50

F-TS-675-RE 55

ABOVE: Three-view drawing of the alternative tailless P 1079 layout illustrates its gullwing design. In plan view the aircraft is not dissimilar to some of Alexander Lippisch's schemes produced some three years earlier.

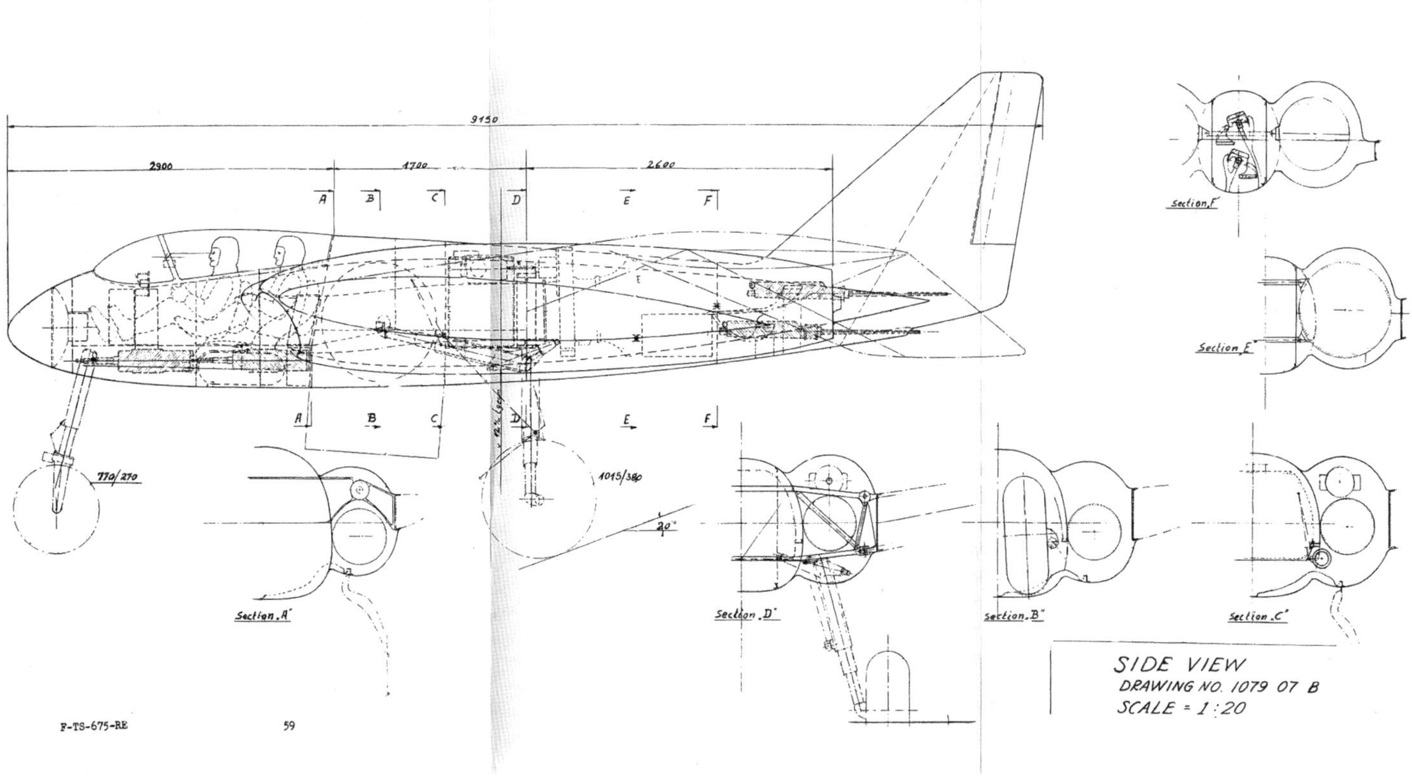

ABOVE: The tailless version of the P 1079 would, according to Heinkel chief designer Siegfried Günter, outperform its conventionally laid-out counterpart. Whether it was originally designed during the war or only afterwards is debatable.

i.e. that the designs have to be processed again, which means that small improvements in flight performance are also to be expected. In a first approximation, however, an assessment of the individual designs can already be made on the basis of the values now calculated."

Specific comments on the individual designs were added to the report on March 25 by Focke-Wulf engineer Herbert Wolff: "About Arado I: The critical point is the poor engine intake, which should cause a thrust loss of at least 4% ... Because of the large wing depth on the fuselage there is a strong centre effect. The cabin structure is strikingly wide, the total surface is relatively large. Regarding Arado II: Despite the smaller wing area, the total surface is exactly as large as that of Arado I. Note the high interference resistance of the engine nacelles.

"Regarding Blohm + Voss P 215: The thick, short fuselage causes a large Mach number influence; the long intake duct on the jet unit (6m) causes a thrust loss of 4%.

"On Dornier P 256: The unswept wing of the Do 335 is carried over, which together with the unfavourable engine nacelle arrangement results in poor Mach number behaviour and high interference (see Arado II). The fuselage is short but the tail units are strikingly large due to the small wing-tail spacing.

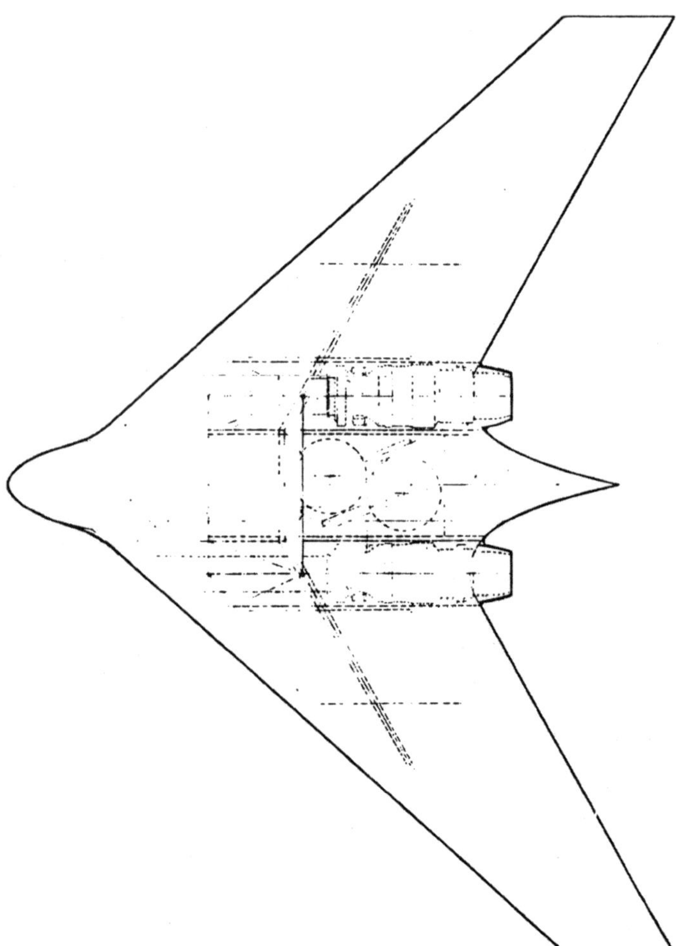

ABOVE: While only one tailless version of the P 1079 was included in the report produced for the Americans at Penzing, it would appear that a second was also prepared. The only known drawing of it is this faded off image from British report 'German Aircraft: New and Projected Types'.

"About Gotha P-60 C: In contrast to the tailless projects, this is the only pure wing design here, which has the smallest surface of all projects. A large centre effect could not be avoided due to the greatly increased wing depth. The engine inlets are endangered due to the large take-off run.

"Regarding Fw II and III: Both designs differ essentially in the different wing sizes of 50 and 40m². The smaller aircraft results in excessive wing loading and therefore high landing speed. The thrust loss is expected to be low (2%)."

In terms of sheer performance, the P-60 C was calculated to be the fastest with 975km/h reached at 3km altitude. The two Focke-Wulf projects were expected to reach 900km/h. The Blohm & Voss P 215.02 was expected to encounter problems due to its thin wingroot profile while both Arado projects would be hampered by high drag resistance but worst of all "the Dornier project stands out due to the unswept wing and the unfavourable engine suspension". Climbing performance was roughly the same for all designs but Focke-Wulf Entwurf III and the P-60 C were the best. And all projects required a take-off run of between 1,000 and 1,200m.

Similarly all projects had an endurance of 1.7 hours at full throttle at 10km. Throttled back to 65%, "all projects except Arado II achieve flight times of 2.8 to 3 hours; Arado II only reaches 2.2h at 80% throttling".

The war diary of the Chef TLR noted: "2 TL bad-weather fighters: EHK meeting in Bad Eilsen from 20 to 24 March clearly showed that the demands of General der Jagdflieger cannot be met, especially for the duration of the flight and defence. The demands of the weapon side are based on the current state of night fighting and do not take into account the development of equipment. The decisive factor for the use of high-quality jet fighter aircraft is the development of new types of armament procedures that result in shooting without having to adjust the speed.

"Working together through company representatives showed that the original TLR requirement of January 27, 1945 can be realised. The claims subsequently revised by TLR will be released on April 2 and distributed. In contrast to TLR, EHK also considers an optimal Otto Jaeger necessary."

A new requiremen[41] was issued in April, though the precise date is unclear, which reverted some of the spec to that of January 27 but also added some new demands.

The first alteration was to the engine section. Any number of Jumo 004s, BMW 003s or HeS 011s could still be used and they still had to provide a top speed of 900km/h, but they now had to do it at an altitude of 10km. And it was made clear that where HeS 011s were specified, it would be necessary to show "arithmetically calculated" performance levels.

Geheime Kommandosache!

Einzelwiderstände

Projekt	Arado I	Arado II	Blohm+Voß	Dornier	Gotha	Focke-Wulf II	Focke-Wulf III
Ausführung	schwanzlos	normal	schwanzlos	normal	Nurflügel	normal	normal
Anordnung d. TL-Geräte	unterm Flügel	unterm Flügel	im Rumpf	unterm Flügel	unter u. über Flügelmitte	im Rumpf	im Rumpf
Flügel							
Größe (m^2)	75/66	50/42,7	55/46,4	41/36,7	54,5	50/43	40/33,6
Spannweite (m)	18,4	15,0	14,4	15,5	13,5	15,8	14,1
Streckung	4,5	4,5	4,1	5,8	3,3	5,0	5,0
Oberfläche (m^2)	145	94	102	79	120	95	74
Pfeilung in t/4	30°	35°	30°	10°	44°	30°	30°
Profildicke innen/außen (%)	12/10	12/10	8/12	12/10	12/10	12/10	12/10
c_{wpo}	0,0075	0,0075	0,0071	0,0071	0,0071	0,0071	0,0071
$f_{wF} = c_{wpo} \cdot F$ (m^2)	0,495	0,320	0,330	0,260	0,387	0,305	0,240
Leitwerke							
Größe (m^2)	5,0 (nur SLW)	14,0	14,3	12,0/10,7	2,0 (nur SLW)	12,5/11,5	10,8/9,8
Oberfläche (m^2)	10,5	28,5	30,0	22,5	4,1	24,0	21,0
Pfeilung in t/4	30°	35°	30°	0/20°	15°	30°	30°
Profildicke (%)	8	12/10	10	12/10	10	10	10
c_{wpo}	0,0078	0,0080	0,0080	0,0080	0,0080	0,0080	0,0080
$f_{wLW} = c_{wpo} F_{LW}$ (m^2)	0,039	0,112	0,114	0,086	0,016	0,092	0,078
Rumpf							

ABOVE: A page from Focke-Wulf's assessment of the March 20-24 jet night fighter designs – showing the competitors.

There was a big change to armament, since defensive weaponry was no longer required unless the oblique firing guns could be used for defence. The latter also had to be capable of firing straight up at 90°, where previously 80° was the maximum angle needed. The equipment list now included the FuG 139 ground-to-air data link but clarified that either Bremen airborne intercept radar or FuG 280 passive search was to be fitted, rather than both systems at the same time.

The alteration most likely to please the aircraft manufacturers was a reduction in flight duration from climb to 10km plus two hours of level flight at full throttle to two hours at full throttle including climb to 10km. This would have significantly reduced the required fuel load and therefore made it possible to have a smaller airframe overall. The requirement for landing speed, formerly 180km/h, was lowered to 165km/h.

The most significant change of all was in the construction details section. In January, the requirement had simply been to keep costs down and use wood as much as possible. In February this had been expanded to incorporate easy maintenance and the use of low quality fuels. Now, in April, it specified: "The front part of the cabin must be equipped with flat armoured windows for reasons of visibility. Engine system and entire fuselage must be set up for the use of primitive fuels (e.g. tank drainage, heating of the tanks, pipes, filters etc. as well as the possibility of flushing with starting fuel before stopping the engine).

"The use of the aircraft as an interceptor (high rate of climb, but as far as possible unchanged horizontal flight time after climbing to operating altitude) with additional special engines (e.g. rocket, Lorin duct) is intended and should be taken into account in the design."

It is unknown whether any manufacturer actually received this final requirement, let alone whether anyone attempted to design an aircraft to it, but it was entirely academic now since Focke-Wulf's Bad Eilsen offices, where the Sonderkommission Schlechtwetter und Nachtjäger was based, would be overrun by British forces on April 8 – bringing a permanent halt to the night fighter competition.

Chapter 13

Pulsejet fighters

Revival of the Argus jet tube 1944–1945

The success of the Fieseler Fi 103/V-1 flying bomb in 1944 and efforts to create a manned suicide attack version - the 'Reichenberg' - brought about renewed interest in the potential of the simple tubular pulsejets made by Argus for use in manned aircraft.

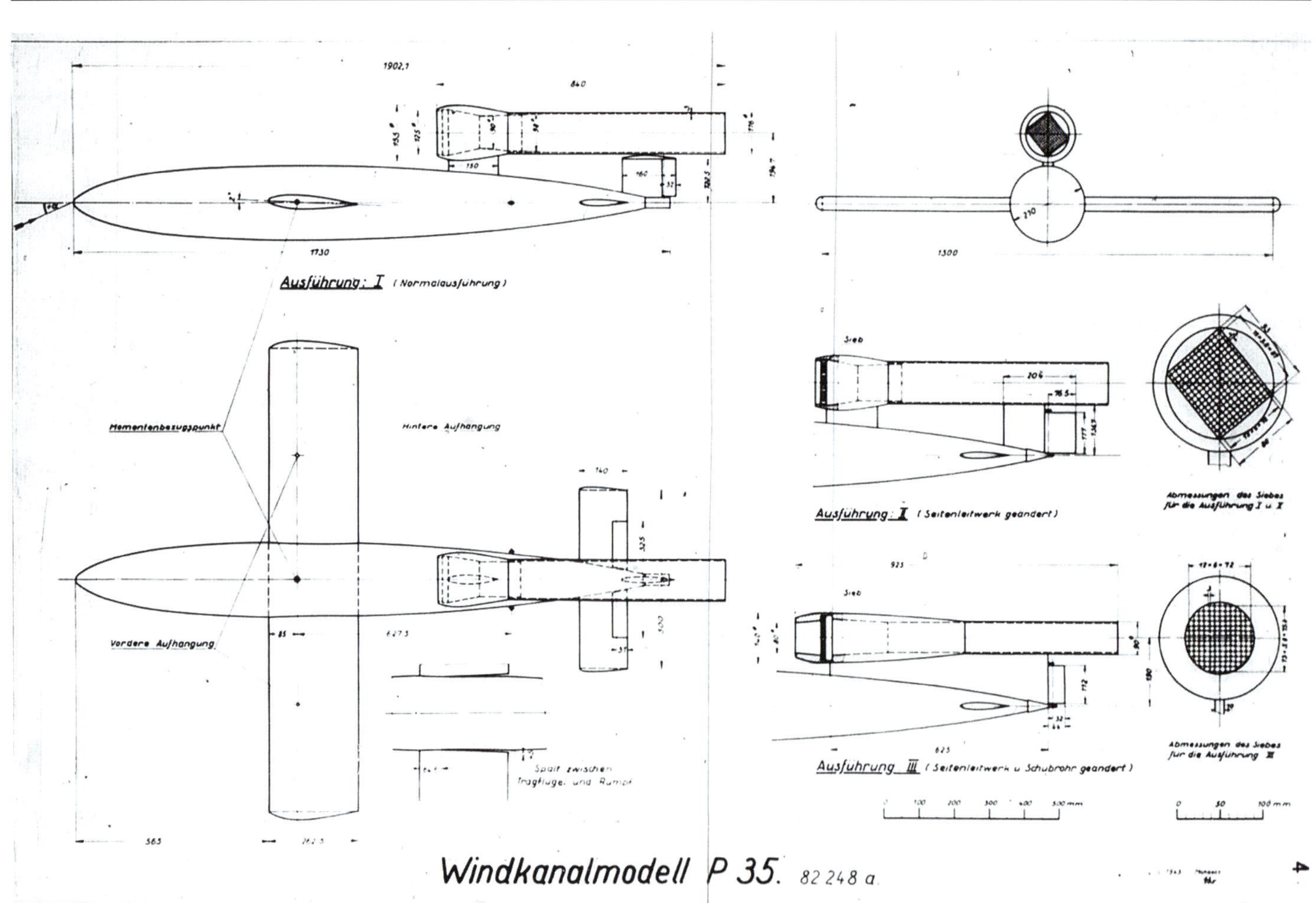

ABOVE: Before it became the Fi 103, aka the V1, the famous pulsejet-powered flying bomb was the Fieseler P 35. This drawing of a P 35 wind tunnel model comes from an LFA report dated October 26, 1943.

Heinkel had defeated Blohm & Voss with its P 1073 design and by October was already proposing a variant of the Julia target defence interceptor fitted with an Argus pulsejet mounted on its back in a similar configuration to that of the Fi 103 flying bomb. The new design, very similar in most respects to the seated version of Julia, was given the name 'Romeo' – presumably drawn from the Shakespeare play *Romeo and Juliet*, which is known as *Romeo and Julia* in Germany.

The Romeo drawing, dated October 15, 1944, was included with the Julia report of November 16, 1944,[1] but no reference was made to it in the text. At around the same time, Junkers was working on its new Fi 103-based ground-attack aircraft, the EF 126 'Elli' – another design with its pulsejet mounted on its back.

JUNKERS EF 126 'ELLI'

Although it was not designed as a fighter, it is worth mentioning what Junkers called the "Infanterieflugzeug 'Elli' (EF 126)" here. The design appears to have been proposed earlier than December 1945, although exactly how much earlier is unclear. The EF 126, later developed into the rocket-powered EF 127 'Walli' detailed earlier, was based on the Fi 103 configuration and some of its components. It was powered by a single Argus As 014 or As 044 engine[2] – a more powerful version of the ubiquitous As 014 pulsejet – and had a wingspan of 6.34m (compared to the Fi 103's 5.37m) with a wing area of 8.9m².

Armament options were: two MG 151/20s and two AB 250 submunition dispensers, two MG 213s or 12 Panzerblitz rockets. The design was initially proposed with a tricycle wheeled undercarriage but a landing skid was later also offered. Both options were on the table during December 1944 and remained under discussion right up to the point when the type was 'declined', with no firm decision apparently made either way.

According to a Chef TLR Fl-E 2 management report[3] of December 21, 1944, under the heading 'Projekt EF 126

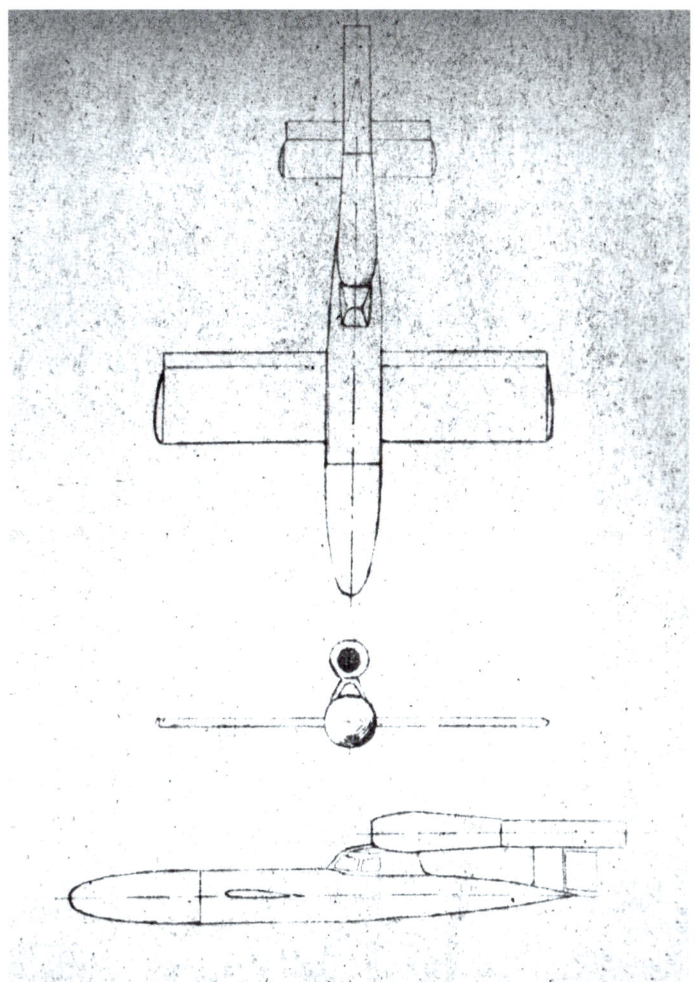

RIGHT: A sketchy drawing of the 'Reichenberg' – a manned version of the Fi 103 – from a postwar American intelligence report. The success of the Fi 103 and proposals for the 'Reichenberg' appear to have prompted a reassessment of the As 014 pulsejet as powerplant for manned aircraft.

BELOW: Heinkel's Romeo utilised the basic airframe of the Julia rocket-propelled interceptor to create a pulsejet-powered light fighter. This drawing was dated October 15, 1944.

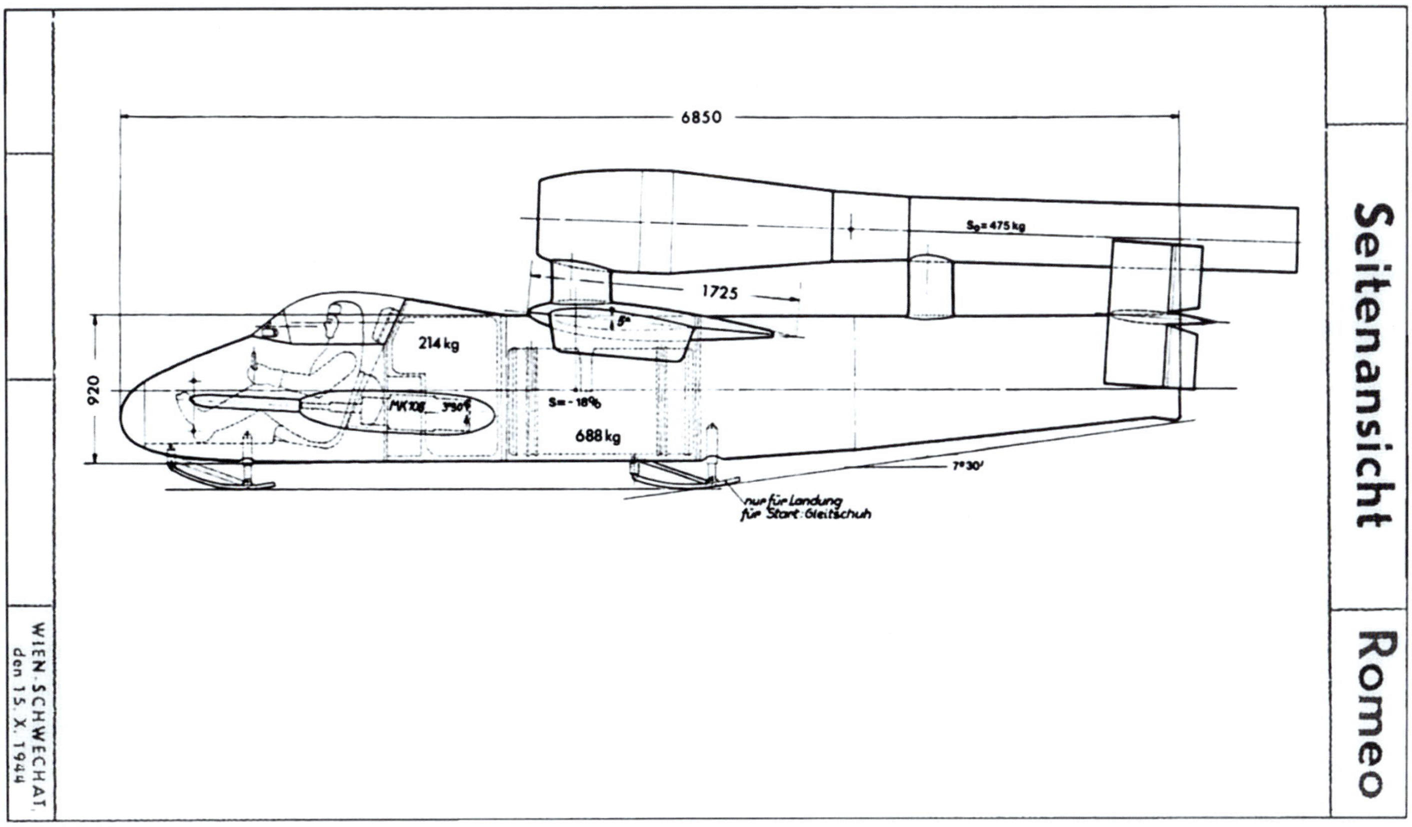

von Junkers': "Device with 450kg thrust cannot be used without launch rockets. Currently checking the function of the Kemm's double pipe and possibly using it for EF 126. Failures due to airframe damage caused by the engine as with V1 are intolerable for manned equipment. Currently review all of these questions. EF 126 would only be of use if it were available in certain numbers in the summer of 1945.

"Mock-up was inspected on 12.12 by General der Schlachtflieger Oberst Hitzschold [General of ground-attack pilots Oberst Hubertus Hitzschhold]. Consideration is being given to converting a project to a landing skid in order to make execution even easier and cheaper. G.d.S. [General der Schlachtflieger] wants to make a decision shortly whether there is interest in this device."

The same report also noted: "EF 126: Changes desired by G.d.S. were planned and G.d.S. visited on December 16. G.d.S. agrees and will make a demand."

The Chef TLR war diary[4] for the week of January 15 to January 21, 1945, noted that: "After another request by the G.d.S. (stopped by EHK) project values offered by Junkers are to be checked by DVL. Comparison with 8-162 (2 x Argus tube) makes Junkers performance questionable. G.d.S. calls for 'Elli' with a skid and winch start instead of launch rockets. Winch for multiple towing operated with BMW 801. Apart from inefficiency, immobility and tactical commitment are considered unbearable."

A note on 'Elli' from the diary entry for the period from March 5 to March 15, 1945,[5] stated that the "project is also declined by the General staff". This appears to have been the last word on 'Elli' before the end of the war. The Soviets continued work on it afterwards.

BLOHM & VOSS P 213.01-01

Back in October 1944, Richard Vogt, chief designer at Blohm & Voss, seems to have been asked to design a very cheap fighter powered by a pulsejet. He wrote to Argus at the end of the month to request details of the As 014 and the company's Hermann Teichgraeber, based at Greiffenberg

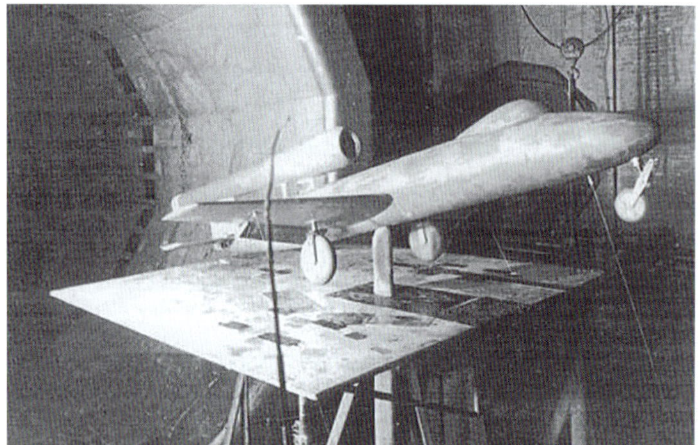

ABOVE: The original design Junkers put forward for the **EF 126** 'Elli' ground-attack aircraft featured a wheeled undercarriage. **LEFT:** Junkers **EF 126** wind tunnel model.

ABOVE: Junkers later developed a variant of the **EF 126** fitted with a skid undercarriage.

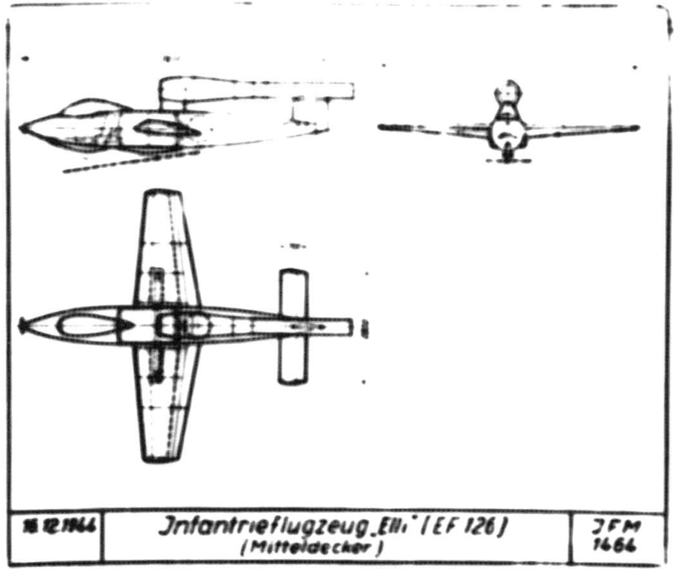

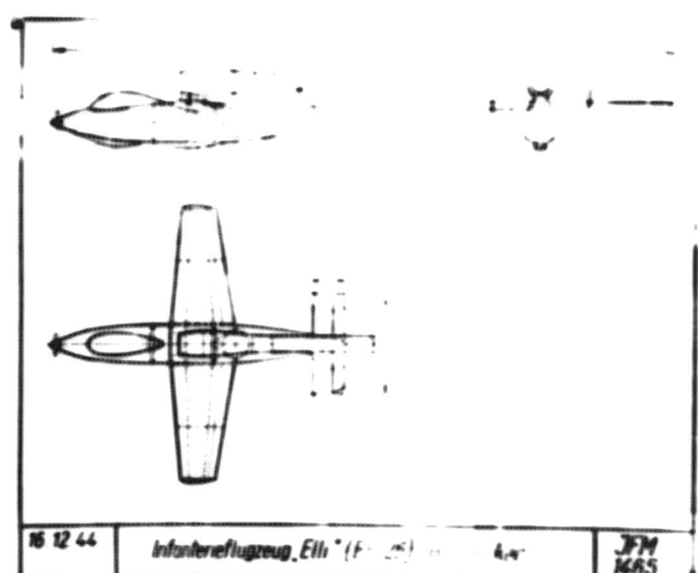

ABOVE: Two variants of the EF 126 'Elli' were showcased during a Junkers presentation on December 16, 1944 – one with a mid-wing layout and the other with the wings positioned on top of the fuselage.

in Silesia, replied on November 3:[6] "We hurry to send you the requested documents about our engine by courier and would be grateful if you would send us a simple overview sketch as soon as possible, from which we will explain the planned installation of our engine in your airframe. We assume that we can give you some useful hints based on the already extensive experience."

Just a week later on November 10, 1944, Blohm & Voss's designers to came up with a three-view sketch of the P 213.01-01 'Miniatur-Jäger'. This tiny fighter, resting on a 500 x 180 mainwheel and 260 x 85 nosewheel tricycle undercarriage, was 6.2m long with a wingspan of 6m and a wing area of 5m². It had a single MK 108 in its nose, just above the tubular intake for its engine. Rather than a conventional fin and tailplanes it had an unusual inverted V-tail.

Vogt wrote back to Teichgraeber the same day, enclosing the sketch,[7] and said: "We thank you for the prompt delivery of the installation documents of the As 014. This enabled us to work out the enclosed project sketch at short notice. In contrast to the external suspension, we have tried to achieve an organic and aerodynamic high-quality installation, without affecting the cooling.

"We ask you to check the installation based on your experience and to give us a statement. We are not worried about air vibrations in the inlet channel. In any case, some experiments in the Göttinger wind tunnel have shown this."

Five days later, Blohm & Voss had completed a report on the P 213.[8] The introduction read: "The examination of the proposal to design a very small fighter with the Argus-Schmidt jet tube As 014 has shown that with a single large-caliber weapon, with a 350kg fuel reserve and with a minimum set of equipment (radio has been completely omitted) can come to a starting weight from 1,280kg.

"That is about half the size of the Kleinstjäger [presumably a reference to the Volksjäger] in the start-up. If one takes into account not only the comparative value of the airframes, but also the manufacturing costs of the two engines – in one case a high-quality compressor and turbine unit and in the other a jet tube that is only required for a few working hours – then the pursuit of this line of thought seems quite profitable.

"Again we were guided by the design ideas of our P 211 Kleinstjäger and mounted all the fuselage equipment on the inlet pipe in order to put the fuselage shell over it. The smaller the internal dimensions, the more advantageous this structure is. What is new in the project described here is the position of the wing on the fuselage boom."

Under 'fuselage', it said: "The air intake pipe is the fuselage carrier and at the same time the assembly frame for all the equipment.

"A beam extends out of this pipe support, under which the jet tube hangs and which at the same time supports the tail unit with its extension. This beam is welded from sheet steel and holds 420 litres of fuel. This type of

ABOVE: Diagram from a Junkers report dated February 14, 1945, showing bending moments for the EF 126. The project would continue for another month after this point before being cancelled just before the end of the war.

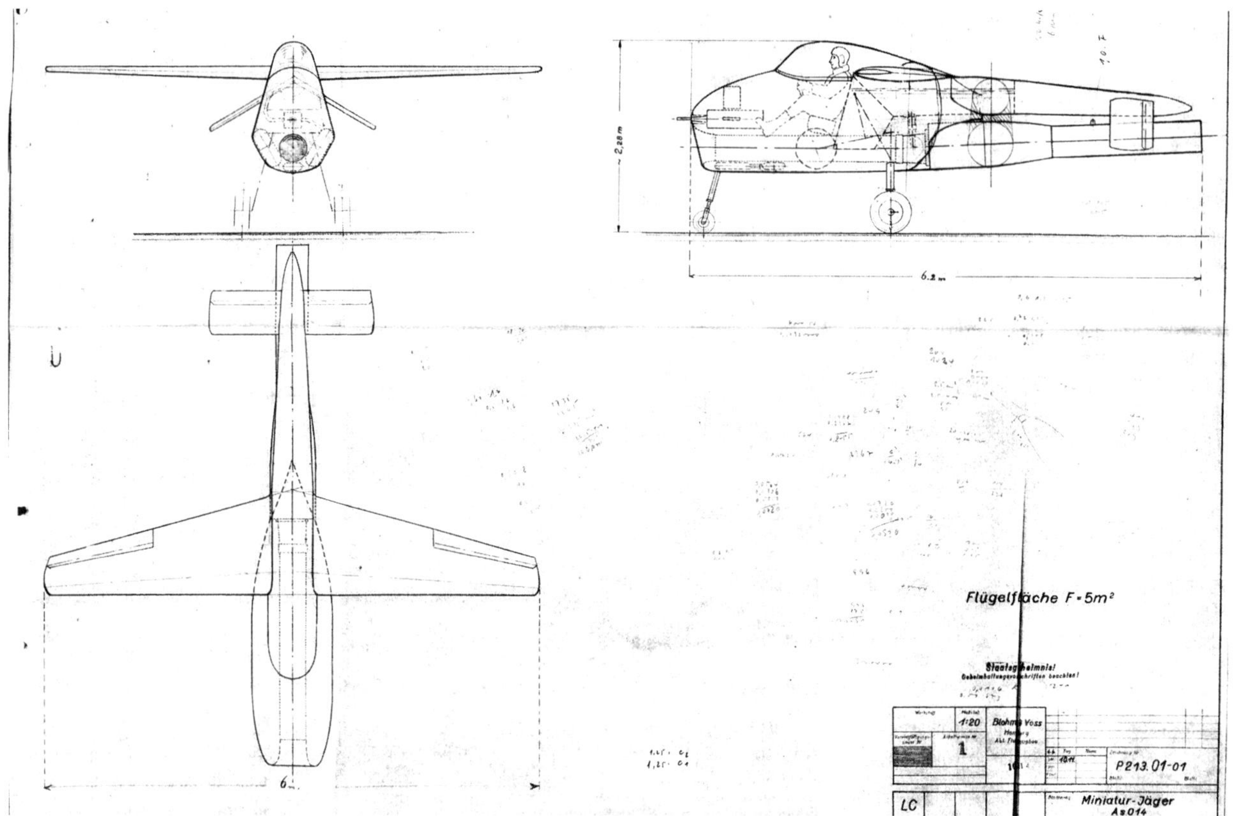

LEFT: The Blohm & Voss P 213.01-01 Miniatur-Jäger. The company believed that a simple rubber sleeve would be enough to alleviate the As 014's hazardous vibrations.

accommodation is much less problematic than the other existing possibility of storage in the impregnated wooden wing. The actual fuselage skin is slipped on from the front as a closed component and screwed to a form frame."

The most important part of the design, given the As 014's history of dangerous vibrations, was the engine installation. The report stated: "The Argus As 014 engine hangs behind the fuselage, organically in the slipstream. At the front it is attached to the normal fork belonging to the engine. This fork is just pivoted upwards. At the back, the support fitting is welded onto the tube as a hanger. A pendulum strut extends into the fuselage. Between the fuselage inlet pipe and the engine inlet, a sufficiently wide rubber sleeve compensates for the engine vibrations."

Also on November 15, Vogt wrote to Blohm & Voss's Berlin representative, W. Bürkner, concerning the P 213.[9] He wrote: "Attached you will receive two copies of the design for Mr Scheibe about the Miniatur-Jäger with the Argus jet tube. One copy is for you.

"Please note that the top speed is relatively good on the ground, but that it drops off immediately in contrast to the turbojet aircraft. Since the range is also very short, there is a considerable difference between the Argus tube and the turbojet. You shouldn't particularly push this project forward with Mr Scheibe. We would only like to offer it if other tasks should not be carried out by us, or if this small machine could also be used as an additional job – as a flying mock-up, so to speak."

Little did Vogt realise that Argus was about to kill off the P 213 less than a month after work on it first began. The company's Dr Goslar sent Vogt a letter[10] on the 15th which Blohm & Voss marked as received on November 21. He wrote: "Dear Dr Vogt! We have just received your message from 10/11/44 with the project sketch of the Miniatur-Jäger. We find the solution proposed by you very interesting and would like to comment briefly on your request.

"First of all, it is not entirely clear to us where you want to put the fuel. After all, approx. 700 liters are required, and unfortunately we do not find any indication in your sketch where this fuel supply should be placed. Regarding the intake pipe for our engine, we unfortunately cannot share your optimism about the vibrations. The vibrations of the inlet pipe are not only caused by cross-blowing, but are also excited by the pulsating suction of our engine. The way the system is shown in the sketch is not possible.

"As with acoustic filters, the inlet pipe would either have to be provided with numerous side holes or be given a shape that deviates from the cylindrical shape. Based on our extensive experience, we have to say that our engine is extremely sensitive at the inlet end and that we can only count on a quick success if the supply conditions of the air to the pipe are not touched if possible.

"Allow us to point out that the pushers that we specify in our installation folder are actually useful pushers and, in addition, the drag of the engine no longer needs to be taken into account for the free moving pipe. All of the other projects currently available to us have taken over the installation of the pipe above the tail unit, which has proven itself in the V1 [Fi 103 flying bomb]. We fully understand the intentions underlying your design; but you will certainly want to get the job as soon as possible, and the less risk the project contains, the more likely it will be.

"We do not want to say any more about your design at the moment because we believe that considering the size of the fuel tank, small changes will still be necessary. Perhaps you can also take our concerns about the engine intake into account and send us a correspondingly modified sketch."

Having had its design so thoroughly dismissed, Blohm & Voss appears to have abandoned the P 213 immediately after receiving this letter.

ZEPPELIN PULSEJET FIGHTER

Less well known than the DVL, the AVA or the LFA, the Forschungsanstalt Graf Zeppelin (FGZ) or 'Graf Zeppelin research institute' was nevertheless a reasonably large and well-equipped organisation which carried out numerous experiments in aerodynamics and other areas of aviation research.

Not to be confused with Luftschiffbau Zeppelin, the airship maker, the FGZ grew out of the Flugtechnischen Institut of the Technischen Hochschule Stuttgart in 1937. In the beginning, it consisted of just six people but by January 1944 it had 400–500 staff including four professors, 17 doctors and 22 qualified engineers.

Its primary areas of interest were bomb aerodynamics, parachutes, underwater blast physics, take-off and landing aids, and the aerodynamics of aircraft with additional exterior structural units – such as troop-carrying pods attached to the wings of Junkers Ju 87 Stukas.

During the early part of 1944, the FGZ worked on launch ramps for the V1 flying bomb and became intimately familiar with the workings of the As 014 in the process. And while the FGZ had been working on ramps, the nearby Forschungsinstitut für Kraftfahrwesen und Fahrzeugmotoren Stuttgart (FKFS) had been working on a solution to the pulsejet vibration problem.

According to a postwar report[11] produced by Ulrich W Hütter, who joined the institute in early 1944, the FGZ and FKFS had worked closely together during the autumn of 1944 on "an anti-aircraft device consisting of an unmanned missile with explosives controlled from the ground via fine steel wires towards approaching enemy formations. The drive was two Argus As 014 pipes.

"In the FKFS, experiments had shown that when these pulsejet tubes were run in parallel operation, so that the inlets of both tubes were parallel and only a small distance from each other, the operation became calmer and smoother than when a single pipe was operated on its own."

Evidently the key to cancelling out the As 014's vibrations was simply to operate a second one right next to it. Hütter said that where a BMW 801 with propeller and other auxiliary machinery took 8,000–9,000 man hours to build,

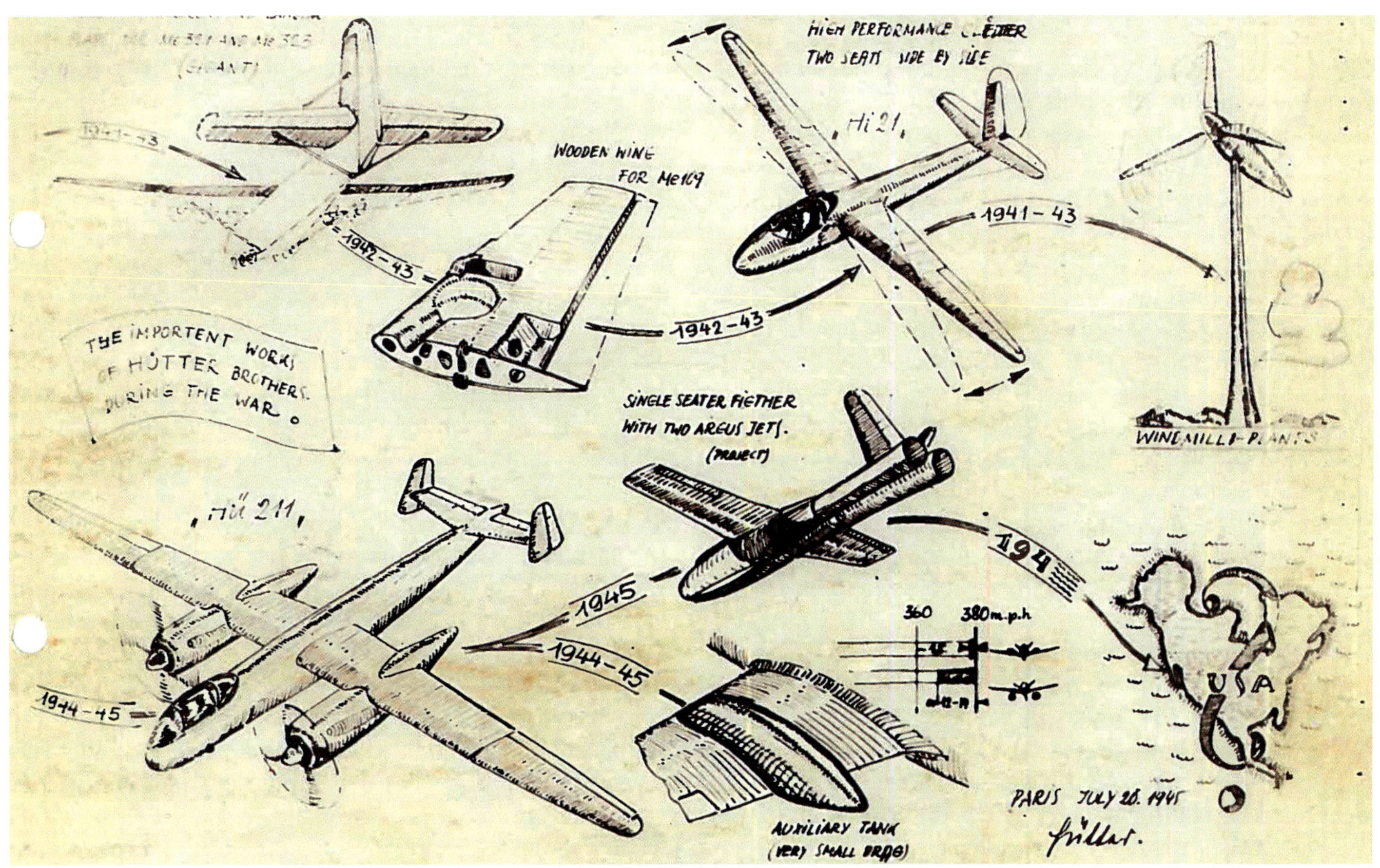

ABOVE: Wolfgang Hütter included a rear view of the Zeppelin twin-pulsejet fighter in a July 20, 1945, report on projects involving himself and his brother Ulrich.

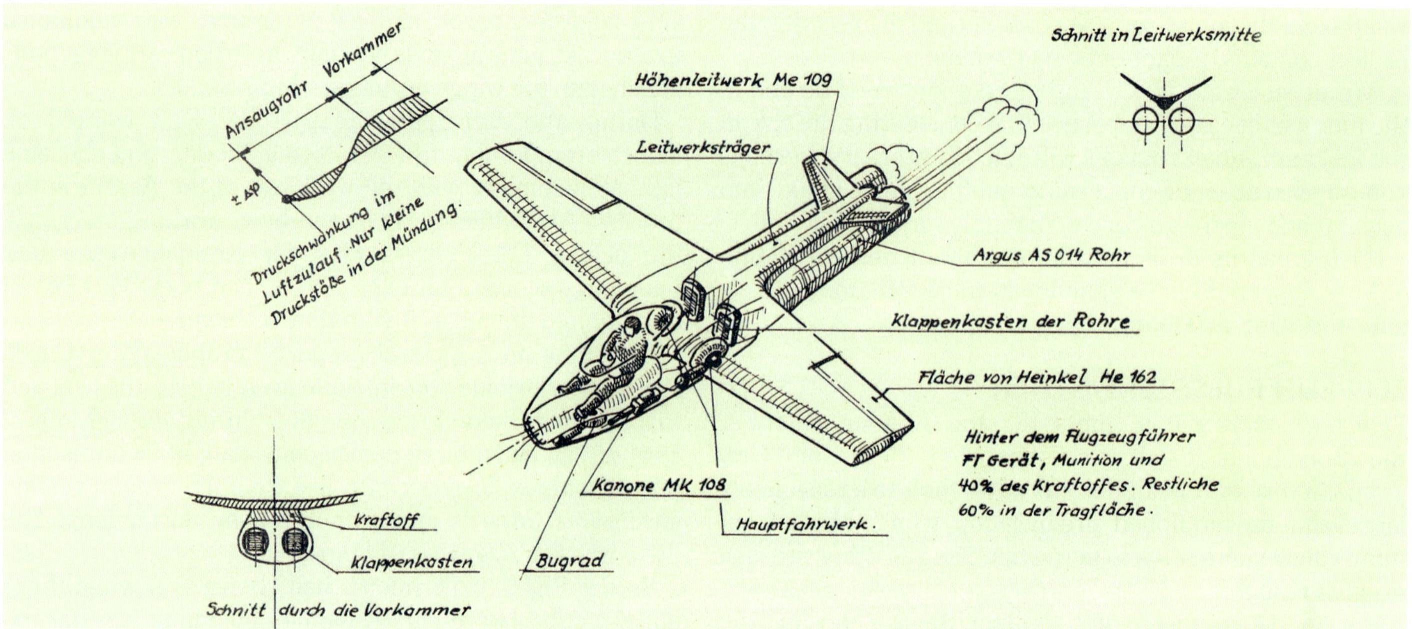

ABOVE: The Zeppelin pulsejet fighter as it appears in Ulrich Hütter's report on the work of the FGZ.

two As 014s took just 500–600 man hours. If a simple fuselage could also be used, the saving in terms of both time and cost would be considerable.

However, the FKFS also discovered that two tubes running side by side only produced a little more thrust than one tube on its own. Aerodynamic drag was blamed for this anomaly, and it was thought that a suitably shaped fuselage to cover the engines' front end would result in big performance gains. This was where Zeppelin came in – its aerodynamic experience was applied to the problem and a fuselage was designed which could house "large amounts of fuel, the explosive, a control system, the large cavities for the intake of the two tubes and the attachment point for the wings".

The FGZ built a test rig and "the double pipes ran quite satisfactorily on the stand, even with large plywood sheaves in the shape of the projected fuselage nose".

There were concerns that the pulsejet missile's guidance system might be difficult to procure, so a manned version was studied in parallel. But "the design difficulties were considerably greater, since the landing gear, radio equipment, weapons etc. had to be accommodated with a larger fuel supply". The aircraft was to have an all-up weight of two tons and a top speed of over 900km/h in horizontal flight at 6,000m.

Hütter noted that several attempts had previously been made to power an aircraft with pulsejets including a "manned Fi 103 with enlarged wings, Junkers 'Lilli' [Elli] device with 500mm tube and DFS/Messerschmitt 235 with two free-floating As 014 pipes". The latter appears to be a reference to the Me 328, since Hütter goes on to say: "However, they repeatedly failed because neither man nor the more sensitive parts of the aircraft are able to cope with the heavy vibrations that occur during tolerable periods of time. There is damage to clothing and skin, to control parts and instruments, the pilot's canopy glazing becomes hard to see through and memory and concentration of the pilot seriously suffer."

Diagrams appended to Hütter's report showed the Zeppelin pulsejet fighter with wings taken from an He 162 and tails surfaces from a Bf 109 arranged in a V at the end of a stubby tail boom above the twin As 014s. It was to be armed with a pair of MK 108 30mm cannon and had a tricycle undercarriage.

Work on the fighter was suspended when it became clear that the He 162 was eating up all available production capacity and it would not be possible to make the Zeppelin pulsejet the "fast solution" required. In addition, there were still question marks over the vibrations produced by the jet tubes at high airspeeds – although with a tuned diffuser "and fully elastic suspension of the entire engine, it is conceivable that these difficulties may have been mastered. Until then, however, a lot of development work was still required". It never happened.

HE 162 WITH ARGUS TUBES

The last known pulsejet project of the war was devised by Heinkel as a continuation of He 162 development. It was outlined in a report[12] published on March 30, 1945 – two day before the Soviet army began their offensive against Vienna where Heinkel's design staff were based. Given the mention of an He 162 powered by pulsejets in connection with Junkers 'Elli' in January however, it would appear that the project had been under way for several months beforehand.

The short description stated: "The occasion for investigation of the 162 with Argus tubes was the considerably lower cost of producing the Argus tubes in comparison with the turbojet power unit. The drawbacks are the great fuel consumption, low altitude output and the necessity for supplementary starting aid. The velocities at lower altitudes, however, are only slightly inferior to the value attainable with turbojet units.

"We have the choice between different-size Argus tubes, the As 014 tube with 335kg static thrust and the As 044 with 500kg static thrust. We investigated the employment of two As 014 tubes above the fuselage, and also of one As 044 tube similarly placed. In contrast with report 114/44, we are nevertheless suggesting the installation of one As 044 tube, owing to the increased thrust which has meanwhile been developed. The arrangement with one tube gives a considerably better rear field of view than with two tubes. For the installation of two strong As 044 tubes, allowing for at least 20 min duration of full-throttle flight at ground level, the body of 162 is too small (approximately 2.2 tons of fuel are required).

"As a result of the pronounced falling-off in the power unit's thrust with altitude (more than with air density) the maximum speed is attained at ground level. For the same reason flight performance is poor at greater altitudes, so that operational flight can only be considered at low altitudes. Take-off must be effected by having recourse to auxiliary starting aid. If the take-off position is not to be betrayed by the smoke of the starting rockets it will be necessary to employ a catapulting device. The amount of fuel required may be completely housed in the fuselage and wing, making external tanks superfluous. The supply of auxiliary power for the radio, fuel and hydraulic systems, is provided by a generating set on board the aircraft."

Two drawings were appended to the report: 162.01-43 and 162.01-44, both dated March 30, 1945. The former showed the He 162 with the large As 044 mounted on its back while the latter showed the aircraft with a pair of As 014s side by side on its back – though not right next to each other, suggesting that Heinkel was unaware of the FGZ's work.

The drawings also show the He 162 with a fuselage length of 9.25m, although this change from the design finalised in mid-October 1944 (fuselage length of 9.05m) does not appear to have been to accommodate more fuel since the As 044 design had a fuselage tank of 470 litres and the twin As 014 design had a tank of 650 litres, both tank capacities having been previously planned for the standard He 162 A-2. It is also uncertain how Heinkel intended to fit 900 litres of fuel into the wings on each design since the standard He 162's wing tanks could only take 280 litres.

Armament was a pair of MK 108s in each case but a third, unnumbered, drawing exists showing the As 044 design equipped with a pair of much longer-barrelled weapons – possibly MK 103s – with their barrels protruding from the sides of the aircraft's fuselage.

It is unclear why Heinkel bothered with this project since the vibrations produced by the Argus pulsejets were seemingly incurable despite the work carried out by the FGZ. None of the pulsejet projects of 1944/45 amounted to anything useful.

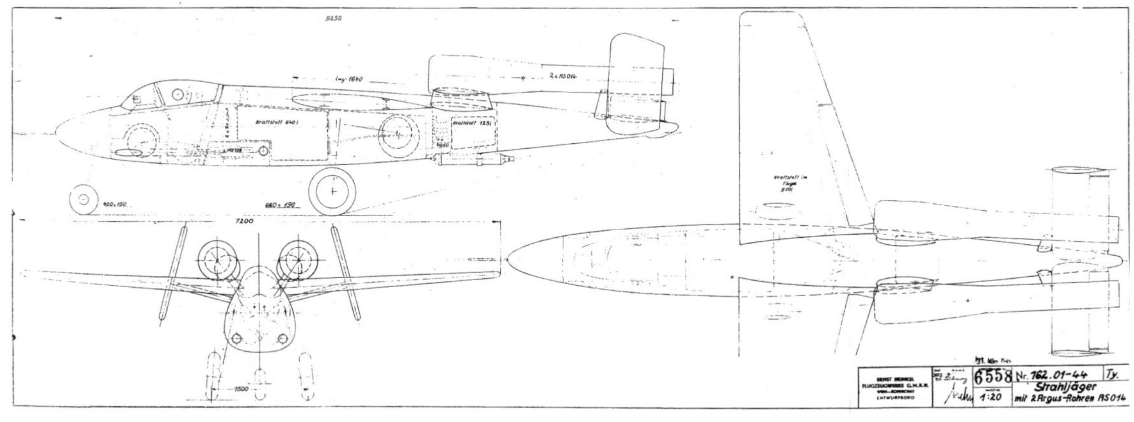

RIGHT: The twin-As 014 pulsejet He 162. Heinkel preferred the single As 044 option.

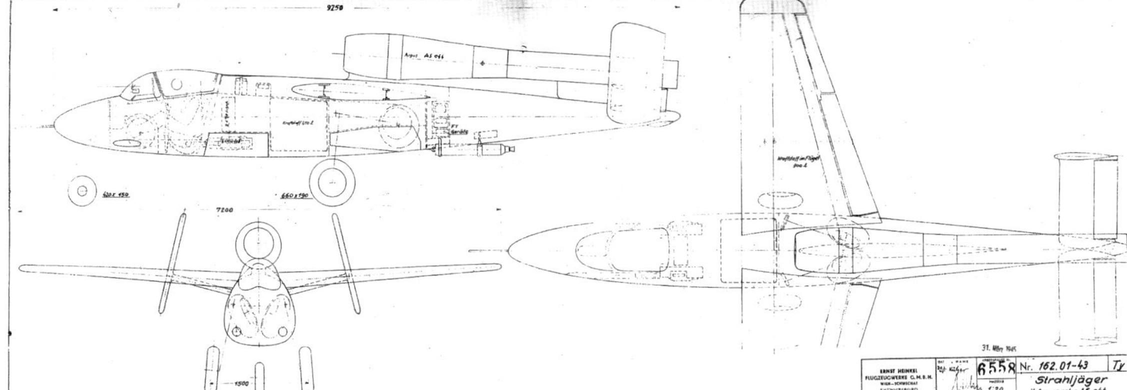

RIGHT: Heinkel's design for an He 162 powered by a single large As 044 pulsejet, as it appeared in drawing 162.01-43.

Chapter 14

Miscellaneous jet fighters

Various projects 1940–1945

Throughout the war but particularly towards the end companies and individual engineers worked on jet fighter projects without any obvious mandate from the RLM. Some of these designs appear as though they might fit this or that set of requirements but are not known to have been part of any competition. In many cases their background remains obscure.

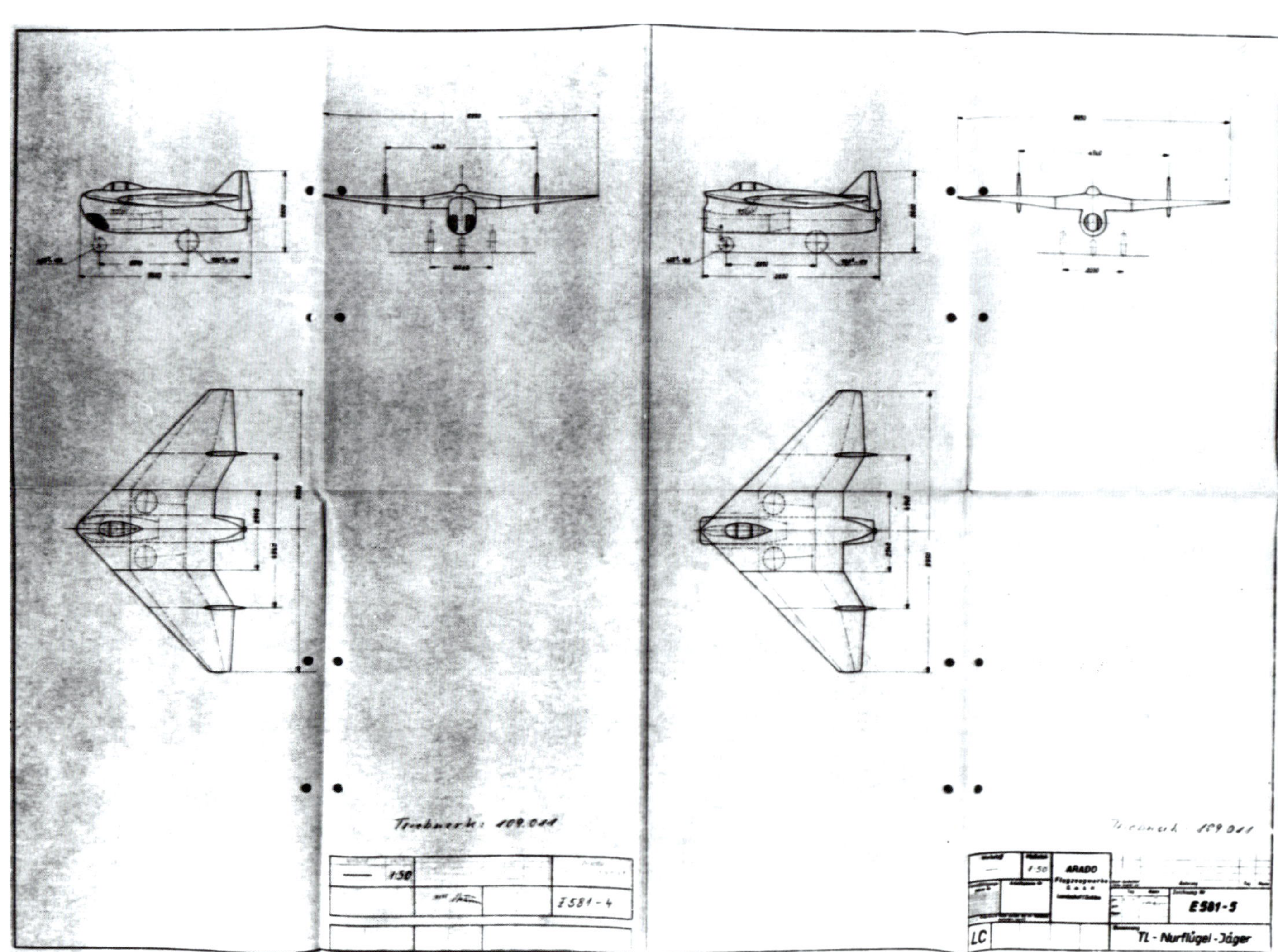

ABOVE: The small single-jet Arado **E 581-4** and **E 581-5** fighters. While they look 'fat' when viewed side-on, from above each design is something like an arrow-shaped flying wing.

ARADO E 581-4 AND -5

Little is known about Arado's outlandish single-jet flying-wing fighter series – E 581. The series seems to have begun in mid-November 1944 and the earliest known version is the E 581-2, which was to be powered by the BMW 003. It had a wing sweepback of 46°, mid-wing fins for control, a 740-litre fuel tank and two MK 108 cannon. Its wingspan was 8.48m with a wing area of 22.5m². By January 1945, the design had evolved into the E 581-4 and E 581-5, each of which was to be powered by a single HeS 011 – putting the small fighter into the same category as the 1-TL-Jäger designs. Both were 5.65m long with a wingspan of 8.95m.

However, no evidence has yet come to light suggesting that the E 581 was ever submitted or considered for the 1-TL-Jäger competition. The only surviving period documentation appears to be a bundle of notes[1] and three-view drawings of the latter two designs, produced by Arado designer Braun at Landeshut and dated March 1945.

BLOHM & VOSS NURFLÜGEL-TL-JÄGER

Blohm & Voss designer Thieme produced a drawing on sheet number Ae 607 on February 5, 1945, showing a very odd-looking aircraft.[2] It was labelled "Nurflügel-TL-Jäger" or 'all-wing jet fighter' and had an 8m wingspan with a 45° sweepback. It also had an offset cockpit which allowed a very narrow wing profile. This also meant that there could be a direct and uninterrupted flow of air from the circular intake at the tip of the wing to the HeS 011 engine at the rear. Small forward-swept whisker-like canards at the front of the fuselage would have presumably lowered the landing speed to an acceptable level, and the four-wheel undercarriage, with two large wheels at the front, two smaller ones at the rear, would have kept the engine intake away from any surface debris in a rough landing area.

The Blohm & Voss Nurflügel-TL-Jäger was probably intended to showcase ideas for solving particular problems facing designers when deciding on a layout for fighters and never received a 'P' number.

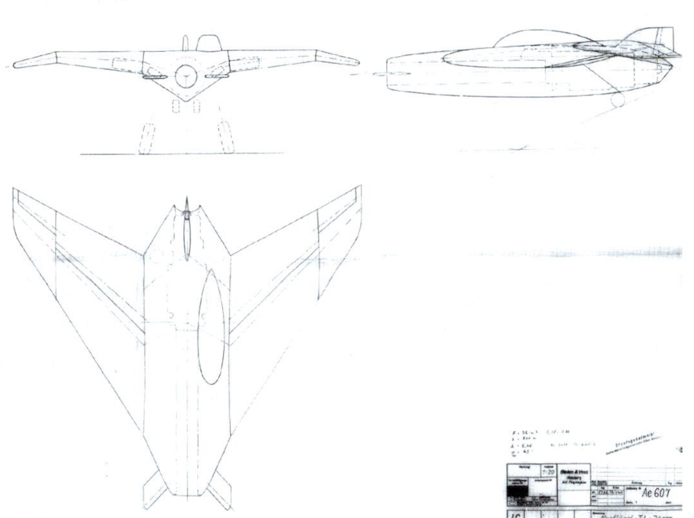

ABOVE: For years it was assumed that this outlandish Blohm & Voss design was pure fantasy – until an original drawing was discovered which proves its authenticity. The aircraft would have had two ordinary-sized undercarriage mainwheels at the front, and another two very small ones to the rear. It also had small 'whisker'-type canards beside its nose intake.

BLOHM & VOSS P 175

The tiny P 175 Bordjager or parasite aircraft had a wingspan of just 6.2m, compared for example to the Me 163 B's 9.3m, a wing area of only 6m² and no undercarriage. Its single Jumo turbojet was mounted ventrally, it had a twin-fin tail, a cockpit canopy that slid rearwards, rather than opening outwards, and a highly unusual split-spike device on its nose – presumably for hooking onto its host aircraft.

The only known drawing of it is dated March 22, 1943 – which coincides with a series of meetings to discuss whether or not to build the Luftwaffe a new long-range maritime reconnaissance aircraft.[3] Blohm & Voss had four aircraft designs involved in this process. Under the heading of 'Atlantik-Aufklärer', a meeting on March 22 discussed and compared the Fw 200, Ju 290, BV 222 and 238, Me 264 and Ju 390.

By the meeting[4] on April 27, 1943, the heading had changed to 'Atlantik-Aufklärer / Fernkampfflugzeug' and the only Blohm & Voss project in contention was the BV 250 – the wheeled undercarriage version of the BV 238 flying boat. B&V documents from this period show that a small-fuselage version of the BV 250, the P 173, was also being heavily worked on during this time.

No direct evidence is known to exist linking the P 175 to any of these projects, yet the timing fits. It would appear that the P 175 was intended to be carried on the back of or underneath a BV 238 or BV 250. But like many B&V projects it was short-lived and went no further.

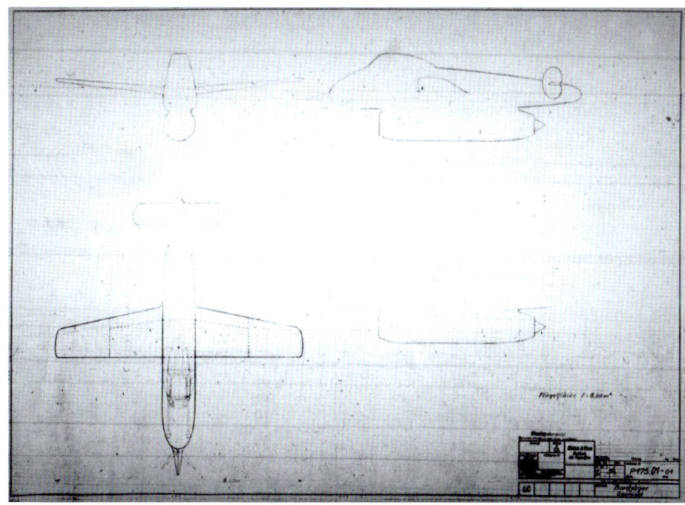

ABOVE: This faded drawing is the only known surviving image of Blohm & Voss's P 175 Bordjäger.

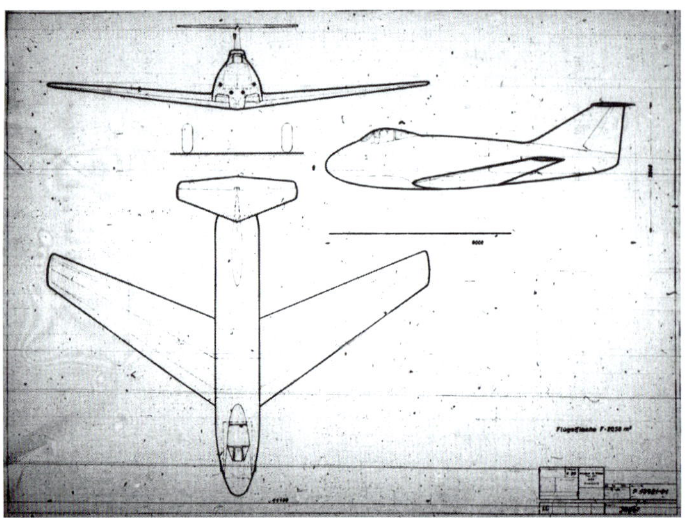

ABOVE: The Blohm & Voss P 197 was a surprisingly modern-looking fighter powered by a pair of Jumo 004s.

BLOHM & VOSS P 197

This advanced-looking single-seat fighter – probably dating from April to May 1944 – was powered by two Jumo 004s fitted side by side in the rear end of the fuselage. This arrangement was possible thanks to the sweepback of the low wing. It had a wingspan of 11.1m with a wing area of 20.5m² and was 9m long. According to British intelligence report 'German Aircraft: New and Projected Types': "Apart from the fact that the tailplane is mounted at the top of the fin the front elevation is rather like that of the Me 262." Armament consisted of two MK 103s underneath the cockpit and an MG 151/20 on either side. The intakes for the jet units were in the nose just forward of the leading edge. Unusually, they were positioned facing slightly downwards and were therefore partially masked when looked at directly from the front. Estimated maximum speed was 660mph. Little else is known about the P 197.

BLOHM & VOSS P 198

Powered by a single very large BMW 018, the P 198 was apparently designed as a high-altitude fighter with a 15m wingspan and 33.5m² wing area. It had a relatively conventional layout except for a very deep fuselage to accommodate the turbojet and bulges on the underside of the wingroots accommodate its undercarriage mainwheels when retracted. The armoured cabin was in the aircraft's nose and the armament was a single MK 412 and a pair of MG 151/20s. It probably dated from mid-1944 but no other data about it has yet been discovered.

BLOHM & VOSS P 202

The twin-jet single-seat P 202 fighter had a high wing designed to be swivelled in flight to give the effect of variable sweep-back without altering the aircraft's centre of lift. According to British intelligence report 'German Aircraft: New and Projected Types': "When the wing is displaced from its central position the velocity of air flow normal to the leading-edge is progressively reduced in relation to the forward speed of the aircraft.

"Since it is the velocity of normal flow which determines the onset of compressibility effects at high speed it is theoretically possible in this way to obtain an increasingly high speed for a given thrust. It is difficult to visualise the possible effects of this particular arrangement upon the flying characteristics but presumably the designers thought that no insuperable problems would arise. The maximum displacement of the wing is 35°.

"For take-off and landing the wing is in its normal position, indeed it is only in this position that the flaps and main undercarriage can be operated. Thus one objection to the normal swept-back wing, namely difficulties of control at low speeds, is overcome. The two jet units (probably BMW 003) are enclosed in a bulge in the lower portion of the fuselage with a common intake in the nose. The empennage is conventional. Armament consists of one MK 103 and two MG 151/20s."

The P 202's overall length was 10.56m and wingspan was 12m in unswivelled configuration or 10m at maximum swivel. Wing area was 20m². The design's precise date is difficult to pin down but Blohm & Voss's numerical design

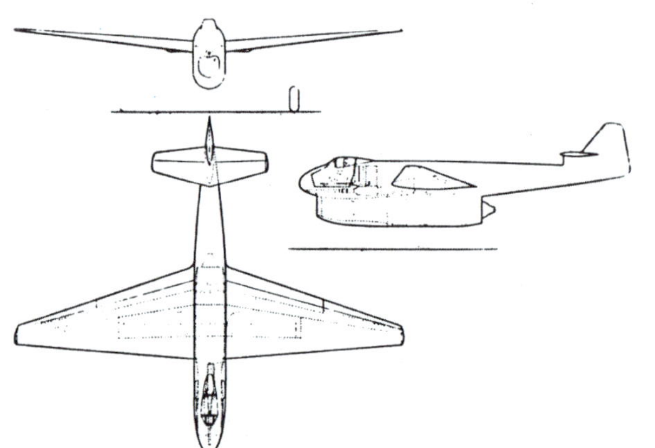

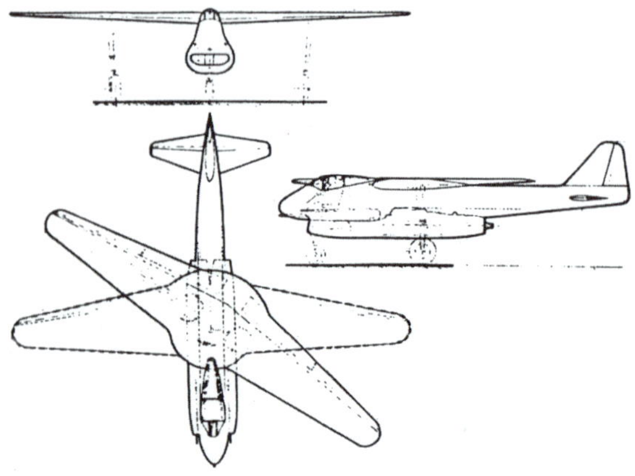

ABOVE LEFT: The enormous BMW 018 turbojet-powered Blohm & Voss P 198 high-altitude fighter. The position of the wings relative to the underside of the fuselage meant that the aircraft would have needed to have extremely long undercarriage legs.
ABOVE RIGHT: The Blohm & Voss P 202 jet fighter with swivelling wing.

series was chronologically sequential and the official company project description for the P 203 was issued in June 1944, so the P 202 probably pre-dates that by several months.

BLOHM & VOSS P 214

Among the most enigmatic of Blohm & Voss's many projects was the P 214.

The only known evidence of it is a series of five sheets: two text and three graphs[5] dated November 20–24, 1944. These describe the project as a 'Bemannte Fla – Bombe', with the 'Fla' evidently being short for 'Flugabwehr' meaning 'air defence', itself meaning 'anti-aircraft'. So the P 214 was, in effect, a 'Manned anti-aircraft bomb'.

Given its size – a wingspan of 7m, similar to that of the He 162 by way of comparison, and a wing area of 10m^2, only slightly smaller than that of the He 162 – it wasn't exactly a missile. It weighed 3,600kg, of which 1,700kg was fuel.

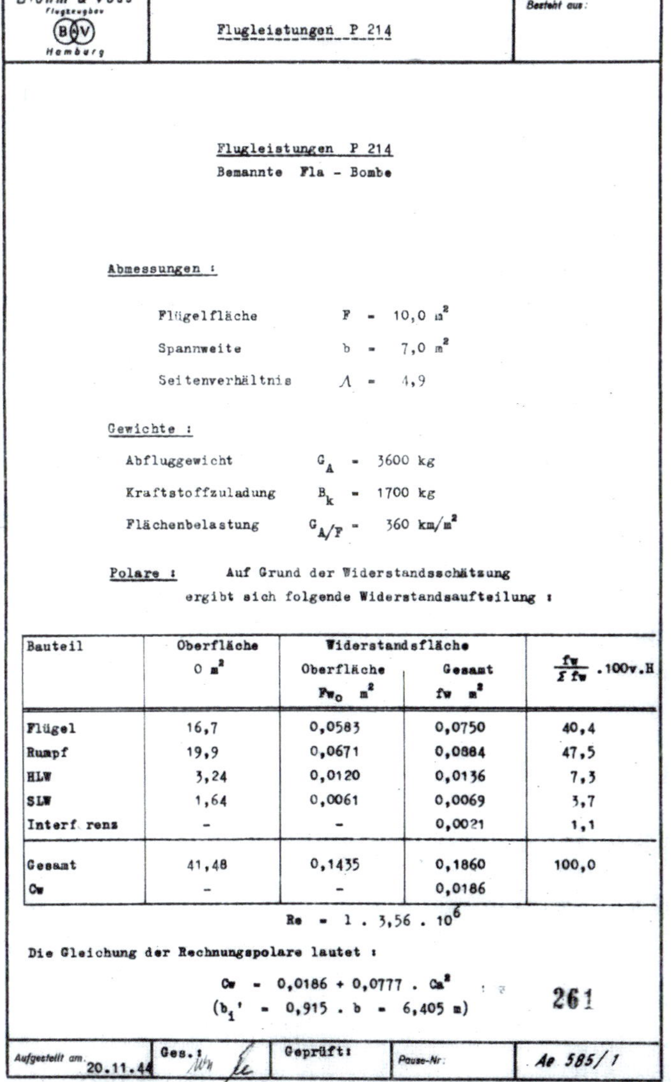

Blohm & Voss Flugzeugbau Hamburg

Flugleistungen P 214

Besteht aus:

Flugleistungen P 214

Bemannte Fla - Bombe

Abmessungen :

Flügelfläche $F = 10{,}0\ m^2$

Spannweite $b = 7{,}0\ m^2$

Seitenverhältnis $\Lambda = 4{,}9$

Gewichte :

Abfluggewicht $G_A = 3600\ kg$

Kraftstoffzuladung $B_k = 1700\ kg$

Flächenbelastung $G_{A/F} = 360\ km/m^2$

Polare : Auf Grund der Widerstandsschätzung ergibt sich folgende Widerstandsaufteilung :

Bauteil	Oberfläche $O\ m^2$	Widerstandsfläche Oberfläche $Fw_o\ m^2$	Widerstandsfläche Gesamt $fw\ m^2$	$\frac{fw}{\Sigma fw}$ .100v.H
Flügel	16,7	0,0583	0,0750	40,4
Rumpf	19,9	0,0671	0,0884	47,5
HLW	3,24	0,0120	0,0136	7,3
SLW	1,64	0,0061	0,0069	3,7
Interf. renz	-	-	0,0021	1,1
Gesamt	41,48	0,1435	0,1860	100,0
Cw	-	-	0,0186	

$Re = 1 \cdot 3{,}56 \cdot 10^6$

Die Gleichung der Rechnungspolare lautet :

$$Cw = 0{,}0186 + 0{,}0777 \cdot Ca^2$$

$$(b_1' = 0{,}915 \cdot b = 6{,}405\ m)$$

261

Aufgestellt am: 20.11.44 | Ges.: | Geprüft: | Pause-Nr: | Ae 585/1

ABOVE: The first of five known pages detailing the Blohm & Voss P 214.

Its empty weight of 1,900kg was almost exactly the same as the empty weight of the Me 163, though the latter carried just over 2,000kg of fuel and oxidiser for its rocket engine.

No engine type is given but the remainder of the text states: "Flight route: The course of the flight is assumed as follows: The aircraft is accelerated on the ground until the speed has reached 700 or 800km/h, with which it should rise below 45°. The thrust for acceleration on the ground should be equal to the initial thrust when climbing."

This certainly suggests rocket thrust. The accompanying graphs show the P 214 climbing to an altitude of 15km (49,212ft) or 17km (55,774ft), depending on whether take-off speed was 700km/h or 800km/h, in just under two minutes. Descending flight then began, with the total flight time including climb expected to last around eight minutes.

It has been suggested that one or more of the late-war designs of Karl Stöckel, an employee of the Deutsche Versuchsanstalt für Luftfahrt (DVL) or 'German aviation research institute', is actually the P 214 – but none of these matches the weights and dimensions of the P 214 given in the Blohm & Voss paperwork, particularly his MGRP which consisted of a 5m wingspan glider attached to the back of a very large rocket, giving a take-off weight of 10 tons. For now it would appear that the P 214's intended appearance is simply unknown.

BMW SINGLE JET-FIGHTERS

As one of the three biggest aero engine manufacturers in Germany during the Second World War, BMW was constantly in contact with the aircraft constructors – liaising over existing models and potential future designs.

In addition to the ongoing development of its war-critical BMW 801 piston engine and piston engines it hoped would soon reach series production such as the BMW 802 and 803, the company was also at the forefront of developing both turbojet and rocket engines.

Towards the end of the war, the BMW 003 turbojet became increasingly important in the minds of those planning for the Luftwaffe's requirements and the troubled BMW 018 was also perceived to have great potential.

The company had previously come up with a series of bomber designs to showcase its products and with EZS-Bericht Nr. 58 of November 3, 1944, it appears to have done the same with fighters – presenting four configurations which leaned on the latest aerodynamic forms being reported on by experimental institutions such as the AVA and LFA. While the four look quite different from one another at first glance, the first three all have exactly the same straight wing shape with swept leading edge and are all powered by the BMW 003. The second and fourth designs have a very similar fuselage and tail design, and all four have a nose intake and a tricycle undercarriage.

The first design, known as 'Strahljäger I', has a twin-tail with the high-position turbojet exhausting above it, 'Strahljäger II' has a conventional tail with the low-

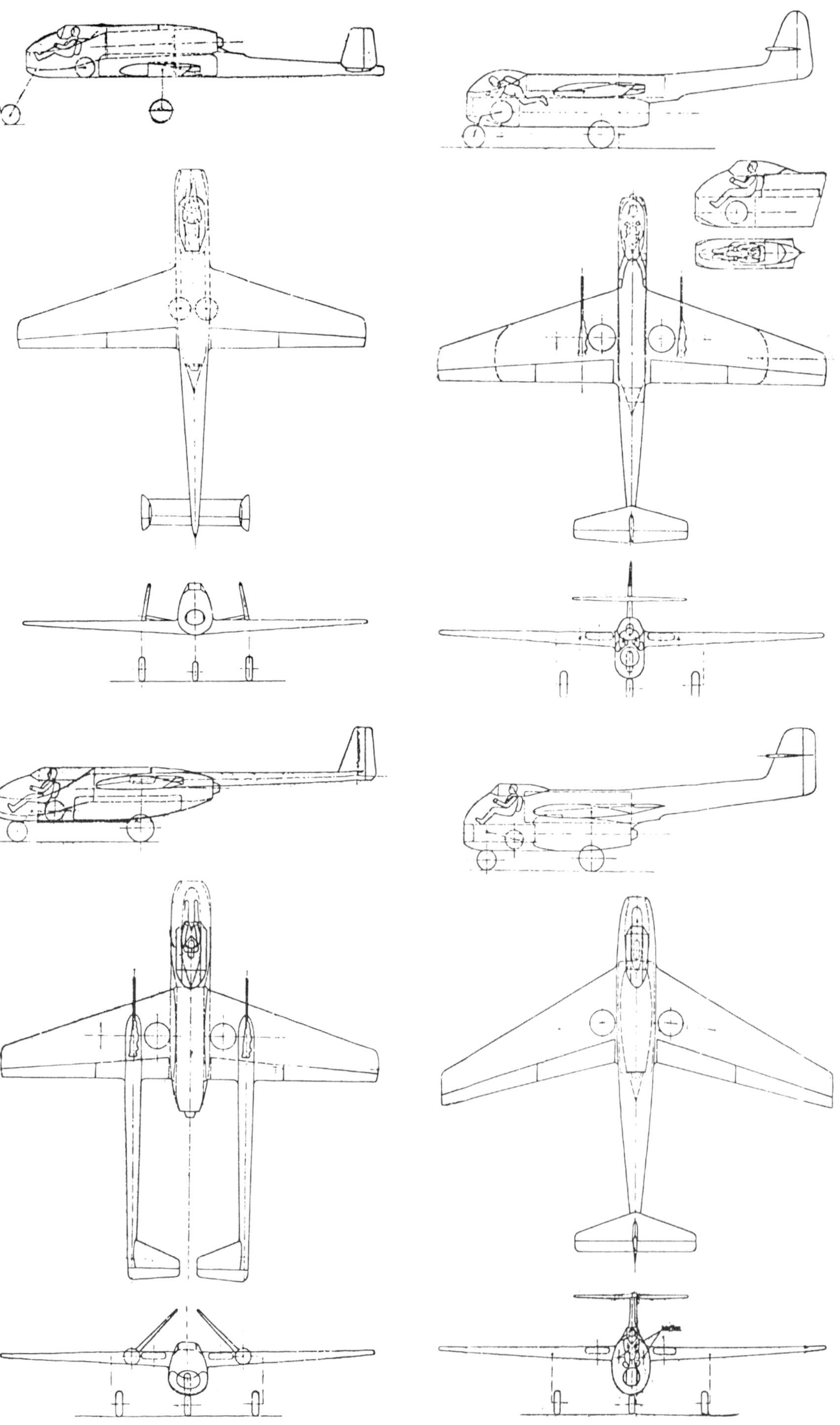

TOP LEFT: BMW's Strahljäger I design had its BMW 003 turbojet installed within the central section of the fuselage, fed by a nose intake. Dating from November 3, 1944, after the He 162 had already been chosen as the Volksjäger, it seems unlikely anyone ever expected it to be built.

TOP RIGHT: The BMW Strahljäger II looks somewhat different from the Strahljäger I but the two are actually quite similar – they both had a BMW 003 engine, both had a nose intake and their wings were largely the same. The Strahljäger II differed in having a more conventional tail, its turbojet in the lower part of the central fuselage and a prone pilot position. A conventional seated pilot position is shown as an option on the original drawing.

BOTTOM LEFT: A centrally positioned turbojet and twin-boom tail are the main differences between the Strahljäger III and the two other BMW 003 designs from the November 3, 1944 report. The Strahljäger III also had an extremely short nosewheel leg, with the wheel itself barely clearing the bottom of the fuselage.

BOTTOM RIGHT: Similar in some ways to the Strahljäger III, the Strahljäger IV differed in having a much larger BMW 018 turbojet in its lower fuselage and in having swept wings. Overall, it was a substantially larger aircraft.

position turbojet exhausting below it and 'Strahljäger III' has a centrally positioned turbojet exhausting between the fins of a twin-boom tail. The final design, known as 'Strahljäger IV', was powered by a BMW 018 and was somewhat larger than the other three.

None of these designs went any further than BMW's report.

DVL JET FIGHTER

Shortly after the war, while documenting a set of high-speed measurement graphs dated January 1944 found among papers taken from the Deutsche Versuchsanstalt für Luftfahrtforschung (DVL), the German institute for aviation research, based in Berlin, US personnel discovered some handwritten notes on their reverse side alongside four sketches.

Three of the sketches show what appears to be a tandem arrangement of a tailless and engineless but highly aerodynamic fighter aircraft at the front attached to a winged pod containing the engine at the rear. Exactly what advantage this was intended to give is unclear, but it may have been an attempt to avoid having a long intake tube, or to provide the engine with air already 'smoothed out' by its passage over the aircraft shape in front.

In the first sketch, the pod is attached via two long narrow booms. In the second more tentative sketch, the pod is attached at a central point and one of the wingtips has sprouted a shorter boom with a little winglet on the end. In the third drawing, the central attachment point and wingtip booms have become better established, with the aircraft at the front shrinking somewhat and the pod growing in size.

In the fourth drawing, the aircraft section and engine pod have effectively been combined into a single structure. The cockpit, which previously had the pilot seated in a reclined position, has been shrunk to a tiny chamber above the gaping engine intake where the pilot is shown lying in a prone position. The aircraft's weapons, presumably a pair of MK 108s, are crammed into spaces within its chin.

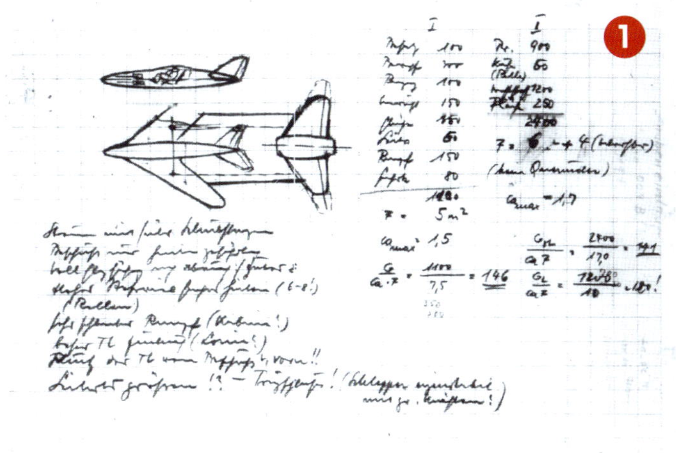

ABOVE: The first drawing in the sequence shows design for an engineless aircraft attached to a winged engine pod by two slender booms.

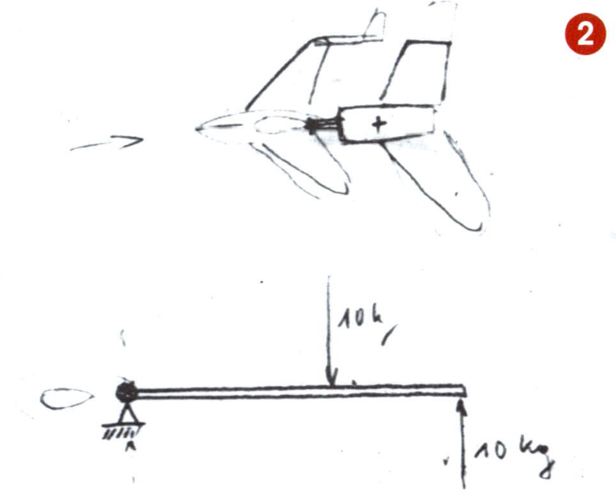

ABOVE: Next, the nameless designer revised the arrangement to feature a central attachment point for the engine pod and wingtip control surfaces instead of full booms.

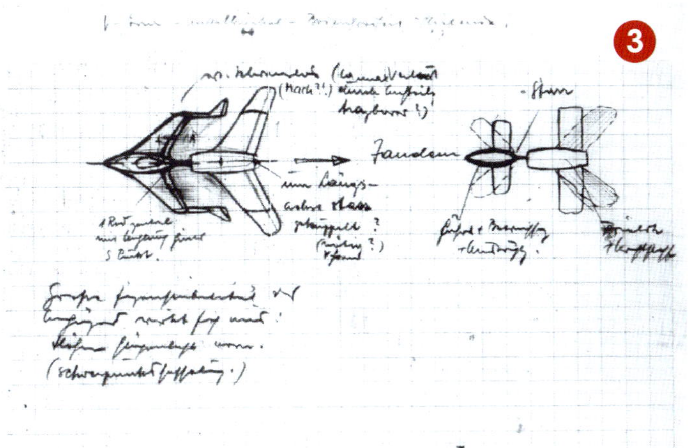

ABOVE: The third design shows the fighter now with full wingtip booms and a central attachment point for the engine pod.

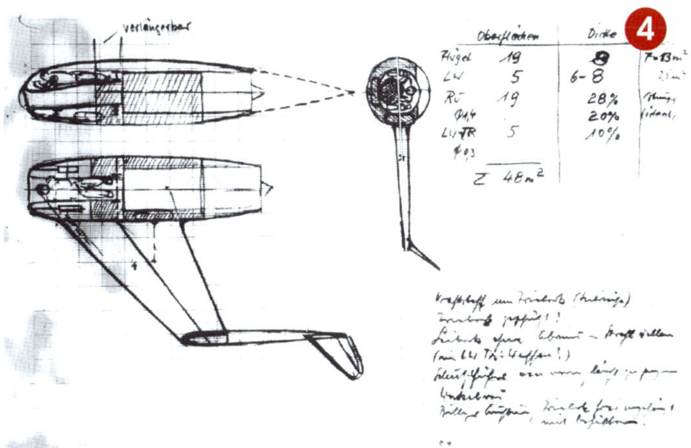

ABOVE: Finally, the aircraft and engine pod are merged to create a one-piece tailless fighter. The cockpit is reduced to a narrow space above the engine intake.

Exactly how the aircraft would have landed is unclear, but it would most likely have used a belly skid.

This design is significant because it appears to be an early precursor to a series of designs that would be produced by Blohm & Voss later in the year. The idea of using wingtip booms was not new – having evidently been the subject of several patents during the mid-to-late 1930s – but the idea of using a single angled surface on the tip of a boom, rather than a full tailfin and tailplanes, appears to have had no precedent at this point.

DORNIER 'CLOVER LEAF'

French forces captured Dornier's facility at Manzell in the Friedrichshafen area but were evidently reluctant to share the material they found there with their British and American allies. As a result, the Americans sent a specialist to interrogate company owner Claude Dornier and his son Peter, who were in French custody in Paris, about it.

The result was an insightful though rather brief document on Dornier's fighter development programme[6] by Alfred V Verville, an aviation pioneer and aircraft designer who worked for Curtiss, Aeromarine, the Thomas-Morse Airplane Company, the General Aeroplane Company and the Fisher Body Corporation, before joining the army as a civilian specialist then starting his own aircraft company.

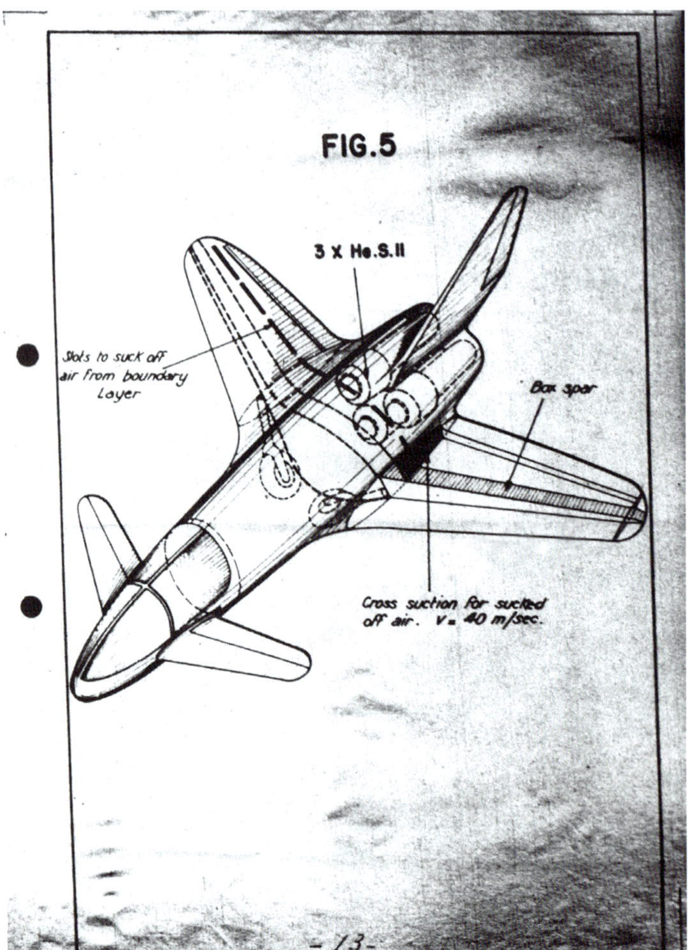

ABOVE: The unnamed triple-jet canard design described to US investigator Alfred V Verville by Claude Dornier and his son Peter.

Verville noted that the Dorniers were "asked to outline the pertinent features and advantages of the Do 335 design and to give a general review of current and projected pursuit plane designs, for the future". After discussions concerning the Do 335 and other piston-engined projects, the conversation turned to a "multiple jet pursuit plane". Verville's report stated: "This design study was carried out by Dornier with a view of developing a high-performing jet fighter of a concept opposite to that of the tailless type. It is of the so-called canard tail-first type using both swept-back wings and swept-back tail surfaces.

"The three Heinkel-Hirth 011 jet engines are arranged in clover leaf fashion. The combustion air is taken in through the fuselage and from wing boundary layer suction slots. The pilot is situated in the fuselage nose. Dornier claims it is possible to increase range by stopping two engines for cruising and without having additional flow drag increase."

The Dornier 'clover leaf' jet fighter discussed by Claude and Peter Dornier with Alfred V Verville is otherwise completely unknown. It does not seem to be referred to in any other known document and there are no other known drawings of it except for the astonishing example featured in Verville's report. Elsewhere it has been suggested that this design might even have been given an official RLM designation as the Do 350. The only known mention of the Do 350 comes from the March 16, 1945 to April 4, 1945 section of the war diary of the Chef TLR. The bullet point labelled 'Nachtjägerprojekt' or 'night fighter project' says: "Discussion with company DW [Dornier-Werke] about new night fighter project and issuing a preliminary ruling. The pattern is named Do 350."

Rather than being the radical triple-jet canard fighter, this sounds much more like Dornier's very conservative 'safe' night fighter design, the P 256/1-01. The timing fits well and a conservative buildable design, as opposed to the host of radical designs competing against the P 256/1-01, would certainly have suited the thinking of the time.

As for Dornier's triple-jet canard design, unless further evidence emerges to provide more background, it must remain one of the war's more mysterious designs.

FOCKE-WULF STRAHLJÄGER (ENTE)

On December 1, 1942, Focke-Wulf worked on calculations for a new tail-first or 'canard' fighter powered by the BMW P 3303 turbojet.[7]

Work on the P 3303 had evidently begun in 1940 or earlier and between April 21 and June 1, 1942, it was given the official designation BMW 018. Clearly, Focke-Wulf did not get this message and even by January 20, 1943, was still referring to it as the P 3303.

A BMW report[8] entitled 'Einbaumappe 109-018', dated June 1, 1942, gives the engine's overall length as 4.75m and

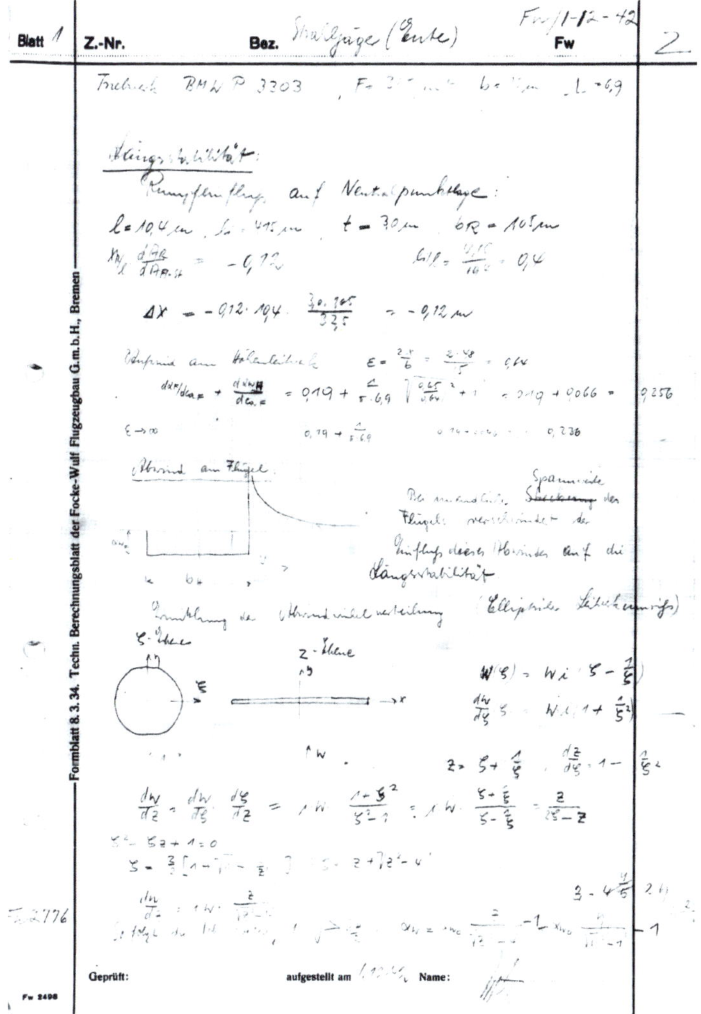

Blatt 1 | Z.-Nr. | Bez. Strahljäger (Ente) | Fw/1-12-42 Fw | 2

Triebwerk BMW P 3303

Formblatt 8.3.34. Techn. Berechnungsblatt der Focke-Wulf Flugzeugbau G.m.b.H., Bremen

Geprüft: | aufgestellt am | Name:

ABOVE: The first page of the three-page document on the Strahljäger (Ente).

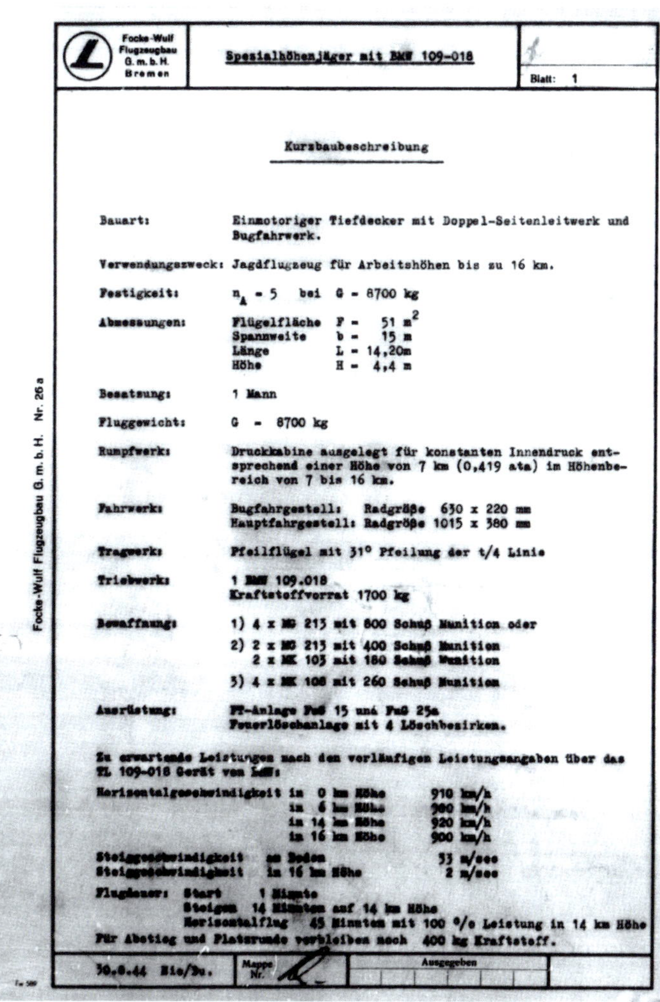

Focke-Wulf Flugzeugbau G.m.b.H. Bremen

Spezialhöhenjäger mit BMW 109-018

Blatt: 1

Kurzbaubeschreibung

Bauart: Einmotoriger Tiefdecker mit Doppel-Seitenleitwerk und Bugfahrwerk.

Verwendungszweck: Jagdflugzeug für Arbeitshöhen bis zu 16 km.

Festigkeit: n_A = 5 bei G = 8700 kg

Abmessungen:
Flügelfläche F = 51 m²
Spannweite b = 15 m
Länge L = 14,20m
Höhe H = 4,4 m

Besatzung: 1 Mann

Fluggewicht: G = 8700 kg

Rumpfwerk: Druckkabine ausgelegt für konstanten Innendruck entsprechend einer Höhe von 7 km (0,419 ata) im Höhenbereich von 7 bis 16 km.

Fahrwerk:
Bugfahrgestell: Radgröße 630 x 220 mm
Hauptfahrgestell: Radgröße 1015 x 380 mm

Tragwerk: Pfeilflügel mit 31° Pfeilung der t/4 Linie

Triebwerk: 1 BMW 109.018
Kraftstoffvorrat 1700 kg

Bewaffnung:
1) 4 x MG 213 mit 800 Schuß Munition oder
2) 2 x MG 213 mit 400 Schuß Munition
2 x MK 103 mit 180 Schuß Munition
3) 4 x MK 108 mit 260 Schuß Munition

Ausrüstung: FT-Anlage FuG 15 und FuG 25a
Feuerlöschanlage mit 4 Löschbezirken.

Zu erwartende Leistungen nach den vorläufigen Leistungsangaben über das TL 109-018 Gerät von BMW:

Horizontalgeschwindigkeit in 0 km Höhe 910 km/h
in 6 km Höhe 900 km/h
in 14 km Höhe 920 km/h
in 16 km Höhe 900 km/h

Steiggeschwindigkeit am Boden 33 m/sec
Steiggeschwindigkeit in 16 km Höhe 2 m/sec

Flugdauer: Start 1 Minute
Steigen 14 Minuten auf 14 km Höhe
Horizontalflug 45 Minuten mit 100 % Leistung in 14 km Höhe
Für Abstieg und Platzrunde verbleiben noch 400 kg Kraftstoff.

30.8.44 Bie/Du. | Mappe Nr. | Ausgegeben

Focke-Wulf Flugzeugbau G.m.b.H. Nr. 26a

ABOVE: The only known description of Focke-Wulf's BMW 018-powered high-altitude fighter of August 1944.

its weight as 1,740kg. By comparison, the Jumo 004 was 3.86m long and weighed only 719kg.

With these specs in mind, the Strahljäger (Ente) must have been a fairly substantial aircraft to accommodate it. However, there is no drawing and beyond these basic facts nothing more is known about it.

FOCKE-WULF SPEZIALHÖHENJÄGER MIT BMW 109-018

Focke-Wulf was working on a different fighter powered by a single BMW 018 in August 1944. The only outline description of the design known to exist,[9] dated August 30, 1944, stated that it was a single-seater 14.2m long with a wingspan of 15m and a wing area of 51m² – which would make it 2m longer than a Bf 110 but with wings that were slightly shorter and much broader. It was described as having a twin-rudder low-wing configuration, also like the Bf 110. Unlike the Bf 110, its wings had a 31° sweepback, it had a tricycle undercarriage with a 630 x 220mm nosewheel and 1015 x 380mm mainwheels and it weighed a lot more at 8,700kg.

According to the description, the aircraft was intended to operate at altitudes up to 16km – the BMW 018 having been designed for high-altitude operations. There were three possible armament options: four MG 213s, two MG 213s and two MK 103s or four MK 108s. And equipment consisted of FuG 15 and FuG 25a plus a fire extinguishing system which operated in four areas.

The design appears on a list of projects[10] drawn up by the RLM's Technisches Amt on May 12, 1944, in connection with its BMW 018 engine, so it may be said with certainty that the project was worked on for at least three months. However, there is no known drawing of the aircraft and besides some performance data, little else is known about it.

FOCKE-WULF SUPER-LORIN AND SUPER-TL

Often included in roundups of Focke-Wulf wartime projects, the so-called Super-Lorin and Super-TL appear only ever to have existed as two very rough sketches. Both designs were apparently drawn by Heinz von Halem, the engineer who created the form of the Triebflügeljäger.

The evidence for this is '(v. H.)' appearing on the original drawings and redrawn postwar versions, evidently dating from the 1960s, having '(Konstr. : v. Halem)' written on them. Von Halem's name is not known to appear in connection with any of the fighter projects created by Focke-Wulf's Entwurf abteilung, headed by Chefing. Ludwig Mittelhuber and his deputies Obering. Quenzer and Obering. Hans Multhopp; nor those created by the company's two experimental divisions, Flugmechanik E, headed by Chefing. Gotthold

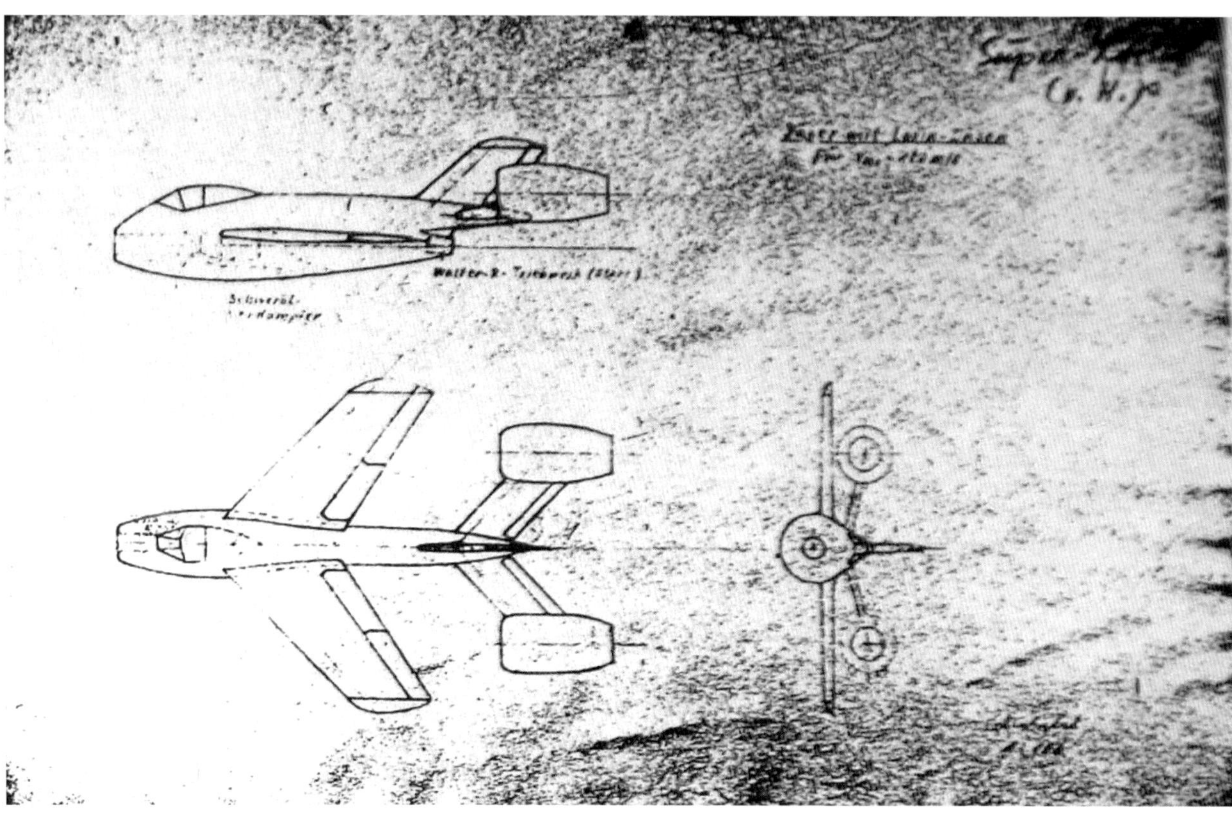

LEFT: The original drawing showing the Super-Lorin is extremely sketchy and offers few details. There is no date and even 'Focke-Wulf' is missing.

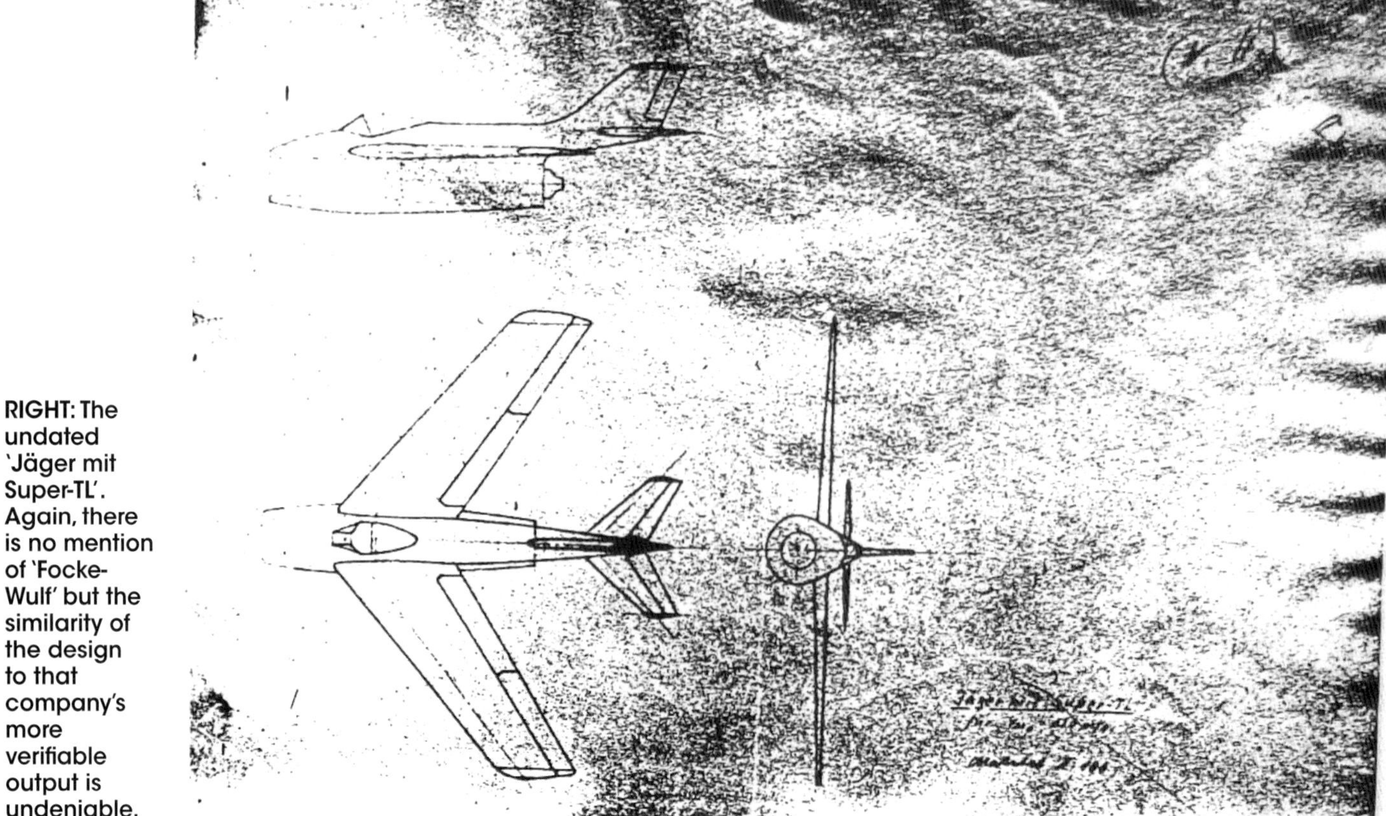

RIGHT: The undated 'Jäger mit Super-TL'. Again, there is no mention of 'Focke-Wulf' but the similarity of the design to that company's more verifiable output is undeniable.

Mathias, and Flugmechanik L, headed by Obering. Herbert Wolff.

It is possible that he worked instead under Pabst in Focke-Wulf's Gas Dynamik Versuchs Abteilung and therefore perhaps had no direct connection to the company's three projects offices.

This would likely explain the very basic nature of the Super-Lorin and Super-TL sketches – which do not even bear the name 'Focke-Wulf'. The Super-Lorin was vaguely similar in outline appearance to Focke-Wulf's other jet fighter projects with the exception of two ramjet tubes attached to its tailplanes. A Walter rocket motor was fitted in the rear of its fuselage for take-off and the turbojet appears to have been positioned centrally within its fuselage, exhausting beneath the aircraft.

The original sketch gives no dimensions but the 1960s redrawing gives length as 11.6m and wingspan as 7.6m. Similarly, no thought appears to have been given originally to either landing gear or armament but the redrawing helpfully suggests positions for a nosewheel and main gear plus two dots on the nose seemingly indicating locations for cannon.

The Super-TL appears to have originally been captioned 'Jäger mit Super-TL', suggesting that only the engine fitted was a 'Super-TL' rather than the fighter as a whole. Again, no dimensions are given in the original but the redrawing offers a length of 10m and a wingspan of 11m. Even the redrawn Super-TL offers no clues as to landing gear position, though it does again add a couple of dots to indicate weaponry positions.

Of all the aircraft designs featured in this book, the evidence supporting the Focke-Wulf Super-Lorin and Super-TL appears to be the flimsiest. Von Halem was a peripheral figure in aircraft design for Focke-Wulf, although his Triebflügeljäger certainly caught the attention of the Allies after the war, and these two sketches, whether they were genuinely drawn up during the war or afterwards, appear to have had no significance.

HENSCHEL P 90

Few aircraft projects worked on by Berlin-based Henschel are known about in any detail due to the loss of that company's papers at the end of the war. Among the more mysterious projects worked on was the P 90. Models of this canard-type pulsejet-powered design were tested by the DVL during 1943 and a report[11] on the outcome was produced on November 24 of that year. The report stated that work on the P 90 had followed on from a series of successful tests on the canard type Henschel P 87 but no other details about the type are revealed.

There has been some speculation about the design – that it was intended for ground-attack or that the Soviets intended to produce it using captured Henschel documents – but the truth is that we know little for certain about it except for its project number, its layout and the fact that

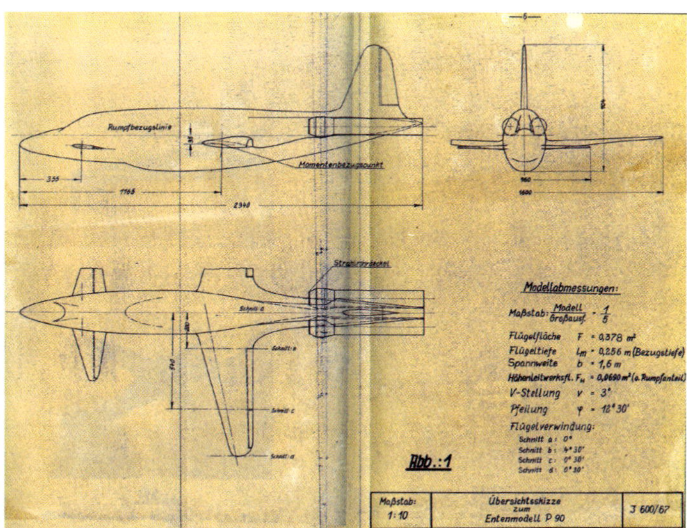

ABOVE: Very little is known about the Henschel P 90. This image is from a report compiled by the DVL, which tested models of it under contract to Henschel.

ABOVE: Wind tunnel model of Henschel's P 90 as it was tested – suspended upside down.

it was to be powered by a pair of pulsejets. This in itself is unusual since the only other pulsejet projects known to have been ongoing in 1943 were the Me 328 and Fi 103.

The twin-pulsejet configuration and the shape of the forward fuselage certainly suggest a manned aircraft – and intriguingly the positioning of the pulsejets also suggests that Henschel's design prefigured the FGZ's conclusion that two pulsejets positioned right next to one another would cancel out each other's vibrations. But nothing is known about the role intended for the aircraft.

LIPPISCH P 15 'DIANA'

At the end of February 1945 Alexander Lippisch was facing up to the fact that his long-cherished ramjet fighter concept, the P 13, was unlikely ever to be built.

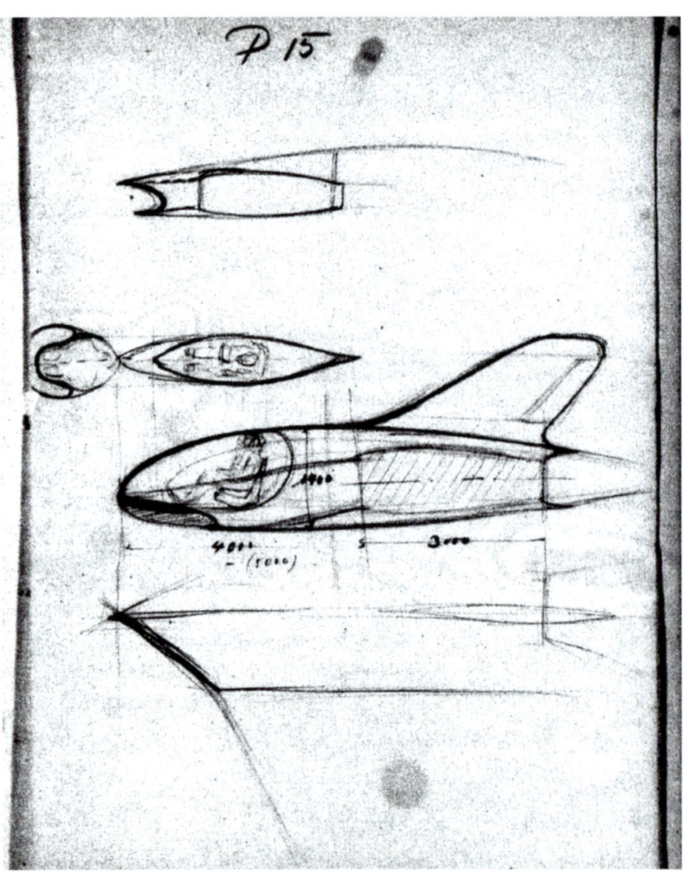

ABOVE: This was evidently Alexander Lippisch's first attempt at the P 15.

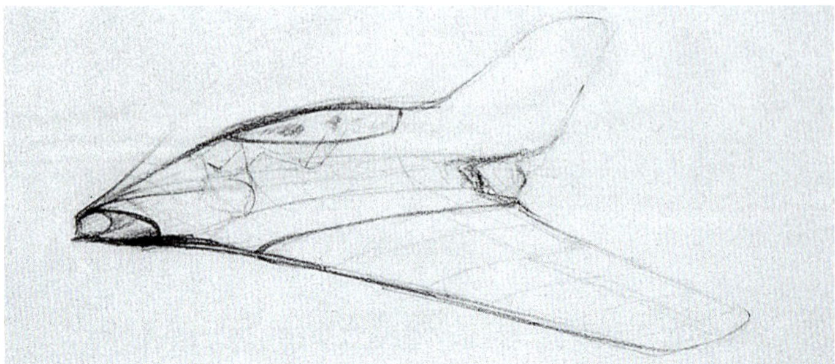

ABOVE: Lippisch's second P 15 exhibits an odd, almost beak-like, split-nose intake.

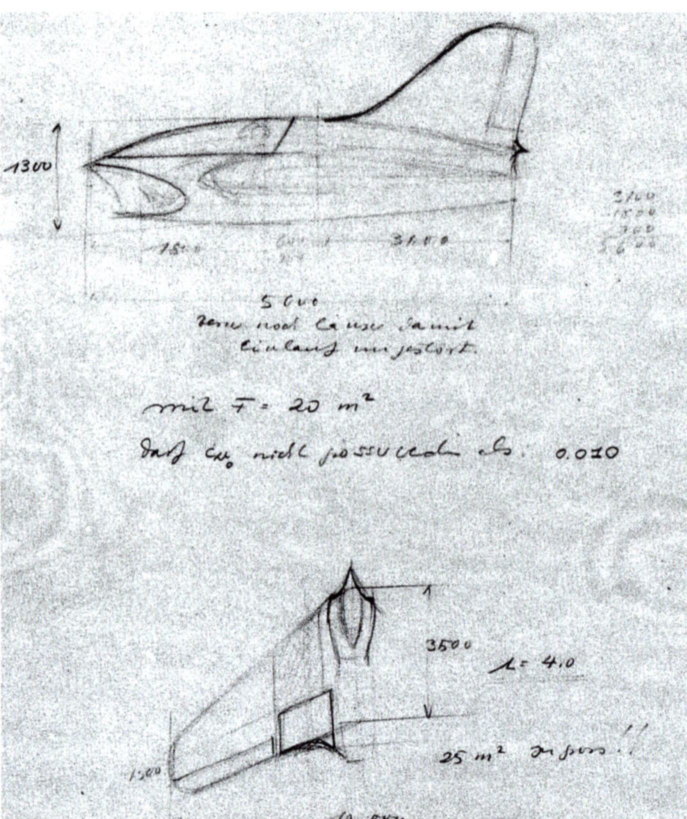

ABOVE: The plan view of this P 15 shows a very sharply pointed nose and gaping side intakes.

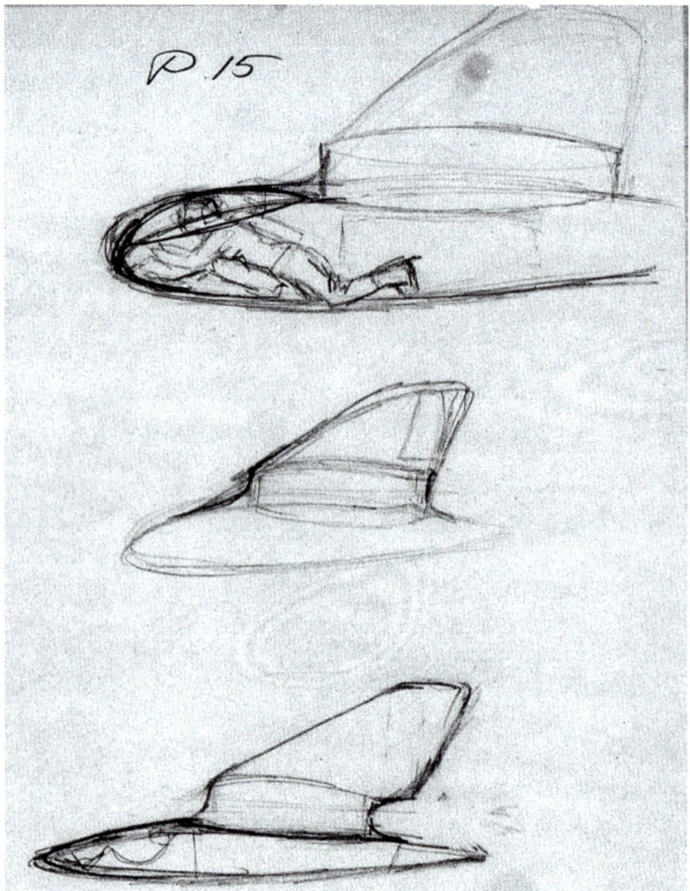

ABOVE: Trio of dorsal-engined P 15s.

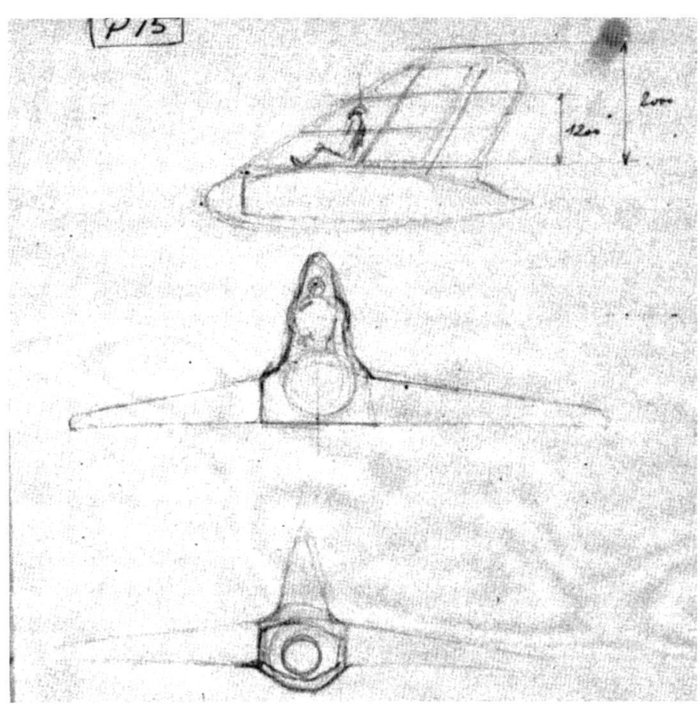

ABOVE: The third type of P 15, after the conventional fin and dorsal fin types, was something resembling Lippisch's P 13a – a very short flying wing housing the turbojet with a fin-cockpit built on top.

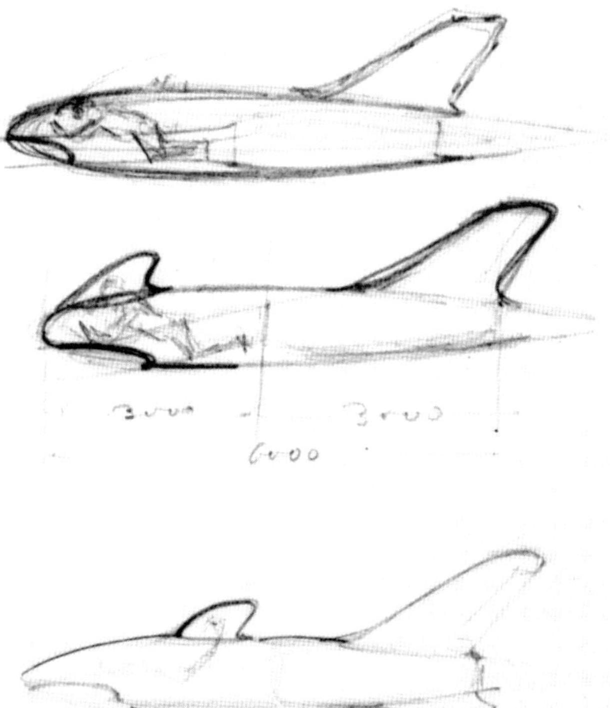

ABOVE: Lippisch's most shark-like designs. Two of these boast highly unusual fin-like bubble cockpits.

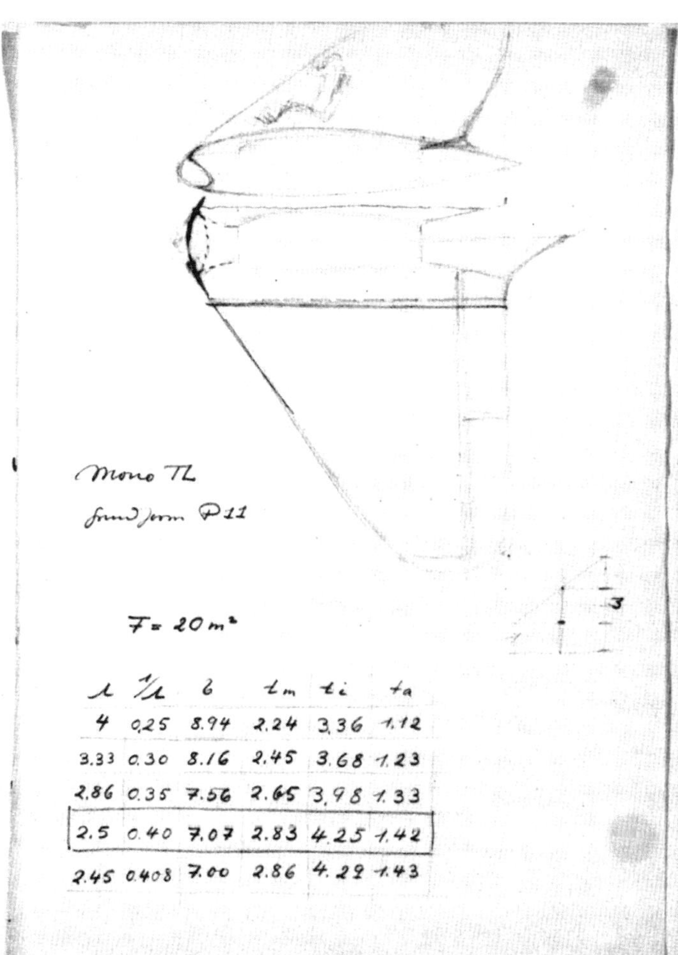

ABOVE: Plan view of the P 13a-style P 15.

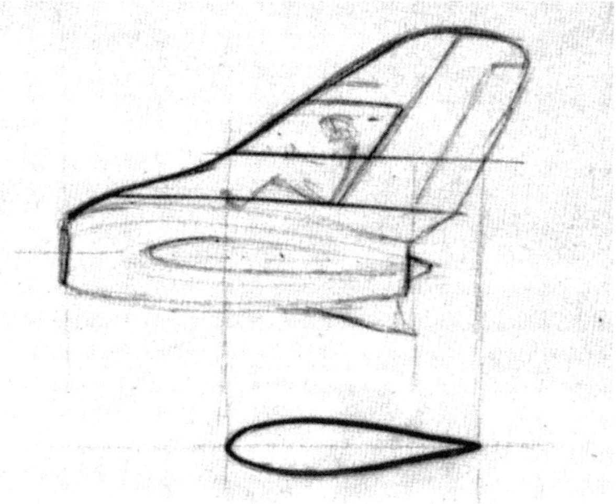

ABOVE: This tiny P 15 is little more than a turbojet mounted in a small flying wing with a small cockpit-fin bolted on top. Whether it would have even been capable of flight is questionable.

His thoughts were turning towards yet more conceptual designs – the elongated slender delta P 14 for high-speed flight – when in early March 1945 he was visited at the LFW in Vienna by Oberst Siegfried Knemeyer. He asked Lippisch if he could design a new single-jet aircraft based on proven aerodynamics.

According to Lippisch's autobiography: "After the 'Blitzaktion' Volksjäger He 162 had unreasonable flight characteristics for the young pilot, at the suggestion of Colonel Knemeyer we designed a jet fighter P 15 'Diana' consisting of existing components: the proven Me 163 airframe, the cockpit and armament of the He 162, and

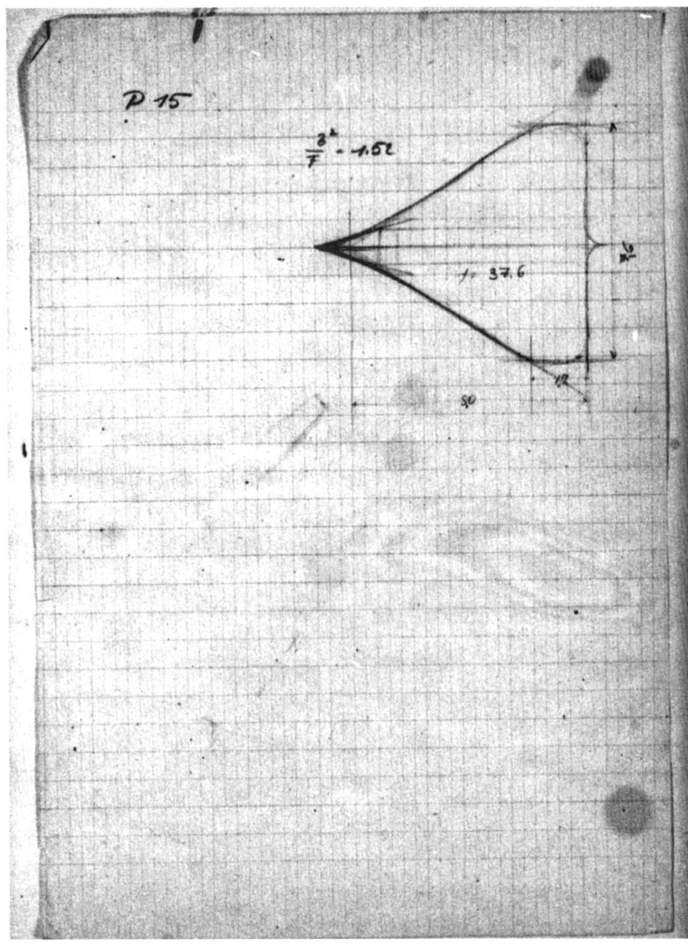

ABOVE: The final drawing of Lippisch's P 15 sequence is a plan view, presumably related to the earlier sketches.

the modified hull of a Me 163 derivative (the Ju 248 = Me 263) with built-in jet engine. The project did not go beyond sketches."

But it seems that these were not Lippisch's first thoughts for the P 15. Initially, he began making sketches showing entirely new designs.[12] These appear to have been based on aspects of various designs from earlier in his career, with shades of his early jet fighter work for Messerschmitt showing through in some designs, and others depicting what amounted to a P 13a fitted with a turbojet in place of its ramjet combustion chamber.

In all, there were around 10 different designs loosely based on three different configurations. The first configuration was something akin to his Messerschmitt P 01 designs, with the pilot seated in the tailless aircraft and the turbojet's intake passing him to feed the engine, mounted in the rear fuselage. There was a single large tailfin and no undercarriage, presumably implying a trolley take-off and skid landing.

The second configuration had the pilot seated centrally in the aircraft with a sharp beak-like mouth intake in the nose.

For the third design, the pilot sat in a very narrow cockpit in the nose with gaping whale shark-like intakes on either side of him. For this design, the fin was only gently swept.

Lippisch then changed tack, moving the turbojet out of the fuselage and mounting it on the aircraft's back where the fin had previously been. The fin itself was positioned above the turbojet. Three designs were drawn up in this layout, the first with the pilot lying prone in the aircraft's nose, unhindered by engine intakes, the second had no pilot shown but the fin was more sharply swept and the third shows the pilot seated – almost reclined on his back – within the aircraft and a strongly swept fin mounted on the turbojet.

The last of the three P 15 configurations shows a broad but very short delta aircraft with the turbojet positioned centrally in the wing and the pilot seated in a cockpit built into an oversize fin above it. This design concept was clearly intended to rely on the extensive aerodynamic work already carried out by Lippisch for the P 13a. A second version with this layout has a planform apparently based on that of the P 11, while the third appears to have been shorter still, with a smaller fin making conditions somewhat cramped for the pilot. A small under-fin is also a feature of this design. This type's wing form is unclear from the side view but it would most likely have been a flying wing – the wing itself providing space for fuel, weapons and landing skids.

Perhaps the most striking of Lippisch's P 15 designs emerged from a return to the first layout with the pilot in the nose and the turbojet to the rear beneath a central fin. A single sheet of paper shows three different designs in this mode – the first has the pilot lying prone in the tiny nose cockpit above the turbojet's intake, the second is similar but introduces a fin-shaped bubble canopy for the pilot. The third retains the oddly fin-like bubble canopy but sees the pilot moved to a seated position in the centre of the aircraft. The fin is strongly swept back and an under-fin is also present.

Lippisch's shark-like original P 15 drawings do not appear to have met Knemeyer's expectations – primarily because he was looking for something that could be built quickly and each of the new sketches would have required a ground-up new design. Therefore Lippisch came up with an alternative that utilised the components mentioned as the P 15 'Diana' instead.

He wrote a brief two-page outline of the project on March 4, 1945, entitled 'Flugtechnische Grundlagen für Projekt P 15 'Diana' or 'Aeronautical Fundamental for Project P 15 'Diana''.[13] This began: "The highest speeds were met with the tailless aircraft Me 163 (constructed by Dr Lippisch). Investigations in high-speed wind tunnels clearly show the superiority of this design in the high-speed range. The flight characteristics are significantly better in all flight areas than those of the most normal modern aircraft. This finding has been consistently confirmed by all testing stations.

"The development of high-speed fighter aircraft based on the normal aircraft does not lead to the necessary tactical speeds (1,000km/h). It therefore seems obvious to use the results already available and the existing tailless designs for these new developments. The present draft is determined by demands resulting from the need for the shortest development time and the fastest production start. Therefore, the task was set so that from the components of the model Me 163 B and C, He 162 and Ju 248 and 263 could be to put together in a new pattern.

"The demands are: Flight weight 3.6 tons, flight time 45 minutes, maximum speed 1,000km/h, armament and equipment such as He 162 and Me 163 B, nosewheel chassis like the He 162, engine HeS 011 (BMW). The following modules are to be used unchanged: Me 163 C swept wing, cockpit He 162, main undercarriage He 162 or Me 109 G, controls Me 163 B, weapons installation He 162 or Me 163 B."

The P 15 was to have a wingspan of 10m and a wing area of 20m² but an increase in wing area in a ratio of 1:1.8 was possible, as was an increase in wingspan by 1:1.4. However, "compared to the He 162 or similar designs based on normal

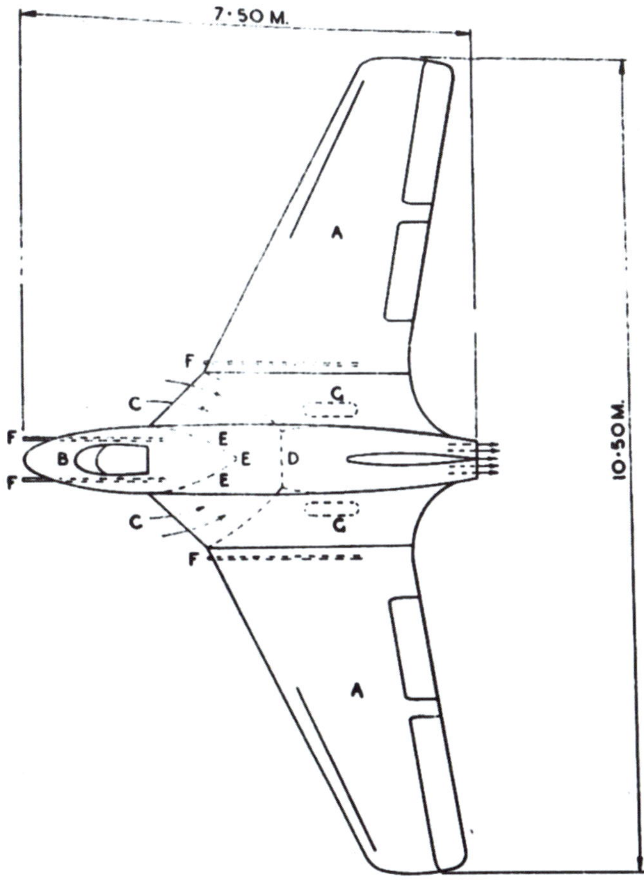

ABOVE: Lippisch claimed that all drawings of the P 15's final form were lost and this sketch was produced during his interrogation by the British to indicate how the aircraft was meant to be put together.

aircraft, a substantial increase in fast flight performance is to be expected, because 1) the swept-wing is superior in the range of high Mach numbers, and because 2) the drag is reduced to a minimum by a closed design while avoiding all overlaps and dead angles.

"The wing is thoroughly tested and has over all other patterns a clear superiority in flight characteristics. It is therefore proposed to create this pattern as quickly as possible, which can be achieved in a very short time using parts and assemblies that are already being manufactured."

According to historian Walter Schick:[14] "The LFA i:n Vienna constructed a 1:25 scale model, after which it was planned for the Wiener Neustadt Aircraft Works (WNF) to commence series production as quickly as possible." Whether any of this actually happened is debatable. Wiener Neustadt was overrun by the Soviets on April 2, 1945, and Vienna itself capitulated on April 16. However, even if the LFA did construct the model and the WNF did begin gearing up to produce the type, none of the German government's aircraft development plans recorded between March 4 and April 16 mentions the P 15. Almost everything that is known about it comes from Lippisch himself and a handful of surviving documents. Lippisch later contended that no drawings of the 'bitsa' P 15 had survived the war – only his initial sketches. Incidentally, none of these bear the name 'Diana', though his project description does. Exactly where the name came from and its significance, if any, remains a mystery.

MESSERSCHMITT P 12

It is possible to determine the origin and purpose of nearly every design worked on by Messerschmitt's Abteilung L during its more than four years in existence – with the exception of the P 12.

This extremely bulky fighter, appearing in a drawing dated September 30, 1942, was designed around BMW's enormous P 3303 turbojet project, later to be renamed the BMW 018, at a time when few other manufacturers were considering it.

At that point, Abteilung L was locked into an increasingly bitter disagreement with Professor Messerschmitt over tailless projects. As department head Alexander Lippisch and his two lieutenants Walter Stender and Dr Hermann Wurster argued among themselves and with Messerschmitt, the remainder of Abteilung L seems to have simply got on with its projects work.

Handrick produced the P 11 on September 13, a fast bomber intended for the long-running Schnellbomber competition which was ongoing at that time, and one of Lippisch's longest-serving staff members, Josef Hubert, produced the P 12 some 17 days later.

Loosely resembling the earlier P 01 designs, it was offered with two different wing forms – either an 11m wingspan or 14.8m – and two different tailfins, a relatively conventional short one which gave the aircraft a length of 7m and a

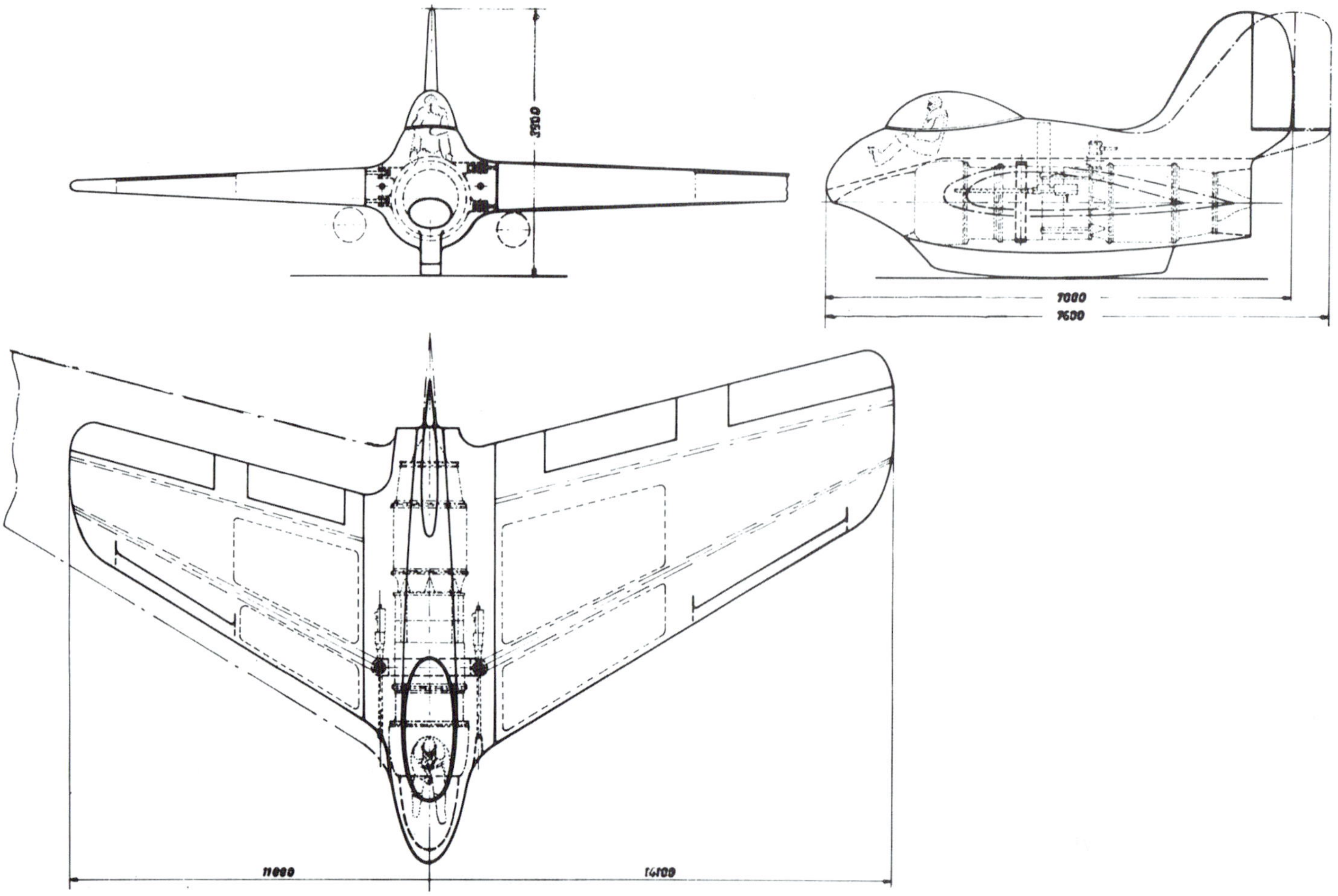

ABOVE: Abteiling L's P 12 was to have been powered by a BMW P 3303 – later to become the BMW 018. But where BMW's own 018-powered design was drawn up in November 1944, the P 12 dates from more than two years earlier – September 1942. Had it been built, the P 12 would have been a very bulky aircraft and landing it on a belly skid might have been tricky.

swept one which increased the overall length by 60cm. The pilot sat beneath a bubble canopy and above a huge sharply angled intake for the massive engine.

Armament was a pair of long-barrelled MG 151s or possibly MK 103s and the drawing shows a cylindrical object under each wing – probably a bomb, although these are only shown in the forward view and not the side or top views. Large as it is for an Abteilung L design, the aircraft was intended to land on a skid and presumably would have had to take off using a catapult or trolley undercarriage since it lacks any other sort of undercarriage.

Exactly what the P 12 was intended to demonstrate or what requirement it was intended to meet is unclear. Lippisch would later return to the P 11 and rework it extensively but the P 12 was left on the shelf – perhaps because the BMW P 3303 never materialised as a viable production model, even after its redesignation as the BMW 018.

MESSERSCHMITT P 1070

Known from just a single drawing, the P 1070 (P 70) was evidently designed in parallel to the P 65 as a smaller alternative.[15] The original drawing was dated '1940' and dimensions given were a wingspan of 8.3m and a length of 7.2m. The engines were BMW P 3304s and armament was two MG 151s and one MG 17.

The original drawing, believed to have been of very poor quality, was passed to British author Eddie Creek by German writer and historian Karl Ries[16] in 1970. Creek made his own clean line drawing of it and gave the original drawing back to Ries. Its current whereabouts are unknown.

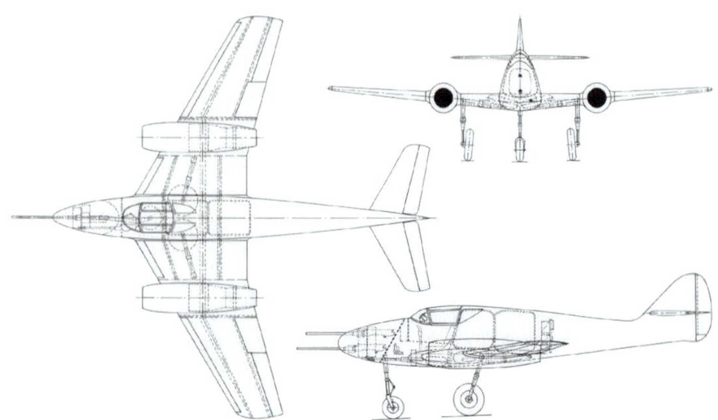

ABOVE: Eddie Creek's drawing of the Messerschmitt P 1070, based on an original drawing now lost. *EN-ARCHIVE*

MESSERSCHMITT P 1090

The Schnellstbomber competition, held at the end of 1942 and into January 1943, resulted in Dornier being given a development contract for its push-pull piston-engine P 231 design – which would become the Do 335. Messerschmitt's entry for the competition, and Dornier's chief rival, had been the twin-fuselage Me 109 Z. This was an expression of Willy Messerschmitt's latest design obsession – reducing the number of different aircraft in production and creating new types by combining existing components in new ways instead.

For example, why build separate and incompatible day fighters, night fighters, fast bombers, torpedo bombers and reconnaissance aircraft when all of them could be built around the same proven fuselage, a selection of just three engines and a standard set of wings? The required aircraft could be made a single-seater or two-seater with the addition of a fuselage insert, longer wings could be fitted for high-altitude operations and a pair of fuselages could even be combined in parallel to create a set of twin-engine aircraft.

This approach is explained in an undated and untitled draft document.[17] It outlines the aircraft shortages facing Germany and points to the fact that 53 different and incompatible aircraft types were in production at that time. It then suggests combining just a few standard Messerschmitt components to create 32 different modular aircraft patterns, 21 of which involve double fuselages.

The Me 109 Z, known initially as the Me 109 Zw or Zwilling 'Twin', was an attempt to create a twin-engine fast bomber almost entirely out of existing Me 109 components – thereby preventing any significant disruption to production lines and dramatically reducing the usual development time. Yet despite the simplicity of its design and the ease with which it could have been built, it was defeated by Dornier's P 231 – soon to be the Do 335 – at the Schellstbomber designs presentation meeting on January 19, 1943.

Nevertheless, just 18 days later, on February 6, Messerschmitt tried again with a slightly different slant on the modular idea in a document entitled 'Denkschrift über die Vereinheitlichung der zur Fertigung zusulassenden Flugzeug-Frontmuster' or 'Memorandum on the standardisation of the front-pattern aircraft to be added for production'.

The document begins by examining the options for creating generic single- and twin-engine aircraft using various existing Messerschmitt components, then goes into more detail on the twin engine designs.

It says: "As an example of the feasibility of this unification of the warplanes, the basic 'twin-engine two-seater aircraft' has been singled out and examined in series as the Me 410, as well as the project P 1090, which was already designed under these provisions, then examined in detail in the following paper. Further investigation of how far the Me 109 can be used in the twin fuselage arrangement as a basic pattern for the twin-engine aircraft is not yet completed."

The P 1090 itself was loosely based on project work already carried out around the Me 210/410, such as fitting it with a V-tail, and its modular nature was supposed to allow it to be configured for a wide variety of roles including night fighter, ground-attack, single-seat and two-seat heavy fighter, high-altitude fighter, fast bomber, three-seat bomber, torpedo bomber and reconnaissance aircraft. For most of these roles it would be powered by a pair of Daimler-Benz DB 603 G piston engines – the engines favoured for the Schnellstbomber – but for the high-altitude fighter it would use two DB 628s. Future development, it was suggested, would allow it to become a twin-jet aircraft with Jumo 004s capable of being altered to suit the same wide variety of roles.

The P 1090's construction description explicitly states that its performance will be "approximately equal to the twin-hull aircraft (e.g., Me 109 Z with DB 603 G) and the single-fuselage aircraft with the two engines in the fuselage" but with the added advantages of "better visibility for the crew and good usability as a basic starting pattern for many aircraft variations with different military tasks".

It was to have a wing area of 28m² that could be expanded to 31m² or 36m² through the use of inserts if necessary and the fuselage was made up of compatible components which could be swapped to provide single-seat or two-seat options, with the former providing an additional fuel tank in place of the second seat. Inside the fuselage there was space for two armoured fuel tanks and a bomb bay capable of holding a 500kg bomb load.

The V-tail could also be enlarged with inserts and the main landing gear consisted of double wheels retracting backwards into the engine nacelles. In addition "for higher flight weight interchangeability with larger wheels is taken into account".

For propulsion, "the installation of the DB 603 G engine is anticipated. It is possible to install other air-cooled engines. The normal fuel system consists of two protected and centrally installed easily replaceable containers".

As a two-seat night fighter the P 1090 would have two DB 603 Gs, four fixed forward-firing MK 108s and two upwards-firing MK 108s with limited adjustment possible, 250kg of armour protection for the crew and two fuel tanks with 1,400 litres and 800 litres. Equipment was FuG 16 ZE, FuG 10 KK and FuG 25a, plus night fighter equipment. Maximum take-off weight was 11 tons. Wing area was 36m².

As a two-seat ground-attack aircraft, the P 1090 would have the same engines and basic equipment but the two fuel tanks would only contain 1,400 litres between them. Fixed forward armament was four MK 108s and two MK 103s plus "4-5 Giesskannen". The Giesskanne was an external gun pod containing six MG 81s linked in pairs. Fixed rearward firing were two MG 151s and bombs could be carried under the wings in place of the 'Giesskannen', or 500kg of bombs could be carried internally.

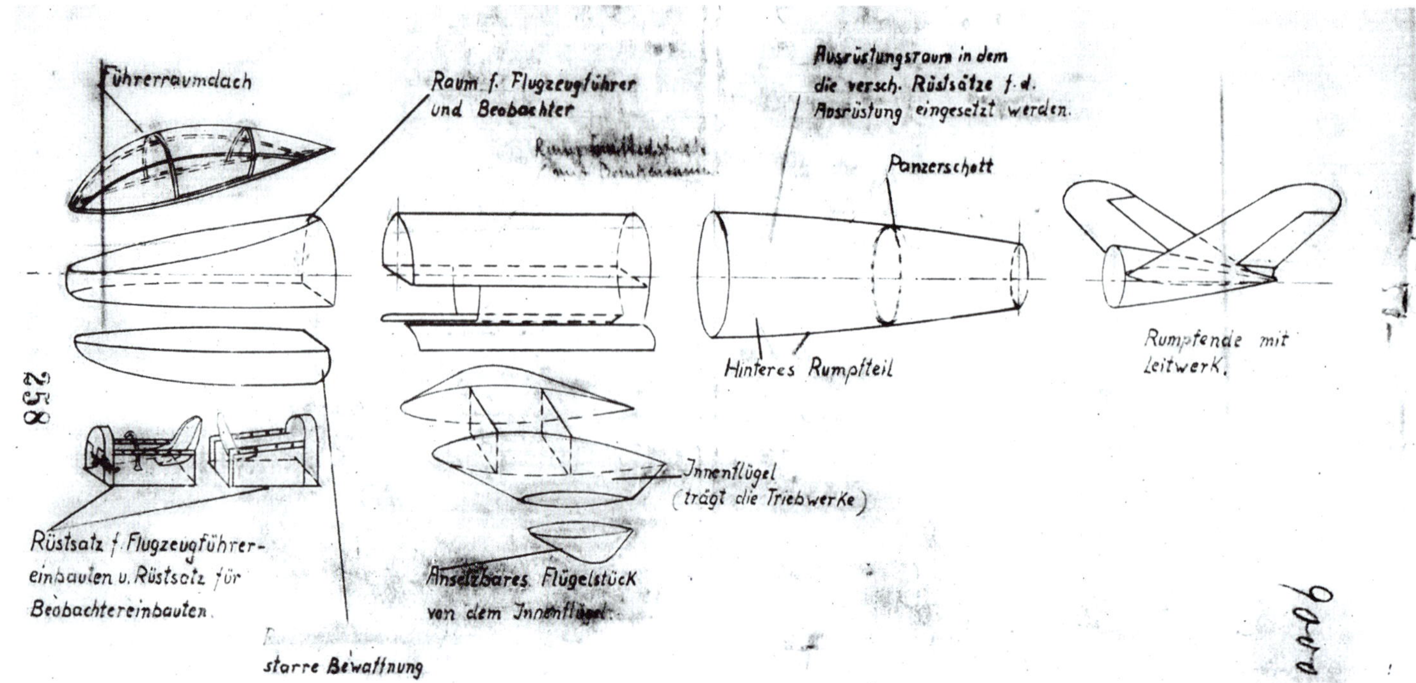

ABOVE: Messerschmitt's modular P 1090 design. Its stated aim was to replace most other German twin-engine aircraft types then in production – but in reality it was a swipe at Dornier's Do 335, which had just defeated the Me 109 Z in the Schnellstbomber competition.

Crew and engine armour amounted to 450kg and overall maximum starting weight was 11 tons. Wing area was 31m².

The two-seat or single-seat heavy fighter version had the same engines and equipment, except for a FuG 26a in place of the 25a. Fuel tankage was the same as the ground-attack version, except with the option of an additional tank if fitted out for single-seat operation, while forward and rearward armament was also the same except with the option of carrying two MK 103s or four MK 108s as a forward-firing gun pack in place of the bomb bay.

If the belly pack wasn't fitted, the bomb bay could be retained to carry a single 500kg bomb, two 250kg bombs or containers for fragmentation bombs. Armour was 250kg. Max take-off weight was the same but wing area was just 28m².

In addition to its two DB 628s, the single-seat high-altitude fighter P 1090 would be equipped with a FuG 16 ZB and FuG 25a. The two fuel tanks contained the usual 1,400 litres but another tank could also be fitted in the bomb bay. Forward-firing weaponry was four MK 108s and rearward one MG 151. A bomb load was 'possible' but not specified. Armour was 250kg and overall weight remained 11 tons. Wing area was, needless to say, the 36m² option.

As a single- or two-seat fast bomber, armament was stripped back to just one forward-firing and one rearward-firing MG 151. Internal bomb load remained 500kg but heavier loads were 'possible', as were external fuel tanks. Wing area was 28m².

Other potential uses for the piston-engined P 1090 included high-altitude fast bomber, torpedo bomber and reconnaissance aircraft.

A more extreme conversion saw the P 1090 become a three-seater medium bomber. The engines would be "two piston engines with up to 2000 PS", fuel tankage was 1,400 litres with the possibility of external tanks or putting a tank in the bomb bay. The bomb load of up to four 500kg bombs could be carried on four wing hard points. There would be upper and lower defensive turrets each mounting an MG 151 Z and a single forward-firing MK 108. Overall weight was 14 tons with a wing area of 36m².

Another conversion would see the P 1090 become a single-seat light jet fighter or two-seat heavy fighter with a pair of Jumo 004s. Three fuselage fuel tanks would carry a total of 2,000 litres or 2,500 litres in 'heavy' form and forward-firing armament was four MK 108s and two MK 103s. The jet version could also become a fighter-bomber, long-range reconnaissance aircraft, ground-attack aircraft, high-altitude fighter, night fighter or torpedo bomber with various alterations of armament and wing area.

It would appear that the P 1090 went no further than this technical description, since the Do 335 had already been chosen as the Luftwaffe's new twin-engine fast bomber. In addition, there appears to have been insufficient support for Messerschmitt's grand plan of standardisation – which would have meant a dramatic re-ordering of Germany's aircraft industry from top to bottom. He seems to have dropped the concept at this point.

MESSERSCHMITT P 1101 DRAWING XVIII/108

As mentioned previously, Messerschmitt's P 1101 series initially encompassed a wide range of experimental multi-engine designs. One of the more unusual[18] was shown in

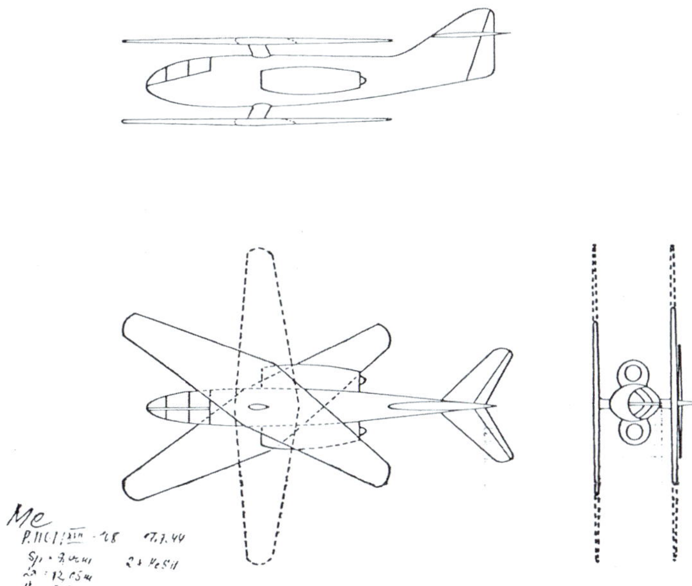

ABOVE: Messerschmitt drawing XVIII/108 showing perhaps the most bizarre known member of the P 1101 family of jet aircraft designs.

drawing XVIII/108 of July 17, 1944. This aircraft operated on a similar principle to the Blohm & Voss P 202 but with two sets of swivelling wings – one above the fuselage and one below. Its two HeS 011 turbojets were attached to the sides of its fuselage and it had a conventional swept tail. Fuselage length is given as 12.05m and wingspan appears to be 8.4m.

The date of the drawing places it after Blohm & Voss's P 202 so this perhaps represents Messerschmitt's own take on the idea, though it appears to have been more conceptual and less practical than Blohm & Voss's design, since there is no clear way of providing the aircraft with an undercarriage.

The XVIII/108 drawing number places this member of the P 1101 family between XVIII/99 – a four-jet P 1101 of spring 1944 – and XVIII/113 of August 30, one of the earliest known 1-TL-Jäger versions of the series.

STÖCKEL RAM-FIGHTERS

During the summer of 1944, when the effects of Allied bombing were becoming increasingly significant, DVL employee Karl Stöckel evidently began thinking of specialised aircraft that might be able to tackle the incoming enemy bombers.

He typed up a brief single-page report[19] on the subject of 'Rammrakete' on August 31, 1944. Whether this was inspired by the Lippisch/Karlson paper on 'Arrows of death' of September 1943 or projects such as the Blohm & Voss BV 40, Bachem Natter, Messerschmitt P 1104 and Heinkel Julia, all ongoing by mid-August 1944, is unknown.

Included with the report were three drawings. The first, dated the day after the typed report, September 1, showed a 5.2m-long aircraft labelled 'Rammrakete' or 'ram rocket' with a 4m wingspan. Its cockpit consisted of an armoured spike with a small canopy of bulletproof glass and a metal plate immediately behind the pilot separating him from the fuel tanks and a rocket motor positioned within a ramjet tube. Beneath the cockpit was a small landing skid and under each triangular wing was a small bomb. The tail consisted of two downturned tailplanes and a single fin.

According to Stöckel's description: "The ramming rocket mainly consists of an armoured cockpit and an engine carrier, to which the tail and structure are attached. The armoured tip is provided with ram edges and a steel cover, it is pushed onto the engine carrier so that it can be separated. A liquid rocket of 1.5 tons thrust or a corresponding launcher and a ramjet are used as the engine.

"The tailplane is designed as a landing skid. The aircraft has a normal control system. Landing aids are left out for the sake of simplicity. Engine carriers, supporting structure and tail unit are made of lightweight material.

"Operational use: The aircraft is intended for ramming. It can do both wing and projectile launch. It is accelerated to almost the speed of sound with rocket and Lorin propulsion and brought up on a steep climb. Horizontal long-range flight at altitude is made using the ramjet. By flying over the bomber briefly, one or more bombs or towed bombs can be brought to impact, and finally the aircraft is piloted into the vertical tailplane of a bomber."

On impact, the wings and engine would tear away from the armoured cockpit or be blasted away using explosive bolts. The pilot could then escape using his parachute. Alternatively, if the aircraft remained intact, a skid landing could be made.

Stöckel listed eight 'tactical considerations': "1) The shockproof and shatterproof placement of the pilot in an armoured case enables a targeted approach of the aircraft to a bomber formation. 2) The high approach speed leaves no time to fight back, even single fighters can attack a bomber formation. 3) The rocket propulsion gives high flight performance, the Lorin propulsion lower fuel consumption. 4) The engine is a device already in series production. 5) The high-quality armoured tip is reusable. 6) The rammer does not require weapons or instrumentation. 7) The training is simple, the manufacture and use of the device must be completed in the shortest possible time. 8) The combination of rocket and ramjet propulsion offers approaches for economical long-distance flight, especially with supersonic speed!"

The two other drawings, both dated August 25, 1944, each appears to show an earlier version of Stöckel's concept. One is labelled 'Raketenjäger mit zusätzlichem Lorin-Abtrieb' or 'Rocket fighter with additional Lorin engine' and is described as being for ramming or aerial bombing. It is 6.8m long with a 6.6m wingspan and has loosely the same layout as the 'Rammrakete' except for a lack of landing skids and with a twin- rather than single-fin.

The final drawing appears to show a similar aircraft fitted with an 'air jet' – presumably a turbojet – and two small rocket motors in place of the rocket motor and ramjet combination. Interestingly, the aircraft is also fitted with 'Nachverbrennung' – 'afterburning'.

The DVL apparently thought enough of Stöckel's concept to make a patent application for it on September 15, 1944, as "DVL Nr. 5130 and 5131 (secret)"[20] but these documents were destroyed shortly before the end of the war – only to be recreated for the Allies by Stöckel himself.

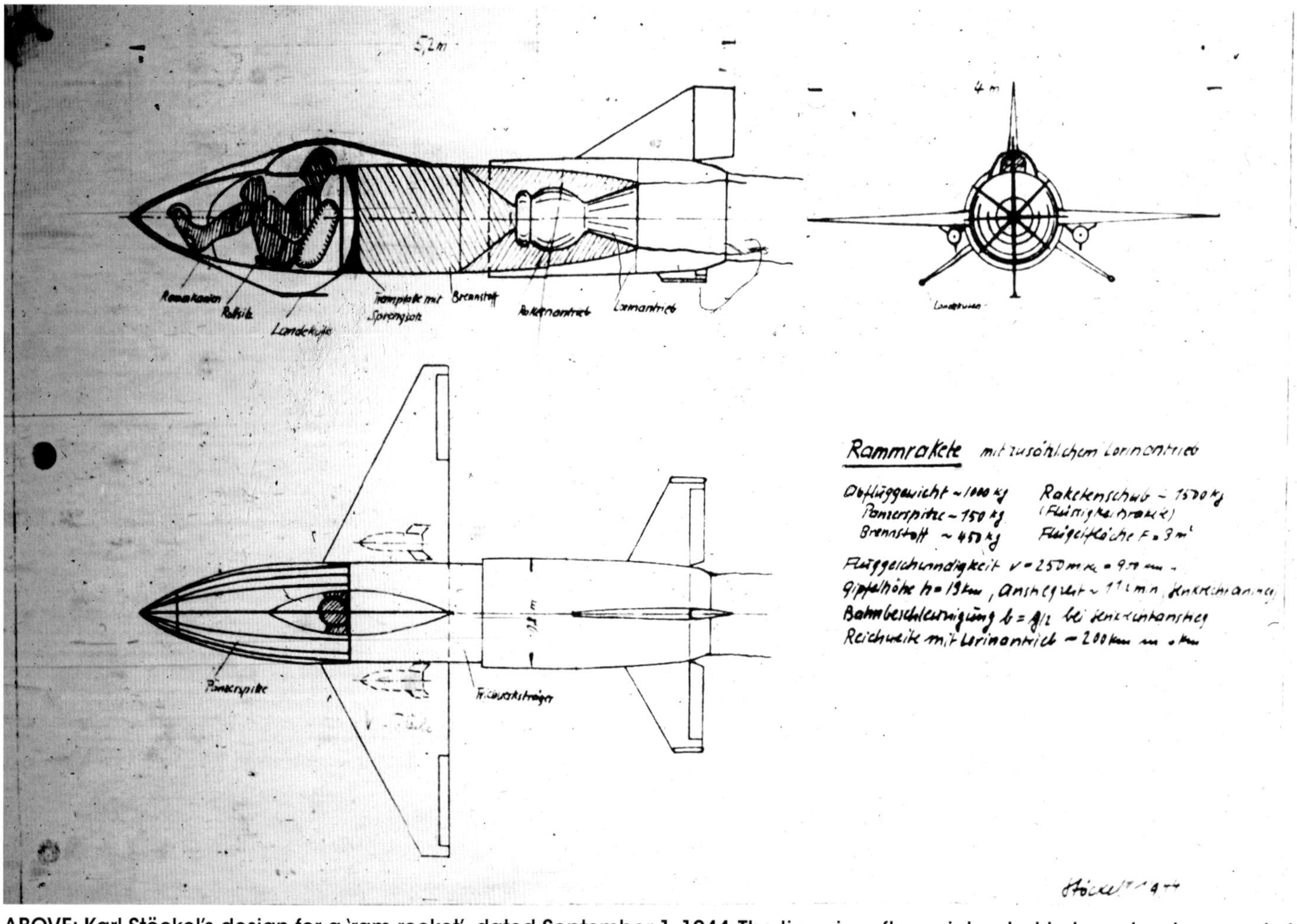

ABOVE: Karl Stöckel's design for a 'ram rocket', dated September 1, 1944. The tiny aircraft was intended to launch using a rocket motor before its ramjet took over. It would climb towards enemy bombers before overshooting them and dropping its bombs on them. The pilot would then turn it around and smash it through a bomber's tail before bailing out.

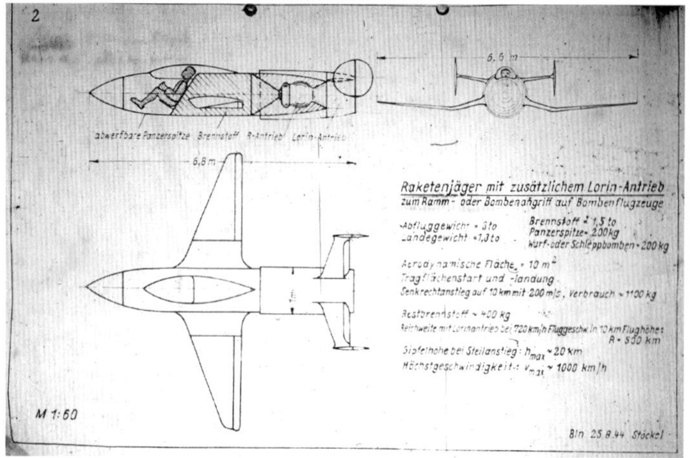

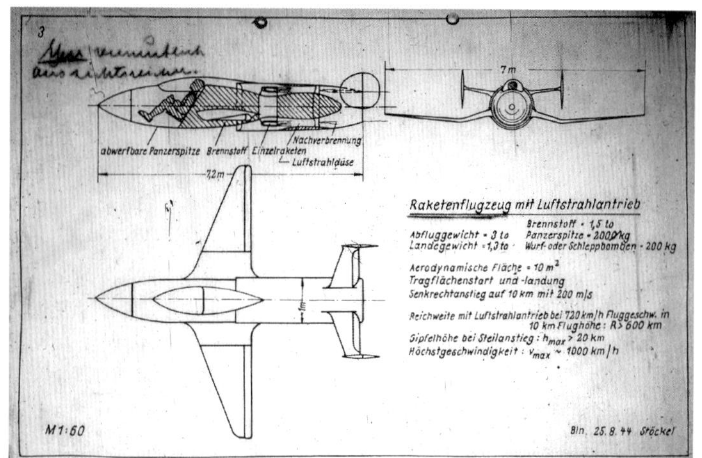

ABOVE LEFT: An earlier version of Stöckel's ram rocket – a 'rocket fighter with additional ramjet engine'. It was slightly larger than the final design but lacked any form of landing skid.
ABOVE RIGHT: This version of Stöckel's ram rocket appears to have had a turbojet engine with afterburning.

ZIPPERMAYR PFEIL FLUGZEUG

Italian-born Austrian scientist Mario Zippermayr carried out industrial and aeronautical research at his laboratory in Vienna until August 1939 when he was drafted into the Luftwaffe.

Having served with Luftgaukommando XVII, the air defence organisation of Vienna, until April 1942, he was then asked by the RLM to re-establish his lab to investigate and find answers to a number of aviation-related engineering problems.

To begin with, he was supplied with equipment by various firms in the city and received military personnel to act as his staff. His initial work was on developing more reliable brushes for aircraft electricity generators and devising a means of tracking enemy bomber formations using radio equipment.

However, at the beginning of 1943 he was able to recruit experienced scientists to his staff and was tasked with developing a means of stabilising air-launched torpedoes during their flight from an aircraft until they hit the water. In addition, he had to find a way of increasing the speed, altitude and angle at which torpedoes could be launched without suffering damage when they impacted on the water's surface.

In ordinary circumstances, torpedoes had to be launched at low level, close to the intended target and with the aircraft flying slowly and horizontally. Zippermayr's twofold solution to this was to equip torpedoes with both gyroscopes and very deep but also very narrow wings – extremely low aspect ratio. The result was a torpedo that was apparently automatically – and extraordinarily – stable when launched, no matter how fast or how high the launch aircraft was flying.

According to BIOS Interrogation Report No. 267: "The device was designated as L40 and provisions were made for it in the emergency programme. In the testing stage, a torpedo thus equipped was dropped from aircraft flying at speeds up to 720km/h and from heights above 1,000m. It was intended to use such torpedoes from the Arado 234. This work was carried out from January 1944 to the end of the war."

In fact, the L40 achieved such good results that "the wings of the L40 device made it desirable to utilise the knowledge gained for the construction of a high-speed aircraft. The ultimate intention in this development was to evolve a plane that would approach or even exceed the speed of sound".

Zippermayr stated during his interrogation that the aerodynamics institute of the Technische Hochschule, Hannover, had carried out wind tunnel tests on the L40 design "and passed the opinion that the structure was sound and suitable and could be applied successfully to actual aircraft".

He was then given orders to develop a jet-propelled aircraft using his torpedo wings. "It was the intention to utilise the new principle first in the development of a high-speed fighter. At the same time experiments were to continue to determine the value of the new design in the construction of fast bombers and transport planes. Dr Zippermayr claims that plans were begun to investigate the possibilities of constructing gigantic high-speed airliners, plans which are said to have been successful as far as they were able to progress before Germany's defeat."

Furthermore: "It was decided that the first experimental model was to be a glider that would be released from a tow or carrier-plane at high altitude. Models were constructed and equipped with instruments to register all possible flying and diving qualities. The results of these experiments were then used in the construction of the first full-scale glider model.

"It was with this model about three months short of the first test-flight that work stopped with the end of the war. This model, about three-fourths complete, is now stored at the Hagen carpenter shop in Lofer. After test-flights of this model were completed work was to begin on a jet-propelled craft. Dr Porsche, located at Schüttgut near Zell-am-See had already been assigned the task of supplying the jet-units."

Zippermayr claimed that his Pfeil Flugzeug or 'Arrow Aircraft' would have several clear advantages over more conventional designs: "Most outstanding, of course, is its tremendous speed. If jets of sufficient power can be devised its speed potential, it is claimed, would approach and possibly exceed the speed of sound. The small, but extremely strong wings have much less drag than the conventional variety, making it possible to realise a much greater efficiency from the engine thrust.

"The craft is expected to be extremely stable, especially on the transverse axis, making the plane easy to control. The torpedo prototype flew without any automatic steering device and even at speeds of 80 to 200m per second was perfectly stable. Even at an incidence angle of 52 degrees the torpedoes showed no tendency to stall nor were there any irregularities in the flight direction.

"It is also claimed that the wing loading for such a craft is much less than for conventional types. It is therefore expected that the landing speed will be reasonably low. It is intended that the highest speeds will not be attempted at less than 10,000m to take advantage of the lesser air density. It is expected that the construction of the plane should make mass production even more economical than that of present-day craft.

"Materials required for completion: The full-scale test model is complete except for the construction of the tail assembly. Panel instruments and other instruments to record the test-flight data are also required. If these things are obtainable, Dr Zippermayr claims that the first test-flight could be made in about three months' time. A carrier plane would also be required. It is Zippermayr's hope that American aviation authorities will interest themselves sufficiently in the craft to make such a carrier available when the model is declared ready to fly.

"Estimate of test flight performance: it is intended that the plane should be released from the carrier-craft at a height of about 6,000-7,000m. The test model (without jets) is then expected to dive at terrific speed until it is approximately 1,000m from the ground. Here the pilot begins to pull the craft out of its dive until the nose has risen into the air and the angle of approach is as much as 52°. In this position, the plane is expected to maintain its straight line of approach until it is about 20ft from the ground. Here it levels off and makes a regular three-point landing. Zippermayr has two highly competent pilots working with him on the plane, both of whom are anxious to make the initial flight.

"Present condition of project: the nearly finished full-scale model and the plans and drawings required to complete the craft are located at the Carpenter Shop Hagen in Lofer, where they are being guarded by workmen employed by Zippermayr. There are no American guards. Some documents and models have been removed by American agencies. The only name Dr Zippermayr remembers of Americans who have already exploited the target is that of an 'Engineer Bramford of SHAEF'. At present, no work is going on, pending instructions from American authorities as to disposition of the work already accomplished, and of the plans, drawings and models."

Zippermayr's work on the Pfeil Flugzeug came to nothing and his original design drawings have never resurfaced but may well still exist in American archives.

ABOVE: This photograph purports to show the mock-up of Zippermayr's Pfeil Flugzeug. Its unusual shape is such that it is not at first obvious in the picture. It is in fact the framework behind the man's head, with the cockpit on the right-hand side of the image and the very deep but also very short wing stretching from the upper rear part of the cockpit towards the tail end, to the left of the image.

ABOVE: A page of sketches accompanied BIOS Interrogation Report No. 267 on the activities of Dr Mario Zippermayr. These illustrate the remarkable shape and size of the Austrian's proposed Pfeil Flugzeug with its incredibly low aspect ratio wings.

REORGANISATION OF THE RLM IN 1944

The way in which decisions were made about new aircraft developments in Germany changed dramatically during the course of 1944. The establishment of the famous Jägerstab on March 1, 1944 (which resulted in the equally famous and impressive increase in German fighter production for the final year of the war – at least on paper) was just the first step in a radical transformation of the Reichsluftfahrtministerium.

First, a little background. The RLM was created in April 1933 with Hermann Göring as Air Minister. From May 1933 it consisted of two large departments – the military Luftschutzamt and the civilian Allgemeines Luftamt. Göring put Erhard Milch, formerly a director of Lufthansa, in charge of the Luftschutzamt, effectively making him his deputy.

In September 1933 the ministry was reorganised into six departments, one of which was the Technisches Amt or Technical Office, which dealt with aircraft design and development. A seventh department was created in 1934 known as Luftzeugmeister, which was in charge of equipment and logistics. The Luftwaffe itself was formally created on February 26, 1935.

Crucially, in 1936, Ernst Udet was appointed to lead the Technical Office under the new title Generalluftzeugmeister. This was far more than just a title, however. The office of Generalluftzeugmeister gave Udet sweeping powers to organise and decide on procurement and production of all equipment for the Luftwaffe, side-lining Milch. Udet essentially became the head of the RLM, responsible only to Göring.

After Udet took his own life on November 17, 1941, Milch took his place as Generalluftzeugmeister and took responsibility for procuring, producing and supplying the Luftwaffe's equipment. He decided what aircraft types were required, which types should be built, how many should be built and what priority they received. This included dictating the allocation and reallocation of resources between different privately run aircraft companies.

Towards the end of February 1944, it became clear to Milch that the RLM's traditional systems for managing aircraft production, and the personnel managing those systems, were simply unable to produce the increases in fighter aircraft production demanded by the Luftwaffe. He therefore approached Albert Speer, head of the Reichsministerium für Rüstung und Kriegsproduktion – Ministry for Equipment and War Production (hereafter referred to as Speer's ministry) – for assistance.[1]

By Milch's own account, within six hours an agreement was reached that Speer's people would apply their expertise to the problem. The idea was quickly run past Göring and then Hitler, who both approved, and the Jägerstab was formed at the beginning of March. It would be jointly led by Milch and Speer with Karl-Otto Saur as their deputy and chief of staff. A planning group was then established which included Hans Kammler of the Waffen-SS. The Jägerstab successfully managed to improve production of fighter aircraft by dispersing production – although by exactly how much is debatable – but it also significantly weakened Milch's position politically. Although he was joint leader of the Jägerstab, most of the credit for its achievements went to Speer and Saur.

The Jägerstab would continue to operate as a separate entity to the Generalluftzeugmeister and Speer's ministry until June 20, 1944 (the last known meeting of the Jägerstab was on June 30, 1944), when Göring ordered that the office of Generalluftzeugmeister be dissolved and that responsibility for equipping the Luftwaffe should be concentrated within Speer's ministry.[2]

This brought to an end the system that had been in place since Milch took over from Udet in 1941. Milch himself was effectively sacked and dropped out of the picture, with Speer and Saur now taking the reins of not only aircraft production but most of the RLM's other functions too. The following month it was decided that the remains of Milch's Generalluftzeugmeister organisation within the RLM would be reformed into the Chef der Technischen Luftrüstung (Chef TLR for short).[3] This would carry out the practical aircraft design, development and testing functions and would answer to the Oberkommando der Luftwaffe (OKL) or Luftwaffe High Command.

Decision-making power would remain with Speer's ministry but that authority would be delegated to a new committee of technical specialists. This was known as the Entwicklungshauptkommission Flugzeuge (EHK) or Main Development Commission for Aircraft.

The Chef TLR was launched on August 1, led by Oberst Ulrich Diesing, and the EHK was officially formed on September 15, chaired by Luftwaffe chief engineer Roluf Lucht with Heinkel director Karl Frydag as his deputy (Göring himself had been filling in on decision-making meetings in the meantime).[4]

The Development Commission concept had been pioneered by Saur three years earlier in July 1941 with the Panzerkommission for tanks and 11 others had followed, including the Munitions Commission, Motor Transport Commission, Electro-Technical Commission, Shipbuilding Commission and so on with the EHK being the last.

According to Saur, a Development Commission had two main tasks: to ensure that development requirements were drawn up only in consultation with production specialists and to avoid overly elaborate designs. Explaining the commissions to Allied interrogators immediately after the war, he said: "It was found that if 100% of effort was required to meet a specification 100%, only 30% of effort was needed to meet 90%."[5]

Following a meeting on December 20–21, 1944, it was decided that the technical aspects of the EHK's work would be divided up into nine special commissions.[6] Willy Messerschmitt and his company were made responsible for assessing and reporting back on day fighter developments; Kurt Tank and Focke-Wulf were made responsible for night fighters; Heinrich Hertel and Junkers got bombers; training aircraft were given to Friedrich Fecher and Siebel; Robert Lusser and Fieseler got 'special aircraft' including rocketplanes and flying bombs; Günter Bock and the DVL got airframe construction; Walter Schilo and BMW got engines; planning and installation of equipment went to Rudolf Stüssel and Deutsche Lufthansa, while planning and installation of armament went to Walter Blume and Arado.

Richard Vogt of Blohm & Voss was given personnel management and Focke-Wulf director Georg Manigold was to manage the EHK's head office.

This meant that from December 22, 1944 onwards, responsibility for assessing day fighter projects was given to Messerschmitt, Tank and his team would assess night fighter projects, Hertel's team would assess bomber designs and so on. In practice, competing companies had to send their project proposals and designs to the relevant special commission and hope that they would be fairly assessed.

Decisions about which designs to proceed with were then meant to be taken collectively during full meetings of the EHK. Unfortunately, this system seems to have quickly broken down, with too many vested interests and rivalries at play, and decision-making ground to a halt.

An order from Adolf Hitler on March 27 effectively stripped the EHK, and thereby Speer's ministry, of its decision- making power and handed it to Kammler[7] but the end of the war was now so close that this made no real difference.

[1] IWM Milch Vol. 29/9579 Stenografischer Bericht über die St/GL – Besprechung unter dem Vorsitz des Generalfeldmarschalls Milch am Freitag, dem 31. März 1944, 11.30 Uhr im Reichsluftfahrtministerium
[2] BA-MA R3-1749/0380 Nachrichten des Reichsministers für Rüstung und Kriegsproduktion, Der Reichsminister der Luftfahrt und Oberbefehlshaber der Luftwaffe. Berlin, den 20. Juni 1944
[3] BA-MA RL3-2570 Der Reichsmarschall des Grossdeutschen Reiches und Oberbefehlshaber der Luftwaffe Nr. /44 geheim (Gen. Qu. 2. Abt. (I)) H. Qu., den .Juli 44 Betr.: Luftwaffen-Rüstungsproduktion
[4] BA-MA R3-1749/0380 Nachrichten des Reichsministers für Rüstung und Kriegsproduktion, Der Reichsminister für Rüstung und Kriegsproduktion TAE-Nr. 13300/44. Berlin, den 15. Sept. 1944
[5] CIOS Evaluation Report 163a 4 July 1945 Interrogation of Saur, Head of Technical Dept. in the Speer Ministry, 11 June 1945
[6] BA-MA R3-1749/126 Entwicklungshauptkommission Flugzeuge An Chef TLR Betrifft: Organisation der EHK. Berlin SW 29, den 22.12.1944. See also BA-MA R3-1749/156 Entwicklungshauptkommission Flugzeuge Aktenvermerk Betr.: Bildung der Entwicklungs-Hauptkommission Flugzeuge. Berlin, 25.11.44
[7] BA-MA RL3-2567 KTB der Chef TLR Berichtsraum 16.3./4.4.1945

A SELECTION OF DOCUMENTS CONCERNING JET FIGHTER DEVELOPMENT IN GERMANY

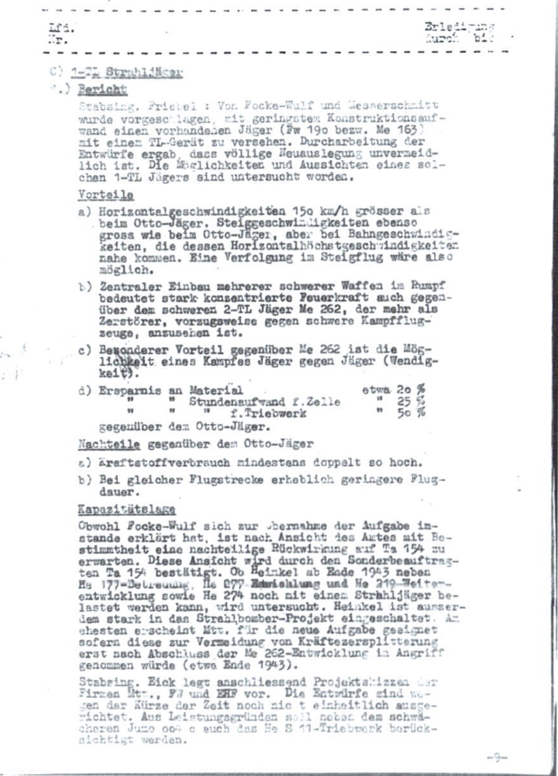

- 8 -

Lfd. Nr. — Erledigung durch bis

C) 1-TL Strahljäger

1.) Bericht

Stabsing. Friebel : Von Focke-Wulf und Messerschmitt wurde vorgeschlagen, mit geringstem Konstruktionsaufwand einen vorhandenen Jäger (Fw 19o bezw. Me 163) mit einem TL-Gerät zu versehen. Durcharbeitung der Entwürfe ergab, dass völlige Neuauslegung unvermeidlich ist. Die Möglichkeiten und Aussichten eines solchen 1-TL Jägers sind untersucht worden.

Vorteile

a) Horizontalgeschwindigkeiten 15o km/h grösser als beim Otto-Jäger. Steiggeschwindigkeiten ebenso gross wie beim Otto-Jäger, aber bei Bahngeschwindigkeiten, die dessen Horizontalhöchstgeschwindigkeiten nahe kommen. Eine Verfolgung im Steigflug wäre also möglich.

b) Zentraler Einbau mehrerer schwerer Waffen im Rumpf bedeutet stark konzentrierte Feuerkraft auch gegenüber dem schweren 2-TL Jäger Me 262, der mehr als Zerstörer, vorzugsweise gegen schwere Kampfflugzeuge, anzusehen ist.

c) Besonderer Vorteil gegenüber Me 262 ist die Möglichkeit eines Kampfes Jäger gegen Jäger (Wendigkeit).

d) Ersparnis an Material etwa 2o %
" " Stundenaufwand f.Zelle " 25 %
" " " f.Triebwerk " 5o %
gegenüber dem Otto-Jäger.

Nachteile gegenüber dem Otto-Jäger

a) Kraftstoffverbrauch mindestens doppelt so hoch.

b) Bei gleicher Flugstrecke erheblich geringere Flugdauer.

Kapazitätslage

Obwohl Focke-Wulf sich zur Übernahme der Aufgabe imstande erklärt hat, ist nach Ansicht des Amtes mit Bestimmtheit eine nachteilige Rückwirkung auf Ta 154 zu erwarten. Diese Ansicht wird durch den Sonderbeauftragten Ta 154 bestätigt. Ob Heinkel ab Ende 1943 neben He 177-Betreuung, He 277-Entwicklung und He 219-Weiterentwicklung sowie He 274 noch mit einem Strahljäger belastet werden kann, wird untersucht. Heinkel ist ausserdem stark in das Strahlbomber-Projekt eingeschaltet. Am ehesten erscheint Mtt. für die neue Aufgabe geeignet sofern diese zur Vermeidung von Kräftezersplitterung erst nach Abschluss der Me 262-Entwicklung in Angriff genommen würde (etwa Ende 1943).

Stabsing. Eick legt anschliessend Projektskizzen der Firmen Mtt., FW und EHF vor. Die Entwürfe sind wegen der Kürze der Zeit noch nicht einheitlich ausgerichtet. Aus Leistungsgründen soll neben dem schwächeren Jumo 004 c auch das He S 11-Triebwerk berücksichtigt werden.

-9-

Lfd. Nr. — Erledigung durch bis

2.) Durchsprache

Generalmajor Galland: Die Notwendigkeit des 1-TL Jägers auch neben der (voraussichtlich 1 Jahr früher kommenden) Me 262 steht ausser Zweifel, da nur mit ihm örtlich und zeitlich an bestimmten Frontabschnitten eine Luftüberlegenheit wird erzielt werden können. Im Interesse der Wendigkeit muss dieses Flugzeug sehr klein werden. Als Gipfelhöhe sind 13 km zu fordern.

K.d.E. weist auf die Gefahr einer Schädigung des Schwanzes durch den heissen Strahl hin. - Für Auftriebserhöhung und Herabsetzung der Flächenbelastung zum Langsamflug wäre ein grosser "taktischer Fowler" von Vorteil.

C-E Chef erwartet, dass die Gefährdung des Schwanzes durch die Abgase beherrschbar sein wird, sodass eine schwanzlose Bauweise aus diesem Grunde nicht unerlässlich ist, zumal mit Abkipp- und Schwerpunktsschwierigkeiten gerechnet werden muss.

C-E 2 : Die sehr hohen Festigkeitsansprüche eines taktischen Fowlers lassen den Stahlflügel wünschenswert erscheinen. Die Unterbringung des Fowlers wird bei den dünnen Profilen allerdings Schwierigkeiten machen.

Prof. Dr. Seewald hebt die Bedeutung auch des Kolbenmotors für Geschwindigkeiten bis 9oo km/h trotz hoher Gewichte für Luftschrauben etc. hervor. Da ein entsprechender Motor bei uns aber z.Zt. nicht vorhanden ist, steht die Notwendigkeit des Einsatzes der TL-Geräte auch vom Otto-Standpunkt aus ausser Zweifel.

3.) Entscheid

Der Generalfeldmarschall befiehlt, die Projekte des 1-TL Jägers mit grösstem Nachdruck weiter zu verfolgen. In der Auslegung sind die beiden Versionen a) grösster Geschwindigkeit, b) einer Dienstgipfelhöhe von mindestens 13 km zu berücksichtigen. Terminmöglichkeit und Leistung legen Verwendung des He S 11-Triebwerks nahe. Mit der Durchführung der Aufgabe, über die zu einem späteren Zeitpunkt entschieden werden wird, wäre zweckmässigerweise eine Jäger-Firma - wie Mtt. oder vielleicht auch Focke-Wulf - zu betrauen. Auch Fa. Heinkel ist aber an den Projektarbeiten zunächst noch weiter zu beteiligen. - Nächster Vortrag über 1-TL Strahljäger am 25.6.43. — C-E 2 / C-E 3 / C-E 25.6.43

D) Auslandsnachrichten (GL/A-Rü)

1.) Flugzeuge

a) Republic Thunderbolt P-47 C (Bewaffnung, Schutz, Leistung).

Bewaffnung von 8 x 12,7 mm-Flügel-MG mit je etwa 3oo Schuss ist durch Beutetrümmer bestätigt. Aus-

-10-

LEFT: Extract from minutes of the RLM Entwicklungsbesprechung of May 28, 1943, discussing the advantages of a single-jet fighter or '1-TL-Strahljäger' and plans to design one. Source: ADIK 158/142

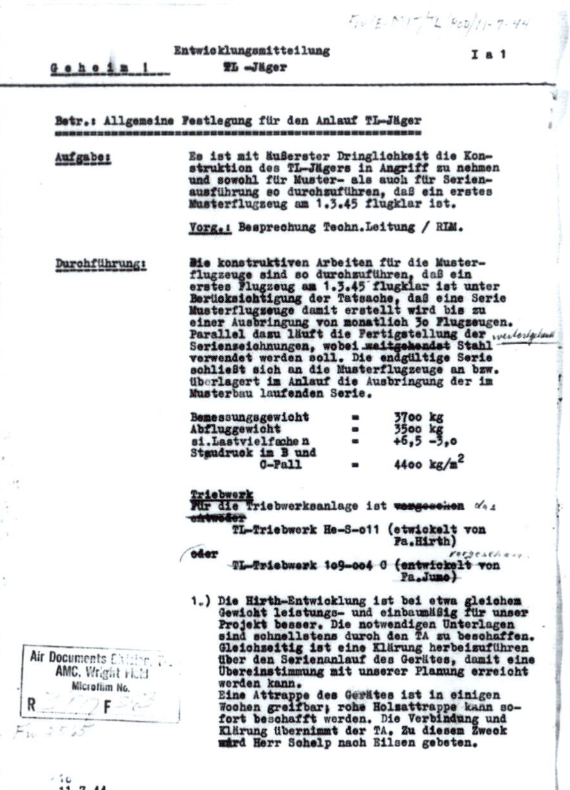

Fw/E-MIT/TL Rod/11-7-44

Geheim!

Entwicklungsmitteilung
TL-Jäger

I a 1

Betr.: Allgemeine Festlegung für den Anlauf TL-Jäger

Aufgabe: Es ist mit äußerster Dringlichkeit die Konstruktion des TL-Jägers in Angriff zu nehmen und sowohl für Muster- als auch für Serienausführung so durchzuführen, daß ein erstes Musterflugzeug am 1.3.45 flugklar ist.

Vorg.: Besprechung Techn.Leitung / RLM.

Durchführung: Die konstruktiven Arbeiten für die Musterflugzeuge sind so durchzuführen, daß ein erstes Flugzeug am 1.3.45 flugklar ist unter Berücksichtigung der Tatsache, daß eine Serie Musterflugzeuge damit erstellt wird bis zu einer Ausbringung von monatlich 3o Flugzeugen. Parallel dazu läuft die Fertigstellung der Serienzeichnungen, wobei ~~weitgehendst~~ weitestgehend Stahl verwendet werden soll. Die endgültige Serie schließt sich an die Musterflugzeuge an bzw. überlagert im Anlauf die Ausbringung der im Musterbau laufenden Serie.

Bemessungsgewicht	=	37oo kg
Abfluggewicht	=	35oo kg
si.Lastvielfachen	=	+6,5 -3,o
Staudruck im B und C-Fall	=	44oo kg/m²

Triebwerk
Für die Triebwerksanlage ist ~~vorgesehen~~ das ~~entweder~~
TL-Triebwerk He-S-o11 (etwickelt von Fa.Hirth)
~~oder~~ vorgesehen.
~~TL-Triebwerk 1o9-oo4 C (entwickelt von Fa.Jumo)~~

1.) Die Hirth-Entwicklung ist bei etwa gleichem Gewicht leistungs- und einbaumäßig für unser Projekt besser. Die notwendigen Unterlagen sind schnellstens durch den TA zu beschaffen. Gleichzeitig ist eine Klärung herbeizuführen über den Serienanlauf des Gerätes, damit eine Übereinstimmung mit unserer Planung erreicht werden kann.
Eine Attrappe des Gerätes ist in einigen Wochen greifbar; rohe Holzattrappe kann sofort beschafft werden. Die Verbindung und Klärung übernimmt der TA. Zu diesem Zweck wird Herr Schelp nach Eilsen gebeten.

11.7.44

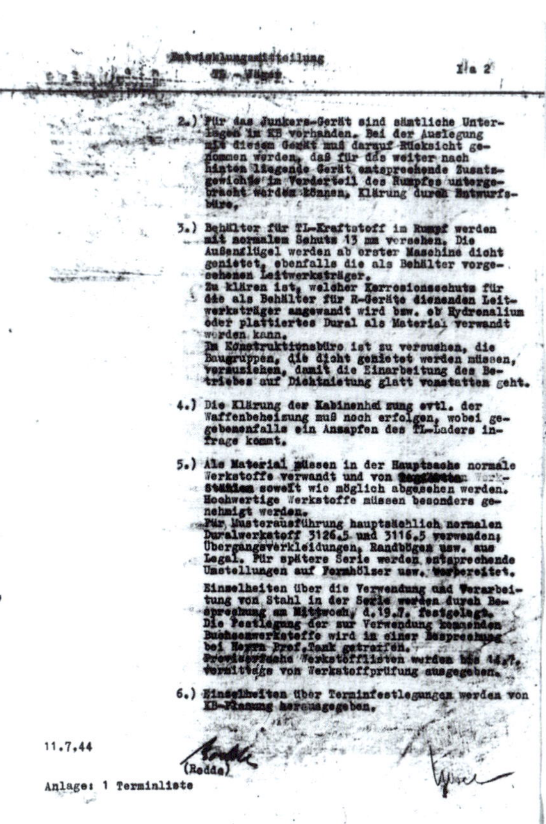

Entwicklungsmitteilung
TL-Jäger

I a 2

2.) Für das Junkers-Gerät sind sämtliche Unterlagen im KB vorhanden. Bei der Auslegung mit diesem Gerät muß darauf Rücksicht genommen werden, daß für das weiter nach hinten liegende Gerät entsprechende Zusatzgewichte im Vorderteil des Rumpfes untergebracht werden können. Klärung durch Entwurfsbüro.

3.) Behälter für TL-Kraftstoff im Rumpf werden mit normalem Schutz 13 mm versehen. Die Außenflügel werden ab erster Maschine dicht genietet, ebenfalls die als Behälter vorgesehenen Leitwerksträger.
Zu klären ist, welcher Korrosionsschutz für die als Behälter für R-Geräte dienenden Leitwerksträger angewandt wird bzw. ob Hydronalium oder plattiertes Dural als Material verwandt werden kann.
Im Konstruktionsbüro ist zu versuchen, die Baugruppen, die dicht genietet werden müssen, vorzuziehen, damit die Einarbeitung des Betriebes auf Dichtnietung glatt vonstatten geht.

4.) Die Klärung der Kabinenheizung evtl. der Waffenbeheizung muß noch erfolgen, wobei gegebenenfalls ein Anzapfen des TL-Laders infrage kommt.

5.) Als Material müssen in der Hauptsache normale Werkstoffe verwandt und von [illegible] Werkstätten soweit wie möglich abgesehen werden. Hochwertige Werkstoffe müssen besonders genehmigt werden.
Für Musterausführung hauptsächlich normalen Duralwerkstoff 3126.5 und 3116.5 verwenden; Übergangsverkleidungen, Randbögen usw. aus Legal. Für spätere Serie werden entsprechende Umstellungen auf Formhölzer usw. vorbereitet.
Einzelheiten über die Verwendung und Verarbeitung von Stahl in der Serie werden durch Besprechung am Mittwoch, d.19.7. festgelegt.
Die Festlegung der zur Verwendung kommenden Buchsenwerkstoffe wird in einer Besprechung bei Herrn Prof.Tank getroffen.
Provisorische Werkstofflisten werden bis 14.7. vormittags von Werkstoffprüfung ausgegeben.

6.) Einzelheiten über Terminfestlegungen werden von KB-Planung herausgegeben.

11.7.44

(Redde)

Anlage: 1 Terminliste

LEFT: Focke-Wulf document dated July 11, 1944, outlining a specification for a new jet fighter which needs to be ready for its first flight on March 1, 1945. A single turbojet is implied but not overtly stated. Source: ADRC/T-2 3997/583

RIGHT: The first two pages of a report on a comparison of HeS 011 single-jet fighter designs dated September 10, 1944. On the same day a specification would be issued for a new single-jet fighter powered by a BMW 003. Source: ADRC/T-2 2414/0527

Protokoll Nr. 1 (893/44) 1

10.9.44.

Vergleich der TL-Jäger-Entwürfe von Focke-Wulf, Heinkel und Messerschmitt
Arbeitstagung der Firmen B.V., F.W., EHAG u.Mtt.

B.u.V. 2 x
F.W. 2 x
Heinkel 2 x
GL/C-E Chef 2 x
GL/C-E 2/III 2 x
Mtt. A.G.:
TDM 1 x
H.Wackerle 1 x
H.Hornung 1 x

Antmann B.u.V.
Mittelhuber F.W.
Begandt "
Nieten "
Voigtberger "
Tielcke "
Wolff "
Hobach Heinkel
Schulz "
Eberhard "
Voigt Mtt.A.G.
Hornung "
Wackerle "
Horn "
Puffert "
Narr "

Geheime Kommandosache!

Auf Veranlassung und Anregung des GL/R-E Chefs traten die Firmen B.u.V., EHAG, F.W. und Mtt. zu einer Arbeitstagung in Oberammergau zusammen. Aufgabe war, Vergleichbare Grundlagen für die Leistungsrechnung zu schaffen und soweit als möglich einen Vergleich der vorliegenden TR-Entwürfe durchzuführen.

1.) Zusammengefaßte Kurzbeschreibungen der vorgelegten einsitzigen Ein-TL-Jäger mit HeS 11 Schubgerät (Auszug aus den Baubeschreibungen).
a) B.u.V.: Kein Entwurf vorgelegt.
b) F.W.:
Aufgabenstellung:
1.) Ein-TL-Jäger mit Zusatz-R-Antrieb zum schnellen Steigen auf große Höhen
2.) TL-Jäger mit Zusatz-R-Gerät, aber weniger R-Kraftstoff, ausreichend für schnelles Steigen auf mittlere Flughöhen.
3.) Reine TL-Jäger ohne R-Gerät, mit ungefähr 2 Stunden Flugdauer in ungefähr 10 - 11 km Höhe.

Ausrüstung:
Bewaffnung: im Rumpf 2 x MK 103
im Flügel je 1 x MG 151
Panzerung: gegen Beschuß von vorne mit 140 kg.

S.O. Archiv Bad Eilsen 8/140

12. Febr.

- 2 -

Protokoll vom 10.9.44. 2

FT: FuG 15 ZY und FuG 25

Abmaße:
Startgewicht bei 1) G = 4900 kg
Startgewicht bei 3) G = 4350 kg
Bemessungsgewicht G = 3700 kg mit n_a = 6,5

TL-Kraftstoff: 1250 kg, davon ~50 % geschützt im Rumpf

F = 17 m^2
B = 8 m
L = 10,5 m
Pfeilwinkel φ = 30°, bezogen auf Vorderkante (= 27° in 1/4 l)
ungepfeiltes Höhenleitwerk
Doppelseitenleitwerk
Haupträder 740 x 240
Druckkabine vorhanden.

c) EHAG:
Aufgabenstellung:
Reiner TL-Jäger

Ausrüstung:
Bewaffnung: im Rumpf 2 x MG 213
oder 2 x MK 108
oder 1 x MK 103 + 1 x 20 mm Waffe
Panzerung: zu 100 kg angenommen
FT: FuG 25 ZY und FuG 25

Abmaße:
Gewicht mit 2 x MG 213 u.1250 kg Kraftstoff G = 4060 kg
~ 60 % des Kraftstoffes geschützt im Rumpf

F = 14 m^2
B = 8 m
L = 9,3 m
Pfeilwinkel φ = 35°, bezogen auf 0,25 t
gepfeiltes V-Leitwerk.
Hauptfahrwerk 710 x 175
Druckkabine vorhanden.

d) Mtt.A.G.:
Aufgabenstellung:
1.) Kleinster TL-Jagdeinsitzer mit höchstmöglichen Flugleistungen
2.) TLR-Jäger mit R-Kraftstoff für schnellstes Steigen oder als Kampfhilfe.+)
3.) Erweiterungsmöglichkeit auf 2 Mann Besatzung durch Einsetzen eines Rumpfzwischenstückes, wobei der Raum für 2.Mann auch für Zusatzwaffen, Zusatzkraftstoff oder Zusatzantrieb ausgenutzt werden kann.

+) (Erweiterungsmöglichkeit dch.Einsetzen eines Rumpfzwischenstücks)

- 3 -

RIGHT: The Volksjäger requirement sent out to companies at 11.38am on September 10, 1944. This is Blohm & Voss's copy but Focke-Wulf's is also known to survive. Source: ADRC/T-2 2455/751

10. Sep. 1944 · 12

590

t
blohmwerft hmb
reichsluft bln
b
+ berlin rlm nr 2267 10.9. 1138=

25
Fi 200

an fa blohm u voss, hamburg z hd herrn dr. voigt =

- - sehr - dringend - -

- - streng vertraulich - -

betr.: projekt 1 tl-jaeger.-
im auftrag von tlr/ fl-e chef werden sie gebeten, projektuntersuchungen entsprechend nachstehenden richtlinien mit aeuszerster dringlichkeit durchzufuehren und unterlagen in spaetestens 3-5 tagen einzureichen.-
1 tl-jaeger billigster bauweise mit bmw-geraet 003. weitgehende verwendung von holz und stahl. v-max in bodennaehe 750 km/h. startrollstrecke nicht ueber 500 m. bei der fahrwerksauslegung i st rechnung zu tragen, dasz landung auf schlechten plaetzen erfolgen musz. flugzeeit in boddennaehe 30 min. mit ~~1000xxx~~ 100 proz.
schub.-
ausruestung.-
----
2 x mk 108 mit je 80-100 schusz (variante 2 mg 151/20 mit je 200-250 schusz). instrumentierung fuer schoenwettereinsatz, fug 16 zy bezw. fug 15.-
panzerung:
----
schutz des flugzeugfuehrers und der 3 cm munition gegen beschusz von vorn gegen 13 mm munition 0 grad schuszwinkel. schutz des flugzeugfuehrers gegen beschusz von hinten gegen 20 mm munition schuszkegel 10 grad. normaler behaelterschutz.-
es ist zu pruefen, welche verkuerzung der startrollstrecke zu erreichen ist, wenn bewaffnung auf 1 x mk 108 reduziert und auf panzerung gegen beschusz von vorn verzichtet wird.-
eventuelle mitnahme von flugzeuggebundenen formauszenbehaeltern zur erhoehung der reichweite unter erhoehung der startrollstrecke ist zu projektieren. (15 min. in bodennaehe).-
eventuelle rueckfragen apparat- nr. 1738 bezw. 3845.=

reichsluft fl. e 2 nr. 14496/44 (roem drei a)+

reichsluft bln
blohmwerft hmb

FAR RIGHT: A note on the organisation of the Entwicklunghauptkommission Flugzeuge dated December 22, 1944, showing the personnel appointed to lead its nine special committees. Source: BA-MA R3-1749/126

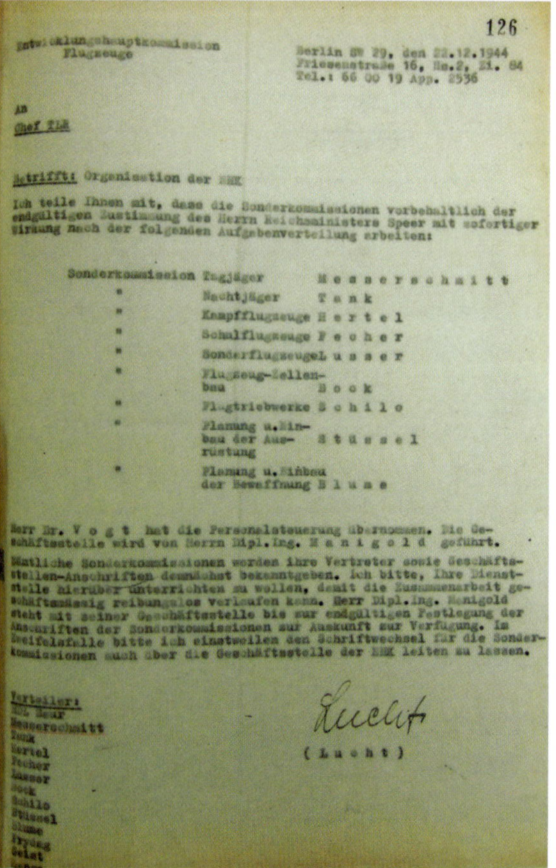

126

Entwicklungshauptkommission Flugzeuge

Berlin SW 29, den 22.12.1944
Friesenstraße 16, Hs.2, Zi. 84
Tel.: 66 00 19 App. 2536

An
Chef TLR

Betrifft: Organisation der EHK

Ich teile Ihnen mit, dass die Sonderkommissionen vorbehaltlich der endgültigen Zustimmung des Herrn Reichsministers Speer mit sofortiger Wirkung nach der folgenden Aufgabenverteilung arbeiten:

Sonderkommission	Tagjäger	Messerschmitt
"	Nachtjäger	Tank
"	Kampfflugzeuge	Hertel
"	Schulflugzeuge	Fecher
"	Sonderflugzeuge	Lusser
"	Flugzeug-Zellenbau	Bock
"	Flugtriebwerke	Schilo
"	Planung u. Einbau der Ausrüstung	Stüssel
"	Planung u. Einbau der Bewaffnung	Blume

Herr Dr. Vogt hat die Personalsteuerung übernommen. Die Geschäftsstelle wird von Herrn Dipl.Ing. Manigold geführt.

Sämtliche Sonderkommissionen werden ihre Vertreter sowie Geschäftsstellen-Anschriften demnächst bekanntgeben. Ich bitte, Ihre Dienststelle hierüber unterrichten zu wollen, damit die Zusammenarbeit geschäftsmässig reibungslos verlaufen kann. Herr Dipl.Ing. Manigold steht mit seiner Geschäftsstelle bis zur endgültigen Festlegung der Anschriften der Sonderkommissionen zur Auskunft zur Verfügung. Im Zweifelsfalle bitte ich einstweilen den Schriftwechsel für die Sonderkommissionen auch über die Geschäftsstelle der EHK leiten zu lassen.

Verteiler:
Messerschmitt
Tank
Hertel
Fecher
Lusser
Bock
Schilo
Stüssel
Blume
Frydag
Geist

Lucht
(Lucht)

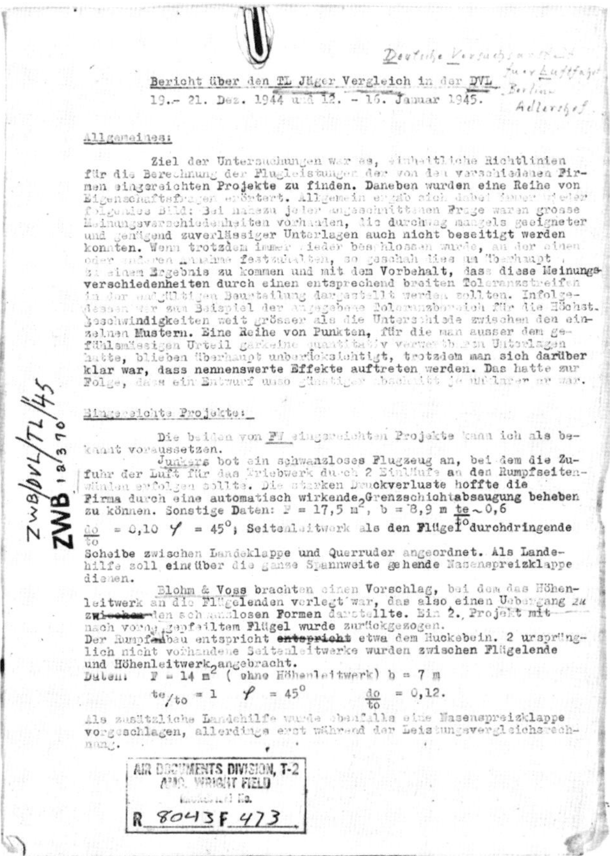

Bericht über den TL Jäger Vergleich in der DVL
19.– 21. Dez. 1944 und 12. – 16. Januar 1945.

Allgemeines:

Ziel der Untersuchungen war es, einheitliche Richtlinien für die Berechnung der Flugleistungen der von den verschiedenen Firmen eingereichten Projekte zu finden. Daneben wurden eine Reihe von Eigenschaftsfragen erörtert. Allgemein ergab sich dabei immer wieder folgendes Bild: Bei diesen jeder angeschnittenen Frage waren grosse Meinungsverschiedenheiten vorhanden, die durchweg mangels geeigneter und genügend zuverlässiger Unterlagen auch nicht beseitigt werden konnten. Wenn trotzdem immer wieder bestimmten Werte, an der einen oder anderen Annahme festzuhalten, so geschah dies um überhaupt zu einem Ergebnis zu kommen und mit dem Vorbehalt, dass diese Meinungsverschiedenheiten durch einen entsprechend breiten Toleranzstreifen in der endgültigen Beurteilung dargestellt werden sollten. Infolgedessen war das Beispiel der angegebene Toleranzbereich für die Höchstgeschwindigkeiten meist grösser als die Unterschiede zwischen den einzelnen Mustern. Eine Reihe von Punkten, für die man ausser dem gefühlsmässigen Urteil keinerlei quantitativ verwertbaren Unterlagen hatte, blieben überhaupt unberücksichtigt, trotzdem man sich darüber klar war, dass nennenswerte Effekte auftreten werden. Das hatte zur Folge, dass ein Entwurf umso günstiger abschnitt je unklarer er war.

Eingereichte Projekte:

Die beiden von FW eingereichten Projekte kann ich als bekannt voraussetzen.

Junkers bot ein schwanzloses Flugzeug an, bei dem die Zufuhr der Luft für das Triebwerk durch 2 Einläufe an den Rumpfseitenwänden erfolgen sollte. Die starken Druckverluste hoffte die Firma durch eine automatisch wirkende Grenzschichtabsaugung beheben zu können. Sonstige Daten: F = 17,5 m², b = 8,9 m te ~ 0,6

do/to = 0,10 φ = 45°; Seitenleitwerk als den Flügel durchdringende Scheibe zwischen Landeklappe und Querruder angeordnet. Als Landehilfe soll eine über die ganze Spannweite gehende Nasenspreizklappe dienen.

Blohm & Voss brachten einen Vorschlag, bei dem das Höhenleitwerk an die Flügelenden verlegt war, das also einen Übergang zu zwischen den schwanzlosen Formen darstellte. Ein 2. Projekt mit nach vorn gepfeiltem Flügel wurde zurückgezogen.
Der Rumpfaufbau entspricht etwa dem Huckebein. 2 ursprünglich nicht vorhandene Seitenleitwerke wurden zwischen Flügelende und Höhenleitwerk angebracht.
Daten: F = 14 m² (ohne Höhenleitwerk) b = 7 m

te/to = 1 φ = 45° do/to = 0,12.

Als zusätzliche Landehilfe wurde ebenfalls eine Nasenspreizklappe vorgeschlagen, allerdings erst während der Leistungsvergleichsrechnung.

Deutsche Versuchsanstalt für Luftfahrt Berlin-Adlershof

ZWB/DVL/TL/45
ZWB 13.3.70

AIR DOCUMENTS DIVISION, T-2
AMC, WRIGHT FIELD
Microfilm No.
R 8043 F 473

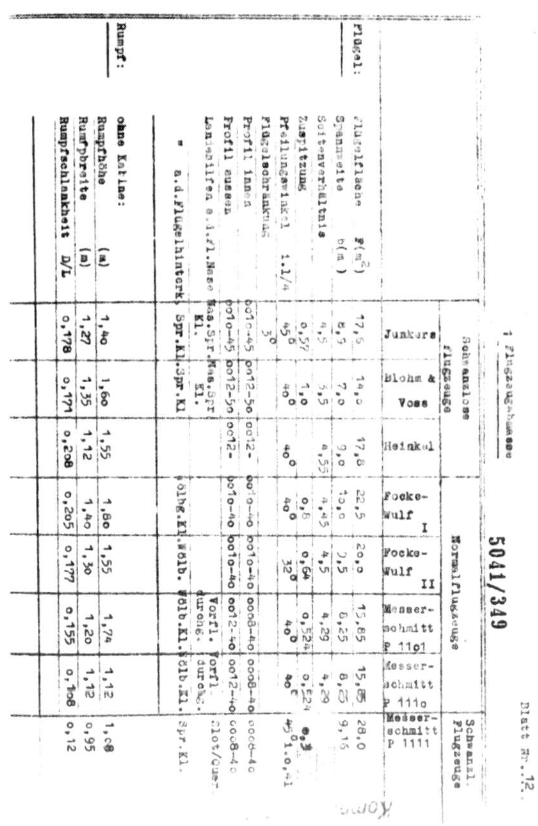

5041/349

TL 1945 Blatt Nr. 12

1 Flugzeugabmessungen

	Schwanzlose Flugzeuge			Normalflugzeuge				Schwanzl. Flugzeuge
	Junkers	Blohm & Voss	Heinkel	Focke-Wulf I	Focke-Wulf II	Messerschmitt P 1101	Messerschmitt P 1110	Messerschmitt P 1111
Flügel:								
Flügelfläche F(m²)	17,5	14,0	17,8	22,5	20,0	15,85	15,85	28,0
Spannweite b(m)	8,9	7,0	9,0	10,0	9,5	8,25	8,25	9,16
Seitenverhältnis	4,5	3,5	4,55	4,45	4,5	4,29	4,29	
Zuspitzung	0,57	1,0		0,8	0,64	0,524	0,524	0,3
Pfeilungswinkel 1.1/4	45°	40°	40°	40°	32°	40°	40°	45° 1.0,41
Flügelschränkung	3°							
Profil innen	0010-45	0012-50	0012-	0010-40	0010-40	0008-40	0008-40	0008-40
Profil aussen	0010-45	0012-50	0012-	0010-40	0010-40	0012-40	0012-40	0008-40
Landehilfen a.d.Fl.Nase	Nas.Spr. Kl.	Nas.Spr. Kl.				Vorfl. durchg.	Vorfl. durchg.	Slot/Quer-
" a.d.Flügelhinterk.	Spr.Kl.	Spr.Kl		Wölbg.Kl.	Wölb.Kl.	Wölb.Kl.	Wölb.Kl.	Spr.Kl.
Rumpf: ohne Kabine:								
Rumpfhöhe (m)	1,40	1,60	1,55	1,80	1,55	1,74	1,12	1,08
Rumpfbreite (m)	1,27	1,35	1,12	1,40	1,30	1,20	1,12	0,95
Rumpfschlankheit D/L	0,178	0,171	0,208	0,205	0,177	0,155	0,108	0,12

FAR LEFT: First page of an assessment report on single-jet fighter designs produced by the DVL in January 1945. Source: ADRC/T-2 8043/473

LEFT: Page from the pre-meeting report for the single-jet fighter comparison meeting held at the end of February 1945, showing details of the eight competing designs. Source: ADRC/MAP 5041/349

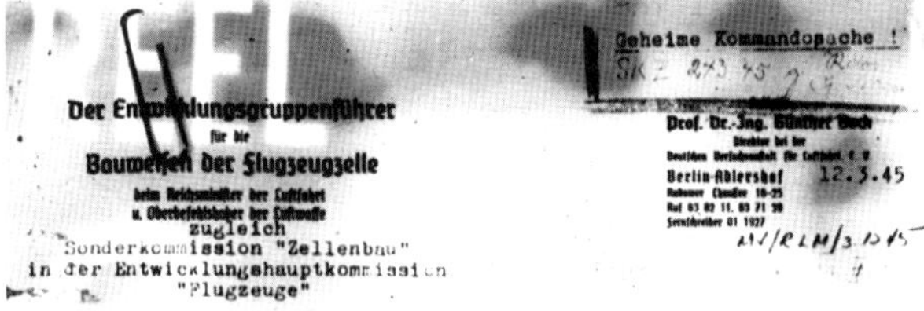

Der Entwicklungsgruppenführer
für die
Bauweisen der Flugzeugzelle
beim Reichsminister der Luftfahrt
u. Oberbefehlshaber der Luftwaffe
zugleich
Sonderkommission "Zellenbau"
in der Entwicklungshauptkommission
"Flugzeuge"

Geheime Kommandosache !

Prof. Dr.-Ing. Günther Bock
Direktor bei der
Deutschen Versuchsanstalt für Luftfahrt, E. V.
Berlin-Adlershof 12.3.45

Herrn
Dipl.-Ing. W e h r s e
Focke-Wulf Flugzeugbau G.m.b.H.
(21) B a d E i l s e n

Air Documents Division, T-2
R 3282 F 270

Betr.: Fertigungsprogramm und Entwicklungsarbeiten

1.) Führer-Notprogramm

Für die Fertigung von Flugzeugen ist ein Führer-Notprogramm aufgestellt worden. Dieses sieht mit verminderten Ausbringungszahlen gegenüber dem bisherigen Programm Baureihen folgender Flugzeugmuster vor:

8-88, 8-109, 8-152, 8-162, 8-190, 8-231, 8-262, 8-335, 8-396 und Mistel.

Wegen der veränderten Rohstoff- und Arbeitslage wird sich das Programm jedoch nicht einhalten lassen, weshalb dieses gekürzt wurde und als "gekürztes Notprogramm" nur noch Arbeiten für die Muster

8-88, 8-152/190, 8-162, 8-234, 8-262, 8-396

vorsieht; es gilt für die Fertigung ab Oktober 45. Noch nicht geklärt ist hierbei die Stellung des Musters 8-335.

Soweit die Materiallage und Energieversorgung usw. es zulassen, sollen jedoch über die in den Notprogrammen vorgesehenen Stückzahlen hinaus noch möglichst viel Flugzeuge fertiggestellt werden.

Die Arbeiten der Sonderkommission "Flugzeugzellenbau" gelten, soweit sie die oben angegebenen Flugzeugmuster des angekürzten bezw. gekürzten Fertigungsnotprogramms betreffen, als ebenfalls unter das Notprogramm fallend.

2.) Entwicklungsarbeiten

Auf der letzten Sitzung der Entwicklungshauptkommission "Flugzeuge" am 1.3.45 wurde die Richtung der Entwicklungsarbeiten besprochen. Hierbei ergab sich entsprechend dem Verwendungszweck der Flugzeugmuster folgendes Bild:

a) Tagjäger

Mit Entwicklungsarbeiten für Tagjäger mit TL-Triebwerken HeS 109.011 A ist bei den Firmen Focke-Wulf und Messerschmitt mit je einem Entwicklungsauftrag zu rechnen. Ob darüber hinaus auch noch die Firmen Junkers oder Blohm & Voss für die Entwicklung eingeschaltet werden, ist noch unbestimmt.

b) Nacht- und Schlechtwetterjäger

Für Nacht- und Schlechtwetterjäger mit je 2 TL-Triebwerken HeS 109.011 A sind Projekte bei Arado und Blohm & Voss in Bearbeitung. Eine Auftragserteilung an beide Firmen ist wahrscheinlich. Ein weiteres Projekt "Jägervariante Horten IX" ist bei der Gothaer Waggonfabrik in der Durcharbeitung.

c) Langstreckenbomber

Mit Entwicklungsarbeiten für Langstreckenbomber mit je 4 TL-Triebwerken HeS 109.011 A ist die Firma Junkers betraut, und zwar sind in diesem Zusammenhang dort die Arbeiten an der 8-287 wieder angelaufen. Im Anschluss an diese Arbeiten sind bei der gleichen Firma Projektarbeiten für einen Langstreckenbomber in Nur-Flügel-Bauweise vorgesehen.

Aus vorstehenden Angaben werden Sie sich ein ungefähres Bild über die Arbeiten auf dem Entwicklungsgebiet, soweit es Ihre eigenen Arbeiten berührt, machen können. Ich bitte Sie, die Leiter Ihrer Arbeitsgruppen bezw. Ihre Mitarbeiter in geeigneter Form zu unterrichten, damit sie sich auf ihrem Spezialgebiet rechtzeitig bei den einzelnen Projekten und Bauvorhaben einschalten können.

Heil Hitler !

Ddr.:
Entwicklungshauptkommission "Flugzeuge
Oberstingr. Dr.-Ing. Kötzschke.

LEFT: Letter dated March 12, 1945, from Günther Bock, leader of the EHK's Sonderkommission 'Zellenbau', to Oberingenieur Friedrich Wehrse at Focke-Wulf concerning the Führer-Notprogramm and development work on single-jet day fighters, twin-jet night- and bad-weather fighters and the four-jet 'Langstreckenbomber'. Source: ADRC/T-2 3282/270

NOTES

Introduction

[1] IWM Milch Vol. 64/6700 Stenographische Neiderschrift über die Konstrukteurbesprechung beim Reichsmarschall am 25. Mai 1944, 11 Uhr im Führerspeiseraum der SS-Kaserne Obersalzberg
[2] ADRC/T-2 3201/986 Messerschmitt descriptive reports for stepping up German Aircraft Production circa 1943
[3] Army Air Force Scientific Advisory Group Technical Intelligence Supplement by H S Tsien, H L Dryden, F L Wattendorf, Lt Col F W Williams, F Zwicky and W H Pickering. Published by Headquarters Air Materiel Command, Publications Branch, Intelligence T-2, Wright Field, Dayton Ohio, May 1946
[4] *Arming the Luftwaffe: The German Aviation Industry in World War II by Daniel Uziel*, McFarland, 2012, p101
[5] TNA AIR 40/164/42 Air P/W Interrogation Unit Main Headquarters Second T.A.F. Trends of Development in the G.A.F. Part IV – The Story of the 8-162. July 12, 1945
[6] BA-MA RL3-2567 KTB der Chef TLR Berichtswoche 16.3/4.4.1945
[7] Although in his weary A Scientist's Analysis of the German Defeat, Interrogation Report No. 210, of October 8, 1946, Professor Walter Georgii, director of the German Aeronautical Research Council, wrote: "Finally propaganda came out with its last miracle, the new technical weapons. This type of propaganda was given considerable circulation in the entire military service. To those who had anything to do with technical development or research, it was clear that V-weapons could not be decisive in changing the course of the war. Any new V-weapons which would actually have been important were not known."

1. Early developments

[1] *Max Valier, Ein Vorkampfer der Weltraumfahrt 1895-1939* by I. Essers 1968, p158
[2] *Ein Dreieck Fliegt* by Alexander Lippisch 1976, p23
[3] ibid. p31
[4] ibid. p27
[5] *Erinnerungen* by Alexander Lippisch 1976, p170
[6] *Ein Dreieck Fliegt* by Alexander Lippisch 1976, p44
[7] *Erinnerungen* by Alexander Lippisch 1976, p176
[8] *Ein Dreieck Fliegt* by Alexander Lippisch 1976, p48
[9] *Erinnerungen* by Alexander Lippisch 1976, p177
[10] ibid. p182
[11] *Heinkel Raketen- und Strahlflugzeuge* by Volker Koos p9
[12] ibid. p10
[13] ibid. p12
[14] ibid. p26
[15] ibid. p18
[16] idid. p20
[17] *Me 163 Rocket Interceptor Volume One* by Stephen Ransom and Hans-Hermann Cammann 2001, p67
[18] *Erinnerungen* by Alexander Lippisch 1976, p186
[19] ibid. p192
[20] ibid. p196
[21] *Heinkel Raketen- und Strahltriebwerk Flugzeuge* by Volker Koos 2008, p29
[22] ibid. p112
[23] BA–MA RL3/780 LC 7/III Gegenwärtiger Stand und künftige Entwicklungsarbeit auf dem Gebiete des Schnellfluges mit Strahltriebwerk 14.10.38
[24] *Heinkel Raketen- und Strahlflugzeuge* by Volker Koos 2008, p38
[25] ADRC/MAP 5028/476 Junkers EF Designs file
[26] *Heinkel Raketen- und Strahlflugzeuge* by Volker Koos 2008, p41
[27] ibid. p122
[28] ibid. p123
[29] ibid. p42
[30] ibid. p128
[31] ibid. p116
[32] ibid. p129
[33] ibid. p130
[34] Messerschmitt Abschrift Anlage zum Schreiben LC 7 Nr. 461/38 (III) g. Kdos. vom 4.1.1939 Vorläufige Technische Richtlinien für schnelle Jagdflugzeuge mit Strahltriebwerk
[35] Iowa State University Library Special Collections and University Archive MS-243 Box 17 Folder 40 Projekte 1939
[36] Iowa State University Library Special Collections and University Archive MS-243 Box 17 Folder 37 Entwurf P 01 12.4.1939 and P 01.116 13.4.1939
[37] *Me 262 Volume One* by J. Richard Smith and Eddie Creek 1997 p58
[38] *Erinnerungen* by Alexander Lippisch 1976, p198
[39] NARA T 177-16/0395 RLM Flugzeugentwicklingsprogramm vom 1.10.36
[40] *Ein Dreieck Fliegt* by Alexander Lippisch 1976, p60
[41] ADRC/T-2 2070/931 Messerschmitt Projektbaubeschreibung P 01-112 February 1940
[42] *Erinnerungen* by Alexander Lippisch 1976, p199
[43] ibid.
[44] ibid. p201
[45] ADRC/T-2 2013/427 Messerschmitt Projektbaubeschreibung P 01-114 August 1940
[46] IWM FD4355/45/164 Professor Willy Messerschmitt Vol. 5
[47] *Ein Dreieck Fliegt* by Alexander Lippisch 1976, p60

2. Abteilung L

[1] Lippisch himself refers to this design as the P 03-Me 263, although the AVA reports go from being marked simply 'P 01' briefly to 'Me 263 (P 01)' then finally to 'Me 263' without P 03 appearing. The only known surviving evidence of the P 03 is a set of 12 handwritten sheets on graph paper in Lippisch's personal collection of papers at Iowa State University, which offer few clues about the actual design. A very faint sketch on one sheet shows a design with tailplanes similar to those appearing on the AVA wind tunnel model drawings. In fact, a document sold through online auction site eBay in 2017 dated November 1941 includes an entry which reads: "Me 327 (Früher 263 = P01) z. Zt. gestoppt wegen Vordringlichkeit Me 163 A und B." This suggests that before it was cancelled due to the urgency with which the Me 163 A and B were required, the P 01-114/P 03/Me 263 had finally received an official type number from the RLM: Me 327
[2] *Ein Dreieck Fliegt* by Alexander Lippisch 1976, p60
[3] The aircraft which eventually bore the designation Me 263 in 1944 – itself originally designated Ju 248 – was unrelated to the P 01-derived 1941 design
[4] ADRC/T-2 2250/311 AVA Bericht 41/14/22 Untersuchung eines Gesamtmodells mit Pfeilflügel (P-01) 23.9.1941
[5] *Ein Dreieck Fliegt* by Alexander Lippisch 1976, p60
[6] *Me 163 Rocket Interceptor Volume One* by Stephen Ransom and Hans-Hermann Cammann 2001, p80
[7] Iowa State University Library Special Collections and University Archive MS-243 Box 17 Folder 37 Messerschmitt blueprints
[8] At this stage there was no 'Me 163 A' since there was no 'B' – it was simply known as the Me 163
[9] Iowa State University Library Special Collections and University Archive MS-243 Box 17 Folder 42 Messerschmitt Projektbaubeschreibung Li P 05 27. August 1941
[10] IWM FD4355/45/649 Professor Willy Messerschmitt Vol. 5 Aktenvermerk Besprechung mit dem RLM 29. August 1941
[11] ibid. /650
[12] IWM Milch Vol. 14/560 GL/C-E 2 Nr. 2976/42 geh. Vortrag C-E 2 Chef über Fernstflugzeuge am 12.5.42 vor Chef Genst.

NOTES CONTINUED

[13] *Ein Dreieck Fliegt* by Alexander Lippisch 1976, p62
[14] *Me 163 Rocket Interceptor Volume One* by Stephen Ransom and Hans-Hermann Cammann 2001, p84
[15] *Ein Dreieck Fliegt* by Alexander Lippisch 1976, p48-49
[16] Iowa State University Library Special Collections and University Archive MS-243 Box 17 Folder 27 Patentanmeldung der Messerschmitt A.G., Augsburg Flugzeug mit nach vorne gepfeilten Flügel 14.11.41
[17] Iowa State University Library Special Collections and University Archive MS-243 Box 17 Folder 28 Patentanmeldung der Messerschmitt A.G., Augsburg Höhenleitwerk für Flugzeuge 14.11.41
[18] Iowa State University Library Special Collections and University Archive MS-243 Box 17 Folder 31 Patentanmeldung der Messerschmitt A.G., Augsburg Pfeilflügel mit drehbaren Enden 14.11.41
[19] *Ein Dreieck Fliegt* by Alexander Lippisch 1976, p74
[20] ibid. p75
[21] *Erinnerungen* by Alexander Lippisch 1976, p204, p211
[22] ADRC/T-2 2344/804 Lippisch Me 163 B + C (Nahaufklärer) sowie M 163 Super Leistungsrechnung Jan-Apr 1944
[23] ADRC/T-2 2645/260 Messerschmitt Flugleistungen P 20 19.4.1943
[24] *Me 163 Rocket Interceptor Volume One* by Stephen Ransom and Hans-Hermann Cammann 2001, p94-95

3. The first competition
[1] *Me 262 Volume 1* by J. Richard Smith and Eddie Creek 1997, p62
[2] ADRC/T-2 2050/1219 Messerschmitt Projektbaubeschreibung P 1065 Nr. 3 März 1940
[3] IWM Speer 76/746 C-Amts-Programm Lieferplan 1.7.40
[4] *Heinkel Raketen- und Strahlflugzeuge* by Volker Koos 2008, p129
[5] *Me 262 Volume 1* by J. Richard Smith and Eddie Creek 1997, p68
[6] *Heinkel Raketen- und Strahlflugzeuge* by Volker Koos 2008, p130
[7] ibid. p131
[8] *Me 262 Volume 1* by J. Richard Smith and Eddie Creek 1997, p72
[9] ibid.
[10] *Heinkel Raketen- und Strahlflugzeuge* by Volker Koos 2008, p134
[11] ibid. p135
[12] *Me 262 Volume 1* by J. Richard Smith and Eddie Creek 1997, p88
[13] *Heinkel Raketen- und Strahlflugzeuge* by Volker Koos 2008, p139
[14] ibid. p140
[15] NARA T321-59/807339 GL/C-Nr. 845/42 g.K. (E/J) St/GL – Befehl Nr. 480 Betrifft: Entwicklungs- und Beschaffungsprogramm "Vulkan" 10. Dez. 1942
[16] ADIK 158/0248 St/GL/C-E 2 Nr. 5223/43 g.Kdos Bericht Nr. 15 über die Entwicklungsbesprechung am 8.1.1943
[17] ADIK 158/0235 St/GL/C-E 2 Nr. 5514/43 g.Kdos Bericht Nr. 17 über die Entwicklungsbesprechung am 21.1.1943
[18] *Heinkel Raketen- und Strahlflugzeuge* by Volker Koos 2008, p147
[19] ADIK 158/0195 St/GL/C-E 2 Nr. 6482/43 g.Kdos. Bericht Nr. 23 über die Entwicklungsbesprechung am 19.3.43
[20] ADIK 158/0190 St/GL/C-E 2 Nr. 6629/43 g.Kdos. Bericht Nr. 24 über die Entwicklungsbesprechung am 22.3.43
[21] ADRC/T-2 2645/156 Messerschmitt Kurz-Projektbaubeschreibung Me 262 25. März 1943
[22] IWM GDC 10/5512 DFS Messungen mit dem Flugzeug He 280 V7 UM 3530/1
[23] *The First and the Last* by Adolf Galland, 1954 (2018 edition) p350
[24] IWM GDC 15/7 Messerschmitt Me 262 Verwendungsmöglichkeiten. Anhang zur Baubeschreibung vom 10.8.43 11.9.43

4. Light fighter
[1] ADRC/T-2 2217/585 Messerschmitt
[2] ADRC/MAP 5041/81 RLM GL/CE-2 – Festigkeitsprüfstelle Nr. 108/12.41 gKdos. Protokoll. Betr.: Lastannahmenvorbesprechung Me P 1079. Vorg.: Besprechung am 17.12.1941 Bln.-Adlershof, d. 30.12.41
[3] IWM Speer FD4355/45/493 GL/C-E 2 Nr. 1272/42 (I) geh. An die Firma Messerschmitt A.G. z.Hd. von Herrn Prof. Messerschmitt Betr.: P 1079 17.2.1942
[4] IWM Speer FD4355/45492 Messerschmitt An das RLM Herrn Generaling. Reidenbach P 1079 24.2.42
[5] ADRC/MAP 5012/56 Messerschmitt Flugzeug mit Strahlrohren 31.3.42
[6] ADRC/T-2 3216/75 Messerschmitt Me 328 als Bordflugzeug, unter dem Gesichtspunkt der Leistungssteigerung des Flugzeuges, sowie wirtschaftlicher und sparstoffscarender Möglichkeiten für den Tragwerksbau 7.7.1942
[7] Counterintuitively, it would later be given project number P 1085, then simply referred to as 'Me 264 6-motorig'
[8] TNA AIR 40/206 Messerschmitt Me 328 B Leichtes Schnellkampfflugzeug 15 Dezember 1942
[9] TNA AIR 40/2426 A.D.I.(K) Report No. 1/1946 The Messerschmitt Organisation and its Work During the War 7. Jan. 1946

5. 1-TL-Jäger part I
[1] TNA AIR 40/91 BIOS Interrogation Report No. 311, Interrogation of Dr. Ing. Otto Ernst Pabst, Wind Tunnel Supervisor and High Speed Aerodynamicist, for the Focke-Wulf Organisation at Bremen, on 11th June, 1945
[2] IWM GDC 16/2102 Focke-Wulf Jagdeinsitzer various presentation cards November 1942
[3] ADRC/T-2 2008/279 Focke-Wulf Bericht Nr. 03028 Grundlagen zum Entwurf eines Jägers mit Turbinentriebwerk by J. Rotta 4.1.43
[4] ADRC/T-2 2264/387 Focke-Wulf Kurzbeschreibung Jagdflugzeug mit Turbinentriebwerk BMW-P3302 10.3.43
[5] ADRC/T-2 2047/844 Focke-Wulf Kurzbeschreibung Jagdflugzeug mit Turbinentriebwerk Jumo 109 004 C 10.3.43
[6] There is no known forward view of the '1. Entwurf', meaning that the precise position of its undercarriage is open to speculation
[7] ADIK 94/0083 Focke-Wulf Betrachtung zum Entwurf eines einmotorigen Jagdflugzeuges mit TL-Triebwerk 15.8.44
[8] IWM GDC 14/205 Folder containing loose papers from Arado Brandenburg, dates ranging from February 18, 1943 to April 15, 1943
[9] IWM GDC 14/200 Vorschlag für die Weiterentwicklung schneller Zweisitzer, Arado Flugzeugwerke, Brandenburg/Havel, August 11, 1943
[10] ADIK 158/142 Der Staatssekretär der Luftfahrt und Generalinspekteur der Luftwaffe St/GL/C-E 2 Nr. 7696/43 g.Kdos. Bericht Nr. 28 über die Entwicklungsbesprechung am 28.5.1943
[11] IWM GDC 18/125 Heinkel list of project aircraft by number
[12] ADRC/T-2 2012/99 Focke-Wulf Baubeschreibung Nr. 264 Jäger mit Junkers Turbinentriebwerk 9.6.43
[13] ADRC/T-2 2414/0649 Focke-Wulf Plan 126 Pause 02 Strahl-Jäger 26.7.43
[14] ADIK 94/0083 Focke-Wulf Betrachtung zum Entwurf eines einmotorigen Jagdflugzeuges mit TL-Triebwerk 15.8.44
[15] T-2 Translation Report No. F-TS-460-RE Description of Construction No. 272 Single-Engine Jet Fighter With Rocket Unit November 11, 1946
[16] ADIK 109/369 Focke-Wulf drawing number 0310 226-92
[17] ADRC/T-2 2414/0496 RLM GL/C-E 2 Protokoll Festigkeitsvorschriften FW-Strahljäger Berlin-Adlershof, den 14.6.1944
[18] ADRC/T-2 2009/87 Messerschmitt Technischer Bericht Nr. 90/43 Leistungsvergleich von TL-Jägern 3.7.43
[19] ADRC/T-2 2604/658 Messerschmitt / JSF drawings of the Me 328 T and C September 1943
[20] IWM Speer FD4355/45/426 Messerschmitt Herrn. Dipl.-Ing. Seitz – Probü – Ihre "Stellungsnahme der Entwicklungsleitung Me 328 zum Stoppbefehl vom 3.9.43" 29.9.43
[21] ADRC/T-2 4064/61 Messerschmitt P 1095 Vorbemerkung undated

NOTES CONTINUED

6. 1,000 x 1,000 x 1,000

[1] ADRC/T-2 2256/643 Luftfahrt-Forschung Wien Projektbaubeschreibung Versuchsflugzeug für Hochgeschwindigkeit 29. Mai 1943
[2] *Conversations with Reimar Horten Volume #2* by David Myhra 2016, p162
[3] *Conversations with Reimar Horten Volume #3* by David Myhra 2016, p219
[4] *Conversations with Reimar Horten Volume #2* by David Myhra 2016, p164
[5] *Conversations with Reimar Horten Volume #1* by David Myhra 2016, p27
[6] IWM Milch Volume 62/5598 Besprechungsnotiz Nr. 90/43 g.Kdos. vom 28.9.1943 Reichsjägerhof
[7] *Conversations with Reimar Horten Volume #3* by David Myhra 2016, p225
[8] ibid. p207
[9] ibid. p218
[10] *Conversations with Reimar Horten Volume #2* by David Myhra 2016, p168
[11] *Conversations with Reimar Horten Volume #3* by David Myhra 2016, p271-272
[12] USSAFE Technical Intelligence Report No. I-68 Notes on the Interrogation of Dr Rudolf Goethert with particular reference to the Gotha P-60 Jet Propelled Flying Wing 7 July 1945
[13] TNA AIR 40/3129 A.I.2(G) Report No. 2383 German Aircraft: New and Projected Types, January 1946
[14] ADRC/T-2 2809/689 Gothaer Waggonfabrik Vorläufige Baubeschreibung 22.11.44
[15] *Conversations with Reimar Horten Volume #3* by David Myhra 2016, p246-7
[16] TNA AIR 40/175 109 Entwicklungs-Sonderkommission Flugzeugzellenbau Prof. Dr.-Ing. G. Bock Langstreckenbomber Projekt-Vergleich 2. Bericht 25.2.45
[17] ADRC/T-2 2809/34 Horten Flugzeugmuster 8-229 Zerstörer 1.3.45
[18] ADRC/T-2 3282/270 Prof. Dr. Ing. Günther Bock – Herrn. Dipl.-Ing. Wehrse, Focke-Wulf Flugzeugbau 12.3.45
[19] ADRC/T-2 4069/556 Junkers TA.-E- Mitteilung (Sondertriebwerk) Triebwerkeinbau in Go 229 (Horten) (V3 + V5) 7. März 1945
[20] BA-MA RL3-2567 KTB der Chef TLR Berichtsraum 16.3/4.4.1945
[21] IWM GDC 2817T Glossary of German Aeronautical Codes, Models, Project Numbers, Abbreviations, Etc. Final Edition November 1947
[22] ADRC/T-2 4007/393 Gothaer Waggonfabrik Go P – 60 Hochgeschwindigkeitsflugzeug 11 März 1945
[23] DSIR 23/14733 Aeronautical Research Council Performance Sub-Committee Research and Aeroplane Development Work done by Dr Alexander Lippisch. Extracts from Technical Intelligence Report A.424 13th August, 1945
[24] *Erinnerungen* by Alexander M. Lippisch 1976, p216
[25] TNA AIR 40/3129 A.I.2(G) Report No. 2383 German Aircraft: New and Projected Types, January 1946
[26] ADRC/T-2 2011/341 OMW-Kobü Sondertriebwerke Aktenvermerk Betr.: Einbau des 004 Gerätes in Delta VI 2 2.12.1944
[27] ADRC/T-2 2908/392 Arado file of notes and sketches by Hans Walland circa March 1943 to September 1943
[28] Prüfbericht über 3- und 6-Komponentenmessungen an der Zuspitzungsreihe von Flügeln kleiner Stockung (Teilbericht: Dreieckflügel) by Lange and Wacke, UM 1023/5, 27.9.1943 via NACA Technical Memorandum No. 1176 Translation May 1948

7. Messerschmitt Me 262 later developments

[1] ADRC/T-2 2263/872 Messerschmitt Baubeschreibungen der Weiterentwicklungen der Me 262 (mit neuem vergrössertem Rumpf) P 1099 (mehrsitzige schwere Jäger) P 1100 (mehrsitzige Bomber) 22.3.44
[2] ibid.
[3] ADRC/T-2 2251/1036 Messerschmitt P 1099 (Me 262 mit vergrössertem Rumpf) Baubeschreibung für Jäger (zweisitzig) 22.3.44
[4] ADRC/MAP 5026/121 Messerschmitt Büro-Darmstadt Aktenvermerk Betrifft: Hochgeschwindigkeitsentwicklung Me 262 Bericht für die Zeit vom 10.1. bis 17.3.1944
[5] ADRC/MAP 5026/230 Sketch signed by Willy Messerschmitt 27.3.44
[6] ADRC/MAP 5026/122 Messerschmitt Oberammergau Aktenvermerk Betrifft: Hochgeschwindigkeitsentwicklung Me 262 Bericht für die Zeit vom 18.3. bis 5.4.1944
[7] ADRC/MAP 5026/282 Messerschmitt Me 262 Projektübergabe V (Hochgeschwindigkeitsentwicklung) 3.4.44
[8] ADRC/MAP 5026/122 Messerschmitt Oberammergau Aktenvermerk Betrifft: Hochgeschwindigkeitsentwicklung Me 262 Bericht für die Zeit vom 18.3. bis 5.4.1944
[9] ADRC/MAP 5026/123 Messerschmitt Aktenvermerk Betr.: Hochgeschwindigkeitsentwicklung Me 262. Bericht für die Zeit vom 6.4. bis 17.4.44
[10] ADRC/MAP 5026/118 Messerschmitt Aktenvermerk Betr.:
[11] TNA AIR 40/2426 ADI(K) Report No. 1/1946 The Messerschmitt Organisation and its Work During the War

8. 1-TL-Jäger part II

[1] ADRC/T-2 115/0221 Focke-Wulf Baubeschreibung Nr. 280 Einmotorige TL-Jäger mit R-Gerät 5.7.44
[2] ADRC/T-2 3997/583 Focke-Wulf Entwicklungsmitteilung TL-Jäger Betr.: Allgemeine Festlegung für den Anlauf TL-Jäger, Rodde 11.7.44
[3] ADIK 94/0084 Focke-Wulf Betrachtung zum Entwurf eines einmotorigen Jagdflugzeuges mit TL-Triebwerk 15.8.44
[4] ADRC/T-2 2008/187 Focke-Wulf Baubeschreibung Nr. 281 Einmotoriger Jäger mit PTL-Gerät 021 18.8.44
[5] NARA T321-59/807004 Der Reichsminister für Rüstung und Kriegsproduktion Jägerstab Punkte aus Besprechung beim Herrn Reichsmarschall am 1.7.44
[6] *Luftfahrt International* Nr. 24 Nov.-Dez. 1977 3691 Heinkel P 1073 Schneller Strahljäger 10. Juli 1944
[7] *Die Strahlflugzeuge der Ernst Heinkel Flugzeugwerke* by Volker Koos 2008, p178
[8] ADRC/T-2 2319/758 Messerschmitt handwritten notes, calculations and tables on the P 1101 signed by Hans Hornung
[9] *Die Strahlflugzeuge der Ernst Heinkel Flugzeugwerke* by Volker Koos 2008, p178
[10] ADRC/T-2 2414/0527 GL/C-E Protokoll Nr. 1 (893/44) Vergleich der TL-Jäger-Entwürfe von Focke-Wulf, Heinkel und Messerschmitt Arbeitstagung der Firmen B.V., F.W., EHAG u. Mtt. 10.9.44
[11] ADRC/T-2 3569/68 Chef TLR Fl. E 2 Nr. 41343/44 g.Kdos (III 3a) Betr.: 1 TL-Jäger mit HES 011. 4.12.44
[12] BA-MA RL3-2567 KTB der Chef TLR Berichtswoche 13.12.-19.12.1944
[13] ADRC/T-2 2151/239 Focke-Wulf Baubeschreibung Nr. 279 Einmotoriger TL-Jäger mit He S 011 8.11.44
[14] ADRC/MAP 5015/493 Messerschmitt Aktennotiz Betrifft: Ausstattung P 1101 Interne Probü-Besprechung am 31.10.44
[15] ADRC/MAP 5015/497 Messerschmitt Vorläufiger Ausstattungsplan, P 1101-Serienausführung P 1101 2.11.44
[16] IWM GDC 15/16 Messerschmitt Projektübergabe P 1101 10.11.44
[17] IWM GDC 15/265 Messerschmitt Versuchs-Bericht Nr. 262 28 L 44 Standschub Jumo TL mit 3m Einlauf 23.11.44
[18] ADRC/T-2 2905/146 DVL Untersuchungen und Mitteilungen Nr. 1448 Abgleich der Jäger-Projekte mit HeS11 TL 9.1.45
[19] ibid.
[20] ADRC/T-2 4004/632 Blohm & Voss Kurzbeschreibung – Schnellst-Jäger mit HeS 011 P 209 November 1944
[21] IWM GDC 23/16 Blohm & Voss Kurzbeschreibung – Schnellst-Jäger mit HeS 011 P 212 – Schwanzlos mit Randleitwerken November 1944
[22] BA-MA R3/1749(FD4955/45)/127 Entwicklungshauptkommission Flugzeuge Br.Nr. 407/44 g.Kdos. Ergebnisse der Sitzungen am 19. und 20.12.1944
[23] ADRC/T-2 2905/146 DVL Untersuchungen und Mitteilungen Nr. 1448 Abgleich der Jäger-Projekte mit HeS11 TL 1. Teil 9.1.45
[24] ADRC/MAP 5032/24 Oberkommando der Luftwaffe Generalquartiermeister Nr. 13215/44 g.Kdos. (6.Abt III T) 2. Ang. Betr.: Entwicklungsforderung für Jagdflugzeuge 11.1.45
[25] BA-MA RL3-2567 KTB der Chef TLR Berichtsraum 8.1.-14.1.1945
[26] ADRC/T-2 8043/473 DVL Bericht über den TL Jäger Vergleich in der DVL 19.-21. Dez. 1944 und 12.-16. Januar 1945
[27] ADRC/MAP 5041/345 Obb. Forschungsanstalt Technischer Bericht TB-Nr. 139/45 B 5 Kurzbeschreibung des Projektes der Fa. Focke-Wulf II (nach Kurzbaubeschreibung Nr. 30) 26.2.45
[28] ADIK 94/0221 Fw-Bezeichnungen für rechnerische Unterlagen Strahljäger 11.12.44, amended 15.2.45
[29] ADIK 111/0500 DVL Untersuchungen und Mitteilungen Nr. 1448 Abgleich der Jäger-Projekte mit HeS11 TL 2. Teil 26.2.45
[30] ADIK 124/0555 Heinkel Bericht-Nr. 114/45
[31] BA-MA Reichsministerium für Rüstung und Kriegsproduktion R3/1749 FD4955/45 219 2. Arbeitstagung der Entwicklungshauptkommission Flugzeuge im Fährhaus

Tornow, Potsdam 23.1.45 / 24.1.45
[32] ADIK 111/0500 DVL Untersuchungen und Mitteilungen Nr. 1448 Abgleich der Jäger-Projekte mit HeS11 TL 2. Teil 26.2.45
[33] ADRC/T-2 2433/935 Messerschmitt Projektbüro Technischer Bericht TB Nr. 139/45 Vergleich der der Entwicklungs-Hauptkommission zur entscheidung eingereichten Ein- TL-Jäger (Vorarbeit für die Sitzung v.27.u.28.2.45)
[34] IWM GDC 15/835 Messerschmitt Projektbüro Technischer Bericht TB Nr. 139/45 Vergleich der der Entwicklungs-Hauptkommission zur entscheidung eingereichten Ein- TL-Jäger (Vorarbeit für die Sitzung v.27.u.28.2.45)
[35] ADRC/T-2 3282/270 Prof. Dr. Ing. Günther Bock – Herrn. Dipl.-Ing. Wehrse, Focke-Wulf Flugzeugbau 12.3.45
[36] BA-MA RL3-2567 KTB der Chef TLR Berichtswoche 26.2.-4.3.1945
[37] BA-MA RL3-2567 KTB der Chef TLR Berichtswoche 5.3.-15.3.1945
[38] ADIK 94/0210 Focke-Wulf Technische Zentrale Mitteilung Betr.: TL-Jäger, Bad Eilsen, den 14.4.1945
[39] ADIK 91/0385 Fw-Bezeichnungen für rechnerische Unterlagen Strahljäger 8-183 17.3.45
[40] GDC 735 M.O.S. (A) Völkenrode Reports and Translations No. 730 Bericht über die Entwicklung der Henschel Flugzeugwerke Nicolaus October 1946
[41] ADRC/T-2 3429/1027 Messerschmitt 1 TL-Jäger Allgemein
[42] TNA AIR 40/3129 A.I.2(G) Report No. 2383 German Aircraft: New and Projected Types January 1946
[43] TNA DSIR 23/15140 Interrogation report No. 9 Investigation of Oberbayerische Forschungsanstalt, Oberammergau, June 8, 9 and 10, 1945
[44] TNA AIR 40/2426 A.D.I.(K) Report No. 1/1946 The Messerschmitt Organisation and its Work During the War January 1946
[45] BA-MA RL3-2567 KTB der Chef TLR Berichtswoche 16.3.-4.4.1945
[46] *He. 1000* by Ernst Heinkel, 1956, p276
[47] IWM GDC 18/200T T-2 Translation Report No. F-TS-670-RE Single Place Fighter With Turbo-jet HeS-11 P 1078 (Heinkel Report) Jagdeinsitzer mit Strahltriebwerk HeS-11 by Eichner and Hohbach September 14, 1945

9. Volksjäger
[1] ADIK 117/0886 Focke-Wulf Einstrahlige Jägerprojekte 20.9.44. See also ADRC/T-2 2455/0773 Blohm & Voss Aktenvermerk über Projektarbeiten für den Volksjäger und die geführten Besprechungen, Dr Vogt 12.10.44. There is evidence which suggests the requirement may also have been sent to Dornier – the company was copied in on the minutes of the first round of discussions concerning the projects, referenced in note 9 below.
[2] ADRC/T-2 2455/751 Berlin RLM Nr. 2267 Betr.: Projekt 1 TL-Jäger 10.9.44
[3] *Luftfahrt International* Nr. 24 Nov.-Dez. 1977 3697 Heinkel Kleinst-Jäger P 1073 mit BMW 003 11.9.44
[4] ADRC/T-2 2455/0773 Blohm & Voss Aktenvermerk über Projektarbeiten für den Volksjäger und die geführten Besprechungen, Dr Vogt 12.10.44
[5] ADRC/T-2 2441/61 Blohm & Voss Kurzbeschreibung BV P 210 und P 211 Jäger mit BMW 109-003. 12.9.44
[6] *Die Entwicklung der deutschen Jagdflugzeuge* by Rüdiger Kosin 1983, p195
[7] IWM GDC 14/409 Arado Jagdflugzeug E 580 12.9.44
[8] ADRC/T-2 2455/0773 Blohm & Voss Aktenvermerk über Projektarbeiten für den Volksjäger und die geführten Besprechungen, Dr Vogt 12.10.44
[9] ADRC/T 2 2455/0749 TLR/Fl-E 2 Nr 9776/44 (IIIA) g.Kos. Besprechungsniederschrift Betr.: Jagdeinsitzer 15.9.44
[10] ADIK 91/433 Focke-Wulf Aktenvermerk Betr. Rollstrecken beim "Volksflugzeug" und "Volksflitzer". 21.9.44
[11] ADRC/T-2 2395/0210 Focke-Wulf Flitzer mit BMW 003-A1 15.9.44
[12] ADRC/T-2 2395/0152 Focke-Wulf project notes 16.9.44
[13] ADIK 117/0886 Focke-Wulf Einstrahlige Jägerprojekte 20.9.44
[14] IWM FD3049/49 Folder No. 3 Führer Protokolle 1942-1945
[15] ADRC/T-2 2455/0743 Berlin RLM Nr. 6553 29/9 1905= Betr.: Kleinstjäger
[16] ADRC/T-2 2455/0773 Blohm & Voss Aktenvermerk über Projektarbeiten für den Volksjäger und die geführten Besprechungen, Dr Vogt 12.10.44
[17] ADRC/T-2 2287/0169 Blohm & Voss P 211 Volksjäger Baubeschreibung 29.9.44

10. Objektschutzjäger
[1] ADRC/T-2 2438/638 RLM GL/C-E 2 Nr. 13546/43 (III C) geh. Betr.: Gleitjäger 19.8.43
[2] ADRC/MAP 5032/84 General der Jagdflieger Brb.Nr. 2145/43 gKdos (T2) Btr.: Rammen feindlicher 4-motoriger Kampfflugzeuge 13.11.43
[3] ADRC/T-2 2438/629 Blohm & Voss Gleitjäger P 186 Aktenvermerk St 14 Lastannahmen-Entwurf 29.11.43
[4] ADRC/T-2 2438/624 Blohm & Voss note Büro 25 Herrn Dibbern 15.12.43
[5] ADRC/T-2 2438/589 Blohm & Voss Fernschreiben Betr.: Gleitjäger 16.3.44
[6] ADRC/T-2 2438/586 Blohm & Voss FS Nr. 215 20.4.44
[7] ADRC/T-2 2138/435 Ramm-Raketen nach Dr Alexander Lippisch, Wien 9.9.43
[8] ADRC/T-2 2154/50 AVA Bericht 42/W/57 Windkanalmessungen an einem Pfeilflugzeug Me 163 C (mit unverwundenem Flügel) 12.2.1943
[9] ADRC/T-2 2156/782 AVA Bericht 42/W/60 Untersuchung eines Gesamtmodells Me 163 B mit verlängertem Rumpf 9.2.1943
[10] *Me 163 Rocket Interceptor Volume Two* by Stephen Ransom and Hans-Hermann Cammann, Classic Publications 2003, p364
[11] ADRC/T-2 BMW Terminablaufplan HS/Sk 225b 22.9.43
[12] *Me 163 Rocket Interceptor Volume One* by Stephen Ransom and Hans-Hermann-Cammann, Classic Publications 2002, p97
[13] ADIK 158/0615 GL/C-Nr. 6723/44 gKdos Entwicklung von kriegswichtigen und Schwerpunktsgeräten 12.5.44
[14] *Natter – Manned Missile of the Third Reich* by Brett Gooden, 2019, p23
[15] CIOS XXX-107 Natter Interceptor Project by Dr C. B. Millikan NavTecMisEu July 1945
[16] ADRC/T-2 2691/455 Projekt Natter (BP-20/Barak I), Bachem-Werk 9.8.44
[17] ADRC/T-2 4003/29 Projekt Natter (BP-20/Barak I), Bachem-Werk 20.9.44/5.10.44
[18] ADRC/T-2 2256/664 Messerschmitt Rechnung am Projekt P 1104, circa September 1944
[19] IWM FD4355/45 Vol. 1/367 Willy Messerschmitt to Erich Bachem 6.10.44
[20] *Heinkel Raketen- und Strahlflugzeuge* by Volker Koos, Aviatic Verlag, 2008, p244
[21] *Me 163 Rocket Interceptor Volume One* by Stephen Ransom and Hans-Hermann-Cammann, Classic Publications 2002, p102
[22] *Me 163 Rocket Interceptor Volume Two* by Stephen Ransom and Hans-Hermann-Cammann, Classic Publications 2003, p366
[23] ADRC/T-2 2012/36 Junkers Kurzbaubeschreibung 248 3.11.44
[24] ADIK 149/739 Junkers Baubeschreibung 263 15.12.44
[25] NavTecMisEu Serial 00276 Technical Report #342-45 Enclosure 1: Heinkel Bericht Nr. 111/44 'Julia' 16.11.44
[26] ADRC/T-2 3415/716 Heinkel Allgemeine Geräteliste (AG Liste) zu 5035 Rüstzustand M1 u. M2 24.11.44/16.1.45
[27] A.I.2(G) Report No. 2383 German Aircraft: New and Projected Types by H. F. King January 1946
[28] ADRC/T-2 3429/672 Junkers Unterlagen über EF 126, 127, He 162, Me 163 B, 248, Julia und Natter 16.12.44
[29] BA-MA R3/1749 Entwicklungshauptkommission Flugzeuge Br. Nr. 407/44 g.Kdos. Ergebnisse der Sitzungen am 19. und 20.12.1944
[30] Chef TLR Fl-E 2 Nr. 16834/44 geh. Lagebericht der Fachabteilung Fl-E 2 21.12.1944
[31] BA-MA RL3-2567 KTB der Chef TLR Berichtswoche 27.12.44-7.1.45
[32] IWM FD3049/49 Volume 1/039 Niederschrift des am 7. Juni 1945 erfolgten Interview zwischen Vertretern des Alliierten Oberkommandos und Herrn Saur
[33] *Natter: Manned Missile of the Third Reich* by Brett Gooden, 2019, p332
[34] *Me 163 Rocket Interceptor Volume Two* by Stephen Ransom and Hans-Hermann-Cammann, Classic Publications 2003, p370
[35] ADIK 124/0652 Heinkel Aktenvermerk Nr. 56/44 Bericht über wie wichtigsten Entwicklungsfragen in Wien (Stand Jahresende 1944) 29.12.44
[36] ADIK 124/0490 Heinkel Frankce Gespräch mit Berliner Büro 27.1.1945
[37] ADIK 124/0359 Heinkel Nr. 69/45 Franche an Herrn Jost – Herrn Benz 16.2.45
[38] ADIK 124/0354 Heinkel Francke Gespräch mit Berliner Büro am 17.2.45
[39] ADIK 124/0330 Heinkel Nr. 77/45 Francke an Herrn Jost 21.2.45

NOTES CONTINUED

[40] ADIK 124/0327 Heinkel Nr. 76/45 Francke an Huffziger 21.2.45
[41] ADIK 124/0265 Heinkel Nr. 93/45 Francke an Braunn 3.3.45
[42] ADIK 124/0214 Heinkel Besprechung am 5.3.45 Julia 6.3.45
[43] ADIK 124/0139 Heinkel TD Dir Francke an Herrn Prof. Dr. Heinkel 28.3.1945

11. Jäger mit Lorinantrieb
[1] ADIK 99/0160 Luftfahrtforschungsanstalt Hermann Göring Bericht Nr. 996 Wirkungsgrade und Grössenverhältnisse von Lorinmotoren by E. Sänger August 1938
[2] CIOS XXXII-66 Deutsche Forschungsanstalt für Segelflug Ainring by Lt Colonel John A. O'Mara section B.4 June 10, 1945
[3] ADIK 89/24 DFS Über einen Lorinantrieb für Strahljäger by E. Sänger und I. Bredt, Ainring (UM 3509) Oktober 1943
[4] ADIK 114/0050 Kurzbeschreibung des vorgeschlagenen Jabo bzw. Jägers mit Lorinantrieb 15.3.44
[5] J. Richard Smith collection ref. A0118 Vorschlag zur Kombination eines 20000-PS-Lorin rohres mit der Lippisch-Delta VI-Zelle, signed by Sänger 18.4.44
[6] *Erinnerungen* by Alexander M. Lippisch, Luftfahrtverlag Axel Zuerl, 1976
[7] Iowa State University Library Special Collections and University Archive MS 243 Box 19, Folder 23
[8] Ibid.
[9] A.I.2(g) Report No. 2383 German Aircraft: New and Projected Types by H.F. King January 1946
[10] TNA AIR 40/92 USSAFE Technical Intelligence Report No. I-82 Interrogation of Dr Alexander Lippisch 28.7.45
[11] *Erinnerungen* by Alexander M. Lippisch, Luftfahrtverlag Axel Zuerl, 1976 and Ein Dreieck Fliegt by Alexander Lippisch, Motorbuch-Verlag, 1976
[12] TNA AIR 40/2001 ADIK Report No. 14/1946 Tailless aircraft development by Lippisch
[13] Ibid.
[14] A.I.2(G) Report No. 2383 German Aircraft: New and Projected Types by H. F. King January 1946
[15] *Ein Dreieck Fliegt* by Alexander Lippisch, Motorbuch-Verlag, 1976, p104
[16] Chef TLR Fl-E 2 Nr. 16834/44 geh. Lagebericht der Fachabteilung Fl-E 2 21.12.1944
[17] A.I.2(G) Report No. 2383 German Aircraft: New and Projected Types by H. F. King January 1946
[18] Iowa State University Library Special Collections and University Archive MS 243 Box 19, Folder 21
[19] ADRC/T-2 3563/628 Luftfahrtforschung Wien drawing SK-P13-6/001 February 1945
[20] *Ein Dreieck Fliegt* by Alexander Lippisch, Motorbuch-Verlag, 1976, p98
[21] *Erinnerungen* by Alexander M. Lippisch, Luftfahrtverlag Axel Zuerl, 1976, p215 to p221
[22] CIOS Evaluation Report 68, Flugtechnisches Fachgruppe (Target No. 5/49) by H. C. Ashley and Major C. F. White, June 7, 1945
[23] TNA AIR 40/91 USSAF Interview 21.5.45 Dr Kurt Tank, Otto Pabst & H. Multhopp
[24] TNA AIR 40/91 ADIK No. 376 – 21.8.45 Propulsive Ducts (Lorin Ducts)
[25] ADIK 198/113 Focke-Wulf 2 Strahlrohr Jäger note 11.2.44
[26] ADIK 198/112 Focke-Wulf Strahlrohr Jäger note 11.2.44
[27] ADIK 198/100 Focke-Wulf Strahlrohr nach Pabst note 24.4.44
[28] ADIK 198/84 Focke-Wulf Strahlrohrjäger X5 note
[29] ADIK 198/40-70 Focke-Wulf Strahlrohrjäger X6 notes
[30] ADRC/T-2 2011/1076 Focke-Wulf Baubeschreibung Nr. 283 4.8.44
[31] ADRC/T-2 3683/1000 Focke-Wulf Lorin-Rohr als Flugzeugtrieb
[32] ADIK 198/131 Focke-Wulf Strahlrohr für Fw 190 A-10 23.6.44
[33] *Helicopters of the Third Reich* by Steve Coates with Jean-Christophe Carbonel, Classic Publications, 2002, p192
[34] Der Triebflügel, E. von Holst, D. Küchemann, K. Solf, Jahrbuch der Lufo 1942, S. I 435
[35] ADIK 89/0002 Focke-Wulf Bericht Nr. 09 042 Leistungsblätter zum Lorin-Triebwerk für das Triebflügelflugzeug by Pabst, March 1944
[36] ADRC/T-2 2008/275
[37] ADRC/T-2 2008/273
[38] A.I.2(G) Report No. 2383 German Aircraft: New and Projected Types by H. F. King January 1946
[39] IWM GDC 10/5519 Deutsche Forschungsanstalt für Segelflug E.V. „Ernst Udet", z. Zt. Ainring, Leistungssteigerung der Me 262 durch Lorin-Zusatzantrieb, by E. Sänger, 31.1.45
[40] CIOS Evaluation Report 118 Enclosure Deutsche Forschungsanstalt für Segelflug, Institute, Ainring (25/7, 5/49, 25/74), by Major R. L. Sadler, F/O Price and F/Lt Block 6th Army Group, Item Group IV, May 25, 1945
[41] ADRC/T-2 2211/10 Skodawerke Baubeschreibung Skoda-Jagdflugzeugentwurf SK P 14 mit Saenger-Lorin-Strahlrohr / 1.5m Brennkammerdurchmesser 23.3.45
[42] ADRC/T-2 2211/1 Skodawerke Zusammenstellung Betrifft: Skoda-Projekt SK P14 24.3.1945
[43] IWM GDC 18/193T T-2 Translation Report No. F-TS-674-RE 15.7.1946 Single Place Fighter with Lorin-Jet Heinkel Report by S. Günter

12. Night and bad-weather fighters
[1] Although it has been suggested that Arado itself proposed a night fighter version of the Ar 234 in mid-1943. See *Arado – History of an Aircraft Company* by Jörg Armin Kranzhoff, Schiffer, 1997, p150
[2] Rechlin Protokoll Nr. 2612 über die Besprechung vom 13.12.43 in Rechlin betr: Ar 234. Rechlin 15.12.43 via Luftfahrt dokumente LD 21 Arado Ar 234 1976 compiled by Karl R. Pawlas p57
[3] Arado Protokoll Nr. 2023 (lfd. Nr. 29) über die Besprechung am 3.6.44 in Landeshut via *Luftfahrt dokumente LD 21 Arado Ar 234* 1976 compiled by Karl R. Pawlas p141
[4] ADRC/T-2 2101/0185 Messerschmitt Abt. Flugerprobung Gruppe Leistungen Versuchs-Bericht Nr. 262 14 L 44 21.7.44
[5] ADIK 117/0731 Focke-Wulf Vorläufige techn. Bedingungen für einen Schlechtwetter- Tag- und Nachtjäger 4.8.44
[6] ADIK 117/0909 FL E 2 Nr. 14352/44 (IIIB) geh. Berlin Betr.: Projekt neuer Zerstörer 28.8.44
[7] ADRC/T-2 2766/95 Focke-Wulf Protokoll Vergleich der Entwürfe des Hochleistungsjägers mit Jumo 222 E/F von den Firmen: Blohm u. Voss, Dornier-Werke, Focke-Wulf 20.9.44
[8] ADRC/T-2 2277/753 Focke-Wulf Kurzbeschreibung Nr. 16 Hochleistungsjäger mit As 9-413 3.10.44
[9] ADRC/T-2 3607/1143 Focke-Wulf Nacht- und Schlechtwetterjäger Besuch von Herrn Emmert am 10.11.44
[10] ADRC/T-2 2022/412 Focke-Wulf Kurzbeschreibung Nr. 21 Nachtjäger und Schlechtwetterzerstörer Jumo 222 E/F und 2 x BMW 003 23.11.44
[11] ADIK 111/0548 Focke-Wulf Kurzbeschreibung Nr. 23 Zweimotoriges TL-Jagdflugzeug mit HeS 109-011 23.11.44
[12] See Chapter 14
[13] ADRC/T-2 2011/203 Messerschmitt Projektübergabe Me 262 Nachtjäger (Einbau Schulflugzeug Stufe I) 5.10.44
[14] *Me 262 Volume Two* by J. Richard Smith and Eddie Creek, Classic Publications, 1998, p294
[15] ADIK 102/0094 OKL Chef TLR F2-EN IV 20.2.45
[16] For the Me 262 B2, Messerschmitt appears to have dropped the usual hyphen 'between the 'B' and '2'. Why this should have been the case is unclear but documents do usually refer to the 'B2' rather than the 'B-2'
[17] ADRC/T-2 2094/0516 Messerschmitt FT-Lageplan für Nachtjäger Me 262-Serie 20.10.44
[18] ADRC/T-2 2009/2 Messerschmitt Me 262 B2 Nachtjäger Projektbaubeschreibung vom 18.1.1945
[19] ADRC/T-2 2660/359 Arado Gewichtsübersicht – Nachtjäger 8-234 19.1.1945
[20] Tank's Sonderkommission appears in period documents variously as the Sonderkommission Nachtjagd-Flugzeuge, Sonderkommission Nachtjäger, Entwicklungssonderkommission „Schlechtwetter- u.Nachtjagd", E.S.K. Nachtjagd and more. Evidently the special commissions system had been so rushed into existence it lacked even the basic formality of a consistent name
[21] Oberkommando der Luftwaffe Generalquartiermeister Nr. 13215/44 g.Kdos. (6.Abt.III T) 2. Ang. Betr.: Entwicklungsforderung für Jagdflugzeuge 11.1.45
[22] ADIK 75/0146 Dornier Baubeschreibung Nr. 1600 Nachtjäger Do 335 A-6 mit 1 x DB 603 E und 1 x DB 603 QE 20.11.1944
[23] ADRC/T-2 2011/244 Focke-Wulf Kurzbeschreibung Nr. 26 Nachtjäger mit 2 x HeS 011 27.1.45

NOTES CONTINUED

[24] ADRC/MAP 5023/22 Fl. E F 1 Nr. 12376/45 (III E) g. Kdos. Betr.: Technische Richtlinien für einen Schlechtwetter- und Nachtjäger 27.1.1945
[25] ADRC/T-2 2438/0495 Blohm & Voss Gewichtsaufstellung P 215.01 22.2.45
[26] ADIK 75/0114 Messerschmitt Me 262 Nachtjäger mit HeS 011-Triebwerken Projektbeschreibung 12.2.1945
[27] ADRC/T-2 2017/0448 Focke-Wulf Kurzbeschreibung Nr. 27 Nachtjäger mit 2 x HeS 011 7.2.45
[28] ADRC/T-2 Focke-Wulf Nachtjäger mit 2 x HeS 011 26.2.45
[29] ADRC/T-2 2965/1079 Entwicklungssonderkommission „Schlechtwetter- u. Nachtjagd" Arbeitsgruppe Flugmechanik Betr.: Me 262 – Nachjäger, B u. V. P 21501, Do P 252/1 26.2.45
[30] ADRC/T-2 4007/422 Dornier Baubeschreibung Nr. 1608 Do P 252/1 Zerstörer für Tag- und Nachteinsatz 27.1.45
[31] BA-MA RL3-2567 KTB der Chef TLR Berichtswoche 26.2.-4.3.1945
[32] ADIK 117/0908 Abschrift dH/5 1.3.45 Der Chef der Techn. Luftrüstung Fl. E Nr. 12376/45 (E 2 III B) g.Kdos. 2. Angel. Betr.: Technische Richtlinien für einen Schlechtwetter- und Nachtjäger
[33] ADIK 75/0007 Dornier Baubeschreibung Nr. 1615 Jäger für Schlechtwetter- und Nachteinsatz Do P 256/1 mit 2 TL-Geräten 109-011.
[34] IWM GDC 16/110 Nachtjäger mit HeS 011 Entwurf II-V 19.3.45
[35] ADRC/T-2 2041/111 Focke-Wulf Flugzeugdaten Entwurf VI 20.3.45
[36] ADRC/T-2 4007/393 Gothaer Waggonfabrik Go P-60 Hochgeschwindigkeitsflugzeug 11.3.45
[37] ADRC/T-2 2904/945 Messerschmitt Baubeschreibung Me 262 B2 Nachtjäger mit 2 x HeS 011 Triebwerken 17.3.45
[38] IWM GDC 15/647 Messerschmitt Projektbaubeschreibung 3 sitziger Nachtjäger mit 2 x HeS 011 Triebwerken 17.3.45
[39] IWM GDC 16/42 Focke-Wulf Entwürfe für Nacht- u. Schlechtwetterjagd 23-25.3.45
[40] ADRC/T-2 8037/1049 Night Fighter and Destroyer Fitted With Two HeS-011 Power Plants P 1079
[41] BA-MA RL3-2568 Der Chef der Technischen Luftrüstung Fl. E Nr. 12376/45 (2/III) g.Kdos. 3. Ang. Betr.: Technische Richtlinien für einen Schlechtwetter- und Nachtjäger April 1945

13. Pulsejet fighters
[1] NavTecMisEu Serial 00276 Technical Report #342-45 Enclosure 1: Heinkel Bericht Nr. 111/44 'Julia' 16.11.44
[2] ADRC 8051/132 Entwicklungs-Notprogramm chart, February 1945
[3] Fl-E 2 Nr. 16834/44 geh. Lagebericht der Fachabteilung Fl-E 2 21.12.44
[4] BA-MA RL3-2567 KTB der Chef TLR Berichtswoche 15.1.-21.1.1945
[5] BA-MA RL3-2567 KTB der Chef TLR Berichtswoche 5.3.-15.3.1945
[6] ADRC 2438/0488 Hermann Teichgraeber – Werk Schosdorf 3.11.44
[7] ADRC 2438/0486 Richard Vogt Betrifft: Miniatur-Jäger As 014 P 213.01-01 10.11.44
[8] IWM GDC 23/54 Blohm & Voss Kurzbeschreibung Miniatur-Jäger mit As 014 P 213 November 1944
[9] ADRC 2348/0482 Blohm & Voss Vertretung Berlin Herrn Dipl.-Ing. W. Buerkner 15.11.44
[10] ADRC 2348/0481 Argus Motoren Gesellschaft M.B.H. Greiffenberg Schles. 15.11.44
[11] IWM GDC 1668 Volkenrode Monograph – Organisation, Arbeitsgebiete und Berichte der Forschungsanstalt Graf Zeppelin (FGZ) Stuttgart Ruit by U. Hütter November 1946
[12] IWM GDC 18/59 Heinkel Bericht Nr. 114a/45

14. Miscellaneous jet fighters
[1] ADRC/T-2 3422/424 Arado Brandenburg notes circa November 1944 to March 1945
[2] ADRC/T-2 4097/194 Blohm & Voss Ae file circa 1943-45
[3] ADIK 158/0190 St/GL/C-E 2 Nr. 6629/43 g. Kdos. Bericht Nr. 24 über die Entwicklungsbesprechung am 22.3.1943
[4] ADIK 158/0167 St/GL/C-E 2 Nr. 7200/43 g. Kdos Bericht Nr. 26 über die Entwicklungsbesprechung am 27.4.1943
[5] ADRC/T-2 4097/261 Blohm & Voss Flugleistungen P 214 Bemannte Fla-Bombe Ae 585 20.11.44
[6] NavTecMisEu Technical Report No. 524-45 – Report on Dornier Pursuit Plane Development Program by Alfred V Verville 13.10.45
[7] ADRC/T-2 3607/388 Focke-Wulf Strahljäger (Ente) notes 1.12.42
[8] ADRC/T-2 2418/482 BMW Einbaumappe 109-018 1.6.42
[9] ADRC/T-2 2414/686 Focke-Wulf Spezialhöhenjäger mit BMW 109-018 30.8.44
[10] ADIK 158/0638 RLM Technisches Amt Amtsgruppe Entwicklung GL/C-Nr. 6724/44 g. Kdos. (E 3) Triebwerks-Entwicklung 12.5.44
[11] IWM GDC 10/5699 DVL 3- und 6-Komponentenmessung am Entenmodell P 90. 24.11.43
[12] ADRC/T-2 2718/769 P 15 sketches Lippisch
[13] ADRC/T-2 8147/440 Flugtechnische Grundlagen für Projekt P 15 "Diana" 4.3.45
[14] *Luftwaffe Secret Projects – Fighters 1939-1945* by Walter Schick and Ingolf Meyer, Midland Publishing, 1997 p73
[15] *Me 262 Volume One* by J. Richard Smith and Eddie J. Creek, Classic Publications, 1997, p71
[16] Correspondence Dan Sharp / Eddie Nielinger November 13/14/2019
[17] ADRC/T-2 3201/986 Messerschmitt Me/Ge/640 22 (undated)
[18] J. Richard Smith collection ref. A0057
[19] ADRC/T-2 4096/1 Rammrakete by Stöckel 31.8.44
[20] IWM GDC 10/14191 Patente über Düsenflugzeuge K. Stöckel 12.12.45

BIBLIOGRAPHY

The vast majority of original source material used in this book was taken from archival documents and documents kept on ADRC/T-2 microfilm. These comprise hundreds of thousands of pages of reports, letters, file notes, cables and teleprint messages produced, sent and received by the German aircraft manufacturers, test centres and RLM in addition to dozens of Allied intelligence reports concerning German aircraft development and related technology. Those documents used are cited in the endnotes to each chapter.

PUBLISHED BOOKS

Amtmann, Hans H. *The Vanishing Paperclips*, Monogram, Boylston, 1988
Coates, Steve with Carbonel, Jean-Christophe *Helicopters of the Third Reich*, Classic Publications, Hersham, 2002
Conradis, Heinz *Forschen und Fliegen: Weg und Werk von Kurt Tank* Musterschmidt Verlag, Göttingen 1955
Ebert, Hans J., Kaiser, Johann B. and Peters Klaus *Willy Messerschmitt – Pioneer of Aviation Design*, Schiffer, 1999
Forsyth, Robert and Creek, Eddie J. *Heinkel He 162 Spatz*, Classic Publications, Hersham, 2008
Galland, Adolf *The First and the Last*, Methuen, London, 1955
Gooden, Brett *Natter: Manned Missile of the Third Reich*, 2019
Griehl, Manfred *Jet Planes of the Third Reich, The Secret Projects Volume One*, Monogram, Sturbridge, 1998
Griehl, Manfred *Jet Planes of the Third Reich, The Secret Projects Volume Two*, Monogram, Sturbridge, 2004
Griehl, Manfred *Last Days of the Luftwaffe*, Frontline Books, Barnsley, 2009
Heinkel, Ernst *He 1000*, Hutchinson & Co., London, 1956
Herwig, Dieter and Rode, Heinz *Luftwaffe Secret Projects Ground Attack & Special Purpose Aircraft*, Midland Publishing, Hinckley, 2003
Horten, Reimar and Selinger, Peter F., *Nurflügel*, H. Weishaupt Verlag, Graz, 1982
Koos, Volker *Heinkel Raketen- und Strahlflugzeuge*, Aviatic Verlag, Oberhaching, 2008
Kosin, Rüdiger *Die Entwicklung der deutschen Jagdflugzeuge*, Bernard & Graefe Verlag, Koblenz, 1983
Kranzhoff, Jörg Armin Arado – History of an Aircraft Company, Schiffer, Atglen, 1997
Lippisch, Alexander *Ein Dreieck Fliegt*, Motorbuch Verlag, Stuttgart, 1976
Lippisch, Alexander Erinnerungen, Luftfahrtverlag Axel Zuerl, Steinebach-Wörthsee, 1976
Masters, David *German Jet Genesis*, Jane's, London, 1982
Myhra, *David Conversations with Reimar Horten Volumes 1-5*, RCW Ebook Publishing, 2016
Nowarra, Heinz J. *Die deutsche Luftrüstung 1933-1945 Band 1-4*, Bernard & Graefe Verlag, Koblenz, 1993
Pabst, Martin *Willy Messerschmitt Zwölf Jahre Flugzeugbau im Führerstaat*, Aviatic Verlag, Oberhaching, 2018
Pabst, Otto E. *Kurzstarter und Senkrechtstarter*, Bernard & Graefe Verlag, Koblenz, 1984
Pohlmann, Hermann *Chronik eines Flugzeugwerkes 1932-1945*, Motorbuch Verlag, Stuttgart, 1979
Radinger, Willy and Schick, Walter *Messerschmitt Geheimprojekte*, Aviatic Verlag, Oberhaching, 1991
Ransom, Stephen and Cammann, Hans-Hermann *Me 163 Rocket Interceptor Volume One*, Classic Publications, Crowborough, 2002
Ransom, Stephen and Cammann, Hans-Hermann *Me 163 Rocket Interceptor Volume Two*, Classic Publications, Hersham, 2003
Schick, Walter and Meyer, Ingolf *Luftwaffe Secret Projects Fighters 1939-1945*, Midland Publishing, Hinckley, 1997
Shepelev, Andrei and Ottens, Huib *Horten Ho 229 Spirit of Thuringia*, Classic Publications, Hersham, 2006/2014
Smith, J. Richard and Creek, Eddie J. *Me 262 Volume One*, Classic Publications, Burgess Hill, 1997
Smith, J. Richard and Creek, Eddie J. *Me 262 Volume Two*, Classic Publications, Burgess Hill, 1998
Smith, J. Richard and Creek, Eddie J. *Me 262 Volume Three*, Classic Publications, Crowborough, 2000
Smith, J. Richard and Creek, Eddie J. *Me 262 Volume Four*, Classic Publications, Crowborough, 2000
Smith, J. Richard and Creek, Eddie J. with Rolotschek, Gerhard, *Dornier Do 335 Pfeil*, Classic Publications, Hersham, 2007
Speer, Albert *Inside the Third Reich*, Weidenfeld & Nicolson, London, 1970
Speer, Albert *The Slave State – Heinrich Himmler's Masterplan for SS Supremacy*, Weidenfeld & Nicolson, London, 1981
Vann, Frank *Willy Messerschmitt*, Patrick Stephens, Sparkford, 1993
Vogt, Richard *Weltumspannende Memoiren eines Flugzeug-Konstrukteurs*, Luftfahrt-Verlag Walter Zuerl, Steinebach-Wörthsee, 1976

INDEX

INDEX CONTINUED